Robert Babuška and Frans C.A. Groen (Eds.)

Interactive Collaborative Information Systems

Studies in Computational Intelligence, Volume 281

Editor-in-Chief
Prof. Janusz Kacprzyk
Systems Research Institute
Polish Academy of Sciences
ul. Newelska 6
01-447 Warsaw
Poland
E-mail: kacprzyk@ibspan.waw.pl

Further volumes of this series can be found on our homepage: springer.com

Vol. 260. Edward Szczerbicki and Ngoc Thanh Nguyen (Eds.)
Smart Information and Knowledge Management, 2009
ISBN 978-3-642-04583-7

Vol. 261. Nadia Nedjah, Leandro dos Santos Coelho, and Luiza de Macedo de Mourelle (Eds.)
Multi-Objective Swarm Intelligent Systems, 2009
ISBN 978-3-642-05164-7

Vol. 262. Jacek Koronacki, Zbigniew W. Ras, Slawomir T. Wierzchon, and Janusz Kacprzyk (Eds.)
Advances in Machine Learning I, 2009
ISBN 978-3-642-05176-0

Vol. 263. Jacek Koronacki, Zbigniew W. Ras, Slawomir T. Wierzchon, and Janusz Kacprzyk (Eds.)
Advances in Machine Learning II, 2009
ISBN 978-3-642-05178-4

Vol. 264. Olivier Sigaud and Jan Peters (Eds.)
From Motor Learning to Interaction Learning in Robots, 2009
ISBN 978-3-642-05180-7

Vol. 265. Zbigniew W. Ras and Li-Shiang Tsay (Eds.)
Advances in Intelligent Information Systems, 2009
ISBN 978-3-642-05182-1

Vol. 266. Akitoshi Hanazawa, Tsutom Miki, and Keiichi Horio (Eds.)
Brain-Inspired Information Technology, 2009
ISBN 978-3-642-04024-5

Vol. 267. Ivan Zelinka, Sergej Celikovský, Hendrik Richter, and Guanrong Chen (Eds.)
Evolutionary Algorithms and Chaotic Systems, 2009
ISBN 978-3-642-10706-1

Vol. 268. Johann M.Ph. Schumann and Yan Liu (Eds.)
Applications of Neural Networks in High Assurance Systems, 2009
ISBN 978-3-642-10689-7

Vol. 269. Francisco Fernández de de Vega and Erick Cantú-Paz (Eds.)
Parallel and Distributed Computational Intelligence, 2009
ISBN 978-3-642-10674-3

Vol. 270. Zong Woo Geem
Recent Advances In Harmony Search Algorithm, 2009
ISBN 978-3-642-04316-1

Vol. 271. Janusz Kacprzyk, Frederick E. Petry, and Adnan Yazici (Eds.)
Uncertainty Approaches for Spatial Data Modeling and Processing, 2009
ISBN 978-3-642-10662-0

Vol. 272. Carlos A. Coello Coello, Clarisse Dhaenens, and Laetitia Jourdan (Eds.)
Advances in Multi-Objective Nature Inspired Computing, 2009
ISBN 978-3-642-11217-1

Vol. 273. Fatos Xhafa, Santi Caballé, Ajith Abraham, Thanasis Daradoumis, and Angel Alejandro Juan Perez (Eds.)
Computational Intelligence for Technology Enhanced Learning, 2010
ISBN 978-3-642-11223-2

Vol. 274. Zbigniew W. Raś and Alicja Wieczorkowska (Eds.)
Advances in Music Information Retrieval, 2010
ISBN 978-3-642-11673-5

Vol. 275. Dilip Kumar Pratihar and Lakhmi C. Jain (Eds.)
Intelligent Autonomous Systems, 2010
ISBN 978-3-642-11675-9

Vol. 276. Jacek Mańdziuk
Knowledge-Free and Learning-Based Methods in Intelligent Game Playing, 2010
ISBN 978-3-642-11677-3

Vol. 277. Filippo Spagnolo and Benedetto Di Paola (Eds.)
European and Chinese Cognitive Styles and their Impact on Teaching Mathematics, 2010
ISBN 978-3-642-11679-7

Vol. 278. Radomir S. Stankovic and Jaakko Astola
From Boolean Logic to Switching Circuits and Automata, 2010
ISBN 978-3-642-11681-0

Vol. 279. Manolis Wallace, Ioannis E. Anagnostopoulos, Phivos Mylonas, and Maria Bielikova (Eds.)
Semantics in Adaptive and Personalized Services, 2010
ISBN 978-3-642-11683-4

Vol. 280. Chang Wen Chen, Zhu Li, and Shiguo Lian (Eds.)
Intelligent Multimedia Communication: Techniques and Applications, 2010
ISBN 978-3-642-11685-8

Vol. 281. Robert Babuška and Frans C.A. Groen (Eds.)
Interactive Collaborative Information Systems, 2010
ISBN 978-3-642-11687-2

Robert Babuška and Frans C.A. Groen (Eds.)

Interactive Collaborative Information Systems

Springer

Dr. Robert Babuška
Delft Center for Systems and Control
Delft University of Technology
Mekelweg 2
2628 CD Delft
The Netherlands
E-mail: r.babuska@tudelft.nl

Dr. Frans C.A. Groen
Faculty of Science
Informatics Institute
Science Park 107
1098 XG, Amsterdam
The Netherlands
E-mail: F.C.A.Groen@uva.nl

ISBN 978-3-642-11687-2 e-ISBN 978-3-642-11688-9

DOI 10.1007/978-3-642-11688-9

Studies in Computational Intelligence ISSN 1860-949X

Library of Congress Control Number: 2010920828

Typeset & Cover Design: Scientific Publishing Services Pvt. Ltd., Chennai, India.

Printed on acid-free paper

9 8 7 6 5 4 3 2 1

springer.com

Preface

The increasing complexity of our world demands new perspectives on the role of technology in decision making. Human decision making has its limitations in terms of information-processing capacity. We need new technology to cope with the increasingly complex and information-rich nature of our modern society. This is particularly true for critical environments such as crisis management and traffic management, where humans need to engage in close collaborations with artificial systems to observe and understand the situation and respond in a sensible way. We believe that close collaborations between humans and artificial systems will become essential and that the importance of research into Interactive Collaborative Information Systems (ICIS) is self-evident.

Developments in information and communication technology have radically changed our working environments. The vast amount of information available nowadays and the wirelessly networked nature of our modern society open up new opportunities to handle difficult decision-making situations such as computer-supported situation assessment and distributed decision making. To make good use of these new possibilities, we need to update our traditional views on the role and capabilities of information systems.

The aim of the *Interactive Collaborative Information Systems* project is to develop techniques that support humans in complex information environments and that facilitate distributed decision-making capabilities. ICIS emphasizes the importance of building actor-agent communities: close collaborations between human and artificial actors that highlight their complementary capabilities, and in which task distribution is flexible and adaptive. To fulfill such a prospect, we need intelligent systems that observe their environment, interpret and fuse information, have learning and decision-making capabilities, and have the ability to work in teams. It also means that we need to study the interaction of humans with their artificial counterparts in such settings and how their information needs can be met. Research within the ICIS projects helps create such views. ICIS combines research from information technology, artificial intelligence and human sciences to obtain a multidisciplinary foundation from which innovative actor-agent systems for critical environments can emerge.

This book focuses on the employment of innovative agent technology, advanced machine learning techniques, and cognition-based interface technology for the use in collaborative decision support systems. It consists of five parts: *Reinforcement learning*, *Collaborative decision making*, *Computer-human interaction modeling*, *Architectures for distributed agent-actor communities*, and *Case studies and applications*.

Reinforcement Learning is the main subject of the first part of the book. This type of learning plays an important role in developing intelligent systems. As the systems (agents) have to realize a common goal, the question is which actions an agent should take in its environment to contribute to that common goal. Reinforcement learning is an established way to implement this process. The core research questions within this subject are the representation of the environment's state, the representation of the actions that an agent can take in a given state, and the long-term reward representing the common goal. Finding the optimal sequence of actions often becomes intractable, especially when the number of agents increases. As a result, approximate solutions are needed. These are discussed in the chapter "*Approximate dynamic programming and reinforcement learning.*" This chapter focuses on an approximate solution for reinforcement learning given a representation with continuous states. The next chapter "*Learning with whom to communicate using relational reinforcement learning*" exploits relational structures to come up with strategies for multi-agent systems. The last chapter "*Switching between representations in reinforcement learning*" investigates when to switch online between feature sets representing the states.

The second part of the book addresses **Collaborative Decision Making**. Decision-theoretic models are given to describe cooperation between multiple agents under uncertainty. In these cases the state of the environment given the agent's observations is uncertain and can be described by, for instance, a probability distribution over the states. Approximate solution methods for these cases are presented in the chapter "*A decision-theoretic approach to collaboration: principal description methods and efficient heuristic approximations*". The next chapter "Efficient Methods for Near-Optimal Sequential Decision Making Under Uncertainty" discusses both Bayesian and distribution-free algorithms for sequential decision making when costs are known. A completely different approach is presented in the chapter on "*Ant colony learning algorithm for optimal control*". In this case an optimization heuristic is used and a novel algorithm is introduced in which the artificial agents (ants) work together to collectively learn optimal control policies. The chapter on "*Map-based support for effective collaboration in micro-mobile virtual teams*" presents collaboration in geo-spatial support systems where the maps aid the distributed decision-making process.

The topic of the third part of the book is **Computer-Human Interaction Modeling**. The first chapter of this part "*Affective dialogue management using factored POMDPs*" shows that partially observable Markov decision processes (POMDPs) are appropriate for this purpose and presents a novel approach to develop an affective dialogue model. The next chapter, "*Context-aware multimodal human computer interaction*," presents multimodal interaction techniques including speech, lip movement, facial expression, and text and visual communication. The chapter "*Design issues for pen-centric interactive maps*" focuses on pen-input recognition systems, in particular on new features for classifying iconic gestures. The response of users to intelligent systems showing adaptive behavior is studied in the chapter "*Interacting with adaptive systems*". Human poses are important for both interaction and situational awareness in human-inhabited environments. The last chapter, "*Example-based human pose recovery under predicted partial occlusions*," deals with that topic also in the case when partial occlusion occurs.

The architecture of intelligent decision-making system is an important glue that lets all the parts work together. This is the topic of the fourth part of the book, **Architectures for Distributed Agent-Actor Communities**. Chapter "*Agility and adaptive autonomy in networked organizations*" addresses the tradeoff in a multi-actor environment between global coordination of activities and respecting the autonomy of the actors involved. The underlying principles of multi-agents organizations that are not only hierarchical, but that can also adapt their structure are the topic of the chapter "*Adaptive hierarchical multi-agent organizations*". In the chapter "*Method for designing networking adaptive interactive hybrid systems*" the various architectures for this type of systems are given and a top-down design methodology is introduced. The tradeoff in a multi-actor environment between global coordination of activities and respecting the autonomy of the actors involved forms the topic studied in the chapter "*Agility and adaptive autonomy in networked organizations*".

Case Studies and Applications are discussed in the last part of the book. Crisis management is the topic of the first two chapters. In the first chapter, "*A call for sensemaking support systems in crisis management*", the information challenges in crisis management are explored and three case studies are investigated. In the next chapter, "*A distributed approach to gas detection and source localization using heterogeneous information*", a system for early detection of gaseous substances and coarse source estimation is presented by using heterogeneous sensor measurements and human reports. Mobility is the application discussed in the next two chapters. The chapter "*Traffic light control by multiagent reinforcement learning systems*" discusses extensions to improve a basic setting of multiple local controllers (agents), each responsible for the optimization of traffic lights around a single junction using reinforcement learning. The chapter "*Fusing heterogeneous and unreliable data from traffic sensors*" deals with traffic data fusion from a variety of traffic sensors using conservation laws and Poisson statistics. The last chapter, "*Bayesian networks for expert systems, theory and practical applications*", shows the strength of Bayesian modeling approaches in three different applications: medical diagnosis support, petrochemical decision support and victim identification.

Acknowledgement. The work described in this book is part of the ICIS project – Interactive Collaborative Information Systems, see http://www.icis.decis.nl/. This project is one of nine ICT research projects funded by the BSIK program (SenterNovem, grant no. BSIK03024) of the Dutch government. The ICIS project started in the fall of 2003 and finished in 2010. The ICIS project is hosted by the D-CIS Lab, an open research partnership of Thales Nederland, the Delft University of Technology, the University of Amsterdam and the Netherlands Foundation of Applied Scientific Research (TNO). The ICIS consortium involves leading IT-driven industries, academic research institutions, technology institutes and high-tech small and medium enterprizes. The financial support of SenterNovem is gratefully acknowledged.

Amsterdam, Delft,
November 2009

Frans C.A. Groen
Robert Babuška

Contents

Part I: Reinforcement Learning

Part II: Collaborative Decision Making

Part IV: Architectures for Distributed Agent-Actor Communities

List of Contributors

Jelmer Marinus van Ast
Delft Center for Systems and Control, Delft University of Technology, Mekelweg 2, 2628 CD Delft, The Netherlands
j.m.vanast@tudelft.nl

Robert Babuška
Delft Center for Systems and Control, Delft University of Technology, Mekelweg 2, 2628 CD Delft, The Netherlands
r.babuska@tudelft.nl

Bram Bakker
Informatics Institute, University of Amsterdam, Science Park 107, 1098 XG Amsterdam, The Netherlands
p.b.bakker@uva.nl

Guido te Brake
TNO Defense, Security and Safety, PO Box 21, 3769 ZG, Soesterberg, The Netherlands
guido.tebrake@tno.nl

Lucian Buşoniu
Delft Center for Systems and Control, Delft University of Technology, Mekelweg 2, 2628 CD Delft, The Netherlands
i.l.busoniu@tudelft.nl

Trung H. Bui
Center for the Study of Language and Information, Stanford University, 210 Panama St, Stanford, CA 94305, USA
thbui@stanford.edu

Willem Burgers
SNN Adaptive Intelligence, Geert Grooteplein 21, 6525 EZ Nijmegen, The Netherlands
w.burgers@science.ru.nl

Alin G. Chiţu
Man Machine Interaction, Delft University of Technology, Mekelweg 4 2628 CD, The Netherlands
a.g.chitu@tudelft.nl

Henriette Cramer
Human-Computer Studies, University of Amsterdam, Science Park 107, 1098 XG Amsterdam, The Netherlands
hcramer@science.uva.nl

Tom Croonenborghs
KH Kempen University College, Kleinhoefstraat 4, 2440 Geel, Belgium
tom.croonenborghs@khk.be

Dragoş Datcu
Man Machine Interaction, Delft University of Technology, Mekelweg 4 2628 CD, The Netherlands
d.datcu@tudelft.nl

Bart De Schutter
Delft Center for Systems and Control & Marine and Transport Technology Department, Delft University of Technology, Mekelweg 2, 2628 CD Delft, The Netherlands
b@deschutter.info

Christos Dimitrakakis
Informatics Institute, University of Amsterdam, Science Park 107, 1098 XG Amsterdam, The Netherlands
christos.dimitrakakis@gmail.com

Kurt Driessens
DTAI, Katholieke Universiteit Leuven, Celestijnenlaan 200A, B-3001 Heverlee, Belgium
kurt.driessens@cs.kuleuven.be

Vanessa Evers
Human-Computer Studies, University of Amsterdam, Science Park 107, 1098 XG Amsterdam, The Netherlands
v.evers@uva.nl

Siska Fitrianie
Man Machine Interaction, Delft University of Technology, Mekelweg 4, 2628 CD, The Netherlands
s.fitrianie@tudelft.nl

Marcel van Gerve
Institute for Computing & Information Sciences, Radboud University Nijmegen, The Netherlands
marcelge@cs.ru.nl

Mattijs Ghijsen
Human Computer Studies Group, Informatics Institute, Universtiteit van Amsterdam, Science Park 107, 1098 XG Amsterdam, The Netherlands
m.ghijsen@uva.nl

Rianne Gouman
D-CIS Lab / Thales Research & Technology Netherlands, P.O. Box 90, 2600 AB, Delft, The Netherlands
rianne.gouman@icis.decis.nl

Frans C.A. Groen
Informatics Institute, University of Amsterdam, Science Park 107, 1098 XG Amsterdam, The Netherlands
f.c.a.groen@uva.nl

Jaap van den Herik
Tilburg centre for Creative Computing, Tilburg University, Warandelaan 2, PO Box 90153, 5000 LE Tilburg, The Netherlands
h.j.vdnherik@uvt.nl

Prof. Dr. S.P. Hoogendoorn
Transport & Planning, Faculty of Civil Engineering and Geosciences, Delft University of Technology
s.p.hoogendoorn@tudelft.nl

Wouter N.H. Jansweijer
Human Computer Studies Group, Informatics Institute, Universtiteit van Amsterdam, Science Park 107, 1098 XG Amsterdam, The Netherlands
w.n.h.jansweijer@uva.nl

Michiel Kamermans
Thales Research & Technology Netherlands, Delftechpark 24, 2628 XH, Delft, The Netherlands
Michiel.Kamermans@icis.decis.nl

Marten Kampman
University of Utrecht, Information & Computing Sciences
martenkampman@gmail.com

Bert Kappen
Radboud University Nijmegen, Donders Institute for Brain, Cognition and Behaviour, Geert Grooteplein 21, 6525 EZ Nijmegen, The Netherlands
b.kappen@science.ru.nl

Masja Kempen
D-CIS Lab / Thales Research & Technology Netherlands, P.O. Box 90, 2600 AB, Delft, The Netherlands
masja.kempen@icis.decis.nl

Leon Kester
TNO Defence, Security and Safety, Oude Waalsdorperweg 63, 2597 AK The Hague, The Netherlands
leon.kester@tno.nl

Rick van der Kleij
TNO Defense, Security and Safety, PO Box 21, 3769 ZG, Soesterberg, The Netherlands
rick.vanderkleij@tno.nl

Dr. J.W.C. Van Lint
Transport & Planning, Faculty of Civil Engineering and Geosciences, Delft University of Technology
j.w.c.vanlint@tudelft.nl

Willem J. Muhren
Department of Information Systems and Management, Tilburg University, PO Box 90153, 5000 LE Tilburg, The Netherlands
w.j.muhren@uvt.nl

Martijn Neef
TNO Defense, Security and Safety, Oude Waalsdorperweg 63, 2597 AK The Hague, The Netherlands
martijn.neef@tno.nl

Ralph Niels
Donders Institute for Brain, Cognition & Behavior, Radboud University Nijmegen, The Netherlands
r.niels@donders.ru.nl

Anton Nijholt
Human Media Interaction Group, University of Twente, PO Box 217, 7500 AE Enschede, The Netherlands
anijholt@cs.utwente.nl

Frans A. Oliehoek
Intelligent System Laboratory Amsterdam, Science Park 107, 1098 XG Amsterdam, The Netherlands
F.A.Oliehoek@uva.nl

Q. Ou
Transport & Planning, Faculty of Civil Engineering and Geosciences, Delft University of Technology
Q.Ou@tudelft.nl

Patrick de Oude
Intelligent Autonomous Systems Group, Informatics Institute, Faculty of Science, University of Amsterdam, Science Park 107, 1098 XG Amsterdam, The Netherlands
P.deOude@uva.nl

Gregor Pavlin
Thales Research & Technology
Netherlands, Delftechpark 24,
2628 XH, Delft, The Netherlands
Gregor.Pavlin@icis.decis.nl

Mannes Poel
Human Media Interaction Group,
University of Twente, PO Box 217,
7500 AE Enschede, The Netherlands
mpoel@cs.utwente.nl

Marc Ponsen
Department of Knowledge
Engineering, Maastricht University,
Mindebroedersberg 6a, 6211 LK,
Maastricht, The Netherlands
m.ponsen@maastrichtuniversity.nl

Ronald Poppe
Human Media Interaction Group,
University of Twente, Enschede,
The Netherlands
poppe@ewi.utwente.nl

Eric Postma
Tilburg centre for Creative
Computing, Tilburg University,
Warandelaan 2, P.O. Box 90153,
5000 LE Tilburg, The Netherlands
e.o.postma@uvt.nl

Jan Ramon
DTAI, Katholieke Universiteit
Leuven, Celestijnenlaan 200A,
B-3001 Heverlee, Belgium
jan.ramon@cs.kuleuven.be

Leon J.M. Rothkrantz
Man Machine Interaction, Delft
University of Technology, Mekelweg
4 2628 CD, The Netherlands and
Netherlands Defence Academy,
Faculty of Technical Sciences, Den
Helder, The Netherlands
l.j.m.rothkrantz@tudelft.nl

Harm van Seijen
TNO Defense, Security and Safety,
Oude Waalsdorperweg 63, 2597 AK
The Hague, The Netherlands
harm.vanseijen@tno.nl

Maarten van Someren
Human-Computer Studies,
University of Amsterdam, Science
Park 107, 1098 XG Amsterdam,
The Netherlands
M.W.vanSomeren@uva.nl

Karl Tuyls
Department of Knowledge
Engineering, Maastricht University,
Mindebroedersberg 6a, 6211 LK,
Maastricht, The Netherlands
k.tuyls@maastrichtuniversity.nl

Bob van der Vecht
TNO Defence, Security and Safety,
Oude Waalsdorperweg 63, 2597 AK
The Hague, The Netherlands
bob.vandervecht@tno.nl

Arnoud Visser
Intelligent System Laboratory
Amsterdam, Science Park 107,
1098 XG Amsterdam,
The Netherlands
A.Visser@uva.nl

Louis Vuurpijl
Donders Institute for Brain,
Cognition & Behavior, Radboud
University Nijmegen,
The Netherlands
l.vuurpijl@donders.ru.nl

Bartel Van de Walle
Department of Information Systems
and Management, Tilburg
University, PO Box 90153, 5000 LE
Tilburg, The Netherlands
bartel@uvt.nl

Shimon Whiteson
Informatics Institute, University of Amsterdam, Science Park 107, 1098 XG Amsterdam, The Netherlands
s.a.whiteson@uva.nl

Wim Wiegerinck
SNN Adaptive Intelligence, Geert Grooteplein 21, 6525 EZ Nijmegen, The Netherlands
w.wiegerinck@science.ru.nl

Bob J. Wielinga
Human Computer Studies Group, Informatics Institute, Universtiteit van Amsterdam, Science Park 107, 1098 XG Amsterdam, The Netherlands
B.J.Wielinga@uva.nl

Niek Wijngaards
D-CIS Lab / Thales Research & Technology Netherlands, P.O. Box 90, 2600 AB, Delft, The Netherlands
niek.wijngaards@icis.decis.nl

Don Willems
Donders Institute for Brain, Cognition & Behavior, Radboud University Nijmegen
d.willems@donders.ru.nl

Zhenke Yang
Man Machine Interaction, Delft University of Technology, Mekelweg 4 2628 CD, The Netherlands
z.yang@tudelft.nl

Job Zwiers
Human Media Interaction Group, University of Twente, PO Box 217, 7500 AE Enschede, The Netherlands
zwiers@cs.utwente.nl

Part I
Reinforcement Learning

Approximate Dynamic Programming and Reinforcement Learning

Lucian Buşoniu, Bart De Schutter, and Robert Babuška

Abstract. Dynamic programming (DP) and reinforcement learning (RL) can be used to address problems from a variety of fields, including automatic control, artificial intelligence, operations research, and economy. Many problems in these fields are described by continuous variables, whereas DP and RL can find exact solutions only in the discrete case. Therefore, approximation is essential in practical DP and RL. This chapter provides an in-depth review of the literature on approximate DP and RL in large or continuous-space, infinite-horizon problems. Value iteration, policy iteration, and policy search approaches are presented in turn. Model-based (DP) as well as online and batch model-free (RL) algorithms are discussed. We review theoretical guarantees on the approximate solutions produced by these algorithms. Numerical examples illustrate the behavior of several representative algorithms in practice. Techniques to automatically derive value function approximators are discussed, and a comparison between value iteration, policy iteration, and policy search is provided. The chapter closes with a discussion of open issues and promising research directions in approximate DP and RL.

1 Introduction

Dynamic programming (DP) and reinforcement learning (RL) can be used to address important problems arising in a variety of fields, including, e.g., automatic

Lucian Buşoniu
Delft Center for Systems and Control, Delft University of Technology, Mekelweg 2, 2628CD Delft, The Netherlands
e-mail: i.l.busoniu@tudelft.nl

Bart De Schutter
Delft Center for Systems and Control & Marine and Transport Technology Department, Delft University of Technology, Mekelweg 2, 2628CD Delft, The Netherlands
e-mail: b.deschutter@tudelft.nl

Robert Babuška
Delft Center for Systems and Control, Delft University of Technology, Mekelweg 2, 2628CD Delft, The Netherlands
e-mail: r.babuska@tudelft.nl

R. Babuška & F.C.A. Groen (Eds.): Interactive Collaborative Information Systems, SCI 281, pp. 3–44.
springerlink.com

control, artificial intelligence, operations research, and economy. From the perspective of automatic control, the DP/RL framework comprises a nonlinear and stochastic optimal control problem [9]. Moreover, RL can be seen as adaptive optimal control [75, 83]. Algorithms that solve such a problem in general would be extremely useful for optimal control. From the perspective of artificial intelligence, RL promises a methodology to build an artificial agent that learns how to survive and optimize its behavior in an unknown environment, without requiring prior knowledge [74]. Developing such an agent is a central goal of artificial intelligence. Because of this mixed inheritance from optimal control and artificial intelligence, two sets of equivalent names and notations are used in DP and RL: e.g., 'controller' has the same meaning as 'agent,' and 'process' has the same meaning as 'environment.' In this chapter, we will mainly use control-theoretical terminology and notations.

The DP/RL problem can be formalized as a Markov decision process (MDP). In an MDP, at each discrete time step, the controller (agent) measures the state of the process (environment) and applies an action, according to a control (behavior) policy. As a result of this action, the process transits into a new state. A scalar reward is sent to the controller to indicate the quality of this transition. The controller measures the new state, and the whole cycle repeats. State transitions can generally be nonlinear and stochastic. This pattern of interaction is represented in Fig. 1. The goal is to find a policy that maximizes the cumulative reward (the return) over the course of interaction [9, 11, 74].

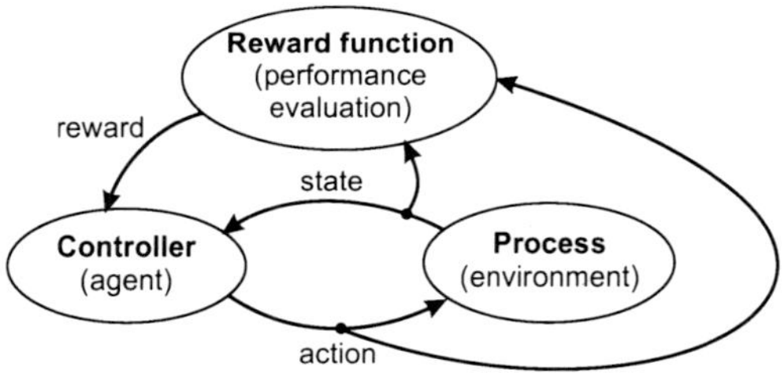

Fig. 1 The basic elements of DP and RL and their flow of interaction

As a conceptual example, consider a garbage-collecting robot. This robot measures its own position and the positions of the surrounding objects; these positions are the state variables. The software of the robot is the controller (the agent). Note that in DP and RL, the process (environment) also includes the physical body of the robot. The controller receives the position measurements and sends motion commands to the motors; these commands are the actions. The dynamics describe the rule according to which positions (states) change as a result of the commands (actions). The reward signal can, e.g., be positive at every time step in which the robot picks up trash, and zero otherwise. In this case, the goal is to pick up as much trash as possible, because this corresponds to accumulating as much reward as possible.

DP algorithms are model based: they require a model of the MDP, in the form of the transition dynamics and the reward function [9, 11]. DP algorithms typically work offline, producing a policy which is then used to control the process. Usually, analytical expressions for the dynamics and the reward function are not required. Instead, given a state and an action, the model is only required to generate a next state and the corresponding reward. RL algorithms are model free [74], and use transition and reward data obtained from the process. RL is useful when a model is difficult or costly to derive. So, RL can be seen as model-free, sample-based, or trajectory-based DP, and DP can be seen as model-based RL. Some RL algorithms work offline, using data collected in advance. Other RL algorithms work online: they compute a solution while simultaneously controlling the process. Online RL is useful when it is difficult or costly to obtain data in advance. Online RL algorithms must balance the need to collect informative data with the need to control the process well.

DP and RL algorithms can be classified by the path they take to search for an optimal policy. *Value iteration* algorithms search for the optimal value function, i.e., the maximal returns as a function of the state and possibly of the control action. The optimal value function is then used to compute an optimal policy. *Policy iteration* algorithms iteratively improve policies. In each iteration, the value function of the current policy is found (instead of the optimal value function), and this value function is used to compute a new, improved policy. *Policy search* algorithms use optimization techniques to directly search for an optimal policy.[1]

Classical DP and RL algorithms require exact representations of the value functions and policies. When some of the variables have a very large or infinite number of possible values (e.g., when they are continuous), exact representations are no longer possible. Instead, value functions and policies need to be approximated. Since many problems of practical interest have large or continuous state and action spaces, approximation is essential in DP and RL. Two main types of approximators can be identified: parametric and nonparametric approximators. *Parametric* approximators are functions of a set of parameters; the form of the function is given a priori, and does not depend on the data. The parameters are tuned using data about the target value function or policy. A representative example is a linear combination of a fixed set of basis functions (BFs). In contrast, the form and number of parameters of a *nonparametric* approximator are derived from the available data. For instance, kernel-based approximators can also be seen as representing the target function with a linear combination of BFs, but, unlike parametric approximation, they define one BF for each data point.

This chapter provides an in-depth review of the literature on approximate DP and RL in large or continuous-space, infinite-horizon problems. Approximate value iteration, policy iteration, and policy search are presented in detail and compared. Model-based (DP), as well as online and batch model-free (RL) algorithms are discussed. Algorithm descriptions are complemented by theoretical guarantees on their performance, and by numerical examples illustrating their behavior in practice. We focus mainly on parametric approximation, but also mention some important nonparametric approaches. Whenever possible, we discuss the general case of

[1] A fourth category of (model-based) algorithms is *model predictive control* [8, 22], which we do not discuss in this chapter.

nonlinearly parameterized approximators. Sometimes, we delve in more detail about linearly parameterized approximators, e.g., because they allow to derive better theoretical guarantees on the resulting approximate solutions.

The remainder of this chapter is structured as follows. After a brief introduction to classical, exact DP and RL in Sect. 2, the need for approximation in DP and RL is explained in Sect. 3. This is followed by an in-depth discussion of approximate value iteration in Sect. 4 and of approximate policy iteration in Sect. 5. Techniques to automatically derive value function approximators are reviewed in Sect. 6. Approximate policy search is discussed in Sect. 7. A representative algorithm from each class (value iteration, policy iteration, and policy search) is applied to an example involving the optimal control of a DC motor: respectively, grid Q-iteration in Sect. 4, least-squares policy iteration in Sect. 5, and pattern search policy optimization in Sect. 7. While not all of the algorithms used in the examples are taken directly from the literature, and some of them are designed by the authors, they are all straightforward instantiations of the class of techniques they represent. Approximate value iteration, policy iteration, and policy search are compared in Sect. 8. Section 9 closes the chapter with a discussion of open issues and promising research directions in approximate DP and RL.

2 Markov Decision Processes; Exact Dynamic Programming and Reinforcement Learning

This section formally describes MDPs and characterizes their optimal solution. Then, exact algorithms for value iteration and policy iteration are described. Because policy search is not typically used in exact DP and RL, it is not described in this section; instead, it will be presented in the context of approximation, in Sect. 7.

2.1 Markov Decision Processes and Their Solution

A MDP is defined by its state space X, its action space U, its transition probability function $\tilde{f} : X \times U \times X \to [0,\infty)$, which describes how the state changes as a result of the actions, and its reward function $\tilde{\rho} : X \times U \times X \to \mathbb{R}$, which evaluates the quality of state transitions. The controller behaves according to its control policy $h : X \to U$.

More formally, at each discrete time step k, given the state $x_k \in X$, the controller takes an action $u_k \in U$ according to the policy h. The probability that the resulting next state x_{k+1} belongs to a region $X_{k+1} \subset X$ of the state space is $\int_{X_{k+1}} \tilde{f}(x_k, u_k, x')\mathrm{d}x'$. For any x and u, $\tilde{f}(x,u,\cdot)$ is assumed to define a valid probability density of the argument '$\cdot$'. After the transition to x_{k+1}, a reward r_{k+1} is provided according to the reward function: $r_{k+1} = \tilde{\rho}(x_k, u_k, x_{k+1})$. The reward evaluates the immediate effect of action u_k, namely the transition from x_k to x_{k+1}. We assume that

$\|\tilde{\rho}\|_\infty = \sup_{x,u,x'} |\tilde{\rho}(x,u,x')|$ is finite.[2] Given $\tilde{f}$ and $\tilde{\rho}$, the current state x_k and action u_k are sufficient to determine the probability density of the next state x_{k+1} and of the reward r_{k+1}. This is the so-called Markov property, which is essential in providing theoretical guarantees about DP/RL algorithms.

Note that, when the state space is countable (e.g., discrete), the transition function can also be given as $\bar{f} : X \times U \times X \to [0,1]$, where $\bar{f}(x_k, u_k, x')$ is the probability of reaching x' after taking u_k in x_k. The function $\tilde{f}$ is a generalization of $\bar{f}$ to uncountable (e.g., continuous) state spaces; in such spaces, the probability of reaching a given point x' in the state space is generally 0, making a description of the form $\bar{f}$ inappropriate. Additionally, the individual rewards themselves can be stochastic; if they are, to simplify the notation, we take $\tilde{\rho}$ equal to the *expected* rewards.

Developing an analytical expression for the transition probability function $\tilde{f}$ is generally a difficult task. Fortunately, most DP algorithms do not require such an analytical expression. Instead, given any state–action pair, the model is only required to generate a corresponding next state and reward. Constructing this generative model is usually easier.

The expected infinite-horizon discounted return of a state x_0 under a policy h accumulates the rewards obtained by using this policy from x_0:[3]

$$R^h(x_0) = \lim_{K\to\infty} \mathrm{E}_{x_{k+1}\sim \tilde{f}(x_k,h(x_k),\cdot)} \left\{ \sum_{k=0}^{K} \gamma^k \tilde{\rho}(x_k, h(x_k), x_{k+1}) \right\} \tag{1}$$

where $\gamma \in [0,1)$ is the discount factor and the expectation is taken over the stochastic transitions. The notation $x_{k+1} \sim \tilde{f}(x_k, h(x_k), \cdot)$ means that the random variable x_{k+1} is drawn from the density $\tilde{f}(x_k, h(x_k), \cdot)$ at each step k. The goal is to find an optimal policy h^* that maximizes the expected return (1) from every initial state. So, the long-term performance (return) must be maximized using only feedback about the immediate, one-step performance (reward). This is challenging because actions taken in the present potentially affect rewards achieved far into the future, and the immediate reward provides no information about these long-term effects. This is the problem of delayed reward [74]. When the infinite-horizon discounted return is used and under certain technical assumptions on the elements of the MDP, there exists at least one stationary deterministic optimal policy [10].

The discount factor can intuitively be seen as a way to encode an increasing uncertainty about rewards that will be received in the future. From a mathematical point of view, discounting ensures that, given bounded rewards, the returns will always be bounded. Choosing γ often involves a tradeoff between the quality of the solution and the convergence rate of the DP/RL algorithm. Some important DP/RL

---

[2] As already mentioned, control-theoretic notations are used instead of artificial intelligence notations. For instance, in the artificial intelligence literature on DP/RL, the state is usually denoted by s, the state space by S, the action by a, the action space by A, and the policy by π.

[3] We assume that the MDP and the policies h have suitable properties to ensure that the return and the Bellman equations in the sequel are well defined. See, e.g., [10] and Appendix A of [9] for a discussion of these properties.

algorithms have a rate of convergence proportional to γ, so they converge faster when γ is smaller (this is the case, e.g., for model-based value iteration, see Sects. 2.2 and 4.1). However, if γ is too small, the solution may be unsatisfactory because it does not sufficiently take into account rewards obtained after a large number of steps.

Instead of discounting the rewards, they can also be averaged over time, or they can simply be added together without weighting [35]. It is also possible to use a finite-horizon return, in which case optimal policies and the optimal value function depend on the time step k. Only infinite-horizon discounted returns, leading to time-invariant optimal policies and value functions, are considered in this chapter.

Policies can be conveniently characterized using their value functions. Two types of value functions exist: state–action value functions (Q-functions) and state value functions (V-functions). The Q-function of a policy h is the return when starting in a given state, applying a given action, and following the policy h thereafter:

$$Q^h(x,u) = \mathrm{E}_{x' \sim \tilde{f}(x,u,\cdot)} \left\{ \tilde{\rho}(x,u,x') + \gamma R^h(x') \right\} \tag{2}$$

The optimal Q-function is defined as the best Q-function that can be obtained by any possible policy:[4]

$$Q^*(x,u) = \max_h Q^h(x,u) \tag{3}$$

A policy that selects for every state an action with the largest optimal Q-value:

$$h^*(x) = \arg\max_u Q^*(x,u) \tag{4}$$

is optimal (it maximizes the return). A policy that maximizes a Q-function in this way is said to be *greedy* in that Q-function. Here, as well as in the sequel, if multiple maximizing actions are encountered when computing greedy policies for some state, any of these actions can be chosen. So, finding an optimal policy can be done by first finding Q^*, and then computing a greedy policy in Q^*.

A central result in DP and RL is the *Bellman optimality equation*:

$$Q^*(x,u) = \mathrm{E}_{x' \sim \tilde{f}(x,u,\cdot)} \left\{ \tilde{\rho}(x,u,x') + \gamma \max_{u'} Q^*(x',u') \right\} \tag{5}$$

This equation gives a recursive characterization of Q^*: the optimal value of taking action u in state x is the expected sum of the immediate reward and of the discounted optimal value achievable in the next state. The Q-function Q^h of a policy h is also characterized by a Bellman equation, given as follows:

[4] Note that, for the simplicity of notation, we implicitly assume that the maximum exists in (3) and in similar equations in the sequel. When the maximum does not exist, the 'max' operator should be replaced by 'sup', and the theory remains valid. For the computation of greedy actions in (4) and in similar equations in the sequel, the maximum must exist to ensure the existence of a greedy policy; this can be guaranteed under certain technical assumptions.

$$Q^h(x,u) = \mathrm{E}_{x' \sim \tilde{f}(x,u,\cdot)} \left\{ \tilde{\rho}(x,u,x') + \gamma Q^h(x', h(x')) \right\} \tag{6}$$

which states that the value of taking action u in state x under the policy h is the expected sum of the immediate reward and of the discounted value achieved by h from the next state.

The V-function $V^h : X \to \mathbb{R}$ gives the return when starting from a particular state and following the policy h. It can be computed from the Q-function: $V^h(x) = Q^h(x, h(x))$. The optimal V-function is defined as $V^*(x) = \max_h V^h(x)$ and can be computed from the optimal Q-function: $V^*(x) = \max_u Q^*(x,u)$. The V-functions V^* and V^h satisfy Bellman equations similar to (5) and (6). The optimal policy can be computed from V^*, but the formula to do so is more complicated than (4): it requires a model of the MDP and computing expectations over the stochastic transitions. This hampers the computation of control policies from V-functions, which is a significant drawback in practice. Therefore, in the sequel we will prefer using Q-functions. The disadvantage of Q-functions is that they are more costly to represent, because in addition to x they also depend on u.

In *deterministic* problems, the transition probability function $\tilde{f}$ is replaced by a simpler transition function, $f : X \times U \to X$. This function is obtained from the stochastic dynamics by using a degenerate density $\tilde{f}(x,u,\cdot)$ that assigns all the probability mass to the state $f(x,u)$. The deterministic rewards are completely determined by the current state and action: $\rho(x,u) = \tilde{\rho}(x,u,f(x,u))$. All the formalism given in this section can be specialized to the deterministic case. For instance, the Bellman optimality equation for Q^* becomes

$$Q^*(x,u) = \rho(x,u) + \gamma \max_{u'} Q^*(f(x,u),u') \tag{7}$$

and the Bellman equation for Q^h becomes

$$Q^h(x,u) = \rho(x,u) + \gamma Q^h(f(x,u), h(f(x,u))) \tag{8}$$

2.2 *Exact Value Iteration*

Value iteration techniques use the Bellman optimality equation to iteratively compute an optimal value function, from which an optimal policy is then derived. As an illustrative example of a DP (model-based) value iteration algorithm, we describe Q-iteration. Let the set of all Q-functions be denoted by $\mathscr{Q}$. Define the Q-iteration mapping $T : \mathscr{Q} \to \mathscr{Q}$, which computes the right-hand side of the Bellman optimality equation (5) for an arbitrary Q-function:

$$[T(Q)](x,u) = \mathrm{E}_{x' \sim \tilde{f}(x,u,\cdot)} \left\{ \tilde{\rho}(x,u,x') + \gamma \max_{u'} Q(x',u') \right\} \tag{9}$$

In the deterministic case, the right-hand side of the deterministic Bellman optimality equation (7) should be used instead. It can be shown that T is a contraction with factor $\gamma < 1$ in the infinity norm, i.e., for any pair of functions Q and Q', it is true that

$\|T(Q) - T(Q')\|_\infty \le \gamma \|Q - Q'\|_\infty$. The Q-iteration algorithm starts from an arbitrary Q-function Q_0 and in each iteration ℓ updates it using

$$Q_{\ell+1} = T(Q_\ell) \tag{10}$$

Because T is a contraction, it has a unique fixed point, and from (7), this point is Q^*. This implies that Q-iteration asymptotically converges to Q^* as $\ell \to \infty$. A similar V-iteration algorithm can be given that computes the optimal V-function V^*.

RL (model-free) techniques like Q-learning [87] and Dyna [73] either learn a model or do not use an explicit model at all. For instance, Q-learning starts from an arbitrary initial Q-function Q_0 and updates it online, using observed transitions $(x_k, u_k, x_{k+1}, r_{k+1})$ [86, 87]. After each transition, the Q-function is updated with

$$Q_{k+1}(x_k, u_k) = Q_k(x_k, u_k) + \alpha_k [r_{k+1} + \gamma \max_{u'} Q_k(x_{k+1}, u') - Q_k(x_k, u_k)] \tag{11}$$

where $\alpha_k \in (0, 1]$ is the learning rate. The term between brackets is the temporal difference, i.e., the difference between the current estimate $Q_k(x_k, u_k)$ of the optimal Q-value of (x_k, u_k) and the updated estimate $r_{k+1} + \gamma \max_{u'} Q_k(x_{k+1}, u')$. This new estimate is actually a single sample of the expectation on the right-hand side of the Q-iteration mapping (9), applied to Q_k in the state–action pair (x_k, u_k). In this sample, $\tilde{f}(x_k, u_k, x')$ is replaced by the observed next state x_{k+1}, and $\tilde{\rho}(x_k, u_k, x')$ is replaced by the observed reward r_{k+1}. In the discrete-variable case, Q-learning asymptotically converges to Q^* as the number of transitions k approaches infinity, if $\sum_{k=0}^{\infty} \alpha_k^2$ is finite, $\sum_{k=0}^{\infty} \alpha_k$ is infinite, and if all the state–action pairs are (asymptotically) visited infinitely often [29, 87].

The third condition can be satisfied if, among other things, the controller has a nonzero probability of selecting any action in every encountered state; this is called exploration. The controller also has to exploit its current knowledge to obtain good performance, e.g., by selecting greedy actions in the current Q-function. A classical way to balance exploration with exploitation is the ε-greedy policy, which selects actions according to

$$u_k = \begin{cases} \arg\max_u Q(x_k, u) & \text{with probability } 1 - \varepsilon_k \\ \text{a uniformly random action in } U & \text{with probability } \varepsilon_k \end{cases} \tag{12}$$

where $\varepsilon_k \in (0, 1)$ is the exploration probability at step k. Usually, ε_k diminishes over time, so that asymptotically, as $Q_k \to Q^*$, the policy used also converges to a greedy, and therefore optimal, policy.

2.3 Exact Policy Iteration

Policy iteration techniques iteratively evaluate and improve policies [9, 11]. Consider a policy iteration algorithm that uses Q-functions. In every iteration ℓ, such an algorithm computes the Q-function Q^{h_ℓ} of the current policy h_ℓ; this step is called policy evaluation. Then, a new policy $h_{\ell+1}$ that is greedy in Q^{h_ℓ} is computed; this step is called policy improvement. Policy iteration algorithms can be either model based or model-free; and offline or online.

To implement policy evaluation, define analogously to (9) a *policy evaluation mapping* $T^h : \mathscr{Q} \to \mathscr{Q}$, which applies to any Q-function the right-hand side of the Bellman equation for Q^h. In the stochastic case, the right-hand side of (6) is used, leading to

$$[T^h(Q)](x,u) = \mathrm{E}_{x' \sim \tilde{f}(x,u,\cdot)} \left\{ \tilde{\rho}(x,u,x') + \gamma Q(x', h(x')) \right\} \tag{13}$$

whereas in the deterministic case, the right-hand side of (8) should be used instead. Like the Q-iteration mapping T, T^h is a contraction with a factor $\gamma < 1$ in the infinity norm. A model-based policy evaluation algorithm can be given that works similar to Q-iteration. This algorithm starts from an arbitrary Q-function Q_0^h and in each iteration τ updates it using[5]

$$Q_{\tau+1}^h = T^h(Q_\tau^h) \tag{14}$$

Because T^h is a contraction, this algorithm asymptotically converges to Q^h. Other ways to find Q^h include online sample-based techniques similar to Q-learning, and directly solving the linear system of equations provided by (6) or (8), which is possible when X and U are discrete and the cardinality of $X \times U$ is not very large [9].

Policy iteration starts with an arbitrary policy h_0. In each iteration ℓ, policy evaluation is used to obtain the Q-function Q^{h_ℓ} of the current policy. Then, an improved policy is computed which is greedy in Q^{h_ℓ}:

$$h_{\ell+1}(x) = \arg\max_u Q^{h_\ell}(x,u) \tag{15}$$

The Q-functions computed by policy iteration asymptotically converge to Q^* as $\ell \to \infty$. Simultaneously, the policies converge to h^*.

The main reason for which policy iteration algorithms are attractive is that the Bellman equation for Q^h is linear in the Q-values. This makes policy evaluation easier to solve than the Bellman optimality equation (5), which is highly nonlinear due to the maximization in the right-hand side. Moreover, in practice, offline policy iteration algorithms often converge in a small number of iterations [45, 74], possibly smaller than the number of iterations taken by offline value iteration algorithms. However, this does not necessarily mean that policy iteration is less computationally costly than value iteration. Even though policy evaluation is generally less costly than value iteration, *every single* policy iteration requires a complete policy evaluation.

Model-free variants of policy iteration can also be given. SARSA is an online model-free policy iteration algorithm proposed in [69] as an alternative to the value iteration-based Q-learning. SARSA starts with an arbitrary initial Q-function Q_0 and updates it using tuples $(x_k, u_k, x_{k+1}, u_{k+1}, r_{k+1})$, as follows:

5 A different iteration index τ is used for policy evaluation because policy evaluation runs in the inner loop of every policy iteration ℓ.

$$Q_{k+1}(x_k,u_k) = Q_k(x_k,u_k) + \alpha_k[r_{k+1} + \gamma Q_k(x_{k+1},u_{k+1}) - Q_k(x_k,u_k)] \tag{16}$$

In contrast to Q-learning (11), which uses the maximal Q-value in the next state to compute the temporal difference, SARSA uses the Q-value of the action actually taken in the next state. This means that SARSA performs online model-free policy evaluation. To select actions, a greedy policy is combined with exploration, using, e.g., the ε-greedy strategy (12). Using a greedy policy means that SARSA implicitly performs a policy improvement at every time step; hence, SARSA is a type of online policy iteration.

Actor–critic algorithms [76] also belong to the class of online policy iteration techniques; they will be presented in Sect. 5.4. The 'actor' is the policy and the 'critic' is the value function.

3 The Need for Approximation in Dynamic Programming and Reinforcement Learning

When the state and action spaces of the MDP contain a large or infinite number of elements, value functions and policies cannot be represented exactly. Instead, approximation must be used. Consider, e.g., the algorithms for exact value iteration of Sect. 2.2. They require to store distinct estimates of the return for every state (in the case of V-functions) or for every state–action pair (Q-functions). When some of the state variables have a very large or infinite number of possible values (e.g., they are continuous), exact storage is no longer possible. Large or continuous action spaces make the representation of Q-functions additionally challenging.

Approximation in DP/RL is not only a problem of representation. Consider, e.g., the Q-iteration algorithm of Sect. 2, which iteratively applies the Q-iteration mapping: $Q_{\ell+1} = T(Q_\ell)$. This mapping would have to be implemented as follows:

$$\textbf{for every } (x,u) \textbf{ do: } Q_{\ell+1}(x,u) = \mathrm{E}_{x' \sim \tilde{f}(x,u,\cdot)} \left\{ \tilde{\rho}(x,u,x') + \gamma \max_{u' \in U} Q_\ell(x',u') \right\} \tag{17}$$

When the state–action space contains an infinite number of elements, it is impossible to loop over all the state–action pairs in a finite time. Instead, an approximate update has to be used that only considers a finite number of state–action samples. Additionally, the expectation on the right-hand side of (17) cannot be computed exactly, but has to be estimated from a finite number of samples, using Monte Carlo methods. Note that, in many RL algorithms, the Monte Carlo approximation does not appear explicitly, but is performed implicitly while processing samples. Q-learning (11) is one such algorithm.

The maximization over the action variable in (17) must be solved for every sample used in the Monte Carlo estimation. In large or continuous action spaces, this maximization is a potentially difficult nonconcave optimization problem, which can only be solved approximately. To simplify this problem, many algorithms discretize the action space in a small number of values, compute the value function for all the discrete actions, and find the maximum among these values using enumeration.

Similar difficulties are encountered in policy iteration algorithms; there, the maximization problem has to be solved at the policy improvement step. Policy search algorithms also need to estimate returns using a finite number of samples, and must find the best policy in the class considered, which is a potentially difficult optimization problem. However, this problem only needs to be solved once, unlike the maximization over actions in value iteration and policy iteration, which must be solved for every sample considered. In this sense, policy search methods are less affected from the maximization difficulties than value iteration or policy iteration.

In deterministic MDPs, the Monte Carlo estimation is not needed, but sample-based updates and approximate maximization are still required.

4 Approximate Value Iteration

For value iteration in large or continuous-space MDPs, the value function has to be approximated. Linearly parameterized approximators make it easier to analyze the theoretical properties of the resulting DP/RL algorithms. Nonlinearly parameterized approximators like neural networks have better representation power than linear parametrizations; however, the resulting DP/RL algorithms are more difficult to analyze.

Consider for instance a linearly parameterized approximator for the Q-function. Such an approximator has n basis functions (BFs) $\phi_1, \ldots, \phi_n : X \times U \to \mathbb{R}$, and is parameterized by a vector[6] of n parameters $\theta \in \mathbb{R}^n$. Given a parameter vector θ, approximate Q-values are computed with

$$\widehat{Q}(x,u) = \sum_{l=1}^{n} \phi_l(x,u)\theta_l = \phi^{\mathrm{T}}(x,u)\theta \qquad (18)$$

where $\phi(x,u) = [\phi_1(x,u), \ldots, \phi_n(x,u)]^{\mathrm{T}}$. The parameter vector θ thus provides a compact (but approximate) representation of a Q-function. Examples of linearly parameterized approximators include crisp discretization [7, 80] (see Example 1), multilinear interpolation [14], Kuhn triangulation [55], and Gaussian radial BFs [51, 80] (see Example 2).

In this section, we describe algorithms for model-based and model-free approximate value iteration. Then, we describe convergence guarantees for approximate value iteration, and apply a representative algorithm to an example.

4.1 Approximate Model-Based Value Iteration

This section describes the approximate Q-iteration algorithm with a general parametric approximator, which is an extension of the exact Q-iteration algorithm in Sect. 2.2. Recall that exact Q-iteration starts from an arbitrary Q-function Q_0 and

[6] All the vectors used in this chapter are column vectors.

in each iteration ℓ updates the Q-function using $Q_{\ell+1} = T(Q_\ell)$, where T is the Q-iteration mapping (9).

Approximate Q-iteration parameterizes the Q-function using a parameter vector $\theta \in \mathbb{R}^n$. It requires two other mappings in addition to T. The *approximation* mapping $F : \mathbb{R}^n \to \mathscr{Q}$ produces an approximate Q-function $\widehat{Q} = F(\theta)$ for a given parameter vector θ. This Q-function is used as an input to the Q-iteration mapping T. The *projection* mapping $P : \mathscr{Q} \to \mathbb{R}^n$ computes a parameter vector θ such that $F(\theta)$ is as close as possible to a target Q-function Q, e.g., in a least-squares sense. Projection is used to obtain a new parameter vector from the output of the Q-iteration mapping. So, approximate Q-iteration starts with an arbitrary (e.g., identically 0) parameter vector θ_0, and updates this vector in every iteration ℓ using the composition of the mappings P, T, and F:

$$\theta_{\ell+1} = (P \circ T \circ F)(\theta_\ell) \tag{19}$$

Of course, the results of F and T cannot be fully computed and stored. Instead, $P \circ T \circ F$ can be implemented as a single entity, or sampled versions of the F and T mappings can be applied. Once a satisfactory parameter vector θ^* (ideally, a fixed point of the composite mapping $P \circ T \circ F$) has been found, the following policy can be used:

$$\widehat{h}^*(x) = \arg\max_u [F(\theta^*)](x,u) \tag{20}$$

Figure 2 illustrates approximate Q-iteration and the relations between the various mappings, parameter vectors, and Q-functions considered by the algorithm.

We use the notation $[F(\theta)](x,u)$ to refer to the Q-function $F(\theta)$ evaluated at the state–action pair (x,u). For instance, a linearly parameterized approximator (18) would lead to $[F(\theta)](x,u) = \phi^{\mathrm{T}}(x,u)\theta$. The notation $[P(Q)]_l$ refers to the lth component in the parameter vector $P(Q)$.

A similar formalism can be given for approximate V-iteration, which is more popular in the literature [19, 25–27, 55, 80]. Many results from the literature deal with the discretization of continuous-variable problems [19, 25, 27, 55]. Such dis-

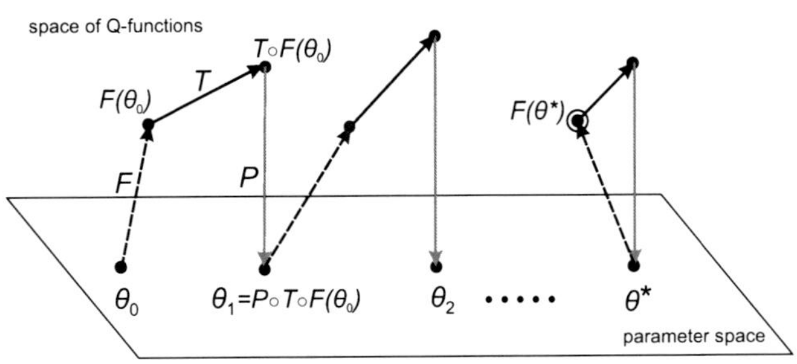

Fig. 2 A conceptual illustration of approximate Q-iteration. In every iteration, F is applied to the current parameter vector to obtain an approximate Q-function, which is then passed through T. The result of T is projected back onto the parameter space with P. Ideally, the algorithm converges to a fixed point θ^*, which leads back to itself when passed through $P \circ T \circ F$. The solution of approximate Q-iteration is the Q-function $F(\theta^*)$.

cretizations are not necessarily crisp, but can use interpolation procedures, which lead to linearly parameterized approximators of the form (18).

4.2 Approximate Model-Free Value Iteration

Approximate model-free value iteration has also been extensively studied [21, 23, 28, 30, 52, 60, 71, 77, 79]. The Q-learning algorithm is the most popular, and has been combined with a variety of approximators, which includes

- linearly parameterized approximators, encountered under several names such as interpolative representations [77] and soft state aggregation [71];
- fuzzy rule bases [23, 28, 30], which can also be linear in the parameters; and
- nonlinearly parameterized approximators such as neural networks [43] and self-organizing maps [79].

The most straightforward way to integrate approximation in Q-learning is by using gradient updates of the parameter vector [74]:

$$\theta_{k+1} = \theta_k + \alpha_k \left[r_{k+1} + \gamma \max_{u'} \widehat{Q}_k(x_{k+1}, u') - \widehat{Q}_k(x_k, u_k) \right] \frac{\partial}{\partial \theta_k} \widehat{Q}_k(x_k, u_k)$$

where the Q-function is parameterized by θ and the term in brackets is an approximation of the temporal difference (see again (11)). With linearly parameterized approximation (18), this update simplifies to

$$\theta_{k+1} = \theta_k + \alpha_k \left[r_{k+1} + \gamma \max_{u'} \left(\phi^{\mathrm{T}}(x_{k+1}, u')\theta_k \right) - \phi^{\mathrm{T}}(x_k, u_k)\theta_k \right] \phi(x_k, u_k)$$

Some algorithms for approximate model-free value iteration work offline and require a batch of samples collected in advance. A good example is fitted Q-iteration, which uses ensembles of regression trees (a nonparametric approximator) to represent the Q-function [21]. Fitted Q-iteration belongs to the class of approximate Q-iteration algorithms. It replaces the exact Q-iteration mapping T in (19) by an approximation derived from the available samples, and the projection mapping P by a process that derives a new ensemble of regression trees in every iteration, to best approximate the current Q-function. Neural-fitted Q-iteration is a similar algorithm, but it approximates the Q-function using neural networks instead of ensembles of regression trees [67].

4.3 Convergence and the Role of Nonexpansive Approximators

An important question in approximate DP/RL is whether the approximate solution computed by the algorithm converges, and, if it does converge, how far the convergence point is from the optimal solution. Convergence is important because it makes the algorithm more amenable to analysis and meaningful performance guarantees.

The convergence proofs for approximate value iteration often rely on contraction mapping arguments. Consider for instance approximate Q-iteration, given by (19). The Q-iteration mapping T is a contraction in the infinity norm with factor $\gamma < 1$, as already mentioned in Sect. 2.2. If the composite mapping $P \circ T \circ F$ of approximate Q-iteration is also a contraction, i.e., $\|(P \circ T \circ F)(\theta) - (P \circ T \circ F)(\theta')\|_\infty \leq \gamma' \|\theta - \theta'\|_\infty$ for all θ, θ' and for a $\gamma' < 1$, then approximate Q-iteration asymptotically converges to a unique fixed point, which we denote by θ^*.

One way to make $P \circ T \circ F$ a contraction is to ensure that F and P are nonexpansions, i.e., that $\|F(\theta) - F(\theta')\|_\infty \leq \|\theta - \theta'\|_\infty$ for all θ, θ' and that $\|P(Q) - P(Q')\|_\infty \leq \|Q - Q'\|_\infty$ for all Q, Q' [26]. In this case, the contraction factor of $P \circ T \circ F$ is the same as that of T: $\gamma' = \gamma < 1$. Under these conditions, as we will describe next, suboptimality bounds can be derived on the solution obtained.

Denote by $\mathscr{F}_{F \circ P} \subset \mathscr{Q}$ the set of fixed points of the composite mapping $F \circ P$ (this set is assumed nonempty). Define $\sigma_{\mathrm{QI}}{}^* = \min_{Q' \in \mathscr{F}_{F \circ P}} \|Q^* - Q'\|_\infty$, the minimum distance between Q^* and any fixed point of $F \circ P$.[7] This distance characterizes the representation power of the approximator; the better the representation power, the closer the nearest fixed point of $F \circ P$ will be to Q^*, and the smaller $\sigma_{\mathrm{QI}}{}^*$ will be. The convergence point θ^* of approximate Q-iteration satisfies the following suboptimality bounds [26, 80]:

$$\|Q^* - F(\theta^*)\|_\infty \leq \frac{2\sigma_{\mathrm{QI}}{}^*}{1-\gamma} \tag{21}$$

$$\|Q^* - Q^{\widehat{h}^*}\|_\infty \leq \frac{4\gamma\sigma_{\mathrm{QI}}{}^*}{(1-\gamma)^2} \tag{22}$$

where $Q^{\widehat{h}^*}$ is the Q-function of a policy $\widehat{h}^*$ that is greedy in $F(\theta^*)$ (20). Equation (21) gives the suboptimality bound of the approximately optimal Q-function, whereas (22) gives the suboptimality bound of the resulting approximately optimal policy. The latter may be more relevant in practice. The following relationship between the policy suboptimality and the Q-function suboptimality was used to obtain (22), and is also valid in general:

$$\|Q^* - Q^h\|_\infty \leq \frac{2\gamma}{(1-\gamma)}\|Q^* - Q\|_\infty \tag{23}$$

where the policy h is greedy in the (arbitrary) Q-function Q.

To take advantage of these theoretical guarantees, P and F should be nonexpansions. When F is linearly parameterized (18), it is fairly easy to ensure its nonexpansiveness by normalizing the BFs ϕ_l, so that for every x and u, we have $\sum_{l=1}^n \phi_l(x,u) = 1$. Ensuring that P is nonexpansive is more difficult. For instance, the most natural choice for P is a least-squares projection:

[7] If the minimum does not exist, then $\sigma_{\mathrm{QI}}{}^*$ should be taken as small as possible so that there still exists a $Q' \in \mathscr{F}_{F \circ P}$ with $\|Q^* - Q'\|_\infty \leq \sigma_{\mathrm{QI}}{}^*$.

$$P(Q) = \arg\min_{\theta} \sum_{l_s=1}^{n_s} |Q(x_{l_s}, u_{l_s}) - [F(\theta)](x_{l_s}, u_{l_s})|^2 \tag{24}$$

for some set of samples $\{(x_{l_s}, u_{l_s}) \mid l_s = 1, \ldots, n_s\}$, where ties in the 'arg min' can be broken arbitrarily. Unfortunately, such a projection can in general be an expansion, and examples of divergence when using it have been given [80, 88]. One way to make P nonexpansive is to choose exactly $n_s = n$ samples (for instance, the centers of the BFs), and require that $\phi_{l_s}(x_{l_s}, u_{l_s}) = 1$ and $\phi_{l_s'}(x_{l_s}, u_{l_s}) = 0$ for $l_s \neq l_s'$. Then, the projection mapping (24) simplifies to an assignment that associates each parameter with the Q-value of the corresponding sample:

$$[P(Q)]_{l_s} = Q(x_{l_s}, u_{l_s}) \tag{25}$$

This mapping is clearly nonexpansive. More general (but still restrictive) conditions on the BFs under which convergence and near optimality are guaranteed are given in [80].

In the area of approximate model-free value iteration (which belongs to approximate RL), many approaches are heuristic and do not guarantee convergence [23, 28, 30, 52, 79]. Those that do guarantee convergence use linearly parameterized approximators [21, 60, 71, 77], and often employ conditions related to the nonexpansiveness properties above, e.g., for Q-learning [71, 77], or for sample-based batch V-iteration [60].

Another important theoretical property of algorithms for approximate DP and RL is consistency. In model-based value iteration, and more generally in DP, an algorithm is consistent if the approximate value function converges to the optimal one as the approximation accuracy increases [19, 25, 70]. In model-free value iteration, and more generally in RL, consistency is usually understood as the convergence to a well-defined solution as the number of samples increases. The stronger result of convergence to an optimal solution as the approximation accuracy also increases is proven in [60, 77].

Example 1 (Grid Q-iteration for a DC motor). Consider a second-order discrete-time model of a DC motor:

$$\begin{aligned} x_{k+1} &= f(x_k, u_k) = Ax_k + Bu_k \\ A &= \begin{bmatrix} 1 & 0.0049 \\ 0 & 0.9540 \end{bmatrix}, \quad B = \begin{bmatrix} 0.0021 \\ 0.8505 \end{bmatrix} \end{aligned} \tag{26}$$

This model is obtained by discretizing a continuous-time model of the DC motor, which was determined by first-principles modeling of the real DC motor. The discretization is performed with the zero-order hold method, using a sampling time of $T_s = 0.005$ s. Using saturation, the position $x_{1,k} = \alpha$ is bounded to $[-\pi, \pi]$ rad, the velocity $x_{2,k} = \dot{\alpha}$ to $[-16\pi, 16\pi]$ rad/s. The control input u_k is constrained to $[-10, 10]$ V. A quadratic regulation problem has to be solved, which is described by the following reward function:

$$r_{k+1} = \rho(x_k, u_k) = -x_k^{\mathrm{T}} Q_{\mathrm{rew}} x_k - R_{\mathrm{rew}} u_k^2$$
$$Q_{\mathrm{rew}} = \begin{bmatrix} 5 & 0 \\ 0 & 0.01 \end{bmatrix}, \quad R_{\mathrm{rew}} = 0.01 \tag{27}$$

With this reward function, a good policy will drive the state (close) to 0 while also minimizing the magnitude of the states along the trajectory and the control effort. The discount factor is chosen $\gamma = 0.95$, which is sufficient to produce a good control policy. Figure 3 presents a near-optimal solution to this problem, computed using the convergent and consistent fuzzy Q-iteration algorithm [14] with an accurate approximator.

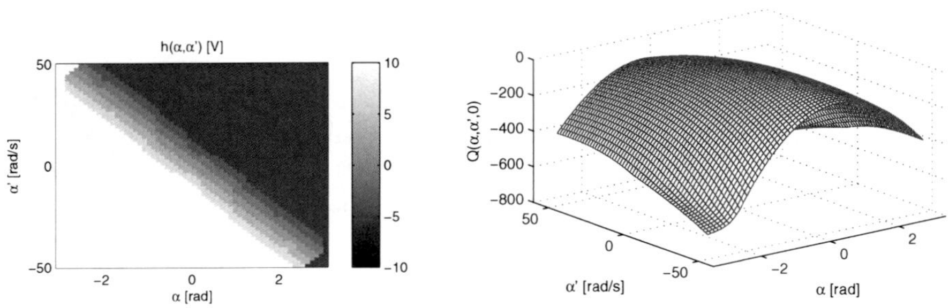

(a) A near-optimal policy. (b) Slice through a near-optimal Q-function, for $u = 0$.

Fig. 3 A near-optimal solution for the DC motor

As an example of approximate value iteration, we develop a Q-iteration algorithm that relies on a gridding of the state space and on a discretization of the action space into a set of finitely many values, $U_{\mathrm{d}} = \{u_1, \ldots, u_M\}$. For this problem, three discrete actions are sufficient to find an acceptable stabilizing policy, $U_{\mathrm{d}} = \{-10, 0, 10\}$. The state space is gridded (partitioned) into a set of N disjoint rectangles. Let X_i be the surface of the ith rectangle in this partition, with $i = 1, \ldots, N$. The Q-function approximator represents distinct slices through the Q-function, one for each of the discrete actions. For a given action, the approximator assigns the same Q-values for all the states in X_i. This corresponds to a linearly parameterized approximator with binary-valued (0 or 1) BFs over the state-discrete action space $X \times U_{\mathrm{d}}$:

$$\phi_{[i,j]}(x,u) = \begin{cases} 1 & \text{if } x \in X_i \text{ and } u = u_j \\ 0 & \text{otherwise} \end{cases} \tag{28}$$

where $[i,j]$ denotes the single-dimensional index corresponding to i and j, and can be computed with $[i,j] = i + (j-1)N$. Note that because the rectangles are disjoint, exactly one BF is active at any point of $X \times U_{\mathrm{d}}$.

To derive the projection mapping P, the least-squares projection (24) is used, taking as state–action samples the cross product of the sets $\{x_1, \ldots, x_N\}$ and U_{d}, where x_i denotes the center of the ith rectangle X_i. These samples satisfy the conditions to simplify P to an assignment of the form (25)

$$[P(Q)]_{[i,j]} = Q(x_i, u_j) \tag{29}$$

Using the linearly parameterized approximator (18) with the BFs (28) and the projection (29) yields the grid Q-iteration algorithm. Because F and P are nonexpansions, this algorithm is convergent.

To apply grid Q-iteration to the DC motor problem, two different grids over the state space are used: a coarse grid, with 20 equidistant bins on each axis (leading to $20^2 = 400$ rectangles), and a fine grid, with 100 equidistant bins on each axis (leading to $100^2 = 10{,}000$ rectangles). The algorithm is considered convergent when the maximum amount by which any parameter changes between two consecutive iterations does not exceed $\varepsilon_{\text{QI}} = 0.001$. For the coarse grid, convergence occurs after 160 iterations, and for the fine grid, after 118. This shows that the number of iterations to convergence is not monotonously increasing with the number of parameters. The finer grid may help convergence because it can achieve a better accuracy. Representative state-dependent slices through the resulting Q-functions (obtained by setting the action argument u to 0), together with the corresponding policies computed with (20), are given in Fig. 4. The accuracy in representing the Q-function is worse for the coarse grid, in Fig. 4(b), than for the fine grid, in Fig. 4(d). The structure of the policy is more clearly visible in Fig. 4(c). Axis-oriented artifacts appear for both grid sizes, due to the limitations of the

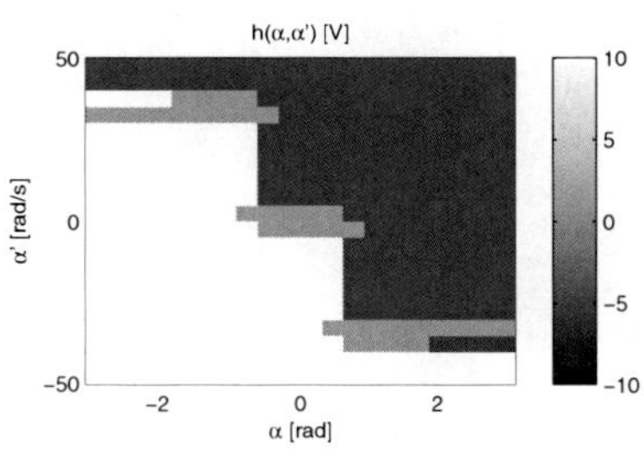

(a) Coarse-grid policy.

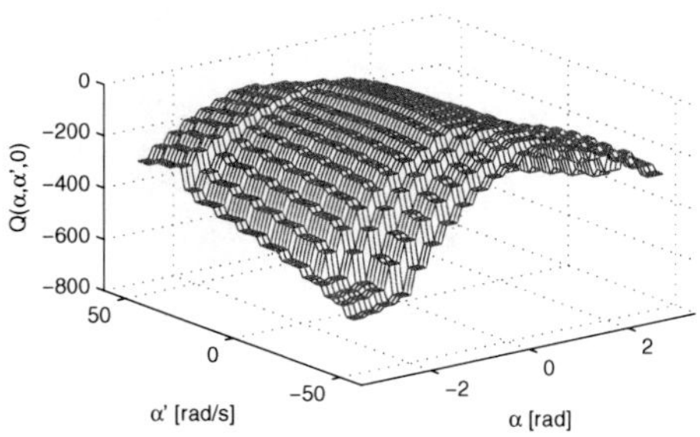

(b) Slice through coarse-grid Q-function, for $u = 0$.

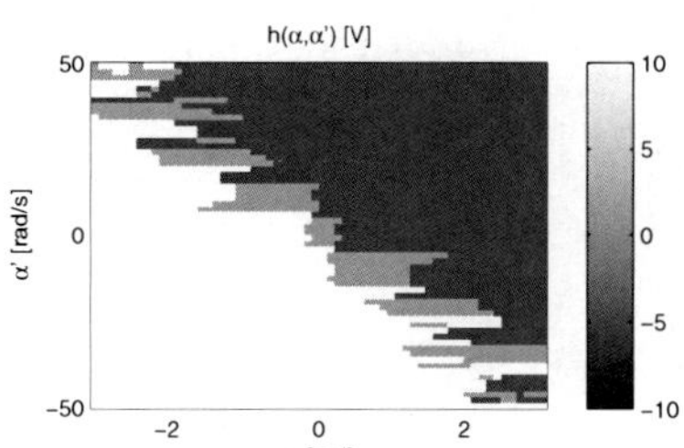

(c) Fine-grid policy.

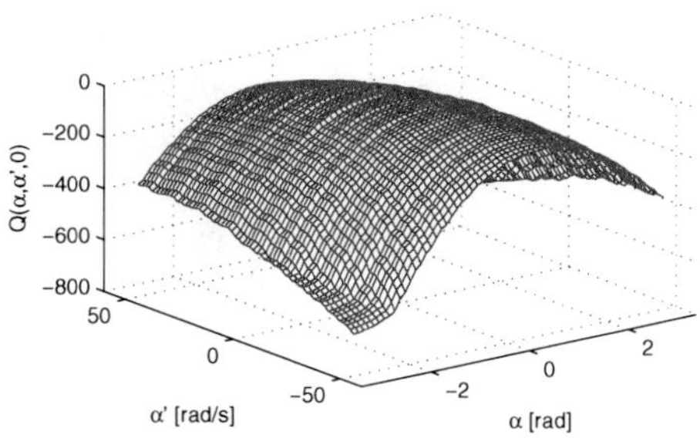

(d) Slice through fine-grid Q-function, for $u = 0$.

Fig. 4 Grid Q-iteration solutions for the DC motor

chosen approximator. For instance, the piecewise-constant nature of the approximator is clearly visible in Fig. 4(b).

5 Approximate Policy Iteration

Recall that policy iteration techniques compute in each iteration the value function of the current policy. Then, they compute a new, improved policy, which is greedy in the current value function, and repeat the cycle (see Sect. 2.3). Approximating the value function is always necessary to apply policy iteration in large and continuous spaces. Sometimes, an explicit representation of the policy can be avoided by computing improved actions on demand from the current value function. Alternatively, the policy can be represented explicitly, in which case policy approximation is generally required. In this case, solving a classical supervised learning problem is necessary to perform policy improvement.

Like in approximate value iteration, the convergence of approximate policy evaluation can be guaranteed more easily with linearly parameterized approximators of the value function. Nonlinearly parameterized approximators, especially neural networks, are also used often in actor–critic algorithms, an important subclass of approximate policy iteration.

Policy evaluation algorithms are discussed first, followed by policy improvement and the resulting policy iteration algorithms. Theoretical guarantees on the solutions obtained are given and a representative algorithm is applied to the DC motor problem of Example 1. Finally, actor–critic techniques are presented.

5.1 Approximate Policy Evaluation

Some of the most powerful algorithms for approximate policy evaluation combine linearly parameterized approximators of the value function with model-free least-squares techniques to compute the parameters [9]. We therefore focus on model-free least-squares policy evaluation in this section. In particular, we discuss the least-squares temporal difference for Q-functions (LSTD-Q) [40] and the least-squares policy evaluation for Q-functions (LSPE-Q) [57]. Model-based approximate policy evaluation can be derived along the same lines as model-based approximate value iteration, see Sect. 4.1.

Assume for now that X and U have a finite number of elements. LSTD-Q and LSPE-Q solve a projected form of the Bellman equation (6): [8]

$$\widehat{Q}^h = P^w(T^h(\widehat{Q}^h)) \tag{30}$$

where P^w performs a weighted least-squares projection onto the space of representable Q-functions, i.e., the space $\left\{\phi^\mathrm{T}(x,u)\theta \mid \theta \in \mathbb{R}^n\right\}$ spanned by the BFs. Here, $w : X \times U \to [0,1]$ is the weight function, which is always interpreted as

[8] A multistep version of this equation can also be given. In this chapter, we only consider the single-step case.

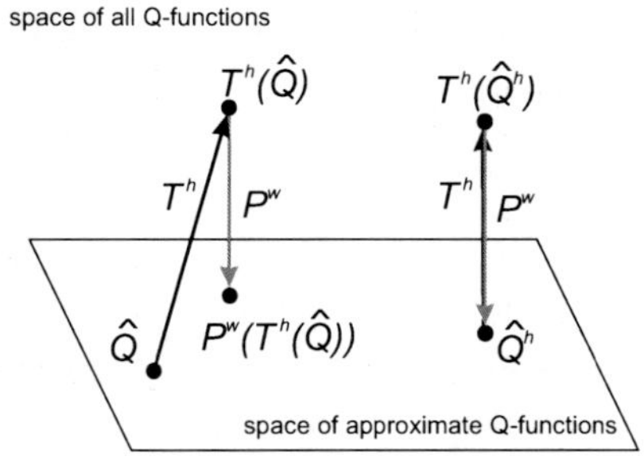

Fig. 5 A conceptual illustration of the projected Bellman equation. Applying T^h and then P^w to an ordinary approximate Q-function leads to a different point in the space of approximate Q-functions (left). In contrast, applying T^h and then P^w to the fixed point $\widehat{Q}^h$ of the projected Bellman equation leads back into the same point (right).

a probability distribution over the state–action space and must therefore satisfy $\sum_{x,u} w(x,u) = 1$. Figure 5 illustrates the projected Bellman equation.

The projection P^w is defined by

$$[P^w(Q)](x,u) = \phi^{\mathrm{T}}(x,u)\theta^{\ddagger}, \text{ where}$$
$$\theta^{\ddagger} = \arg\min_{\theta} \sum_{(x,u)\in X\times U} w(x,u) \left|\phi^{\mathrm{T}}(x,u)\theta - Q(x,u)\right|^2$$

The function w controls the distribution of the error between a Q-function and its projection, and therefore indirectly controls the accuracy of the solution $\widehat{Q}^h$ of the projected Bellman equation, via (30).

By writing the projected Bellman equation (30) in a matrix form, it can eventually be transformed into a linear equation in the parameter vector (see [9, 40] for details):

$$\Gamma\theta^h = \gamma\Lambda\theta^h + z \tag{31}$$

where $\Gamma, \Lambda \in \mathbb{R}^{n\times n}$ and $z \in \mathbb{R}^n$. The original high-dimensional Bellman equation (6) has thus been replaced by the low-dimensional linear system (31). A solution θ^h of this system can be used to find an approximate Q-function with (18).

The matrices Γ, Λ and the vector z can be estimated from transition samples. Consider a set of samples $\{(x_{l_s}, u_{l_s}, x'_{l_s} \sim f(x_{l_s}, u_{l_s}, \cdot), r_{l_s} = \rho(x_{l_s}, u_{l_s}, x'_{l_s})) \,|\, l_s = 1, \ldots, n_s\}$ constructed by drawing state–action samples (x,u) and then computing corresponding next states and rewards. The probability distribution of the state–action samples is given by the weight function w. The estimates of Γ, Λ, and z are updated with

$$\begin{aligned}
&\Gamma_0 = 0, \quad \Lambda_0 = 0, \quad z_0 = 0\\
&\Gamma_{l_s} = \Gamma_{l_s-1} + \phi(x_{l_s}, u_{l_s})\phi^{\mathrm{T}}(x_{l_s}, u_{l_s})\\
&\Lambda_{l_s} = \Lambda_{l_s-1} + \phi(x_{l_s}, u_{l_s})\phi^{\mathrm{T}}(x'_{l_s}, h(x'_{l_s}))\\
&z_{l_s} = z_{l_s-1} + \phi(x_{l_s}, u_{l_s})r_{l_s}
\end{aligned} \tag{32}$$

LSTD-Q processes the entire batch of samples using (32) and then solves the equation

$$\frac{1}{n_s}\Gamma_{n_s}\widehat{\theta}^h = \gamma\frac{1}{n_s}\Lambda_{n_s}\widehat{\theta}^h + \frac{1}{n_s}z_{n_s} \tag{33}$$

to find an approximate parameter vector $\widehat{\theta}^h$. Asymptotically, as $n_s \to \infty$, it is true that $\frac{1}{n_s}\Gamma_{n_s} \to \Gamma$, $\frac{1}{n_s}\Lambda_{n_s} \to \Lambda$, and $\frac{1}{n_s}z_{n_s} \to z$. Therefore, $\widehat{\theta}^h \to \theta^h$ when $n_s \to \infty$. The parameter vector $\widehat{\theta}^h$ obtained by LSTD-Q gives an approximate Q-function via (18), which can then be used to perform policy improvement. Note that the division by n_s, although not necessary from a formal point of view, helps to increase the numerical stability of the algorithm.

An alternative to LSTD-Q is LSPE-Q, which starts with an arbitrary initial parameter vector θ_0 and updates it with

$$\begin{aligned} \theta_{l_s} &= \theta_{l_s-1} + \alpha(\theta^{\ddagger}_{l_s} - \theta_{l_s-1}), \text{ where} \\ \frac{1}{l_s}\Gamma_{l_s}\theta^{\ddagger}_{l_s} &= \gamma\frac{1}{l_s}\Lambda_{l_s}\theta_{l_s-1} + \frac{1}{l_s}z_{l_s} \end{aligned} \tag{34}$$

and where α is a step size parameter. To ensure its invertibility, Γ can be initialized to a small multiple of the identity matrix.

The linear systems in (33) and (34) can be solved in several ways, e.g., (i) by matrix inversion, (ii) by Gaussian elimination, or (iii) by incrementally computing the inverse with the Sherman–Morrison formula. Although for the derivation above it was assumed that X and U are countable, the updates (32), together with LSTD-Q and LSPE-Q, can be applied also in uncountable (e.g., continuous) state–action spaces. Note also that when the BF vector $\phi(x,u)$ is sparse,[9] the computational efficiency of the updates (32) can be improved by exploiting this sparsity.

To guarantee the asymptotical convergence of LSPE-Q to θ^h, the weight (probability of being sampled) of each state–action pair, $w(x,u)$, must be identical to the steady-state probability of this pair along an infinitely long trajectory generated with the policy h [9]. In contrast, LSTD-Q (33) may have meaningful solutions for many weight functions w, which can make it more robust in practice. An advantage of LSPE-Q over LSTD-Q stems from the incremental nature of LSPE-Q, which means it can benefit from a good initial value of the parameters.

Analogous least-squares algorithms can be given to compute linearly parameterized approximate V-functions [9]. Recall, however that, as explained in Sect. 2.1, when V-functions are used, it is more difficult to compute greedy policies, and therefore to perform policy improvements.

Gradient-based versions of policy evaluation can also be given, using linearly parameterized approximators [72, 81] or nonlinearly parameterized approximators such as neural networks [1, 20]. When combined with linearly parameterized

[9] The BF vector is sparse, e.g., when the discrete-action approximator described in the upcoming Example 2 is used. This is because the BF vector contains zeros for all the discrete actions that are different from the current discrete action.

approximators, gradient-based algorithms usually require more samples than least-squares algorithms to achieve a similar accuracy [36, 91].

Note that the only requirement imposed on the approximator by the convergence guarantees of approximate policy evaluation is its linearity in the parameters. In contrast, approximate value iteration imposes additional requirements to ensure that the approximate value iteration mapping is a contraction (Sect. 4.3).

5.2 *Policy Improvement: Approximate Policy Iteration*

To obtain a policy iteration algorithm, a method to perform policy improvement is required in addition to approximate policy evaluation. Consider first the case in which the policy is not explicitly represented. Instead, greedy actions are computed on demand from the value function, for every state where a control action is required, using, e.g., in the case of Q-functions:

$$h_{\ell+1}(x) = \arg\max_{u} \widehat{Q}^{h_\ell}(x,u) \tag{35}$$

where ℓ is the iteration index. The policy is then implicitly defined by the value function. If only a small, discrete set of actions is considered, the maximization in the policy improvement step can be solved by enumeration. In this case, policy improvement is exact. For instance, the algorithm obtained by combining exact policy improvement with policy evaluation by LSTD-Q is least-squares policy iteration (LSPI) [39, 40].

Policies can also be approximated, e.g., using a parametric approximator with a parameter vector $\vartheta \in \mathbb{R}^{\mathscr{N}}$. For instance, a linearly parameterized policy approximator uses a set of state-dependent BFs $\varphi_1, \ldots, \varphi_{\mathscr{N}} : X \to \mathbb{R}$ and approximates the policy with[10]

$$\widehat{h}(x) = \sum_{i=1}^{\mathscr{N}} \varphi_i(x)\vartheta_i = \varphi^{\mathrm{T}}(x)\vartheta \tag{36}$$

where $\varphi(x) = [\varphi_1(x), \ldots, \varphi_{\mathscr{N}}(x)]^{\mathrm{T}}$. For simplicity, the parametrization (36) is only given for scalar actions, but it can easily be extended to the case of multiple action variables. For this policy parametrization, approximate policy improvement can be performed by solving the linear least-squares problem

$$\vartheta_{\ell+1} = \arg\min_{\vartheta} \sum_{i_s=1}^{\mathscr{N}_s} \left\| \varphi^{\mathrm{T}}(x_{i_s})\vartheta - \arg\max_{u} \phi^{\mathrm{T}}(x_{i_s},u)\theta_\ell \right\|_2^2 \tag{37}$$

[10] Calligraphic notation is used to differentiate variables related to policy approximation from variables related to value function approximation. So, the policy parameter is ϑ and the policy BFs are denoted by φ, whereas the value function parameter is θ and the value function BFs are denoted by ϕ. Furthermore, the number of policy parameters and BFs is $\mathscr{N}$, and the number of samples for policy approximation is $\mathscr{N}_s$.

to find an improved policy parameter vector $\vartheta_{\ell+1}$, where $\{x_1,\dots,x_{\mathcal{N}_s}\}$ is a set of samples for policy improvement. In this formula, $\arg\max_u \phi^{\mathrm{T}}(x_{i_s},u)\theta_\ell = \arg\max_u \widehat{Q}^{h_\ell}(x_{i_s},u)$ is an improved greedy action for the sample x_{i_s}; notice that the policy $\widehat{h}_\ell$ is now also an approximation.

In sample-based policy iteration, instead of waiting with policy improvement until a large number of samples have been processed and an accurate approximation of the Q-function for the current policy has been obtained, policy improvements can also be performed after a small number of samples. Such a variant is sometimes called *optimistic* policy iteration [9, 11]. In the extreme, fully optimistic case, a policy that is greedy in the current value function is applied at every step. If the policy is only improved once every several steps, the method is partially optimistic. One instance in which optimistic updates are useful is when approximate policy iteration is applied online, since in that case the policy should be improved often. Optimistic variants of approximate policy iteration can be derived, e.g., using gradient-based policy evaluation [20] and least-squares policy evaluation [31, 32]. For instance, approximate SARSA is in fact a type of optimistic policy iteration [33] (see also Sect. 2.3).

Instead of using the Bellman equation to compute the value function of a policy, this value function can also be estimated by Monte Carlo simulations. This is the approach taken in [41], where Q-functions obtained by Monte Carlo policy evaluation are used to obtain improved policies represented as support vector classifiers.

5.3 Theoretical Guarantees

As long as the policy evaluation and improvement errors are bounded, approximate PI algorithms eventually produce policies with a bounded suboptimality. We formalize these convergence results, which apply to general (possibly nonlinearly parameterized) approximators.

Consider the general case in which both the value functions and the policies are approximated for *nonoptimistic* policy iteration. Consider also the case in which Q-functions are used. Assume that the error in every policy evaluation step is bounded by ε_Q:

$$\|\widehat{Q}^{\widehat{h}_\ell} - Q^{\widehat{h}_\ell}\|_\infty \le \varepsilon_Q, \quad \text{for any } \ell \ge 0$$

and the error in every policy improvement step is bounded by ε_h, in the following sense:

$$\|T^{\widehat{h}_{\ell+1}}(\widehat{Q}^{\widehat{h}_\ell}) - T(\widehat{Q}^{\widehat{h}_\ell})\|_\infty \le \varepsilon_h, \quad \text{for any } \ell \ge 0$$

where $T^{\widehat{h}_{\ell+1}}$ is the policy evaluation mapping for the improved (approximate) policy, and T is the Q-iteration mapping (9). Then, approximate policy iteration eventually produces policies with performances that lie within a bounded distance from the optimal performance [40]:

$$\limsup_{\ell\to\infty} \left\| \widehat{Q}^{\widehat{h}_\ell} - Q^* \right\|_\infty \le \frac{\varepsilon_h + 2\gamma\varepsilon_Q}{(1-\gamma)^2} \tag{38}$$

For an algorithm that performs exact policy improvements, such as LSPI, $\varepsilon_h = 0$ and the bound is tightened to

$$\limsup_{\ell \to \infty} \|\widehat{Q}^{h_\ell} - Q^*\|_\infty \leq \frac{2\gamma\varepsilon_Q}{(1-\gamma)^2} \tag{39}$$

where $\|\widehat{Q}^{h_\ell} - Q^{h_\ell}\|_\infty \leq \varepsilon_Q$, for any $\ell \geq 0$. Note that computing ε_Q (and, when approximate policies are used, computing ε_h) may be difficult in practice, and the existence of these bounds may require additional assumptions on the MDP.

These guarantees do not imply the convergence to a fixed policy. For instance, both the value function and policy parameters might converge to limit cycles, so that every point on the cycle yields a policy that satisfies the bound. Similarly, when exact policy improvements are used, the value function parameter may oscillate, implicitly leading to an oscillating policy. This is a disadvantage with respect to approximate value iteration, which can be guaranteed to converge monotonically to its fixed point Sect. 4.3.

Similar results hold when V-functions are used instead of Q-functions [11].

Optimistic policy iteration improves the policy before an accurate value function is available. Because the policy evaluation error can be large, the performance guarantees given above are not useful in the optimistic case. The behavior of optimistic policy iteration has not been properly understood yet, and can be very complicated. It can, e.g., exhibit a phenomenon called chattering, whereby the value function converges to a stationary function, while the policy sequence oscillates, because the limit of the value function parameter corresponds to multiple policies [9, 11].

Example 2 (LSPI for the DC motor). In this example, LSPI will be applied to the DC motor problem of Example 1. The action space is again discretized into the set $U_\mathrm{d} = \{-10, 0, 10\}$, which contains $M = 3$ actions. Only these discrete actions are allowed into the set of samples. A number of N normalized Gaussian radial basis functions (RBFs) $\bar{\phi}_i : X \to \mathbb{R}$, $i = 1, \ldots, N$ are used to approximate over the state space. The RBFs are defined as follows:

$$\bar{\phi}_i(x) = \frac{\phi_i'(x)}{\sum_{i'=1}^{N} \phi_{i'}'(x)}, \qquad \phi_i'(x) = \exp\left[-\frac{(x_1 - c_{i,1})^2}{b_{i,1}^2} - \frac{(x_2 - c_{i,2})^2}{b_{i,2}^2}\right] \tag{40}$$

where ϕ_i' are (nonnormalized) Gaussian axis-parallel RBFs, $(c_{i,1}, c_{i,2})$ is the center of the ith RBF, and $(b_{i,1}, b_{i,2})$ is its radius. The centers of the RBFs are arranged on an equidistant 9×9 grid in the state space. The radii of the RBFs along each dimension are taken identical to the distance along that dimension between two adjacent RBFs; this yields a smooth interpolation of the Q-function over the state space. The RBFs are replicated for every discrete action, and to compute the state-discrete action BFs, all the RBFs that do not correspond to the current discrete action are taken equal to 0. Approximate Q-values can then be computed with $\widehat{Q}(x, u_j) = \phi^\mathrm{T}(x, u_j)\,\theta$ for the state–action BF vector

$$\phi(x,u_j) = [\underbrace{0,\ldots,0}_{u_1},\ldots,0,\underbrace{\bar{\phi}_1(x),\ldots,\bar{\phi}_N(x)}_{u_j},0,\ldots,\underbrace{0,\ldots,0}_{u_M}]^{\mathrm{T}} \in \mathbb{R}^{NM}$$

and a parameter vector $\theta \in \mathbb{R}^n$ with $n = NM = 3N$.

First, LSPI with exact policy improvements is applied, starting from an initial policy h_0 that is identically equal to -10 throughout the state space. The same set of $n_s = 7500$ samples is used in every LSTD-Q policy evaluation. The samples are random, uniformly distributed over the state-discrete action space $X \times U_d$. To illustrate the results of LSTD-Q, Fig. 6 presents the first improved policy found by LSPI, h_1, and its approximate Q-function, computed with LSTD-Q.

The complete LSPI algorithm converged in 11 iterations. Figure 7 shows the resulting policy and Q-function, which are good approximations of the near-optimal solution in Fig. 3. Compared to the results of grid Q-iteration in Fig. 4, LSPI needed fewer BFs (9×9 rather than 400 or 10,000) and was able to find a better approximation of the policy. This is mainly because the Q-function is largely smooth (see Fig. 3(b)), which means it can be represented well using the wide RBFs considered. In contrast, the grid BFs give a discontinuous approximate Q-function, which is less appropriate for this problem. Although certain types of continuous BFs can

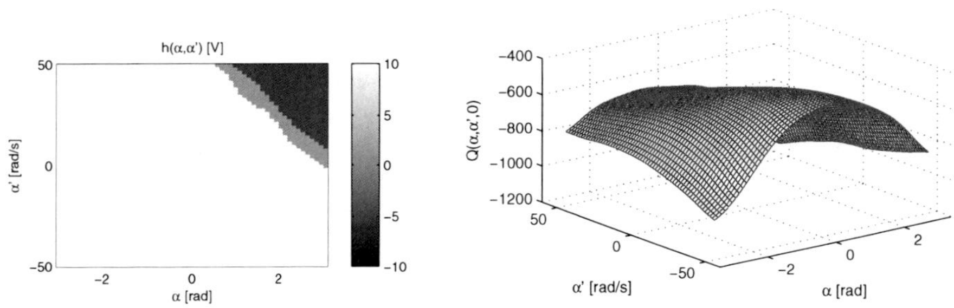

(a) Policy. (b) Slice through the Q-function, for $u = 0$.

Fig. 6 An early policy and its approximate Q-function, for LSPI with exact policy improvements

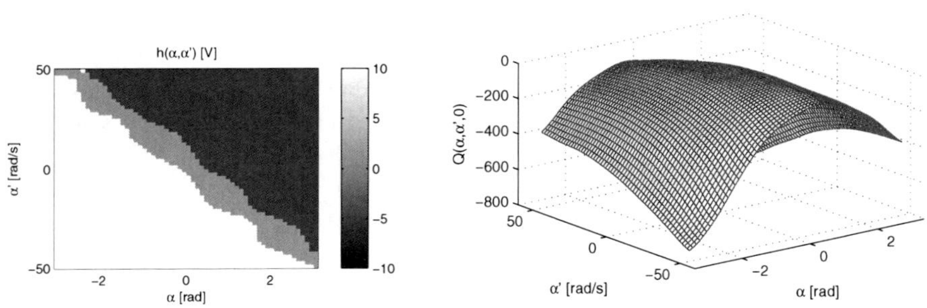

(a) Policy. (b) Slice through the Q-function, for $u = 0$.

Fig. 7 Results of LSPI with exact policy improvements on the DC motor be used with Q-iteration, using wide RBFs such as these is unfortunately not possible, because they do not satisfy the assumptions for convergence, and indeed lead to divergence when they are too wide. A disadvantage of these wide RBFs is that they fail to properly identify the policy nonlinearities in the top-left and bottom-right corners of Fig. 3(a), and the corresponding changes in the Q-function.

Another observation is that LSPI converges in a significantly fewer iterations than grid Q-iteration did in Example 1 (namely, 11 iterations for LSPI, instead of the 160 iterations taken by grid Q-iteration with the coarse grid, and of the 118 iterations taken with the fine grid). Such a convergence rate advantage of policy iteration over value iteration is often observed in practice. Note, however, that while LSPI did converge faster, it was actually more computationally intensive than grid Q-iteration: it required approximately 105 s to run, whereas grid Q-iteration only required 0.5 s for the coarse grid and 6 s for the fine grid.[11] This is mainly because the cost of policy evaluation with LSTD-Q is at least quadratic in the number $n = NM$ of state–action BFs (see the updates (32)). In contrast, the cost of every grid Q-iteration is only $O(n \log(N))$, when binary search is used to locate the position of a state on the grid.

Next, LSPI with *approximate policies* and approximate policy improvements is applied. The policy approximator is (36) and uses the same RBFs as the Q-function approximator ($\varphi_i = \bar{\phi}_i$). As before, $N_s = 7500$ samples are used for policy evaluation. Note that the approximate policy produces continuous actions, which must be quantized (into discrete actions belonging to U_d) before performing policy evaluation, because the Q-function approximator only works for discrete actions. Approximate policy improvement is performed with (37), using a set of $\mathcal{N}_s = 2500$ random, uniformly distributed state samples. The same set is used in every iteration.

In this experiment, both the Q-functions and the policies *oscillate* in the steady state of the algorithm, with a period of 2 iterations. The execution time until the oscillation was detected was 104 s. The differences between the 2 distinct policies and Q-functions on the limit cycle are too small to be noticed in a figure. Instead, Fig. 8 shows the evolution of the policy parameter that changes the most in steady state.

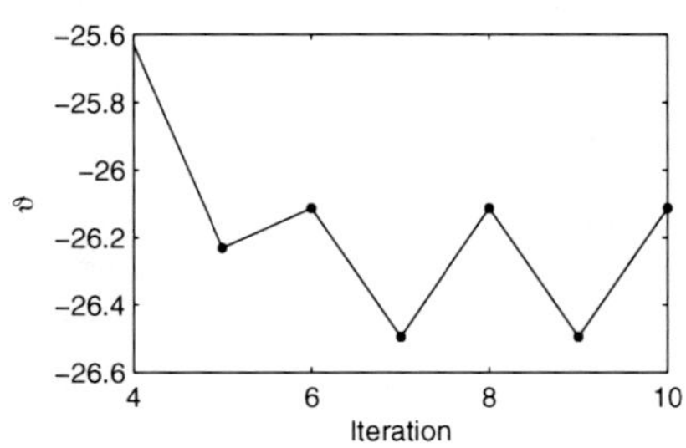

Fig. 8 The variation of one of the policy parameters, starting with the fourth iteration and until the oscillation was detected

[11] For all the experiments in this chapter, the algorithms are run in MATLAB 7, on a PC with an Intel T2400 CPU and 2 GB of RAM.

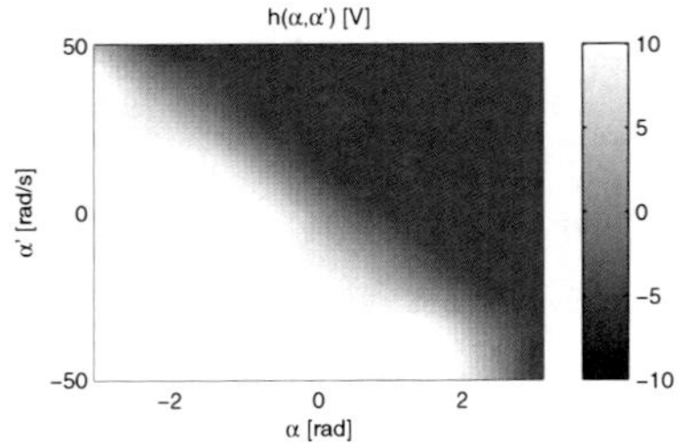

(a) Policy.

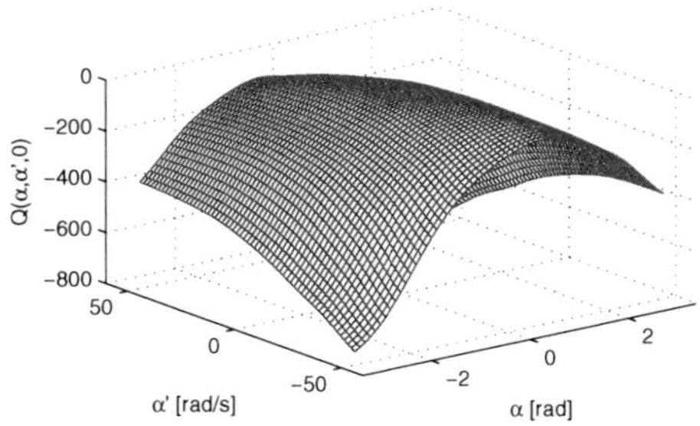

(b) Slice through the Q-function, for $u = 0$.

Fig. 9 Results of LSPI with approximate policy improvement on the DC motor

Its oscillation is clearly visible. The appearance of oscillations may be related to the fact that the weaker suboptimality bound (38) applies when approximate policies are used, rather than the stronger bound (39), which applies to the experiment with exact policy improvements.

Figure 9 presents one of the two policies from the limit cycle, and one of the Q-functions. The policy and Q-function have a similar accuracy to those computed with exact policy improvements, even though in this experiment the solution oscillated instead of converging to a stationary value. The approximate policy has the added advantage that it produces continuous actions.

5.4 Actor–Critic Algorithms

Actor–critic algorithms are a special case of online approximate policy iteration. They were introduced in [3] and have been investigated often since then [5, 6, 12, 38, 56, 76]. In typical actor–critic methods, both the policy and the value function are approximated using differentiable approximators (often neural networks [44, 61, 64]) and updated using gradient rules. The critic is the approximate value function (typically a V-function), and the actor is the approximate policy.

Next, a typical actor–critic algorithm is formalized. Denote by $\widehat{V}(x;\theta)$ the approximate V-function, parameterized by $\theta \in \mathbb{R}^N$, and by $\widehat{h}(x;\vartheta)$ the approximate policy, parameterized by $\vartheta \in \mathbb{R}^{\mathscr{N}}$. We use the notation $\widehat{V}(x;\theta)$ (and respectively, $\widehat{h}(x;\vartheta)$) to make explicit the dependence of the parameter vector θ (and respectively, ϑ). Because the algorithm does not distinguish between the value functions of the different policies, the value function notation is not superscripted by the policy. After each transition from x_k to x_{k+1}, the temporal difference $\delta_{\mathrm{TD},k} = r_{k+1} + \gamma\widehat{V}(x_{k+1};\theta_k) - \widehat{V}(x_k;\theta_k)$ is computed. This temporal difference is analogous to the temporal difference for Q-functions, used, e.g., in Q-learning (11). It is a sample of the difference between the right-hand and left-hand sides of the Bellman equation for the policy V-function:

$$V(x) = \mathrm{E}_{x' \sim \tilde{f}(x,h(x),\cdot)} \left\{ \rho(x,h(x),x') + V(x') \right\} \tag{41}$$

Since the exact values of the current state, $V(x_k)$, and of the next state, $V(x_{k+1})$, are not available, they are replaced by their approximations.

Once the temporal difference $\delta_{\mathrm{TD},k}$ is available, the policy and V-function parameters are updated with

$$\theta_{k+1} = \theta_k + \alpha_{\mathrm{C}} \frac{\partial \widehat{V}(x_k;\theta_k)}{\partial \theta} \delta_{\mathrm{TD},k} \tag{42}$$

$$\vartheta_{k+1} = \vartheta_k + \alpha_{\mathrm{A}} \frac{\partial \widehat{h}(x_k;\vartheta_k)}{\partial \vartheta} [u_k - \widehat{h}(x_k;\vartheta_k)]\, \delta_{\mathrm{TD},k} \tag{43}$$

where α_{C} and α_{A} are learning rates (step sizes) for the critic and the actor, respectively, and the notation $\frac{\partial \widehat{V}(x_k;\theta_k)}{\partial \theta}$ means that the derivative $\frac{\partial \widehat{V}(x;\theta)}{\partial \theta}$ is evaluated for the state x_k and the parameter θ_k (and analogously in (43)). In the critic update (42), the temporal difference takes the place of the prediction error $V(x_k) - \widehat{V}(x_k;\theta_k)$, where $V(x_k)$ is the exact value of x_k given the current policy. Since this exact value is not available, it is replaced by the estimate $r_{k+1} + \gamma\widehat{V}(x_{k+1};\theta_k)$ offered by the Bellman equation (41), thus leading to the temporal difference. In the actor update (43), the actual action u_k applied at step k can be different from the action $\widehat{h}(x_k;\vartheta_k)$ indicated by the policy. This change of the action indicated by the policy is the form taken by exploration in the actor–critic algorithm. When the exploratory action u_k leads to a positive temporal difference, the policy is adjusted toward this action. Conversely, when $\delta_{\mathrm{TD},k}$ is negative, the policy is adjusted away from u_k. This is because, like in the critic update, the temporal difference is interpreted as a correction of the predicted performance, so that, e.g., if the temporal difference is positive, the obtained performance is considered better than the predicted one.

An important advantage of actor–critic algorithms stems from the fact that their policy updates are incremental and do not require the computation of greedy actions. This means that it is not necessary to solve a difficult optimization problem over the action variable, and continuous actions are easy to handle. The convergence of actor–critic methods is not guaranteed in general. Some actor–critic algorithms employing a value function approximator that is related in a specific way to the policy approximator are provably convergent [6, 37, 38]. Note that, because they use gradient-based updates, all actor–critic algorithms can remain stuck in locally optimal solutions.

6 Finding Value Function Approximators Automatically

Parametric approximators of the value function play an important role in approximate value iteration and approximate policy iteration, as seen in Sects. 4 and 5. Given the functional form of such an approximator, the DP/RL algorithm computes its parameters. There still remains the problem of finding a good functional form, well suited to the problem at hand. In this section, we consider linearly parameterized approximators such as (18), in which case a good set of BFs has to be found. This focus is motivated by the fact that, in the literature, most methods to find value function approximators are given in this linear setting. Also note that many approaches require a discrete and not too large action space, and focus their effort on finding good state-dependent BFs.

The BFs can be designed in advance, in which case two approaches are possible. The first approach is to design the BFs so that a uniform resolution is obtained over the entire state space (for V-functions) or over the entire state–action space (for Q-functions). Unfortunately, such an approach suffers from the curse of dimensionality: the complexity of a uniform-resolution approximator grows exponentially with the number of state (and possibly action) variables. The second approach is to focus the resolution in certain parts of the state(–action) space, where the value function has a more complex shape, or where it is more important to approximate it accurately. Prior knowledge about the shape of the value function or about the importance of certain areas of the state(–action) space is necessary in this case. Unfortunately, such prior knowledge is often nonintuitive and very difficult to obtain without actually computing the value function.

A more general alternative is to find BFs automatically, rather than designing them. Such an approach should provide BFs suited to each particular problem. BFs can be either constructed offline [47, 51] or adapted while the DP/RL algorithm is running [55, 65]. Since convergence guarantees typically rely on a fixed set of BFs, adapting the BFs while running the DP/RL algorithm leads to a loss of these guarantees. Convergence guarantees can be recovered by ensuring that BF adaptation is stopped after a finite number of updates; fixed-BF proofs can then be applied to guarantee asymptotic convergence [21].

In the remainder of this section, we give a brief overview of available techniques to find BFs automatically. Resolution refinement techniques are discussed first, followed by BF optimization, and by other techniques for automatic BF discovery.

6.1 Resolution Refinement

Resolution refinement techniques start with a few BFs (a coarse resolution) and then refine the BFs as the need arises. These techniques can be further classified in two categories:

(1) Local refinement (splitting) techniques evaluate whether a particular area of the state space (corresponding to one or several neighboring BFs) has a sufficient accuracy, and add new BFs when the accuracy is deemed insufficient. Such techniques have been proposed, e.g., for Q-learning [65, 66, 84], for V-iteration [55], for Q-iteration [53, 82], and for policy evaluation [27].
(2) Global refinement techniques evaluate the global accuracy of the representation, and refine the BFs if the accuracy is deemed insufficient. All the BFs can be refined uniformly [19], or the algorithm can decide which areas of the state space require more resolution [55]. For instance, in [19, 55], global refinement is applied to V-iteration, while in [77] it is used for Q-learning.

A variety of criteria are used to decide when the BFs should be refined. In [55], an overview of typical criteria is given, together with a comparison between them in the context of V-iteration. For instance, local refinement in a certain area can be performed:

- when the value function is not (approximately) constant in that area [55];
- when the value function is not (approximately) linear in that area [53, 55];
- when the Bellman error (the error between the left-hand and right-hand sides of the Bellman equation, see the upcoming Sect. 6.2) is large in that area [27];
- or using various other heuristics [65, 82, 84].

Global refinement can be performed, e.g., until a desired solution accuracy is met [19]. The approach in [55] works in discrete-action problems, and refines the areas where the V-function is poorly approximated *and* that affect other areas where the actions dictated by the policy change. This approach globally identifies the areas of the state space that must be approximated more accurately to find a good policy.

Resolution refinement techniques increase the memory and computational expenses of the DP/RL algorithm whenever they increase the resolution. Care must be taken to prevent the memory and computation expenses from becoming prohibitive. This is an important concern both in approximate DP and in approximate RL. Equally important in approximate RL are the restrictions imposed on resolution refinement by the limited amount of data available. Increasing the power of the approximator means that more data will be required to compute an accurate solution, so the resolution cannot be refined to arbitrary levels for a given amount of data.

6.2 Basis Function Optimization

Basis function optimization techniques search for the best placement and shape of a (usually fixed) number of BFs. Consider, e.g., the linear parametrization (18) of the Q-function. To optimize the n BFs, they can be parameterized by a vector of BF parameters ξ that encodes their locations and shapes. For instance, a radial BF is characterized by its center and width. Denote the parameterized BFs by $\varphi_l(x;\xi) : X \times U \to \mathbb{R}$, $l = 1,\dots,n$ to highlight their dependence on ξ. The BF optimization algorithm searches for an optimal parameter vector ξ^* that optimizes a certain criterion related to the accuracy of the value function representation.

Many optimization techniques can be applied to compute the BF parameters. For instance, gradient-based optimization has been used for temporal difference [71] and least-squares temporal difference algorithms [51]. The cross-entropy method has been applied to least-squares temporal difference [51] and Q-iteration algorithms [15].

The most widely used optimization criterion is the Bellman error, also called Bellman residual [51, 71]. This is a measure of the extent to which the approximate value function violates the Bellman equation, which would be precisely satisfied by the exact value function. For instance, the Bellman error for an estimate $\widehat{Q}$ of the optimal Q-function Q^* is derived from the Bellman optimality equation, namely (5) in the stochastic case and (7) in the deterministic case. So, for a deterministic MDP, this Bellman error is

$$\int_X \int_U \left| \widehat{Q}(x,u) - \rho(x,u) - \gamma \max_{u'} \widehat{Q}(f(x,u),u') \right|^2 \mathrm{d}u \, \mathrm{d}x \tag{44}$$

In practice, an approximation of the Bellman error is computed using a finite set of samples. The suboptimality $\|\widehat{Q} - Q^*\|_\infty$ of an approximate Q-function $\widehat{Q}$ is bounded by a constant multiple of the infinity norm of its Bellman error $\|\widehat{Q} - T\widehat{Q}\|_\infty$ [11, 89]. Furthermore, the Q-function suboptimality is related to the policy suboptimality by (23), which means that minimizing the Bellman error is useful in principle. Unfortunately, minimizing the *quadratic* Bellman error (44) may lead to a large *infinity norm* of the Bellman error, so it is unclear whether minimizing (44) leads to a near-optimal approximate Q-function and policy.

Another possible criterion for optimizing the BFs is the performance (returns) of the policy obtained by the DP/RL algorithm [15].

6.3 Other Methods for Basis Function Construction

It is possible to construct BFs using various other techniques that are different from resolution refinement and optimization. For instance, in [46, 47], a spectral analysis of the MDP transition dynamics is performed to find the BFs. These BFs are then used in LSPI. Because the BFs represent the underlying topology of the state transitions, they provide a good accuracy in representing the value function.

Many nonparametric approximators can be seen as generating a set of BFs automatically. The number, location, and possibly also the shape of the BFs are not established in advance, but are determined by the nonparametric regression algorithm. For instance, in [21], regression trees are used to represent the Q-function in every iteration of the fitted Q-iteration algorithm. The method to build the regression trees implicitly determines a set of BFs that represent well the Q-function at the current iteration. In [32, 90], kernel-based approximators are used in LSPI. Originally, kernel-based approximation uses a BF for each sample, but in [90] a kernel sparsification procedure automatically determines a reduced number of BFs, and in [32], BFs are added online only when they improve the accuracy. In [60], kernel-based approximators are combined with value iteration. Least-squares support vector machines are applied to policy evaluation by least-squares temporal difference in [31], and to Q-learning in [85]. Support vector regression is used with SARSA in [33]. Self-organizing maps are combined with Q-learning in [79].

Example 3 (Finding RBFs for LSPI in the DC motor problem). Consider again the DC motor problem of Example 1 and its solution found with LSPI in Example 2. As already discussed, the LSPI solution of Fig. 7(a) does not properly take into account the nonlinearities of the policy seen in the top-left and bottom-right corners of Fig. 3(a). This is because the corresponding variations in the Q-function, seen in Fig. 3(b), are not represented well by the wide RBFs employed. To improve the resolution in the corners where the Q-function is not well approximated, a resolution refinement technique could be applied.

An alternative is to parameterize the RBFs (40) and optimize their locations and shapes. In this case, the RBF parameter vector, denoted by ξ, would contain the two-dimensional centers and radii of all the RBFs: $\xi = [c_{1,1}, c_{1,2}, b_{1,1}, b_{1,2}, \dots, c_{N,1}, c_{N,2}, b_{N,1}, b_{N,2}]^{\mathrm{T}}$. Such an approach is proposed in [51], where the Bellman error is minimized using gradient descent and cross-entropy optimization.

7 Approximate Policy Search

Algorithms for approximate policy search represent the policy approximately, most often using a parametric approximator. An optimal parameter vector is then sought using optimization techniques. In certain special cases (e.g., when the state space is finite and not too large), the parametrization might exactly represent an optimal policy. However, in general, optimal policies can only be represented approximately. We consider in this section policy search techniques that do not employ value functions. Such techniques are useful when it is undesirable to compute value functions, e.g., because value-function-based techniques fail to obtain a good solution.

Denote by $\widehat{h}(x;\vartheta)$ the approximate policy, parameterized by $\vartheta \in \mathbb{R}^{\mathcal{N}}$. Policy search algorithms search for an optimal parameter vector that maximizes the return $R^{\widehat{h}(x;\vartheta)}$ for all $x \in X$. Three additional types of approximation are necessary to implement a general policy search algorithm (see also Sect. 3):

1. When X is large or continuous, computing the return for every state is not possible. A practical procedure to circumvent this difficulty requires choosing a finite set X_0 of representative initial states. Returns are estimated only for states in X_0, and the optimization criterion (score) is the weighted average return over X_0 [49, 54]:
$$s(\vartheta) = \sum_{x_0 \in X_0} w(x_0) R^{\widehat{h}(x;\vartheta)}(x_0) \tag{45}$$
The representative states are weighted by $w : X_0 \to (0,1]$. The set X_0, together with the weight function w, will determine the performance of the resulting policy. For instance, initial states that are deemed more important can be assigned larger weights. Note that maximizing the returns from states in X_0 only results in an approximately optimal policy, because it cannot guarantee that returns from other states in X are maximal.
2. In the computation of the returns, the infinite sum in (1) has to be replaced by a finite sum over K steps. For discounted returns, a value of K that guarantees a maximum absolute error $\varepsilon_{\mathrm{MC}} > 0$ in estimating the returns is [48]
$$K = \left\lceil \log_\gamma \frac{\varepsilon_{\mathrm{MC}}(1-\gamma)}{\|\tilde{\rho}\|_\infty} \right\rceil \tag{46}$$
where $\lceil \cdot \rceil$ produces the smallest integer larger than or equal to the argument (ceiling).
3. Finally, in stochastic MDPs, Monte Carlo simulations are required to estimate the expected returns. This procedure is consistent, i.e., as the number of

simulations approaches infinity, the estimate converges to the correct expectation. Results from Monte Carlo simulation can be applied to bound the approximation error for a finite number of simulations.

If prior knowledge about a (near-)optimal policy is available, an ad-hoc policy parametrization can be designed. For instance, parametrizations that are linear in the state variables can be used, if it is known that a (near-)optimal policy is a linear state feedback. Ad-hoc parametrizations are typically combined with gradient-based optimization [49, 54, 68].

When prior knowledge about the policy is not available, a richer policy parametrization has to be used. In this case, the optimization criterion is likely to have many local optima, and may also be nondifferentiable. This means that gradient-based algorithms are unsuitable, and global gradient-free optimization algorithms are required. Examples of such techniques include evolutionary optimization (genetic algorithms in particular), tabu search, pattern search, and the cross-entropy method. For instance, evolutionary computation has been applied to policy search in [2, 18, 24], Chap. 3 of [17], and cross-entropy optimization in [16, 48]. Chapter 4 of [17] describes a method to find a policy with the model-reference adaptive search, which is closely related to the cross-entropy method.

Example 4 (Approximate policy search for the DC motor). Consider again the DC motor problem of Example 1. First, we derive a policy parametrization based on prior knowledge, and apply policy search to this parametrization. Then, a policy parametrization that does not rely on prior knowledge is given, and the results obtained with these two parametrizations are compared. To optimize the parameters, we use the global gradient-free pattern search optimization [42, 78].[12]

Because the system is linear and the reward function is quadratic, the optimal policy would be a linear state feedback if the constraints on the state and action variables were disregarded [9]. Taking now into account the constraints on the action, we assume that a good approximation of an optimal policy is linear in the state variables, up to the constraints on the action:

$$\widehat{h}(x) = \text{sat}\{\vartheta_1 x_1 + \vartheta_2 x_2, -10, 10\} \tag{47}$$

where 'sat' denotes saturation. In fact, an examination of the near-optimal policy in Fig. 3(a) reveals that this assumption is largely correct: the only nonlinearities appear in the top-left and bottom-right corners of the figure, and they are probably due to the constraints on the state variables. We use the parametrization (47) and search for an optimal parameter vector $\vartheta^* = [\vartheta_1^*, \vartheta_2^*]^{\mathrm{T}}$.

A set X_0 of representative states must be selected. To obtain a uniform performance across the state space, we select a regular grid of representative states: $X_0 = \{-\pi, -2\pi/3, -\pi/3..., \dots, \pi\} \times \{-16\pi, -12\pi, -8\pi, \dots, 16\pi\}$, weighted uniformly by $w(x_0) = \frac{1}{|X_0|}$. We impose a maximum error $\varepsilon_{\text{MC}} = 0.01$ in the estimation of the return. A bound on the reward function (27) can be computed with

[12] We use the pattern search algorithm from the *Genetic Algorithm and Direct Search Toolbox* of MATLAB 7.

$$\begin{aligned}\|\rho\|_\infty &= \sup_{x,u} \left| -x^{\mathrm{T}} Q_{\mathrm{rew}} x - R_{\mathrm{rew}} u^2 \right| \\ &= \left| -[\pi \ 16\pi] \begin{bmatrix} 5 & 0 \\ 0 & 0.01 \end{bmatrix} \begin{bmatrix} \pi \\ 16\pi \end{bmatrix} - 0.01 \cdot 10^2 \right| \\ &\approx 75.61\end{aligned}$$

To find the trajectory length K required to achieve the precision $\varepsilon_{\mathrm{MC}}$, we substitute the values of $\varepsilon_{\mathrm{MC}}$, $\|\rho\|_\infty$, and $\gamma = 0.95$ in (46); this yields $K = 233$. Because the problem is deterministic, simulating multiple trajectories from every initial state is not necessary; instead, a single trajectory from every initial state suffices.

Pattern search is applied to optimize the parameters ϑ, starting with a zero initial value of these parameters. The algorithm is considered convergent when the variation of best score decreases below the threshold $\varepsilon_{\mathrm{PS}} = 0.01$ (equal to $\varepsilon_{\mathrm{MC}}$). The resulting policy is shown in Fig. 10. As expected, it closely resembles the near-optimal policy of Fig. 3(a), with the exception of the nonlinearities in the corners.

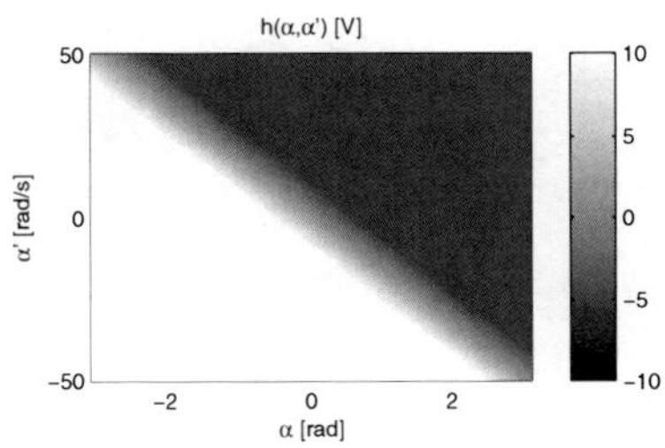

Fig. 10 Results of policy search on the DC motor with the policy parametrization (47). The policy parameter is $\widehat{\vartheta}^* \approx [-16.69, -1]^{\mathrm{T}}$.

The execution time of pattern search was approximately 152 s. This is larger than the execution time of fuzzy Q-iteration in Example 1, which was 6 s for the fine grid and 0.5 s for the coarse grid. It is comparable to the execution time of LSPI in Example 2, which was 105 s when using exact policy improvements, and 104 s with approximate policy improvements. Policy search spends the majority of its execution time estimating the score function (45), which is a computationally expensive operation. For this experiment, the score of 74 different parameter vectors had to be computed until convergence. The complexity can be decreased by taking a smaller number of representative states or larger values for $\varepsilon_{\mathrm{MC}}$ and $\varepsilon_{\mathrm{PS}}$, at the expense of a possible decrease in the control performance.

Consider now the case in which no prior knowledge about the optimal policy is available. In this case, a general policy parametrization must be used. We choose the linear policy parametrization (36), repeated here for easy reference:

$$\widehat{h}(x) = \sum_{i=1}^{\mathcal{N}} \varphi_i(x) \vartheta_i = \varphi^{\mathrm{T}}(x) \vartheta$$

Normalized RBFs (40) are defined, with their centers arranged on an equidistant 7×7 grid in the state space. The radii of the RBFs along each dimension are taken identical to the distance along that dimension between two adjacent RBFs. A total of 49 parameters (for 7×7 RBFs) have to be optimized. This number is larger than for the parametrization (47) derived from prior knowledge, which only had 2 parameters. The same X_0, $\varepsilon_{\mathrm{MC}}$, are used as for the simple parametrization.

The solution obtained by pattern search optimization is shown in Fig. 11. Compared to Fig. 10, the policy varies more slowly in the linear portion; this is because the wide RBFs used lead to a smooth interpolation. The score obtained by the RBF policy of Fig. 11 is -230.69, slightly worse than the score obtained by the policy of Fig. 10, which is -229.25.

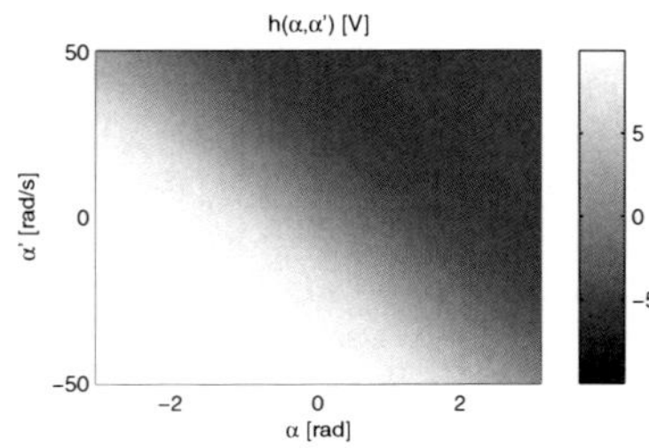

Fig. 11 Results of policy search on the DC motor with the policy parametrization (36)

The algorithm required 30,487 s to converge, and had to compute the score of 11,440 parameter vectors. As expected, the computational cost is much larger than for the simple parametrization because many more parameters have to be optimized. This illustrates the benefits of using a policy parametrization that is appropriate for the problem considered. Unfortunately, deriving an appropriate parametrization requires prior knowledge, which is not always available.

8 Comparison of Approximate Value Iteration, Policy Iteration, and Policy Search

While a definitive comparison between approximate value iteration, approximate policy iteration, and approximate policy search will depend on the particular algorithms considered, some general remarks can be made.

Algorithms for approximate policy iteration often converge in a smaller number of iterations than algorithms for approximate value iteration, as illustrated in Examples 1 and 2. However, approximate policy evaluation is a difficult problem in itself, which must be solved at each single policy iteration. Since the cost of a policy evaluation may be comparable to the cost of value iteration, it is unclear how the entire policy iteration algorithm compares to value iteration from the point of view of computational cost.

The convergence guarantees for approximate policy evaluation impose less restrictive requirements on the approximator than the guarantees of approximate value iteration; this is an advantage for approximate policy iteration. Namely, for policy evaluation, it suffices if the approximator is linearly parameterized (Sect. 5.3), whereas for value iteration, additional properties are required to ensure that the approximate value iteration mapping is a contraction (Sect. 4.3). Moreover, efficient least-squares algorithms such as LSTD-Q can be used to compute a 'one-shot' solution to the policy evaluation problem. These advantages stem from the *linearity* of the Bellman equation for the value function of a given policy, e.g., (8), whereas the Bellman optimality equation, which characterizes the optimal value function, e.g., (7), is *highly nonlinear* due to the maximization in the right-hand side. Note, however, that for value iteration, monotonous convergence to a unique solution is usually guaranteed, whereas policy iteration is generally only guaranteed to converge to a sequence of policies that all provide at least a guaranteed level of performance (see Sect. 5.3).

Approximate policy search is useful in two cases. The first case is when the form of a (near-)optimal policy is known, and only a few parameters need to be determined. In this case, optimization can be used to find a good parameter vector with moderate computational costs. The second case is when, even though prior knowledge is not available, it is undesirable to compute value functions, e.g., because value-function-based techniques fail to obtain a good solution or require too restrictive assumptions. In this case, a general policy parametrization can be defined, and a policy search technique that does not rely on value functions can be used to optimize the parameters. Such techniques are usually free from numerical problems—such as divergence to infinity—even when used with general nonlinear parameterizations, which is not the case for value and policy iteration. However, because of its generality, this approach typically incurs large computational costs.

9 Summary and Outlook

This chapter has described DP and RL for large or continuous-space, infinite-horizon problems. After introducing the necessary background in exact DP and RL, the need for approximation in DP and RL has been explained, and approximate versions for the three main categories of DP/RL algorithms have been discussed: value iteration, policy iteration, and policy search. Theoretical guarantees have been given and practical algorithms have been illustrated using numerical examples. Additionally, techniques to automatically determine value function approximators have been reviewed, and the three categories of algorithms have been compared.

Approximate DP/RL is a young but active and rapidly expanding field of research. Many issues in this field remain open. Some of these issues are specific to approximate DP/RL, while others also apply to exact DP and RL.

Next, we discuss some open issues that are specific to approximate DP and RL.

- Automatically finding good approximators is essential in high-dimensional problems, because approximators that provide a uniform accuracy would

require too much memory and computation. Adaptive value-function approximators are being extensively studied (Sect. 6). In policy search, finding approximators automatically is a comparatively under-explored, but promising idea. Nonparametric approximation is an elegant and powerful framework to derive a good approximator from the data [21, 32, 90].
- Continuous-action MDPs are less often studied than discrete-action MDPs, among others because value iteration and policy improvement are significantly more difficult when continuous actions are considered (Sect. 3). However, for some problems continuous actions are important. For instance, stabilizing a system around an unstable equilibrium requires continuous actions to avoid chattering of the control action, which would otherwise damage the system in the long run. Continuous actions are easier to handle in actor–critic and policy search algorithms [38, 54, 62].
- Owing to their sample efficiency and relaxed convergence requirements, least-squares techniques for policy evaluation are extremely promising in approximate policy iteration. However, they typically work offline and assume that a large number of samples is used for every policy evaluation. From a learning perspective, it would be interesting to study these techniques in the online case, where the policy must be improved once every few samples, before each policy evaluation has converged. Such optimistic least-squares policy iteration algorithms are rarely studied in the literature.

Finally, we present several important open issues that apply to both the exact and approximate cases.

- In practice, it is essential to provide guaranteed performance during online RL. Online RL algorithms should ideally guarantee a monotonous increase in their expected performance. Unfortunately, this is generally impossible because all the online RL algorithms need to explore, i.e., try out actions that may be suboptimal, in order to make sure their performance does not remain stuck in a local optimum. Therefore, weaker requirements could be used, where an overall trend of increased performance is guaranteed, while still allowing for bounded and temporary decreases in performance due to exploration.
- Designing a good reward function is an important and nontrivial step of applying DP and RL. Classical texts on RL recommend to keep the reward function as simple as possible; it should only reward the achievement of the final goal [74]. Unfortunately, a simple reward function often makes online learning very slow, and more information may need to be included in the reward function. Such informative rewards are sometimes called shaping rewards [58]. Moreover, additional higher-level requirements often have to be considered in addition to the final goal. For instance, in automatic control, the controlled state trajectories often have to satisfy requirements on overshoot and the rate of convergence to an equilibrium, etc. Translating such requirements into the 'language' of rewards can be very challenging.
- It is important to address problems in which the state signal cannot be measured directly, because such problems often arise in practice. These problems

are called partially observable in the DP/RL literature. Algorithms for this type of problem are being extensively researched [4, 34, 50, 59, 63].

- RL has become one of the dominating paradigms for learning in distributed multiagent systems [13]. New challenges arise in multiagent RL, as opposed to the single-agent case. Two major new challenges are that the curse of dimensionality is made worse by the multiple agents present in the system, and that the control actions of the agents must be coordinated to reach their intended result. Approximation is an essential, though largely unexplored open issue also in multiagent RL. This means that a good understanding of single-agent approximate RL is required to develop effective multiagent RL algorithms.

References

1. Baddeley, B.: Reinforcement learning in continuous time and space: Interference and not ill conditioning is the main problem when using distributed function approximators. IEEE Transactions on Systems, Man, and Cybernetics—Part B: Cybernetics 38(4), 950–956 (2008)
2. Barash, D.: A genetic search in policy space for solving Markov decision processes. In: AAAI Spring Symposium on Search Techniques for Problem Solving under Uncertainty and Incomplete Information. Palo Alto, US (1999)
3. Barto, A.G., Sutton, R.S., Anderson, C.W.: Neuronlike adaptive elements than can solve difficult learning control problems. IEEE Transactions on Systems, Man, and Cybernetics 13(5), 833–846 (1983)
4. Baxter, J., Bartlett, P.L.: Infinite-horizon policy-gradient estimation. Journal of Artificial Intelligence Research 15, 319–350 (2001)
5. Berenji, H.R., Khedkar, P.: Learning and tuning fuzzy logic controllers through reinforcements. IEEE Transactions on Neural Networks 3(5), 724–740 (1992)
6. Berenji, H.R., Vengerov, D.: A convergent actor-critic-based FRL algorithm with application to power management of wireless transmitters. IEEE Transactions on Fuzzy Systems 11(4), 478–485 (2003)
7. Bertsekas, D.P.: Adaptive aggregation methods for infinite horizon dynamic programming. IEEE Transactions on Automatic Control 34(6), 589–598 (1989)
8. Bertsekas, D.P.: Dynamic programming and suboptimal control: A survey from ADP to MPC. European Journal of Control 11(4-5) (2005); Special issue for the CDC-ECC-05 in Seville, Spain
9. Bertsekas, D.P.: Dynamic Programming and Optimal Control, 3rd edn., vol. 2. Athena Scientific, Belmont (2007)
10. Bertsekas, D.P., Shreve, S.E.: Stochastic Optimal Control: The Discrete Time Case. Academic Press, London (1978)
11. Bertsekas, D.P., Tsitsiklis, J.N.: Neuro-Dynamic Programming. Athena Scientific, Belmont (1996)
12. Borkar, V.: An actor–critic algorithm for constrained Markov decision processes. Systems & Control Letters 54, 207–213 (2005)
13. Buşoniu, L., Babuška, R., De Schutter, B.: A comprehensive survey of multi-agent reinforcement learning. IEEE Transactions on Systems, Man, and Cybernetics 38(2), 156–172 (2008)

14. Buşoniu, L., Ernst, D., De Schutter, B., Babuška, R.: Consistency of fuzzy model-based reinforcement learning. In: Proceedings 2008 IEEE International Conference on Fuzzy Systems (FUZZ-IEEE 2008), Hong Kong, pp. 518–524 (2008)
15. Buşoniu, L., Ernst, D., De Schutter, B., Babuška, R.: Fuzzy partition optimization for approximate fuzzy Q-iteration. In: Proceedings 17th IFAC World Congress (IFAC 2008), Seoul, Korea, pp. 5629–5634 (2008)
16. Buşoniu, L., Ernst, D., De Schutter, B., Babuška, R.: Policy search with cross-entropy optimization of basis functions. In: Proceedings 2009 IEEE International Symposium on Adaptive Dynamic Programming and Reinforcement Learning (ADPRL 2009), Nashville, US, pp. 153–160 (2009)
17. Chang, H.S., Fu, M.C., Hu, J., Marcus, S.I.: Simulation-Based Algorithms for Markov Decision Processes. Springer, Heidelberg (2007)
18. Chin, H.H., Jafari, A.A.: Genetic algorithm methods for solving the best stationary policy of finite Markov decision processes. In: Proceedings 30th Southeastern Symposium on System Theory, Morgantown, US, pp. 538–543 (1998)
19. Chow, C.S., Tsitsiklis, J.N.: An optimal one-way multigrid algorithm for discrete-time stochastic control. IEEE Transactions on Automatic Control 36(8), 898–914 (1991)
20. Coulom, R.: Feedforward neural networks in reinforcement learning applied to high-dimensional motor control. In: Cesa-Bianchi, N., Numao, M., Reischuk, R. (eds.) ALT 2002. LNCS (LNAI), vol. 2533, pp. 403–413. Springer, Heidelberg (2002)
21. Ernst, D., Geurts, P., Wehenkel, L.: Tree-based batch mode reinforcement learning. Journal of Machine Learning Research 6, 503–556 (2005)
22. Ernst, D., Glavic, M., Capitanescu, F., Wehenkel, L.: Reinforcement learning versus model predictive control: a comparison on a power system problem. IEEE Transactions on Systems, Man, and Cybernetics—Part B: Cybernetics 39(2), 517–529 (2009)
23. Glorennec, P.Y.: Reinforcement learning: An overview. In: Proceedings European Symposium on Intelligent Techniques (ESIT 2000), Aachen, Germany, pp. 17–35 (2000)
24. Gomez, F.J., Schmidhuber, J., Miikkulainen, R.: Efficient non-linear control through neuroevolution. In: Fürnkranz, J., Scheffer, T., Spiliopoulou, M. (eds.) ECML 2006. LNCS (LNAI), vol. 4212, pp. 654–662. Springer, Heidelberg (2006)
25. Gonzalez, R.L., Rofman, E.: On deterministic control problems: An approximation procedure for the optimal cost I. The stationary problem. SIAM Journal on Control and Optimization 23(2), 242–266 (1985)
26. Gordon, G.: Stable function approximation in dynamic programming. In: Proceedings 12th International Conference on Machine Learning (ICML 1995), Tahoe City, US, pp. 261–268 (1995)
27. Grüne, L.: Error estimation and adaptive discretization for the discrete stochastic Hamilton-Jacobi-Bellman equation. Numerical Mathematics 99, 85–112 (2004)
28. Horiuchi, T., Fujino, A., Katai, O., Sawaragi, T.: Fuzzy interpolation-based Q-learning with continuous states and actions. In: Proceedings 5th IEEE International Conference on Fuzzy Systems (FUZZ-IEEE 1996), New Orleans, US, pp. 594–600 (1996)
29. Jaakkola, T., Jordan, M.I., Singh, S.P.: On the convergence of stochastic iterative dynamic programming algorithms. Neural Computation 6(6), 1185–1201 (1994)
30. Jouffe, L.: Fuzzy inference system learning by reinforcement methods. IEEE Transactions on Systems, Man, and Cybernetics—Part C: Applications and Reviews 28(3), 338–355 (1998)
31. Jung, T., Polani, D.: Least squares SVM for least squares TD learning. In: Proceedings of 17th European Conference on Artificial Intelligence (ECAI 2006), Riva del Garda, Italy, pp. 499–503 (2006)

32. Jung, T., Polani, D.: Kernelizing LSPE(λ). In: Proceedings 2007 IEEE Symposium on Approximate Dynamic Programming and Reinforcement Learning (ADPRL 2007), Honolulu, US, pp. 338–345 (2007)
33. Jung, T., Uthmann, T.: Experiments in value function approximation with sparse support vector regression. In: Proceedings 15th European Conference on Machine Learning (ECML 2004), Pisa, Italy, pp. 180–191 (2004)
34. Kaelbling, L.P., Littman, M.L., Cassandra, A.R.: Planning and acting in partially observable stochastic domains. Artificial Intelligence 101, 99–134 (1998)
35. Kaelbling, L.P., Littman, M.L., Moore, A.W.: Reinforcement learning: A survey. Journal of Artificial Intelligence Research 4, 237–285 (1996)
36. Konda, V.: Actor–critic algorithms. Ph.D. thesis, Massachusetts Institute of Technology, Cambridge, US (2002)
37. Konda, V.R., Tsitsiklis, J.N.: Actor–critic algorithms. In: Solla, S.A., Leen, T.K., Müller, K.R. (eds.) Advances in Neural Information Processing Systems, vol. 12, pp. 1008–1014. MIT Press, Cambridge (2000)
38. Konda, V.R., Tsitsiklis, J.N.: On actor–critic algorithms. SIAM Journal on Control and Optimization 42(4), 1143–1166 (2003)
39. Lagoudakis, M., Parr, R., Littman, M.: Least-squares methods in reinforcement learning for control. In: Vlahavas, I.P., Spyropoulos, C.D. (eds.) SETN 2002. LNCS (LNAI), vol. 2308, pp. 249–260. Springer, Heidelberg (2002)
40. Lagoudakis, M.G., Parr, R.: Least-squares policy iteration. Journal of Machine Learning Research 4, 1107–1149 (2003)
41. Lagoudakis, M.G., Parr, R.: Reinforcement learning as classification: Leveraging modern classifiers. In: Proceedings 20th International Conference on Machine Learning (ICML 2003), Washington, US, pp. 424–431 (2003)
42. Lewis, R.M., Torczon, V.: Pattern search algorithms for bound constrained minimization. SIAM Journal on Optimization 9(4), 1082–1099 (1999)
43. Lin, L.J.: Self-improving reactive agents based on reinforcement learning, planning and teaching. Machine Learning 8(3/4), 293–321 (1992); Special Issue on Reinforcement Learning
44. Liu, D., Javaherian, H., Kovalenko, O., Huang, T.: Adaptive critic learning techniques for engine torque and air-fuel ratio control. IEEE Transactions on Systems, Man, and Cybernetics—Part B: Cybernetics 38(4), 988–993 (2008)
45. Madani, O.: On policy iteration as a newton s method and polynomial policy iteration algorithms. In: Proceedings 18th National Conference on Artificial Intelligence and 14th Conference on Innovative Applications of Artificial Intelligence AAAI/IAAI 2002, Edmonton, Canada, pp. 273–278 (2002)
46. Mahadevan, S.: Samuel meets Amarel: Automating value function approximation using global state space analysis. In: Proceedings 20th National Conference on Artificial Intelligence and the 17th Innovative Applications of Artificial Intelligence Conference (AAAI 2005), Pittsburgh, US, pp. 1000–1005 (2005)
47. Mahadevan, S., Maggioni, M.: Proto-value functions: A Laplacian framework for learning representation and control in Markov decision processes. Journal of Machine Learning Research 8, 2169–2231 (2007)
48. Mannor, S., Rubinstein, R.Y., Gat, Y.: The cross-entropy method for fast policy search. In: Proceedings 20th International Conference on Machine Learning (ICML 2003), Washington, US, pp. 512–519 (2003)

49. Marbach, P., Tsitsiklis, J.N.: Approximate gradient methods in policy-space optimization of Markov reward processes. Discrete Event Dynamic Systems: Theory and Applications 13, 111–148 (2003)
50. McCallum, A.: Overcoming incomplete perception with utile distinction memory. In: Proceedings 10th International Conference on Machine Learning (ICML 1993), Amherst, US, pp. 190–196 (1993)
51. Menache, I., Mannor, S., Shimkin, N.: Basis function adaptation in temporal difference reinforcement learning. Annals of Operations Research 134, 215–238 (2005)
52. Millán, J.d.R., Posenato, D., Dedieu, E.: Continuous-action Q-learning. Machine Learning 49(2-3), 247–265 (2002)
53. Munos, R.: Finite-element methods with local triangulation refinement for continuous reinforcement learning problems. In: van Someren, M., Widmer, G. (eds.) ECML 1997. LNCS, vol. 1224, pp. 170–182. Springer, Heidelberg (1997)
54. Munos, R.: Policy gradient in continuous time. Journal of Machine Learning Research 7, 771–791 (2006)
55. Munos, R., Moore, A.: Variable-resolution discretization in optimal control. Machine Learning 49(2-3), 291–323 (2002)
56. Nakamura, Y., Moria, T., Satoc, M., Ishiia, S.: Reinforcement learning for a biped robot based on a CPG-actor-critic method. Neural Networks 20, 723–735 (2007)
57. Nedić, A., Bertsekas, D.P.: Least-squares policy evaluation algorithms with linear function approximation. Discrete Event Dynamic Systems 13, 79–110 (2003)
58. Ng, A.Y., Harada, D., Russell, S.: Policy invariance under reward transformations: Theory and application to reward shaping. In: Proceedings 16th International Conference on Machine Learning (ICML 1999), Bled, Slovenia, pp. 278–287 (1999)
59. Ng, A.Y., Jordan, M.I.: PEGASUS: A policy search method for large MDPs and POMDPs. In: Proceedings 16th Conference in Uncertainty in Artificial Intelligence (UAI 2000), Palo Alto, US, pp. 406–415 (2000)
60. Ormoneit, D., Sen, S.: Kernel-based reinforcement learning. Machine Learning 49(2-3), 161–178 (2002)
61. Pérez-Uribe, A.: Using a time-delay actor–critic neural architecture with dopamine-like reinforcement signal for learning in autonomous robots. In: Wermter, S., Austin, J., Willshaw, D.J. (eds.) Emergent Neural Computational Architectures Based on Neuroscience. LNCS (LNAI), vol. 2036, pp. 522–533. Springer, Heidelberg (2001)
62. Peters, J., Schaal, S.: Natural actor–critic. Neurocomputing 71(7-9), 1180–1190 (2008)
63. Porta, J.M., Vlassis, N., Spaan, M.T., Poupart, P.: Point-based value iteration for continuous POMDPs. Journal of Machine Learning Research 7, 2329–2367 (2006)
64. Prokhorov, D., Wunsch, D.C.: Adaptive critic designs. IEEE Transactions on Neural Networks 8(5), 997–1007 (1997)
65. Ratitch, B., Precup, D.: Sparse distributed memories for on-line value-based reinforcement learning. In: Boulicaut, J.-F., Esposito, F., Giannotti, F., Pedreschi, D. (eds.) ECML 2004. LNCS (LNAI), vol. 3201, pp. 347–358. Springer, Heidelberg (2004)
66. Reynolds, S.I.: Adaptive resolution model-free reinforcement learning: Decision boundary partitioning. In: Proceedings 17th International Conference on Machine Learning (ICML 2000), Stanford University, US, pp. 783–790 (2000)
67. Riedmiller, M.: Neural fitted Q-iteration – first experiences with a data efficient neural reinforcement learning method. In: Gama, J., Camacho, R., Brazdil, P.B., Jorge, A.M., Torgo, L. (eds.) ECML 2005. LNCS (LNAI), vol. 3720, pp. 317–328. Springer, Heidelberg (2005)

68. Riedmiller, M., Peters, J., Schaal, S.: Evaluation of policy gradient methods and variants on the cart-pole benchmark. In: Proceedings 2007 IEEE Symposium on Approximate Dynamic Programming and Reinforcement Learning (ADPRL 2007), Honolulu, US, pp. 254–261 (2007)
69. Rummery, G.A., Niranjan, M.: On-line Q-learning using connectionist systems. Tech. Rep. CUED/F-INFENG/TR166, Engineering Department, Cambridge University, UK (1994)
70. Santos, M.S., Vigo-Aguiar, J.: Analysis of a numerical dynamic programming algorithm applied to economic models. Econometrica 66(2), 409–426 (1998)
71. Singh, S.P., Jaakkola, T., Jordan, M.I.: Reinforcement learning with soft state aggregation. In: Tesauro, G., Touretzky, D.S., Leen, T.K. (eds.) Advances in Neural Information Processing Systems, vol. 7, pp. 361–368 (1995)
72. Sutton, R.S.: Learning to predict by the method of temporal differences. Machine Learning 3, 9–44 (1988)
73. Sutton, R.S.: Integrated architectures for learning, planning, and reacting based on approximating dynamic programming. In: Proceedings 7th International Conference on Machine Learning (ICML 1990), Austin, US, pp. 216–224 (1990)
74. Sutton, R.S., Barto, A.G.: Reinforcement Learning: An Introduction. MIT Press, Cambridge (1998)
75. Sutton, R.S., Barto, A.G., Williams, R.J.: Reinforcement learning is adaptive optimal control. IEEE Control Systems Magazine 12(2), 19–22 (1992)
76. Sutton, R.S., McAllester, D.A., Singh, S.P., Mansour, Y.: Policy gradient methods for reinforcement learning with function approximation. In: Solla, S.A., Leen, T.K., Müller, K.R. (eds.) Advances in Neural Information Processing Systems, vol. 12, pp. 1057–1063. MIT Press, Cambridge (2000)
77. Szepesvári, C., Smart, W.D.: Interpolation-based Q-learning. In: Proceedings 21st International Conference on Machine Learning (ICML 2004), Bannf, Canada, pp. 791–798 (2004)
78. Torczon, V.: On the convergence of pattern search algorithms. SIAM Journal on Optimization 7(1), 1–25 (1997)
79. Touzet, C.F.: Neural reinforcement learning for behaviour synthesis. Robotics and Autonomous Systems 22(3-4), 251–281 (1997)
80. Tsitsiklis, J.N., Van Roy, B.: Feature-based methods for large scale dynamic programming. Machine Learning 22(1-3), 59–94 (1996)
81. Tsitsiklis, J.N., Van Roy, B.: An analysis of temporal difference learning with function approximation. IEEE Transactions on Automatic Control 42(5), 674–690 (1997)
82. Uther, W.T.B., Veloso, M.M.: Tree based discretization for continuous state space reinforcement learning. In: Proceedings 15th National Conference on Artificial Intelligence and 10th Innovative Applications of Artificial Intelligence Conference (AAAI 1998/IAAI 1998), Madison, US, pp. 769–774 (1998)
83. Vrabie, D., Pastravanu, O., Abu-Khalaf, M., Lewis, F.: Adaptive optimal control for continuous-time linear systems based on policy iteration. Automatica 45(2), 477–484 (2009)
84. Waldock, A., Carse, B.: Fuzzy Q-learning with an adaptive representation. In: Proceedings 2008 IEEE World Congress on Computational Intelligence (WCCI 2008), Hong Kong, pp. 720–725 (2008)
85. Wang, X., Tian, X., Cheng, Y.: Value approximation with least squares support vector machine in reinforcement learning system. Journal of Computational and Theoretical Nanoscience 4(7-8), 1290–1294 (2007)

86. Watkins, C.J.C.H.: Learning from delayed rewards. Ph.D. thesis, King's College, Oxford (1989)
87. Watkins, C.J.C.H., Dayan, P.: Q-learning. Machine Learning 8, 279–292 (1992)
88. Wiering, M.: Convergence and divergence in standard and averaging reinforcement learning. In: Boulicaut, J.-F., Esposito, F., Giannotti, F., Pedreschi, D. (eds.) ECML 2004. LNCS (LNAI), vol. 3201, pp. 477–488. Springer, Heidelberg (2004)
89. Williams, R.J., Baird, L.C.: Tight performance bounds on greedy policies based on imperfect value functions. In: Proceedings 8th Yale Workshop on Adaptive and Learning Systems, New Haven, US, pp. 108–113 (1994)
90. Xu, X., Hu, D., Lu, X.: Kernel-based least-squares policy iteration for reinforcement learning. IEEE Transactions on Neural Networks 18(4), 973–992 (2007)
91. Yu, H., Bertsekas, D.P.: Convergence results for some temporal difference methods based on least-squares. Tech. Rep. LIDS 2697, Massachusetts Institute of Technology, Cambridge, US (2006)

Learning with Whom to Communicate Using Relational Reinforcement Learning

Marc Ponsen, Tom Croonenborghs, Karl Tuyls, Jan Ramon, Kurt Driessens, Jaap van den Herik, and Eric Postma

Abstract. Relational reinforcement learning is a promising direction within reinforcement learning research. It upgrades reinforcement learning techniques using relational representations for states, actions, and learned value functions or policies to allow natural representations and abstractions of complex tasks. Multi-agent systems are characterized by their relational structure and present a good example of a

Marc Ponsen
Department of Knowledge Engineering, Maastricht University, Mindebroedersberg 6a, 6211 LK, Maastricht, The Netherlands
e-mail: m.ponsen@maastrichtuniversity.nl

Tom Croonenborghs
KH Kempen University College, Kleinhoefstraat 4, 2440 Geel, Belgium
e-mail: tom.croonenborghs@khk.be

Karl Tuyls
Department of Knowledge Engineering, Maastricht University, Mindebroedersberg 6a, 6211 LK, Maastricht, The Netherlands
e-mail: k.tuyls@maastrichtuniversity.nl

Jan Ramon
DTAI, Katholieke Universiteit Leuven, Celestijnenlaan 200A, B-3001 Heverlee, Belgium
e-mail: jan.ramon@cs.kuleuven.be

Kurt Driessens
DTAI, Katholieke Universiteit Leuven, Celestijnenlaan 200A, B-3001 Heverlee, Belgium
e-mail: kurt.driessens@cs.kuleuven.be

Jaap van den Herik
Tilburg centre for Creative Computing, Tilburg University, Warandelaan 2, PO Box 90153, 5000 LE Tilburg, The Netherlands
e-mail: h.j.vdnherik@uvt.nl

Eric Postma
Tilburg centre for Creative Computing, Tilburg University, Warandelaan 2, PO Box 90153, 5000 LE Tilburg, The Netherlands
e-mail: e.o.postma@uvt.nl

R. Babuška & F.C.A. Groen (Eds.): Interactive Collaborative Information Systems, SCI 281, pp. 45–63.
springerlink.com

complex task. In this article, we show how relational reinforcement learning could be a useful tool for learning in multi-agent systems. We study this approach in more detail on one important aspect of multi-agent systems, i.e., on learning a communication policy for cooperative systems (e.g., resource distribution). Communication between agents in realistic multi-agent systems can be assumed costly, limited, and unreliable. We perform a number of experiments that highlight the conditions in which relational representations can be beneficial when taking the constraints mentioned above into account.

1 Introduction

Relational reinforcement learning (RRL) has emerged in the machine learning community as a promising subfield of reinforcement learning (RL) (e.g., [4, 6, 25, 28]). It upgrades RL techniques using relational representations for states, actions, and learned value functions or policies to allow natural representations and abstractions of complex tasks. RRL offers a state-space representation that is much richer than the one used in classical (or propositional) methods. This leads to a serious state-space reduction, allowing the use of generalized prior knowledge and inferring general new knowledge as a result.

So far, most of the work in RRL has focused on single-agent situations, i.e., environments in which only one agent operates and learns. Because multi-agent problems contain, besides a complex environment, also interactions between the agents, it is reasonable to assume that this new paradigm can also contribute to the multi-agent system's (MAS) domain.

An important source of complexity in MAS is the fact that agents only have partial information about the environment. Because of the inherent generalization and abstraction that comes with relational representations, agents can more naturally generalize over unseen parts of the environment. Also, in complex MAS, agents can interfere with each other's actions, plans, or goals. Hence, agents need to take these external influences into account and need to learn which other agents to keep in mind when optimizing ones own behavior. The latter can, for example, be achieved through communication . Through communication, agents can limit the influence of partial observability, and obtain valuable information about other agents. Harmful interactions with agents can be avoided, while on the opposite, agents with common interest can work together to achieve a higher reward. However, communication in complex MAS is not free. Communicating with other agents comes with a cost (e.g., the communication medium may not be free and/or have a limited bandwidth), is often limited (e.g., communication is only possible with agents within a certain range) or unreliable (e.g., messages may not always arrive error free or even not at all).

In this article, we investigate the effect of using a relational representation with respect to agent communication. We define an abstract cooperative MAS, wherein agents can learn to communicate with each other to achieve a higher reward. More precisely, we consider a population of agents that each have to learn how to complete a number of tasks. The agents can have some prior knowledge about tasks, which allows them to complete the task more easily. Additionally, agents may ask other agents for advice. Hence, they can learn to complete a task not only by finding a solution themselves but also by asking an agent who already knows how to complete it. Therefore, it is important that agents have a good idea of who is good at which task, and as such obtain a good impression of the agent relationships. We perform a number of experiments that illustrate the benefit of using RRL for agent communication in MAS. More specifically, we empirically validate if (relational) communication is beneficial compared to no communication, and whether relational representations can cope with issues such as limited and unreliable communication. We also test whether our relational communication approach is scalable in terms of agents and task complexity. Our results illustrate that when using a relational communication approach, agents can learn to complete tasks faster through advice seeking.

The remainder of this paper is structured as follows: in Section 2 and 3 we explain RL and RRL, respectively. The application of RRL in MAS, along with existing work, is discussed in Section 4. In Section 5, we describe our experiments that must empirically validate the advantages of using relational communication in MAS. We conclude in Section 6.

2 Reinforcement Learning

Most RL research is based on the framework of Markov decision processes (MDP) [19]. MDP are sequential decision making problems for fully observable worlds. They are defined by a tuple (S,A,T,R). Starting in an initial state s_0 at each discrete time step $t = 0,1,2,\ldots$ an adaptive agent observes an environment state s_t contained in a finite set of states $S = \{s_1,s_2,\ldots,s_n\}$, and executes an action a from a finite set of admissible actions $A = \{a_1,a_2,\ldots,a_m\}$. The agent receives an immediate reward $R : S \rightarrow \mathbb{R}$, that assigns a value or reward for being in that state, and moves to a new state s' depending on a probabilistic transition function $T : S \times A \times S \rightarrow [0,1]$. The probability for ending up in state s' after doing action a in state s is denoted as $T(s,a,s')$. For all actions a, and all states s and s', we have that $T(s,a,s') \geq 0$ and $\sum_{s' \in S} T(s,a,s') = 1$. An MDP respects the *Markov property*: the future dynamics, transitions, and rewards fully depend on the current state, i.e., $T(s_{t+1}|s_t,a_t,s_{t-1},a_{t-1},\ldots) = T(s_{t+1}|s_t,a_t)$ and $R(s_t|s_{t-1},\ldots) = R(s_t)$. The transition function T and reward function R together are often referred to as the model of the environment. The learning task in MDP is to find a policy $\pi : S \rightarrow A$ for selecting actions with maximal expected (discounted) future reward. The quality of a policy is indicated by a *value function* V^{π}. The value $V^{\pi}(s)$ specifies the total amount of reward which an agent may expect to accumulate over the future, starting from state s and following the policy π. Informally, the value function indicates the long-term desirability of states after taking into account the states that are likely to follow and the rewards available in those states. In a discounted infinite horizon MDP, the expected cumulative reward (i.e., the value function) is defined as

$$V^{\pi}(s) = E\left[\sum_{t=0}^{\infty} \gamma^t R(S_t) | s_0 = s\right] \tag{1}$$

A discount factor $\gamma \in [0,1\rangle$ is introduced to ensure that the rewards returned are bounded (finite) values. The variable γ determines the relevance of future rewards. Setting γ to 0 results in a *myopic* view on rewards (i.e., only the immediate reward is optimized), whereas values closer to 1 will increase the contribution of future rewards to the sum.

The value for a given policy π, expressed by Equation 1, can be iteratively computed by the so-called *Bellman Equation* [1]. In value iteration approaches, one starts with arbitrarily chosen value function V_0^{π} and, during each iteration t, for each state $s \in S$ the value function is updated based on the immediate reward and the current estimates of V^{π}:

$$V_{t+1}^{\pi}(s) = R(s) + \gamma \sum_{s' \in S} T(s,\pi(s),s') V_t^{\pi}(s') \tag{2}$$

The process of updating state value functions based on current estimates of successor state values is referred to as *bootstrapping*. The depth of successor states considered in the update can be varied, i.e., one can perform a shallow update where one only looks at immediate successor states or a deep update where successors of successors are also considered. The value function estimates of successor states are used to update the value function estimate of the current state. This is called a *backup* operation. Different algorithms use different backup strategies, e.g., sample backups (sample a single successor state) or full backups (sample all successor states).

The objective of an MDP is to find the optimal policy, i.e., the policy that receives the most reward. The optimal policy $\pi^*(s)$ is such that $V^{\pi^*}(s) \geq V^{\pi}(s)$ for all $s \in S$ and all policies π. It is proven that the optimal value function, often abbreviated as V^*, satisfies the following Bellman optimality criterion:

$$V^*(s) = R(s) + \gamma \max_{\alpha \in A} \left[\sum_{s' \in S} T(s,a,s') V^*(s')\right] \tag{3}$$

Solving Equation 3 can be done in an iterative manner, similar to the computation of the value function for a given policy such as expressed in Equation 2. The Bellman optimality criterion is turned into an update rule:

$$V_{t+1}^{\pi}(s) = R(s) + \gamma \max_{\alpha \in A} \left[\sum_{s' \in S} T(s,a,s') V_t^{\pi}(s')\right] \tag{4}$$

The optimal action can then be selected as follows:

$$\pi^*(s) = \arg\max_a \left[R(s) + \gamma \sum_{s' \in S} T(s,a,s') V^*(s') \right] \tag{5}$$

Besides learning statevalues, one can also define state-action value functions, or so-called *Q-functions*. Q-functions map state-action pairs to values, $Q : S \times A \rightarrow \mathbb{R}$. They reflect the long-term desirability of performing action a in state s, and then performing policy π thereafter. The Q-function is defined as follows:

$$Q^*(s,a) = R(s,a) + \gamma \sum_{s' \in S} T(s,a,s') V^*(s') \tag{6}$$

The optimal policy π^* selects the action which maximizes the optimal-action value function $Q^*(s,a)$ for each state $s \in S$:

$$\pi^*(s) = \arg\max_a Q^*(s,a) \tag{7}$$

Solutions for and properties of state value functions can be straightforwardly translated to state-action values.

When an environment's model (i.e., transition T function and reward R function) is known, the optimal policy can be computed using a dynamic programming approach, as for example with *policy iteration* or *value iteration*. When the model of the environment is unknown, as it usually is, we can use sampling-based techniques such as *temporal difference learning* as an alternative. These techniques do not depend on a model but rather collect samples of interactions with the environment and update the value estimates based on these interactions. An important issue is the used sampling strategy, better known in the RL literature as the exploration–exploitation dilemma, i.e., when to explore the environment and when to exploit acquired knowledge. Various exploration and exploitation strategies exist, such as ε-greedy and Boltzmann exploration. For a thorough overview, we refer to [30]. Temporal difference learning methods such as Q-learning [29] and SARSA [20] are model-free solution methods. The algorithms are described in detail in [23]. The update rule for the most popular algorithm, *one-step* Q-learning, is denoted as

$$Q(a,s) \leftarrow (1-\alpha)Q(a,s) + \alpha \left[r + \gamma \max_{a'} Q(a',s') \right] \tag{8}$$

where α is the step-size parameter, and γ the discount rate. Both algorithms are proven to converge to optimal policies under loose conditions, given that no abstractions have been applied. Unfortunately for complex, real-world problems, solving MDPs is impractical and complexity must be reduced to keep learning tractable. In the following section, we will discuss the benefit of relational reinforcement learning for generalizing over the state, action, and policy space.

3 Relational Reinforcement Learning

Relational reinforcement learning (RRL) combines the RL setting with relational learning or inductive logic programming (ILP) [14] to represent states, actions, and policies using the structures and relations that identify them. These structural representations allow abstraction from and generalization over specific goals, states, and actions. Because RRL algorithms try to solve problems at an abstract level, the solutions will often carry to different instantiations of that abstract problem. For example, resulting policies learned by an RRL system often generalize over domains with varying number of existing objects.

A straightforward example is the blocks world. A number of blocks with different properties (size, color, and so on) are placed on each other or on the floor. It is assumed that an infinite number of blocks can be put on the floor and that all blocks are neatly stacked onto each other, e.g., a block can only be on one other block at the same time. The possible actions consist of moving one clear block (e.g., a block with no other block on top of it) onto another clear block or onto the floor. It is impossible to represent such blocks world states with a propositional representation without an explosion of the number of states. Consider as an example the right-most state in Figure 1. In first-order logic (FOL), this state can be represented, presuming that this state is called *s*, by the conjunction

$$\{on(s,c,floor) \wedge clear(s,c) \wedge on(s,d,floor) \wedge on(s,b,d) \wedge on(s,a,b) \wedge clear(s,a) \wedge on(s,e,floor) \wedge clear(s,e)\},$$

which is obtained through executing the move-action (indicated by the arrow), noted as *move(r,s,a,b)*, in the previous state *r*.

One of the most important benefits of the relational learning approach is that one can generalize over states, actions, objects selectively. One can abstract away only those parts of the statespace that are less important. As an example, in the blocks world scenario, one can say "there exists a small block which is on a large block" ($\exists B1,B2 : block(B1,small,\_), block(B2,large,\_), on(B1,B2)$). Objects $B1$ and $B2$ are free variables, and can be instantiated by any block in the environment. Hence, RRL can generalize over blocks, and learn policies for a variable number of objects without causing an explosion of the number of states.

Although, RRL is a relatively new domain, several approaches have been proposed during the past few years. One of the first methods developed within RRL was

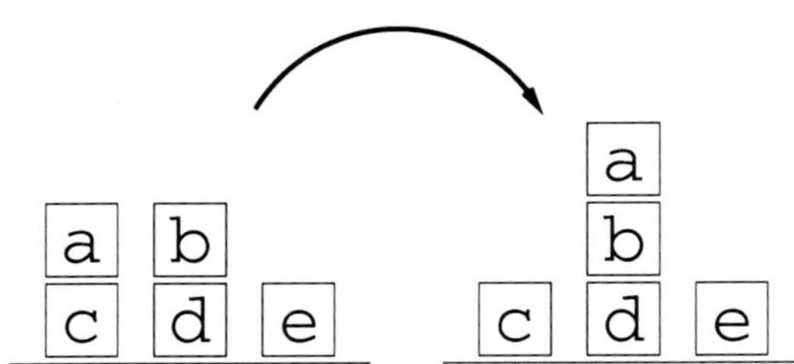

Fig. 1 The blocks world

Algorithm 1 The Relational Reinforcement Learning Algorithm

Require: initialize $Q(s,a)$ arbitrarily
Require: $e \leftarrow 0$
  repeat
Require: initialize s_0 arbitrarily
    $Examples \leftarrow \oslash$
    $i \leftarrow 0$
    repeat
      take a for s using policy $\pi(s)$ and observe r and s'
      $Q(a,s) \rightarrow (1-\alpha)Q(a,s) + \alpha \left[r + \gamma \max_{a'} Q(a',s') \right]$
      $Examples \leftarrow Examples \cup \{a,s,Q(s,a)\}$
      $i \leftarrow i+1$
    until (s_i is terminal)
    Update $\hat{Q}_e$ using *Examples* and a relational regression algorithm to produce $\hat{Q}_{e+1}$
    $e \leftarrow e+1$
  until (no more episode)

relational Q-learning [6]. A high level description of the algorithm is given in Table 1. Relational Q-learning behaves similar to standard Q-learning, with the exception of the agent's representation of the states S, actions A, and learned value function. In relational reinforcement learning, this representation contains structural or relational information about the environment. Furthermore, it employs a relational regression algorithm to generalize over the value function (and policy) space. Learning examples stored as a tuple $< a,s,Q(s,a) >$ are processed by an incremental relational regression algorithm to produce a relational value function (or policy) as a result.[1] So far, a number of different relational regression learners have been developed.[2]

In the present work, we employ the incremental first-order regression tree algorithm $RRL-TG$ [5] as our chosen regression algorithm. The algorithm is denoted in Table 2. Given a set of examples, the algorithm incrementally builds up a tree (*using top down induction*). It starts with an empty tree with one leaf, which contains all stored examples. Then, when the algorithm reaches a leaf node, candidate tests defined in the so-called *language bias* are evaluated. The tree may be expanded by candidate tests that reduce variance among Q-values sufficiently. The best among these candidate tests is then selected to expand the tree.

4 Multi-agent Relational Reinforcement Learning

During the 1990s multi-agent systems (MAS) have become a popular approach in solving computational problems of distributed nature as for instance load balancing or distributed planning systems. They are a conceptually proved solution method for problems of this nature. However, designing a cooperative MAS with both a global

[1] An example of such a relational Q-function is illustrated in Figure 3.
[2] A thorough discussion and comparison can be found in [4].

Algorithm 2 TG Algorithm

Require: input is a collection of examples $\langle a,s,Q(s,a)\rangle$
Require: initialize tree with single leaf with empty statistics
  $i \leftarrow 0$
  repeat
    sort down $Example_i$ in tree until it reaches leaf and update statistics
    if statistics in leaf indicate new split **then**
      generate new internal node using indicated test and grow 2 new leafs with empty statistics
    end if
    $i \leftarrow i+1$
  until (no more examples)

high reward for the system and high individual rewards for the different agents is still a difficult problem, see, e.g., [9, 17, 18, 26]. The general goal is therefore to have a good system performance where all individual agents in the MAS contribute to some part of the collective through their individual actions.

Several approaches have been proposed, ranging from joint action learners [10] to individual local Q-learners. All of these approaches have their own merits as drawbacks in a multi-agent context. In the first approach, i.e., the joint action space approach, the state and action space are defined as the Cartesian product of the agent's individual state and action spaces. This implies that the state information is shared among the agents and actions are taken and evaluated synchronously. It is obvious that this approach leads to very big state-action spaces, and assumes instant communication between the agents. Clearly, this approach is in contrast with the basic principles of many contemporary multi-agent applications such as distributed control, asynchronous actions, incomplete information, cost of communication, etc. In the local or selfish Q-learners setting, the presence of the other agents is totally neglected, and agents are considered to be selfish reinforcement learners. The effects caused by the other agents also acting in that same environment are considered as noise. In between these approaches we can find examples which try to overcome the drawbacks of the joint action approach [2, 13, 15, 21, 22, 27]. There has also been quite some effort to extend these RL techniques to partially observable MDPs and non-Markovian settings [11].

The complexity of designing a MAS and modeling the behavior of agents within stems from three important sources. First, partial observability is all but unavoidable in MAS because even if the environment is fully observable, the intentions and plans of other agents cannot be assumed to be known, especially when multiple agents in the MAS are updating their behavior through learning. Local decision making, and therefore inherently partial observability, is a key requirement to make MAS expandable both in the number of agents and in the size of the environment they work in. Agents can more naturally generalize over unseen parts of the environment using a rich relational language. Second, agents in a MAS can interfere with each other's actions, plans, or goals. This is yet another reason why joint-action learners and individual Q-learners mentioned previously are not appropriate for MAS. One needs to take external influences into account when optimizing locally. The problem then translates into learning which other agents to keep in mind when optimizing one's own behavior. Relational representations can be of great help when trying to define (and eventually learn) important relations between agents in MAS, for example, in learning with whom to communicate.

Communication is major component of MAS, which can potentially overcome the previously mentioned complexities. Namely, through communication agents can limit the influence of partial observability, and obtain valuable information about other agents. Harmful interactions with agents can be avoided, while on the opposite, agents with common interest can work together to achieve a higher reward. However, communication in MAS is often assumed to be either *limited*, *unreliable*, or *costly*. Consider, for example, the situation where agents can only communicate with agents that are located within a certain range or the cost could be dependent on the distance between agents. Hence, deciding when to communicate with whom adds a layer of complexity to the behavior of agents in a MAS. Therefore, we believe that multi-agent relational reinforcement learning (MARRL) applied to the problem of efficient communication can be a valuable contribution to creating scalable, complex MAS.

So far, all the studies on RRL have focused on the single-agent case. To our knowledge, there is almost no existing work on applying RRL in a MAS. Earlier, van Otterlo et al. [16] already mentioned the possible benefits of using RRL in MAS. They state that cognitive and sapient agents especially need a learning component where the RL paradigm is the most logical choice for this. Since these agents are (usually) logic-based, RRL is indicated as a very suitable learning method for intelligent agents. Letia and Precup [12] present a system where multiple agents, represented by GOLOG programs, act in the same environment. The GOLOG programs can supply initial plans and prior knowledge. The agents however cannot communicate. There also exist a game-theoretic extension of GOLOG that can be used to implement different agents and compute Nash policy pairs [7]. Hernández et al. [8] use logical decision trees to add a learning factor to BDI (belief, desire, and intension) agents. Only Croonenborghs et al. [3] present an approach of using RRL in MAS. They show that when using a relational learner, agents can induce knowledge by observing other agents. However, they do not consider the issue of communication between agents. As we consider communication to be one of the crucial aspects of MARRL, we will focus in this article on using RRL to learn with whom to communicate. In the next section, we describe experiments in an cooperative MAS wherein agents learn an communication policy under several conditions.

5 Empirical Evaluation

In this section, we will empirically investigate the benefits of using RRL for communication between agents in an abstract cooperative MAS. More specifically, we will investigate the influence on the learning performance when allowing agents to seek advice, both in a relational and propositional system. Furthermore, we will vary the quality of the prior knowledge of agents (i.e., agents may not always have the right answer), limit the communication capabilities of agents and finally vary the number of agents and task difficulty in our system. We will also apply relational communication to a challenging multi-state task.

Since we would like to study the effects and benefits of efficient communication, we do not consider the aspects of agent action interference and partial observability in our test environment. Hence, all agents are working on their own task without the possibility to interfere with each other. Not only is this still a challenging problem, an isolated study of the relations between the agents and methods that gain maximally from this, is necessary to handle more complex problems.

5.1 Learning Task

The learning task of the agents consists of learning how to complete a number of different tasks. Tasks and agents can be characterized by a number of properties (e.g., color, size, etc.). Every agent is assigned to a random task in the beginning of an episode. An agent receives a positive reward if it completes his task and the episode is ended when all agents completed their tasks. The number of possible actions (henceforth called *primitive actions*) an agent can execute for solving a certain task can vary from task to task. To complete a task, the agent has to execute the *optimal* action, a single action from the possible primitive actions for that task.

Some agents have a priori knowledge about certain tasks; these agents are called *experts* for that task. A *pure expert* only has access to the task's optimal action, and therefore has no need for exploring the action space for this particular task. Besides pure experts we also introduce different levels of expertise: agents may have some knowledge about the task at hand, i.e., they know that the optimal action is in some subset of all primitive actions for that task and consequently need to explore a constrained part of the action space. Non-experts must learn a policy to complete the task. Determining whether an agent is expert for a task can be done in many ways. For example, we can define an agent as expert if a certain property, or all, match for both task and agent. Seeking advice from an expert immediately completes the task, since its action space only contains the optimal action. However, it may also prove useful seeking advice from an agent who has a high degree of expertise for a task, since it can be expected that this agent learns to complete the task faster.

To evaluate whether communication can improve learning performance, we will run a series of experiments with and without communication. For the first, agents can seek advice from other agents by asking them which action they would execute according to their current policy in the current task of the inquirer. Therefore, the action space of an agent for some task consists of the number of the primitive actions for this task and the *advice actions*, one for every agent in the environment.

5.2 *Experimental Results*

All agents are relational Q-learners who use the incremental relational tree learner RRL-TG [5] to learn a generalized Q-function (we refer to Section 3 for details). To guide this tree building a declarative bias is specified. This declarative bias contains a language bias to specify the predicates that can be used as tests to partition the tree. For all agents this language bias contains predicates that identify the task and the action taken by the agent. Furthermore, it contains candidate tests that check certain properties of the task at hand and properties of the other agents. The candidate tests are listed in Table 1.

Table 1 The candidate tests collected in the language bias for building up the tree

Candidate test	Explanation
`primitive_action(A,B)`	primitive action A has id B
`advice_action(A,B)`	advice action A is suggested by agent B
`task_has_id(A,B)`	task A has id B
`task_has_property(A,B)`	task A has a certain property B
`agent_has_property(A,B)`	agent A has a certain property B

We use two versions for the language bias: one to mimic a propositional learner (i.e., standard Q-learning) and a relational variant. The propositional variant straightforwardly uses the first three candidate tests, wherein agents try out every instantiation of primitive and advice actions until they find the optimal one (e.g., the one that produces a reward of 1). The relational variant of the language bias includes all tests, thereby allowing agents to generalize over properties for the agents and tasks. Given these properties, relational learning agent might discover for a certain task it is smart to seek advice from an agent with certain properties.

During learning, the agents follow a Boltzmann exploration policy [24]. Each agent is maximally allowed 100 actions per episode. Each agent receives a reward of 1.0 if he completes the task and neither reward nor punishment otherwise. Every 100 episodes, the learned policies are tested by freezing the Q-function approximation following a greedy policy on 100 testepisodes to check for optimality. Unless mentioned otherwise, all results are averaged over 5 runs. We assume that the transition function is fully deterministic, and that agents have full observability (e.g., agents can see all the other agents and their properties). Although this is a restricted setting in terms of RL, this is not important for our learning task since we are investigating the effect of communication under several conditions.

We want to investigate whether RRL can generalize over agent and task properties, and therefore can quickly discover experts in the population (if any), so they can communicate (i.e, seek advice) with these experts. We expect this can potentially accelerate the learning process. To validate this claim empirically, we compare the average learning performance for agents in two situations, namely, (1) learning with and (2) learning without means for communication. The learning task without advice boils down to selecting the optimal *primitive action* for a task. If we add *advice actions* to the action space, agents do not necessary need to learn to locate the optimal *primitive action* for each task, it may be much better to seek advice from an expert. This is especially true in our experimental setup, since we do not yet consider communication costs or constraints (although this is certainly possible in our task, and we will elaborate on this in our future work).

5.2.1 Seeking Advice

Consider the following experimental setting: we use a MAS containing 10 agents and 10 different tasks. Both agents and tasks are described by one property with a number of possible values (e.g., the property is shape and possible values are circle, square, etc.). The tasks all have a different difficulty. In the first task there are only 5 primitive actions available, for the second task these are doubled to 10 primitive actions and so forth until 2560 actions for the tenth task. In this setting, each agent is an expert for a certain task if he shares the property with that task. The experiment is set up in such a way that every agent is expert for exactly one different task and hence there is exactly one expert for every task.

Figure 4 presents the result for the basic setting. It clearly shows the gain that agents with the capability of communication have over agents without this capability. Learning performance is increased by communication, both using a propositional and relational language bias. An example policy learned with the propositional language bias is shown in Figure 2. As we can see, this approach mimics a standard Q-learner where for each instantiation of a task, the optimal action must be located (albeit through personal exploration or by asking a specific agent for advice). No benefit is taking from the fact that certain agents (with a matching property with the task) are experts for solving this particular task. Figure 3 illustrates the learned policy using the relational language bias. This learned policy concisely and elegantly illustrates the optimal policy: for a particular task, one should seek advice from its corresponding expert; otherwise, the agent itself is expert. From

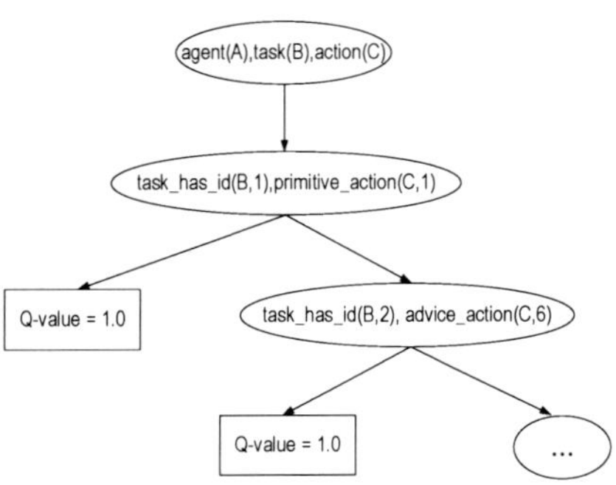

Fig. 2 Example policy learned using the propositional language bias

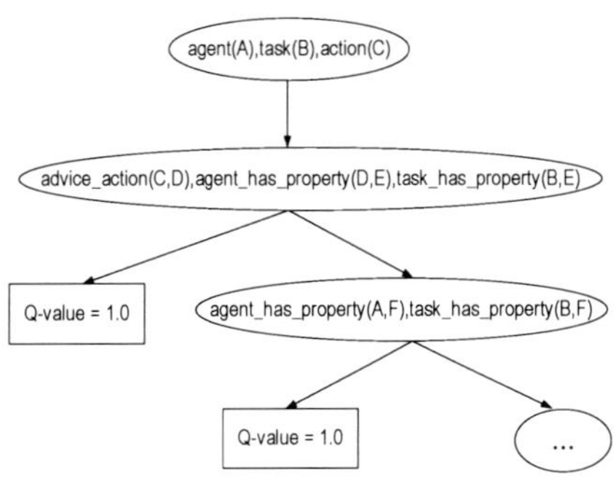

Fig. 3 Example policy learned using the relational language bias

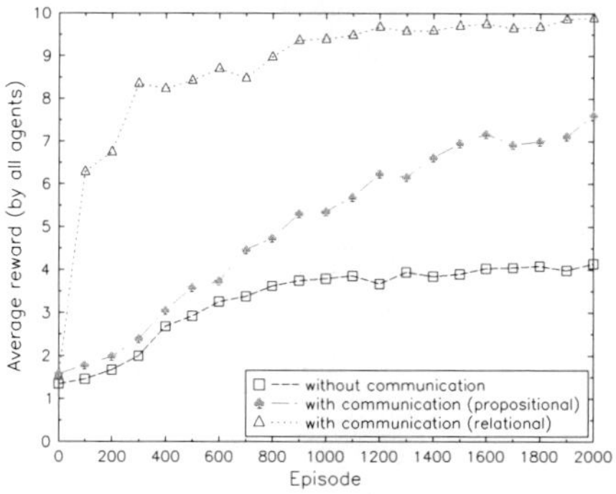

Fig. 4 Communication versus no Communication

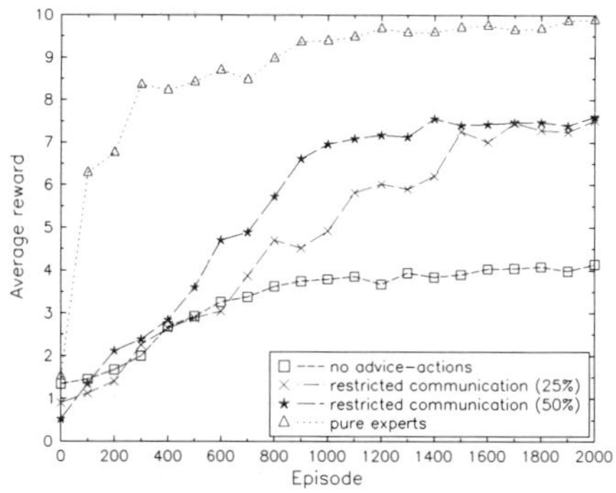

Fig. 5 Influence of restricted communication

Figure 4, we may conclude that the generalization of relational learning is essential to take optimal advantage of the added communication.

5.2.2 Influence of Restricted Communication

We also investigate the influence of restricted communication on the performance of the agents. Figure 5 shows the result when agents can no longer communicate with every other agent. The percentages indicated in the labels reflect the percentage of possible advice actions at every step. The setting *no advice-actions* corresponds to the setting *without communication* in the previous experiments. In the setting *pure experts*, agents may communicate with all other agents. The uncertainty of communication causes a drop in performance, but due to the generalization in the agents' policy, the possibility of communication still has a beneficial effect.

5.2.3 Influence of Prior Knowledge

In these experiments, we want to evaluate the influence of the quality of prior knowledge. To investigate this, we extended the possible actions the agents can execute in a task when they are expert for that task. Figure 6 shows the influence of the prior knowledge that agents have[3]. As previously explained, in the setting *pure experts* the action space of an expert only includes the optimal action. As a result experts always give optimal advice. This is clearly the most ideal setting in terms of agent communication. In the setting *optimal + advice*, the experts can also execute advice actions and hence they still have to learn to select the primitive action instead of asking for advice. In the worst possible setting *no experts*, the agents have no prior knowledge, i.e., the possible actions for the agents are all primitive actions and advice actions. We also investigated two intermediate settings where a subset of all primitive actions is selected, namely, 50% and 75% of the possible primitive actions.

[3] The results of the pure experts and the agents without communication are repeated for clarity.

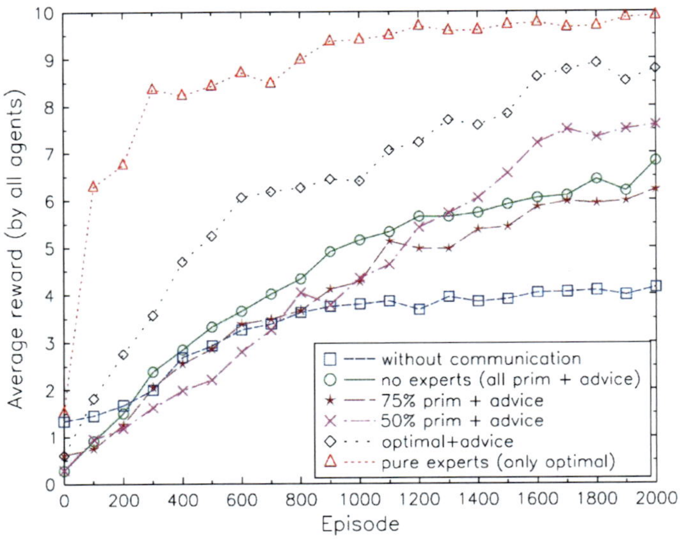

Fig. 6 The influence of prior knowledge

We may conclude from the graph that when using relational learning communication, learning performance is improved even when prior knowledge is limited. In the early stages of learning, the extra possibility of seeking advice does not hurt the agent; only in the very beginning the performance is slightly worse due to the bigger action space. Once some agents learned to complete some tasks, this knowledge is distributed through the population resulting in a performance gain.

5.2.4 Influence of Task Difficulty and Number of Agents

To determine the influence of the task difficulty and the number of agents, we sampled tasks with varying difficulty (i.e., varying the size of action space) restricted by a maximum value. Furthermore, agent and task property values are now randomly sampled and unlike the previous experiments we lose the guarantee that an expert is present in the population. We performed experiments with an increasing maximum value for the action space (this shifts the balance between advice and primitive actions) and with an increasing number of agents compared to tasks (this way we increase the likelihood of finding an expert in the population, but also increases the number of advice actions).

In Figure 7, we see that for learning without communication (the experiments denoted with 'NC', the number following this represents the maximum number of actions for this task), agents are steadily moving toward an optimal policy as the number of episodes increases. The convergence rate is determined by the difficulty of the randomly generated tasks. We can also see that for the experiments with advice (denoted with 'C'), agents perform better, in particular in situations where agents are faced with more complex tasks. It seems that when tasks are relatively

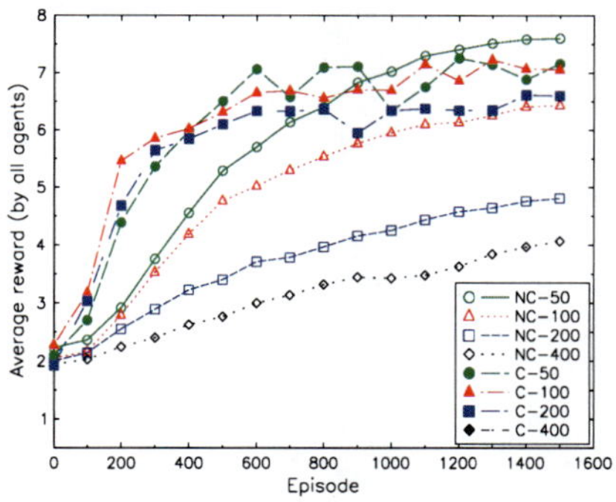

Fig. 7 Influence of the task difficulty

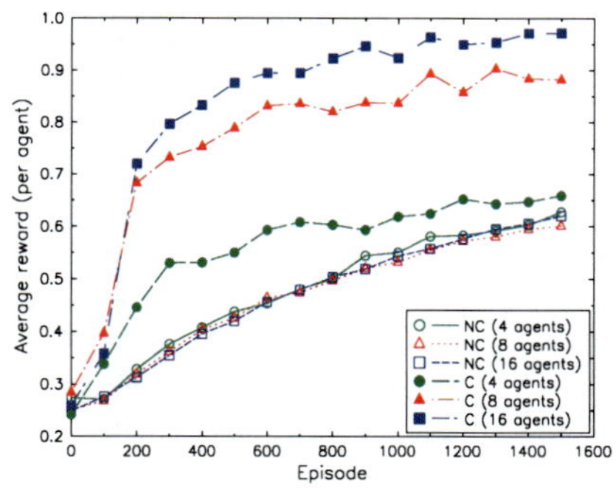

Fig. 8 Influence of the number of agents

easy (i.e., see 'NC-50' and 'C50' in Figure 7), it can be at least as efficient to learn to complete the task individually, rather than seeking advice. Only when generating complex tasks, does communication significantly improve learning performance.

Figure 8 shows the results for different sizes of the agent population using the same setting as in the previous paragraph with the maximal number of actions set to 200. The number of agents is denoted by the number behind 'NC' (no communication) or 'C' (communication). The plot shows that the number of agents is irrelevant to the learning performance of the setting without communication. This was expected since agents are learning individually and cannot benefit from the knowledge of others. When we allow agents to seek advice, we see much faster learning in particular when many agents are present in the system. A likely explanation is that more agents increase the likelihood of an expert in the population, no matter what task is randomly generated.

5.2.5 Application to Complex Learning Task

In the last series of experiments, we will show that the earlier discussed results also apply for more difficult environments. We extended the number of properties: we provide both agent and task with 3 properties, with, respectively, 4, 3, and 2 possible values. Therefore, a total of 24 different agent types and task types can be assembled. More precisely, agents and tasks all have a color (with values green, red, yellow, or blue), shape (square, cube, or sphere) and size (small or large). The property values for tasks have a weight value associated with them that determine the difficulty of a task. For example, a blue task is four times more difficult compared to a yellow task (i.e., has four times more possible actions). We vary the task difficulty by setting a minimum and a maximum number of allowed actions.

Additionally, we limit both the quality of prior knowledge of agents and move to a multi-state task. Now tasks have to be completed by sequentially solving a number of subtasks (an analogy can be made with an agent moving through a 1-dimensional grid-world to reach a certain goal location). When all subtasks are successfully completed, the agent is rewarded with a reward of 1.0. The latter can be seen as a first

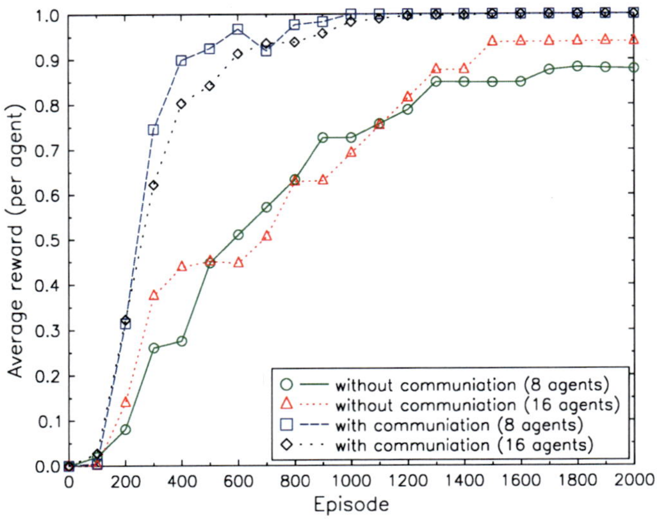

Fig. 9 Multi-state task

step toward multi-state dispersion and cooperation games and is a similar environment as the one used in [3]. We sample random tasks and agents out of the 24 types uniformly. Whereas in the previous settings, pure experts were defined by a single property; now all three properties are taken into account. A pure expert (i.e., an agent that can immediately complete the task) must have all properties matched with those of the task. If only a subset of the properties match, the agent will have a certain degree of expertise, depending on the number of matching properties and their type (e.g., a matching color property may have a larger influence than a matching shape property). Given a set of agent and task properties, a degree of expertise can be calculated. If an agent is, let us say 80% expert of a task, the agent only needs to explore 20% of the action space. Of course, if no properties match, the agent has no expertise at all and must explore the whole-action space for this task. Again, in this setting it is possible that no, or very few, experts are generated for the tasks at hand.

Figure 9 presents the result obtained for a multi-state task consisting of four subtasks. It can be noted that when using RRL, communication clearly helps the agent to learn a better policy faster.

6 Conclusions

In this article, we introduced the novel idea of cross-fertilization between relational reinforcement learning and multi-agent systems to perform well complex multi-state dynamic planning tasks. We proposed to use a relational representation of the state space in multi-agent reinforcement learning as this has many proved benefits over the propositional one, as, for instance, handling large state spaces, a rich relational language, modeling of other agents without a computational explosion, and generalization over new derived knowledge.

We started this article with a short introduction to relational reinforcement learning and multi-agent learning. Furthermore, we investigated the positive effects of relational reinforcement learning applied to the problem of agent communication in MAS. More precisely, we investigated the learning performance of RRL given some communication constraints. On the basic of our empirical results in our abstract co-operative MAS we make the following conclusions:

1. Communication in general pays off in MAS, and in particular when employing a relational representation of the MAS (see Figure 4).
2. Communication is still beneficial to the learning performance, even when communication is limited (see Figure 5) or unreliable (see Figure 6).
3. Communication is in particular useful for MAS with large number of agents (see Figure 8), and complex tasks (see Figure 7). Therefore, our approach scales well to more complex MAS.
4. Rapid convergence is achieved in more complex multi-state problem (see Figure 9).

In our future work, we plan to continue along this track and gradually extend our approach to even more complex and realistic settings, e.g., including more elaborate communication possibilities, a cost model, and interference between agents since this is an important part of every multi-agent system. Furthermore, a thorough theoretical study of the described properties and settings will be part of our future research.

References

1. Bellman, R.: Dynamic Programming. Princeton University Press, Princeton (1957)
2. Claus, C., Boutilier, C.: The dynamics of reinforcement learning in cooperative multi-agent systems. In: Proceedings of the 15th International Conference on Artificial Intelligence, pp. 746–752 (1998)
3. Croonenborghs, T., Tuyls, K., Ramon, J., Bruynooghe, M.: Multi-agent relational reinforcement learning. In: Tuyls, K., 't Hoen, P.J., Verbeeck, K., Sen, S. (eds.) LAMAS 2005. LNCS (LNAI), vol. 3898, pp. 192–206. Springer, Heidelberg (2006), `http://www.cs.kuleuven.ac.be/cgi-bin-dtai/publ_info.pl?id=41977`
4. Driessens, K.: Relational reinforcement learning. Ph.D. thesis, Department of Computer Science, Katholieke Universiteit Leuven (2004), `http://www.cs.kuleuven.be/publicaties/doctoraten/cw/CW2004_05.abs.html`
5. Driessens, K., Ramon, J., Blockeel, H.: Speeding up relational reinforcement learning through the use of an incremental first order decision tree learner. In: Flach, P.A., De Raedt, L. (eds.) ECML 2001. LNCS (LNAI), vol. 2167, pp. 97–108. Springer, Heidelberg (2001)
6. Džeroski, S., De Raedt, L., Driessens, K.: Relational reinforcement learning. Machine Learning 43, 7–52 (2001)

7. Finzi, A., Lukasiewicz, T.: Game theoretic golog under partial observability. In: AAMAS 2005: Proceedings of the fourth international joint conference on Autonomous agents and multiagent systems, pp. 1301–1302. ACM, New York (2005), http://doi.acm.org/10.1145/1082473.1082743
8. Guerra-Hernández, A., Fallah-Seghrouchni, A.E., Soldano, H.: Learning in BDI multiagent systems. In: Dix, J., Leite, J. (eds.) CLIMA 2004. LNCS (LNAI), vol. 3259, pp. 218–233. Springer, Heidelberg (2004)
9. Hoen, P., Tuyls, K.: Engineering multi-agent reinforcement learning using evolutionary dynamics. In: Proceedings of the 15th European Conference on Machine Learning (2004)
10. Hu, J., Wellman, M.P.: Experimental results on Q-learning for general-sum stochastic games. In: ICML 2000: Proceedings of the Seventeenth International Conference on Machine Learning, pp. 407–414. Morgan Kaufmann Publishers Inc., San Francisco (2000)
11. Kaelbling, L., Littman, M., Moore, A.: Reinforcement learning: A survey. Journal of Artificial Intelligence Research (1996)
12. Letia, I.A., Precup, D.: Developing collaborative golog agents by reinforcement learning. International Journal on Artificial Intelligence Tools 11(2), 233–246 (2002)
13. Littman, M.: Markov games as a framework for multi-agent reinforcement learning. In: Proceedings of the Eleventh International Conference on Machine Learning, pp. 157–163 (1994)
14. Muggleton, S., De Raedt, L.: Inductive logic programming: Theory and methods. Journal of Logic Programming 19,20, 629–679 (1994)
15. Nowé, A., Parent, J., Verbeeck, K.: Social agents playing a periodical policy. In: Proceedings of the 12th European Conference on Machine Learning, Freiburg, pp. 382–393 (2001)
16. van Otterlo, M.: A characterization of sapient agents. In: International Conference Integration of Knowledge Intensive Multi-Agent Systems (KIMAS 2003), Boston, Massachusetts (2003)
17. Panait, L., Luke, S.: Cooperative multi-agent learning: The state of the art. Autonomous Agents and Multi-Agent Systems 11(3), 387–434 (2005)
18. Panait, L., Tuyls, K., Luke, S.: Theoretical advantages of lenient learners: An evolutionary game theoretic perspective. Journal of Machine Learning Research 9, 423–457 (2008)
19. Puterman, M.: Markov decision processes: Discrete stochastic dynamic programming. John Wiley and Sons, New York (1994)
20. Rummery, G.A., Niranjan, M.: On-line Q-learning using connectionist systems. Tech. rep., Cambridge University Engineering Department (1994)
21. Sen, S., Airiau, S., Mukherjee, R.: Towards a Pareto-optimal solution in general-sum games. In: The Proceedings of the Second Intenational Joint Conference on Autonomous Agents and Multiagent Systems, Melbourne, Australia, July 2003, pp. 153–160 (2003)
22. Stone, P.: Layered learning in multi-agent systems. MIT Press, Cambridge (2000)
23. Sutton, R., Barto, A.: Reinforcement Learning: an introduction. MIT Press, Cambridge (1998)
24. Sutton, R., Barto, A.: Reinforcement Learning: An Introduction. The MIT Press, Cambridge (1998)
25. Tadepalli, P., Givan, R., Driessens, K.: Relational reinforcement learning: An overview. In: Proceedings of the ICML 2004 Workshop on Relational Reinforcement Learning (2004)

26. Tumer, K., Wolpert, D.: COllective INtelligence and Braess' Paradox. In: Proceedings of the Sixteenth National Conference on Artificial Intelligence, pp. 104–109 (2000)
27. Tuyls, K., Verbeeck, K., Lenaerts, T.: A selection-mutation model for Q-learning in Multi-Agent Systems. In: The second International Joint Conference on Autonomous Agents and Multi-Agent Systems. ACM Press, Melbourne (2003)
28. van Otterlo, M.: The logic of adaptive behavior: Knowledge representation and algorithms for the Markov decision process framework in first-order domains. Ph.D. thesis, Department of Computer Science, University of Twente, Enschede, The Netherlands, p. 512 (May 2008)
29. Watkins, C.: Learning with delayed rewards. Ph.D. thesis, Cambridge University (1989)
30. Wiering, M.: Explorations in efficient reinforcement learning. Ph.D. thesis, Universiteit van Amsterdam (1999)

Switching between Representations in Reinforcement Learning

Harm van Seijen, Shimon Whiteson, and Leon Kester

Abstract. This chapter presents and evaluates an online representation selection method for factored Markov decision processes (MDPs). The method addresses a special case of the feature selection problem that only considers certain subsets of features, which we call *candidate representations*. A motivation for the method is that it can potentially deal with problems where other structure learning algorithms are infeasible due to a large degree of the associated dynamic Bayesian network. Our method uses switch actions to select a representation and uses off-policy updating to improve the policy of representations that were not selected. We demonstrate the validity of the method by showing for a contextual bandit task and a regular MDP that given a feature set containing only a single relevant feature, we can find this feature very efficiently using the switch method. We also show for a contextual bandit task that switching between a set of relevant features and a subset of these features can outperform each of the individual representations. The reason for this is that the switch method combines the fast performance increase of the small representation with the high asymptotic performance of the large representation.

1 Introduction

In *reinforcement learning* (RL) [7, 13], an agent seeks an optimal control policy for a sequential decision problem. When the sequential decision problem is modeled as a *Markov decision process* (MDP) [2], the agent's policy can be represented as a mapping from each state it may encounter to a probability distribution over the available actions.

Harm van Seijen and Leon Kester
TNO Defence, Security and Safety, Oude Waalsdorperweg 63, 2597 AK The Hague, The Netherlands
e-mail: harm.vanseijen@tno.nl, leon.kester@tno.nl

Shimon Whiteson
University of Amsterdam, Science Park 107, 1098 XG Amsterdam, The Netherlands
e-mail: s.a.whiteson@uva.nl

R. Babuška & F.C.A. Groen (Eds.): Interactive Collaborative Information Systems, SCI 281, pp. 65–84.
springerlink.com

RL suffers from what is often called the 'curse of dimensionality': the state space grows exponentially as function of the number of problem parameters and it becomes quickly infeasible to solve. For the subdomain of factored MDPs, where the state space is described through a set of features, one can often considerably reduce the complexity of the problem by making use of its structure. Boutilier et al. [3] showed for the planning case that by representing the transition and reward functions as a dynamic Bayesian network (DBN) and representing the associated conditional probability tables (CPTs) as a tree, it becomes possible to avoid the explicit enumeration of states. This allows the computation of value estimates and policy to be performed on partitions instead of individual states. They later extended this work by using an even more compact way to represent the CPTs, algebraic decision diagrams [6], and by combining it with approximate dynamic programming techniques [11].

The success of this approach inspired other people to use the idea of representing the transition and reward function as a DBN to speed up learning of the model in model-based RL. Although these methods still use explicit state enumeration, the model is more efficiently learned and the sample efficiency is also improved. The early work assumed the structure of the DBN is known, and only focussed on learning the values of the decision trees [8]. Strehl et al. [12] were able to relax the prior knowledge by showing how the structure of the DBN could be efficiently learned with only the degree of the DBN as prior knowledge. This work recently got extended by combining it with the KWIK framework [9], resulting in a method with better bounds [4].

Although these recent methods can learn the structure efficiently when the degree is small, the representation size grows exponentially in the degree N of the DBN [1, 4, 12]; therefore, for large values of N, learning the DBN model becomes infeasible.

We present in this chapter an alternative approach that can potentially better deal with cases, where the current structure-learning algorithms are infeasible. Our approach is different in two major ways. First, our method searches for the feature set that has currently the best performance, instead of searching for the feature set that has the highest performance in the limit. The second difference is that our method relies on knowledge in the form of relevant feature combinations. The advantage of this form of prior knowledge is that it allows for representing more specific domain knowledge than just the degree of the DBN, effectively reducing the problem size.

The general principle behind the algorithm is a very intuitive one. Each subset of features forms a *candidate representation*. At the start of each episode, the agent selects one candidate representation and uses its policy to perform action–selection and policy updating till the end of the episode. If the episode is finished, the value of the representation as a whole is updated.

Selection of the representation suffers from the same exploration-exploitation dilemma as regular action selection, i.e. the agent can choose either the representation that has the highest current estimated value, or it can choose a suboptimal representation to improve its policy and its estimate of that policy. Due to this exploration cost there is a trade-off between the performance gain by using a more compact representation and the performance decrease due to the exploration of suboptimal representations. To reduce the exploration cost, we defined a number of conditions under which off-policy updating is possible of the policies of the unselected representations and of the value of the representation as a whole. This can increase the performance considerably.

The rest of this chapter is organized as follows. In Sect. 2, we give a formal definition of a factored MDP. In Sect. 3, we define different types of features and relate these to valid candidate representations. Section 4 describes the algorithm for the case of a contextual bandit problem, a variation to the classical multi-armed bandit problem where the outcome of pulling an arm depends on context information. In Sect. 5 we extend the ideas from Sect. 4 to the case of an MDP. In Sect. 6, we provide empirical results for the contextual bandit case and the full MDP case. We finish this chapter with a conclusion and a future work section.

2 Background on Factored MDP

In this section, we will describe the framework of a factored MDP. We will adopt the notation from [5]. An MDP is formally defined as a 4-tuple $\langle X, A, T, R\rangle$ where

- X is the set of all states the agent can encounter
- A is the set of all actions available
- $T(x,a,x') = P(x'|x,a)$ is the transition function
- $R(x,a) = E(r|x,a)$, is the reward function.

An MDP satisfies the equations for the *Markov property*:

$$P(x_{t+1},r_{t+1}|x_t,a_t) = P(x_{t+1},r_{t+1}|x_t,a_t,r_t,x_{t-1},a_{t-1},\ldots,r_1,x_0,a_0) \tag{1}$$

for all x_{t+1},r_{t+1} and all possible values of $x_t,a_t,r_t,\ldots,r_1,x_0,a_0$.

For a factored MDP, the state space is described using a set of state variables or *features*: $X = \{X_1,\ldots,X_n\}$, where each X_i takes on values in some domain $Dom(X_i)$. We will only consider discrete and finite domains throughout this chapter. We will use the notation $\mathscr{D}$ to indicate the set of values with non-zero probability. As an example, consider a feature set X consisting of two identical features X_1 and X_2, for the size of X the following holds:

$$|Dom(X)| = |Dom(X_1)| \cdot |Dom(X_2)|$$

while

$$|\mathscr{D}(X)| = |\mathscr{D}(X_1)| = |\mathscr{D}(X_2)|$$

A state x defines a value $x(i)$ for each variable X_i. We will refer to the set of variables that describes a state space as a *representation*. For an instantiation $y \in \mathscr{D}(Y)$ and a subset of these variables $Z \subset Y$, we use $y[Z]$ to denote the value of the variables Z in the instantiation y.

The goal of a RL agent is to find an optimal policy $\pi^* = P(a|x)$, which maximizes the expected discounted return $\langle R_t\rangle$ for state s_t, with R_t defined as

$$R_t = r_{t+1} + \gamma r_{t+2} + \gamma^2 r_{t+3} + \cdots = \sum_{k=0}^{\infty} \gamma^k r_{t+k+1} \qquad (2)$$

where γ is a discount factor with $0 \leq \gamma \leq 1$.

3 Feature Selection in Factored MDPs

In this section, we define several classes of features and show how these relate to constructing a *valid candidate representation*.

3.1 Feature Types

In this section we consider features from the feature set X, where X describes the set of all states for some MDP $M = \langle X, A, T, R \rangle$.

Definition 1. A feature $X_k \subseteq X$ is *irrelevant* with respect to Y if the following holds for all x^-_{t+1}, y^+_t, a_t

$$P(x^-_{t+1}, r_{t+1} | y^-_t, a_t) = P(x^-_{t+1}, r_{t+1} | y^+_t, a_t) \qquad (3)$$

with

$$\begin{aligned} x^-_{t+1} &\in \mathscr{D}(X \setminus X_k) \\ y^+_t &\in \mathscr{D}(Y \cup X_k) \\ y^-_t &= y^+_t[Y \setminus X_k] \end{aligned}$$

Informally spoken, an irrelevant feature is a feature who's feature value neither affect the next value of any other feature (except potentially its own value) nor the reward received when a representation is used with all features from $Y \cup X_k$ are present.

The complement class is the class of relevant features:

Definition 2. A feature $X_k \subseteq X$ is *relevant* with respect to Y if it is not irrelevant with respect to Y.

We can divide the irrelevant feature class into three subclasses: constant, empty, and redundant features.

Definition 3. A *constant* feature $X_k \subseteq X$ is a feature with $|\mathscr{D}(X_k)| = 1$.

A constant feature is a feature that never changes value. Note that a feature that stays constant during an episode but changes value in between episodes is not constant according to this definition nor is it irrelevant. The difference between the two is that a feature with $|\mathscr{D}(X_k)| = 1$ can be removed from a representation, without affecting the Markov property (as we will prove in Sect. 3.2), while removing a feature of the second type can cause a violation of the Markov property. The Markov property can be violated since the history might reveal something about the current value of the feature and therefore affect the transition probability.

Definition 4. An *empty* feature $X_k \subseteq X$ is a feature for which the following holds

$$|\mathscr{D}(X_k)| > 1$$

$$P(x^-_{t+1}, r_{t+1}) = P(x^-_{t+1}, r_{t+1} | x(k)_t)$$

for all $x^-_{t+1} \in \mathscr{D}(X \setminus X_k), r_{t+1}, x(k)_t$.

An empty feature is a feature that contains no useful information, i.e., it is irrelevant with respect to the empty set.

Definition 5. A feature $X_k \subseteq X$ is *redundant* with respect to Y if it is irrelevant with respect to Y and it is not an empty or a constant feature.

A redundant feature does contain useful information; however, this information is shared with some other features in Y. By removing some feature set Z from Y, feature X_i can become a relevant feature with respect to $Y \setminus Z$.

Apart from the relevant/irrelevant classification, we define two other classifications: dependent and independent features.

Definition 6. A feature $X_k \subseteq X$ is *independent* if for all x_{t+1}, x_t, a_t the following holds

$$P(x_{t+1}(k)) = P(x_{t+1}(k) | x^-_{t+1}, r_{t+1}, x_t, a_t) \tag{4}$$

with

$$\begin{aligned} x_{t+1}(k) &= x_{t+1}[X_k] \\ x^-_{t+1} &= x_{t+1}[X \setminus X_k] \end{aligned}$$

So the value of an independent feature does not depend on the previous state or current values of other features or the reward just received. Note that this does not prevent an independent feature to influence the next feature value of other features or the next reward. An independent feature is unique in the sense that it can contain relevant information, but leaving the feature out still gives a Markov representation as we will prove in the next subsection. Therefore, the regular methods still converge, although the resulting policy is not optimal in X. However, since we are primarily interested in the best online performance instead of the optimal performance, omitting independent features can play an important role in finding the best representation.

For the sake of completeness, we also define the counterpart of an independent feature:

Definition 7. A feature $X_k \subseteq X$ is *dependent* if it is not independent.

3.2 Candidate Representation

A subset of features $Y \subseteq X$ forms a candidate representation. We will now define what we mean by a *valid* candidate representation.

Definition 8. Consider the MDP $M = \langle X, A, T, R \rangle$. A subset of features $Y \subseteq X$ is a *valid candidate representation* if the Markov property applies to it, i.e., if the following conditions hold:

$$P(y_{t+1}, r_{t+1} | y_t, a_t) = P(y_{t+1}, r_{t+1} | y_t, a_t, r_t, y_{t-1}, a_{t-1}, \ldots, r_1, y_0, a_0) \qquad (5)$$

for all $y_{t+1} \in \mathscr{D}(Y), r_{t+1}$ and all possible values of $y_t, a_t, r_t, \ldots, r_1, y_0, a_0$.

The full feature set X is per definition Markov. The following theorem shows how different feature sets can be constructed from X that maintain the Markov property, i.e., that are valid candidate representations. We will only consider valid candidate representations in the rest of this chapter.

Theorem 1. *Consider the MDP $M = \langle X, A, T, R \rangle$. A subset of features $Y \subset X$ is a valid candidate representation if for the set of missing features the following holds*

$$\forall X_i \notin Y : X_i \text{ is irrelevant w.r.t. } Y \text{ or } X_i \text{ is an independent feature.}$$

Proof. To prove the above theorem it suffices to prove that if for an arbitrary set Z with $Y \subseteq Z \subseteq X$, the Markov property holds; it also holds for $Z \setminus Z_k$ if Z_k is irrelevant w.r.t. Y or independent.

For the proof, we will make use of the following formulas, which can be easily deduced from the Bayesian statistics rules:

$$P(a|b,c) \cdot P(b|c) \quad = \quad P(a,b|c) \qquad (6)$$
$$P(a|b) = P(a|b,c,d) \quad \Rightarrow \quad P(a|b) = P(a|b,c) \qquad (7)$$
$$P(a,b_i|c) = P(a,b_i|c,d) \quad \text{for all } i \quad \Rightarrow \quad P(a|c) = P(a|c,d) \qquad (8)$$

with $P(b_i, b_j) = 0$ for all $i \neq j$ and $\sum_i P(b_i) = 1$.

First, we will prove that the Markov property is conserved if Z_k is an irrelevant feature.

$$P(z_{t+1}, r_{t+1} | z_t, a_t) = P(z_{t+1}, r_{t+1} | z_t, a_t, r_t, z_{t-1}, a_{t-1}, \ldots, r_1, z_0, a_0)$$
$$P(z^-_{t+1}, r_{t+1} | z_t, a_t) = P(z^-_{t+1}, r_{t+1} | z_t, a_t, r_t, z_{t-1}, a_{t-1}, \ldots, r_1, z_0, a_0) \quad \text{using rule (8)}$$
$$P(z^-_{t+1}, r_{t+1} | z^-_t, a_t) = P(z^-_{t+1}, r_{t+1} | z_t, a_t, r_t, z_{t-1}, a_{t-1}, \ldots, r_1, z_0, a_0) \quad \text{using rule (3)}$$
$$P(z^-_{t+1}, r_{t+1} | z^-_t, a_t) = P(z^-_{t+1}, r_{t+1} | z^-_t, a_t, r_t, z^-_{t-1}, a_{t-1}, \ldots, r_1, z^-_0, a_0) \quad \text{using rule (7)}$$

Now, for an independent feature:

$$P(z_{t+1}, r_{t+1} | z_t, a_t) = P(z_{t+1}, r_{t+1} | z_t, a_t, r_t, z_{t-1}, a_{t-1}, \ldots, r_1, z_0, a_0)$$
$$P(z^-_{t+1}, r_{t+1} | z_t, a_t) = P(z^-_{t+1}, r_{t+1} | z_t, a_t, r_t, z_{t-1}, a_{t-1}, \ldots, r_1, z_0, a_0) \quad \text{using rule (8)}$$

We now multiply both sides with $P(z(k)_t)$ and rewrite the left hand side as

$$P(z^-_{t+1}, r_{t+1} | z_t, a_t) \cdot P(z(k)_t)$$
$$= P(z^-_{t+1}, r_{t+1} | z(k)_t, z^-_t, a_t) \cdot P(z(k)_t | z^-_t, a_t) \quad \text{using rule (4)}$$
$$= P(z^-_{t+1}, r_{t+1}, z(k)_t | z^-_t, a_t) \quad \text{using rule (6)}$$

and the right hand side as

$$P(z^-_{t+1}, r_{t+1} | z_t, a_t, r_t, z_{t-1}, a_{t-1}, \ldots, r_1, z_0, a_0) \cdot P(z(k)_t)$$
$$= P(z^-_{t+1}, r_{t+1} | z(k)_t, z^-_t, a_t, r_t, z_{t-1}, a_{t-1}, \ldots, r_1, x_0, a_0)$$
$$\cdot P(z(k)_t | z^-_t, a_t, r_t, z_{t-1}, a_{t-1}) \quad \text{using rule (4)}$$
$$= P(z^-_{t+1}, r_{t+1} | z(k)_t, z^-_t, a_t, r_t, z_{t-1}, a_{t-1}, \ldots, r_1, x_0, a_0)$$
$$\cdot P(z(k)_t | z^-_t, a_t, r_t, z_{t-1}, a_{t-1}, \ldots, r_1, z_0, a_0) \quad \text{using rule (5)}$$
$$= P(z^-_{t+1}, r_{t+1}, z(k)_t | z^-_t, a_t, r_t, z_{t-1}, a_{t-1}, \ldots, r_1, z_0, a_0)$$

Combining these results gives

$$P(z^-_{t+1}, r_{t+1}, z(k)_t | z^-_t, a_t) = P(z^-_{t+1}, r_{t+1}, z(k)_t | z^-{}_t, a_t, r_t, z_{t-1}, a_{t-1}, \ldots, r_1, z_0, a_0)$$
$$P(z^-_{t+1}, r_{t+1} | z^-_t, a_t) = P(z^-_{t+1}, r_{t+1} | z^-_t, a_t, r_t, z^-_{t-1}, a_{t-1}, \ldots, r_1, z^-_0, a_0) \text{ using (7),(8)} \quad \square$$

4 Representation Selection for a Contextual Bandit

In this section, we describe our representation selection method for a contextual bandit, a subclass of MDP problems where only a single action has to be taken. We will start by introducing an example contextual bandit task that will serve as a reference for the rest of this section. In Sect. 4.2, we will explain how formally the selection problem can be seen as solving a derived MDP. Section 4.3 shows how a representation can be evaluated. Section 4.4 describes how the performance of our method can be increased by off-policy updating the Q-values of the representations and of the switch actions. Note that for a contextual bandit, each subset of features forms a valid candidate representation, since there is no history to consider and, therefore, the Markov property is never violated.

4.1 A Contextual Bandit Example

The standard multi-armed bandit problem represents a class of RL tasks based on a traditional slot machine but with multiple arms. Pulling an arm results in a reward drawn from a distribution associated with that arm. The goal of an agent is to maximize its reward over iterative pulls. The agent has no prior knowledge about the distributions associated with each arm. The dilemma the agent faces when performing this task is to either pull the arm that has currently the highest expected reward or to improve the estimates of the other arms. This is often referred to as the exploration versus exploitation dilemma. The contextual bandit problem is a variation of

(a)

X	Y	R
0	0	+4
0	1	+2
1	0	-2
1	1	-4

(b)

X	R
0	+3
1	-3

(c)

Y	R
0	+1
1	-1

Fig. 1 Expected return of arm a_1 for the full representation (a) and representation S_X (b) and S_Y (c) assuming a prior probability of 0.5 for the features values 0 and 1 of features X and Y

the multi-armed bandit problem where the agent observes context information that can affect the reward distributions of the arms. We can connect the bandit problem to an MDP by interpreting the arms as actions and the context as the state. When the context is described through features, we can interpret the problem as a factored MDP where episodes have length 1.

We will now describe an example of a contextual bandit that will be used as a reference for the next sections. Consider a contextual bandit with two arms: a_0 and a_1. The context is described by two binary features: X and Y. The expected reward for a_0 is 0 for all context states, while the expected reward for a_1 can be either positive or negative depending on the context state. Figure 1(a) shows the expected reward for arm a_1 for all feature value combinations. The agent is given two candidate representations to choose from on this task. Representation S_X consists of only feature X, while representation S_Y consists of only feature Y. Both representations partition the original state space into two abstract states, one corresponding to each feature value. We can calculate the expected reward for these states from the full table in Fig. 1(a) and the prior probabilities of the feature values, which we assume to be 0.5 for both values of X and Y. Figure 1(b) and (c) shows the expected reward of arm a_1 for representation S_X and S_Y, respectively. Note that the optimal policy of representation S_X has an expected reward of 1.5 (action a_0 will be chosen when $X = 1$), while the optimal policy for representation S_Y has an expected reward of 0.5.

In the next section, we will show for this example task how the concept of representation selection can be interpreted as solving a derived MDP.

4.2 Constructing the Switch Representation

Formally, the representation is one of the elements that defines an MDP task. Therefore, an agent that can choose between multiple representations actually chooses between multiple MDP tasks, each of them having a different state space X, transition function T, and reward function R, but with equal action set A. We can combine these multiple MDP tasks into a single 'switch' MDP task by defining *switch actions*, actions that have no effect on the environment but merely select the subtask the agent will try to solve.

In Fig. 2, we can see how this switch MDP is constructed from the individual MDPs for the example task described earlier. The agent can choose between an

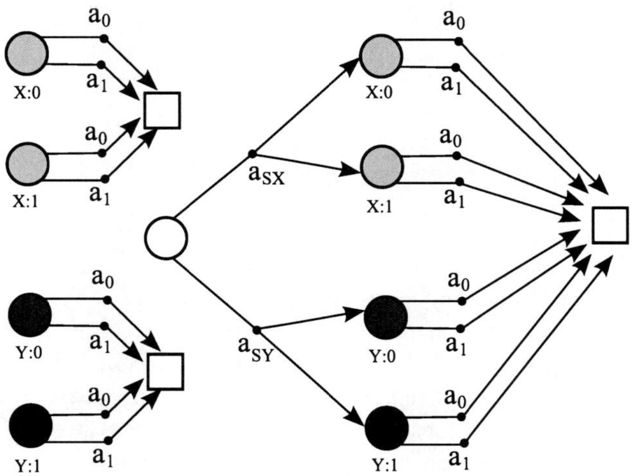

Fig. 2 The individual MDPs for the contextual bandit example and the derived switch MDP. The circles indicate specific states, the squares indicate the end state, and the black dots indicate actions. Stochastic actions have multiple arrows.

MDP that uses representations S_X, based on feature X, and S_Y, which is based on feature Y. Switch actions are defined that let the agent use either representation S_X or S_Y. The state space of the switch MDP consists of 6 states, a single start state from which the representation selection occurs, the non-terminal states from representations S_X and S_Y, and one terminal state. Note that by including switching actions, the derived MDP has episodes of length 2. The switch actions are stochastic actions with 0 reward.

Defining a single start state from which switch actions are selected guarantees that the Q-values of the switch actions are set independent from the feature values. We will motivate this approach by demonstrating what happens for our bandit example if the representation is selected based on the Q-values of the candidate representations. Assume that the current state is defined by features values (x = 1, y = 0). From Fig. 1(a), we can see that the expected reward is -2. Representation S_X, which only sees feature X, predicts a reward of -3 in this case, while representation S_Y predicts a reward of +1. Therefore, if the representation selection would be based on the Q-values of the representations under consideration, S_Y would be selected in this case, and arm a_1 would be pulled which, despite the prediction of +1, would result in a reward drawn from a distribution with a mean of -2. On the other hand, if we had selected S_X, action a_0 would be the optimal action, which has an expected reward of 0. This illustrates that using the Q-values of the individual representations can lead to incorrect behavior. By defining switch actions with specific Q-values this problem is avoided, since these Q-values represent the average performance of a representation, which is 1.5 for S_X and 0.5 for S_Y.

The switch MDP is an MDP with special properties. One of its properties is that not all the actions are external actions. Although the MDP framework does not distinguish between internal or external actions, for our performance criteria it does matter, since we want to optimize only with respect to the external actions. In the following sections, we will see how this property can be exploited to increase the online performance.

4.3 Evaluation of a Representation

In the example bandit task, the agent has to choose between only two equally sized representations. In the general case, however, there can be many candidate representations of different sizes and information content. The policy of a representation is not the same at each time step, but it will improve over time until all of its Q-values have converged. Since small representations converge faster, a good online strategy can be to use a small representations for the early phase of learning and switch to a larger representations with more information content, at a later stage. To be able to switch from one representation to another at the right moment, the agent needs to know the *current expected return* for each representation, which we define as follows:

Definition 9. The *current expected return of a representation* is the expected return of the start state of that representation, assuming the current policy of the representation is used. If the start state is drawn from a probability distribution of start states, then the current expected return refers to the weighted average of the expected returns of the start states, weighted according to their probability.

The Q-value of a switch action gives an estimate of the current expected return of the corresponding representation. To get a meaningful estimate, this Q-value cannot be updated by bootstrapping from the Q-values of the representation, since these converge too slowly and do not give an accurate estimate for the current policy. Instead, we use a Monte Carlo update, which uses the complete return. Since the policy of representations changes during learning, the most recent return gives the least biased estimate for determining the *current* expected return. However, since the variance on the return is very high, it is necessary to average over multiple returns to get an accurate estimate. Tuning the learning rate is therefore essential for a good trade-off between the error due to variance and the error due to bias. The learning rate of the switch actions is therefore set independently from the learning rate for the representations.

4.4 Improving Performance by Off-Policy Updating

Since the agent has to consider multiple representations, the exploration cost increases. Fortunately, the exploration cost can be reduced considerably by using off-policy updating techniques to simultaneously improve the policy of all representations. We can reduce the exploration cost further by also using off-policy updating for the switch actions. Off-policy updating means that the policy used to generate the behavior, i.e., the *behavior policy*, is different from the policy we try to evaluate, i.e., the *evaluation policy*.

4.4.1 Updating of the Unselected Representations

At the start of an episode, the agent selects via a switch action the representation it will use to base its external action selection on. If we assume that the agent can observe the full feature set, it can also observe what state it would end up in had it taken a different switch action. Therefore, parallel to the real experience sequence from the selected representation, an experience sequence for the other candidate representations can be constructed. These sequences can be used to perform extra updates.

Given our example task, assume that the context is described by feature values $x = 0$ and $y = 1$ and assume that representation S_X is selected for action selection. A possible experience sequence is then

$s_{start} \rightarrow a_{S_X} \rightarrow s_{x:0} \rightarrow a_1 \rightarrow r, s_{end}$

With this sequence we can update $Q(s_{x:0},a_1)$. By observing feature Y, we can create an experience sequence for S_Y

$s_{start} \rightarrow a_{S_Y} \rightarrow s_{y:1} \rightarrow a_1 \rightarrow r, s_{end}$

This parallel sequence can be used to update $Q(s_{y:1},a_1)$. We have to take into account, however, that the update is biased since the action selection is based on representation S_X.

Underlying this bias is that state $s_{y:1}$ is actually an aggregation of two states from the full features set: $(x = 0, y = 1)$ and $(x = 1, y = 1)$. The expected reward for state–action pair $(s_{y:1},a_1)$ is a weighted average of the expected rewards of these two underlying states. Since representation S_X aggregates the states from the full features set in a different way, the two states underlying $s_{y:1}$ correspond to two different states in representation S_X. Therefore, if the selection probability of a_1 for those states is different, the rewards are not properly weighted to estimate the expected reward for $(s_{y:1},a_1)$.

This bias can be avoided by only updating an unselected representation under certain conditions.

Theorem 2. *Assume that action selection occurs according to representation S_1 and we want to update representation S_2 based on a parallel experience sequence. We can perform an unbiased update of S_2 if one of the following two conditions hold:*

1. *If the feature set of S_1 is a subset of the feature set of S_2*
2. *If the action was an exploratory action under an exploration scheme that does not depend on the specific state, like an ε-greedy exploration scheme*

Proof (sketch). Under the first condition, a single state of S_2 always corresponds to a single state of S_1, and therefore the states aggregated by S_2 never get incorrectly weighted. For the second case, remember that both S_1 and S_2 are constructed from X by removing irrelevant and independent features. Therefore, all the features that are in S_1 but not in S_2 are either irrelevant or independent. If they are irrelevant, the one-step model is the same regardless of the feature value and therefore using a different weighting of the aggregated states does not matter. If the feature is an independent feature, the one-step model should be weighted according to the feature value probability. When taking an exploration action, the selection probability of an action is the same regardless of the underlying state, and therefore this guarantees that we correctly weigh the aggregated states.

By off-policy updating of the unselected representations, we can improve the policy of a representation even if we do not select it, resulting in an overall better performance for the switching method. Besides updating the Q-values of the unselected representations, we can also update the Q-values of the switch actions corresponding to the unselected representations. This is discussed in the next section.

4.4.2 Updating the Unselected Switch Actions

Since the switch actions are updated using Monte Carlo updates, we cannot apply the conditions from the previous section for off-policy updating of the switch actions. Off-policy Monte Carlo updating can be achieved using a technique called importance sampling [10].

To understand the difference with regular (on-policy) Monte Carlo updates, consider that we determine the Q-value of a state–action pair (x,a) by simply taking the average of all returns seen so far:

$$Q(x,a) = \frac{\sum_{i=1}^{N} R_i(x,a)}{N} \tag{9}$$

where $R_i(x,a)$ is the return followed by the ith visit of state–action pair (x,a) and N is the total number of returns observed for (x,a). A similar off-policy version can then be made by taking the weighted average:

$$Q(x,a) = \frac{\sum_{i=1}^{N} w_i(x,a) \cdot R_i(x,a)}{\sum_{i=1}^{N} w_i(x,a)} \tag{10}$$

where $w_i(x,a)$ is the weight assigned to the ith return for (x,a). The value of $w_i(x,a)$ is computed as follows. Let $p(x,a)$ be the probability of the state action sequence following (x,a) occurring under the estimation policy π and $p'(x,a)$ be the probability of it occurring under the behavior policy π'. Then the weight $w_i(x,a)$ is equal to the relative probability of the observed experience sequence of occurring under π and π', i.e., by $p(x,a)/p'(x,a)$. These probabilities can be expressed in their policy probabilities by

$$w(x_t,a_t) = \frac{p(x_t,a_t)}{p'(x_t,a_t)} = \prod_{k=t+1}^{T-1} \frac{\pi(x_k,a_k)}{\pi'(x_k,a_k)} \tag{11}$$

For a deterministic evaluation policy π, the weight w is non-zero only when all actions taken under π' match the action that would have been taken under π. If this is the case, the above equation simplifies to

$$w(x_t,a_t) = \prod_{k=t+1}^{T-1} \frac{1}{\pi'(x_k,a_k)} \qquad \text{if } \pi(x_k) = a_k \text{ for all } k \geq t+1 \tag{12}$$

where $\pi(x_k)$ refers to the action the agent would take at timestep k (with probability 1) when following this deterministic policy.

We will now consider again our example contextual bandit task and the state–action sequence from the unselected representation:

$s_{start} \rightarrow a_{S_Y} \rightarrow s_{y:1} \rightarrow a_1 \rightarrow r, s_{end}$

Since the state–action pair that requires off-policy updating, (s_{start}, a_{S_Y}), is followed by only a single action, and since we use a deterministic evaluation policy, the weight expression is reduced even further to

$$w = \frac{1}{\pi'(s_{x:0}, a_1)} \qquad \text{if } \pi(s_{x:0}) = a_1 \tag{13}$$

Given that we use an ε-greedy behavior policy and the condition that $\pi(s_{x:0}) = a_1$, the weight w is a fixed value and can be scaled to 1 in this case.

If the external action is an optimal action, we can perform the following update of the switch action

$$Q(s_0, a_{S_Y}) = (1 - \alpha)\, Q(s_0, a_{S_Y}) + \alpha \cdot r \tag{14}$$

To update the switch action of an unselected representation S_i, two conditions have to hold

1. The reward should be unbiased with respect to S_i, i.e., one of the two conditions of Sect. 4.4 should hold.
2. The selected action is an optimal action according to S_i.

5 Representation Selection for an MDP

Although the framework introduced in the previous section is not strictly limited to the bandit case, applying it to the more general MDP task brings up some complications. As explained in the previous section, for a contextual bandit problem, any subset of features from the total feature set forms a valid candidate representations. This is not generally the case for a full MDP, since partial observability often leads to non-Markov behavior. For a valid candidate representation, Theorem 1 should hold. We will only consider valid candidate representations in this section. In the following subsections, we explain how off-policy updating works in the case of an MDP.

5.1 Off-Policy Updating of the Unselected Representations

The method to off-policy update of an unselected representation in the MDP case is very similar to the bandit case. The only difference is that for the bandit case, the update target consists only of the reward, while for the MDP case, the update target consists of a reward and next state. For updating an unselected candidate representations, this next state is the state according to that candidate representations. So if we have to consider the representations S_X and S_Y and select S_X as the representation for action selection, we use the sequence

$\cdots \rightarrow r_t, s_{x_t} \rightarrow a_t \rightarrow r_{t+1}, s_{x_{t+1}} \rightarrow \cdots$

to perform the update

$$Q(s_{x_t},a_t) = (1-\alpha)\,Q(s_{x_t},a_t) + \alpha\,(r_{t+1} + \gamma \max_a Q(s_{x_{t+1}},a)) \tag{15}$$

And, if one of the conditions for unbiased updates from Sect. 4.4 holds, we use the parallel sequence of S_Y

$\cdots \rightarrow r_t, s_{y_t} \rightarrow a_t \rightarrow r_{t+1}, s_{y_{t+1}} \rightarrow \cdots$

to perform the update

$$Q(s_{y_t},a_t) = (1-\alpha)\,Q(s_{y_t},a_t) + \alpha\,(r_{t+1} + \gamma \max_a Q(s_{y_{t+1}},a)) \tag{16}$$

5.2 *Off-Policy Updating of the Unselected Switch Actions*

For the contextual bandit case, it was possible to perform very efficiently off-policy Monte Carlo updates of the unselected switch actions. Therefore, exploration of the switch actions is not necessary for a contextual bandit problem.

Off-policy Monte Carlo updates for the MDP case are much less efficient. For a deterministic evaluation policy π, all the actions of an episode generated using the behavior policy π' have to be equal to π to get a non-zero weight (see (13)). Since for our switch method the behavior policy is based on a different representation, this is rarely the case. If an stochastic evaluation policy would be used, the weights will always be non-zero, but the difference in weight values will be very large, also causing the off-policy updates to be inefficient. Therefore, for the MDP case, we only perform updates of switch action corresponding to the selected representation and use exploration to ensure all switch actions are updated.

6 Experimental Results and Discussion

6.1 *Contextual Bandit Problem*

In the first experiment, we compare the performance of the switching method against the performance of the full representation given the prior knowledge that only one feature from a set of features is a relevant feature, while the others are empty features. The feature values are randomly drawn from their domain values after each pull of an arm. We make this comparison for a feature set of 3 and of 4 features. Each feature has 8 feature values. The bandit has two arms with opposite expected reward. Depending on the context, one has an expected reward of +1, while for the other arm it is -1. The reward is drawn from a normal distribution with a standard deviation of 2. For half of the feature values of the information-carrying feature, the first arm has the +1 expected reward. Therefore, without this feature the expected reward is 0.

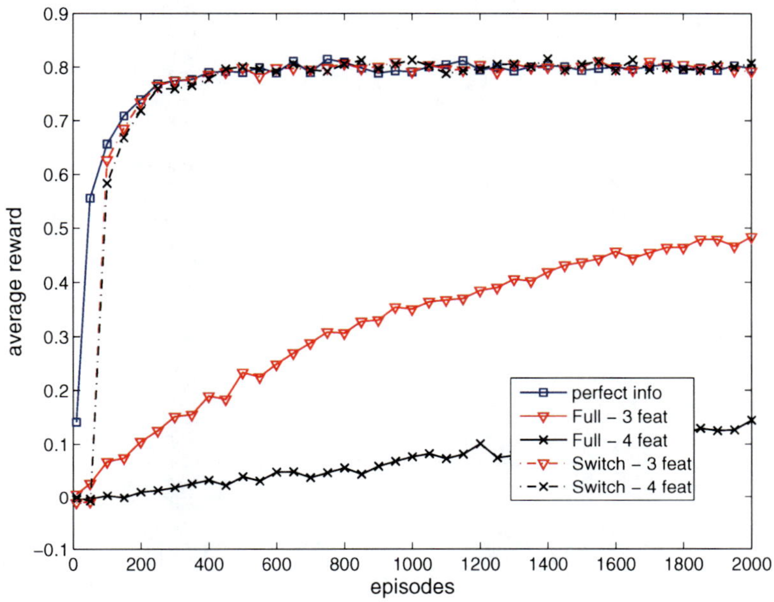

Fig. 3 Comparison of the switch method with the full feature set for a feature set of size 3 and size 4 for a contextual bandit problem. The perfect info graph shows what the performance would have been had the agent known beforehand what the relevant feature would be.

The candidate representations for the switch algorithm consist, in this case, of one representation per feature. We used a learning rate of 0.01 for the representation evaluation and selected the representation greedy with respect to this evaluation. The exploration scheme is ε-greedy with $\varepsilon = 0.2$ for all methods. To kick start the switch method, we used only exploration for the first 50 episodes. Since all candidate representations are updated during this exploration phase, this has a positive effect on the total reward.

Figure 3 shows the results. To get an upper bound, we also plotted the performance of a representation that consists only of the relevant feature. All results are averaged over 10,000 independent runs and smoothed.

The advantage of the switching method is very clear for this task. While the full representation grows exponentially with the number of features, with 512 context states for a set of 3 features and 4096 for a set of 4, the total number of states for the switching methods grows linear, with 24 states for a set of 3 features and 32 states a set of 4. The simultaneous updating of the representations on exploration increases the performance even further and brings it very close to the representation containing only the information-carrying feature.

For our second experiment, we consider a different contextual bandit task. For this bandit task, we have three information-containing features, but one of them contains the most information. We will demonstrate with this task that using a small incomplete representation for the initial learning phase can improve the overall

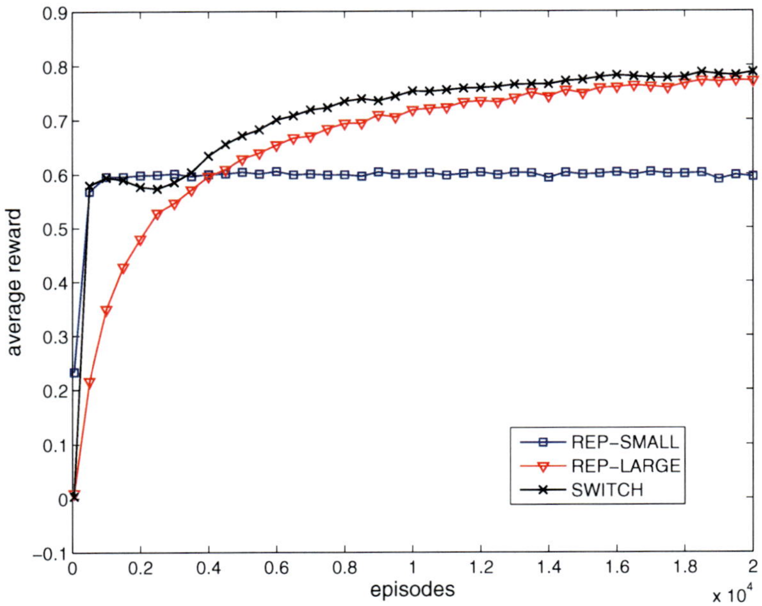

Fig. 4 Average reward for a contextual bandit when switching between a large and a small representation

performance for certain tasks. We compared in this case the performance of a representation containing only the most important feature (REP-SMALL), with the representation containing all three relevant features (REP-LARGE) and the switch method that has both representations as candidate set (SWITCH). We used a learning rate of 0.001 for the representation evaluation and greedy representation selection. The exploration scheme is again ε-greedy with $\varepsilon = 0.2$ for all methods. The switch method uses only exploration for the first 100 episodes. The results are averaged over 10,000 independent runs and smoothed.

Figure 4 shows the average reward for the first 20,000 episodes. The performance of representation REP-SMALL is about 3/4 of that of representation REP-LARGE. After about 500 episodes, REP-LARGE has seen enough samples to outperform REP-SMALL. After the initial exploration phase, the switch method catches up very quickly with REP-SMALL, then the performance shows a small ditch before it starts climbing to new performance levels. We explain the ditch as following: while REP-SMALL is quickly recognized as the better representation, we keep improving the Q-values of REP-LARGE as well as the Q-values of the switch actions by off-policy updating. Once the Q-value of the selection action for REP-LARGE approaches the Q-value of the selection action for REP-SMALL, for some of the runs, the estimates will prematurely indicate REP-LARGE as the better one, causing a small ditch in the performance. Then, when the Q-values of REP-LARGE further improve, it will really outperform REP-SMALL causing the climb in performance. Interesting about the performance of the switch method is that it outperforms REP-LARGE at each point during learning. We explain this as following; the exact point where REP-LARGE outperforms REP-SMALL is different for each run. Since the switch method uses an up-to-date estimate of the expected reward for each representation, it simply makes longer use of REP-SMALL for those runs where the Q-values of REP-LARGE are improved more slowly than average. Therefore, once REP-SMALL has been properly learned its performance forms a lower boundary for the remaining episodes. This lower boundary is not present when using only REP-LARGE (which performance boundary is 0) and therefore on average the switch method will do better at each point during learning.

6.2 MDP Task

The task we will consider is an episodic navigational task, where the agent has to move through a narrow corridor while avoiding bumping into the walls, see Fig. 5. The agent can take any of four movement actions: up, down, left, and right. On top of this, the agent moves an additional step in either the up or the down direction (think of the corridor as being on a rocking ship). The information about the direction of the additional step is useful, since if the additional step would cause the agent to hit the wall, it can prevent this by making a step in the opposite direction. Whether the additional step is in the up or down direction, which depends on the context, is specified by a set of independent features that change value at each timestep. Only one of those features contains useful information, but the agent does not know which one. This creates an MDP equivalent of the first contextual bandit problem. Each step results in a reward of -1 plus an additional reward of -10 if the agent hits a wall. The environment is stochastic causing the agent to make, with a probability of 20%, a step in a random direction instead of the direction corresponding to its action. The discount factor is 0.95.

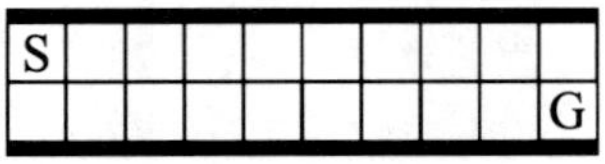

Fig. 5 The corridor task. The agent has to move from S to G while avoiding bumping into the wall (thick black line).

We compare the online performance of the switch method against the performance of the full feature set for a feature set that consists, besides the position feature, of 3 or 4 features describing the direction of the additional step. The candidate representations for the switch method consist of the position feature and one additional feature. We used a learning rate of 0.05 for all states and an ε-greedy policy with $\varepsilon = 0.1$ for the switch actions as well as regular actions.

Figure 6 shows the online performance for the first 10,000 episodes for a set of 3 features and a set of 4. As a reference, we also show the performance of a representation consisting of only the position feature and the relevant additional feature. The results are averaged over 100 independent runs and smoothed.

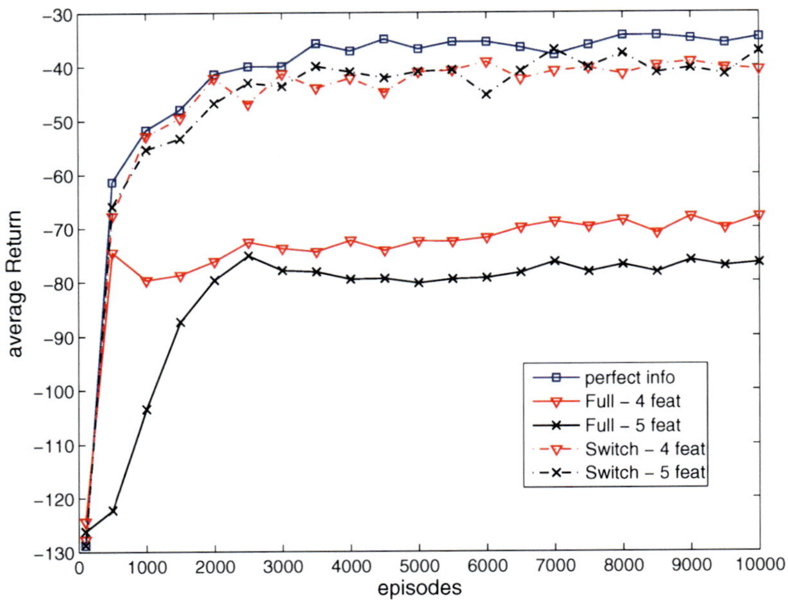

Fig. 6 Comparison of the switch method with the full representation on the corridor task for a feature set size of 4 and 5. The perfect info graph shows what the performance would have been had the agent known beforehand what the relevant feature would be.

The performance of the full representations increases fast during the initial learning phase, but after that increases only very slowly. As expected, the switch method performs a lot better and is close to the performance of the reference representation. That the switch method converge to a performance slightly lower than the reference representation is due to the additional exploration of the switch actions, causing suboptimal representations to be selected.

7 Conclusions

We have presented a method for online representation selection for factored MDPs. The method addresses a special case of the feature selection problem, that only considers certain subsets of features, called *candidate representations*. The problem of representation selection is formalized by defining switch actions that select the representation to be used. The switch actions combine the MDP tasks corresponding to the candidate representations into a single switch MDP, which is then solved using standard RL techniques. Since the agent can observe the full feature set, parallel experience sequence can be constructed corresponding to the unselected representations. These parallel sequences can be used to off-policy update the Q-values of unselected representations. To update the Q-values of the switch actions, Monte

Carlo updates are used since this gives a better estimate of the current performance of a representation than bootstrapping from the representations Q-values.

We demonstrated the validity of the method by demonstrating for a contextual bandit task and a regular MDP that given a feature set containing only a single relevant feature, we can find this feature very efficiently using the switch method. We also showed for a contextual bandit task that switching between a set of relevant features and a subset of these features, our method can outperform both individual representations, since it combines the fast performance increase of the small representation with the high asymptotic performance of the large representation.

8 Future Work

Having a set of candidate representations means it also contains some structure information. For example, the feature set size of the largest candidate representation gives an upper limit for the degree of a DBN structure representation. In this way, prior knowledge about features can be translated into structural parameters. We would like to compare our feature-based algorithm with structure-based learning algorithms on this translation. We expect that for problems with limited learning time, our algorithm has a higher online performance, since from the start of an episode the exploration–exploitation dilemma is taken into account, whereas structure-learning algorithms typically perform exploration until they have an accurate model of the environment. We also would like to see if we can combine some of our ideas about online estimation of the current value of a feature set with structure learning.

Throughout this chapter we only considered valid candidate representations, i.e., representations for which the Markov property holds. We expect, however, that the method will also perform well if there are some candidate representations among the total set of candidate representations that are non-Markov, since their lower average performance will refrain the agent from selecting them. We would like to test this intuition on a number of benchmark problems.

Acknowledgments. The authors thank Christos Dimitrakakis for discussions about this chapter.

References

1. Abbeel, P., Koller, D., Ng, A.: Learning factor graphs in polynomial time and sample complexity. The Journal of Machine Learning Research 7, 1743–1788 (2006)
2. Bellman, R.E.: A Markov decision process. Journal of Mathematical Mechanics 6, 679–684 (1957)
3. Boutilier, C., Dearden, R., Goldszmidt, M.: Exploiting structure in policy construction. In: International Joint Conference on Artificial Intelligence, vol. 14, pp. 1104–1113 (1995)
4. Diuk, C., Li, L., Leffler, B.: The adaptive k-meteorologists problem and its application to structure learning and feature selection in reinforcement learning. In: Proceedings of the 26th Annual International Conference on Machine Learning. ACM, New York (2009)

5. Guestrin, C., Koller, D., Parr, R., Venkataraman, S.: Efficient solution algorithms for factored mdps. Journal of Artificial Intelligence Research 19, 399–468 (2003)
6. Hoey, J., St-Aubin, R., Hu, A., Boutilier, C.: Spudd: Stochastic planning using decision diagrams. In: Proceedings of the Fifteenth Conference on Uncertainty in Artificial Intelligence, pp. 279–288. Morgan Kaufmann, San Francisco (1999)
7. Kaelbling, L.P., Littman, M.L., Moore, A.P.: Reinforcement learning: A survey. Journal of Artificial Intelligence Research 4, 237–285 (1996)
8. Kearns, M., Koller, D.: Efficient reinforcement learning in factored mdps. In: International Joint Conference on Artificial Intelligence, vol. 16, pp. 740–747 (1999)
9. Li, L., Littman, M., Walsh, T.: Knows what it knows: a framework for self-aware learning. In: Proceedings of the 25th international conference on Machine learning, pp. 568–575. ACM, New York (2008)
10. Siegmund, D.: Importance sampling in the monte carlo study of sequential tests. Annals of Statistics 4, 673–684 (1976)
11. St-Aubin, R., Hoey, J., Boutilier, C.: Apricodd: Approximate policy construction using decision diagrams. In: Proceedings of Advances in Neural Information Processing Systems, pp. 1089–1095. MIT Press, Cambridge (2000)
12. Strehl, A., Diuk, C., Littman, M.: Efficient structure learning in factored-state mdps. In: Proceedings of the Twenty-Second National Conference on Artificial Intelligence, vol. 22, p. 645. AAAI Press/MIT Press, Menlo Park/Cambridge (2007)
13. Sutton, R.S., Barto, A.G.: Reinforcement Learning: An Introduction. MIT Press, Cambridge (1998)

Part II
Collaborative Decision Making

A Decision-Theoretic Approach to Collaboration: Principal Description Methods and Efficient Heuristic Approximations

Frans A. Oliehoek and Arnoud Visser

Abstract. This chapter gives an overview of the state of the art in decision-theoretic models to describe cooperation between multiple agents in a dynamic environment. Making (near-) optimal decisions in such settings gets harder when the number of agents grows or the uncertainty about the environment increases. It is essential to have compact models, because otherwise just representing the decision problem becomes intractable. Several such model descriptions and approximate solution methods, studied in the Interactive Collaborative Information Systems project, are presented and illustrated in the context of crisis management.

1 Introduction

Making decisions is hard. Even though we humans make thousands of decisions a day, some decisions, especially important ones, are difficult to make. This is even more true for decision making in complex dynamic environments. This chapter focuses on such complex decision problems and particularly on situations where there are multiple decision makers or *agents* that have to cooperate.

When compared to computer systems, humans perform extremely well in making most decisions. Still, there is a growing need for the development of intelligent decision support systems and implementing cognitive abilities in autonomous artificial agents, because human decision making has its limitations. For instance, human situation awareness is characterized by structural biases and humans are

Frans A. Oliehoek
Intelligent System Laboratory Amsterdam, Science Park 107, NL 1098 XG Amsterdam, The Netherlands
e-mail: F.A.Oliehoek@uva.nl

Arnoud Visser
Intelligent System Laboratory Amsterdam, Science Park 107, NL 1098 XG Amsterdam, The Netherlands
e-mail: A.Visser@uva.nl

R. Babuška & F.C.A. Groen (Eds.): Interactive Collaborative Information Systems, SCI 281, pp. 87–124.
springerlink.com

conservative estimators as compared to applying for example Bayesian statistics in handling uncertainties [14]. As such, human decision making may be substantially improved when assisted by intelligent decision support systems [21], providing an important social and economic incentive to develop such systems with intelligent behavior.

The Interactive Collaborative Information Systems (ICIS) project focuses on the development of intelligent techniques and methods that can be applied to decision support systems. It particularly aims to improve overall quality of information processing and decision making under conditions of stress, risk and uncertainty. For systems to be effective under such circumstances, several requirements must be met, including are the capabilities to:

1. Reach a set of (predetermined) goals by influencing the environment in which the system operates.
2. Cope with changes in the dynamic environment.
3. Support reasoning with uncertainty, reasoning with risks and reasoning under a lack of knowledge, necessary because of the nondeterministic nature of the real world.

The systems under concern are connected to the real world and can observe and influence this world. Also, such systems need to make use of their experience of previous interactions with the world. That is, such a system needs capabilities of tackling the problem of *sequential decision making*, making a series of decisions over time. This chapter is concerned with methods to realize such capabilities. In particular, it focuses on decision-theoretic methods for dynamic planning, collaboration and optimization.

Following the ICIS project, the work in this chapter is illustrated by the application to crisis management situations. These results, however, can be used in other domains of application that can be characterized by decision making in complex, dynamic and nondeterministic environments.

An example of a challenging environment is the RoboCup Rescue League [40]. Inspired on the earthquake in Kobe in 1995 [75], the agent competition RoboCup Rescue simulates an earthquake in an urban environment (see Fig. 1). In this scenario, buildings collapse causes roads to get blocked and people to get trapped in the debris, damages to gas pipes cause fires to break out all over the city and parts of the communication infrastructure fails.

In this chaotic setting, teams of firefighters, police officers and ambulances have to make decisions locally, based on only limited information. The objective of these emergency response services is to minimize the casualties and structural damage. To this end, it is important to make effective plans to deal with the situation, but this is difficult due to the uncertainties. For instance, it is hard to estimate how fast the fire will spread or how many firefighting units one should allocate to a particular fire site, and how long it will take them to control the fire. Moreover, it is also important to try to improve the situational awareness, and these two goals may compete with each other. For instance, it may be necessary to trade off human capacity between fighting fire at a particular site and reconnaissance. Not performing such information-gaining

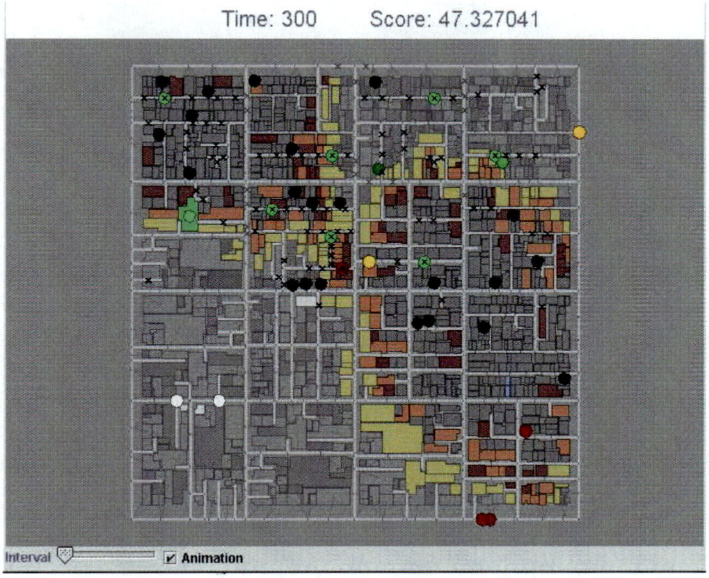

Fig. 1 Robocup Rescue simulates an earthquake in an urban environment. This is a top view of a city center, with gray buildings and white roads. Circles indicate agents; a black circle for instance indicates an agent which died due to a fire or a collapse.

activities allows allocating more agents to deal with the current situation, but may impose severe risks, e.g., a yet unknown fire source may spread out of control.

1.1 Forms of Uncertainty

The example of RoboCup Rescue illustrates how various forms of uncertainty complicate effectively resolving situations. Here, we formalize these different types of uncertainty as considered in this chapter.

The first type of uncertainty we consider is *outcome uncertainty*: the outcome or effects of actions may be uncertain. In particular, we will assume that the possible outcomes of an action are known, but that each of those outcomes is realized with some probability. This means that the state transitions are stochastic.

In the real world an agent might not be able to determine what the state of the environment exactly is, which means that there is *state uncertainty*. In such cases, we say that the environment is *partially observable*. Partial observability results from noisy and/or limited sensors. Because of *sensor noise* an agent can receive faulty or inaccurate observations. When sensors are limited the agent is unable to observe the differences between states that cannot be detected by the sensor, e.g., the presence or absence of an object outside a laser range-finder's field of view. When the same sensor reading might require different action choices, this phenomenon is referred to as *perceptual aliasing*.

Another complicating factor is the presence of multiple agents that each make decisions that influence the environment. Such an environment in which multiple agents operate is called a *multiagent system (MAS)* [70, 72, 78, 79]. The difficulty in MASs is that each agent can be uncertain regarding the actions of other agents. This is obvious in MASs with self-interested agents, where agents may be unwilling to share information. In a cooperative setting, the agents have the same goal and therefore are willing to share information and coordinate their actions. Still, it is non-trivial how such coordination should be performed. Especially when communication capabilities are limited or absent, *how* the agents should coordinate their actions becomes problematic. This problem is magnified in partially observable environments: as the agents are not assumed to observe the state—each agent only knows its own observations received and actions taken—there is no common signal they can condition their actions on. Note that this problem is in addition to the problem of partial observability, and not instead of it; even if the agents could freely and instantaneously communicate their individual observations, the joint observations would not disambiguate the true state.

1.2 Decision-Theoretic Approach to MASs

Many approaches to multiagent systems share the drawback of not having a measure of quality of the generated plans. To overcome this problem, we build upon the field of decision theory (DT).

Decision theory describes how a decision maker, or agent, should make a decision given its preferences over alternatives. Let us for the moment assume that an agent has preferences over states s of the environment. If the agents' preferences satisfy the axioms of utility theory, they can be conveniently described by a utility function u that maps each state to a real number $u(s)$ [61]. This number indicates how the agent values that state: if A is preferred to B, then $u(A) > u(B)$.

If the agent has a model of how its actions a influence the environment, i.e. when the probabilities $\Pr(s|a)$ are available, the agent can compute the *expected* utility of each action as

$$u(a) \equiv E\left[u(s)|a\right] = \sum_{s} u(s)\Pr(s|a). \tag{1}$$

A rational agent should select the action that maximizes this expected utility.

Decision-theoretic planning (DTP) extends DT to sequential decision problems and has roots in control theory and operations research. In control theory, one or more controllers control a stochastic system with a specific output as goal. Operations research considers tasks related to scheduling, logistics and work flow and tries to optimize the concerning systems. Many decision-theoretic planning problems can be formalized as *Markov decision processes* (MDPs). In the last decades, the MDP framework has gained in popularity in the AI community as a model for planning under uncertainty [8, 36].

The MDP is a framework for sequential decision making at predetermined points in time, i.e., it is a discrete-time model. The extension of the MDP to continuous time is called a *semi Markov decision process (SMDP)* [56]. Also in control theory much research has considered continuous time settings [66]. In order to solve such continuous time settings, however, time is discretized (again leading to a regular MDP formulation), or special assumptions, such as linear dynamics and quadratic costs, are required [5]. As such, most approaches to DTP for multiagent systems have focused on extensions of the MDP, a notable exception is presented by Van der Broek et al. [9].

MDPs can be used to formalize a discrete-time planning task of a single agent in a stochastically changing environment, on the condition that the agent can observe the state of the environment. The environment is observed at discrete *time steps*, also referred to as *stages* [8]. The number of time steps the agent will interact with the environment is called the *horizon* of the decision problem, and will be denoted by h. Every time step the state changes stochastically, but the agent chooses an action that selects a particular transition function: i.e., taking an action from a particular state at time step t induces a probability distribution over states at time step $t+1$. The probabilities of state transitions are specified by the model. The goal of planning for such an MDP is to find a *policy* that is optimal with respect to the desired behavior. This desired behavior is specified by the reward model: for each action taken from each possible state of the environment, a reward is issued. The agent has to maximize the expected long-term reward signal.

When the agent knows the probabilities of the state transitions, i.e., when it knows the model, it can contemplate the expected transitions over time and compute a plan that is most likely to reach a specific goal state, minimizes the expected costs or maximizes the expected reward. This stands in contrast to reinforcement learning (RL) [71], where the agent does not have a model of the environment, but has to learn good behavior by repeatedly interacting with the environment. Reinforcement learning can be seen as the combined task of learning the model of the environment *and* planning, although in practice often it is not necessary to explicitly recover the environment model.

In order to deal with the introduced sensor uncertainty, a *partially observable Markov decision process (POMDP)* extends the MDP model by incorporating observations and their probability of occurrence conditional on the state of the environment [39]. A POMDP, however, only considers one agent in an environment. We consider the setting where there are multiple agents that each may influence the state of this environment, as illustrated in Fig. 2. To incorporate the influence of multiple agents, the models need to be adapted to consider the joint effect of the individual actions, i.e., transition probabilities depend on *joint actions* and similarly for observations.

The effects of uncertainty with respect to other agents may be mitigated through communication. Under the stringent assumptions of instantaneous, cost- and noise-free communication, these effects can be discarded altogether, and the problem reduces to a POMDP [57]. However, in general, these assumptions are too strong and deciding *when* to communicate *what* becomes part of the problem.

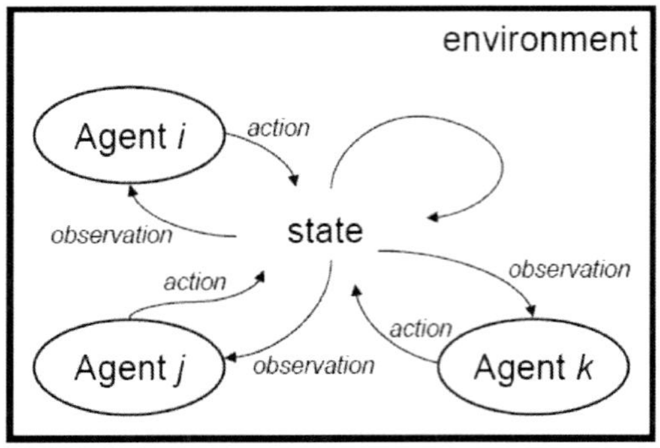

Fig. 2 A schematic representation of multiple agents in a dynamic environment. The state is influence by the combined actions of all agents.

There are two perspectives on MASs. One option is to consider each agent separately, and have each such agent maintain an explicit model of the other agents, we refer to this as the *subjective perspective* of a MAS. This is the approach as chosen in the *recursive modeling method (RMM)* [29] and the *Interactive POMDP (I-POMDP)* framework [28]. On the other hand, the *decentralized partially observable Markov decision process (Dec-POMDP)* [4] is a generalization to multiple agents of a POMDP and can be used to model a team of cooperative agents that are situated in a stochastic, partially observable environment. It presents an *objective perspective* of the MAS, in which we try to find plans for all agents at the same time.

Some other models that we do not consider in details stem from game theory. In particular, extensive games [52] can model problems of sequential decision making in uncertain environment. However, the game trees needed to model complex environments are extremely large. The more recently introduced MAIDs [44] and NIDs [25] can be more compact than extensive games, but this is especially the case with respect to the structure of the variables of influence, not with respect to decisions made over multiple stages.

1.3 Overview

This chapter gives an overview of some different decision-theoretic models for multiagent planning. First, in Section 2 we treat objective models, in particular the Dec-POMDP. Next, Section 3 presents the interactive POMDP, the most well-known subjective model. As illustrated in Section 4, there still is a gap between the state of the art in DTP and many real-world applications. Our research tries to close this gap from two sides. On the one hand, we try to scale up the problems decision-theoretic methods can handle as explained in the remainder of Section 4. On the other hand, Section 5 shows how we employ efficient heuristic methods to large realistic tasks. Finally, Section 6 concludes.

2 The Objective Approach: Dec-POMDPs

This section introduces the objective approach to planning for MASs. In particular it focuses on the decentralized POMDP [4], which is a model for objective sequential decision making for a team of cooperative agents. It is roughly equivalent to the *multiagent team decision problem (MTDP)* [57].

2.1 Decentralized POMDPs

Here, we formally define the Dec-POMDP model and its components.

Definition 1. A *decentralized partially observable Markov decision process (Dec-POMDP)* is defined as a tuple $\langle n,S,A,P_T,R,O,P_O,h,b^0 \rangle$ where:

- n is the number of agents.
- S is a finite set of states.
- A is the set of joint actions.
- P_T is the transition function.
- R is the immediate reward function.
- O is the set of joint observations.
- P_O is the observation function.
- h is the horizon of the problem.
- $b^0 \in \mathscr{P}(S)$ is the initial state distribution at time $t = 0$.[1]

The Dec-POMDP model extends single-agent (PO)MDP models by considering *joint* actions and observations. In particular, $A = \times_i A_i$ is the set of *joint actions*. Here, A_i is the set of actions available to agent i which can be different for each agent. Every time step, one joint action $\mathbf{a} = \langle a_1,...,a_n \rangle$ is taken. In a Dec-POMDP, agents only know their own individual action; they do not observe each other's actions. Similar to the set of joint actions, $O = \times_i O_i$ is the set of joint observations, where O_i is a set of observations available to agent i. Every time step the environment emits one joint observation $\mathbf{o} = \langle o_1,...,o_n \rangle$, from which each agent i only observes its own component o_i, as illustrated in Fig. 3.

Actions and observations are the interface between the agents and their environment. The Dec-POMDP framework describes this environment by its *states* and *transitions*. A Dec-POMDP specifies an environment model simply as the set of possible states $S = \{s_1,...,s_{|S|}\}$ together with the probabilities of state transitions. In particular, the transition from some state to another depends stochastically on the past states and actions. This probabilistic dependence models *outcome uncertainty*: the fact that the outcome of an action cannot be predicted with full certainty as discussed in Section 1.1.

An important characteristic of Dec-POMDPs is that the states possess the *Markov property*. That is, the probability of a particular next state depends on the current state and joint action, but not on the whole history:

[1] $\mathscr{P}(\cdot)$ denotes the set of probability distributions over $(\cdot)$.

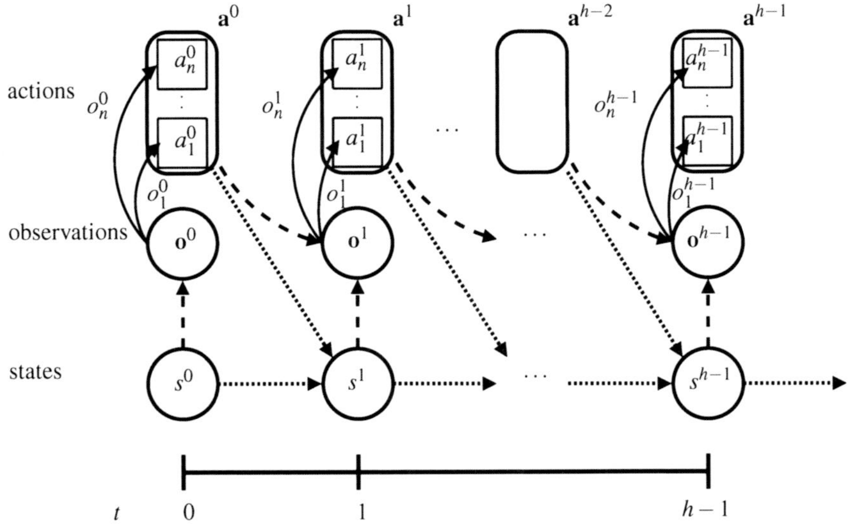

Fig. 3 An illustration of the dynamics of a Dec-POMDP. At every stage the environment is in a particular state. This state emits a joint observation, of which each agent observes its individual observation. Then each agent selects an action forming the joint action.

$$\Pr(s^{t+1}|s^t,\mathbf{a}^t,s^{t-1},\mathbf{a}^{t-1},...,s^0,\mathbf{a}^0) = \Pr(s^{t+1}|s^t,\mathbf{a}^t). \tag{2}$$

Also, we will assume that the transition probabilities are *stationary*, meaning that they are independent of the stage t.

In a way similar to how the transition model P_T describes the stochastic influence of actions on the environment, the observation model describes how the state of the environment is perceived by the agents. Formally, P_O is a mapping from joint actions and successor states to probability distributions over joint observations: $P_O : A \times S \rightarrow \mathscr{P}(O)$, i.e., it specifies

$$\Pr(\mathbf{o}^t|\mathbf{a}^{t-1},s^t). \tag{3}$$

The Dec-POMDP is truly decentralized in the sense that during execution the agents are assumed to act based on their individual observations only and no additional communication is assumed. This does not mean that Dec-POMDPs cannot handle communication: communication can be modeled implicitly through the regular actions and observations as will be illustrated in Section 4.1.

The reward function $R : S \times A \rightarrow \mathbb{R}$ is used to specify the goal of the agents. In particular, a desirable sequence of joint actions should correspond to a high 'long-term' reward, formalized as the *return*.

Definition 2. Let the *return* or *cumulative reward* of a Dec-POMDP be defined as total of the rewards received during an execution:

$$\sum_{t=0}^{h} R(s^t, \mathbf{a}^t) \tag{4}$$

where $R(s^t, \mathbf{a}^t)$ is the reward received at time step t.

We consider as optimality criterion the *expected cumulative reward*, where the expectation refers to the expectation over sequences of states and executed joint actions. The planning problem is to find a tuple of policies, called a *joint policy* that maximizes the expected cumulative reward.

Note that, in contrast to reinforcement learning settings [71], in a Dec-POMDP, the agents are assumed not to observe the immediate rewards. Observing the immediate rewards could convey information regarding the true state which is not present in the received observations. This is undesirable as all information available to the agents should be modeled in the observations. When planning for Dec-POMDPs the only thing that matters is the *expectation* of the cumulative future reward, not the actual reward obtained.

The assumption in this text is that planning takes place in an off-line phase, after which the plans are executed in the on-line phase. In the decentralized setting, however, this statement can use some clarification. The on-line execution phase is completely decentralized: each agent only knows the joint policy as found in the planning phase and its individual history of actions and observations. The planning phase, however, is centralized: a single centralized computer computes a joint plan and consequently distributes these plans to the agents, who then merely execute the plans on-line.

Example 1 (The FIREFIGHTING *problem).* As an example, we consider a benchmark problem in which a team of n fire fighters have to extinguish fires in a row of N_H houses. Each house H is characterized by an integer parameter fl_H, or fire level. It indicates to what degree a house is burning, and it can have n_f different values, $0 \leq fl_H < n_f$. Its minimum value is 0, indicating the house is not burning.

At every time step, the agents receive a reward of $-fl_H$ for each house and each agent can choose to move to any of the houses to fight fires at that location. That is, each agent has actions $H_1 \ldots H_{N_H}$ If a house is burning ($fl_H > 0$) and no fire fighting agent is present, its fire level will increase by one point with probability 0.8 if any of its neighboring houses are burning, and with probability 0.4 if none of its neighbors are on fire. A house that is not burning can only catch fire with probability 0.8 if one of its neighbors is on fire. When two agents are in the same house, they will extinguish any present fire completely, setting the house's fire level to 0. A single agent present at a house will lower the fire level by one point with probability 1 if no neighbors are burning, and with probability 0.6 otherwise. Each agent can only observe whether there are flames (F) or not (N) at its location. Flames are observed with probability 0.2 if $fl_H = 0$, with probability 0.5 if $fl_H = 1$, and with probability 0.8 otherwise. Initially, the agents start outside any of the houses, and the fire level fl_H of each house is drawn from a uniform distribution.

In case there is only one agent, the Dec-POMDP reduces to a standard POMDP. In a POMDP, the agent still cannot observe the state, so it is not possible to specify a policy as a mapping from states to actions as is done in an MDP. However, it turns out that maintaining a probability distribution over states, called *belief*, $b \in \mathscr{P}(S)$, is a sufficient statistic:

$$\forall_{s^t} \quad b^t(s^t) \equiv \Pr(s^t | o^t, a^{t-1}, o^{t-1}, \ldots, a^0, o^0). \tag{5}$$

As a result, a single agent in a partially observable environment can specify its policy as a series of mappings from the set of beliefs to actions.

Unfortunately, in the general Dec-POMDP case no such simplifications are possible. Even though the transition and observation model can be used to compute a *joint* belief, this computation requires knowledge of the joint actions and observations. During execution, the agents have no access to this information and thus can not compute such a joint belief.

2.2 Histories and Policies

The goal of planning in a Dec-POMDP is to find a (near-) optimal tuple of policies, and these policies specify for each agent how to act in a specific situation. A Dec-POMDP specifies h time steps or stages $t = 0,...,h-1$. At each of these stages, there is a state s^t, joint observation $\mathbf{o}^t$ and joint action $\mathbf{a}^t$. Therefore, when the agents have to select their k-th actions (at $t = k-1$), the history has the following form:

$$\left(s^0, \mathbf{o}^0, \mathbf{a}^0, s^1, \mathbf{o}^1, \mathbf{a}^1, \ldots, s^{k-1}, \mathbf{o}^{k-1}\right). \tag{6}$$

Here s^0 is the initial state, drawn according to the initial state distribution b^0. The initial joint observation $\mathbf{o}^0$ is assumed to be the empty joint observation: $\mathbf{o}^0 = \mathbf{o}_\emptyset$.

From this history of the process, the states remain unobserved and agent i can only observe its own actions and observations. Therefore, an agent will have to base its decision regarding which action to select on the sequence of actions and observations observed up to that point.

Definition 3. We define *the action-observation history (AOH) for agent i*, $\vec{\theta}_i$, as the sequence of actions taken by and observations received by agent i. At a specific time step t, this is

$$\vec{\theta}_i^t = \left(o_i^0, a_i^0, o_i^1 \ldots, a_i^{t-1}, o_i^t\right). \tag{7}$$

The *joint action-observation history*, $\vec{\boldsymbol{\theta}}$, is the action-observation history for all agents: $\vec{\boldsymbol{\theta}}^t = \langle \vec{\theta}_1^t, \ldots, \vec{\theta}_n^t \rangle$. Agent *i*s set of possible AOHs at time t is $\vec{\Theta}_i^t$ and the set of all its AOHs is $\vec{\Theta}_i = \cup_{t=0}^{h-1} \vec{\Theta}_i^t$.[2] Finally, the set of all possible *joint* AOHs is given by $\vec{\Theta} = \cup_{t=0}^{h-1} (\vec{\Theta}_1^t \times \ldots \times \vec{\Theta}_n^t)$. At $t = 0$, the action-observation history is empty, denoted by $\vec{\boldsymbol{\theta}}^0 = \vec{\boldsymbol{\theta}}_\emptyset$.

[2] In a particular Dec-POMDP, it may be the case that not all of these histories can actually be realized, because of the probabilities specified by the transition and observation model.

Definition 4. The *observation history (OH) for agent i*, $\vec{o}_i$, is the sequence of observations an agent has received. At a specific time step t, this is

$$\vec{o}_i^t = \left(o_i^0, o_i^1, \ldots, o_i^t\right). \tag{8}$$

The *joint observation history*, $\vec{\mathbf{o}}$, is the OH for all agents: $\vec{\mathbf{o}}^t = \langle \vec{o}_1^t, \ldots, \vec{o}_n^t \rangle$. Similar to the notation for AOHs, the set of OHs for agent i at time t is denoted $\vec{O}_i^t$. We also use $\vec{O}_i$ and $\vec{O}$ and the empty observation history is denoted $\vec{\mathbf{o}}_\emptyset$.

Definition 5. The *action history (AH) for agent i*, $\vec{a}_i$, is the sequence of actions an agent has performed. At a specific time step t, we write

$$\vec{a}_i^t = \left(a_i^0, a_i^1, \ldots, a_i^{t-1}\right). \tag{9}$$

Notation for joint action histories and sets are analogous to those for observation histories. Finally we note that, clearly, a (joint) action-observation history consists of a (joint) action- and a (joint) observation history: $\vec{\boldsymbol{\theta}}^t = \langle \vec{\mathbf{o}}^t, \vec{\mathbf{a}}^t \rangle$.

The action-observation history of an agent specifies all the information the agent has when it has to decide upon an action. For the moment we assume that an individual policy π_i for agent i is a deterministic mapping from action-observation histories to actions. However, the number of such histories grows exponentially with the horizon of the problem: e.g., at the last time step $h-1$, there are $(|A_i|\,|O_i|)^{h-1}$ action-observation histories for agent i. The number of policies for agent i is exponential in this number, and thus doubly exponential in the horizon h.

It is possible to reduce the number of policies under consideration by realizing that a lot of policies specify the same behavior. This is illustrated by the left side of Fig. 4, which clearly shows that under a *deterministic* policy only a subset of possible action-observation histories are reached. Policies that only differ with respect to an action-observation history that is not reached in the first place, manifest the same behavior. The consequence is that to specify a deterministic policy, the observation history suffices: when an agent takes its action deterministically, he will be able to infer what action he took from only the observation history as illustrated by the right side of Fig. 4.

Definition 6. A *pure* or *deterministic policy*, π_i, for agent i in a Dec-POMDP is a mapping from observation histories to actions, $\pi_i : \vec{O}_i \to A_i$. The set of pure policies of agent i is denoted Π_i.

Note that also for pure policies we write $\pi_i(\vec{\theta}_i)$. In this case we mean the action that π_i specifies for the observation history contained in $\vec{\theta}_i$. We use $\pi = \langle \pi_1, \ldots, \pi_n \rangle$ to denote a *joint policy* Also we use $\pi_{\neq i} = \langle \pi_1, \ldots, \pi_{i-1}, \pi_{i+1}, \ldots, \pi_n \rangle$, to denote a profile of policies for all agents but i.

Apart from pure policies, it is also possible to have the agents execute *randomized policies*, i.e., policies that do not always specify the same action for the same situation, but in which there is an element of chance that decides which action is performed. We will not consider such policies here, since in a Dec-POMDP, there always is an optimal pure joint policy [48].

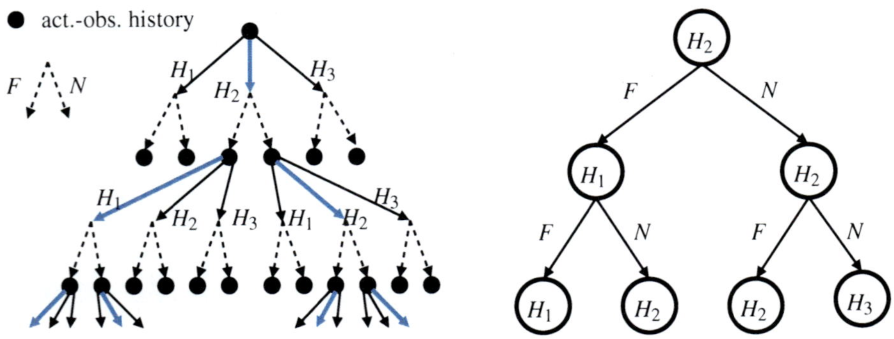

Fig. 4 Left: a tree of action-observation histories $\vec{\theta}_i$ for one of the agents in the $\langle N_H = 3, n_f = 3\rangle$ FIREFIGHTING problem. An arbitrary deterministic policy π_i is highlighted. Clearly shown is that π_i only reaches a subset of histories $\vec{\theta}_i$. ($\vec{\theta}_i$ that are not reached are not further expanded.) Right: The same policy can be shown in a simplified policy tree.

A common way to represent the temporal structure in a policy is to split it in *decision rules* δ_i that specify the policy for each stage. An individual policy is then represented as a sequence of decision rules $\pi_i = (\delta_i^0, \ldots, \delta_i^{h-1})$. In the case of a deterministic policy, the form of the decision rule for stage t is a mapping from length-t observation histories to actions $\delta_i^t : \vec{O}_i^t \to A_i$. A joint decision rule $\delta^t = \langle \delta_1^t, \ldots, \delta_n^t \rangle$ specifies a decision rule for each agent.

We will also consider policies that are partially specified with respect to time. Formally, $\varphi^t = (\delta^0, \ldots, \delta^{t-1})$ denotes the past joint policy at stage t, which is a partial joint policy π specified for stages 0,...,$t-1$.

Clearly, policies differ in how much reward they can expect to accumulate. We consider the expected cumulative reward of a joint policy, also referred to as its *value*.

Definition 7. The *value* $V(\pi)$ of a joint policy π is defined as

$$V(\pi) \equiv E\left[\sum_{t=0}^{h-1} R(s^t, \mathbf{a}^t) \middle| \pi, b^0\right], \tag{10}$$

where the expectation is over states and observations.

In particular, we can calculate this expectation recursively using

$$V_\pi^t(s^t, \vec{\mathbf{o}}^t) = R\left(s^t, \pi(\vec{\mathbf{o}}^t)\right) + \sum_{s^{t+1} \in S} \sum_{\mathbf{o}^{t+1} \in O} \Pr(s^{t+1}, \mathbf{o}^{t+1} | s^t, \pi(\vec{\mathbf{o}}^t)) V_\pi^{t+1}(s^{t+1}, \vec{\mathbf{o}}^{t+1}). \tag{11}$$

The value of joint policy π is then given by

$$V(\pi) = \sum_{s^0 \in S} V_\pi(s^0, \vec{\boldsymbol{\theta}}_\emptyset) b^0(s^0). \tag{12}$$

Example 2 (Optimal policies for the FIREFIGHTING *problem).* As an example, Fig. 5 shows an optimal joint policy for horizon 3 of the $\langle N_H = 3, n_f = 3 \rangle$ FIREFIGHTING problem. One agent initially moves to the middle house to fight fires there, which helps prevent fire from spreading to its two neighbors. The other agent moves to house 3, and stays there if it observes fire, and moves to house 1 if it does not observe flames. As well as being optimal, such a joint policy makes sense intuitively speaking.

no observation → go house 3	no observation → go house 2
flames → go house 3	flames → go house 2
no flames → go house 1	no flames → go house 2
flames, flames → go house 1	flames, flames → go house 1
flames, no flames → go house 1	flames, no flames → go house 1
no flames, flames → go house 2	no flames, flames → go house 1
no flames, no flames → go house 2	no flames, no flames → go house 1

Fig. 5 An optimal joint policy for FIREFIGHTING $\langle N_H = 3, n_f = 3 \rangle$, horizon 3. On the left the policy for the first agent, on the right the second agent's policy.

2.3 *Solving Dec-POMDPs*

The fact that the number of pure joint policies is doubly exponential in the horizon h provides some intuition about how hard the problem is. This intuition is supported by the result that the problem of finding the optimal solution for a finite-horizon Dec-POMDP with $n \geq 2$ is NEXP-complete [4]. Moreover, Dec-POMDPs cannot be approximated efficiently: even finding an ε-approximate solution is NEXP-hard [59]. Also, the problem of solving over an infinite horizon is undecidable, which is a direct result of the undecidability of (single-agent) POMDPs over an infinite horizon [45].

In the rest of this section, we provide background on the two main approaches of solving finite-horizon Dec-POMDPs: the forward view and the backward view. For a more complete overview of finite-horizon solution methods we refer to [48] and for an overview including infinite-horizon approaches to [64].

2.3.1 The Forward View: Heuristic Search

The forward view of Dec-POMDPs is given by the combination of multiagent A* [73] and the method proposed by Emery-Montemerlo [22] and the works derived from those methods. In particular, both approaches can be unified in a generalization of MAA*, creating a new view that may be called the forward perspective to solving Dec-POMDPs [48].

Multiagent A* (MAA*)

Szer et al. introduced a heuristically guided policy search method called *multiagent A* (MAA*)* [73]. It performs a guided A*-like search over partially specified joint policies, pruning joint policies that are guaranteed to be worse than the best (fully specified) joint policy found so far by an admissible heuristic.

In particular MAA* considers joint policies that are partially specified with respect to time: a partial joint policy $\varphi^t = (\delta^0, \delta^1, \ldots, \delta^{t-1})$ specifies the joint decision rules for the first t stages. For such a partial joint policy φ^t a heuristic value $\widehat{V}(\varphi^t)$ is calculated by taking $V^{0\ldots t-1}(\varphi^t)$, the actual expected reward φ^t achieves over the first t stages, and adding $\widehat{V}^{t\ldots h-1}$, a heuristic value for the remaining $h-t$ stages. Clearly, when $\widehat{V}^{t\ldots h-1}$ is an *admissible heuristic*—a guaranteed overestimation—so is $\widehat{V}(\varphi^t)$.

MAA* starts by placing the completely unspecified joint policy φ^0 in an open list. Then, it proceeds by selecting partial joint policies $\varphi^t = (\delta^0, \delta^1, \ldots, \delta^{t-1})$ from the list and 'expanding' them: generating all $\varphi^{t+1} = (\delta^0, \delta^1, \ldots, \delta^{t-1}, \delta^t)$ by appending all possible joint decision rules δ^t for next time step (t). The left side of Fig. 6 illustrates the expansion process. After expansion, all created children are heuristically evaluated and placed in the open list. Any partial joint policies φ^{t+1} with $\widehat{V}(\varphi^{t+1})$ less than the expected value $V(\pi)$ of some earlier found (fully specified) joint policy π, can be pruned. The search ends when the list becomes empty, at which point an optimal fully specified joint policy has been found.

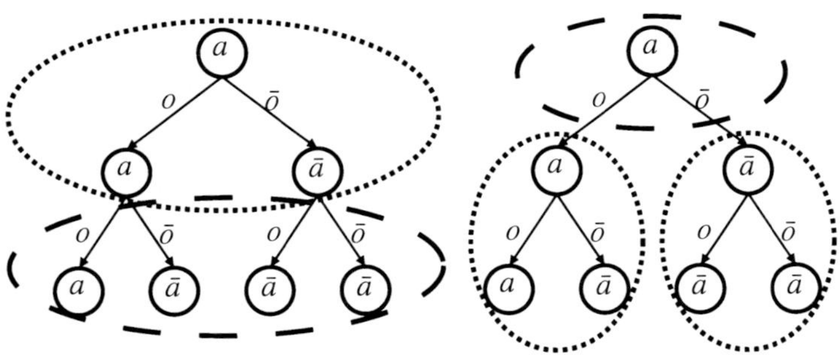

Fig. 6 Difference between policy construction in MAA* (left) and dynamic programming (right) for an agent with actions a,$\bar{a}$ and observations o,$\bar{o}$. The dashed components are newly generated, dotted components result from the previous iteration. MAA* 'expands' a partial policy from the leaves, while dynamic programming backs up a set of 'sub-tree policies' forming new ones.

Dec-POMDPs as series of Bayesian games

The MAA* algorithm can be generalized by interpreting a Dec-POMDP as a series of Bayesian games. A Bayesian game (BG) [52] is an extension of a normal form game in which the agents can hold some private information which is expressed by their *type*. BGs can be used to approximate Dec-POMDPs [22]. In this method, agents construct and solve a BG for each stage of the process in an on-line fashion. Such modeling is exact when using an optimal payoff function for the BGs [48]. Modeling Dec-POMDPs through BGs is appealing because it decomposes the recursive Dec-POMDP problem into a conceptually simpler problem for each stage.

It allows for efficient approximations, since local optima for BGs can be computed efficiently using solution methods from game theory. Also, it opens a new branch of related work and techniques that may be used, some of which are discussed in Section 4.

The crucial difficulty in making a decision at some stage t in a Dec-POMDP is that the agents lack a common signal on which to condition their actions and must rely instead on their individual (action-)observation histories. Given b^0 and φ^t, the joint policy followed for stages $0 \dots t-1$, this situation can be modeled as a BG with identical payoffs. Such a game $BG(b^0,\varphi^t)$ consists of the set of agents $\{1 \dots n\}$, their joint actions A, the set of their joint types Θ, a probability distribution over these joint types $\Pr(\cdot)$ and a payoff function u that maps a joint type and action to a real number $u(\boldsymbol{\theta},\mathbf{a})$. A joint type $\boldsymbol{\theta} \in \Theta$ specifies a type for each agent $\boldsymbol{\theta} = \langle \theta_1, \dots, \theta_n \rangle$. Since the type of an agent represents the private information it holds, types are AOHs $\theta_i \equiv \vec{\theta}_i^t$ and joint types correspond to joint AOHs $\boldsymbol{\theta} \equiv \vec{\boldsymbol{\theta}}^t$. Given φ^t and b^0, the probability distribution over joint AOHs is welldefined and the payoff function is given by $u(\boldsymbol{\theta},\mathbf{a}) \equiv Q^*(\vec{\boldsymbol{\theta}}^t,\mathbf{a})$, the optimal Q-value function of the Dec-POMDP. Although Q^* is intractable to compute, heuristic Q-value functions $\widehat{Q}$ can be computed, for instance using the underlying MDP's value function.

We can solve such a BG by computing the expected heuristic value $\widehat{V}$ for all joint BG-policies $\boldsymbol{\beta}^t = \langle \beta_1, \dots, \beta_n \rangle$, where an individual BG-policy maps types to actions $\beta_i(\vec{\theta}_i^t) = a_i^t$. This valuation is given by

$$\widehat{V}(\boldsymbol{\beta}^t) = \sum_{\vec{\boldsymbol{\theta}}^t} \Pr(\vec{\boldsymbol{\theta}}^t | \varphi^t, b^0) \widehat{Q}(\vec{\boldsymbol{\theta}}^t, \boldsymbol{\beta}^t(\vec{\boldsymbol{\theta}}^t)), \tag{13}$$

where $\boldsymbol{\beta}^t(\vec{\boldsymbol{\theta}}^t) = \langle \beta_i(\vec{\theta}_i^t) \rangle_{i=1\dots n}$ denotes the joint action that results from application of the individual BG-policies to the individual AOHs $\vec{\theta}_i^t$ specified by $\vec{\boldsymbol{\theta}}^t$. The solution $\boldsymbol{\beta}^{t,*}$ is the joint BG-policy with the highest expected value. Note that if φ^t is deterministic, the probability of $\vec{\boldsymbol{\theta}}^t = \langle \vec{\mathbf{a}}^t, \vec{\mathbf{o}}^t \rangle$ is non-zero for only one $\vec{\mathbf{a}}^t$ per $\vec{\mathbf{o}}^t$ ($\vec{\mathbf{a}}^t$ can be reconstructed from $\vec{\mathbf{o}}^t, \varphi^t$). Therefore, in effect the BG-policies reduce to decision rules: mappings from observation histories to actions $\beta_i(\vec{o}_i^t) = a_i^t$.

Generalized MAA*

The modeling of a stage of a Dec-POMDP as a BG as outlined above can be applied in a heuristic policy search scheme called Generalized MAA* (GMAA*) [48], which generalizes MAA* and the BG-method of [22]. Algorithm 3 shows GMAA*, which maintains an open list P of partial joint policies φ^t and their heuristic values $\widehat{V}(\varphi^t)$. Every iteration the highest ranked φ^t is selected and expanded, i.e., the Bayesian game $BG(\varphi^t,b^0)$ is constructed and all joint BG-policies $\boldsymbol{\beta}^t$ are evaluated. Consequently, these joint BG-policies are used to construct a new set of partial policies

$$\Phi_{\texttt{Next}} := \{\varphi^{t+1} = (\varphi^t, \boldsymbol{\beta}^t)\} \tag{14}$$

Algorithm 3 GMAA*

$\underline{v}^\star \leftarrow -\infty$
$\mathrm{P} \leftarrow \{\varphi^0 = ()\}$
repeat
  $\varphi^t \leftarrow$ SelectHighestRankedPartialJPol(P)
  $\Phi_{\text{new}} \leftarrow$ ConstructAndSolveBG(φ^t, b^0)
  if Φ_{new} contains full policies $\Pi_{\text{new}} \subseteq \Phi_{\text{new}}$ **then**
    $\pi' \leftarrow \arg\max_{\pi \in \Pi_{\text{new}}} V(\pi)$
    if $V(\pi') > \underline{v}^\star$ **then**
      $\underline{v}^\star \leftarrow V(\pi')$ {found new lower bound}
      $\pi^\star \leftarrow \pi'$
      $\mathrm{P} \leftarrow \{\varphi \in \mathrm{P} \mid \widehat{V}(\varphi) > \underline{v}^\star\}$ {prune P}
    end if
    $\Phi_{\text{new}} \leftarrow \Phi_{\text{new}} \setminus \Pi_{\text{new}}$ {remove full policies}
  end if
  $\mathrm{P} \leftarrow (\mathrm{P} \setminus \varphi^t) \cup \{\varphi \in \Phi_{\text{new}} \mid \widehat{V}(\varphi) > \underline{v}^\star\}$ {remove processed/add new partial policies}
until P is empty

and their heuristic values. When the heuristic values are an upper bound to the true values, any lower bounds $\underline{v}^\star$ (i.e., full joint policies) that are found can be used to prune P. When P becomes empty, the optimal policy has been found.

GMAA* as outlined here is MAA* reformulated to work on BGs. The BG-method of [22] is similar, but does not backtrack, i.e., rather than constructing all new partial policies $\forall_{\boldsymbol{\beta}^t}\ \varphi^{t+1} = (\varphi^t, \boldsymbol{\beta}^t)$ only the best-ranked partial policy $(\varphi^t, \boldsymbol{\beta}^{t,*})$ is constructed and the open list P will never contain more than 1 partial policy. A generalization k-GMAA* constructs the k best-ranked partial policies, allowing to trade off computation time and solution quality.

2.3.2 The Backward View: Dynamic Programming

GMAA* incrementally builds policies from the first stage $t = 0$ to the last $t = h-1$. Dynamic programming (DP) for Dec-POMDPs [37] constructs policies the other way around: starting with a set of '1-step policies' (actions) that can be executed at the last stage, they construct a set of 2-step policies to be executed at $h-2$, etc.

It should be stressed that the policies maintained are quite different from those used by GMAA*. In particular, a partial policy in GMAA* has the form $\varphi^t = (\delta^0, \delta^1, \ldots, \delta^{t-1})$. The policies maintained by DP do not have such a correspondence to decision rules. We define the *time-to-go* τ at stage t as

$$\tau = h - t. \tag{15}$$

Now $q_i^{\tau=k}$ denotes a k-steps-to-go *sub-tree policy* for agent i. That is, $q_i^{\tau=k}$ is a policy tree that has the same form as a full policy for the horizon-k problem. Within the original horizon-h problem $q_i^{\tau=k}$ is a candidate for execution starting at stage $t = h-k$. The set of k-steps-to-go sub-tree policies maintained for agent i is denoted $Q_i^{\tau=k}$. Dynamic programming for Dec-POMDPs is based on backup operations: constructing $Q_i^{\tau=k+1}$ from $Q_i^{\tau=k}$. For instance, the right side of Fig. 6 shows how $q_i^{\tau=3}$, a 3-steps-to-go sub-tree policy, is constructed from two $q_i^{\tau=2} \in Q_i^{\tau=2}$. Also illustrated is the difference between this process and MAA* expansion (on the left side).

Dynamic programming consecutively constructs $Q_i^{\tau=1}, Q_i^{\tau=2}, \ldots, Q_i^{\tau=h}$ for all agents i. However, the size of the set $Q_i^{\tau=k+1}$ is given by

$$|Q_i^{\tau=k+1}| = |A_i| \, |Q_i^{\tau=k}|^{|O_i|}, \tag{16}$$

and as a result the sizes of the maintained sets grow doubly exponential with k. To counter this source of intractability, Hansen et al. [37] propose to eliminate dominated sub-tree policies. The expected reward of a particular sub-tree policy $q_i^{\tau=k}$ depends on the probability over states when $q_i^{\tau=k}$ is started (at stage $t = h - k$) as well as the probability with which the other agents $j \neq i$ select their sub-tree policies $q_j^{\tau=k} \in Q_j^{\tau=k}$. If we let $q_{\neq i}^{\tau=k}$ denote a sub-tree profile for all agents but i, and $Q_{\neq i}^{\tau=k}$ the set of such profiles, we can say that $q_i^{\tau=k}$ is dominated if it is not maximizing at any point in the *multiagent belief* space: the simplex over $S \times Q_{\neq i}^{\tau=k}$. Hansen et al. test for dominance over the entire multiagent belief space by linear programming. Removal of a dominated sub-tree policy $q_i^{\tau=k}$ of an agent i may cause a sub-tree policy $q_j^{\tau=k}$ of an other agent j to become dominated. Therefore, they propose to iterate over agents until no further pruning is possible, a procedure known as *iterated elimination of dominated policies* [52].

Finally, when the last backup step is completed the optimal policy can be found by evaluating all joint policies $\pi \in Q_1^{\tau=h} \times \cdots \times Q_n^{\tau=h}$ for the initial belief b^0.

2.4 Special Cases and Generalization

Because of the negative complexity results for Dec-POMDPs, much research has focused on special cases of Dec-POMDPs. This section reviews some of these. For a more comprehensive overview of special cases, the reader is referred to [32, 57, 64].

2.4.1 Factored Dec-POMDPs

A *factored* Dec-POMDP has a state space $S = X_1 \times \ldots \times X_{|X|}$ that is spanned by $X = \{X_1, \ldots, X_{|X|}\}$ a set of state variables, or *factors*. A state corresponds to an assignment of values for all factors $s = \langle x_1, \ldots, x_{|X|} \rangle$. In a factored Dec-POMDP, the transition and observation model can be compactly represented by exploiting conditional independence between variables. In particular, the transition and observation model can be represented by a dynamic Bayesian network (DBN) [8].

Even though factored Dec-POMDPs can be represented more compactly, solving them is still hard. However, by adding additional assumptions to the factored Dec-POMDP model, more efficient solutions are sometimes possible [2, 32, 64].

2.4.2 Degrees of Observability

The literature has identified different categories of observability [32, 57]. When the observation function is such that the individual observation for each of the agents will always uniquely identify the true state, the problem is considered *fully-* or *individually observable*. In such a case, a Dec-POMDP effectively reduces to a *multi-agent Markov decision process (MMDP)* [7].

In this setting a (joint) action can be selected based on the state without considering the history, because the state is Markovian. Moreover, because each agent can observe the state, there is an effective way to coordinate. One can think of the situation as a regular MDP with a 'puppeteer' agent that selects joint actions. For this 'underlying MDP' an optimal solution π^* can be found efficiently (P-complete [53]) with standard dynamic programming techniques [56]. Such a solution $\pi^* = (\delta^0, \ldots, \delta^{h-1})$ specifies a mapping from states to joint actions for each stage $\forall_t\ \delta^t : S \rightarrow A$ and can be split into individual policies $\pi_i = \left(\delta_i^0, \ldots, \delta_i^{h-1}\right)$ with $\forall_t\ \delta_i^t : S \rightarrow A_i$ for all agents.

The other extreme is when the problem is *non-observable*, meaning that none of the agents observes any useful information. This is modeled by the fact that agents always receive a null-observation, $\forall_i\ O_i = \{o_{i,\emptyset}\}$. Under non-observability agents can only employ an open-loop plan. A result of this is that non-observable setting is easier from a complexity point of view (NP-complete [57]).

Between these two extremes, there are partially observable problems. One more special case has been identified, namely, the case where not the individual, but the joint observation identifies the true state. This case is referred to as *jointly-* or *collectively observable*.

Definition 8. A jointly observable Dec-POMDP is referred to as a *Dec-MDP*.

Even though all observation together identify the state in a Dec-MDP, each agent still has a partial view. As such Dec-MDPs are a non-trivial sub-class of Dec-POMDPs for which the NEXP-completeness result holds [3].

2.4.3 Explicit Communication

The Dec-POMDP model does not explicitly define communication. Of course, it is possible to include communication actions and observations in the regular set of actions and observations, so a Dec-POMDP can handle communication implicitly. The Dec-POMDP, however, has been extended to incorporate explicitly communication actions and observations. The resulting model the Dec-POMDP-Com [31, 32] additionally includes a set of messages Σ that can be sent by each agent and a cost function C_Σ that specifies the cost of sending each message. The MTDP has a similar extension, called the Com-MTDP [57], which is equivalent to the Dec-POMDP-Com.

Although the Dec-POMDP-Com model itself could allow different communication models, studies so far have considered noise-free instantaneous broadcast communication. That is, at a stage in the process each agent broadcasts its message and receives the messages sent by all other agents instantaneously and without errors.

In the most general case, the goal in a Dec-POMDP-Com is to

> "find a joint policy that maximizes the expected total reward over the finite horizon. Solving for this policy embeds the *optimal meaning* of the messages chosen to be communicated" — Goldman and Zilberstein [31]

That is, in this perspective the semantics of the communication actions become part of the optimization problem. This problem is considered in [31, 67, 80].

One can also consider the case where messages have fixed semantics. In such a case the agents need a mechanism to process these semantics (i.e., to allow the messages to affect their beliefs). For instance, when the agents share their local observations, each agent maintains a joint belief and performs an update of this joint belief, rather than maintaining the list of observations. It was shown that under cost-free communication, a joint communication policy that shares the local observation at each stage is optimal [57].

Models with explicit communication seem more general than the models without, but (in the absence of special mechanisms to interpret the messages) it is possible to transform the former to the latter. That is, a Dec-POMDP-Com can be transformed to a Dec-POMDP [32, 64].

2.4.4 Generalization: Partially Observable Stochastic Games (POSGs)

The generalization of the Dec-POMDP is the *partially observable stochastic game (POSG)*. It has the same components as a Dec-POMDP, except that it specifies not a single reward function, but a collection of reward functions, one for each agent. This means that a POSG assumes self-interested agents that each want to maximize their individual expected cumulative reward.

The consequence of this is that there is no longer an optimal joint policy, simply because 'optimality' no longer defined. Rather the joint policy to suggest should be a (Bayesian) Nash Equilibrium, and preferably a Pareto optimal NE. However, there is no clear way to identify the 'best' one. Moreover, such a Pareto optimal NE is only guaranteed to exist in randomized policies (for a finite POSG), which means that it is no longer possible to perform brute-force policy evaluation. Also search methods based on alternating maximization are no longer guaranteed to converge for POSGs.

3 The Subjective Approach

The Dec-POMDP presents an objective perspective in which the goal is to find a plan, a (near-) optimal joint policy, for all agents simultaneously. In the subjective perspective, we reason for one particular protagonist agent. That is, this agent tries to take the best actions by predicting the other agents in the environment.

This perspective has been computationally formalized in the *recursive modeling method (RMM)* [29, 30]. The limitation of the RMM framework, however, is that the agents interact in the setting of repeated strategic games, and such games do not take into account the environment, or observations of agents. As such it is difficult to model many realistic settings, where agents have different information regarding the environment.

The *interactive POMDP (I-POMDP)* is a different framework that overcomes these limitations [28]. It is closely related to the POMDP in how it models the environment. The I-POMDP focuses on the decision making process of a single self-interested agent, situated in an environment with other agents. Gradually, dependent on the so-called strategy level, the influence of other agents is incorporated into the decisions of the protagonistic agent. Because of this, the framework also admits non-cooperative and competitive MASs.

3.1 Interactive POMDPs

When planning from the perspective of a protagonist agent in a POMDP-like environment that includes other agents, an obvious first approach is to simply ignore the other agents. The problem with this approach is that the result of its action and observations are influenced by other agents.

It is possible to treat this influence as noise, but that this approach has two main drawbacks: first, this approximation decreases the value of the optimal policy. Second, as these approximations are incorporated in the transition- and observation model, they are the same for each time-step. During execution, however, the other agent's policy might be non-stationary, thus the probability of another agent performing some action might depend on the timestep. This non-stationarity is actually guaranteed to take place when the other agent learns of its experience, and especially when it tries to learn and predict the behavior of the protagonist agent.

Therefore, instead of approximating other agents' influence through noise, Gmytrasiewicz and Doshi [28] propose to predict the behavior of other agents. To do this, they maintain an *interactive belief* over worldstates and models of other agents. A model m_i for an agent i from the set of possible models M_i describes the internal state and probabilities of future actions of agent i.

Definition 9. Formally, each model $m_i \in M_i$ for agent i is defined as a triple $m_i = \langle \vec{o}_i, \pi_i, P_{O_i} \rangle$, where $\vec{o}_i$ is the history of observations of agent i, π_i is the policy (referred to as 'function') of agent i and P_{O_i} is agent i's observation function.

Furthermore, Gmytrasiewicz and Doshi divide the set of models into two classes: *sub-intentional* and *intentional* models. The former being simpler, the latter being more complex by attributing beliefs, preferences and rationality to the agent. The intentional model that is considered in I-POMDPs is an agent's *type*:

Definition 10. A *type* θ_i of an agent i that participates in a 'POMDP-like' environment is a tuple $\langle b_i, \hat{\theta}_i \rangle$, where $b_i \in \mathscr{P}(S)$ is the belief over states of agent i and $\hat{\theta}_i = \langle A_i, P_{T_i}, R_i, O_i, P_{O_i}, OC_i \rangle$ is called agent i's *frame*. This frame in turn consists of $A_i, P_{T_i}, R_i, O_i, P_{O_i}$ the actions, transition- and reward function, observations and observation function for agent i, and OC_i, the optimality criterion of agent i: usually the cumulative (discounted) reward. A type is an element of the space of intentional models: $\theta_i \in M_i^{intentional}$.

Given these definitions, we can give the definition of an I-POMDP as follows:

Definition 11. Formally, an *interactive-POMDP (I-POMDP)* of agent i is a tuple $\langle IS_i, A, P_{T_i}, R_i, O_i, P_{O_i} \rangle$, where:

- IS_i is the set of *interactive states*, defined as $IS_i \equiv S \times (\times_{j \neq i} M_j)$.
- A is the set of joint actions.
- $P_{T_i}, R_i, O_i, P_{O_i}$ are the transition- and reward function, observations and observation function for agent i. These are defined over joint actions, i.e. $P(s'|s,\mathbf{a})$ and $R(s,\mathbf{a})$, but over individual observations, i.e. $P(o_i|\mathbf{a},s')$.

An I-POMDP is a POMDP defined over interactive states, which include models of the other agents. This means that an interactive state fully specifies the future behavior of the other agents. As such, it is possible to transform the POMDP belief update to the I-POMDP setting [28]. Since it is rather complex, we do not repeat this transformed belief update here, but just mention that to predict the actions of the other agents, it uses probabilities $\forall_j \, P(a_j|\theta_j)$ given by the model m_j.

When considering intentional models (types), the formal definition of I-POMDPs as above leads to an infinite hierarchy of beliefs, because an I-POMDP for agent i defines its belief over models and thus types of other agents, which in turn define a belief over the type of agent i, etc. To overcome this problem one can define *finitely nested I-POMDPs* [28]. Here, a 0-th level belief for agent i, $b_{i,0}$, is a belief over world-states S. An k-th level belief $b_{i,k}$ is defined over world-states and models consisting of types that admit beliefs of (up to) level $k-1$. The actual number of levels that the finitely nested I-POMDP admits is called the *strategy level*.

Example 3 (The FIREFIGHTING *problem as an I-POMDP).* Here we illustrate how the FIREFIGHTING problem may be represented as a finitely nested I-POMDP. To ease the following discussion we assume that there are two agents: i and j.

Suppose we define two sub-intentional models for agent j: one model m_{H_1} specifies that agent j will always perform action H_1, the other m_{rand} assumes that agent j will act random; i.e.,will select its actions with uniform probability. Together these models form the set $M_{j,0}$ of 0th-level models[3] of agent j and this set can be used to construct a 1st-level I-POMDP model for agent i. In this I-POMDP $IP_{i,1}$,

- the interactive state space $IS_{i,1} = S \times M_{j,0}$ is the product of the set of world states S, which is defined in the same way as in the Dec-POMDP case (i.e., the state specifies the fire levels of each of the houses), and the possible models,
- the set of joint actions is equal to the Dec-POMDP case: it is the cross product of the sets of individual actions $\{H_1 \ldots H_{N_H}\}$,
- the transition, observation and reward function are also defined in the same way as in the Dec-POMDP case.

Because each of the possible models $m \in M_{j,0}$ fully specifies the behavior of agent j, i.e., it specifies $\Pr(a_j|m)$, the interactive state possesses the Markov property and $IP_{i,1}$ can be solved.

[3] Clearly, this is a very simple example. A more realistic scenario might include more sub-intentional models.

Now in turn $IP_{i,1}$ can be interpreted as a model for agent i. In particular, given an interactive belief for $b_{i,1} \in IS_{i,1}$ the behavior of agent i is completely specified. This means that we can use it as a model $m_{i,1} = \langle b_{i,1}, IP_{i,1} \rangle$ for agent i in the interactive 2nd level state space $IS_{j,2} = S \times M_{i,1}$ for agent j, etc.

A final note in this example is that by updating the belief over interactive states in the belief update not only the belief over states is updated, but also the belief over the models of the other agent. For instance, consider a FIREFIGHTING problem where the observation spaces are augmented to include the observation of the other agent if they happen to fight fire at the same house. In this case, when agent i observes agent j at a house different than house 1, it learns that the model m_{H_1} is incorrect and will update its beliefs to reflect this, thereby improving future predictions of agent j.

A graphical model formulation of I-POMDPs called *interactive dynamic influence diagram (I-DID)* has also been proposed [20]. This model allows for a more compact representation of sequential decision problems by making explicit the structure of state variables and their influences. Like the I-POMDP itself, it gives a subjective view from a particular agent.

Such a subjective approach is very appropriate for systems with self-interested agents, but can also be applied to cooperative agents. The subjective perspective also allows for flexibility, for instance in systems where agents may come and go. Also, the subjective approach can be used to allow efficient approximations, such as demonstrated in Section 5.2.

3.2 Solving I-POMDPs

As mentioned in Section 3.1, an I-POMDP actually considers reasoning from one agent's perspective, which is why it is very similar to single agent planning.

For a finitely nested I-POMDP, the reasoning the agent performs in fact is of the form "what would the other agents do if they think that I think that...". A finitely nested I-POMDP can be solved bottom up [28]: An I-POMDP of strategic level 0 for agent i is a regular POMDP that can be solved. This gives a level 0 policy. Now, assuming that agent i can only have certain particular beliefs (over states, since it is a level-0 model), this induces a probability distribution over actions, which can be used in the level 1 I-POMDP of agent j, effectively transforming the latter to a POMDP as well. The solution (a level 1 I-POMDP policy π_j) can then be used for the level 2 I-POMDP for the agent i, etc.

The problem in the above procedure lies in the assumption that the modeled agent's belief must be a member of some particular finite set of beliefs, instead of the infinite set of beliefs. Even though one can argue that both agents have the same initial belief b^0 and that therefore the number of (0-th level) beliefs over states is finite, the number of possible models of other agents is infinite. Effectively this means that an arbitrary discretization of the interactive belief space has to be made. To overcome this problem, it is possible to group models in behavioral equivalence classes [60]. This induces a partition of the infinite interactive state space and thus leads to a finite representation of the interactive state space that allows for exact solutions.

3.3 *The Complexity of Solving I-POMDPs*

For finitely nested I-POMDPs, similar results hold as for POMDPs: value iteration is guaranteed to converge and the value function is piecewise linear and convex [28]. The authors also state that solving a finitely nested I-POMDP of strategy level l is PSPACE-hard, because it is equivalent to solving $O(M^l)$ POMDPs, *assuming that the number of models considered at each level is bounded by a number* M. However, the space of models considered can be very large or even infinite. Moreover, when updating the belief of another agent's model, we have to consider what observations the other agent may have received and introduce a new resulting model for each of them. This means that the number of considered models grows exponentially with the planning horizon.

4 Application of Decision-Theoretic Models and the Need to Scale up

In this section, we give an example of the application of decision-theoretic models to a realistic crisis management task, namely, RoboCup Rescue. The resulting model is huge and current methods are still far away from solving such a model. We argue that this is typical for the application of decision-theoretic tools to real-world problems and that it motivates the search for methods that scale better.

Consequently, we address some different methodologies that contribute to realizing better scaling for MASs under outcome and state uncertainty. Most of these methods have been proposed for Dec-POMDPs. Most work on I-POMDPs has focused on making the I-POMDP belief update more tractable [17, 18], clustering models [60, 81], or extending single-agent POMDP techniques [19].

4.1 *An Example: RoboCup Rescue as a Dec-POMDP*

Here, we present an example that shows how firefighting in RoboCup Rescue can be modeled as a factored Dec-POMDP and that the resulting model is intractable.[4] It would also be possible to model it as an I-POMDP, but similar scaling problems would result.

4.1.1 State Description

Here, we describe a formal model of the state space for the task of fire fighting in RoboCup Rescue. It is based on the dynamic factors in the RoboCup Rescue world, restricted to the factors relevant for firefighting. Static factors, i.e. factors that do not change throughout the simulation, will not have to be incorporated in the state space. In RoboCup Rescue, the world is given by a map, consisting of buildings,

[4] This is a simplified summary of the description in [50].

roads and nodes, which act as the glue between roads and buildings. A typical map consists of approximately 1000 buildings and the same number of roads and nodes.

This world is populated by civilians, rescue agents (platoons) and the immobile rescue centers. The rescue agents (both mobile and center agents) can in turn be divided into fire, police and ambulance agents. The center agents function as communication centers. We only consider (about 15) mobile firefighting agents here.

The mobile agents can be located on roads, nodes or in buildings. So the number of valid positions on a map is the sum of these elements (i.e., typically around 3000). Also these mobile agents have a particular health, expressed in health points (HPs). Fire-brigade agents have a particular amount of water left.

The earthquake that takes place at the beginning of the simulation has a large effect on the state: buildings collapse and catch fire, these collapses can cause roads to get blocked (in either direction) and civilians to get trapped in debris. Starting from this initial state, fires will spread if left unattended. The fire simulator is based on heat energy as primary concept. Fire propagation's main component is heat radiation, which is simulated by dividing the open (non-building) area of the map in (approx. 20000) cells for which the air temperature is computed. Other factors that determine the spread of fire are properties of buildings. The dynamic factors are the heat of a building, whether it is burning, how much fuel is left and how much water is put in by extinguish actions. Static properties like the size and composition of a particular building (wood, steel or reinforced concrete) and how close it is to surrounding buildings also influence the spreading of fire. However, because these properties are static, they do not need to be incorporated in the state description. Instead the probability of a particular building i catching fire given that a neighboring building j is on fire is modeled through the transition model.

Table 1 summarizes the state factors. Note that, as this example focuses on the firefighting aspect, certain factors (e.g. the position, health and other properties of other agents) are ignored.

4.1.2 Actions

The actions for an agent i in RoboCup Rescue can be divided into domain level actions A_i^d and communication actions A_i^c. A mobile agent can perform both a domain level action as communication within one timestep (e.g. a fire-brigade agent can move/extinguish and communicate). This means that the set of actions A_i for a mobile agent i is the Cartesian product of all domain level and communication actions $A_i = A_i^d \times A_i^c$. In this section, we will discuss the domain level actions. Section 4.1.4 will deal with communication actions.

All mobile agents can perform the *move* action. The argument of this *move* action is a path along which the agent should move. Clearly, the *move* actions are dependent on the current position of the agent. Also, there is a maximum distance that an agent can travel in one timestep (333m). This means that two paths that deviate only after this point lead to the same action. Fire-brigades have two specialized actions: *extinguish* and *refill*. The *extinguish* action specifies a building and the amount of water (in liters) to direct to that building.The *refill* action restores the

Table 1 State variables or *factors* for a particular RoboCup Rescue world

factor	values
for all fire agents in the world (±15)	
current position	valid positions
health	0–9999HPs
amount of water	0–15000l
for all roads (±1000)	
blocked? (2 directions)	blocked/free
for all buildings (±1000)	
heat energy	0–10^6 GJ
state	(not) burning
fuel left	percent(%)
amount of water	0–150000l
for all air cells (±20000)	
temperature	20-10000°C

water supply of the brigade and can only be performed at 'refuges'; these are special buildings where agents can find shelter.

4.1.3 Observations

As with actions we specify the set of observations for agent i as the Cartesian product of domain and communication observations $O_i = O_i^d \times O_i^c$. Here we treat the domain observations, communication observations are treated in section 4.1.4.

At each timestep, only objects within a range of 10 m are seen, except for fiercely burning buildings, which can be observed from a larger distance. When an agent executed a *move* action, only observations of the new position are received (i.e. no observations are made 'en route'). On average 4–6 static objects (building and roads) can be visually observed during a timestep [54].

Observing an object means that the agent receives the object ID, its type and properties. For a road this property is whether or not it is blocked (in both ways), for a building, the so-called fieriness, is observed. This fieriness factor is a direct function of the amount of fuel and water left in the building and determines the part of the area counted as damaged.

4.1.4 Communication

Communication consists of both an action (by the sender) and an observation (for the receiver). In RoboCup Rescue there are two forms of communication actions: *say* and *tell*. The *say* messages are directly transferred (i.e., shouted), the latter transmitted by radio. Both types of communication are broadcast: *say* messages can be picked up by agents of all types within 30 m, *tell* messages can be received by all agents of the same type regardless of the distance. The restrictions that are posed on communication vary per competition. We assume that platoon agents can send 4 *tell* messages and one *say* message and that all agents can receive all messages. Restrictions on the number of received messages can also be incorporated [50].

In a Dec-POMDP, we can model communication by introducing communication actions and observations. The basic idea is that for each joint communication action $\mathbf{a}^c$ one joint communication observation $\mathbf{o}^c$ can be introduced that for each agent contains the messages sent by the other agents. Restrictions with respect to communication distance can be modeled by making communication dependent on the (next) state s'. That is, it is possible to specify a communication model of the form $\Pr(\mathbf{o}^c|\mathbf{a}^c,s')$.

The complete observation model is then given as the product of this communication model and the regular, domain observation model:

$$\Pr(\langle \mathbf{o}^d,\mathbf{o}^c\rangle|\langle \mathbf{a}^d,\mathbf{a}^c\rangle,s') = \Pr(\mathbf{o}^c|\mathbf{a}^c,s')\cdot\Pr(\mathbf{o}^d|\mathbf{a}^d,s'). \tag{17}$$

In a pure planning framework, messages have no a priori semantics. Instead the planning process should embed the 'optimal meaning' in each communication action as explained in Section 2.4.3 In RoboCup Rescue all messages are 256 bytes. When an agent can send 4 *tell* and 1 *say* message, it has $8\cdot256\cdot5 = 10240$ bits to encode its communication action and thus that $|A_i^c| = 2^{10240}$. This means that the number of joint communication actions $|A^c| = 2^{10240n}$. Clearly this way of treating communication is intractable.

4.1.5 Transition, Observation and Reward Model

The transition model of a factored Dec-POMDP can be compactly described by a two-stage DBN. Because the state description is the same as used by the simulator components, these structures and probabilities for this DBN can be found by analyzing the code of the simulation system. For the (domain) observation model we can make a similar argument.

The reward function is easily derived from the scoring function. A typical scoring function is

$$Score(s) = (P + S/S_0)\cdot\sqrt{B/B_0}, \tag{18}$$

where P is the number of living agents, S_0 is the total sum of health points (HPs) at start of the simulation, S is the remaining sum of HPs, B_0 is the total area of houses and B is the area of houses that remained undamaged.

This gives us the reward function of the Dec-POMDP in the following way:

$$R(s,\mathbf{a},s') = R(s,s') = Score(s') - Score(s). \tag{19}$$

The horizon is finite in the RoboCup Rescue competition (300 time-steps). However, in the real-life setting we will typically want to plan for a varying horizon (until all fire is extinguished and all trapped people are either rescued or dead). This can be accomplished by treating the problem as one of infinite horizons.

4.1.6 Complexity

Here, we will give a brief argument of the complexity of the Dec-POMDP representation. By ignoring some factors of the statespace, we effectively performed a first abstraction to reduce the state space. However, the state space as presented in Table 1 is still huge. When there are $n = 15$ fire-brigade agents and 3000 valid positions this already leads to 3000^{15} different configurations. When considering only the first four factors, we already get a state space of

$$\begin{aligned} |nr\_pos|^{15} \cdot |HPs|^{15} \cdot |water|^{15} \cdot 2^{|2 \cdot nr\_roads|} &= \\ 3000^{15} \cdot 10000^{15} \cdot 15000^{15} \cdot 2^{2000} &\approx \\ 10^{52} \cdot 10^{60} \cdot 10^{62} \cdot 10^{602} &= 10^{776} \end{aligned}$$

and this is not even including the state of each of the 1000 buildings. We already saw that the number of joint communication actions is prohibitively large and the same holds for domain actions and observations. Therefore, this is clearly an intractable problem.

In the above, we modeled RoboCup Rescue as a Dec-POMDP and concluded that the resulting model is intractable to solve. Even though this is just an example, it is exemplary for the current state of affairs: when modeling a realistic problem using decision-theoretic tools, the result is typically an intractable problem. On the one hand, this motivates the need for decision-theoretic methods that scale better. On the other hand, some research tries to directly find heuristic solutions that perform well, thereby trying to closer the gap between principled solutions and real-life applications from both sides. In Section 5, we will provide some examples of our work in the latter category. The rest of this section focuses on ways to scale up decision-theoretic approaches. In particular, we give an overview of hierarchical decompositions, exploiting independence between agents and compressions of policies. We focus on aspects of MASs, but stress that scaling up methods for single-agent (PO)MDPs is still a very important area of research [36].

4.2 Aggregation and Hierarchical Decompositions

Some common sense reveals that the intractability of the flat description of the previous section is no surprise: In the real world, a fireman who is extinguishing some building typically will not care about properties of a building beyond the zone of potential fire spreading. Nor will he consider what the exact position of a colleague at the other side of town is. Effectively, in order for a firefighter to perform his job he needs to focus on those state variables that are relevant for his current task. States that do not differ on relevant variables can be grouped or *aggregated*. In such a way state aggregations reduce the size of the state space [15, 24, 27, 55].

Aggregation is closely related to hierarchical methods [1, 16, 68]. That is, these above insights may be used to provide a hierarchical decomposition of tasks, bringing leverage to the decision process, by constraining the number of possible policies. An example how such a decomposition might look for the RoboCup Rescue

Dec-POMDP is given in [50]: At the lowest level each group of burning buildings (fire zone) may be considered separately, coordination between different fire zones is performed at a higher district level and the highest level considers the entire city. So far, most other hierarchical approaches have focussed on single-agent and fully observable settings, typically using the framework of SMDPs. An important direction of future research is the examination of more principled hierarchical methods for partially observable MASs.

4.3 Modeling and Exploiting Independence between Agents

In a complex MAS, not all agents may need to consider each other's actions. A hierarchical decomposition may make this assumption explicit by having some agents participate in different parts of the decomposition (e.g. different fire zones in the example above). On a smaller scale, however, there may also be independence between agents. For instance consider a slightly modified version of FIREFIGHTING, called FIREFIGHTINGGRAPH, illustrated in Fig. 7. This version of the problem restricts each agent to only go to any of two fixed houses that are assigned to it. In such a setting it is conceivable that agent 1 only has to directly coordinate with agent 2, but not with agent 3.

In the last decade, much research has been devoted to exploiting such independence between agents. However, many of these approaches make very restrictive assumptions. For instance, models with more assumptions on observability and/or communication have been considered [35, 42]. These approaches try to exploit independence between agents, but they either require the agents to observe the state of the system (as in a multiagent MDP) or to observe a subset of state variables (as in a factored Dec-MDP) and communicate at every stage. For the particular case of transition and observation independent Dec-POMDPs [2], it has been suggested to exploit independence [47, 76]. However, the assumption of transition and

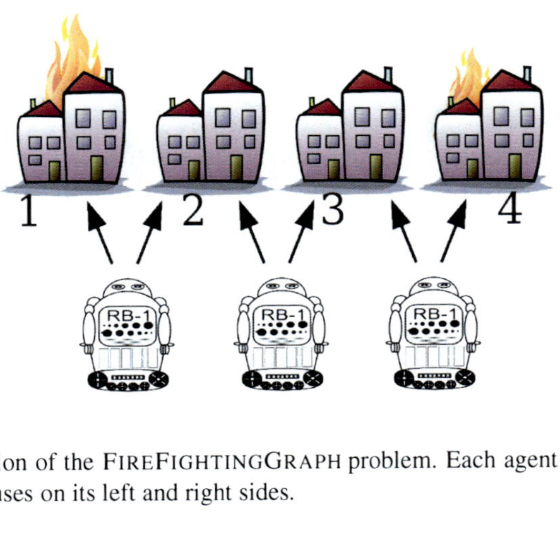

Fig. 7 An illustration of the FIREFIGHTINGGRAPH problem. Each agent is restricted to go fight fire at the houses on its left and right sides.

observation independence (TOI) severely limits the applicability of these models. In particular, TOI implies that the agents have disjoint sets of individual states and observations, and that one agent cannot influence the state transition or the observation of another agent. In practice, this means that many interesting tasks, such as two robots carrying an object, cannot be modeled.

A more general analysis of locality of interaction in factored Dec-POMDPs shows that the last stage in the process contains the highest degree of independence but, when moving back in time (towards the initial stage $t = 0$), the scope of dependence grows [49]. Still, the authors are able to exploit independence between agents in the optimal solution of factored Dec-POMDPs, because most of the independence is located where it is most needed: the overwhelming majority of the computational effort of solving a Dec-POMDP is spent in the last stage [73]. The method of [49] is an extension of GMAA*, which replaces the BGs by *collaborative graphical BGs (CGBGs)*, which are less computationally expensive to solve in later stages where independence is high. Even though they can only exploit independence in the last stage to guarantee optimality (GMAA* needs to return *all* partial policies for intermediate stages to guarantee optimality), an experimental evaluation shows a speedup of two orders of magnitude.

4.4 Compression of the Considered Policy Space

A third category of techniques to facilitate better scaling tries to compress directly the space of policies that is considered. The first example of this type of approach is presented in [22] that proposes to prune low-probability joint types (i.e., joint action-observation histories with a low probability) from the BGs constructed for each stage. Subsequently, the smaller BGs are easier to solve, since there are a lot less joint BG-policies (i.e., decision rules for the stage for which the BG is constructed) to consider. As such, the search space of joint policies is effectively trimmed by pruning in the space of histories. This approach is refined by clustering histories with similar (heuristic) payoff profiles, rather than pruning [23].

Even though these pruning and clustering techniques proved to be quite effective, they are a form of lossy compression of the BG and no quality guarantees are available. Recently, a criterion has been identified that *guarantees* that two individual histories have the same optimal value, allowing *lossless clustering* and therefore faster optimal solutions of Dec-POMDPs [51]. The authors experimentally demonstrate that in several well-known test problems, the proposed method allows for the optimal solution of significantly longer horizons.

In the backward view of solving Dec-POMDPs similar techniques have been employed. Probably the most successful method here is memory-bounded dynamic programming (MBDP) [63]. Rather than pruning dominated sub-tree policies $q_i^{\tau=k}$, MBDP prunes all sub-tree policies except a few in each iteration. More specifically, for each agent m sub-tree policies that perform well on a set of heuristically sampled multiagent belief points are retained. Effectively this means that at every iteration the search space of policies is trimmed to a fixed size. As a result, MBDP has only linear space and time complexity with respect to the horizon. The MBDP algorithm still depends on the exhaustive generation of the sets $Q_i^{\tau=k+1}$ which now contain $|A_i|\,m^{|O_i|}$ sub-tree policies. Moreover, in each iteration all $\left(|A_*|\,m^{|O_*|}\right)^n$ joint sub-tree policies have to be evaluated for each of the sampled belief points. To counter this growth, an extension is proposed that limits the considered observations during the backup step to the *maxObs* most likely observations [62]. MBDP with observation compression (MBDP-OC) [10] improves upon this by not pruning low-probability observations, but rather clustering observations in a manner that minimizes the expected loss.

MBDP-OC is able to bound the error introduced relative to MBDP without observation compression. The error introduced by MBDP itself, however, is unbounded. Recently, an optimal algorithm based on DP that performs a lossless compression of the sub-tree policies was introduced [6]. The idea is that sub-tree policies are represented in a smaller more compact form, similar to *sequence form* [43] and especially the so-called *tests* from the work on predictive state representations [65]. Policy compression results in significant improvement in performance compared to regular DP for Dec-POMDPs.

5 Efficient Heuristic Approaches for Teams of Agents

The decision-theoretic models covered in this chapter have some desirable properties. They are well defined and allow for quantitative comparisons of solutions. Unfortunately, the solution methods scale poorly. The previous section described some techniques that can improve scalability. Even though this is a step in the right direction, pure decision-theoretic solutions are still impractical for most real-world settings.

Our strategy has been to tackle the gap between decision-theory and real-world applications from two sides. This section describes work that approaches it from the application side using efficient heuristic methods. First, we describe a body of work that employs a limited form of decision-theoretic reasoning to allocate heuristically defined roles using variants of the Dec-POMDP. Second, we describe how a heuristic variant of the subjective approach to decision theory can be employed for the task of exploration in emergency sites.

5.1 Allocating Pre-specified Roles Sensibly

Over the last decades, many heuristic approaches to multiagent systems have been proposed and employed with considerable success in (near-) real-life settings [38, 58, 69, 74]. Of these approaches, many are based on the belief-desire-intention (BDI) paradigm [26] and extensions for MASs [11–13, 33, 34]. In such 'BDI teams' agents reason over their beliefs, desires and intentions and that of the team to select a plan, or *role* from a library of pre-compiled plans.

These heuristic approaches have been able to scale to larger problems than the decision-theoretic approaches treated in earlier sections. Still, the latter have some desirable properties. As such, some research has tried to integrate DTP within these BDI-teams to improve performance, by applying it to substitute only a part of the reasoning. In particular, the task of role allocation has been considered.

For instance, decision-theoretic reasoning has been applied to assign roles to agents in the context of the RoboCup Soccer simulation league [41]. Their approach is based on using coordination graphs to indicate which agents should coordinate when select a role. Nair and Tambe [46] consider role allocation for TOP plans [58]. They define the role-based multiagent team decision problem (RMTDP) which is an extension of the Dec-POMDP to include role taking and role executing actions. The RMTDP is constructed and solved more efficiently by exploiting properties of the TOP plan. For instance, the features tested in the TOP plan correspond to the set of states S of the RMTDP. More important, the organization hierarchy of the TOP plan is used to calculate upper bounds for the expected reward when a role is chosen. This upper bound is used to prune less attractive role allocations. For this algorithm, experimental results on a simplified RoboCup Rescue scenario are presented.

5.2 Frontier Selection in Exploration

Multi-robot exploration is a challenging application area in which multiple robots have to explore an area as efficiently as possible. An instance of this problem can be found in the RoboCup Rescue League: inside the Virtual Robot competition multiple robots have to coordinate their actions to explore jointly an as large as possible area after a disaster. At the beginning, the robots have no knowledge of the environment, but by driving around and observing, it can be explored. Exploration can be formally defined as the task to maximize the knowledge about the environment. When the environment is modeled with an occupancy grid map, the exploration problem is the problem of maximizing the cumulative information of each grid cell.

One of the difficulties in multi-robot exploration is that information about the environment is distributed; each robot has a partial and incomplete view represented by the occupancy grid map it individually retains. Most of the knowledge about the environment is directly derived from the sensors of the robot, only occasionally augmented with some additional (older) observations from the other robots when they are relayed over a multi-hop communication network. However, communication may not be available at all times. Also, for a setting like this it is infeasible to pre-compute a plan describing what to do in all possible settings, since this would entail finding an exploration plan for every possible world. For these reasons, a subjective approach as outlined in Section 3 is more suitable for this type of problems.

In particular, the exploration problem as introduced can be formalized as an I-POMDP. In this representation, each robot maintains its private occupancy grid map. The full state vector contains both s_p the true state of the physical environment (which includes the agent positions) as well as s_i the maintained grid of each agent i. Domain-level actions of robots typically consist of low-level actuations. However, in this setting it is possible to use a higher abstraction level by defining 'exploration frontiers'. Such frontiers are the boundaries between explored and unexplored areas as specified by each individual occupancy grid. Action a_{ij} of agent i dictates that it goes to frontier f_j, the j-th frontier specified by its grid. Selecting a frontier means a decision on the action to take. Given s_p the state of the physical environment and the actions selected by the agents, a new physical state s'_p results stochastically. Each agent makes a local observation of this new state s'_p which allows them to update (their belief of) their individual maps s'_i, as well as their beliefs over the maps maintained by other agents. Any received communication may be incorporated in these beliefupdates. The reward function can be defined as the difference in cumulative information between two states s and s'.

The above defines all necessary components for an I-POMDP, except the intentional models. A sensible level-0 sub-intentional model for an agent is to go to the nearest frontier on its map m_{near}. We do not define alternative sub-intentional models, so $M_{j,0} = \{m_{near}\}$ for each agent j. As before, the interactive states of a level-1 I-POMDP for agent i can now be defined as $IS_{i,1} = S \times M_{j,0}$. Here, a $s \in S$ is a full state (specifies the physical states and the grid of each agent). Each interactive belief of such an agent i fully specifies its behavior and therefore serves as a model for a level-2 I-POMDP for agent j, etc.

In theory, for each agent i we could build such an I-POMDP model and the resulting behavior should be coordinated. Of course, such an I-POMDP representation is intractable; the number of interactive states even for an I-POMDP of level 1 is given by product of the number of physical states s_p (i.e., possible worlds) and the number of possible maps the other agent. Still the I-POMDP formulation as presented can serve as a guideline for efficient heuristic subjective methods. In particular, the number of interactive states can be greatly reduced by simply dropping the maps of other agents j from the state description for an I-POMDP of agent i. A side effect is that the sub-intentional model of an agent j is no longer welldefined: it specifies to go to the nearest frontier according to agent j's grid map, but the state no longer includes this map of agent j. Therefore, agent i assumes that agent j has the same map as itself. This approximation may work well, because the agents that need to coordinate closely are those that are physically near, which in turn means that they are likely to have similar maps.

A second issue is the infinite number of physical worlds that the robots could explore. At the end, the reasoning over hidden world states performed in an (interactive) POMDP has as its goal to derive expected future reward values for each action, which can then be used to select the best action. Now the idea is to more directly use a heuristic function to valuate each of the possible individual actions. The heuristic which used is based on the physical properties of the observation process. The difference between explored and unexplored areas is a gradual one. Some observations beyond the frontiers are already available, which can be used the estimate the future reward [77]. In particular, we use an estimate of the information gained by each possible individual action by looking at the area visible beyond the frontier.

By predicting the other agents' actions (by assuming that they go to the frontier nearest to them) each individual action a_i leads to a joint action. This joint action leads to an expected increase in information, by adding up the size of the areas beyond those frontiers. Such an approximation may work well, because in a

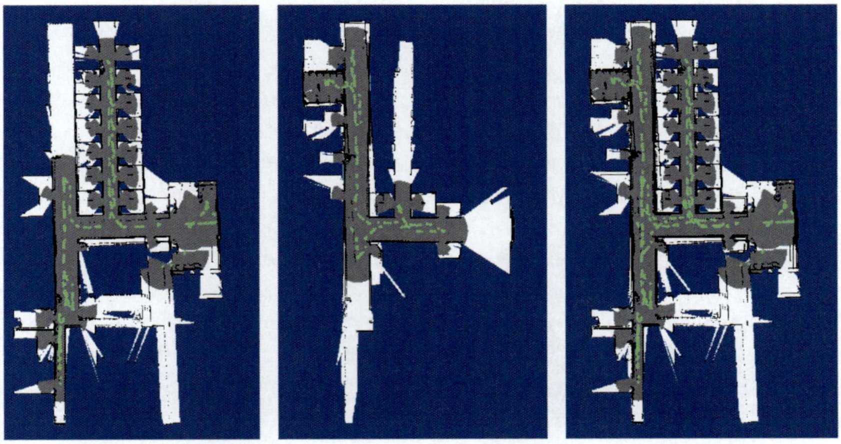

(a) Map generated by one robot (b) Map generated by another robot (c) combined Map

Fig. 8 Map resulting from an autonomous exploration in the RoboCup Rescue 2006 Competition

exploration setting, each non-explored world typically is equally likely, i.e., there typically is no good Bayesian prior that gives a better estimate of the value of each action than the sparse observations beyond the frontier.

Note that assuming the other agents go to the nearest frontier may result in miscoordination since in fact they will employ similar reasoning. To reduce the probability of such miscoordinations, prediction of the other agents is enhanced by assuming that the other agents also trade off information gain and travel cost. Frontiers are allocated to robots in order, where the combination which has the highest estimated information gain / travel cost balance is selected first.

The result is a heuristic exploration method inspired by the I-POMDP formulation given above. Experimental evaluations demonstrate that the method results in coordinated behavior where each robot explores a different part of the environment. An example of such cooperation is given in Fig. 8, where the exploration effort of two robots is indicated with their individual maps and the shared map. Gray areas indicate fully explored areas, and white areas indicate to be explored areas. The boundaries between white and gray areas are the frontiers. It is clear that the overlap in the exploration effort of both robots is small.

6 Conclusions

In this chapter, an overview is given of a number of decision-theoretic tools and models to allow reasoning with uncertainty, reasoning with risks and reasoning under a lack of knowledge. The decentralized POMDP framework is described as a model for objective sequential decision making for a team of cooperative agents and compared with an alternative subjective approach; interactive POMDP. For both models we described solution methods and the known complexity results.

We also modeled a more realistic task as a Dec-POMDP and illustrated that the resulting model is prohibitively large. Therefore, we argue that there still is a big gap between decision-theoretic methods and real-life applications. In our opinion, this gap should be closed from both sides: by scaling up principled solutions and identifying efficient heuristics. In the former category we discussed some techniques that may allow for the solution of larger Dec-POMDPs. In the latter, we gave an example of how Dec-POMDPs can be applied to tackle a smaller part of the problem (role assignment), and a heuristic approach for the task of exploration. Since the approach to exploration takes a subjective perspective, it can be interpreted as an approximation to an I-POMDP. Future work should investigate whether such approximations are possible for a broader class of partially observable multiagent problems.

Acknowledgements. We wish to thank our colleagues: Julian de Hoog, Julian Kooij, Matthijs Spaan, Nikos Vlassis and Shimon Whiteson.

References

1. Barto, A.G., Mahadevan, S.: Recent advances in hierarchical reinforcement learning. Discrete event dynamic systems: Theory and applications 13, 343–379 (2003)
2. Becker, R., Zilberstein, S., Lesser, V., Goldman, C.V.: Solving transition independent decentralized Markov decision processes. Journal of Artificial Intelligence Research 22, 423–455 (2004)
3. Bernstein, D.S., Givan, R., Immerman, N., Zilberstein, S.: The complexity of decentralized control of Markov decision processes. Mathematics of Operations Research 27(4), 819–840 (2002)
4. Bernstein, D.S., Zilberstein, S., Immerman, N.: The complexity of decentralized control of Markov decision processes. In: Proc. of Uncertainty in Artificial Intelligence, pp. 32–37 (2000)
5. Bertsekas, D.P.: Dynamic Programming and Optimal Control, 3rd edn., vol. I. Athena Scientific, Belmont (2005)
6. Boularias, A., Chaib-draa, B.: Exact dynamic programming for decentralized POMDPs with lossless policy compression. In: Proc. of the International Conference on Automated Planning and Scheduling (2008)
7. Boutilier, C.: Planning, learning and coordination in multiagent decision processes. In: Proc. of the 6th Conference on Theoretical Aspects of Rationality and Knowledge, pp. 195–210 (1996)
8. Boutilier, C., Dean, T., Hanks, S.: Decision-theoretic planning: Structural assumptions and computational leverage. Journal of Artificial Intelligence Research 11, 1–94 (1999)
9. van den Broek, B., Wiegerinck, W., Kappen, B.: Graphical models inference in optimal control of stochastic multi-agent systems. Journal of Artificial Intelligence Research 32, 95–122 (2008)
10. Carlin, A., Zilberstein, S.: Value-based observation compression for DEC-POMDPs. In: Proc. of the International Joint Conference on Autonomous Agents and Multi Agent Systems, pp. 501–508 (2008)

11. Cohen, P.R., Levesque, H.J.: Intention is choice with commitment. Artificial Intelligence 42(3), 213–261 (1990)
12. Cohen, P.R., Levesque, H.J.: Confirmations and joint action. In: Proc. of the International Joint Conference on Artificial Intelligence, pp. 951–957. Morgan Kaufmann, San Francisco (1991)
13. Cohen, P.R., Levesque, H.J.: Teamwork. Nous 25(4) (1991)
14. Dawes, R.M.: Rational Choice in an Uncertain World. Hartcourt Brace Jovanovich (1988)
15. Dean, T., Givan, R.: Model minimization in Markov decision processes. In: Proc. of the National Conference on Artificial Intelligence, pp. 106–111 (1997)
16. Dietterich, T.G.: Hierarchical reinforcement learning with the MAXQ value function decomposition. Journal of Artificial Intelligence Research 13, 227–303 (2000)
17. Doshi, P.: Approximate state estimation in multiagent settings with continuous or large discrete state spaces. In: Proc. of the International Joint Conference on Autonomous Agents and Multi Agent Systems, p. 13 (2007)
18. Doshi, P., Gmytrasiewicz, P.J.: A particle filtering based approach to approximating interactive POMDPs. In: Proc. of the National Conference on Artificial Intelligence, pp. 969–974 (2005)
19. Doshi, P., Perez, D.: Generalized point based value iteration for interactive POMDPs. In: Proc. of the National Conference on Artificial Intelligence, pp. 63–68 (2008)
20. Doshi, P., Zeng, Y., Chen, Q.: Graphical models for interactive POMDPs: representations and solutions. Autonomous Agents and Multi-Agent Systems 18(3), 376–416 (2008)
21. Druzdzel, M.J., Flynn, R.R.: Decision Support Systems. In: Encyclopedia of Library and Information Science. The Taylor & Francis, Inc., New York (2003)
22. Emery-Montemerlo, R., Gordon, G., Schneider, J., Thrun, S.: Approximate solutions for partially observable stochastic games with common payoffs. In: Proc. of the International Joint Conference on Autonomous Agents and Multi Agent Systems, pp. 136–143 (2004)
23. Emery-Montemerlo, R., Gordon, G., Schneider, J., Thrun, S.: Game theoretic control for robot teams. In: Proc. of the IEEE International Conference on Robotics and Automation, pp. 1175–1181 (2005)
24. Feng, Z., Hansen, E.: An approach to state aggregation for POMDPs. In: AAAI 2004 Workshop on Learning and Planning in Markov Processes – Advances and Challenges, pp. 7–12 (2004)
25. Gal, Y., Pfeffer, A.: Networks of influence diagrams: A formalism for representing agents' beliefs and decision-making processes. Journal of Artificial Intelligence Research 33, 109–147 (2008)
26. Georgeff, M.P., Pell, B., Pollack, M.E., Tambe, M., Wooldridge, M.: The belief-desire-intention model of agency. In: Rao, A.S., Singh, M.P., Müller, J.P. (eds.) ATAL 1998. LNCS (LNAI), vol. 1555, pp. 1–10. Springer, Heidelberg (1999)
27. Givan, R., Dean, T., Greig, M.: Equivalence notions and model minimization in Markov decision processes. Artificial Intelligence 147(1-2), 163–223 (2003)
28. Gmytrasiewicz, P.J., Doshi, P.: A framework for sequential planning in multi-agent settings. Journal of Artificial Intelligence Research 24, 49–79 (2005)
29. Gmytrasiewicz, P.J., Durfee, E.H.: A rigorous, operational formalization of recursive modeling. In: Proc. of the International Conference on Multiagent Systems, pp. 125–132 (1995)
30. Gmytrasiewicz, P.J., Noh, S., Kellogg, T.: Bayesian update of recursive agent models. User Modeling and User-Adapted Interaction 8(1-2), 49–69 (1998)

31. Goldman, C.V., Zilberstein, S.: Optimizing information exchange in cooperative multi-agent systems. In: Proc. of the International Joint Conference on Autonomous Agents and Multi Agent Systems, pp. 137–144 (2003)
32. Goldman, C.V., Zilberstein, S.: Decentralized control of cooperative systems: Categorization and complexity analysis. Journal of Artificial Intelligence Research 22, 143–174 (2004)
33. Grosz, B.J., Kraus, S.: Collaborative plans for complex group action. Artificial Intelligence 86(2), 269–357 (1996)
34. Grosz, B.J., Sidner, C.: Plans for discourse. In: Intentions in Communication. MIT Press, Cambridge (1990)
35. Guestrin, C., Koller, D., Parr, R.: Multiagent planning with factored MDPs. In: Advances in Neural Information Processing Systems, vol. 14, pp. 1523–1530 (2002)
36. Guestrin, C., Koller, D., Parr, R., Venkataraman, S.: Efficient solution algorithms for factored MDPs. Journal of Artificial Intelligence Research 19, 399–468 (2003)
37. Hansen, E.A., Bernstein, D.S., Zilberstein, S.: Dynamic programming for partially observable stochastic games. In: Proc. of the National Conference on Artificial Intelligence, pp. 709–715 (2004)
38. Jennings, N.R.: Controlling cooperative problem solving in industrial multi-agent systems using joint intentions. Artificial Intelligence 75(2), 195–240 (1995)
39. Kaelbling, L.P., Littman, M.L., Cassandra, A.R.: Planning and acting in partially observable stochastic domains. Artificial Intelligence 101(1-2), 99–134 (1998)
40. Kitano, H., Tadokoro, S., Noda, I., Matsubara, H., Takahashi, T., Shinjoh, A., Shimada, S.: RoboCup rescue: Search and rescue in large-scale disasters as a domain for autonomous agents research. In: Proc. of the International Conference on Systems, Man and Cybernetics, pp. 739–743 (1999)
41. Kok, J.R., Spaan, M.T.J., Vlassis, N.: Non-communicative multi-robot coordination in dynamic environments. Robotics and Autonomous Systems 50(2-3), 99–114 (2005)
42. Kok, J.R., Vlassis, N.: Collaborative multiagent reinforcement learning by payoff propagation. Journal of Machine Learning Research 7, 1789–1828 (2006)
43. Koller, D., Megiddo, N., von Stengel, B.: Fast algorithms for finding randomized strategies in game trees. In: Proc. of the 26th ACM Symposium on Theory of Computing, pp. 750–759 (1994)
44. Koller, D., Milch, B.: Multi-agent influence diagrams for representing and solving games. Games and Economic Behavior 45(1), 181–221 (2003)
45. Madani, O., Hanks, S., Condon, A.: On the undecidability of probabilistic planning and infinite-horizon partially observable Markov decision problems. In: Proc. of the National Conference on Artificial Intelligence, pp. 541–548 (1999)
46. Nair, R., Tambe, M.: Hybrid BDI-POMDP framework for multiagent teaming. Journal of Artificial Intelligence Research 23, 367–420 (2005)
47. Nair, R., Varakantham, P., Tambe, M., Yokoo, M.: Networked distributed POMDPs: A synthesis of distributed constraint optimization and POMDPs. In: Proc. of the National Conference on Artificial Intelligence, pp. 133–139 (2005)
48. Oliehoek, F.A., Spaan, M.T.J., Vlassis, N.: Optimal and approximate Q-value functions for decentralized POMDPs. Journal of Artificial Intelligence Research 32, 289–353 (2008)
49. Oliehoek, F.A., Spaan, M.T.J., Whiteson, S., Vlassis, N.: Exploiting locality of interaction in factored POMDPs. In: Proc. of the International Joint Conference on Autonomous Agents and Multi Agent Systems, December 2008, pp. 517–524 (2008)

50. Oliehoek, F.A., Visser, A.: A hierarchical model for decentralized fighting of large scale urban fires. In: Proc. of the AAMAS 2006 Workshop on Hierarchical Autonomous Agents and Multi-Agent Systems (H-AAMAS), pp. 14–21 (2006)
51. Oliehoek, F.A., Whiteson, S., Spaan, M.T.J.: Lossless clustering of histories in decentralized POMDPs. In: Proc. of the International Joint Conference on Autonomous Agents and Multi Agent Systems, pp. 577–584 (2009)
52. Osborne, M.J., Rubinstein, A.: A Course in Game Theory. The MIT Press, Cambridge (1994)
53. Papadimitriou, C.H., Tsitsiklis, J.N.: The complexity of Markov decision processes. Mathematics of Operations Research 12(3), 441–451 (1987)
54. Post, S., Fassaert, M.: A communication and coordination model for 'RoboCupRescue' agents. Master's thesis, University of Amsterdam (2004)
55. Poupart, P.: Exploiting structure to efficiently solve large scale partially observable Markov decision processes. Ph.D. thesis, Department of Computer Science, University of Toronto (2005)
56. Puterman, M.L.: Markov Decision Processes—Discrete Stochastic Dynamic Programming. John Wiley & Sons, Inc., Chichester (1994)
57. Pynadath, D.V., Tambe, M.: The communicative multiagent team decision problem: Analyzing teamwork theories and models. Journal of Artificial Intelligence Research 16, 389–423 (2002)
58. Pynadath, D.V., Tambe, M.: An automated teamwork infrastructure for heterogeneous software agents and humans. Autonomous Agents and Multi-Agent Systems 7(1-2), 71–100 (2003)
59. Rabinovich, Z., Goldman, C.V., Rosenschein, J.S.: The complexity of multiagent systems: the price of silence. In: Proc. of the International Joint Conference on Autonomous Agents and Multi Agent Systems, pp. 1102–1103 (2003)
60. Rathnasabapathy, B., Doshi, P., Gmytrasiewicz, P.: Exact solutions of interactive POMDPs using behavioral equivalence. In: Proc. of the International Joint Conference on Autonomous Agents and Multi Agent Systems, pp. 1025–1032 (2006)
61. Russell, S., Norvig, P.: Artificial Intelligence: A Modern Approach, 2nd edn. Pearson Education, London (2003)
62. Seuken, S., Zilberstein, S.: Improved memory-bounded dynamic programming for decentralized POMDPs. In: Proc. of Uncertainty in Artificial Intelligence (2007)
63. Seuken, S., Zilberstein, S.: Memory-bounded dynamic programming for DEC-POMDPs. In: Proc. of the International Joint Conference on Artificial Intelligence, pp. 2009–2015 (2007)
64. Seuken, S., Zilberstein, S.: Formal models and algorithms for decentralized decision making under uncertainty. Autonomous Agents and Multi-Agent Systems 17(2), 190–250 (2008)
65. Singh, S., James, M.R., Rudary, M.R.: Predictive state representations: a new theory for modeling dynamical systems. In: Proc. of Uncertainty in Artificial Intelligence, pp. 512–519 (2004)
66. Sontag, E.D.: Mathematical control theory: deterministic finite dimensional systems, 2nd edn. Textbooks in Applied Mathematics. Springer, New York (1998)
67. Spaan, M.T.J., Gordon, G.J., Vlassis, N.: Decentralized planning under uncertainty for teams of communicating agents. In: Proc. of the International Joint Conference on Autonomous Agents and Multi Agent Systems, pp. 249–256 (2006)
68. Stone, P.: Layered Learning in Multiagent Systems: A Winning Approach to Robotic Soccer. MIT Press, Cambridge (2000)

69. Stone, P., Veloso, M.: Task decomposition, dynamic role assignment, and low-bandwidth communication for real-time strategic teamwork. Artificial Intelligence 110(2), 241–273 (1999)
70. Stone, P., Veloso, M.: Multiagent systems: A survey from a machine learning perspective. Autonomous Robots 8(3), 345–383 (2000)
71. Sutton, R.S., Barto, A.G.: Reinforcement Learning: An Introduction. The MIT Press, Cambridge (1998)
72. Sycara, K.P.: Multiagent systems. AI Magazine 19(2), 79–92 (1998)
73. Szer, D., Charpillet, F., Zilberstein, S.: MAA*: A heuristic search algorithm for solving decentralized POMDPs. In: Proc. of Uncertainty in Artificial Intelligence, pp. 576–583 (2005)
74. Tambe, M.: Towards flexible teamwork. Journal of Artificial Intelligence Research 7, 83–124 (1997)
75. Tierney, K.J., Goltz, J.D.: Emergency response: Lessons learned from the kobe earthquake. Tech. rep., Disaster Research Center (1997), `http://dspace.udel.edu:8080/dspace/handle/19716/202`
76. Varakantham, P., Marecki, J., Yabu, Y., Tambe, M., Yokoo, M.: Letting loose a SPIDER on a network of POMDPs: Generating quality guaranteed policies. In: Proc. of the International Joint Conference on Autonomous Agents and Multi Agent Systems (2007)
77. Visser, A., Xingrui-Ji, van Ittersum, M., González Jaime, L.A., Stancu, L.A.: Beyond frontier exploration. In: Visser, U., Ribeiro, F., Ohashi, T., Dellaert, F. (eds.) RoboCup 2007: Robot Soccer World Cup XI. LNCS (LNAI), vol. 5001, pp. 113–123. Springer, Heidelberg (2008)
78. Vlassis, N.: A Concise Introduction to Multiagent Systems and Distributed Artificial Intelligence. In: Synthesis Lectures on Artificial Intelligence and Machine Learning. Morgan & Claypool Publishers, San Francisco (2007)
79. Wooldridge, M.: An Introduction to MultiAgent Systems. Wiley, Chichester (2002)
80. Xuan, P., Lesser, V., Zilberstein, S.: Communication decisions in multi-agent cooperation: Model and experiments. In: Proc. of the International Conference on Autonomous Agents (2001)
81. Zeng, Y., Doshi, P., Chen, Q.: Approximate solutions of interactive dynamic influence diagrams using model clustering. In: Proc. of the National Conference on Artificial Intelligence, pp. 782–787 (2007)

Efficient Methods for Near-Optimal Sequential Decision Making under Uncertainty

Christos Dimitrakakis

Abstract. This chapter discusses decision making under uncertainty. More specifically, it offers an overview of efficient Bayesian and distribution-free algorithms for making near-optimal sequential decisions under uncertainty about the environment. Due to the uncertainty, such algorithms must not only learn from their interaction with the environment but also perform as well as possible while learning is taking place.

1 Introduction

It could be argued that automated decision making is the main application domain of artificial intelligence systems. This includes tasks from selecting moves in a game of chess, choosing the best navigation route in a road network responding to questions posed by humans, playing a game of poker and exploring other planets of the solar system. While chess playing and navigation both involve accurate descriptions of the problem domain, the latter problems involve uncertainty about both the nature of the environment and its current state. For example, in the poker game, the nature of the other players (i.e. the strategies the use) is not known. Similarly, the state of the game is not known perfectly—i.e. their cards are not known, but it might be possible to make an educated guess.

This chapter shall examine acting under uncertainty in environments with state, wherein a *sequence* of decisions must be made. Each decision made has an effect on the environment, thus changing the environment's state. One type of uncertainty arises when it is not known how the environment works, i.e. the agent can observe the state of the environment but is not certain what is the effect of each possible action in each state. The decision-making agent must therefore explore the environment, but not in a way that is detrimental to its performance. This balancing act is

Christos Dimitrakakis
Informatics Institute, University of Amsterdam, Science Park 107, 1098 XG Amsterdam, The Netherlands
e-mail: `christos.dimitrakakis@gmail`

R. Babuška & F.C.A. Groen (Eds.): Interactive Collaborative Information Systems, SCI 281, pp. 125–153.
springerlink.com

commonly referred to as the exploration–exploitation trade-off. This chapter gives an overview of current methods for achieving near-optimal online performance in such cases.

Another type of uncertainty arises when the environment's state cannot be observed directly, and can only be inferred. In that case, the state is said to be hidden or partially observable. When both types of uncertainty occur simultaneously, then the problem's space complexity increases polynomially with time, because we must maintain all the observation history to perform inference. The problem becomes even harder when there are multiple agents acting within the environment. However, we shall not consider this case in this chapter.

The remainder of this chapter is organized as follows. First, we give an introduction to inference and decision making under uncertainty in both the Bayesian and distribution-free framework. Section 3 introduces some sequential decision-making problems under uncertainty. These problems are then formalized within the framework of Markov decision processes in Sect. 4, which can be used to make optimal decisions when the uncertainty is only due to stochasticity, i.e. when the effects of any decisions are random, but arise from a known probability distribution that is conditioned on the agent's actions and the current state. Section 5 discusses the extension of this framework when these probability distributions are not known. It is shown that the result is another Markov decision process with an infinite number of states, and various methods for approximately solving it are discussed. When the states of the environment are not directly observed, the problem becomes much more complex; this case is examined in Sect. 6. Finally, Sect. 7 identifies open problems and possible directions of future research.

Notation

We shall write $\mathbb{I}\{X\}$ for the indicator function that equals 1 when X is true, and 0 otherwise. We consider actions $a \in \mathscr{A}$ and contexts (states, environments or outcomes) $\mu \in \mathscr{M}$. We shall denote a sequence of observations from some set $\mathscr{X}$ as $x^t \triangleq x_1, \dots, x_t$, with $x_k \in \mathscr{X}$.

In general, $\mathbf{P}(X)$ will denote the probability of any event X selected from nature, and $\mathbf{E}$ denotes expectations. When observations, outcomes or events are generated via some specific process μ, we shall explicitly denote this by writing $\mathbf{P}(\cdot|\mu)$ for the probability of events. Frequently, we shall use the shorthand $\mu(\cdot)$ to denote probabilities (or densities, when there is no ambiguity) under the process μ. With this scheme, we make no distinction between the name of the process and the distribution it induces. Thus, the notation $\mu(\cdot)$ may imply a marginalization. For instance, if a process μ defines a probability density $\mu(x,y)$ over observations $x \in \mathscr{X}$, and $y \in \mathscr{Y}$, we shall write $\mu(x)$ for the marginal $\int_{\mathscr{Y}} \mu(x,y)\,dy$. Finally, expectations under the process will be written as $\mathbf{E}_\mu(\cdot)$ or equivalently $\mathbf{E}(\cdot|\mu)$. In some cases, it will be convenient to employ equality relations of the type $\mu(x_t = x)$ to denote the density at x at time t under process μ.

2 Decision Making under Uncertainty

Imagine an agent acting within some environment and let us suppose that it has decided upon a fixed plan of action. Predicting the result of any given action within the plan, or the result of the complete plan, may not be very easy, since there can be many different sources of uncertainty. Thus, it might be hard to *evaluate* actions and plans and consequently to find the *optimal* action or plan. This section will focus on the case where only a single decision must be made.

The simplest type of uncertainty arises when events that take place within the world can be stochastic. Those may be (apparently) truly random events, such as which slit will a photon will pass through in the famous two-slit experiment, or events that can be considered random for all practical purposes, such as the outcome of a die roll. A possible decision problem in that setting would be whether to accept or decline a particular bet on the outcome of one die roll: if the odds are favorable, we should accept, otherwise decline.

The second source of uncertainty arises when we do not know exactly how the world works. Consider the problem of predicting the movement of planets given their current positions and velocities. Modeling their orbits as circular will of course result in different predictions to modeling their orbits as elliptic. In this problem, the decision taken involves the selection of the appropriate model.

Estimating which model, or set of models, best corresponds to our observations of the world becomes harder when there is, in addition, some observation stochasticity. In the given example, that would mean that we would not be able to directly observe the planets' positions and thus it would be harder to determine the best model.

The model selection problems that we shall examine in this section involve two separate, well-defined, phases. The first phase involves collecting data, and the second phase involves making a decision about which is the best model. Many statistical inference problems are of this type, such as creating a classifier from categorical data. The usual case is that the observations have already been collected and now form a fixed *dataset*. We then define a set of classification models and the decision making task is to choose one or more classifiers from the given set of models. A lot of recent work on classification algorithms is in fact derived from this type of decision making framework [10, 54, 55]. A straightforward extension of the problem to online decision making resulted in the boosting algorithm [28]. The remainder of this section discusses making single decisions under uncertainty in more detail.

2.1 Utility, Randomness and Uncertainty

When agents (and arguably, humans [50]) make decisions, they do so on the basis of some preference order among possible outcomes. With perfect information, the rational choice is easy to determine, since the probability of any given outcome given the agent's decision is known. This is a common situation in games of chance. However, in the face of uncertainty, establishing a preference order among actions is no longer trivial.

In order to formalize the problem, we allow the agent to select some action a from a set of $\mathscr{A}$ possible choices. We furthermore define a set of contexts, states or environments $\mathscr{M}$, such that the preferred action may differ depending which is the current context $\mu \in \mathscr{M}$. If the context is perfectly known, then we can simply take the most preferred action in that context.

One way to model this preference is to define a utility function $U : \mathscr{A} \times \mathscr{M} \to \mathbb{R}$ mapping from the set of possible μ and a to the real numbers.

Definition 1 (Utility). For any context $\mu \in \mathscr{M}$, and actions $a_1, a_2 \in \mathscr{A}$, we shall say that we *prefer* a_1 to a_2 and write $a_1 \succ a_2$, iff $U(a_1,\mu) > U(a_2,\mu)$. Similarly, we write that $a_1 = a_2$ iff $U(a_1,\mu) = U(a_2,\mu)$.

The transitivity and completeness axioms of utility theory are satisfied by the above definition, since the utility functions' range are the real numbers. Thus, if both U and μ are known, then the optimal action must exist and is *by definition* the one that maximizes the utility for the given context μ.

$$a^* = \arg\max_{a \in \mathscr{A}} U(a,\mu). \tag{1}$$

We can usually assign a *subjective preference* for each action a in all μ, thus making the function U well defined. It is possible, however, that we have some uncertainty about μ. We are then obliged to use other formalisms for assigning preferences to actions.

2.2 Uncertain Outcomes

We now consider the case where μ is uncertain. This can occur when μ is chosen randomly from some known distribution with density p, as is common in lotteries and random experiments. It may be also chosen by some adversary, which is usual in deterministic games such as chess. In games of chance, such as backgammon, it is some combination of the two. Finally, μ could be neither randomly nor selected by some adversary, but in fact, simply not precisely known: we may only know a set $\mathscr{M}$ that contains μ.

Perhaps the simplest way to assign preferences in the latter case is to select the action with the highest worst-case utility:

Definition 2 (Maximin utility). Our preference $V(a)$ for action a is

$$V(a) \triangleq \inf_{\mu \in \mathscr{M}} U(a,\mu). \tag{2}$$

This is mostly a useful ordering for the adversarial setting. In the stochastic setting, its main disadvantage is that we may avoid actions which have the highest utility for most high-probability outcomes, apart for some outcomes with near-zero probability, whose utility is small. A natural way to take the probability of outcomes into account is to use the notion of expected utility:

Definition 3 (Expected utility). Our preference $V(a)$ for action a is the expectation of the utility under the given distribution with density p of possible outcomes:

$$V(a) \triangleq \mathbf{E}(U|a) = \int_{\mathscr{M}} U(a,\mu)p(\mu)\,d\mu. \tag{3}$$

This is a good method for the case when the distribution from which μ will be chosen is known. Another possible method, which can be viewed as a compromise between expected and maximin utility, is to assign preferences to actions based on how likely they are to be close to the best action. For example, we can take the action which has the highest probability of being ε-close to the best possible action, with $\varepsilon > 0$:

Definition 4 (Risk-sensitive utility).

$$V(a;\varepsilon) \triangleq \int_{\mathscr{M}} \mathbb{I}\left\{U(a,\mu) \geq U(a',\mu) - \varepsilon,\ \forall a' \in \mathscr{A}\right\} p(\mu)\,d\mu. \tag{4}$$

Thus, an action chosen with the above criterion is guaranteed to be ε-close to the actual best action, with probability $V(a;\varepsilon)$. This criterion could be alternatively formulated as the probability that the action's utility is greater than a fixed threshold θ, rather than being ε-close to the utility of the optimal action. A further modification involves fixing a small probability $\delta > 0$ and then solving for ε, or θ to choose the action which has the lowest regret ε, or the highest guaranteed utility θ, with probability $1 - \delta$. Further discussion of such issues including some of the above preference relations is given in [29, 30, 40, 50].

The above definitions are not strictly confined to the case where μ is random. In *Bayesian*, or *subjectivist* viewpoint of probability, we may also assign probabilities to events that are not random. Those probabilities do not represent possible random outcomes, but *subjective beliefs*. It thus becomes possible to extend the above definitions from uncertainty about random outcomes to uncertainty about the environment.

2.3 Bayesian Inference

Consider now that we are acting in one of many possible environments. With knowledge of the true environment and the utility function, it would be trivial, in some sense, to select the utility-maximizing action. However, suppose that we do *not* know which of the many possible environments we are acting in. One possibility is to use the maximin utility rule, but this is usually too pessimistic. An arguably better alternative is to assign a *subjective probability* to each environment, which will represent our belief that it corresponds to reality.[1] It then is possible to use expected utility to select actions.

[1] This is mathematically equivalent to the case where the environment was drawn randomly from a known distribution.

This is not the main advantage of using subjective probabilities, however. It is rather the fact that we can then use standard probabilistic inference methods to *update our belief* as we acquire more information about the environment. With enough data, we can be virtually certain about which is the true environment, and thus confidently take the most advantageous action. When the data are few, a lot of models have a relatively high probability and this uncertainty is reflected in the decision making.

More formally, we define a set of models $\mathscr{M}$ and a prior density ξ_0 defined over its elements $\mu \in \mathscr{M}$. The prior $\xi_0(\mu)$ describes our initial belief that the particular model μ is correct. Sometimes it is unclear how to best choose the prior. The easiest case to handle is when it is known that the environment was randomly selected from a probability distribution over $\mathscr{M}$ with some density ψ: it is then natural (and optimal) to simply set $\xi_0 = \psi$. Another possibility is to use experts to provide information: then the prior can be obtained via formal procedures of prior elicitation [14, 19]. However, when this is not possible either due to the lack of experts or due to the fact that the number of parameters to be chosen is very large, then the priors can be chosen intuitively, according to computational and convenience considerations, or with some automatic procedure; a thorough overview of these topics is given by [7, 32]. For now, we shall assume that we have somehow chosen a prior ξ_0.

The procedure for calculating the belief ξ_t at time t is relatively simple: Let x_t denote our observations at time t. For each model μ in our set, we can calculate the posterior probability $\xi_{t+1}(\mu)$ from their prior $\xi_t(\mu)$. This can be done via the definition of joint densities, which in this form is known as Bayes' rule:

$$\xi_{t+1}(\mu) \triangleq \xi_t(\mu|x_t) = \frac{\mu(x_t)\xi_t(\mu)}{\int_{\mathscr{M}} \mu'(x_t)\xi_t(\mu')\,d\mu'}, \tag{5}$$

where we have used the fact that $\xi_t(x_t|\mu) = \mu(x_t)$, since the distribution of observations for a specific model μ is independent of our subjective belief about which models are most likely.

The advantage of using a Bayesian approach to decision making under uncertainty is that our knowledge about the true environment μ at time t is captured via the density $\xi_t(\mu)$, which represents our belief. Thus, we are able to easily utilize any of the action preferences outlined in the previous section by replacing the density p with ξ_t.

As an example, consider the case where we have obtained t observations and must choose the action that appears best. After t observations, we would have reached a belief $\xi_t(\mu)$ and our *subjective* value for each action would be simply

$$V_t(a) \triangleq \mathbf{E}(U|a,\xi_t) = \int_{\mathscr{M}} U(a,\mu)\xi_t(\mu)\,d\mu. \tag{6}$$

At this point, perhaps some motivation is needed to see why this is a good idea. Let μ^* be the true model, i.e. in fact $\mathbf{E}(U|a) = U(a,\mu^*)$ and assume that $\mu^* \in \mathscr{M}$. Then, under relatively lax assumptions,[2] it can be shown (c.f. [50]) that $\lim_{t\to\infty} \xi_t(\mu) = \delta(\mu - \mu^*)$, where δ is the Dirac delta function. This implies that the probability measure that represents our belief concentrates[3] around μ^*.

There are, of course, a number of problems with this formulation. The first is that we may be unwilling or unable to specify a prior. The second is that the resulting model may be too complicated for practical computations. Finally, although it is relatively straightforward to compute the expected utility, risk-sensitive computations are hard in continuous spaces, since they require calculating the integral of a maximum. In any such situation, distribution-free bounds may be used instead.

2.4 Distribution-Free Bounds

When we have very little information about the distribution, and we do not wish to utilize "uninformative" or "objective" priors [7], we can still express the amount of knowledge acquired through observation by the judicious use of distribution-free concentration inequalities. The main idea is that while we cannot accurately express the possible forms of the underlying distribution, we can always imagine a worst possible case.

Perhaps the most famous such inequality is Markov's inequality, which holds for any random variable X and all $\varepsilon > 0$:

$$\mathbf{P}(|X| \geq \varepsilon) \leq \frac{\mathbf{E}(|X|)}{\varepsilon}. \tag{7}$$

Such inequalities are of the greatest use when applied to estimates of unknown parameters. Since our estimates are functions of our observations, which are random variables, the estimates themselves are also random variables. Perhaps the best way to see this is through the example of the following inequality, which applies whenever we wish to estimate the expected value of a bounded random variable by averaging n observations:

Lemma 1 (Hoeffding inequality). *If $\hat{x}_n \triangleq \frac{1}{n}\sum_{i=1}^n x_i$, with $x_i \in [b_i, b_i + h_i]$ drawn from some arbitrary distribution f_i and $\bar{x}_n \triangleq \frac{1}{n}\sum_i \mathbf{E}(x_i)$, then for all $\varepsilon \geq 0$*

$$\mathbf{P}\left(|\hat{x}_n - \bar{x}_n| \geq \varepsilon\right) \leq 2\exp\left(-\frac{2n^2\varepsilon^2}{\sum_{i=1}^n h_i^2}\right). \tag{8}$$

The above inequality is simply interpreted as telling us that the probability of us making a large estimation error decreases exponentially in the number of samples. Unlike the Bayesian viewpoint, where the mean was a random variable, we now consider the mean to be a fixed unknown quantity and our estimate to be the random

[2] If the true model is not in the set of models, then we may in fact diverge.

[3] To prove that in more general terms is considerably more difficult, but has been done recently by Zhang [58].

variable. So, in practice, such inequalities are useful for obtaining bounds on the performance of algorithms and estimates, rather than expressing uncertainty *per se*.

More generally, such inequalities operate on sets. For any model set $\mathscr{M}$ of interest, we should be able to obtain an appropriate concentration function $\delta(M,n)$ on subsets $M \subset \mathscr{M}$, with $n \in \mathbb{N}$ being the number of observations, such that

$$\mathbf{P}(\mu \notin M) < \delta(M,n), \tag{9}$$

where one normally considers μ to be a fixed unknown quantity and M to be a random estimated quantity. As an example, the right-hand side of the Hoeffding inequality can be seen as a specific instance of a concentration function, where $M = \{x : |\hat{x}_n - x| < \varepsilon\}$. A detailed introduction to concentration functions in much more general terms is given by Talagrand [53].

An immediate use of such inequalities is to calculate high probability bounds on the utility of actions. First, given the deterministic payoff function $U(a,\mu)$, and a suitable $\delta(M,n)$, let

$$\theta(M,a) \triangleq \inf_{\mu \in M} U(a,\mu) \tag{10}$$

be a lower bound on the payoff of action a in set M. It immediately follows that the payoff $U(a)$ we shall obtain for taking action a will satisfy

$$\mathbf{P}[U(a,\mu) < \theta(M,a)] \leq \delta(M,n), \tag{11}$$

since the probability of our estimated set not containing μ is bounded by $\delta(M,n)$. Thus, one may fix a set M and then select a maximizing $\theta(a,M)$. This will give a guaranteed performance with probability at least $1 - \delta(M,n)$.

Such inequalities are also useful to bound the expected *regret*. Let the regret of a procedure that has payoff U relative to some other procedure with payoff U^* be $U^* - U$. Now consider that we want to take an action a that minimizes the expected regret relative to some maximum value U^*, which can be written as

$$\mathbf{E}(U^* - U|a) = \mathbf{E}(U^* - U|a, \mu \in M)\mathbf{P}(\mu \in M) + \mathbf{E}(U^* - U|a, \mu \notin M)\mathbf{P}(\mu \notin M), \tag{12}$$

where the first term of the sum corresponds to the expected regret that we incur when the model is in the set M, while the right term corresponds to the case when the model is not within M. Each of these terms can be bounded with (9) and (10): the first term is bounded by $\mathbf{E}(U^* - U|a, \mu \in M) < U^* - \theta(M,a)$ and $\mathbf{P}(\mu \in M) < 1$, while the second term is bounded by $\mathbf{E}(U^* - U|a, \mu \notin M) < U^* - \inf(U)$ and $\mathbf{P}(\mu \notin M) < \delta(M,n)$. Assuming that the infimum exists, the expected regret for any action $a \in \mathscr{A}$ is bounded by

$$\mathbf{E}(U^* - U|a) \leq (U^* - \theta(M,a)) + (U^* - \inf(U))\delta(M,n). \tag{13}$$

Thus, such methods are useful for taking an action that minimizes a bound on the expected regret, that maximizes the probability, its utility is greater than some threshold, or finally, that maximizes a lower bound on the utility with at least some probability. However, they cannot be used to select actions that maximize expected utility simply because the expectation cannot be calculated as we do not have an explicit probability density function over the possible models. We discuss two upper confidence bound-based methods for bandit problems in Sect. 3.2 and for the reinforcement learning problem in Sect. 5.5.

3 Sequential Decision Making under Uncertainty

The previous section examined two complementary frameworks for decision making under uncertainty. The task was relatively straightforward, as it involved a fixed period of data collection, followed by a single decision. This is far from an uncommon situation in practice, since a lot of real-world decisions are made in such a way.

In a lot of cases, however, decisions may involve future collection of data. For example, during medical trials, data are continuously collected and assessed. The trial may have to be stopped early if the risk to the patients is deemed too great. A lot of such problems were originally considered in the seminal work of Wald [56].

This section will present a brief overview of the three main types of sequential decision-making problems: stopping problems, which are the simplest type, bandit problems, which can be viewed as a generalization of stopping problems, and reinforcement learning, which is a general enough framework to encompass most problems in sequential decision making. The reader should also be aware of the links of sequential decision making to classification [28], optimization [5, 16, 36] and control [1, 8, 9].

3.1 Stopping Problems

Let us imagine an experimenter who needs to make a decision $a \in \mathscr{A}$, where $\mathscr{A}$ is the set of possible decisions. The effect of each different decision will depend on both a and the actual situation in which the decision is taken. However, although the experimenter can quantify the consequences of his decisions for each possible situation, he is not sure what the situation actually is. So, he first must spend some time to collect information about the situation before committing himself to a specific decision. The only difficulty is that the information is not free. Thus, the experimenter must decide at which point he must *stop* collecting data and finally make a decision.

Such *optimal stopping* problems arise in many settings: clinical trials [15], optimization [11], detecting changes in distributions [42], active learning [24, 49] as well as the problem of deciding when to halt an optimization procedure [11]. A general introduction to such problems can be found in [18].

As before, we assume the existence of a *utility function* $U(a,\mu)$ defined for all $\mu \in \mathscr{M}$, where $\mathscr{M}$ is the set of all possible universes of interest. The experimenter knows the utility function, but is not sure in which universe the experiment is taking place. This uncertainty about which $\mu \in \mathscr{M}$ is true is expressed via a subjective distribution $\xi_t(\mu) \triangleq \mathbf{P}(\mu|\xi_t)$, where ξ_t represents the belief at time t.

The expected utility of *immediately* taking an action at time t can then be written as $V_0(\xi_t) = \max_a \sum_\mu U(a,\mu)\xi_t(\mu)$, i.e. the experimenter takes the action which seems best on average. Now, consider that instead of making an immediate decision, he has the opportunity to take k more observations $D^k = (d_1, \ldots, d_k)$ from a sample space S^k, at a cost $c > 0$ per observation[4], thus allowing him to update his belief to

$$\xi_{t+k}(\mu|\xi_t) \triangleq \xi_t(\mu|D_k). \tag{14}$$

What the experimenter must do in order to choose between immediately making a decision a and continuing sampling is to compare the utility of making a decision now with the cost of making k observations plus the utility of making a decision after k timesteps, when the extra data would enable a more informed choice.

The problem is in fact a dynamic programming problem. The utility of making an immediate decision is

$$V_0(\xi_t) = \max_a \int_M U(a,\mu)\xi_t(\mu)\,d\mu \tag{15}$$

Similarly, we denote the utility of an immediate decision at any time $t+T$ by $V_0(\xi_{t+T})$. The utility of taking at most k samples before making a decision can be written recursively as

$$V_{k+1}(\xi_t) = \max\{V_0(\xi_t), \mathbf{E}[V_k(\xi_{t+1})|\xi_t] - c\}, \tag{16}$$

where the expectation with respect to ξ_t is in fact taken over all possible observations under belief ξ_t:

$$\mathbf{E}[V_k(\xi_{t+1})|\xi_t] = \sum_{d_{t+1} \in S} V_k(\xi_{t+1}(\mu|d_{t+1}))\xi_t(d_{t+1}), \tag{17}$$

$$\xi_t(d_{t+1}) = \int_{\mathscr{M}} \mu(d_{t+1})\xi_t(\mu)\,d\mu, \tag{18}$$

where $\xi_t(\mu|d_{t+1})$ indicates the specific next belief ξ_{t+1} arising from the previous belief ξ_t and the observations d_{t+1}.

This indicates that we can perform a backwards induction procedure, starting from all possible terminal belief states ξ_{t+T} to calculate the value of stopping immediately. We would only be able to insert the payoff function U directly when we calculate V_0. Taking $T \to \infty$ gives us the exact solution. In other words, one should stop and make an immediate decision if the following holds for all $k > 0$:

$$V_0(\xi_t) \geq V_k(\xi_t). \tag{19}$$

Note that if the payoff function is bounded we can stop the procedure at $T \propto c^{-1}$.

[4] The case of non-constant cost is not significantly different.

A number of bounds may be useful for stopping problems. For instance, the expected value of perfect information to an experimenter can be used as a surrogate to the complete problem, and was examined by McCall [41]. Conversely, lower bounds on the expected sample size and the regret were examined by Hoeffding [34]. Stopping problems also appear in the context of bandit problems, or more generally, reinforcement learning. For instance, the problem of finding a near-optimal plan with high probability can be converted to a stopping problem [27].

3.2 Bandit Problems

We now consider a generalization of the stopping problem. Imagine that our experimenter visits a casino and is faced with n different bandit machines. Playing a machine at time t results in a random payoff $r_t \in R \subset \mathbb{R}$. The average payoff of the ith machine is μ_i. Thus, at time t, the experimenter selects a machine with index $i \in \{1, \ldots, n\}$ to receive a random payoff with expected value $\mathbf{E}(r_t | a_t = i) = \mu_i$, which is *fixed but unknown*. The experimenter's goal is to leave the casino with as much money as possible. This can be formalized as maximizing the expected sum of discounted future payoffs to time T:

$$\mathbf{E}\left(\sum_{t=1}^{T} \gamma^k r_t\right), \tag{20}$$

where $\gamma \in [0,1]$ is a discount factor that reduces the importance of payoffs far in the future as it approaches 0. The horizon T may be finite or infinite, fixed, drawn from a known distribution or simply unknown. It is also possible that the experimenter is allowed to stop at any time, thus adding stopping to the set of possible decisions and making the problem a straightforward generalization of the stopping problem. If the average payoffs are known, then the optimal solution is obviously to always take action $a^* = \arg\max_i \mu_i$, no matter what γ and T are and is thus completely uninteresting.

This problem was typically studied in a Bayesian setting [15, 31], where computable optimal solutions have been found for some special cases [31]. However, recently, there have been algorithms that achieve optimal regret rates in a distribution-free setting. In particular, the UCB1 algorithm by Auer et al. [3] selects the arm with highest empirical mean plus an upper confidence bound, with an error probability schedule tuned to achieve low regret. More specifically, let the empirical mean of the ith arm at time t be

$$\hat{\mathbf{E}}[r_t | a_t = i] \triangleq \frac{1}{n_i^t} \sum_{k: a_k = i}^{t} r_t, \qquad n_i^t \triangleq \sum_{k=1}^{t} \mathbb{I}\{a_k = i\}, \tag{21}$$

where n_i^t is the number of times arm i has been played until time t. After playing each arm once, the algorithm always selects the arm maximizing:

$$\hat{\mathbf{E}}[r_t|a_t = i] + \sqrt{\frac{2\log t}{n_i^t}}. \tag{22}$$

This guarantees a regret that only scales with rate $O(\log T)$.

The setting has been generalized to continuous time [15], non-stationary or adversarial bandits [2], continuous spaces[1, 5] and to trees [16, 39], while a collection of related results can be found in [13]. Finally, the bandit payoffs can also depend on a context variable. If this variable cannot be affected by the experimenter, then it is sufficient to learn the mean payoffs for all contexts in order to be optimal. However, the problem becomes much more interesting when the experimenter's actions also influence the context of the game. This directly leads us to the concept of problems with state.

3.3 Reinforcement Learning and Control

We now generalize further to problems where the payoffs depend not only on the individual actions that we perform but also on a context, or state. This is a common situation in games. A good example is blackjack, where drawing or stopping (the two possible actions in a game) have expected payoffs that depend on your current hand (the current state, assuming the croupier's hand is random and independent of your own hand).

Both reinforcement learning and control problems are formally identical. Nevertheless, historically, classical control (c.f.[52]) addressed the case where the objective is a known functional of the state s and action a. Reinforcement learning, on the other hand, started from the assumption that the objective function itself is unknown (though its functional *form* is known) and must be estimated. Both discrete time control and reinforcement learning problems can be formalized in the framework of Markov decision processes.

4 Markov Decision Processes

Definition 5 (Markov decision process). A Markov decision process (MDP) is defined as the tuple $\mu = (\mathscr{S}, \mathscr{A}, \mathscr{T}, \mathscr{R})$ comprising a set of states $\mathscr{S}$, a set of actions $\mathscr{A}$, a transition distribution $\mathscr{T}$ conditioning the next state on the current state and action

$$\mathscr{T}(s'|s,a) \triangleq \mu(s_{t+1}{=}s'|s_t{=}s, a_t = a) \tag{23}$$

satisfying the Markov property $\mu(s_{t+1} \mid s_t, a_t) = \mu(s_{t+1} \mid s_t, a_t, s_{t-1}, a_{t-1}, \ldots)$, and a reward distribution $\mathscr{R}$ conditioned on states and actions:

$$\mathscr{R}(r|s,a) \triangleq \mu(r_{t+1}{=}r \mid s_t{=}s, a_t{=}a), \tag{24}$$

with $a \in \mathscr{A}$, $s, s' \in \mathscr{S}$, $r \in \mathbb{R}$. Finally,

$$\mu(r_{t+1}, s_{t+1}|s_t, a_t) = \mu(r_{t+1}|s_t, a_t)\mu(s_{t+1}|s_t, a_t). \tag{25}$$

We shall denote the set of all MDPs as $\mathscr{M}$. We are interested in sequential decision problems where, at each time step t, the agent seeks to maximize the expected utility

$$\sum_{k=1}^{T-t} \gamma^k \mathbf{E}[r_{t+k} \mid \cdot], \tag{26}$$

where r is a stochastic reward and u_t is simply the discounted sum of future rewards. We shall assume that the sequence of rewards arises from a Markov decision process.

In order to describe how we select actions, we define a policy π as a distribution on actions conditioned on the current state $\pi(a_t|s_t)$. If we are interested in the rewards to some horizon T, then we can define a T-horizon value function for an MDP $\mu \in \mathscr{M}$ at time t as

$$V^{\pi}_{\mu,t,T}(s) \triangleq \sum_{k=1}^{T-t} \gamma^k \mathbf{E}[r_{t+k} \mid s_t = s, \pi, \mu], \tag{27}$$

the expected sum of future rewards given that we are at state s at time t and selecting actions according to policy π in the MDP μ. Superscripts and subscripts of V may be dropped when they are clear from context.

The value function of any policy can be calculated recursively by starting from the terminal states.

$$V^{\pi}_{\mu,t,T}(s) = \mathbf{E}[r_{t+1} \mid s_t{=}s, \pi, \mu] + \gamma \sum_{s'} \mu(s_{t+1}{=}s' \mid s_t{=}s, a_t{=}a) V^{\pi}_{\mu,t+1,T}(s'). \tag{28}$$

The optimal value function, i.e. the value function of the optimal policy π^*, can be calculated by a maximizing over actions at each stage:

$$V^{\pi^*}_{\mu,t,T}(s) = \max_{a \in \mathscr{A}} \mathbf{E}[r_{t+1} \mid s_t{=}s, a_t{=}a, \mu] + \gamma \sum_{s'} \mu(s_{t+1}{=}s' \mid s_t{=}s, a_t{=}a) V^{\pi^*}_{\mu,t+1,T}(s'). \tag{29}$$

The recursive form of this equation is frequently referred to as the Bellman recursion and allows us to compute the value of a predecessor state from that of a following state. The resulting algorithm is called *backwards induction* or *value iteration*. When the number of states is finite, the same recursion allows us to compute the value of states when the horizon is infinite, as then, $\lim_{T\to\infty} V^{\pi}_{\mu,t,T} = V^{\pi}_{\mu}$ for all finite t. Finally, note that frequently we shall denote $V^{\pi^*}_{\mu,t,T}$ simply by V^* when the environmental variables are clear from context.

5 Belief-Augmented Markov Decision Processes (BAMDPs)

When the MDP is unknown, we need to explicitly take into account our uncertainty. In control theory, this is referred to as the problem of dual control, see [52], Sect. 5.2. This involves selecting actions (control inputs) such as to improve parameter (and state) estimation in order to hopefully reduce future costs. This behavior is called probing. At the same time, the control strategy must not neglect the minimization of the cost at the current time.

This type of dilemma in its simplest form occurs in the already discussed bandit problems, where we must strike an optimal balance between exploring alternative bandits and exploiting the apparently best bandit. Any optimal solution must take into account the uncertainty that we have about the environment.

A natural idea is to use a Bayesian framework (c.f. [25]) to represent our beliefs. As summarized in Sect. 2.3, this involves maintaining a belief $\xi_t \in \Xi$, about which MDP $\mu \in \mathscr{M}$ corresponds to reality. In a Bayesian setting, $\xi_t(\mu)$ is a subjective probability density over MDPs.

5.1 Bayesian Inference with a Single MDP Model Class

We shall cover the case where it is known that the true MDP μ^* is in some set of MDPs $\mathscr{M}$. For example, it may be known that the MDP has at most K discrete states and that the rewards at each state are Bernoulli, but we know neither the actual transition probabilities nor the reward distributions. Nevertheless, we can use closed-form Bayesian techniques to model our belief about which model is correct: For discrete state spaces, transitions can be expressed as multinomial distributions, to which the Dirichlet density is a conjugate prior. Similarly, unknown Bernoulli distributions can be modeled via a Beta prior. If we assume that the densities are not dependent, then the prior over all densities is a product of Dirichlet priors, and similarly for rewards. Then, we only need a number of parameters of order $O(|\mathscr{S}|^2|\mathscr{A}|)$. The remainder of this section discusses this in more detail.

Let $\mathscr{M}$ be the set of MDPs with unknown transition probabilities and state space $\mathscr{S}$ of size K. We denote our belief at time $t+1$ about which MDP is true as simply our belief density at time t conditioned on the latest observations:

$$\xi_{t+1}(\mu) \triangleq \xi_t(\mu | r_{t+1}, s_{t+1}, s_t a_t) \tag{30a}$$

$$= \frac{\mu(r_{t+1}, s_{t+1} | s_t, a_t)\xi_t(\mu)}{\int_{\mathscr{M}} \mu'(r_{t+1}, s_{t+1} | s_t, a_t)\xi_t(\mu')\, d\mu'}. \tag{30b}$$

We consider the case where $\mathscr{M}$ is an infinite set of MDPs, where each MDP $\mu \in \mathscr{M}$ corresponds to a particular joint probability distribution over the state–action pairs.

We shall begin by defining a belief for the transition of each state–action pair s,a separately. First, we denote by $\tau_{s,a} \in [0,1]^K$ the parameters of the multinomial distribution over the K possible next states, from a specific starting state s and action a. Our belief will be a Dirichlet distribution—a function of $x \in \mathbb{R}^K$ with $\|x\|_1 = 1$ and $x \in [0,1]^K$, with parameters $\psi^{s,a} \in \mathbb{N}^K$. If we denote the parameters of our belief ξ_t at time t by $\psi^{s,a}(t)$, then the Dirichlet density over possible multinomial distributions can be written as

$$\xi_t(\tau_{s,a} = x) = \frac{\Gamma(\psi^{s,a}(t))}{\prod_{i \in \mathscr{S}} \Gamma(\psi_i^{s,a}(t))} \prod_{i \in \mathscr{S}} x_i^{\psi_i^{s,a}(t)}, \tag{31}$$

where $\psi_i^{s,a}$ denotes the ith component of $\psi^{s,a}$. The set of parameters ψ can be written in matrix form as $\Psi(t)$ to denote the $|\mathscr{S}||\mathscr{A}| \times |\mathscr{S}|$ matrix of state–action–state transition counts at time t. The *initial* parameters $\Psi(0)$ form a matrix that defines the parameters of our set *prior* Dirichlet's distributions.

Thus, for any belief ξ_t, the Dirichlet parameters are $\{\psi_i^{j,a}(t) : i, j \in \mathscr{S}, a \in \mathscr{A}\}$. These values are initialized to $\Psi(0)$ and are updated via simple counting:

$$\psi_i^{j,a}(t+1) = \psi_i^{j,a}(t) + \mathbb{I}\{s_{t+1} = i \wedge s_t = j \wedge a_t = a\}, \tag{32}$$

meaning that every time we observe a specific transition s_t, a_t, s_{t+1}, we increment the corresponding Dirichlet parameter by one.

We now need to move from the distribution of a single state–action pair to the set of transition distributions for the whole MDP. In order to do this easily, we shall make the following simplifying assumption:

Assumption 5.1. *For any $s,s' \in \mathscr{S}$ and $a,a' \in \mathscr{A}$,*

$$\xi(\tau_{s,a}, \tau_{s',a'}) = \xi(\tau_{s,a})\xi(\tau_{s',a'}). \tag{33}$$

This assumption significantly simplifies the model but does not let us take into advantage of the case where there may be some dependencies in the transition probabilities. Now, we shall denote the matrix of state–action–state transition *probabilities* for a specific MDP μ as $\mathscr{T}^{\mu}$. Analogously to $\tau_{s,a}$, we denote, for the specific MDP μ, the next state distribution multinomial parameter vector from pair (s,a) to be $\tau_{s,a}^{\mu}$, with $\tau_{s,a,i}^{\mu} \triangleq \mu(s_{t+1} = i \mid s_t{=}s, a_t{=}a)$. Then, we obtain

$$\xi_t(\mu) = \xi_t(\mathscr{T}^{\mu}) = \xi_t(\tau_{s,a} = \tau_{s,a}^{\mu} \forall s \in \mathscr{S}, a \in \mathscr{A}) \tag{34a}$$

$$= \prod_{s\in\mathscr{S}} \prod_{a\in\mathscr{A}} \xi_t(\tau_{s,a} = \tau_{s,a}^{\mu}), \tag{34b}$$

$$= \prod_{s\in\mathscr{S}} \prod_{a\in\mathscr{A}} \frac{\Gamma(\psi^{s,a}(t))}{\prod_{i\in\mathscr{S}} \Gamma(\psi_i^{s,a}(t))} \prod_{i\in\mathscr{S}} \left(\tau_{s,a,i}^{\mu}\right)^{\psi_i^{s,a}(t)}, \tag{34c}$$

where we used Assumption 5.1. This means that the transition counts Ψ are a sufficient statistic for expressing the density over $\mathscr{M}$.

We can additionally model $\mu(r_{t+1}|s_t, a_t)$ with a suitable belief (e.g. a Beta, Normal-Gamma or Pareto prior) and assume independence. This in no way complicates the exposition for MDPs.

5.2 *Constructing and Solving BAMDPs*

In order to optimally select actions in this framework, we need to consider how our actions will affect our future beliefs. The corresponding Bayesian procedure is not substantially different from the optimal stopping procedure outlined in Sect. 3.1. The approach outlined in this section was suggested originally in [6] under the name of Adaptive Control Processes and was investigated more fully in [25, 26].

It involves the creation of an *augmented* MDP, with a state comprising the original MDP's state s_t and our belief state ξ_t. We can then solve the problem *in principle* via standard dynamic programming algorithms such as backwards induction, similarly to Sect. 3.1. We shall call such models BAMDPs.

In BAMDPs, we are at some combined belief and environment state $\omega_t = (\xi_t, s_t)$ at each point in time t, which we call the *hyper-state* . For every possible action a_t, we may observe any $s_{t+1} \in \mathscr{S}$ and any possible reward $r_{t+1} \in R \subset \mathbb{R}$, which would lead to a unique new belief ξ_{t+1} and thus a unique new hyper-state $\omega_{t+1} = (\xi_{t+1}, s_{t+1})$.

More formally, we may give the following definition:

Definition 6 (BAMDP). A *Belief-Augmented MDP* ν (BAMDP) is an MDP $\nu = (\Omega, \mathscr{A}, \mathscr{T}', \mathscr{R}')$ where $\Omega = \mathscr{S} \times \Xi$, where Ξ is an appropriate set of probability measures on $\mathscr{M}$, and $\mathscr{T}', and \mathscr{R}'$ are the transition and reward distributions conditioned jointly on the MDP state s_t, the belief state ξ_t and the action a_t, respectively. Here, the density $p(\xi_{t+1}|\xi_t, r_{t+1}, s_{t+1}, s_t, a_t)$ is singular, since ξ_{t+1} is a deterministic function of $\xi_t, r_{t+1}, s_{t+1}, s_t, a_t$. Thus, we can define the transition

$$\nu(\omega_{t+1}|a_t, \omega_t), \tag{35}$$

where $\omega_t \triangleq (s_t, \xi_t)$.

It should be obvious that s_t, ξ_t jointly form a Markov state in this setting, called the *hyper-state*. In general, we shall denote the components of a future hyper-state ω_t^i as (s_t^i, ξ_t^i). However, occasionally, we will abuse notation by referring to the components of some hyper-state ω as s_ω, ξ_ω. We shall use $\mathscr{M}_B$ to denote the set of BAMDPs.

As in the MDP case, finite horizon problems only require looking at all future hyper-states until the horizon T, where we omit the subscript ν for the value function:

$$V_{t,T}^*(\omega_t) = \max_{a_t} \mathbf{E}[r_{t+1}|\omega_t, a_t, \nu] + \gamma \int_\Omega V_{t+1,T}^*(\omega_{t+1}) \nu(\omega_{t+1}|\omega_t, a_t)\, d\omega_{t+1}. \tag{36}$$

It is easy to see that the size of the set of hyper-states in general grows exponentially with the horizon. Thus, we cannot perform value iteration with bounded memory, as we did for discrete MDPs. One possibility is to continue expanding the belief tree until we are certain of the optimality of an action. As has previously been observed [17, 20], this is possible since we can always obtain upper and lower bounds on the utility of any policy from the current hyper-state. In addition, we can apply such bounds on future hyper-states in order to efficiently expand the tree.

5.3 Belief Tree Expansion

Let the current belief be ξ_t and suppose we observe $x_t^i \triangleq (s_{t+1}^i, r_{t+1}^i, a_t^i)$. This observation defines a unique subsequent belief ξ_{t+1}^i. Together with the MDP state s, this creates a hyper-state transition from ω_t to ω_{t+1}^i.

By recursively obtaining observations for future beliefs, we can obtain an unbalanced tree with nodes $\{\omega_{t+k}^{i} : k = 1, \dots, T; i = 1, \dots\}$. However, we cannot hope to be able to fully expand the tree. This is especially true in the case where observations (i.e. states, rewards or actions) are continuous, where we cannot perform even a full single-step expansion. Even in the discrete case, the problem is intractable for infinite horizons—and far too complex computationally for the finite horizon case. However, had there been efficient tree expansion methods, this problem would be largely alleviated.

All tree search methods require the expansion of leaf nodes. However, in general, a leaf node may have an infinite number of children. We thus need some strategies to limit the number of children. More formally, let us assume that we wish to expand in node $\omega_t^i = (\xi_t^i, s_t^i)$, with ξ_t^i defining a density over $\mathscr{M}$. For discrete state/action/reward spaces, we can simply enumerate all the possible outcomes $\{\omega_{t+1}^{j}\}_{j=1}^{|\mathscr{S}\times\mathscr{A}\times R|}$, where R is the set of possible reward outcomes. Note that if the reward is deterministic, there is only one possible outcome per state–action pair. The same holds if $\mathscr{T}$ is deterministic, in both cases making an enumeration possible. While, in general, this may not be the case, since rewards, states or actions can be continuous, in this chapter, we shall only examine the discrete case.

5.4 Bounds on the Optimal Value Function

Let Ω_T be the set of leaf nodes of the partially expanded belief tree and ν the BAMDP process. If the values of the leaf nodes were known, then we could easily perform the backwards induction procedure, shown in Algorithm 4 for BAMDPs: If one thinks of the BAMDP as a very large MDP, one can see that the algorithm (also called value iteration) is identical to Eq. (29), with the subtle difference that the reward only depends on the next hyper-state.

The main problem is obtaining a good estimate for V_T^*, i.e. the value of leaf nodes. Let $\pi^*(\mu)$ denote the policy such that, for any π,

---

Algorithm 4 Backwards induction action selection

1: **input** Process ν, time t, leaf nodes Ω_T, leaf node values V_T^*.
2: **for** $n = T-1, T-2, \dots, t$ **do**
3: **for** $\omega \in \Omega_n$ **do**
4:

$$\begin{aligned} a_n^*(\omega) &= \arg\max_a \sum_{\omega' \in \Omega_{n+1}} \nu(\omega'|\omega,a)[\mathbf{E}(r|\omega',\omega,\nu) + V_{n+1}^*(\omega')] \\ V_n^*(\omega) &= \sum_{\omega' \in \Omega_{n+1}} \nu(\omega'|\omega,a^*)[\mathbf{E}(r|\omega',\omega,\nu) + V_{n+1}^*(\omega')] \end{aligned} \tag{37}$$

5: **end for**
6: **end for**
7: **return** a_t^*

$$V_{\mu}^{\pi^*(\mu)}(s) \geq V_{\mu}^{\pi}(s), \qquad \forall s \in \mathscr{S}. \tag{38}$$

Furthermore, let the mean MDP arising from the belief ξ at hyper-state $\omega = (s, \xi)$ be $\bar{\mu}_\xi \triangleq \mathbf{E}[\mu|\xi]$.

Proposition 1. *The optimal value function V^* of the BAMPD ν at any hyper-state $\omega = (s, \xi)$ is bounded by the following inequalities*

$$\int V_{\mu}^{\pi^*(\mu)}(s)\xi(\mu)d\mu \geq V^*(\omega) \geq \int V_{\mu}^{\pi^*(\bar{\mu}_\xi)}(s)\xi(\mu)\,d\mu. \tag{39}$$

The proof is given in the appendix. In POMDPs, a trivial lower bound can be obtained by calculating the value of the blind policy [33, 51], which always takes the a *fixed* action, i.e. $a_t = a$ for all t. Our lower bound is in fact the BAMDP analogue of the value of the blind policy in POMDPs if we consider BAMDP policies are POMDP actions. The analogy is due to the fact that for any policy π, which selects actions by considering only the MDP state, i.e. such that $\pi(a_t|s_t, \xi_t) = \pi(a_t|s_t)$, it holds trivially that $V^\pi(\omega) \leq V^*(\omega)$, in the same way that it holds trivially if we consider only the set of policies which always take the same action. In fact, of course $V^*(\omega) \geq V^\pi(\omega)$ for any π, by definition. In our case, we have made this lower bound tighter by considering $\pi^*(\bar{\mu}_\xi)$, the policy that is greedy with respect to the current mean estimate.

The upper bound itself is analogous to the POMDP value function bound given in Theorem 9 of [33]. The crucial difference is that, in our case, both bounds can only be approximated via Monte Carlo sampling with some probability, unless $\mathscr{M}$ is finite.

5.4.1 Leaf Node Lower Bound

A lower bound can be obtained by calculating the expected value of any policy. In order to have a tighter bound, we can perform value iteration in the mean MDP. Let us use $\bar{\mu}_\xi$ to denote the *mean MDP* for belief ξ, with transition probabilities $\mathscr{T}_{\bar{\mu}_\xi}$ and mean rewards $R_{\bar{\mu}_\xi}$:

$$\mathscr{T}_{\bar{\mu}_\xi} \triangleq \bar{\mu}_\xi(s_{t+1}|s_t, a_t) = \mathbf{E}(\mathscr{T}^\mu|\xi_t) \tag{40}$$

$$R_{\bar{\mu}_\xi} \triangleq \bar{\mu}_\xi(s_{t+1}|s_t, a_t) = \mathbf{E}(\mathscr{R}^\mu|\xi_t). \tag{41}$$

Similarly, let $V_{\bar{\mu}_\xi}^\pi$ be the column vector of the value function of the mean MDP to obtain

$$\begin{aligned}
V_{\bar{\mu}_\xi}^\pi &= R_{\bar{\mu}_\xi} + \gamma \int \mathscr{T}_{\bar{\mu}_\xi}^\pi V_{\bar{\mu}_\xi}^\pi \xi(\mu)\,d\mu \\
&= R_{\bar{\mu}_\xi} + \gamma \left(\int \mathscr{T}_{\bar{\mu}_\xi}^\pi \xi(\mu)\,d\mu \right) V_{\bar{\mu}_\xi}^\pi \\
&= R_{\bar{\mu}_\xi} + \gamma \mathscr{T}_{\bar{\mu}_\xi}^\pi V_{\bar{\mu}_\xi}^\pi.
\end{aligned}$$

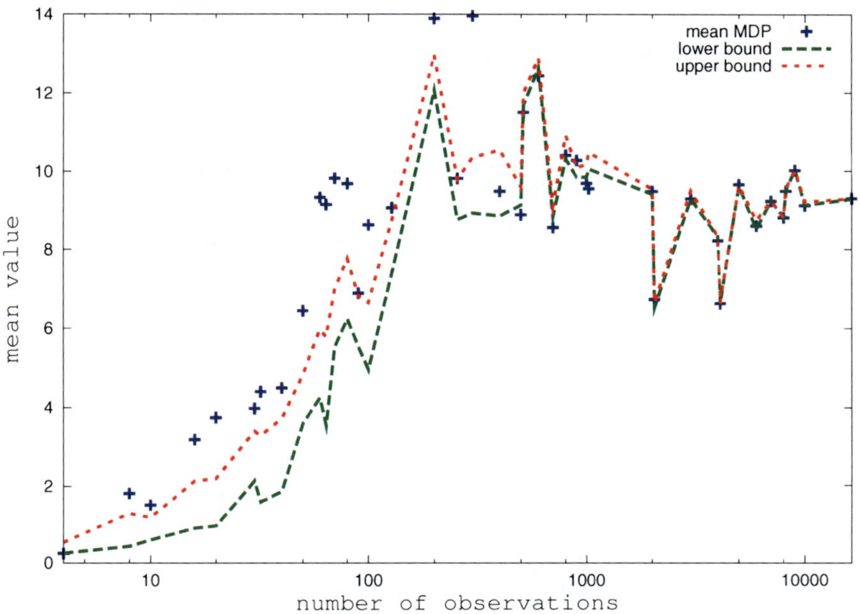

Fig. 1 Illustration of upper and lower bounds on the value function, averaged over all states, as more observations are acquired. It can be seen that the mean MDP value (crosses) is far from the bounds initially. The bounds were calculated by taking the empirical mean of 1000 MDP samples from the current belief.

This is now a standard Bellman recursion, which we can use to obtain the policy $\pi^*_{\bar{\mu}}$ which is optimal with respect to the mean MDP. Unfortunately, the value function of the mean MDP does not generally equal the expected value of the BAMDP:

$$V^{\pi}_{\bar{\mu}_{\xi}} \neq \mathbf{E}[V^{\pi}|\xi]. \tag{42}$$

Nevertheless, the stationary policy that is optimal with respect to the mean MDP can be used to evaluate the right-hand side for $\pi = \pi^*_{\bar{\mu}}$, which will hopefully result in a relatively tight lower bound. This is illustrated in Figure 1, which displays upper and lower bounds on the BAMDP value function as observations are acquired in a simple maze task.

If the beliefs ξ can be expressed in closed form, it is easy to calculate the mean transition distribution and the mean reward from ξ. For discrete state spaces, transitions can be expressed as multinomial distributions, to which the Dirichlet density is a conjugate prior. In that case, for Dirichlet parameters $\{\psi^{j,a}_i(\xi) : i, j \in \mathscr{S}, a \in \mathscr{A}\}$, we have

$$\bar{\mu}_{\xi}(s'|s,a) = \frac{\psi^{s,a}_{s'}(\xi)}{\sum_{i \in \mathscr{S}} \psi^{s,a}_i(\xi)} \tag{43}$$

Similarly, for Bernoulli rewards, the corresponding mean model arising from the beta prior with parameters $\{\alpha^{s,a}(\xi), \beta^{s,a}(\xi) : s \in \mathscr{S}, a \in \mathscr{A}\}$ is $\mathbf{E}[r|s,a,\bar{\mu}_\xi] = \alpha^{s,a}(\xi)/(\alpha^{s,a}(\xi) + \beta^{s,a}(\xi))$. Then, the value function of the mean model can be found with standard value iteration.

5.4.2 Bounds with High Probability

In general, neither the upper nor the lower bounds can be expressed in closed form. However, the integral can be approximated via Monte Carlo sampling.

Let us be in some hyper-state $\omega = (s, \xi)$. We can obtain c MDP samples from the belief at ω: $\mu_1, \ldots, \mu_c \sim \xi(\mu)$. In order to estimate the upper bound, for each μ_k, we can derive the optimal policy $\pi^*(\mu_k)$ and estimate its value function $\tilde{v}_k^* \triangleq V_{\mu_k}^{\pi^*(\mu_k)} \equiv V_{\mu_k}^*$. We may then average these samples to obtain

$$\hat{v}_c^*(\omega) \triangleq \frac{1}{c} \sum_{k=1}^{c} \tilde{v}_k^*(s), \tag{44}$$

where s is the state at hyper-state ω. Let $\bar{v}^*(\omega) = \int_{\mathscr{M}} \xi(\mu) V_\mu^*(s)\, d\mu$. It holds that $\lim_{c \to \infty}[\hat{v}_c] = \bar{v}^*(\omega)$ and that $\mathbf{E}[\hat{v}_c] = \bar{v}^*(\omega)$. Due to the latter, we can apply a Hoeffding inequality

$$\mathbf{P}\left(|\hat{v}_c^*(\omega) - \bar{v}^*(\omega)| > \varepsilon\right) < 2\exp\left(-\frac{2c\varepsilon^2}{(V_{\max} - V_{\min})^2}\right), \tag{45}$$

thus bounding the error within which we estimate the upper bound. For $r_t \in [0,1]$ and discount factor γ, note that $V_{\max} - V_{\min} \leq 1/(1-\gamma)$.

The procedure for the lower bound is identical, but we only need to estimate the value $\tilde{v}_k \triangleq V_{\mu_k}^{\pi^*(\bar{\mu}_\xi)}$ of the mean-MDP-optimal policy $\pi^*(\bar{\mu}_\xi)$ for each one of the sampled MDPs μ_k. Thus, we obtain a pair of high probability bounds for any BAMDP node.

5.5 *Discussion and Related Work*

Employing the above bounds together with efficient search methods [16, 21, 22, 35] is a promising direction. Even when the search is efficient, however, the complexity remains prohibitive for large problems due to the high branching factor and the dependency on the horizon.

Poupart et al. [44] have proposed an analytic solution to Bayesian reinforcement learning. They focus on how to efficiently approximate Eq. (36). More specifically, they use the fact that the optimal value function is the upper envelope of a set of linear segments

$$V_{t,T}^*(\omega_t) = \max_{\alpha \in \Gamma} \alpha(\omega_t), \tag{46}$$

with

$$\alpha(\omega_t) = \int_{\mathscr{M}} \xi_t(\mu)\alpha(s_t,\mu)\,d\mu. \tag{47}$$

They then show that the $k+1$-horizon α-function can be computed from the k-horizon α-function via the following backwards induction:

$$\alpha(\omega_t) = \sum_i \mu(s_{t+1}{=}i \mid s_t, a^*(\omega_t))[\mathbf{E}(R \mid s_{t+1}{=}i, s_t, a^*(\omega_t)) + \gamma\alpha(\omega_{t+1}^{i,a^*(\omega)})], \tag{48}$$

$$V^{k+1}(\omega) = \max_{a \in \Gamma^{k+1}} \alpha(\omega), \tag{49}$$

where ω_{t+1}^i denotes the augmented state resulting from starting at ω_t, taking action $a^*(\omega_t)$ and transiting to the MDP state i. However, the complexity of this process is still exponential with the planning horizon. For this reason, the authors use a projection of the α-functions. This is performed on a set of belief points selected via sampling a default trajectory. The idea is that one may generalize from the set of sample beliefs to other, similar, belief states. In fact, the approximate value function they derive is a lower bound on the BAMDP value function. It is an open question whether or when the bounds presented here are tighter.

At this point, however, it is unclear under what conditions efficient online methods would outperform approximate analytic methods. Perhaps a combination of the two methods would be of interest; i.e. using the online search and bounds in order to see when the point-based approximation has become too inaccurate. The online search could also be used to sample new belief points.

It should be noted that online methods break down when $\gamma \to 1$ because the belief tree cannot, in general, be expanded to the required depth. In fact, the only currently known methods which are nearly optimal in the undiscounted case are based on distribution-free bounds [4]. Similar to the bandit case, it is possible to perform nearly optimally in an unknown MDP by considering upper confidence bounds on the value function. Auer et al. [4] in fact give an algorithm that achieves regret $O(D|\mathscr{S}|\sqrt{|\mathscr{A}|T})$ after T steps for any unknown MDP with diameter[5] D. In exactly the same way as the Bayesian approach, the algorithm maintains counts Ψ over observe state–action–state transitions. In order to obtain an optimistic value function, the authors consider an augmented MDP which is constructed from a set of plausible MDPs M, where M is such that $\mathbf{P}(\Psi|\mu \notin M) < \delta$. Then, instead of augmenting the state space, they augment the action space by allowing the simultaneous choice of actions $a \in \mathscr{A}$ and MDPs $\mu \in M$. The policy is then chosen by performing average value iteration [45] in the augmented MDP.

There has not been much work yet for Bayesian methods in the undiscounted case, with the exception of Bernoulli bandit problems [38]. It may well be that in fact naive look-ahead methods are impractical. In order to achieve $O(D|\mathscr{S}|\sqrt{|\mathscr{A}|T})$ regret with online methods [22] requires an instantaneous regret ε_t such that $\sum_t^T \varepsilon_t < \sqrt{T}, \Rightarrow \varepsilon_t < 1/\sqrt{t}$. If we naively bound our instantaneous regret by discounting, then

[5] Intuitively, the maximin expected time needed to reach any state from any other state, where the max is taken over state pairs and the min over policies. See [45] for details.

that would require our horizon to grow with rate $O(\log_{1/\gamma} \sqrt{t}/(1-\gamma))$, something which seems impractical with current online search techniques.

6 Partial Observability

A useful extension of the MDP model can be obtained by not allowing the agent to directly observe the state of the environment, but an observation variable o_t that is conditioned on the state. This more realistic assumption is formally defined as follows:

Definition 7 (Partially observable Markov decision process). A partially observable Markov decision process μ (POMDP) is the tuple $\mu = (\mathscr{S}, \mathscr{A}, \mathscr{O}, \mathscr{T}, \mathscr{R})$ comprising a set of observations $\mathscr{O}$, a set of states $\mathscr{S}$, a set of actions $\mathscr{A}$, a transition–observation distribution $\mathscr{T}$ conditioned the current state and action $\mu(s_{t+1} = s', o_{t+1} = o | s_t = s, a_t = a)$ and a reward distribution $\mathscr{R}$, conditioned on the state and action $\mu(r_{t+1} = r | s_t = s, a_t = a)$, with $a \in \mathscr{A}$, $s, s' \in \mathscr{S}$, $o \in \mathscr{O}$, $r \in \mathbb{R}$.

We shall denote the set of POMDPs as $\mathscr{M}_P$. For POMDPs, it is often assumed that one of the two following factorizations holds:

$$\mu(s_{t+1}, o_{t+1} | s_t, a_t) = \mu(s_{t+1} | s_t, a_t) \mu(o_{t+1} | s_{t+1}) \tag{50}$$

$$\mu(s_{t+1}, o_{t+1} | s_t, a_t) = \mu(s_{t+1} | s_t, a_t) \mu(o_{t+1} | s_t, a_t). \tag{51}$$

The assumption that the observations are only dependent on a single state or a single state–action pair is a natural decomposition for a lot of practical problems.

POMDPs are *formally identical* to BAMDPs. More specifically, BAMDPs correspond to a special case of a POMDP in which the state is split into two parts: one fully observable dynamic part and one unobservable, continuous, but stationary part, which models the unknown MDP. Typically, however, in POMDP applications, the unobserved part of a state is dynamic and discrete.

The problem of acting optimally in POMDPs has two aspects: the first is state estimation and the second is acting optimally given the estimated state. As far as the first part is concerned, given an initial state probability distribution, updating the belief for a discrete state space amounts to simply maintaining a multinomial distribution over the states. However, the initial state distribution might not be known. In that case, we may assume an initial prior density over the multinomial state distribution. It is easy to see that this is simply a special case of an unknown state transition distribution, where we insert a special initial state which is only visited once. We shall, however, be concerned with the more general case of full exploration in POMDPs, where all state transition distributions are unknown.

6.1 Belief POMDPs

It is possible to create an augmented MDP for POMDP models, by endowing them with an additional belief state, in the same manner as MDPs. However, now the belief state will be a joint probability distribution over $\mathcal{M}_P$ and $\mathcal{S}$. Nevertheless, each (a_t, o_{t+1}) pair that is observed leads to a unique subsequent belief state. More formally, a belief-augmented POMDP is defined as follows:

Definition 8 (Belief POMDP). A *Belief POMDP* ν (BAPOMPD) is an MDP $\nu = (\Omega, \mathcal{A}, \mathcal{O}, \mathcal{T}', \mathcal{R}')$ where $\Omega = \mathcal{G} \times \mathcal{B}$, where $\mathcal{G}$ is a suitable set of probability measures on $\mathcal{S}$, and $\mathcal{B}$ is the set of probability measures on $\mathcal{M}_P$, $\mathcal{T}'$ and $\mathcal{R}'$ are the belief state transition and reward distributions conditioned on the belief state ξ_t and the action a_t, respectively, such that the following factorizations are satisfied for all $\mu \in \mathcal{M}_P$, $\xi_t \in \mathcal{B}$

$$p(s_{t+1}|s_t, a_t, s_{t-1}, \dots, \mu) = \mu(s_{t+1}|s_t, a_t) \quad (52)$$

$$p(o_t|s_t, a_t, o_{t-1}, \dots, \mu) = \mu(o_t|s_t, a_t) \quad (53)$$

$$p(\xi_{t+1}|o_{t+1}, a_t, \xi_t) = \int_{\mathcal{M}_P} p(\xi_{t+1}|\mu, o_{t+1}, a_t, \xi_t)\xi_{t+1}(\mu|o_{t+1}, a_t, \xi_t)\, d\mu \quad (54)$$

We shall denote the set of BAPOMDPs with $\mathcal{M}_{BP}$. Again, Eq. (54) simply assures that the transitions in the belief-POMDP are well defined. The Markov state $\xi_t(\mu, s_t)$ now jointly specifies a distribution over POMDPs and states.[6] As in the MDP case, in order to be able to evaluate policies and select actions optimally, we need to first construct the BAPOMDP. This requires calculating the transitions from the current belief state to subsequent ones according to our possible future observations as well as the probability of those observations. The next section goes into this in more detail.

6.2 The Belief State

In order to simplify the exposition, in the following we shall assume first that each POMDP has the same number of states. Then, $\xi(s_t = s|\mu)$ describes the probability that we are in state s at time t given some belief ξ and assuming we are in the POMDP μ. Similarly, $\xi(s_t = s, \mu)$ is the joint probability given our belief. This joint distribution can be used as a state in an expanded MDP, which can be solved via backward induction, as will be seen later. In order to do this, we must start with an initial belief ξ_0 and calculate all possible subsequent beliefs. The belief at time $t+1$ depends only on the belief time t and the current set of observations r_{t+1}, o_{t+1}, a_t. Thus, the transition probability from ξ_t to ξ_{t+1} is just the probability of the observations according to our current belief, $\xi_t(r_{t+1}, o_{t+1}|a_t)$. This can be calculated by first noting that given the model and the state, the probability of the observations no longer depends on the belief, i.e.

$$\xi_t(r_{t+1}, o_{t+1}, | s_t, a_t, \mu) = \mu(r_{t+1}, o_{t+1} \mid a_t, s_t) = \mu(r_{t+1} \mid a_t, s_t)\mu(o_{t+1} \mid a_t, s_t). \quad (55)$$

[6] The formalism is very similar to that described in [47], with the exception that we do not include the actual POMDP state in the model.

The probability of any particular observation can be obtained by integrating over all the possible models and states

$$\xi_t(r_{t+1}, o_{t+1} \mid a_t) = \int_{\mathscr{M}_P} \int_{\mathscr{S}} \mu(r_{t+1}, o_{t+1} \mid a_t, s_t) \xi_t(\mu, s_t) \, d\mu \, ds_t. \tag{56}$$

Given that a particular observation is made from a specific belief state, we now need to calculate what belief state it would lead to. For this, we need to compute the posterior belief over POMDPs and states. The belief over POMDPs is given by

$$\xi_{t+1}(\mu) \triangleq \xi_t(\mu \mid r_{t+1}, o_{t+1}, a_t,) \tag{57}$$

$$= \frac{\xi_t(r_{t+1}, o_{t+1}, a_t \mid \mu)\xi_t(\mu)}{\xi_t(r_{t+1}, o_{t+1}, a_t)} \tag{58}$$

$$= \frac{\xi_t(\mu)}{Z} \iint_{\mathscr{S}} \mu(r_{t+1}, o_{t+1}, a_t \mid s_{t+1}, s_t) \xi_t(s_{t+1}, s_t \mid \mu) \, ds_{t+1} ds_t \tag{59}$$

where $Z = \xi_t(r_{t+1}, o_{t+1}, a_t)$ is a normalizing constant. Note that $\xi_t(s_{t+1}, s_t \mid \mu) = \mu(s_{t+1} \mid s_t)\xi_t(s_t \mid \mu)$, where $\xi_t(s_t \mid \mu)$ is our belief about the state in the POMDP μ. This can be updated using the following two steps. First, the filtering step

$$\xi_{t+1}(s_t \mid \mu) \triangleq \xi_t(s_t \mid r_{t+1}, o_{t+1}, a_t, \mu) \tag{60}$$

$$= \frac{\mu(r_{t+1}, o_{t+1} \mid s_t, a_t)\xi_t(s_t \mid \mu)}{\xi_t(r_{t+1}, o_{t+1} \mid a_t, \mu)}, \tag{61}$$

where we adjust our belief about the previous state of the MDP based on. Then, we must perform a prediction step

$$\xi_{t+1}(s_{t+1} \mid \mu) = \int_{\mathscr{S}} \mu(s_{t+1} \mid s_t = s)\xi_{t+1}(s_t = s \mid \mu) \, ds, \tag{62}$$

where we calculate the probability over the current states given our new belief concerning the previous states. These predictions can be used to further calculate a new possible belief, since our current belief corresponds to a distribution over $\mathscr{M}_P$. For each possible μ, we determine how our beliefs would change as we acquire new observations. The main difficulty is maintaining the joint distribution over states and POMDPs.

6.3 Belief Compression

Belief-augmented POMDPs in general admit no compact representation of our current belief. This is due to the fact that there is a uncertainty both about which POMDP we are acting in and about the state of each possible POMDP. In fact, the sufficient statistic for such a problem consists of *the complete history of observations*.

This problem is not unique in reinforcement learning, however. A lot of other inference problems admit no compact posterior distributions. Gaussian processes [46], for example, have complexity $O(n^3)$ in the number of observations n.

For the specific case of BAPOMDPs, Ross et al. [47] give a fixed-sample approximation for the value function with a strict upper bound on the error. This work uses the idea of α-vector backups employed in [44] to evaluate the BAMDP value function on a finite POMDP. In addition to the value function approximation, the authors also employ particle filters to represent the belief. In very closely related work, Poupart and Vlassis [43] employ sampling from reachable *beliefs* together with an efficient analytical approximation to the BAPOMDP value function.

7 Conclusion, Future Directions and Open Problems

Online methods for reinforcement learning have now reached a point of relative maturity, especially in the distribution-free framework. Methods for near-optimal reinforcement learning now exist in the discrete case [4]. The continuous case is covered only for bandit problems, however [5, 16, 39]. The continuous case extension of the discrete framework may be relatively straightforward, but it is nevertheless unclear whether a naive extension (by a simple discretization) of the bounds in [4] will be sufficient. Related results in policy learning in continuous spaces [23] show an exponential dependency on the number of dimensions, even though the policy iteration procedure used therein is quite efficient. This, however, is probably due to the generic weakness that interval-based methods have with dealing with high dimensional spaces: they require a number of partitionings of the state space exponential in the number of dimensions. Whether such methods can be further improved remains to be seen.

There are some approaches which use sparse sampling [37] to deal with this problem. These methods have been employed in a Bayesian online settings as well [57], with promising results. Such methods, however, are not useful when rewards are undiscounted: Kearn's sparse sampling [37] method relies on the fact that the discount factor acts as an implicit horizon. It is possible to employ average-reward value iteration (see e.g. [45]) to obtain upper bounds on the optimal value function at the leaf nodes of such a tree. However, the main problem with that approach is that the average-reward value iteration in general diverges; thus, there is no easy way to calculate the value of predecessor states.

Current research to improve the performance in the online case is mostly focused on improved methods for tree search [16, 21, 22, 35, 39, 48, 57]. Offline methods have been explored in the context of point-based approximations to analytic solutions [44] as well as in the context of linear programming [12]. Both approaches could be effective tools to reduce computation, by allowing one to generalize over belief states. This is an approach followed by Ross et al. [47] and Poupart and Vlassis [43] for exploration in POMDPs.

In summary, the following questions should be of interest to researchers in the field: (a) How well do value function approximations on BA(PO)MDPs generalize to unseen belief states? (b) How to perform an effective discretization of the continuous space. Can we go beyond interval-based methods? (c) How can we sample MDP and belief states efficiently? (d) Can Bayesian methods be extended to the undiscounted case?

Acknowledgements. Thanks to Ronald Ortner and Nikos Vlassis for interesting discussions and ideas.

Appendix

Proof (Proposition 1). By definition, $V^*(\omega) \geq V^\pi(\omega)$ for all $\omega = (s, \xi)$, for any policy π. The lower bound follows trivially, since

$$V^{\pi^*(\bar{\mu}_\xi)}(\omega) \triangleq \int V_\mu^{\pi^*(\bar{\mu}_\xi)}(s)\xi(\mu)\,d\mu. \tag{63}$$

The upper bound is derived as follows. First, note that for any function f,

$$\max_x \int f(x,u)\,du \leq \int \max_x f(x,u)\,du.$$

Then, we remark that:

$$V^*(\omega) = \max_\pi \int V_\mu^\pi(s)\xi(\mu)\,d\mu \tag{64a}$$

$$\leq \int \max_\pi V_\mu^\pi(s)\xi(\mu)\,d\mu \tag{64b}$$

$$= \int V_\mu^{\pi^*(\mu)}(s)\xi(\mu)\,d\mu. \tag{64c}$$

□

References

1. Agrawal, R.: The continuum-armed bandit problem. SIAM Journal on Control and Optimization 33(6), 1926–1951 (1995)
2. Auer, P.: Using confidence bounds for exploitation-exploration trade-offs. J. Machine Learning Research 3, 397–422 (2002)
3. Auer, P., Cesa-Bianchi, N., Fischer, P.: Finite time analysis of the multiarmed bandit problem. Machine Learning 47(2/3), 235–256 (2002)
4. Auer, P., Jaksch, T., Ortner, R.: Near-optimal regret bounds for reinforcement learning. In: Proceedings of NIPS 2008 (2008)
5. Auer, P., Ortner, R., Szepesvari, C.: Improved Rates for the Stochastic Continuum-Armed Bandit Problem. In: Bshouty, N.H., Gentile, C. (eds.) COLT. LNCS (LNAI), vol. 4539, pp. 454–468. Springer, Heidelberg (2007)

6. Bellman, R., Kalaba, R.: A mathematical theory of adaptive control processes. Proceedings of the National Academy of Sciences of the United States of America 45(8), 1288–1290 (1959), `http://www.jstor.org/stable/90152`
7. Berger, J.: The case for objective Bayesian analysis. Bayesian Analysis 1(3), 385–402 (2006)
8. Bertsekas, D.: Dynamic programming and suboptimal control: From ADP to MPC. Fundamental Issues in Control, European Journal of Control 11(4-5) (2005); From 2005 CDC, Seville, Spain
9. Bertsekas, D.P.: Dynamic Programming and Optimal Control. Athena Scientific, Belmont (2001)
10. Blumer, A., Ehrenfeuch, A., Haussler, D., Warmuth, M.: Learnability and the Vapnik-Chervonenkis Dimension. Journal of the Association for Computing Machinery 36(4), 929–965 (1989)
11. Boender, C., Rinnooy Kan, A.: Bayesian stopping rules for multistart global optimization methods. Mathematical Programming 37(1), 59–80 (1987)
12. Castro, P.S., Precup, D.: Using linear programming for bayesian exploration in Markov decision processes. In: Veloso, M.M. (ed.) IJCAI, pp. 2437–2442 (2007)
13. Cesa-Bianchi, N., Lugosi, G.: Prediction, Learning and Games. Cambridge University Press, Cambridge (2006)
14. Chen, M.H., Ibrahim, J.G., Yiannoutsos, C.: Prior elicitation, variable selection and Bayesian computation for logistic regression models. Journal of the Royal Statistical Society (Series B): Statistical Methodology 61(1), 223–242 (1999)
15. Chernoff, H.: Sequential Models for Clinical Trials. In: Proceedings of the Fifth Berkeley Symposium on Mathematical Statistics and Probability, vol. 4, pp. 805–812. Univ. of Calif. Press, Berkeley (1966)
16. Coquelin, P.A., Munos, R.: Bandit algorithms for tree search. In: UAI 2007, Proceedings of the 23rd Conference in Uncertainty in Artificial Intelligence, Vancouver, BC, Canada (2007)
17. Dearden, R., Friedman, N., Russell, S.J.: Bayesian Q-learning. In: AAAI/IAAI, pp. 761–768 (1998), `http://citeseer.ist.psu.edu/dearden98bayesian.html`
18. DeGroot, M.H.: Optimal Statistical Decisions. John Wiley & Sons, Chichester (1970)
19. Dey, D., Muller, P., Sinha, D.: Practical nonparametric and semiparametric Bayesian statistics. Springer, Heidelberg (1998)
20. Dimitrakakis, C.: Nearly optimal exploration-exploitation decision thresholds. In: Int. Conf. on Artificial Neural Networks, ICANN (2006); IDIAP-RR 06-12
21. Dimitrakakis, C.: Tree exploration for Bayesian RL exploration. In: Proceedings of the international conference on computational intelligence for modelling, control and automation, CIMCA 2008 (2008)
22. Dimitrakakis, C.: Complexity of stochastic branch and bound for belief tree search in Bayesian reinforcement learning. Tech. Rep. IAS-UVA-09-01, University of Amsterdam (2009)
23. Dimitrakakis, C., Lagoudakis, M.G.: Algorithms and bounds for rollout sampling approximate policy iteration. In: Girgin, S., Loth, M., Munos, R., Preux, P., Ryabko, D. (eds.) EWRL 2008. LNCS (LNAI), vol. 5323, pp. 27–40. Springer, Heidelberg (2008), `http://www.springerlink.com/content/93u40ux345651423`
24. Dimitrakakis, C., Savu-Krohn, C.: Cost-minimising strategies for data labelling: optimal stopping and active learning. In: Hartmann, S., Kern-Isberner, G. (eds.) FoIKS 2008. LNCS, vol. 4932, pp. 96–111. Springer, Heidelberg (2008)

25. Duff, M.O.: Optimal learning computational procedures for Bayes-adaptive Markov decision processes. Ph.D. thesis, University of Massachusetts at Amherst (2002)
26. Duff, M.O., Barto, A.G.: Local bandit approximation for optimal learning problems. In: Mozer, M.C., Jordan, M.I., Petsche, T. (eds.) Advances in Neural Information Processing Systems, vol. 9, p. 1019. The MIT Press, Cambridge (1997), citeseer.ist.psu.edu/147419.html
27. Even-Dar, E., Mannor, S., Mansour, Y.: Action elimination and stopping conditions for the multi-armed and reinforcement learning problems. Journal of Machine Learning Research, 1079–1105 (2006)
28. Freund, Y., Schapire, R.E.: A decision-theoretic generalization of on-line learning and an application to boosting. Journal of Computer and System Sciences 55(1), 119–139 (1997)
29. Friedman, M., Savage, L.J.: The Utility Analysis of Choices Involving Risk. The Journal of Political Economy 56(4), 279 (1948)
30. Friedman, M., Savage, L.J.: The Expected-Utility Hypothesis and the Measurability of Utility. The Journal of Political Economy 60(6), 463 (1952)
31. Gittins, C.J.: Multi-armed Bandit Allocation Indices. John Wiley & Sons, New Jersey (1989)
32. Goldstein, M.: Subjective Bayesian analysis: Principles and practice. Bayesian Analysis 1(3), 403–420 (2006)
33. Hauskrecht, M.: Value-function approximations for partially observable Markov decision processes. Journal of Artificial Intelligence Resesarch, 33–94 (2000)
34. Hoeffding, W.: Lower bounds for the expected sample size and the average risk of a sequential procedure. The Annals of Mathematical Statistics 31(2), 352–368 (1960), http://www.jstor.org/stable/2237951
35. Hren, J.F., Munos, R.: Optimistic planning of deterministic systems. In: Girgin, S., Loth, M., Munos, R., Preux, P., Ryabko, D. (eds.) EWRL 2008. LNCS (LNAI), vol. 5323, pp. 151–164. Springer, Heidelberg (2008)
36. Kall, P., Wallace, S.: Stochastic programming. Wiley, New York (1994)
37. Kearns, M., Singh, S.: Near-optimal reinforcement learning in polynomial time. In: Proc. 15th International Conf. on Machine Learning, pp. 260–268. Morgan Kaufmann, San Francisco (1998), citeseer.ist.psu.edu/kearns98nearoptimal.html
38. Kelly, F.P.: Multi-armed bandits with discount factor near one: The bernoulli case. The Annals of Statistics 9(5), 987–1001 (1981)
39. Kocsis, L., Szepesvári, C.: Bandit based Monte-Carlo planning. In: Fürnkranz, J., Scheffer, T., Spiliopoulou, M. (eds.) ECML 2006. LNCS (LNAI), vol. 4212, pp. 282–293. Springer, Heidelberg (2006)
40. Luce, R.D., Raiffa, H.: Games and Decisions. John Wiley and Sons, Chichester (1957); Republished by Dover in 1989
41. McCall, J.: The Economics of Information and Optimal Stopping Rules. Journal of Business 38(3), 300–317 (1965)
42. Moustakides, G.: Optimal stopping times for detecting changes in distributions. Annals of Statistics 14(4), 1379–1387 (1986)
43. Poupart, P., Vlassis, N.: Model-based bayesian reinforcement learning in partially observable domains. In: International Symposium on Artificial Intelligence and Mathematics, ISAIM (2008)
44. Poupart, P., Vlassis, N., Hoey, J., Regan, K.: An analytic solution to discrete Bayesian reinforcement learning. In: ICML 2006, pp. 697–704. ACM Press New York (2006)

45. Puterman, M.L.: Markov Decision Processes: Discrete Stochastic Dynamic Programming. John Wiley & Sons, New Jersey (1994/2005)
46. Rasmussen, C.E., Williams, C.K.I.: Gaussian Processes for Machine Learning. MIT Press, Cambridge (2006)
47. Ross, S., Chaib-draa, B., Pineau, J.: Bayes-adaptive POMDPs. In: Platt, J., Koller, D., Singer, Y., Roweis, S. (eds.) Advances in Neural Information Processing Systems, vol. 20. MIT Press, Cambridge (2008)
48. Ross, S., Pineau, J., Paquet, S., Chaib-draa, B.: Online planning algorithms for POMDPs. Journal of Artificial Intelligence Resesarch 32, 663–704 (2008)
49. Roy, N., McCallum, A.: Toward optimal active learning through sampling estimation of error reduction. In: Proc. 18th International Conf. on Machine Learning, pp. 441–448. Morgan Kaufmann, San Francisco (2001), `citeseer.ist.psu.edu/roy01toward.html`
50. Savage, L.J.: The Foundations of Statistics. Dover Publications, New York (1972)
51. Smith, T., Simmons, R.: Point-based POMDP algorithms: Improved analysis and implementation. In: Proceedigns of the 21st Conference on Uncertainty in Artificial Intelligence (UAI 2005), pp. 542–547 (2005)
52. Stengel, R.F.: Optimal Control and Estimation, 2nd edn. Dover, New York (1994)
53. Talagrand, M.: A new look at independence. Annals of Probability 24(1), 1–34 (1996)
54. Vapnik, V.N.: The Nature of Statistical Learning Theory. Springer, Heidelberg (2000)
55. Vapnik, V.N., Chervonenkis, A.Y.: On the uniform convergence of relative frequencies of events to their probabilities. Theory of Probability and its Applications 16(2), 264–280 (1971)
56. Wald, A.: Sequential Analysis. John Wiley & Sons, Chichester (1947); Republished by Dover in 2004
57. Wang, T., Lizotte, D., Bowling, M., Schuurmans, D.: Bayesian sparse sampling for online reward optimization. In: ICML 2005, pp. 956–963. ACM, New York (2005)
58. Zhang, T.: From ϵ-entropy to KL-entropy: Analysis of minimum information complexity density estimation. Annals of Statistics 34(5), 2180–2210 (2006)

Ant Colony Learning Algorithm for Optimal Control

Jelmer Marinus van Ast, Robert Babuška, and Bart De Schutter

Abstract. Ant colony optimization (ACO) is an optimization heuristic for solving combinatorial optimization problems and is inspired by the swarming behavior of foraging ants. ACO has been successfully applied in various domains, such as routing and scheduling. In particular, the agents, called ants here, are very efficient at sampling the problem space and quickly finding good solutions. Motivated by the advantages of ACO in combinatorial optimization, we develop a novel framework for finding optimal control policies that we call Ant Colony Learning (ACL). In ACL, the ants all work together to collectively learn optimal control policies for any given control problem for a system with nonlinear dynamics. In this chapter, we discuss the ACL framework and its implementation with crisp and fuzzy partitioning of the state space. We demonstrate the use of both versions in the control problem of two-dimensional navigation in an environment with variable damping and discuss their performance.

1 Introduction

Ant Colony Optimization (ACO) is a metaheuristic for solving combinatorial optimization problems [1]. Inspired by ants and their behavior in finding shortest paths

Jelmer Marinus van Ast
Delft Center for Systems and Control, Delft University of Technology,
Mekelweg 2, 2628 CD Delft, The Netherlands
e-mail: `j.m.vanast@tudelft.nl`

Robert Babuška
Delft Center for Systems and Control, Delft University of Technology,
Mekelweg 2, 2628 CD Delft, The Netherlands
e-mail: `r.babuska@tudelft.nl`

Bart De Schutter
Delft Center for Systems and Control & Marine and Transport Technology,
Delft University of Technology, Mekelweg 2, 2628 CD Delft, The Netherlands
e-mail: `b@deschutter.info`

R. Babuška & F.C.A. Groen (Eds.): Interactive Collaborative Information Systems, SCI 281, pp. 155–182.
springerlink.com

from their nest to sources of food, the virtual ants in ACO aim at jointly finding optimal paths in a given search space. The key ingredients in ACO are the pheromones. With real ants, these are chemicals deposited by the ants and their concentration encodes a map of trajectories, where stronger concentrations represent the trajectories that are more likely to be optimal. In ACO, the ants read and write values to a common pheromone matrix. Each ant autonomously decides on its actions biased by these pheromone values. This indirect form of communication is called stigmergy. Over time, the pheromone matrix converges to encode the optimal solution of the combinatorial optimization problem, but the ants typically do not all converge to this solution, thereby allowing for constant exploration and the ability to adapt the pheromone matrix to changes in the problem structure. These characteristics have resulted in a strong increase of interest in ACO over the last decade since its introduction in the early nineties [2].

The Ant System (AS), which is the basic ACO algorithm, and its variants have successfully been applied to various optimization problems, such as the traveling salesman problem [3], load balancing [4], job shop scheduling [5, 6], optimal path planning for mobile robots [7], and routing in telecommunication networks [8]. An implementation of the ACO concept of pheromone trails for real robotic systems has been described in [9]. A survey of industrial applications of ACO has been presented in [10]. An overview of ACO and other metaheuristics to stochastic combinatorial optimization problems can be found in [11].

This chapter presents the extension of ACO to the learning of optimal control policies for continuous-time, continuous-state dynamic systems. We call this new class of algorithms Ant Colony Learning (ACL) and present two variants, one with a crisp partitioning of the state space and one with a fuzzy partitioning. The work in this chapter is based on [12] and [13], in which we have introduced these respective variants of ACL. This chapter not only unifies the presentation of both versions, it also discusses them in greater detail, both theoretically and practically. Crisp ACL, as we call ACL with crisp partitioning of the state space, is the more straightforward extension of classical ACO to solving optimal control problems. The state space is quantized in a crisp manner such that each value of the continuous-valued state is mapped to exactly one bin. This renders the control problem non-deterministic as state transitions are measured in the quantized domain, but originate from the continuous domain. This makes the control problem much more difficult to solve. Therefore, the variant of ACL in which the state space is partitioned with fuzzy membership functions was developed. We call this variant Fuzzy ACL. With a fuzzy partitioning, the continuous-valued state is mapped to the set of membership functions. The state is a member of each of these membership functions to a certain degree. The resulting Fuzzy ACL algorithm is somewhat more complicated compared to Crisp ACL, but there are no non-deterministic state transitions introduced.

One of the first real applications of the ACO framework to optimization problems in continuous search spaces has been described in [14] and [15]. An earlier application of the ant metaphor to continuous optimization appears in [16], with more recent work like the Aggregation Pheromones System in [17] and the Differential Ant-Stigmergy Algorithm in [18]. Reference [19] is the first work linking ACO to optimal control. Although presenting a formal framework, called *ant programming*, no application, or study of its performance is presented. Our algorithm shares some similarities with the Q-learning algorithm. An earlier work [20] introduced the Ant-Q algorithm, which is the most notable other work relating ACO with Q-learning. However, Ant-Q has been developed for combinatorial optimization problems and not to optimal control problems. Because of this difference, ACL is novel in all major structural aspects of the Ant-Q algorithm, namely the choice of the action selection method, the absence of the heuristic variable, and the choice of the set of ants used in the update. There are only a few publications that combine ACO with the concept of fuzzy control [21–23]. In all three publications, fuzzy controllers are obtained using ACO, rather than developing an actual fuzzy ACO algorithm, as introduced in this chapter.

This chapter has been structured as follows. In Sect. 2, the ACO heuristic is reviewed with special attention paid to the Ant System and the Ant Colony System, which are among the most popular ACO algorithms. Sect. 3 presents the general layout of ACL, with more detail about its crisp and fuzzy version. Sect. 4 presents the application of both ACL algorithms on the control problem of two-dimensional vehicle navigation in an environment with a variable damping profile. The learning performance of Crisp and Fuzzy ACL is studied using a set of performance measures and the resulting policies are compared with an optimal policy obtained by the fuzzy Q-iteration algorithm from [24]. Sect. 5 concludes this chapter and presents an outline for future research.

2 Ant Colony Optimization

This section presents the preliminaries for understanding the ACL algorithm. It presents the framework of all ACO algorithms in Sect. 2.1. The Ant System, which is the most basic ACO algorithm, is discussed in Sect. 2.2 and the Ant Colony System, which is slightly more advanced compared to the Ant System, is discussed in Sect. 2.3. ACL is based on these two algorithms.

2.1 ACO Framework

ACO algorithms have been developed to solve hard combinatorial optimization problems [1]. A combinatorial optimization problem can be represented as a tuple $P = \langle \mathscr{S}, F \rangle$, where $\mathscr{S}$ is the solution space with $s \in \mathscr{S}$ a specific candidate solution and where $F : \mathscr{S} \rightarrow \mathbb{R}_+$ is a fitness function assigning strictly positive values to candidate solutions, where higher values correspond to better solutions. The purpose of the algorithm is to find a solution $s^* \in \mathscr{S}$ or a set of solutions $\mathscr{S}^* \subseteq \mathscr{S}$ that maximize the fitness function. The solution s^* is then called an optimal solution and $\mathscr{S}^*$ is called the set of optimal solutions.

In ACO, the combinatorial optimization problem is represented by a graph consisting of a set of vertices and a set of arcs connecting the vertices. A particular solution s is a concatenation of solution components (i, j), which are pairs of vertices i and j connected by the arc ij. The concatenation of solution components forms a path from the initial vertex to the terminal vertex. This graph is called the *construction* graph as the solutions are constructed incrementally by moving over the graph. How the terminal vertices are defined depends on the problem considered. For instance, in the traveling salesman problem[1], there are multiple terminal vertices, namely for each ant the terminal vertex is equal to its initial vertex, after visiting all other vertices exactly once. For application to control problems, as considered in this chapter, the terminal vertex corresponds to the desired state of the system. Two values are associated with the arcs: a pheromone trail variable τ_{ij} and a heuristic variable η_{ij}. The pheromone trail represents the acquired knowledge about the optimal solutions over time and the heuristic variable provides a priori information about the quality of the solution component, i.e., the quality of moving from vertex i to vertex j. In the case of the traveling salesman problem, the heuristic variables typically represent the inverse of the distance between the respective pair of cities. In general, a heuristic variable represents a short-term quality measure of the solution component, while the task is to acquire a concatenation of solution components that overall form an optimal solution. A pheromone variable, on the other hand, encodes the measure of the long-term quality of concatenating the respective solution components.

2.2 The Ant System

The most basic ACO algorithm is called the Ant System (AS) [25] and works as follows. A set of M ants is randomly distributed over the vertices. The heuristic variables η_{ij} are set to encode the prior knowledge by favoring the choice of some vertices over others. For each ant c, the partial solution $s_{\mathrm{p},c}$ is initially empty and all pheromone variables are set to some initial value $\tau_0 > 0$. We will call each move of the ants over the graph a *step*. In every step, each ant decides, based on some probability distribution, which solution component (i,j) to add to its partial solution $s_{\mathrm{p},c}$. The probability $p_c\{j|i\}$ for an ant c on a vertex i to move to a vertex j within its feasible neighborhood $\mathcal{N}_{i,c}$ is defined as:

$$p_c\{j|i\} = \frac{\tau_{ij}^{\alpha}\eta_{ij}^{\beta}}{\sum_{l\in\mathcal{N}_{i,c}}\tau_{il}^{\alpha}\eta_{il}^{\beta}}, \forall j \in \mathcal{N}_{i,c}, \tag{1}$$

with $\alpha \geq 1$ and $\beta \geq 1$ determining the relative importance of η_{ij} and τ_{ij}, respectively. The feasible neighborhood $\mathcal{N}_{i,c}$ of an ant c on a vertex i is the set of not yet visited vertices that are connected to i. By moving from vertex i to vertex j, ant c adds the associated solution component (i,j) to its partial solution $s_{\mathrm{p},c}$ until it

[1] In the traveling salesman problem, there is a set of cities connected by roads of different lengths and the problem is to find the sequence of cities that takes the traveling salesman to all cities, visiting each city exactly once and bringing him back to its initial city with a minimum length of the tour.

reaches its terminal vertex and completes its candidate solution. The ant is now said to have completed a *trial*.

After all ants have completed their trial, the candidate solutions of all ants are evaluated using the fitness function $F(s)$ and the resulting value is used to update the pheromone levels by:

$$\tau_{ij} \leftarrow (1-\rho)\tau_{ij} + \sum_{s \in \mathscr{S}_{\text{upd}}} \Delta\tau_{ij}(s), \tag{2}$$

with $\rho \in (0,1)$ the evaporation rate and $\mathscr{S}_{\text{upd}}$ the set of solutions that are eligible to be used for the pheromone update, which will be explained further on in this section. This update step is called the *global* pheromone update step. The pheromone deposit $\Delta\tau_{ij}(s)$ is computed as:

$$\Delta\tau_{ij}(s) = \begin{cases} F(s) & \text{, if } (i,j) \in s \\ 0 & \text{, otherwise.} \end{cases}$$

The pheromone levels are a measure of how desirable it is to add the associated solution component to the partial solution. In order to incorporate forgetting, the pheromone levels decrease by some factor. This is called pheromone evaporation in correspondence to the physical evaporation of the chemical pheromones for real ant colonies. By evaporation it can be avoided that the algorithm prematurely converges to suboptimal solutions. Note that in (2) the pheromone level on all vertices is evaporated and only those vertices that are associated with the solutions in the update set receive a pheromone deposit.

In the next trial, each ant repeats the previous steps, but now the pheromone levels have been updated and can be used to make better decisions about which vertex to move to. After some stopping criterion has been reached, the values of τ and η on the graph encode the solution for all (i,j)-pairs. This final solution can be extracted from the graph as follows:

$$(i,j): \quad j = \arg\max_{l}(\tau_{il}^{\alpha}\eta_{il}^{\beta}), \quad \forall i.$$

Note that all ants are still likely to follow suboptimal trajectories through the graph, thereby exploring constantly the solution space and keeping the ability to adapt the pheromone levels to changes in the problem structure.

There exist various rules to construct $\mathscr{S}_{\text{upd}}$, of which the most standard one is to use all the candidate solutions found in the trial. This update set is then called $\mathscr{S}_{\text{trial}}$.[2] This update rule is typical for the Ant System. However, other update rules have shown to outperform the AS update rule in various combinatorial optimization problems. Rather than using the complete set of candidate solutions from the last trial either the best solution from the last trial, or the best solution since the initialization

[2] In ACO literature, this is usually called $\mathscr{S}_{\text{iter}}$, where an iteration has the same meaning as what we call a trial. We prefer to use the word trial in order to stress the similarity between our ACO algorithm for optimal control and reinforcement learning [26].

of the algorithm can be used. The former update rule is called *Iteration Best* in the literature (which could be called *Trial Best* in our terminology), and the latter is called *Best-So-far*, or *Global Best* in the literature [1]. These methods result in a strong bias of the pheromone trail reinforcement towards solutions that have been proven to perform well and additionally reduce the computational complexity of the algorithm. As the risk exists that the algorithm prematurely converges to suboptimal solutions, these methods are only superior to AS if extra measures are taken to prevent this, such as the local pheromone update rule, explained in Sect. 2.3. Two of the most successful ACO variants in practice that implement the update rules mentioned above are the Ant Colony System [27] and the MAX-MIN Ant System [28]. Because of its relation to ACL, we will explain the Ant Colony System next.

2.3 The Ant Colony System

The Ant Colony System (ACS) [27] is an extension to the AS and is one of the most successful and widely used ACO algorithms. There are four main differences between the AS and the ACS:

1. The ACS uses the global-best update rule in the global pheromone update step. This means that only the one solution that has been found since the start of the algorithm that has the highest fitness, called s_{gb}, is used to update the pheromone variables at the end of the trial. This is a form of elitism in ACO algorithms that has shown to significantly speed up the convergence to the optimal solution.
2. The global pheromone update is only performed for the (i,j)-pairs that are an element of the global best solution. This means that not all pheromone levels are evaporated, as with the AS, but only those that also receive a pheromone deposit.
3. The pheromone deposit is weighted by ρ. As a result of this and the previous two differences, the global pheromone update rule is:

$$\tau_{ij} \leftarrow \begin{cases} (1-\rho)\tau_{ij} + \rho\Delta\tau_{ij}(s_{\text{gb}}) & \text{, if } (i,j) \in s_{\text{gb}} \\ \tau_{ij} & \text{, otherwise.} \end{cases}$$

4. An important element from the ACS algorithm that acts as a measure to avoid premature convergence to suboptimal solutions is the local pheromone update step, which occurs for each ant after each step within a trial and is defined as follows:

$$\tau_{ij}(\kappa+1) = (1-\gamma)\tau_{ij}(\kappa) + \gamma\tau_0, \tag{3}$$

where $\gamma \in (0,1)$ is a small parameter similar to ρ, ij is the index corresponding to the (i,j)-pair just added to the partial solution, and τ_0 is the initial value of the pheromone trail. The effect of (3) is that during the trial the visited solution components are made less attractive for other ants to take, in that way promoting the exploration of other, less frequently visited, solution components.

5. There is an explicit exploration–exploitation step with the selection of the next node j, where with a probability of ε, j is chosen as being the node with the highest value of $\tau_{ij}^{\alpha}\eta_{ij}^{\beta}$ (exploitation) and with the probability $(1-\varepsilon)$, a random action is chosen according to (1) (exploration).

The Ant Colony Learning algorithm, which is presented next, is based on the AS combined with the local pheromone update step from the ACS algorithm.

3 Ant Colony Learning

This section presents the Ant Colony Learning (ACL) class of algorithms. First, in Sect. 3.1, the optimal control problem for which ACL has been developed is introduced. Sect. 3.2 presents the general layout of ACL. ACL with crisp state space partitioning is presented in Sect. 3.3 and the fuzzy variant in Sect. 3.4. Regarding the practical application of ACL, Sect. 3.5 discusses the ways of setting the parameters from the algorithm and Sect. 3.6 presents a discussion on the relation of ACL to reinforcement learning.

3.1 The Optimal Control Problem

Assume that we have a nonlinear dynamic system, characterized by a continuous-valued state vector $\boldsymbol{x} = \begin{bmatrix} x_1 & x_2 & \dots & x_n \end{bmatrix}^{\mathrm{T}} \in \mathscr{X} \subseteq \mathbb{R}^n$, with $\mathscr{X}$ the state space and n the order of the system. Also assume that the state can be controlled by an input $\boldsymbol{u} \in \mathscr{U}$ that can only take a finite number of values and that the state can be measured at discrete time steps, with a sample time T_{s} with k the discrete time index. The sampled system is denoted as:

$$\boldsymbol{x}(k+1) = \boldsymbol{f}(\boldsymbol{x}(k),\boldsymbol{u}(k)). \tag{4}$$

The optimal control problem is to control the state of the system from any given initial state $\boldsymbol{x}(0) = \boldsymbol{x}_0$ to a desired goal state $\boldsymbol{x}(K) = \boldsymbol{x}_{\mathrm{g}}$ in at most K steps and in an optimal way, where optimality is defined by minimizing a certain cost function. As an example, take the following quadratic cost function:

$$J(s) = J(\tilde{\boldsymbol{x}},\tilde{\boldsymbol{u}}) = \sum_{k=0}^{K-1} \boldsymbol{e}^{\mathrm{T}}(k+1)\mathbf{Q}\boldsymbol{e}(k+1) + \boldsymbol{u}^{\mathrm{T}}(k)\mathbf{R}\boldsymbol{u}(k), \tag{5}$$

with s the solution found by a given ant, $\tilde{\boldsymbol{x}} = \boldsymbol{x}(1)\dots\boldsymbol{x}(K)$ and $\tilde{\boldsymbol{u}} = \boldsymbol{u}(0)\dots\boldsymbol{u}(K-1)$ the sequences of states and actions in that solution, respectively, $\boldsymbol{e}(k+1) = \boldsymbol{x}(k+1) - \boldsymbol{x}_{\mathrm{g}}$ the error at time $k+1$, and $\mathbf{Q}$ and $\mathbf{R}$ positive definite matrices of appropriate dimensions. The problem is to find a nonlinear mapping from the state to the input $\boldsymbol{u}(k) = \boldsymbol{g}(\boldsymbol{x}(k))$ that, when applied to the system in $\boldsymbol{x}_0$, results in a sequence of state-action pairs $(\boldsymbol{u}(0),\boldsymbol{x}(1))$, $(\boldsymbol{u}(1),\boldsymbol{x}(2))$, …, $(\boldsymbol{u}(K-1),\boldsymbol{x}_{\mathrm{g}})$ that minimizes this cost function. The function $\boldsymbol{g}$ is called the control policy. We make the assumption that

Q and **R** are selected in such a way that $\boldsymbol{x}_{\mathrm{g}}$ is reached in at most K time steps. The matrices **Q** and **R** balance the importance of speed versus the aggressiveness of the controller. This kind of cost function is frequently used in optimal control of linear systems, as the optimal controller minimizing the quadratic cost can be derived as a closed expression after solving the corresponding Riccati equation using the **Q** and **R** matrices and the matrices of the linear state space description [29]. In our case, we aim at finding control policies for nonlinear systems which in general cannot be derived analytically from the system description and the **Q** and **R** matrices. Note that our method is not limited to the use of quadratic cost functions.

The control policy we aim to find with ACL will be a state feedback controller. This is a reactive policy, meaning that it will define a mapping from states to actions without the need of storing the states (and actions) of previous time steps. This poses the requirement on the system that it can be described by a state-transition mapping for a quantized state $\boldsymbol{q}$ and an action (or input) $\boldsymbol{u}$ in discrete time as follows:

$$\boldsymbol{q}(k+1) \sim p(\boldsymbol{q}(k),\boldsymbol{u}(k)), \tag{6}$$

with p a probability distribution function over the state-action space. In this case, the system is said to be a Markov Decision Process (MDP) and the probability distribution function p is said to be the Markov model of the system. Note that finding an optimal control policy for an MDP is equivalent to finding the optimal sequence of state-action pairs from any given initial state to a certain goal state, which is a combinatorial optimization problem. When the state transitions are stochastic, like in (6), it is a stochastic combinatorial optimization problem. As ACO algorithms are especially applicable to (stochastic) combinatorial optimization problems, the application of ACO to deriving control policies is evident.

3.2 General Layout of ACL

ACL can be categorized as a model-free learning algorithm as the ants do not have direct access to the model of the system and must learn the control policy just by interacting with it. Note that as the ants interact with the system in parallel the implementation of this algorithm to a real system requires multiple copies of this system, one for each ant. This hampers the practical applicability of the algorithm to real systems. However, when a model of the real system is available, the parallelism can take place in software. In that case, the learning happens off-line and after convergence the resulting control policy can be applied to the real system. In principle, the ants could also interact with one physical system, but this would require either a serial implementation in which each ant performs a trial and the global pheromone update takes place after the last ant has completed its trial, or a serial implementation in which each ant only performs a single step, after which the state of the system must be reinitialized for the next ant. In the first case the local pheromone update will lose most of its use, while in the second case, the repetitive reinitializations will cause a strong increase of simulation time and heavy load on the system. In either case, the usefulness of ACL will be heavily compromised.

An important consideration for an ACO approach to optimal control is which set of solutions to use in the global pheromone update step. When using the Global Best pheromone update rule in an optimal control problem, all ants have to be initialized to the same state, as starting from states that require less time and less effort to reach the goal would always result in a better Global Best solution. Ultimately, initializing an ant exactly in the goal state would be the best possible solution and no other solution, starting from more interesting states, would get the opportunity to update the pheromones in the global pheromone update phase. In order to find a control policy from *any* initial state to the goal state, the Global Best update rule cannot be used. By simply using all solutions of all ants in the updating, like in the original AS algorithm, the resulting algorithm does allow for random initialization of the ants over the state space and is therefore used in ACL.

3.3 ACL with Crisp State Space Partitioning

For a system with a continuous-valued state space, optimization algorithms like ACO can only be applied if the state space is quantized. The most straightforward way to do this is to divide the state space into a finite number of bins such that each state value is assigned to exactly one bin. These bins can be enumerated and used as the vertices in the ACO construction graph.

3.3.1 Crisp State Space Partitioning

Assume a continuous-time, continuous-state system sampled with a sample time T_s. In order to apply ACO, the state $\boldsymbol{x}$ must be quantized into a finite number of bins to get the quantized state $\boldsymbol{q}$. Depending on the sizes and the number of these bins, portions of the state space will be represented with the same quantized state. One can imagine that applying an input to the system that is in a particular quantized state results in the system to move to a next quantized state with some probability. In order to illustrate these aspects of the quantization, we consider the continuous model to be cast as a discrete stochastic *automaton*. An automaton is defined by the triple $\Sigma = (\mathscr{Q}, \mathscr{U}, \phi)$, with $\mathscr{Q}$ a finite or countable set of discrete states, $\mathscr{U}$ a finite or countable set of discrete inputs, and $\phi : \mathscr{Q} \times \mathscr{U} \times \mathscr{Q} \to [0,1]$ a state transition function.

Given a discrete state vector $\boldsymbol{q} \in \mathscr{Q}$, a discrete input symbol $\boldsymbol{u} \in \mathscr{U}$, and a discrete next state vector $\boldsymbol{q}' \in \mathscr{Q}$, the (Markovian) state transition function ϕ defines the probability of this state transition, $\phi(\boldsymbol{q},\boldsymbol{u},\boldsymbol{q}')$, making the automaton stochastic. The probabilities over all states $\boldsymbol{q}'$ must each sum up to one for each state-action pair $(\boldsymbol{q},\boldsymbol{u})$. An example of a stochastic automaton is given in Figure 1. In this figure, it is clear that, e.g., applying an action $u = 1$ to the system in $q = 1$ can move the system to a next state that is either $q' = 1$ with a probability of 0.2 or $q' = 2$ with a probability of 0.8.

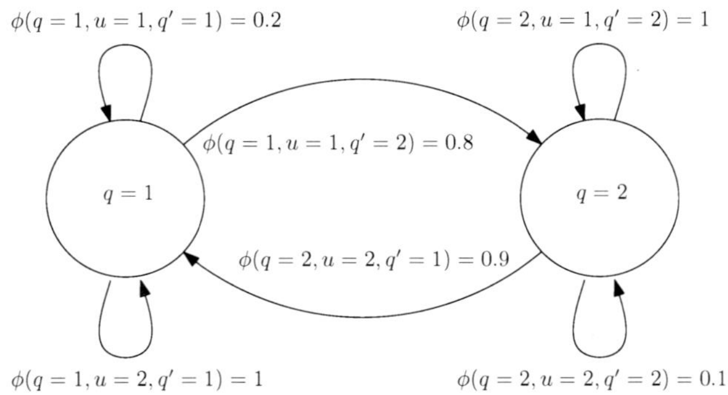

Fig. 1 An example of a stochastic automaton

The probability distribution function determining the transition probabilities reflects the system dynamics and the set of possible control actions is reflected in the structure of the automaton. The probability distribution function can be estimated from simulations of the system over a fine grid of pairs of initial states and inputs, but for the application of ACL this is not necessary. The algorithm can directly interact with the continuous-state dynamics of the system as will be described in the following section.

3.3.2 Crisp ACL Algorithm

At the start of every trial, each ant is initialized with a random continuous-valued state of the system. This state is quantized onto $\mathscr{Q}$. Each ant has to choose an action and add this state-action pair to its partial solution. It is not known to the ant to which state this action will take him, as in general there is a set of next states to which the ant can move, according to the system dynamics.

No heuristic values are associated with the vertices, as we assume in this chapter that no a priori information is available about the quality of solution components. This is implemented by setting all heuristic values equal to one. It can be seen that η_{ij} disappears from (1) in this case. Only the value of α remains as a tuning parameter, now in fact only determining the amount of exploration as higher values of α make the probability higher for choosing the action associated with the largest pheromone level. In the ACS, there is an explicit exploration–exploitation step when the next node j associated with the highest value of $\tau_{ij}^{\alpha}\eta_{ij}^{\beta}$ is chosen with some probability ε (exploitation) and a random node is chosen with the probability $(1-\varepsilon)$ according to (1) (exploration). In our algorithm, as we do not have the heuristic factor η_{ij}, the amount of exploration over exploitation can be tuned by the variable α. For this reason, we do not include an explicit exploration–exploitation step based on ε. Without this and without the heuristic factor, the algorithm needs less tuning and is easier to apply.

Action Selection
The probability of an ant c to choose an action $\boldsymbol{u}$ when in a state $\boldsymbol{q}_c$ is:

$$p_c\{\boldsymbol{u}|\boldsymbol{q}_c\} = \frac{\tau^{\alpha}_{\boldsymbol{q}_c\boldsymbol{u}}}{\sum_{\boldsymbol{l}\in\mathscr{U}_{\boldsymbol{q}_c}} \tau^{\alpha}_{\boldsymbol{q}_c\boldsymbol{l}}}, \tag{7}$$

where $\mathscr{U}_{\boldsymbol{q}_c}$ is the action set available to ant c in state $\boldsymbol{q}_c$.

Local Pheromone Update
The pheromones are initialized equally for all vertices and set to a small positive value τ_0. During every trial, all ants construct their solutions in parallel by interacting with the system until they either have reached the goal state or the trial exceeds a certain pre-specified number of steps K_{max}. After every step, the ants perform a local pheromone update, equal to (3), but in the setting of Crisp ACL:

$$\tau_{\boldsymbol{q}_c\boldsymbol{u}_c}(k+1) \leftarrow (1-\gamma)\tau_{\boldsymbol{q}_c\boldsymbol{u}_c}(k) + \gamma\tau_0. \tag{8}$$

Global Pheromone Update
After completion of the trial, the pheromone levels are updated according to the following global pheromone update step:

$$\begin{aligned}\tau_{\boldsymbol{qu}}(\kappa+1) \leftarrow &(1-\rho)\tau_{\boldsymbol{qu}}(\kappa) \\ &+\rho \sum_{\substack{s\in\mathscr{S}_{\text{trial}};\\ (\boldsymbol{q},\boldsymbol{u})\in s}} J^{-1}(s), \quad \forall(\boldsymbol{q},\boldsymbol{u}) : \exists s \in \mathscr{S}_{\text{trial}}; (\boldsymbol{q},\boldsymbol{u}) \in s,\end{aligned} \tag{9}$$

with $\mathscr{S}_{\text{trial}}$ the set of all candidate solutions found in the trial and κ the trial counter. This type of update rule is comparable to the AS update rule, with the important difference that only the pheromone levels are evaporated that are associated with the elements in the update set of solutions. The pheromone deposit is equal to $J^{-1}(s) = J^{-1}(\tilde{\boldsymbol{q}},\tilde{\boldsymbol{u}})$, the inverse of the cost function over the sequence of quantized state-action pairs in s according to (5).

The complete algorithm is given in Algorithm 5. In this algorithm, the assignment $\boldsymbol{x}_c \leftarrow \text{random}(\mathscr{X})$ in Step 7 selects for ant c a random state $\boldsymbol{x}_c$ from the state space $\mathscr{X}$ with a uniform probability distribution. The assignment $\boldsymbol{q}_c \leftarrow \text{quantize}(\boldsymbol{x}_c)$ in Step 9 quantizes for ant c the state $\boldsymbol{x}_c$ into a quantized state $\boldsymbol{q}_c$ using the prespecified quantization bins from $\mathscr{Q}$. A flow diagram of the algorithm is presented in Figure 2. In this figure, everything that happens in the outer loop is the trial and the inner loops represent the interaction steps with the system for all ants in parallel.

3.4 ACL with Fuzzy State Space Partitioning

The number of bins required to accurately capture the dynamics of the original system may become very large even for simple systems with only two state variables. Moreover, the time complexity of ACL grows exponentially with the number of bins, making the algorithm infeasible for realistic systems. In particular, note that

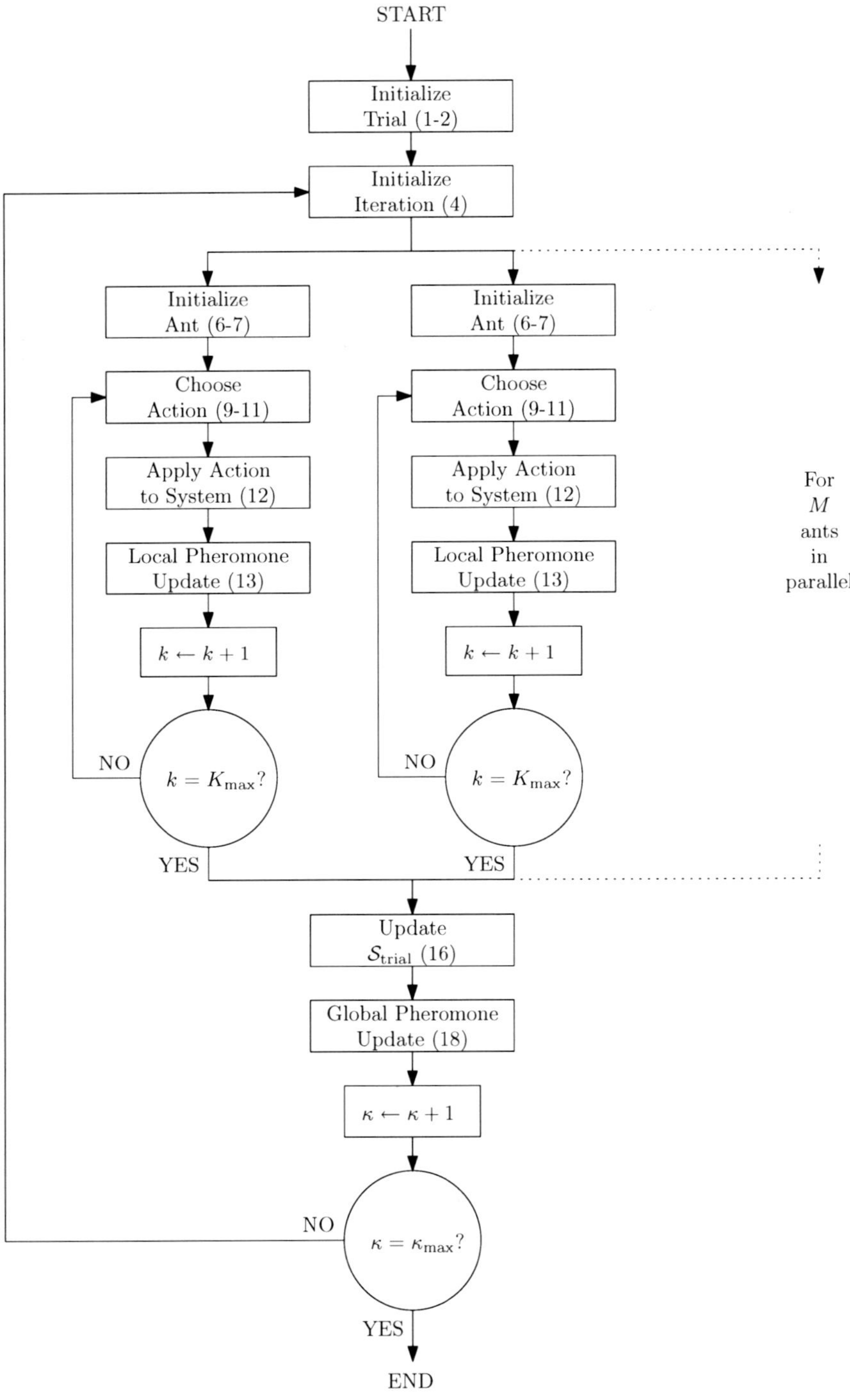

Fig. 2 Flow diagram of the Ant Colony Learning algorithm for optimal control problems. The numbers between parentheses refer to the respective lines of Algorithm 5.

Algorithm 5 The Ant Colony Learning algorithm for optimal control problems

Require: $\mathscr{Q}, \mathscr{U}, \mathscr{X}, \boldsymbol{f}, M, \tau_0, \rho, \gamma, \alpha, K_{\max}, \kappa_{\max}$

$\kappa \leftarrow 0$
$\tau_{\boldsymbol{qu}} \leftarrow \tau_0, \quad \forall(\boldsymbol{q},\boldsymbol{u}) \in \mathscr{Q} \times \mathscr{U}$
repeat
  $k \leftarrow 0$; $\mathscr{S}_{\text{trial}} \leftarrow \emptyset$
  for all ants c in parallel **do**
    $s_{\text{p},c} \leftarrow \emptyset$
    $\boldsymbol{x}_c \leftarrow \text{random}(\mathscr{X})$
    repeat
      $\boldsymbol{q}_c \leftarrow \text{quantize}(\boldsymbol{x}_c)$
      $\boldsymbol{u}_c \sim p_c\{\boldsymbol{u}|\boldsymbol{q}_c\} = \frac{\tau^{\alpha}_{\boldsymbol{q}_c\boldsymbol{u}}}{\sum_{\boldsymbol{l}\in\mathscr{U}_{\boldsymbol{q}_c}} \tau^{\alpha}_{\boldsymbol{q}_c\boldsymbol{l}}}, \quad \forall \boldsymbol{u} \in \mathscr{U}_{\boldsymbol{q}_c}$
      $s_{\text{p},c} \leftarrow s_{\text{p},c} \bigcup \{(\boldsymbol{q}_c,\boldsymbol{u}_c)\}$
      $\boldsymbol{x}_c(k+1) \leftarrow \boldsymbol{f}(\boldsymbol{x}_c(k),\boldsymbol{u}_c)$
      Local pheromone update:
      $\tau_{\boldsymbol{q}_c\boldsymbol{u}_c}(k+1) \leftarrow (1-\gamma)\tau_{\boldsymbol{q}_c\boldsymbol{u}_c}(k) + \gamma\tau_0$
      $k \leftarrow k+1$
    until $k = K_{\max}$
    $\mathscr{S}_{\text{trial}} \leftarrow \mathscr{S}_{\text{trial}} \bigcup \{s_{\text{p},c}\}$
  end for
  Global pheromone update:
  $\tau_{\boldsymbol{qu}}(\kappa+1) \leftarrow (1-\rho)\tau_{\boldsymbol{qu}}(\kappa) + \rho \sum_{\substack{s\in\mathscr{S}_{\text{trial}};\\ (\boldsymbol{q},\boldsymbol{u})\in s}} J^{-1}(s), \quad \forall(\boldsymbol{q},\boldsymbol{u}) : \exists s \in \mathscr{S}_{\text{trial}}; (\boldsymbol{q},\boldsymbol{u}) \in s$
  $\kappa \leftarrow \kappa + 1$
until $\kappa = \kappa_{\max}$

Ensure: $\tau_{\boldsymbol{qu}}, \quad \forall(\boldsymbol{q},\boldsymbol{u}) \in \mathscr{Q} \times \mathscr{U}$

for systems with fast dynamics in certain regions of the state space, the sample time needs to be chosen smaller in order to capture the dynamics in these regions accurately. In other regions of the state space where the dynamics of the system are slower, the faster sampling requires a denser quantization, increasing the number of bins. All together, without much prior knowledge of the system dynamics, both the sampling time and the bin size need to be small enough, resulting in a rapid explosion of the number of bins.

A much better alternative is to approximate the state space by a parameterized function approximator. In that case, there is still a finite number of parameters, but this number can typically be chosen to be much smaller compared to using crisp quantization. The universal function approximator used in this chapter is the fuzzy approximator.

3.4.1 Fuzzy State Space Partitioning

With fuzzy approximation, the domain of each state variable is partitioned using membership functions. We define the membership functions for the state variables to be triangular-shaped such that the membership degrees for any value of the state

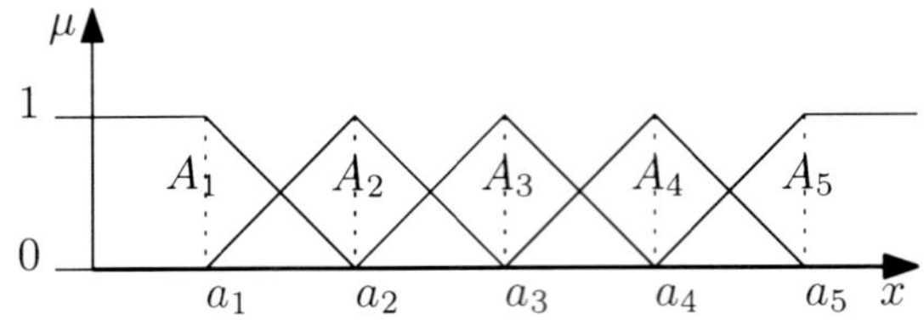

Fig. 3 Membership functions $A_1, \ldots A_5$, with centers $a_1, \ldots, a_5$ on an infinite domain

on the domain always sum up to one. Only the centers of the membership functions have to be stored. An example of such a fuzzy partitioning is given in Figure 3.

Let A_i denote the membership functions for x_1, with a_i their centers for $i = 1, \ldots, N_A$, with N_A the number of membership functions for x_1. Similarly for x_2, denote the membership functions by B_i, with b_i their centers for $i = 1, \ldots, N_B$, with N_B the number of membership functions for x_2. Similarly, the membership functions can be defined for the other state variables in $\boldsymbol{x}$, but for the sake of notation, the discussion in this chapter limits the number to two, without loss of generality. Note that in the example in Sect. 4 the order of the system is four.

The membership degree of A_i and B_i are respectively denoted by $\mu_{A_i}(x_1(k))$ and $\mu_{B_i}(x_2(k))$ for a specific value of the state at time k. The degree of fulfillment is computed by multiplying the two membership degrees:

$$\beta_{ij}(\boldsymbol{x}(k)) = \mu_{A_i}(x_1(k)) \cdot \mu_{B_j}(x_2(k)).$$

Let the vector of all degrees of fulfillment for a certain state at time k be denoted by:

$$\begin{aligned}\boldsymbol{\beta}(\boldsymbol{x}(k)) = [&\beta_{11}(\boldsymbol{x}(k)) \quad \beta_{12}(\boldsymbol{x}(k)) \quad \ldots \quad \beta_{1N_B}(\boldsymbol{x}(k)) \\ &\beta_{21}(\boldsymbol{x}(k)) \quad \beta_{22}(\boldsymbol{x}(k)) \quad \ldots \quad \beta_{2N_B}(\boldsymbol{x}(k)) \\ &\ldots \quad \beta_{N_A N_B}(\boldsymbol{x}(k))]^{\mathrm{T}},\end{aligned} \tag{10}$$

which is a vector containing β_{ij} for all combinations of i and j. Each element will be associated to a vertex in the graph used by Fuzzy ACL. Most of the steps taken from the AS and ACS algorithms for dealing with the $\boldsymbol{\beta}(\boldsymbol{x}(k))$ vectors from (10) need to be reconsidered. This is the subject of the following section.

3.4.2 Fuzzy ACL Algorithm

In the crisp version of ACL, all combinations of these quantized states for the different state variables corresponded to the nodes in the graph and the arcs corresponded to transitions from one quantized state to another. Because of the quantization, the resulting system was transformed into a stochastic decision problem. However, the pheromones were associated to these arcs as usual. In the fuzzy case, the state space is partitioned by membership functions and the combination of the indices to these membership functions for the different state variables corresponds to the nodes in the construction graph. With the fuzzy interpolation, the system remains a deterministic decision problem, but the transition from node to node now does not directly correspond to a state transition. The pheromones are associated to the arcs as usual, but the updating needs to take into account the degree of fulfillment of the associated membership functions.

In Fuzzy ACL, which we introduced in [12], an ant is not assigned to a certain vertex at a certain time, but to all vertices according to some degree of fulfillment at the same time. For this reason, a pheromone τ_{ij} is now denoted as $\tau_{i\boldsymbol{u}}$ with i the index of the vertex (i.e. the corresponding element of $\boldsymbol{\beta}$) and $\boldsymbol{u}$ the action. Similar to the definition of the vector of all degrees of fulfillment in (10), the vector of all pheromones for a certain action $\boldsymbol{u}$ at time k is denoted as:

$$\boldsymbol{\tau}_{\boldsymbol{u}}(k) = \left[\tau_{1\boldsymbol{u}}(k)\ \tau_{2\boldsymbol{u}}(k)\ \dots\ \tau_{N_A N_B \boldsymbol{u}}(k)\right]^{\mathrm{T}}. \tag{11}$$

Action Selection

The action is chosen randomly according to the probability distribution:

$$p_c\{\boldsymbol{u}|\boldsymbol{\beta}_c(k)\}(k) = \sum_{i=1}^{N_A N_B} \beta_{c,i}(k) \frac{\tau_{i,\boldsymbol{u}}^{\alpha}(k)}{\sum_{\boldsymbol{\ell}\in\mathscr{U}} \tau_{i,\boldsymbol{\ell}}^{\alpha}(k)}. \tag{12}$$

Note that when $\boldsymbol{\beta}_c$ contains exactly one 1 and for the rest only zeros, this would correspond to the crisp case, where the state is quantized to a set of bins and (12) then reduces to (7).

Local Pheromone Update

The local pheromone update from (3) can be modified to the fuzzy case as follows:

$$\begin{aligned} \boldsymbol{\tau}_{\boldsymbol{u}} &\leftarrow \boldsymbol{\tau}_{\boldsymbol{u}}(1-\boldsymbol{\beta}) + ((1-\gamma)\boldsymbol{\tau}_{\boldsymbol{u}} + \gamma\tau_0)\boldsymbol{\beta} \\ &= \boldsymbol{\tau}_{\boldsymbol{u}}(1-\gamma\boldsymbol{\beta}) + \tau_0(\gamma\boldsymbol{\beta}), \end{aligned} \tag{13}$$

where all operations are performed element-wise.

As all ants update the pheromone levels associated with the state just visited and the action just taken in parallel, one may wonder whether or not the order in which the updates are done matters when the algorithm is executed on a standard CPU, where all operations are done in series. With crisp quantization the ants may indeed sometimes visit the same state and with fuzzy quantization the ants may very well share some of the membership functions with a membership degree larger than zero. We will show that in both cases the order of updates in series does not influence the final value of the pheromones after the joint update. In the crisp case, the local pheromone update from (8) may be rewritten as follows:

$$\begin{aligned} \tau_{q\boldsymbol{u}}^{(1)} &\leftarrow (1-\gamma)\tau_{q\boldsymbol{u}} + \gamma\tau_0 \\ &= (1-\gamma)(\tau_{q\boldsymbol{u}} - \tau_0) + \tau_0. \end{aligned}$$

Now, when a second ant updates the same pheromone level $\tau_{\boldsymbol{qu}}$, the pheromone level becomes:

$$\begin{aligned}\tau_{\boldsymbol{qu}}^{(2)} &\leftarrow (1-\gamma)(\tau_{\boldsymbol{qu}}^{(1)}-\tau_0)+\tau_0\\ &=(1-\gamma)^2(\tau_{\boldsymbol{qu}}-\tau_0)+\tau_0.\end{aligned}$$

After n updates, the pheromone level is:

$$\tau_{\boldsymbol{qu}}^{(n)} \leftarrow (1-\gamma)^n(\tau_{\boldsymbol{qu}}-\tau_0)+\tau_0, \tag{14}$$

which shows that the order of the update is of no influence to the final value of the pheromone level.

For the fuzzy case a similar derivation can be made. In general, after all the ants have performed the update, the pheromone vector is:

$$\boldsymbol{\tau}_{\boldsymbol{u}} \leftarrow \left(\prod_{c=1}^{M}(1-\gamma\boldsymbol{\beta}_c)\right)(\boldsymbol{\tau}_{\boldsymbol{u}}-\tau_0)+\tau_0, \tag{15}$$

where again all operations are performed element-wise. This result also reveals that the final values of the pheromones are invariant with respect to the order of updates. Furthermore, also note that when $\boldsymbol{\beta}_c$ contains exactly one 1 and for the rest only zeros, corresponding to the crisp case, the fuzzy local pheromone update from either (13) or (15) reduces to the crisp case in respectively (8) or (14).

Global Pheromone Update

In order to derive a fuzzy representation of the global pheromone update step, it is convenient to rewrite (9) using indicator functions:

$$\begin{aligned}\tau_{\boldsymbol{qu}}(\kappa+1) \leftarrow &\left\{(1-\rho)\tau_{\boldsymbol{qu}}(\kappa)+\rho\sum_{s\in\mathscr{S}_{\text{upd}}}J^{-1}(s)\mathscr{I}_{\boldsymbol{qu},s}\right\}\mathscr{I}_{\boldsymbol{qu},\mathscr{S}_{\text{upd}}}\\ &+\tau_{\boldsymbol{qu}}(\kappa)(1-\mathscr{I}_{\boldsymbol{qu},\mathscr{S}_{\text{upd}}}).\end{aligned} \tag{16}$$

In (16), $\mathscr{I}_{\boldsymbol{qu},\mathscr{S}_{\text{upd}}}$ is an indicator function that can take values from the set $\{0,1\}$ and is equal to 1 when $(\boldsymbol{q},\boldsymbol{u})$ is an element of a solution belonging to the set $\mathscr{S}_{\text{upd}}$ and 0 otherwise. Similarly, $\mathscr{I}_{\boldsymbol{qu},s}$ is an indicator function as well and is equal to 1 when $(\boldsymbol{q},\boldsymbol{u})$ is an element of the solutions s and 0 otherwise.

Now, by using the vector of pheromones from (11), we introduce the vector indicator function $\mathscr{I}_{\boldsymbol{u},s}$, being a vector of the same size as $\boldsymbol{\tau}_u$, with elements from the set $\{0,1\}$ and for each state indicating whether it is an element of s. A similar vector indicator function $\mathscr{I}_{\boldsymbol{u},\mathscr{S}_{\text{trial}}}$ is introduced as well. Using these notations, we can write (16) for all states q together as:

$$\boldsymbol{\tau}_{\boldsymbol{u}}(\kappa+1) \leftarrow \left\{ (1-\rho)\boldsymbol{\tau}_{\boldsymbol{u}}(\kappa) + \rho \sum_{s\in\mathscr{S}_{\text{upd}}} J^{-1}(s)\mathscr{I}_{\boldsymbol{u},s} \right\} \mathscr{I}_{\boldsymbol{u},\mathscr{S}_{\text{upd}}} + \boldsymbol{\tau}_{\boldsymbol{u}}(\kappa)(1-\mathscr{I}_{\boldsymbol{u},\mathscr{S}_{\text{upd}}}), \tag{17}$$

where all multiplications are performed element-wise.

The most basic vector indicator function is $\mathscr{I}_{\boldsymbol{u},s^{(i)}}$, which has only one element equal to 1, namely the one for $(\boldsymbol{q},\boldsymbol{u}) = s^{(i)}$. Recall that a solution $s = \{s^{(1)}, s^{(2)}, \ldots, s^{(N_s)}\}$ is an ordered set of solution components $s^{(i)} = (\boldsymbol{q}_i, \boldsymbol{u}_i)$. Now, $\mathscr{I}_{\boldsymbol{u},s}$ can be created by taking the union of all $\mathscr{I}_{\boldsymbol{u},s^{(i)}}$:

$$\mathscr{I}_{\boldsymbol{u},s} = \mathscr{I}_{\boldsymbol{u},s^{(1)}} \cup \mathscr{I}_{\boldsymbol{u},s^{(2)}} \cup \ldots \cup \mathscr{I}_{\boldsymbol{u},s^{(N_s)}} = \bigcup_{i=1}^{N_s} \mathscr{I}_{\boldsymbol{u},s^{(i)}}, \tag{18}$$

where we define the union of indicator functions in the light of the fuzzy representation of the state by a fuzzy union set operator, such as:

$$(A \cup B)(x) = \max[\mu_A(x), \mu_B(x)]. \tag{19}$$

In this operator, where A and B are membership functions, x a variable and $\mu_A(x)$ the degree to which x belongs to A. Note that when A maps x to a crisp domain $\{0,1\}$, the union operator is still valid. Similar to (18), if we denote the set of solutions as the ordered set of solutions $\mathscr{S}_{\text{upd}} = \{s_1, s_2, \ldots, s_{N_S}\}$, $\mathscr{I}_{\boldsymbol{u},\mathscr{S}_{\text{trial}}}$ can be created by taking the union of all $\mathscr{I}_{\boldsymbol{u},s}$:

$$\mathscr{I}_{\boldsymbol{u},\mathscr{S}_{\text{trial}}} = \mathscr{I}_{\boldsymbol{u},s_1} \cup \mathscr{I}_{\boldsymbol{u},s_2} \cup \ldots \cup \mathscr{I}_{\boldsymbol{u},s_{N_S}} = \bigcup_{i=1}^{N_S} \mathscr{I}_{\boldsymbol{u},s_i}. \tag{20}$$

In fact, $\mathscr{I}_{\boldsymbol{u},s_i}$ can be regarded as a representation of the state-action pair $(\boldsymbol{q}_i, \boldsymbol{u}_i)$. In computer code, $\mathscr{I}_{\boldsymbol{u},s_i}$ may be represented by a matrix of dimensions $|\mathscr{Q}| \cdot |\mathscr{U}|$ containing only one element equal to 1 and the others zero.[3]

In order to generalize the global pheromone update step to the fuzzy case, realize that $\boldsymbol{\beta}(\boldsymbol{x}(k))$ from (10) can be seen as a generalized (fuzzy, or graded) indicator function $\mathscr{I}_{\boldsymbol{u},s^{(i)}}$, if combined with an action $\boldsymbol{u}$. The elements of this generalized indicator function take values in the range $[0,1]$, for which the extremes 0 and 1 form the special case of the original (crisp) indicator function. We can thus use the state-action pair where the state has the fuzzy representation of (10) directly to compose $\mathscr{I}_{\boldsymbol{u},s^{(i)}}$. Based on this, we can compose $\mathscr{I}_{\boldsymbol{u},s}$ and $\mathscr{I}_{\boldsymbol{u},\mathscr{S}_{\text{trial}}}$ in the same way as in (18) and (20), respectively. With the union operator from (19) and the definition of the various indicator functions, the global pheromone update rule from (17) can be used in both the crisp and fuzzy variant of ACL.

---

[3] In MATLAB this may conveniently be represented by a sparse matrix structure, rendering the indicator-representation useful for generalizing the global pheromone update rule and while still being a memory efficient representation of the state-action pair.

Note that we need more memory to store the fuzzy representation of the state compared to its crisp representation. With pair-wise overlapping normalized membership functions, like the ones shown in Figure 3, at most 2^d elements are nonzero, with d the dimension of the state space. This means an exponentially growth in memory requirements for increasing state space dimension, but also that it is independent of the number of membership functions used for representing the state space.

As explained in Sect. 3.2, for optimal control problems, the appropriate update rule is to use all solutions by all ants in the trial $\mathscr{S}_{\text{trial}}$. In the fuzzy case, the solutions $s \in \mathscr{S}_{\text{trial}}$ consist of sequences of states and actions and the states can be fuzzified so that they are represented by sequences of vectors of degrees of fulfillment $\boldsymbol{\beta}$. Instead of one pheromone level, in the fuzzy case a set of pheromone levels are updated to a certain degree. It can be easily seen that as this update process is just a series of pheromone *deposits* the final value of the pheromone levels relates to the sum of these deposits and is invariant with respect to the order of these deposits. This is also the case for this step in Crisp ACL.

Regarding the terminal condition for the ants, with the fuzzy implementation, none of the vertices can be identified as being the terminal vertex. Rather one has to define a set of membership functions that can be used to determine to what degree the goal state has been reached. These membership functions can be used to express the linguistic fuzzy term of the state being *close to the goal*. Specifically, this is satisfied when the membership degree of the state to the membership function with its core equal to the goal state is larger than 0.5. If this has been satisfied, the ant has terminated its trial.

3.5 Parameter Settings

Some of the parameters in both Crisp and Fuzzy ACL are similarly initialized as in the ACS. The global and local pheromone trail decay factors are set to a preferably small value, respectively $\rho \in (0,1)$ and $\gamma \in (0,1)$. There will be no heuristic parameter associated to the arcs in the construction graph, so only an exponential weighting factor for the pheromone trail $\alpha > 0$ has to be chosen. Increasing α leads to more biased decisions towards the one corresponding to the highest pheromone level. Choosing $\alpha = 2$ or 3 appears to be an appropriate choice. Furthermore, the control parameters for the algorithm need to be chosen, such as the maximum number of steps per trial, K_{max}, and the maximum number of trials, T_{max}. The latter one can be set to, e.g., 100 trials, where the former one depends on the sample time T_{s} and a guess of the time needed to get from the initial state to the goal optimally, T_{guess}. A good choice for K_{max} would be to take 10 times the expected number of steps, $K_{\text{max}} = 10T_{\text{guess}}/T_{\text{s}}$. Specific to ACL, the number and shape of the membership functions, or the number of quantization bins, and their spacing over the state domain need to be determined. Furthermore, the pheromones are initialized as $\tau_{i\boldsymbol{u}} = \tau_0$ for all $(i,\boldsymbol{u})$ in Fuzzy ACL and as $\tau_{\boldsymbol{q}\boldsymbol{u}} = \tau_0$ for all $(\boldsymbol{q},\boldsymbol{u})$ in Crisp ACL and where τ_0 is a small, positive value. Finally, the number of ants M must be chosen large enough such that the complete state space can be visited frequently enough.

3.6 Relation to Reinforcement Learning

In reinforcement learning [26], the agent interacts in a step-wise manner with the system, experiencing state transitions and rewards. These rewards are a measure of the short-time quality of the taken decision and the objective for the agent is to create a state-action mapping (the policy) that maximizes the (discounted) sum of these rewards. Particularly in Monte Carlo learning, the agent interacts with the system and stores the states that it visited and the actions that it took in those states until it reaches some goal state. Now, its trial ends and the agent evaluates the sequence of state-action pairs over some cost function and based on that updates a value function. This value function basically stores the information about the quality of being in a certain state (or in the Q-learning [30] and SARSA [31] variants, the quality of choosing a particular action in a certain state). Repeating this procedure of interacting with the system and updating its value function will eventually result in the value function converging to the optimal value function, which corresponds to the solution of the control problem. It is said that the control policy has been *learned* because it has been automatically derived purely by interaction with the system.

This procedure shows strong similarities with the procedure of ACL. The main difference is that rather than a single agent, in ACL there is a multitude of agents (called ants) all interacting with the system in parallel and all contributing to the update of the value function (called the set of pheromone levels). For exactly the same reason as with reinforcement learning, the procedure in ACL enables the ants to *learn* the optimal control policy.

However, as all ants work in parallel, the interaction of all these ants with a single real-world system is problematic. You would require multiple copies of the system in order to keep up this parallelism. In simulation this is not an issue. If there is a model of the real-world system available, making many copies of this model in software is very cheap. Nevertheless, it is not completely correct to call ACL for this reason a *model-based* learning algorithm. As the model may be a blackbox model, with the details unknown to the ants, it is purely a way to benefit from the parallelism of ACL.

Because of the parallelism of the ants, ACL could be called a multi-agent learning algorithm. However, this must not be confused with multi-agent reinforcement learning as this stands for the setting in which multiple reinforcement learning agents all act in a single environment according to their own objective and, in order for each of them to behave optimally, they need to take into account the behavior of the other agents in the environment. Note that in ACL the ants do not need to consider, or even notice, the existence of the other ants. All they do is benefit from each other's findings.

4 Example: Navigation with Variable Damping

This section presents an example application of both Crisp and Fuzzy ACL to a continuous-state dynamic system. The dynamic system under consideration is a simulated two-dimensional (2D) navigation problem and similar to the one described in [24]. Note that it is not our purpose to demonstrate the superiority of ACL over any other method for this specific problem. Rather we want to demonstrate the functioning of the algorithm and compare the results for both its versions.

4.1 Problem Formulation

A vehicle, modeled as a point-mass of 1 kg, has to be steered to the origin of a two-dimensional surface from any given initial position in an optimal manner. The vehicle experiences a damping that varies nonlinearly over the surface. The state of the vehicle is defined as $\boldsymbol{x} = \begin{bmatrix} c_1 \; v_1 \; c_2 \; v_2 \end{bmatrix}^{\mathrm{T}}$, with c_1, c_2 and v_1, v_2 the position and velocity in the direction of each of the two principal axes, respectively. The control input to the system $\boldsymbol{u} = \begin{bmatrix} u_1 \; u_2 \end{bmatrix}^{\mathrm{T}}$ is a two-dimensional force. The dynamics are:

$$\dot{\boldsymbol{x}}(t) = \begin{bmatrix} 0 & 1 & 0 & 0 \\ 0 & -b(c_1,c_2) & 0 & 0 \\ 0 & 0 & 0 & 1 \\ 0 & 0 & 0 & -b(c_1,c_2) \end{bmatrix} \boldsymbol{x}(t) + \begin{bmatrix} 0 & 0 \\ 1 & 0 \\ 0 & 0 \\ 0 & 1 \end{bmatrix} \boldsymbol{u}(t),$$

where the damping $b(c_1,c_2)$ in the experiments is modeled by an affine sum of two Gaussian functions:

$$b(c_1,c_2) = b_0 + \sum_{i=1}^{2} b_i \exp\left[- \sum_{j=1}^{2} \frac{(c_j - m_{j,i})^2}{\sigma_{j,i}^2} \right],$$

for which the chosen values of the parameters are $b_0 = 0.5$, $b_1 = b_2 = 8$, and $m_{1,1} = 0$, $m_{2,1} = -2.3$, $\sigma_{1,1} = 2.5$, $\sigma_{2,1} = 1.5$ for the first Gaussian, and $m_{1,2} = 4.7$, $m_{2,2} = 1$, $\sigma_{1,2} = 1.5$, $\sigma_{2,2} = 2$ for the second Gaussian. The damping profile can be seen in, e.g., Figure 7, where darker shading means more damping.

4.2 State Space Partitioning and Parameters

The partitioning of the state space is as follows:

- For the position c_1: $\{-5, -3.75, -2, -0.3, 0, 0.3, 2, 2.45, 3.5, 3.75, 4.7, 5\}$ and for c_2: $\{-5, -4.55, -3.5, -2.3, -2, -1.1, -0.6, -0.3, 0, 0.3, 1, 2.6, 4, 5\}$.
- For the velocity in both dimensions v_1, v_2: $\{-2, 0, 2\}$.

This particular partitioning of the position space is composed of a baseline grid $\{-5, -0.3, 0, 0.3, 5\}$ and adding to it, extra grid lines inside each Gaussian damping region. This partitioning is identical to what is used in [24]. In the crisp case, the crossings in the grid are the centers of the quantization bins, whereas in the fuzzy case, these are the centers of the triangular membership functions. The action set consists of 9 actions, namely the cross-product of the sets $\{-1,0,1\}$ for both dimensions. The local and global pheromone decay factors are respectively $\gamma = 0.01$ and $\rho = 0.1$. Furthermore, $\alpha = 3$ and the number of ants is 200. The sampling time is $T_s = 0.2$ and the ants are randomly initialized over the complete state space at the start of each trial. An ant terminates its trial when its position and velocity in both dimensions are within a bound of ± 0.25 and ± 0.05 from the goal respectively. Each experiment is carried out 30 times. The quadratic cost function from (5) is used with the matrices $\mathbf{Q} = \text{diag}(0.2, 0.1, 0.2, 0.1)$ and $\mathbf{R} = 0$. The cost function thus only takes into account the deviation of the state to the goal. We will run the fuzzy Q-iteration algorithm from [24] for this problem with the same state space partitioning in order to derive the optimal policy to which we can compare the policies derived by both versions of the ACL algorithm.

4.3 *Results*

The first performance measure considered represents the cost of the control policy as a function of the number of trials. The cost of the control policy is measured as the average cost of a set of trajectories resulting from simulating the system controlled by the policy and starting from a set of 100 pre-defined initial states, uniformly distributed over the state space. Figure 4 shows these plots. As said earlier, each experiment is carried out 30 times. The black line in the plots shows the average cost over these 30 experiments and the gray area represents the range of costs for the thirty experiments. From Figure 4 it can be concluded that the Crisp ACL algorithm does not converge as fast and smoothly compared to Fuzzy ACL and that it converges to a larger average cost. Furthermore, the gray region for Crisp ACL is

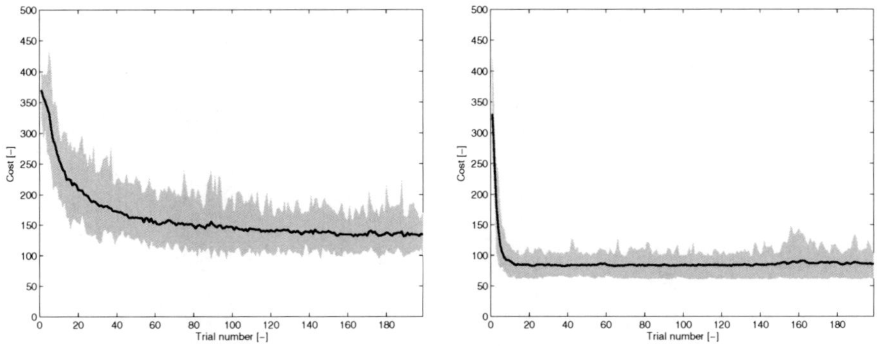

(a) Crisp ACL (b) Fuzzy ACL

Fig. 4 The cost of the control policy as a function of the number of trials passed since the start of the experiment. The black line in each plot is the average cost over 30 experiments and the gray area represents the range of costs for these experiments.

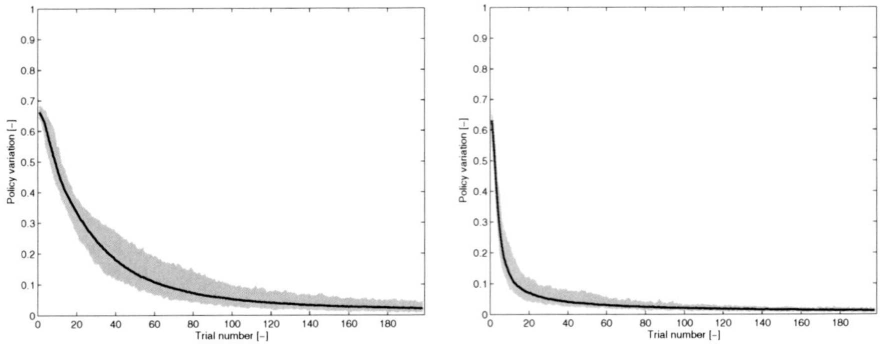

(a) Crisp ACL (b) Fuzzy ACL

Fig. 5 Convergence of the algorithm in terms of the fraction of states (cores of the membership functions or centers of the quantization bins) for which the policy changed at the end of a trial. The black line in each plot is the average policy variation over 30 experiments and the gray area represents the range of policy variation for these experiments.

larger than that of Fuzzy ACL, meaning that there is a larger variety of policies that the algorithm converges to.

The convergence of the algorithm can also be measured by the fraction of quantized states (crisp version), or cores of the membership functions (fuzzy version) for which the policy has changed at the end of a trial. This measure will be called the *policy variation*. The policy variation as a function of the trials for both Crisp and Fuzzy ACL is depicted in Figure 5. It shows that for both ACL versions the policy variation never really becomes completely zero and confirms that Crisp ACL converges slower compared to Fuzzy ACL. It also provides the insight that although

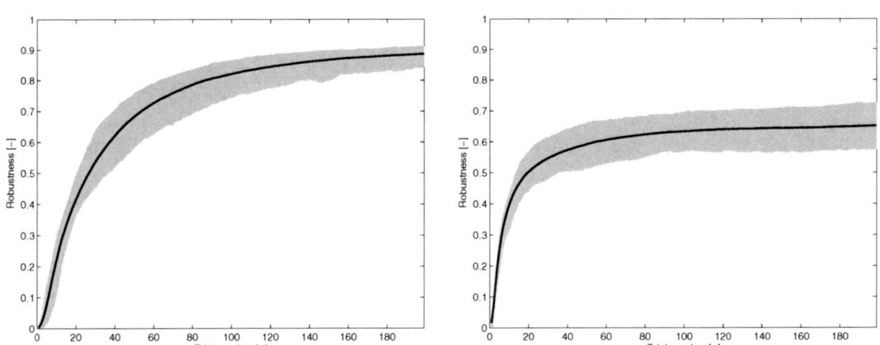

(a) Crisp ACL (b) Fuzzy ACL

Fig. 6 Evolution of the robustness of the control policy as a function of the number of trials. The black line in each plot is the average robustness over 30 experiments and the gray area represents the range of the robustness for these experiments.

for Crisp ACL the pheromone levels do not converge to the same value at every run of the algorithm, they do all converge in about the same way in the sense of policy variation. This is even more true for Fuzzy ACL, where the gray area around the average policy variation almost completely disappears for an increasing number of trials. The observation that the policy variation never becomes completely zero indicates that there are some states for which either action results in nearly the same cost.

A third performance measure that is considered represents the robustness of the control policy. For each state, there is a certain number of actions to choose from. If the current policy indicates that a certain action is the best, it does not say how much better it is compared to the second-best action. If it is only a little better, a small update of the pheromone level for that other action in this state may switch the policy for this state. In that case, the policy is not considered to be very robust. The robustness for a certain state is defined as follows:

$$
\begin{aligned}
\boldsymbol{u}_{\max 1}(\boldsymbol{q}) &= \arg \max_{\boldsymbol{u} \in \mathscr{U}} (\tau_{\boldsymbol{qu}}), \\
\boldsymbol{u}_{\max 2}(\boldsymbol{q}) &= \arg \max_{\boldsymbol{u} \in \mathscr{U} \setminus \{\boldsymbol{u}_{\max 1}(\boldsymbol{q})\}} (\tau_{\boldsymbol{qu}}), \\
\text{robustness}(\boldsymbol{q}) &= \frac{\tau^{\alpha}_{\boldsymbol{q}\boldsymbol{u}_{\max 1}(\boldsymbol{q})} - \tau^{\alpha}_{\boldsymbol{q}\boldsymbol{u}_{\max 2}(\boldsymbol{q})}}{\tau^{\alpha}_{\boldsymbol{q}\boldsymbol{u}_{\max 1}(\boldsymbol{q})}},
\end{aligned}
$$

where α is the same as the α from the Boltzmann action selection rule ((12) and (7)). The robustness of the policy is the average robustness over all states. Figure 6 shows the evolution of the average robustness during the experiments. It also shows that the robustness increases with the number of trials, explaining the decrease in the policy variation from Figure 5. The fact that it does not converge to one explains that there remains a probability larger than zero for the policy to change. This suggests that ACL algorithms are potentially capable of adapting the policy to changes in the cost function due to changing system dynamics or a change of the goal state.

Finally, we present the policies and the behavior of a simulated vehicle controlled by these policies for both Crisp and Fuzzy ACL. In the case of Crisp ACL, Figure 7(a) depicts a slice of the resulted policy for zero velocity. It shows the mapping of the positions in both dimensions to the input. Figure 7(b) presents the trajectories of the vehicle for various initial positions and zero initial velocity. In both figures, the mapping is shown for a grid three times finer than the grid spanned by the cores of the membership functions. For the crisp case, the states in between the centers of the partition bins are quantized to the nearest center. Figure 8 presents the results for Fuzzy ACL. Comparing these results with those from the Crisp ACL algorithm in Figure 7, it shows that for Crisp ACL, the trajectories of the vehicle go widely around the regions of stronger damping, in one case (starting from the top-left corner) even clearly taking a suboptimal path to the goal. For Fuzzy ACL, the trajectories are very smooth and seem to be more optimal; however, the vehicle does not avoid the regions of stronger damping much. The policy of Fuzzy ACL is also much more regular compared to the one obtained by Crisp ACL.

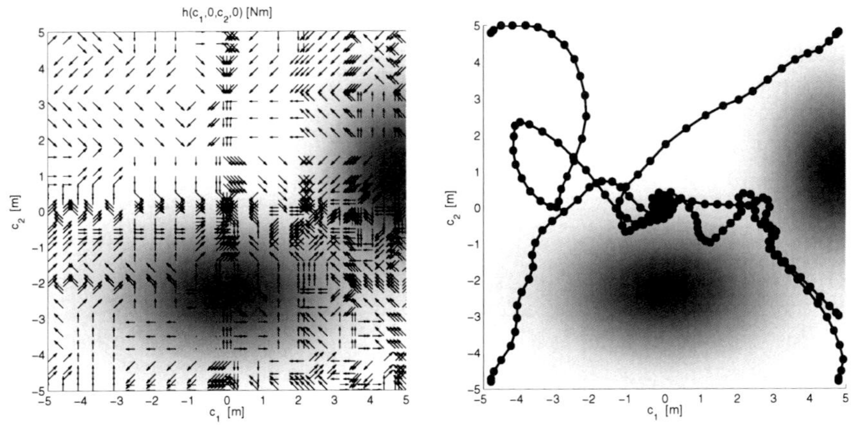

(a) Crisp ACL: policy (b) Crisp ACL: simulation

Fig. 7 (a) A slice of the best resulting policy for zero velocity obtained by Crisp ACL in one of the thirty runs. It shows the control input for a fine grid of positions. The multi-Gaussian damping profile is shown, where darker shades represent regions of more damping. (b) The trajectories of the vehicle under this policy for various initial positions and zero initial velocity. The markers indicate the positions at twice the sampling time.

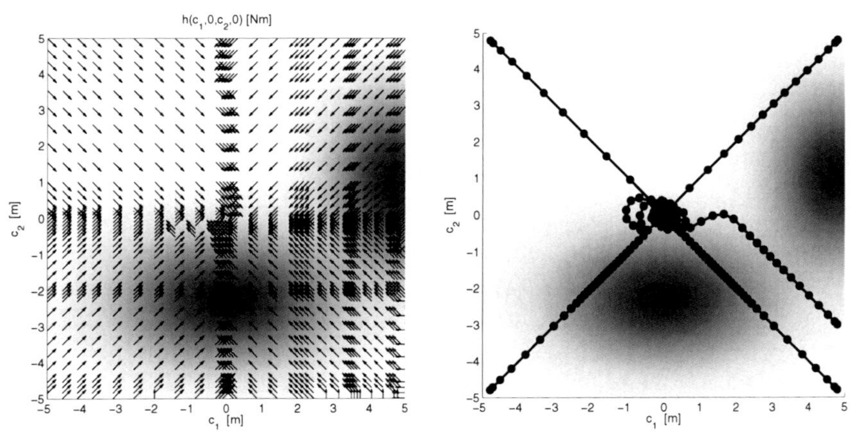

(a) Fuzzy ACL: policy (b) Fuzzy ACL: simulation

Fig. 8 (a) A slice of the best resulting policy for zero velocity obtained by Fuzzy ACL in one of the thirty runs. It shows the control input for a fine grid of positions. The multi-Gaussian damping profile is shown, where darker shades represent regions of more damping. (b) The trajectories of the vehicle under this policy for various initial positions and zero initial velocity. The markers indicate the positions at twice the sampling time.

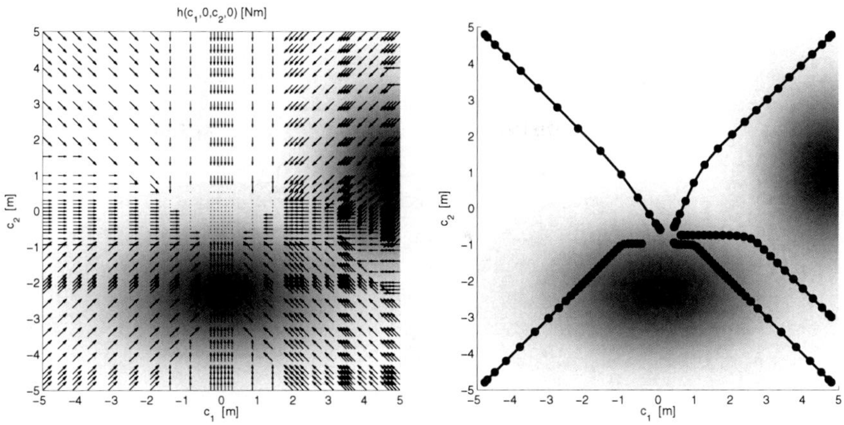

(a) Fuzzy Q-iteration: Policy

(b) Fuzzy Q-iteration: simulation

Fig. 9 The baseline optimal policy derived by the fuzzy Q-iteration algorithm. (a) A slice of the policy for zero velocity. (b) The trajectories of the vehicle under this policy for various initial positions and zero initial velocity. The markers indicate the positions at twice the sampling time.

In order to verify the optimality of the resulting policies, we also present the optimal policy and trajectories obtained by the fuzzy Q-iteration algorithm from [24]. This algorithm is a model-based reinforcement learning method developed for continuous state spaces and is guaranteed to converge to the optimal policy on the partitioning grid. Figure 9 presents the results for fuzzy Q-iteration. It can be seen that the optimal policy from fuzzy Q-iteration is very similar to the one obtained by Fuzzy ACL, but that it manages to avoid the regions of stronger damping somewhat better. Even with the optimal policy, the regions of stronger damping are not completely correctly avoided. Especially regarding the policy around the lower damping region, where it steers the vehicle towards an even stronger damped part of the state space. The reason for this is the particular choice of the cost function in these experiments. The results in [24] show that for a discontinuous cost function, where reaching the goal results in a sudden and steep decrease in cost, the resulting policy avoids the highly damped regions much better. Also note that in those experiments the policy interpolates between the actions resulting in a much smoother change of the policy between the centers of the membership functions. Theoretically, Fuzzy ACL also allows for interpolation of the action set, but this is outside the scope of this chapter. Also, the ACL framework allows for different cost functions than the quadratic cost function used in this example. Designing the cost function carefully is very important for any optimization or control problem and directly influences the optimal policy. It is however not the purpose of this chapter to find the best cost function for this particular control problem.

5 Conclusions and Future Work

Ant Colony Learning (ACL), a novel method for the design of optimal control policies for continuous-state dynamic systems based on Ant Colony Optimization (ACO), has been discussed in this chapter. The partitioning of the state space is a crucial aspect to ACL and two versions have been presented with respect to this. In Crisp ACL, the state space is partitioned using bins such that each value of the state maps to exactly one bin. Fuzzy ACL, on the other hand, uses a partitioning of the state space with triangular membership functions. In this case, each value of the state maps to the membership functions to a certain membership degree. Both ACL algorithms are based on the combination of the Ant System and the Ant Colony System, but have some distinctive characteristics. The action selection step does not include an explicit probability of choosing the action associated with the highest value of the pheromone matrix in a given state, but only uses a random-proportional rule. In this rule, the heuristic variable, typically present in ACO algorithms, is left out as prior knowledge can also be incorporated through an initial choice of the pheromone levels. The absence of the heuristic variable makes ACL more straightforward to apply. Another particular characteristic of our algorithm is that it uses the complete set of solutions of the ants in each trial to update the pheromone matrix, whereas most ACO algorithms typically use some form of elitism, selecting only one of the solutions from the set of ants to perform the global pheromone update. In a control setting, this is not possible as ants initialized closer to the goal will have experienced a lower cost than ants initialized further away from the goal, but this does not mean that they must be favored in the update of the pheromone matrix.

The applicability of ACL to optimal control problems with continuous-valued states is outlined and demonstrated on the nonlinear control problem of two-dimensional navigation with variable damping. The results show convergence of both the crisp and fuzzy versions of the algorithm to suboptimal policies. However, Fuzzy ACL converged much faster and the cost of its resulting policy did not change as much over repetitive runs of the algorithm compared to Crisp ACL. It also converged to a more optimal policy than Crisp ACL. We have presented the additional performance measures of policy variation and robustness in order to study the convergence of the pheromone levels in relation to the policy. These measures showed that ACL converges as a function of the number of trials to an increasingly robust control policy, meaning that a small change in the pheromone levels will not result in a large change in the policy. Compared to the optimal policy derived from fuzzy Q-iteration, the policy from Fuzzy ACL appeared to be close to optimal. It was noted that the policy could be improved further by choosing a different cost function. Interpolation of the action space is another way to get a smoother transition of the policy between the centers of the state space partitioning. Both these issues were not considered in more detail in this chapter and we recommend to look further into this in future research.

The ACL algorithm currently did not incorporate prior knowledge of the system, or of the optimal control policy. However, sometimes prior knowledge of such kind is available and using it could be a way to speed up the convergence of the algorithm and to increase the optimality of the derived policy. Future research must focus on this issue. Comparing ACL to other related methods, like for instance Q-learning, is also part of our plans for future research. Finally, future research must focus on a more theoretical analysis of the algorithm, such as a formal convergence proof.

References

1. Dorigo, M., Blum, C.: Ant colony optimization theory: a survey. Theoretical Computer Science 344(2-3), 243–278 (2005)
2. Colorni, A., Dorigo, M., Maniezzo, V.: Distributed optimization by ant colonies. In: Varela, F.J., Bourgine, P. (eds.) Towards a Practice of Autonomous Systems: Proceedings of the First European Conference on Artificial Life, pp. 134–142. MIT Press, Cambridge (1992)
3. Dorigo, M., Stützle, T.: Ant Colony Optimization. The MIT Press, Cambridge (2004)
4. Sim, K.M., Sun, W.H.: Ant colony optimization for routing and load-balancing: survey and new directions. IEEE Transactions on Systems, Man, Cybernetics, Part A 33(5), 560–572 (2003)
5. Huang, R.H., Yang, C.L.: Ant colony system for job shop scheduling with time windows. International Journal of Advanced Manufacturing Technology 39(1-2), 151–157 (2008)
6. Alaykran, K., Engin, O., Dyen, A.: Using ant colony optimization to solve hybrid flow shop scheduling problems. International Journal of Advanced Manufacturing Technology 35(5-6), 541–550 (2007)
7. Fan, X., Luo, X., Yi, S., Yang, S., Zhang, H.: Optimal path planning for mobile robots based on intensified ant colony optimization algorithm. In: Proceedings of the IEEE International Conference on Robotics, Intelligent Systems and Signal Processing (RISSP 2003), Changsha, Hunan, China, October 2003, pp. 131–136 (2003)
8. Wang, J., Osagie, E., Thulasiraman, P., Thulasiram, R.K.: HOPNET: A hybrid ant colony optimization routing algorithm for mobile ad hoc network. Ad Hoc Networks 7(4), 690–705 (2009)
9. Purnamadjaja, A.H., Russell, R.A.: Pheromone communication in a robot swarm: necrophoric bee behaviour and its replication. Robotica 23(6), 731–742 (2005)
10. Fox, B., Xiang, W., Lee, H.P.: Industrial applications of the ant colony optimization algorithm. International Journal of Advanced Manufacturing Technology 31(7-8), 805–814 (2007)
11. Bianchi, L., Dorigo, M., Gambardella, L.M., Gutjahr, W.J.: Metaheuristics in stochastic combinatorial optimization: a survey. IDSIA, Manno, Switzerland, Tech. Rep. 08 (March 2006)
12. van Ast, J.M., Babuška, R., De Schutter, B.: Novel ant colony optimization approach to optimal control. International Journal of Intelligent Computing and Cybernetics 2(3), 414–434 (2009)
13. van Ast, J.M., Babuška, R., De Schutter, B.: Fuzzy ant colony optimization for optimal control. In: Proceedings of the American Control Conference (ACC 2009), Saint Louis, MO, USA, June 2009, pp. 1003–1008 (2009)
14. Socha, K., Blum, C.: An ant colony optimization algorithm for continuous optimization: application to feed-forward neural network training. Neural Computing & Applications 16(3), 235–247 (2007)
15. Socha, K., Dorigo, M.: Ant colony optimization for continuous domains. European Journal of Operational Research 185(3), 1155–1173 (2008)

16. Bilchev, G., Parmee, I.C.: The ant colony metaphor for searching continuous design spaces. In: Fogarty, T. (ed.) AISB-WS 1995. LNCS, vol. 993, pp. 25–39. Springer, Heidelberg (1995)
17. Tsutsui, S., Pelikan, M., Ghosh, A.: Performance of aggregation pheromone system on unimodal and multimodal problems. In: Proceedings of the 2005 Congress on Evolutionary Computation (CEC 2005), Edinburgh, Scotland, September 2005, pp. 880–887 (2005)
18. Korosec, P., Silc, J., Oblak, K., Kosel, F.: The differential ant-stigmergy algorithm: an experimental evaluation and a real-world application. In: Proceedings of the 2007 Congress on Evolutionary Computation (CEC 2007), Singapore, September 2007, pp. 157–164 (2007)
19. Birattari, M., Caro, G.D., Dorigo, M.: Toward the formal foundation of Ant Programming. In: Proceedings of the International Workshop on Ant Algorithms (ANTS 2002), pp. 188–201. Springer, Brussels (2002)
20. Gambardella, L.M., Dorigo, M.: Ant-Q: A reinforcement learning approach to the traveling salesman problem. In: Prieditis, A., Russell, S. (eds.) Machine Learning: Proceedings of the Twelfth International Conference on Machine Learning, pp. 252–260. Morgan Kaufmann Publishers, San Francisco (1995)
21. Casillas, J., Cordn, O., Herrera, F.: Learning fuzzy rule-based systems using ant colony optimization algorithms. In: Proceedings of the ANTS 2000. From Ant Colonies to Artificial Ants: Second International Workshop on Ant Algorithms, Brussels, Belgium, September 2000, pp. 13–21 (2000)
22. Zhao, B., Li, S.: Design of a fuzzy logic controller by ant colony algorithm with application to an inverted pendulum system. In: Proceedings of the IEEE International Conference on Systems, Man, Cybernetics, Taipei, Taiwan, October 2006, pp. 3790–3794 (2006)
23. Zhu, W., Chen, J., Zhu, B.: Optimal design of fuzzy controller based on ant colony algorithms. In: Proceedings of the IEEE International Conference on Mechatronics and Automation, Luoyang, China, June 2006, pp. 1603–1607 (2006)
24. Buşoniu, L., Ernst, D., De Schutter, B., Babuška, R.: Continuous-state reinforcement learning with fuzzy approximation. In: Tuyls, K., Nowe, A., Guessoum, Z., Kudenko, D. (eds.) ALAMAS 2005, ALAMAS 2006, and ALAMAS 2007. LNCS (LNAI), vol. 4865, pp. 27–43. Springer, Heidelberg (2008)
25. Dorigo, M., Maniezzo, V., Colorni, A.: Ant system: optimization by a colony of cooperating agents. IEEE Transactions on Systems, Man, and Cybernetics, Part B 26(1), 29–41 (1996)
26. Sutton, R.S., Barto, A.G.: Reinforcement Learning: An Introduction. MIT Press, Cambridge (1998)
27. Dorigo, M., Gambardella, L.: Ant Colony System: a cooperative learning approach to the traveling salesman problem. IEEE Transactions on Evolutionary Computation 1(1), 53–66 (1997)
28. Stützle, T., Hoos, U.: MAX MIN Ant System. Journal of Future Generation Computer Systems 16, 889–914 (2000)
29. Åström, K.J., Wittenmark, B.: Computer Controlled Systems—Theory and Design. Prentice-Hall, Englewood Cliffs (1990)
30. Watkins, C., Dayan, P.: Q-learning. Machine Learning 8(3-4), 279–292 (1992)
31. Rummery, G.A., Niranjan, M.: On-line Q-learning using connectionist systems. Cambridge University, Tech. Rep. CUED/F-INFENG/TR166 (1994)

Map-Based Support for Effective Collaboration in Micro-mobile Virtual Teams

Guido te Brake and Rick van der Kleij

1 Introduction

Teamwork is important in many organizations. When a task is assigned to a team rather than to an individual, there are several benefits, hence the appealing maxim "two heads are better than one." Sometimes the job requires bringing people together who are dispersed across different geographical locations to work on a common task using information and communication technologies. These teams are often called virtual teams [36].

Virtual teams have a number of benefits that make them attractive to a wide range of public and private organizations. For example, virtual teams can be composed of the best professionals for the task regardless of their physical location, thus potentially enhancing the quality of the distributed teamwork. Furthermore, global virtual teams can pass work from east to west as the earth rotates to maintain the team's effort or momentum [15, 27]. But not all virtual teams are global in nature. Virtual teams can also be composed of members that are situated in close vicinity, such as police officers individually patrolling the same neighborhood, but in other areas. When a crime is in progress in their neighborhood they need to temporarily team up and work together using communication and information technologies. These teams pose different research challenges, because team members often physically encounter each other during the day, and do not experience cultural or time-zone difficulties. To make these teams more effective and to counter the specific problems that they face, technical, cognitive, and organizational aspects require attention.

Guido te Brake
TNO Defense, Security and Safety, PO Box 21, 3769 ZG, Soesterberg, The Netherlands
e-mail: guido.tebrake@tno.nl

Rick van der Kleij
TNO Defense, Security and Safety, PO Box 21, 3769 ZG, Soesterberg, The Netherlands
e-mail: rick.vanderkleij@tno.nl

R. Babuška & F.C.A. Groen (Eds.): Interactive Collaborative Information Systems, SCI 281, pp. 183–203.
springerlink.com

Many virtual teams, such as the police officers in our example, are highly mobile as well, meaning that they do not work from a fixed location but are constantly on the move. We call these teams micro-mobile virtual teams, with micro meaning that the team members operate in close vicinity of each other. These teams often work according to principles of self-organization with high self-managing capacity, meaning that they are able to dynamically alter their organizational structure to meet the demands of the situation. A trend toward this type of organization can be found in the military domain (where it is referred to as power-to-the-edge or network enabled capability), crisis management, police, and fire brigades, security organizations, rail transport organizations, and so forth. A defining characteristic of micro-mobile virtual teams is that they need to work without a continuous superimposed command and control structure. For these teams, it is essential that team members have a good understanding of the ongoing situation and of the activities and plans of their colleagues. For complex task environments, information and decision support systems that support mobile team members are required.

An important part of the information essential for effective functioning in micro-mobile virtual teams is geospatial in nature: where am I and where are my colleagues? Maps can be used to present geospatial information and are ideal for sharing location-based information about events or resources, and can support collaborative planning activities [28]. Unfortunately, little focus has been put on collaboration using spatial information systems, as stated in 2003 by the National Academy of Sciences [24]: ‘Most of the science and decision making involved in geoinformation is the product of collaborative teams. Current geospatial technologies are a limiting factor because they do not provide any direct support for group efforts’.

This chapter presents an overview of work on map-based collaboration support for micro-mobile virtual teams in the crisis management domain and discusses work that was recently done. The following section provides an introduction to map-based collaboration support and presents a number of recent research efforts. Section 4 addresses how geospatial support can help teams to deal with network breakdowns, a frequently occurring problem in practice. Section 5 describes the effect of differences in map orientation on team effectiveness for communicating geospatial information. We end with a discussion on the validity of lab experiments and draw a number of conclusions.

2 Map-Based Collaboration Support

A look at the recent literature shows that much of the research conducted on maps for mobile devices focuses on single-user navigation and location-based services. Collaboration using map-based devices and support to obtain common ground have been studied but to a much lesser extent. This section gives a brief overview of some of the recent work on map-based collaboration support. First, we will focus on geospatial information for collaboration, and, next, we will look into mobile systems.

2.1 Collaborative Geographic Information System

Work has been done on integrating the realms of geographic information systems (GIS) and computer-supported collaborative work (CSCW), but mainly for teams working in desktop settings. Medeiros et al. [23] present an architecture to support collaborative spatial decision making. Their aim was to integrate GIS, group decision support systems, electronic meeting systems, and workflow concepts in a framework based on coordination premises. The coordination features should aid design team members to cope with their activities supporting the aspects of multi-criteria spatial analysis within a distributed GIS. Ideas presented in this study can be used in mobile settings for dispersed virtual teams, but specific mobility-related issues are not addressed.

Convertino et al. [13] investigated strategies to support knowledge sharing in distributed, synchronous collaboration to achieve good shared awareness. Their goal was to propose, justify, and assess a multiple view approach to support common ground in geocollaboration within multi-role teams. Convertino et al. argue that a collaborative workspace, which includes multiple role-specific views coordinated with a team view, affords a separation between role-specific and shared data, enables the team to filter out role-specific details and share strategic knowledge, and allows learning about knowledge and expertise within the team. Some key issues are discussed that need to be addressed when designing multiple views as a collaborative visualization.

According to the definition of Antunes and André [5], geocollaborative systems address the computational support to situations where people are working in different locations, gathering geographically related data in the field and sharing knowledge. Antunes et al. propose a conceptual framework identifying the design issues that set the stage for eliciting the requirements of geocollaborative systems. The conceptual framework has five elements: places, teams, tasks, artifacts, and georeferenced knowledge. They also highlight two important relationships: (1) a task-artifact relationship, related to the need to increase the organizational decision making abilities through concerted efforts; and (2) an artifact–knowledge relationship, related to the need to support mechanisms for jointly understanding georeferenced data.

2.2 Mobile Collaborative GIS

Alarcón et al. [1] address mobility issues of collaborative GIS. One of the challenges when designing collaborative mobile applications is to consider the context where they will be used. Contextualized applications are easy to adopt by the users; unfortunately, the design of contextualized tools is not evident. Alarcón et al. present a framework of contextual elements to be considered during the conception, analysis, and design phases of mobile collaborative applications. This framework supports developers to identify non-functional requirements and part of the architectural design to get contextualized applications. Their approach does not specifically focus on maps, but can be used when developing map-based collaborative applications.

They tested their framework in developing a mobile collaboration tool for a mobile ad-hoc network (MANET), with a focus on availability, portability, and transparency during disaster relief efforts [2]. Interesting results where found, though the study was rather technologically driven.

Bortenschlager et al. describe the development of a map-based mobile information system for disaster management [9]. Based on a number of high-level requirements such as "Disaster Management systems must provide (near) real-time (geo-)data," and "Disaster Management systems must be able to cope with instability with respect to frequent changes such as topology or network connectivity," they developed a software architecture based on the peer-to-peer paradigm. Their application is robust and promising, but a thorough evaluation of the proposed concepts remains to be done.

Relevant work has been done at the GeoVista institute at Pennsylvania State University, especially within the GeoCollaborative Crisis Management (GCCM) project [10]. This project focused both on emergency operations centers and on first responders in the field and therefore included mobile personnel. Cai et al. state that GIS is indispensable in all stages of crisis management, and its use should focus on teamwork and collaboration. The project goals were twofold:

1. Understanding the roles of geographical information in distributed crisis management activities and
2. Development of computational models, geospatial information technologies, and human–computer systems, to facilitate geocollaborative crisis management

Like some of the approaches described in the previous section, Cai et al. [10] tried to integrate group work with GIS. Cai et al. distinguish two general approaches taken by research groups to integrate these realms:

1. Extending a GIS with some ad-hoc features of group support. This is the approach taken by most geographical information scientists, especially those in a spatial decision-making context. Some interesting extensions include: shared graphics [6], argumentation map [29], and shared interaction and annotation on the map view [21, 30, 33].
2. Extending groupware toolkits with selected GIS functions. This is the approach taken by the CSCW community. For example, GroupArc [12] is an extension of GroupKit [31] to support video conferencing with geographic information. Toucan Navigate (www.infopatterns.net) is an extension of Groove (www.groove.net) to support 'virtual map room' functions.

Cai et al. [10] use a different approach to enable group work with geographical information in geocollaborative crisis management. Their central idea is to use a collaboration agent to combine selected GIS and collaborative functions dynamically (at runtime) according to the need of the ongoing activity. This is based on the belief that a geocollaborative application is likely to need a small subset of functions from GIS and groupware, but what this subset is depends on the kinds of collaborative activities. Their multidisciplinary group consists of GIS experts, information experts, and computer scientists. Topics that are addressed are system architecture, database structure, natural, and efficient multimodal input. What is not covered by their work are the organizational innovations these new technologies enable, such as changes in roles and responsibilities for operations centers and fielded first responders, effects on leadership, or possibilities for self-synchronization. MacEachren and Cai [22] propose a framework for human-GIS-human dialogue modeling, in which maps can play three facilitating roles:

- Objects to talk about (Signifying objects of collaboration). A key goal for visual signification here is to facilitate shared situation assessment by signifying the location of objects of central importance to the team members.
- Objects to think with (Signifying group thinking). In this role, maps (as a type of visual displays) serve as a medium and resource for structuring human thinking. They are tangible representations that prompt mental representations (thus are stimuli for accessing memory, constructing new knowledge, and representing the evolution of that knowledge). For example, maps can create a shared workspace which delineates a common spatial frame of references for the group showing features (roads, township boundaries, water bodies, etc.) that are not directly part of the task, but do provide the necessary background knowledge of the area that supports the discussion of the situation.
- Objects to coordinate perspectives and actions (Signifying group activities). Dealing with complex events involves multiple individuals collaborating on an extended activity. Each member of the team must individually keep track of the status of the collaboration (who has done what, who needs help, what is the progress so far) to decide what to do next, who to talk to, and how to respond. Visual displays can provide mechanisms to directly support awareness of coordinated activities, using colors and showing a history of made annotations.

In the same paper, MacEachren and Cai address group cognition. Following the framework of Scaife and Rogers [32] on externalization by visual representation, MacEachren and Cai suggest the map supports group cognition in three ways:

1. First, the map view supports computational offloading because the maps serve as the group's external memory device where both the resources and the results of the cognitive processes are recorded visually. Threads of cognition directly 'operate' on the objects in the map, and the results of each segment of cognition are immediately available to others. None of the human participants needs to maintain mentally the spatial knowledge encoded in the display.
2. Second, the map view *re-represents* private (or personal) cognition in such a way that it is comprehensible to other members. Re-representation provides the mechanism to 'mesh' together individual cognition into group cognitive processes and ensures that segments of the distributed cognition are well connected. This topic is strongly related to the concept of shared awareness.
3. Third, graphical constraining is provided by the map view since the encoding of all the relevant objects into a common geospatial reference system eliminates the possibility of miscommunication or misunderstanding on the spatial configuration of the problem, and support choosing appropriate actions.

Ashdown and Cummings [7] focus on asymmetry that is often encountered in practice. Teams performing physical tasks are often distributed in space and organized hierarchically. This means that to support collaboration between members systems must account for the asymmetry in physical environment, organizational roles, and available technology. Using urban search and rescue as an example, they first describe the factors that cause this asymmetry. Then, they discuss the way information should be shared and the type of awareness that should be supported. It is suggested that different display and interaction devices for operators must be used at the organizational levels to complement their situations and needs. At the tactical level (operations center), large map displays can be used to show the complete area, providing overview and planning opportunities, whereas on the operational level, small local regions of interest can be shown on portable devices to fielded responders.

3 Geospatial Support of Team Decision Making

Most work on geocollaboration described in the previous section was technologically driven: The goal was the development of software architecture or a collaboration tool for virtual teams. Several important issues were addressed in these papers, such as mobile ad hoc networking, asymmetry in hardware and bandwidth, robustness and network stability, collaborative workspaces, shared awareness, and group decision support. Unfortunately, little attention was given to human and team decision-making processes that need to be supported. Few references were made to literature on cognitive decision making, situation awareness, or teams. We feel that adjusting decision support to these processes is key to develop successful and effective systems.

Figure 1 shows a model of a generic decision-making process. The phases in this model can be found in many different decision-making models, although with varying names. One of the most complex tasks between team members in virtual teams is the development of shared situation awareness (SA). Although numerous definitions of SA have been proposed, Endsley's definition [17], "the perception of elements in the environment within a volume of time and space, the comprehension of their meaning, and the projection of their status in the near future," is firmly established and widely accepted. While some definitions of SA are specific

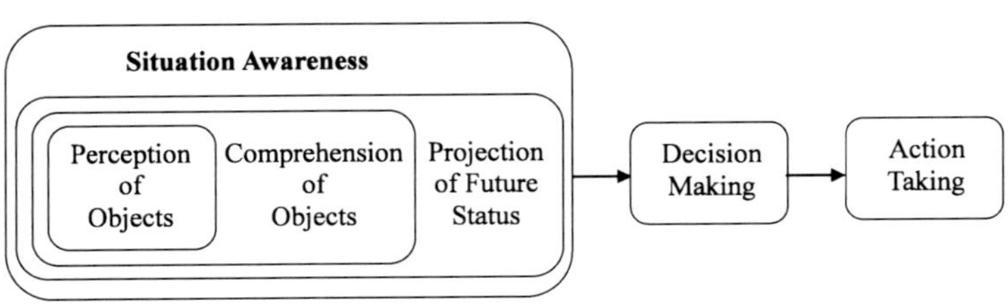

Fig. 1 Endsley's generic decision making process [17]

to the environment from which they were adapted, Endsley's definition is applicable across multiple task domains. Developing situation awareness can already be a challenging task for individuals, but developing shared situation awareness between dispersed team members in micro-mobile teams is even more difficult.

Maintaining an ongoing awareness of events in the work environment and each other's endeavors is essential to achieve the coordination required for collaborative action [20, 35, 37]. People keep up with information about the demand for their work, how particular tasks are progressing, what fellow workers are doing, who is communicating with whom, what equipment is out of order, and so forth. To work effectively, distributed collaborators need to be assured that their partners are 'present' in some sense [11]. Carroll et al. [11] argue that virtual team members need to know what tools and resources their counterparts can access, and what relevant information their collaborators know, as well as their attitudes and goals. Without some sort of knowledge of what other team members are doing, it becomes difficult, or even impossible, to engage in coordinated teamwork. When proper SA is developed, decisions can be made, and actions planned.

Geographic information visualization can support collaboration at different stages of building SA. Maps can facilitate these processes by providing a suitable external visualization that provides a common reference to all team members, supporting SA at the level of perception [17]. This is in line with MacEachren and Cai's [22] visualize objects to talk about role of maps. Second, team members can develop a shared understanding of the situation, called comprehension, by communicating and deliberating on the relevant objects and their temporal and spatial relations, now and in the future (projection). These processes are supported by MacEachren and Cai's second role for maps: visualize objects to think with. Geospatial visualization will structure the reasoning process conducted in the team, especially if interaction with the map and the objects is possible.

Decision making and planning in virtual teams can also benefit from map usage, as plans and actions can be annotated on maps and can easily be discussed and scrutinized by other team members. If time permits complete what-if scenarios can be examined and visualized, but even under time-critical circumstances a quick common visualization of alternatives and options can be provided. Finally, geographical aspects of action execution, such as the movement of team members and vehicles, can be followed realtime by all team members, leading to continuously updated SA.

4 Network Breakdowns and Collaboration

New mobile data communication technologies make it possible for emergency services to have immediate access at any place and any time to information needed to build SA. For example, firemen situated in a crisis area can use a personal digital assistant (PDA) to gather information about the situation and use this information instantly for the execution of their tasks. When information is available to all team members and the situation awareness is shared, the need for explicit communication and coordination may diminish. All relevant information of the activities of their colleagues is available, enabling everyone to adjust their own activities without explicitly discussing details. Teams are then said to be working in a self-organized fashion. Self-organization assumes good network connectivityto provide everybody with reliable and up-to-date information. In practice, however, disconnections of varying lengths are rather common. Reliability of networks is an issue in several of the studies described in Sect. 2 of this chapter. It is obvious that team performance will be low when the quality of the used network is bad. In Sect. 4.1, we will investigate how teams deal with network breakdowns, which strategies they choose, and how they improve team resilience. In Sect. 4.2, we will try to support teams by making uncertainty in the location information explicit.

4.1 Dealing with Unstable Networks

In a lab experiment, we examined the effect of network loss on coordination, strategy, and team performance. A distinction was made between a reliable network that provides stable data communication and a network that suffered from interruptions. The teams of participants had the assignment to rescue victims that were distributed in a synthetic environment that was generated with Unreal Tournament 2004. Unreal Tournament provides a dynamic and to some extent realistic multi-user task environment (see Fig. 2). Victims either had to be saved by one, two or three participants. In this manner, team members were forced to collaborate. Victims could be saved by walking up to them, after which they disappeared both from the virtual world and the map.

A map of the virtual world was presented on a support screen that showed buildings (orange), a church (red), streets (gray), and parks, grass, and trees (green) (shown in Fig. 3). Victims were indicated on the map by a black circle, a clock

Fig. 2 A scene from the synthetic world generated by Unreal Tournament 2004

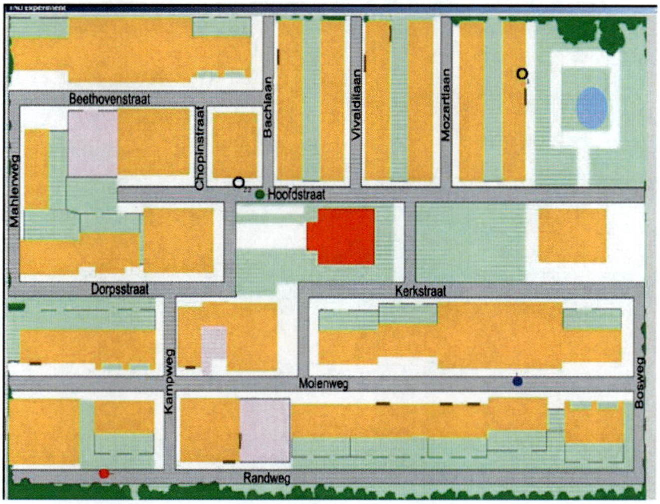

Fig. 3 Example of the map provided as support, indicating the location of the responders (on the screen visible in red, blue, and green), and victims (numbered black circles). Street names are given in Dutch because all participants were Dutch.

showing the time left to rescue a victim, and a number indicating how many participants were needed to rescue it. The participants were shown as a red, blue, and green circle.

The team was provided with visual feedback on the team score, the time that a particular victim could be rescued, the number of team members needed to rescue the victim, and whether a network connection was available. Audio feedback was provided when the connectivity failed or was regained, and when new victims were available. During a period of non-connectivity (lasting 2 minutes), the information on the map was not updated except for the participant's own position, which is provided by GPS for which no communication is required. At that time, of course, the other participants with a network connection no longer received updates of the position of the disconnected participant. Only the data link went down, voice communication remained possible.

Stable and reliable communication proved very important for self-organizing virtual mobile teams. Teams in the reliable networked conditions got significantly higher scores than teams suffering from interruptions in data communication. This study also showed that process and outcome satisfaction are effected negatively by disruptions in data communication, and that the quality of the information exchange between team members is lowered. Virtual mobile teams working under critical conditions can only work optimally if they have the disposition of reliable networks.

In the unreliably connected condition, basically three different behaviors emerged. A strategy chosen by some teams was that people without an up-to-date information picture teamed up with a fellow responder who did have a working data connection. These two-person teams focused on victims that needed two persons to be rescued, the third responder focused on the other victims or participated

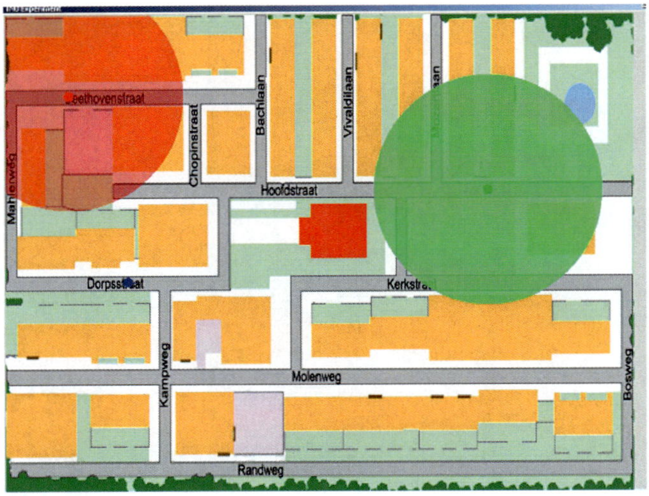

Fig. 4 User interface with supporting information about location uncertainty

when three responders were required. Another strategy that was observed was that a participant with a working data link coached the other using the available voice channel by giving directions and advice. Finally, some teams just wandered around helplessly until the network was up again. A more elaborate description of this experiment and the results can be found in Te Brake et al. [34].

4.2 Network-Aware Support: Blob Interface

Support systems that use the network status and signal this (or the uncertainty of the information that is provided) to the user are called network-aware systems. We evaluated a network-aware support concept to increase understanding of the factors that are important for successful teamwork in mobile virtual teams of first responders [39]. The same virtual environment was used as in the previous sections. Participants performed a simulated search and rescue team task and were equipped with a digitized map and real-time situation updates on the location of other participants in a simulated disaster area. The connection to a server was made deliberately error prone in one condition, leading to occasional losses of network connections. Consequently, participants were not provided with real-time situation updates.

To help them deal with this problem, we equipped team members with a network-aware application that signaled network loss to them and adapted the graphical representation of the location of fellow team members accordingly to the quality of location information present. In accordance with the pedestrian navigation system of Baus et al. [8], we designed an application delivering real-time situation updates, providing information on the location of other responders, and providing information on possible victims on a digitized map of the area. The interface, which we call the blob interface, also involves two levels of uncertainty indicators (Fig. 4). The first level is a simple static visual notification indicating that network connections are lost. Rounded semi-transparent blobs appear on the digitized maps of responders over the last-known locations of their teammates until connections are restored, hence the name blob interface. The second level adds animated detail about characteristics of the uncertainty, specifically the possible location of fellow team members during the period that network connections are lost. The blobs grow in size over time approximating the maximum movement speed of fellow responders until connections are restored. That is, the application adapts the graphical presentations of the locations of fellow first responders on the map according to the quality of location information present. This is hypothesized to assist mobile virtual first responders in making inferences about the possible location of fellow responders during losses of network connections, supporting collaboration and coordination.

Hence, when compared to an application that does not adapt to changes in availability of network components, we expect that the blob interface would improve coordination between interdependent physically distributed first responders, reduces cognitive load, requires less time for response, coordination, and decision making, and eventually allows for more lives to be saved. Interestingly, the opposite may also be true. Presenting uncertainty may also overload the user with information (e.g., display clutter), leading to a higher mental workload, inefficient information exchange, and more stress (cf. [4, 25]). In the latter view, there is a trade-off between the value of presenting additional awareness information and the display clutter and cognitive overload that may result from it.

The research was set up as a detailed empirical study to increase knowledge and understanding of the factors that are important for successful teamwork in mobile virtual teams of first responders. Information is critical to collaboration in mobile virtual teams. However, these teams sometimes have to work in the presence of an unreliable communications infrastructure, leading to information shortages and suboptimal coordination between team members. Our main goal was to conceive and test a solution to this problem; namely a mobile network-aware groupware application that signals network loss to the users and adapts the graphical presentations of the locations of fellow team members according to the quality of location information present. This application was hypothesized to assist mobile virtual first responders in making inferences about the possible location of fellow responders during losses of network connections, helping them to coordinate their collaborative efforts and fight the crisis effectively.

To test the application and investigate the added value of signification of network loss and visualization of geospatial information uncertainty, we compared the network-aware application to three realistic situations: a situation in which participants had complete and reliable information (i.e., there was no need for signification of network loss); a baseline situation in which participants had almost no information at all; and a situation in which participants were connected to an error-prone link, however, without a network-aware application signaling losses of network connections.

In the following test, we discuss the main results for these four conditions. First, we discuss the importance of the real-time mapping of participants' positions on a representation of the environment in the form of a map. Second, we discuss the added value of network-aware support.

To objectively evaluate the team's performance, we calculated the overall team score. The overall team score was based on the total number of people saved. A total of 40 victims appeared during each trial. Some of these victims required only one rescuer to be saved. However, other victims required two cooperating rescuers, which made participants interdependent for successful performance on the task. To make things more complex, we introduced time pressure to the task. The lightly injured victims needed to be rescued within 60 seconds; the more severely injured victims needed to be rescued within 30 seconds. Thus, our task included a total of four different types of victims. Questionnaires on satisfaction, information exchange, and a rating scale on mental effort were administered at the end of each experimental trial. Furthermore, an interview was conducted at the end of the experiment to subjectively evaluate the support conditions and assess the strategies used by the participants.

In general, the results objectively demonstrated the benefits of delivering accurate and real-time location-awareness information to virtual teams (cf. Nova [26]). Teams connected to a server over a fast and reliable link showed superior performance to teams with no network connection. Providing accurate and real-time location-awareness information also increased the quality of information exchange in the team and the amount of satisfaction with team processes and outcomes. Interestingly, these benefits were accomplished at the costs of a higher mental workload. Participants apparently needed to accelerate their cognitive functioning to process the extra information (see also [40]).

As expected, with regard to overall subjective evaluation of the four different support conditions, all teams strongly preferred the reliable network condition, which delivered real-time situation updates, provided information on the location of other responders, and information on the victims. Most teams selected the error-prone condition with network-aware support as second best, although two teams selected this condition as being least preferred. Teams appreciated especially the notification function of the blobs, signaling that network connections were lost. Three teams found the animations useful in assisting them in visualizing the possible location of fellow responders during the loss of network connections. The two teams that preferred the blob interface condition least indicated that the blobs cluttered the screen, were distracting, and led to inertness. The condition without a data network and the error-prone condition without network-aware support were selected by an equal amount of teams as being least preferred.

Interesting to note is that teams reported that there was more interpersonal communication in the reliable network condition than in other conditions. Participants perceived the condition without a data network available as the condition in which the lowest amount of communication had occurred, a perception that was confirmed by our observations. This was rather unexpected. We had hypothesized that additional information on the situation, the location of other responders, and information on the victims would lower the need to communicate: Why communicate when you have all information to perform the task at your disposal? Apparently, the more information teams have, the more there is to communicate about. This explanation was acknowledged by our participants. Teams used information to coordinate their joint efforts. When there was no information at all, teams only communicated their own locations to each other to make sure that they covered as much of the disaster area as possible.

With regard to work strategies, most teams used a combination of two strategies. The first strategy, which was used in all four experimental conditions, was to work separately from three equally sized areas on the map until their joint efforts were needed. Another strategy the teams used, except for teams in the no network condition, was to work with a self-selected dispatcher function coordinating the joint efforts of other members. Usually, this function was fulfilled by the person who was the furthest away from the location of the victim that needed to be rescued.

In addition, we learned from the interviews at the end of the experiment that teams connected to a server over a reliable link communicated more to coordinate their joint efforts. As mentioned, this was contrary to expectations. We hypothesized that the accurate presentation of partners' position in the environment would result in a simplification of communication, i.e., a decrease in the volume of communication. An explanation is that the increased availability of information increased the need for information exchange and communication, again increasing the expenditure of effort.

Participants indicated a preference for the blob interface over the condition without support signaling network loss. Teams appreciated especially the notification function of the blobs, signaling that network connections were lost. Animations visualizing the possible location of fellow responders during loss of network connections were not perceived as useful by the majority of our participants. The blob interface, however, did not reduce the cognitive load of team members, nor did it increase satisfaction or made information exchange between interdependent team members more efficient. More importantly, the blob interface did not help virtual teams to save more 'lives' in our experimental setup. Thus, notwithstanding the theorized benefits of dynamic signaling of network loss and the preference for network-aware support of our participants, presenting uncertainty did not result in more optimal team work. A more elaborate description of this experiment and the results can be found in [39].

5 Map Orientation and Collaboration

In Sect. 3, we saw that objects that are visualized on maps support the development of shared situation awareness in teams at all three levels defined by Endsley. When discussing the location of objects of relevance for the task at hand, it is important that people look at the same picture, otherwise errors are easily made. In the previous sections, we saw that problems can occur when the network has high latency, causing some of the team members to look at old data while others already have an updated picture.

Problems can also occur when the maps used have different reference points. An important design issue for applications using mobile maps is the choice between heading-up and north-up map visualization. North-up maps are presented with north always shown on top. Hence, north-up maps depict the world always in the same way. Heading-up maps (sometimes called head-up, forward-up, track-up, or rotating) adapts to the movement and orientation of the user, rotating the map such that the heading of the user is at the top of the map. Heading-up maps are favored when rapid decisions are required regarding lateral turns or while performing a targeted search [14, 41, 42], and therefore are often used in car navigation systems. Using a heading-up map, no mental rotation is required to align the position of oneself to the map. Thus, targets are generally found more quickly with a heading-up map than with a north-up map and less cognitive effort is required [41]. Although effective for navigation, heading-up maps may interfere with team work. For example, heading-up oriented geographical maps do not allow for shared reference points (i.e., left, up, etc.), making it more difficult to coordinate team efforts.

In a recent experiment, we studied the communication of specific locations on maps between members of a virtual team under two conditions: north-up maps and heading-up maps. Communication of locations is a basic and frequent task among mobile virtual team members while conducting higher level tasks such as coordination and planning, and hence, efficiency is very important. However, miscommunication of locations can have serious consequences for safety and task efficiency. For critical domains, to limit error, it is important to make communication about geospatial information as unequivocal as possible.

In our experimental setup, two participants had to reach a shared understanding through a map-mediated human-to-human dialogue about a specific preset location on digitized maps (see Figs. 5 and 6). Maps team members used varied in orientation and sometimes additional information was added to the maps providing for shared reference points, such as landmarks or compass rose (Fig. 7).

To objectively assess the team's performance, two task-related measures were taken: accuracy and task duration. Accuracy was defined as the distance between the actual location of the crosshair on the computer screen of one of the participants and the location chosen by the other participant. The smaller the difference, the higher the accuracy. Task duration was the time in milliseconds the participants took to complete the task. First responders need to act in concert, and smooth and efficient information exchange is essential to their task. To evaluate the perceived quality of the information exchange during task performance, a summated rating scale was administered at the end of each experimental block. Furthermore, a structured team interview was conducted at the end of the experiment to subjectively evaluate the map versions used throughout the experiment. The results of this experiment showed clearly that heading-up maps interfere with teamwork. Consistent with our expectations, we found that teams using maps with orientations that were identical for both team members (i.e., so called north-up maps), outperformed teams that used maps that varied in orientation between team members. Although no

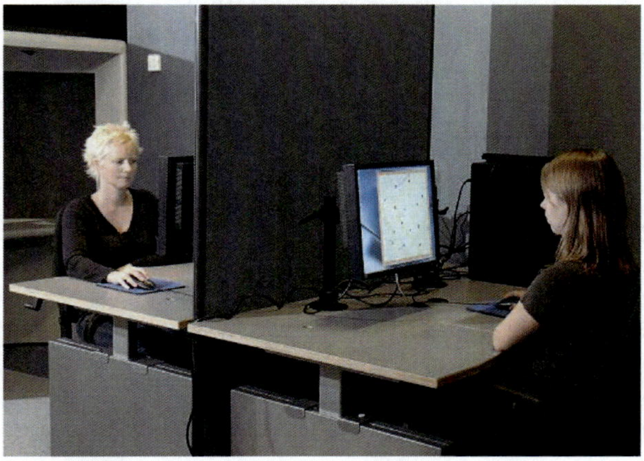

Fig. 5 The experimental setup (staged)

Fig. 6 Left, the map showing the location to the first participant with a crosshair at the lower right. Right, a rotated map shown to the other participant. This example shows a trial in the heading-up condition.

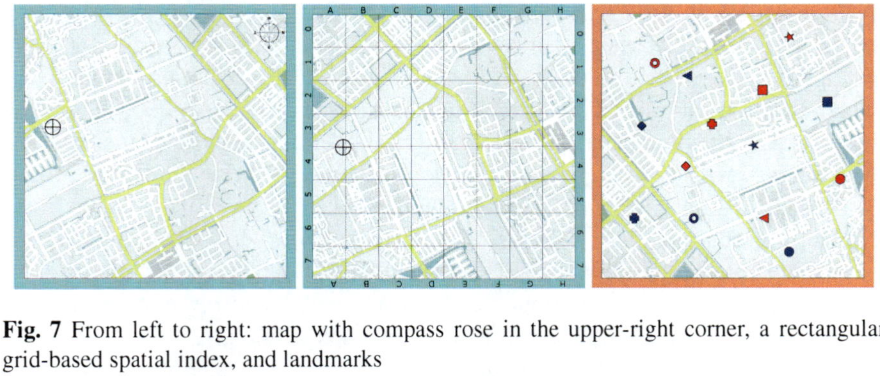

Fig. 7 From left to right: map with compass rose in the upper-right corner, a rectangular grid-based spatial index, and landmarks

difference between map orientations was found for task duration, teams using north-up maps were more accurate than teams in the heading-up condition.

Moreover, it was found that the quality of information exchange was higher in the north-up condition than in the heading-up condition, $F(1,13)=5.094$, $p=.021$. Teams reported higher levels of completeness, speed, and amount of information given, and received in discussions while performing the task in north-up blocks. As expected, with regard to overall subjective evaluation of the map orientation conditions, there was a significant difference between conditions $F(2,26)=15.64$, $p=.00$. Post hoc analyses revealed that participants preferred the maps with extra geospatial information over the base map. Moreover, participants preferred the base map with rectangular grid-based spatial index over maps with compass rose and landmarks. No differences were found between maps with compass rose and landmarks.

For teams in the heading-up condition, it was found that the difference in angle between the two maps was of no importance to our participants in performing the task. It seems that differences between north-up and heading-up conditions on accuracy were solely the result of the extra effort that participants had to undertake in determining whether there was a difference in orientation present, not in the difference per se. To inform each other correctly about the specific preset location on the map, both team members had to realize that there was a misalignment between both maps relative to each other. Then, they had to figure out how it was misaligned and fix the misalignment. There are a variety of ways to accomplish these tasks. We learned from the debriefing sessions with our participants that the strategy most of our teams used was to mentally rotate one of the maps to align both maps before starting the dialogue. Of course, this takes time and demands working memory.

Thus, even though heading-up maps clearly have benefits when it comes to individual navigation tasks, when designing for teamwork maps that are aligned in a fixed orientation, such as north-up maps are the better design option. Implications for practice are that optimal design for team navigation could be to provide both orientations, controllable by the user. This means that fielded first responders can switch between map modes depending on whether they need to coordinate their work. However, because reorienting after a switch in view will take time and mental effort, this needs further research.

Our hypothesis that teams that have the use of landmarks, a compass rose, or a spatial grid will outperform teams without the use of reference points was also supported. In both conditions, both accuracy and time required improved by adding the additional information. Especially the grid was very effective, but the other also improved accuracy and time significantly. Thus, providing additional reference points to teams having map-mediated human-to-human dialogues helps them to determine an exact location on a map. This raises the question whether the use of additional reference points helps teams using heading-up maps to perform as good as teams using north-up maps. An additional analysis was performed revealing no significant differences between both conditions when maps with a grid were considered. So, when it is possible to build the grid into the system, heading-up displays will perform equally well as north-up displays. For a more extensive treatment of the results, refer to [38].

To conclude, our findings have relevant practical implications. The results of studies like these will help practitioners to develop innovative concepts for the organization of virtual teamwork and for the development of groupware to support virtual teams. The knowledge gained in this study is also usable for other domains. Virtual and highly mobile teams, such as patrolling police officers, security on airports, or service employees at railway stations, face similar considerations. Future research should focus on more advanced functions of maps as well, including zooming and panning functions. Because differences in maps between team members will be larger when these functionalities are implemented, coordination could be hampered in these settings.

6 Limitations and Validity

One limitation of the present research we have discussed in this chapter is that the used task setting may limit the generalizability of the results. Laboratory research has met with some criticism over the years. For example, it has been posed that results are unique to the specific test setting and may not generalize to the real world [16]. However, we think that laboratory studies allow us to scrutinize new concepts and to eliminate design flaws in a quick and cost-effective manner before a new system is brought to the field for extensive testing. Moreover, Anderson et al. [3] showed with meta-analysis that laboratory findings actually generalize fairly well, making statements about poor generalizability of lab research to the field misleading and incorrect. Still, this does not mean that experimentation in laboratories and virtual environments can replace field testing. Estimating precise effect sizes will often need to be done in realistic settings and will depend on domain-specific characteristics.

Another possible limitation is the use of a virtual environment in experiments on network-aware decision support. Two concepts describe how realistically a simulator mimics a real-world task: *fidelity* and *validity*. Fidelity is the level of realism that a simulation presents to the user. When a simulator has 100% fidelity it represents the real world in a one-on-one matter. It is needless to say that this 100% fidelity, does not occur and that the development of a simulator toward 100% fidelity is expensive. Validity indicates the degree to which the simulator reflects the underlying construct, that is, whether it simulates what it intends to simulate. The question validity addresses is: did we design the right simulator? When a virtual environment is used to evaluate new concepts, validity of the environment must be high. It may not be required that the virtual world looks very realistic, but the characteristics of the task conducted by the participants in the virtual environment must resemble the situation in the real world (for a more elaborate discussion on experimentation in virtual [game-based] domains, see also [18]).

Finally, university students participated in our experiments. Clearly, students may act differently than professional first responders. To limit generalization problems to a minimum, the tasks that were used throughout our experiments, although representative of the domain, did not require specific knowledge about search and rescue tasks, but focused on the development of situation awareness and the use of uncertain information in planning and decision making.

7 Conclusions

To enable effective decision making in micro-mobile virtual teams is difficult due to a number of reasons. Important prerequisites for enabling virtual operations are the availability of fast and stable networks, information sharing, interoperable systems, and good map-based support. However, attention must be given to the design of individual and team support. Especially establishing shared situation awareness between team members in dynamic task environments and keeping this awareness up to date has proven to be problematic. Visualization of geoinformation on maps facilitates all three levels of situation awareness described by Endsley's model, and enables more effective decision making and action execution. Visualization of network breakdowns and indicating uncertainty were found beneficial by team members in coping with data loss. On the other hand, it is also clear that the blobs are just a first step, requiring further improvement before optimal network-aware support is delivered to the user. In our experiment, user speed was the only variable used to determine the area of uncertainty, but depending on the domain and task, additional information about user direction, goals, and intentions can be used to better estimate the probable location of team members. The experiment on map orientation provides useful guidelines for system developers and information analysts for specifying functional requirements for future mobile collaborative systems. It can then help with deciding between heading-up and north-up visualization, and provides practical tips for improving geospatial understanding of the environment.

Much work still remains to be done, but undoubtedly the availability of cheap smart phones and fast wireless networks will bring map-based support on handhelds to the field in many domains in the very near future. The effectiveness of the applications that support teams will depend on the task support that is provided, the ease of using the maps, and the quality of the information presented.

References

1. Alarcón, R., Guerrero, L.A., Ochoa, S.F., Pino, J.A.: Analysis and design of mobile collaborative applications using contextual elements. Computing and Informatics 22, 1001–1026 (2006)
2. Aldunate, R., Ochoa, S., Pena-Mora, F., Nussbaum, M.: Robust Mobile Ad-hoc Space for Collaboration to Support Disaster Relief Efforts Involving Critical Physical Infrastructure. ASCE Journal of Computing in Civil Engineering 20(1), 13–27 (2006)
3. Anderson, C.A., Lindsay, J.J., Bushman, B.J.: Research in the psychological laboratory: Truth or triviality? Current Directions in Psychological Science 8, 3–9 (1999)

4. Antifakos, S., Schwaninger, A., Schiele, B.: Evaluating the effects of displaying uncertainty in context-aware applications. In: Davies, N., Mynatt, E.D., Siio, I. (eds.) UbiComp 2004. LNCS, vol. 3205, pp. 54–69. Springer, Heidelberg (2004)
5. Antunes, P., André, P.: A Conceptual Framework for the Design of Geo-Collaborative Systems. Group Decision and Negotiation 15, 273–295 (2006)
6. Armstrong, M.P.: Requirements for the development of GIS-based group decision support systems. Journal of the American Society of Information Science 45, 669–677 (1994)
7. Ashdown, M., Cummings, M.L.: Asymmetric Synchronous Collaboration Within Distributed Teams. In: Harris, D. (ed.) HCII 2007 and EPCE 2007. LNCS (LNAI), vol. 4562, pp. 245–255. Springer, Heidelberg (2007)
8. Baus, J., Krüger, A., Wahlster, W.: A resource adaptive mobile navigation system. In: Proceedings of the 2002 International Conference on Intelligent User Interfaces (IUI 2002), pp. 15–22. ACM Press, New York (2002)
9. Bortenschlager, M., Fichtel, T., Leitinger, S., Rieser, H., Steinmann, R.: A Map- and Location-based GeoCollaboration System for Disaster Management in Outdoor Environments. In: 4th International Symposium on LBS and TeleCartography 2007, Hong Kong (2007)
10. Cai, G., MacEachren, A.M., Brewer, I., McNeese, M.D., Fuhrmann, R.S.S.: Map-Mediated GeoCollaborative Crisis Management. In: ISI, Seiten, pp. 429–435 (2005)
11. Carroll, J.M., Rosson, M.B., Convertino, G., Ganoe, C.H.: Awareness and teamwork in computer-supported collaborations. Interacting with Computers 18, 21–46 (2006)
12. Churcher, C., Churcher, N.: Realtime Conferencing in GIS. Transactions in GIS 3(1), 23–30 (1999)
13. Convertino, G., Ganoe, C.H., Schafer, W.A., Yost, B., Carroll, J.M.: A multiple View Approach to Support Common Ground in Distributed and Synchronous Geo-Collaboration. In: Proceedings of the Third International Conference on Coordinated and Multiple Views in Exploratory Visualization, pp. 121–132 (2005)
14. Darken, R.P., Cevik, H.: Map Usage in Virtual Environments: Orientation Issues. In: Proceedings of the IEEE Virtual Reality. VR 1999, Houston, TX, USA, March 13-17, pp. 133–140 (1999)
15. Davis, D.D.: The Tao of Leadership in Virtual Teams. Organizational Dynamics 33, 47–62 (2004)
16. Driskell, J.E., Salas, E.: Can you study real teams in contrived settings? The value of small group research: Toward an understanding of teams. In: Salas, E., Swezey, W. (eds.) Teams: Their training and performance, pp. 101–124. Ablex, Norwood (1992)
17. Endsley, M.R.: Towards a Theory of Situation Awareness. Human Factors 37(1), 32–64 (1995)
18. Frey, A., Hartig, J., Ketzel, A., Zinkernagel, A., Moosbrugger, H.: The use of virtual environments based on a modification of the computer game Quake III Arena ® in psychological experimenting. Computers in Human Behavior 23, 2026–2039 (2007)
19. Johansen, R.: Current user approaches to groupware. In: Johansen, R. (ed.) Groupware: Computer support for business teams, pp. 12–44. Free Press, New York (1988)
20. Kraut, R.E., Fussell, S.R., Brennan, S.E., Siegel, J.: Understanding effects of proximityon collaboration: Implications for technologies to support remote collaborative work. In: Hinds, P.J., Kiesler, S. (eds.) Distributed work, pp. 137–162. Massachusetts Institute of Technology, Cambridge (2002)

21. MacEachren, A.M., Cai, G., Sharma, R., Brewer, I., Rauschert, I.: Enabling collaborative geoinformation access and decision-making through a natural, multimodal interface. International Journal of Geographical Information Science 19(1), 1–26 (2005)
22. MacEachren, A.M., Cai, G.: Supporting group work in crisis management: visually mediated human-GIS-human dialogue. Environment and Planning B: Planning and Design 33, 435–456 (2006)
23. Medeiros, S.P.J., de Souza, J.M., Strauch, J.C.M., Pinto, G.R.B.: Coordination aspects in a spatial group decision support collaborative system. In: Proceedings of the 2001 ACM symposium on Applied computing, pp. 182–186 (2001)
24. Muntz, R.R., Barclay, T., Dozier, J., Faloutsos, C., MacEachren, A.M., Martin, J.L., et al.: IT Roadmap to a Geospatial Future. National Academy of Sciences Press, Washington (2003)
25. Nicholls, D., Battino, P., Marti, P., Pozzi, S.: Presenting uncertainty to controllers and pilots. In: Proc. of the 5th Int. seminar on ATM R&D, FAA and Eurocontrol (2003)
26. Nova, N.: The influences of location awareness on computer-supported collaboration. Doctoral dissertation, École polytechnique fédérale de Lausanne (2007)
27. Priest, H.A., Stagl, K.C., Klein, C., Salas, E.: Virtual teams: Creating context for distributed teamwork. In: Bowers, C., Salas, E., Jentsch, F. (eds.) Creating high-tech teams: Practical guidance on work performance and technology, pp. 185–212. American Psychological Association, Washington (2006)
28. Reichenbacher, T.: Mobile Cartography - Adaptive Visualisation of Geographic Information on Mobile Devices. Verlag Dr. Hut, München (2004)
29. Rinner, C.: Argumentation maps: GIS-based discussion support for online planning. Environment and Planning B-Planning and Design 28, 847–863 (2001)
30. Rogers, Y., Brignull, H., Scaife, M.: Designing Dynamic Interactive Visualisations to Support Collaboration and Cognition. In: First International Symposium on Collaborative Information Visualization Environments, IV 2002, London, pp. 39–50 (2002)
31. Roseman, M., Greenberg, S.: GroupKit: A groupware toolkit for building real time conferencing applications. In: Proceedings of the ACM CSCW Conference on Computer Supported Cooperative Work, Toronto, Canada, pp. 43–50 (1992)
32. Scaife, M., Rogers, Y.: External cognition: how do graphical representations work? International Journal of Human-Computer Studies 45, 185–213 (1996)
33. Shiffer, M.J.: Multimedia GIS for planning support and public discourse. Cartography and Geographic Information Systems 25(2), 89–94 (1998)
34. Te Brake, G., Van der Kleij, R., Cornelissen, M.: Distributed mobile teams: Effects of connectivity and map orientation on teamwork. In: Fiedrich, F., Van de Walle, B. (eds.) Proceedings of the 5th International ISCRAM Conference, Washington, DC, USA, pp. 642–650 (2008)
35. Thompson, L.F., Coovert, M.D.: Teamwork online: The effects of computer conferencing on perceived confusion, satisfaction, and postdiscussion accuracy. Group Dynamics: Theory, Research, and Practice 7, 135–151 (2003)
36. Townsend, A., DeMarie, S., Hendrickson, A.: Virtual teams: Technology and the workplace of the future. Academy of Management Executive 12(3), 17–29 (2007)
37. Van der Kleij, R.: Overcoming distance in virtual teams: Effects of communication media, experience, and time pressure on distributed teamwork. Doctoral dissertation, University of Amsterdam (2007)
38. Van der Kleij, R., Te Brake, G.: Map-Mediated Human-to-Human Dialogues: Effects of Map Orientation Differences and Shared Reference Points on Location-Finding Speed and Accuracy (2009) (submitted for publication)

39. Van der Kleij, R., De Jong, A.M., Te Brake, G., De Greef, T.: Network-aware support for mobile distributed teams. Computers in Human Behavior 25, 940–948 (2009)
40. Van der Kleij, R., Paashuis, R.M., Langefeld, J.J., Schraagen, J.M.C.: Effects of long-term use of video-communication technologies on the conversational process [Special issue]. International Journal of Cognition, Technology and Work 6, 57–59 (2004)
41. Viita, D., Werner, S.: Alignment effects on simple turn decisions in track-up and north-up maps. In: Proceedings of the Human factors and ergonomics society 50th Annual meeting. HFES 2006, San Francisco, CA, USA, October 16- 20, pp. 1519–1522. HFES, Santa Monica (2006)
42. Wickens, C.D., Liang, C.-C., Prevet, T., Olmos, O.: Electronic maps for terminal area navigation: effects of frame of reference and dimensionality. The international journal of aviation psychology 6(3), 241–271 (1996)

Part III
Computer-Human Interaction Modeling

Affective Dialogue Management Using Factored POMDPs

Trung H. Bui, Job Zwiers, Mannes Poel, and Anton Nijholt

Abstract. Partially Observable Markov Decision Processes (POMDPs) have been demonstrated empirically to be good models for robust spoken dialogue design. This chapter shows that such models are also very appropriate for designing affective dialogue systems. We describe how to model affective dialogue systems using POMDPs and propose a novel approach to develop an affective dialogue model using factored POMDPs. We apply this model for a single-slot route navigation dialogue problem as a proof of concept. The experimental results demonstrate that integrating user's affect into a POMDP-based dialogue manager is not only a nice idea but is also helpful for improving the dialogue manager performance given that the user's affect influences their behavior. Further, our practical findings and experiments on the model tractability are expected to be helpful for designers and researchers who are interested in practical implementation of dialogue systems using the state-of-the-art POMDP techniques.

1 Introduction

The HAL 9000 computer character is popular in the speech and language technology research field since his capabilities can be linked to different research topics of the field such as speech recognition, natural language understanding, lip reading, natural language generation, and speech synthesis [20, chap. 1]. This artificial agent is often referred to as a dialogue system, a computer system that is able to talk with humans in a way more or less similar to the way in which humans converse with each other.

Trung H. Bui
Center for the Study of Language and Information, Stanford University,
210 Panama St, Stanford, CA 94305, USA
e-mail: thbui@stanford.edu

Job Zwiers, Mannes Poel, and Anton Nijholt
Human Media Interaction Group, University of Twente, Postbus 217,
7500 AE Enschede, The Netherlands
e-mail: zwiers,mpoel,anijholt@cs.utwente.nl

R. Babuška & F.C.A. Groen (Eds.): Interactive Collaborative Information Systems, SCI 281, pp. 207–236.
springerlink.com

Furthermore, HAL is affective[1]. He is able to recognize the affective states of the crew members through their voice and facial expressions and to adapt his behavior accordingly. HAL can also express emotions, which is explained by Dave Bowman, a crewman in the movie:

> Well, he acts like he has genuine emotions. Of course, he's programmed that way to make it easier for us to talk with him.

Precisely, HAL is an "ideally" Affective Dialogue System (ADS), a dialogue system that has specific abilities relating to, arising from, and deliberately influencing people's emotions [30].

Designing and developing ADSs has recently received much interest from the dialogue research community [2]. A distinctive feature of these systems is affect modeling. Previous work mainly focused on showing the system's emotions to the user in order to achieve the designer's goal such as helping the student to practice nursing tasks [17] or persuading the user to change their dietary behavior [33]. A different and more challenging problem is to infer the interlocutor's affective state (hereafter called "user") and to adapt the system's behavior accordingly*.

Solving this problem could enhance the adaptivity of a dialogue system in many application domains. For example, in the information seeking dialogue domain, if a dialogue system is able to detect the critical phase of the conversation which is indicated by the user's vocal expressions of anger or irritation, it could determine whether it is better to keep the dialogue or to pass it over to a human operator [5]. Similarly, many communicative breakdowns in a training system and a telephone-based information system could be avoided if the computer was able to recognize the affective state of the user and to respond to it appropriately [25]. In the intelligent spoken tutoring dialogue domain, the ability to detect and adapt to student emotions is expected to narrow the performance gap between human and computer tutors [6].

This chapter addresses this problem (see *, this page) by introducing a dialogue management system which is able to act appropriately by taking into account some aspects of the user's affective state. The computational model used to implement this system is called the *affective dialogue model*. Concretely, our system processes two main inputs, namely the observation of the user's action (e.g., dialogue act) and the observation of the user's affective state. It then selects the most appropriate action based on these inputs and the context. In human–computer dialogue, building this sort of system is difficult because the recognition results of the user's action and affective state are ambiguous and uncertain. Furthermore, the user's affective state cannot be directly observed and usually changes over time. Therefore, an affective dialogue model should take into account basic dialogue principles, such as turn-taking and grounding, as well as dynamic aspects of the user's affect.

An intuitive solution is to extend conventional methods for developing spoken dialogue systems such as the Rapid Dialogue Prototyping Methodology (RDPM) [12] by integrating an Affect Recognition (AR) module and define a set of rules for the

[1] We use the terms "emotional" and "affective" interchangeably as adjectives describing either physical or cognitive components of the interlocutor's emotion [30, p. 24].

system to act given the observation of the user's affective state (e.g., using the rules defined by Ortony et al. [28]). However, it is nontrivial to handle uncertainty using a pure rule-based approach such as the RDPM.

From recent literature [11, 49] it follows that Partially Observable Markov Decision Processes (POMDPs) are suitable for use in designing these affective dialogue models for three main reasons. First, the POMDP model allows for realistic modeling of the user's affective state, the user's intention, and other user's hidden state components by incorporating them into the state space. Second, recent results in dialogue management research [49] show that the POMDP-based dialogue manager is able to cope well with *uncertainty* that can occur at many levels inside a dialogue system from the automatic speech recognition (ASR) and natural language understanding (NLU) to the dialogue management. Third, the POMDP environment can be used to create a *simulated* user which is useful for learning and evaluation of competing dialogue strategies [37].

In this chapter, we first introduce the basic components of an ADS. Second, we give an overview of the Partially Observable Markov Decision Process (POMDP) techniques and their applications to the dialogue management problem. Third, we describe our factored POMDP approach to affective dialogue modeling. Finally, we address various technical issues when using a state-of-the-art approximate POMDP solver to compute a near-optimal policy for a single slot route navigation application.

2 Components of an Affective Dialogue System

An ADS is a multimodal dialogue system where the user's affective state might be recognized from speech, facial expression, physiological sensors, or combined multimodal input channels. The system expresses emotions through multimodal output channels such as a talking head or a virtual human. Figure 1 shows an architecture of a speech-based ADS. The speech input is first processed by the Automatic Speech Recognition (ASR) module and the semantic meaning is then derived by the Natural Language Understanding (NLU) module. Similarly, the user's affect is interpreted by the AR module and the result is sent to the Dialogue Manager (DM). The DM processes both inputs from the NLU module and the AR module and produces the system action and the system's affective state which are processed by the natural language generation module and the speech synthesis module before sending to the user.

Based on a general statistical Dialogue System (DS) framework presented in [50] and our prototype DSs [12, 14], the interaction between the system and the user is described by the following cycle. The user has a goal g_u in mind before starting a dialogue session with the system. The user's goal g_u might change during the system–user interaction process. At the beginning, the DM sends an action a_s (e.g., informing that the system is ready or greeting the user) and optionally an affective state e_s (whether it is appropriate to show the system's affective state depends on each particular application). Action a_s usually represents the system's intent and is

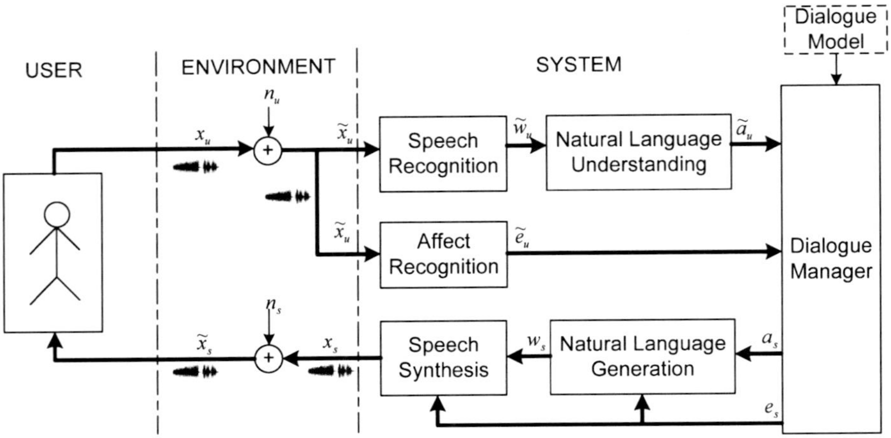

Fig. 1 Components of an affective speech-based dialogue system. Bold arrows show the main flow of the interaction process.

formulated as a sequence of dialogue acts and their associated semantic content. The natural language generation module processes the tuple $< a_s,e_s >$ and realizes a sequence of utterances w_s. The tuple $< w_s,e_s >$ is then processed by the speech synthesis module and the outcome is an audio signal x_s. The signal x_s might be corrupted by the environment noise n_s and the user perceives an audio signal $\tilde{x}_s$. Based on the perceived signal $\tilde{x}_s$, the user infers the system's intent $\tilde{a}_s$ and the affective state $\tilde{e}_s$. The tuple $< \tilde{a}_s,\tilde{e}_s >$ might be different from its counterpart $< a_s,e_s >$ due to a misunderstanding by the user or the corrupted noise n_s from the environment or both. Based on the user's goal g_u, the user forms a communicative action (i.e., intent) a_u. Action a_u might also be influenced by the user's affective state e_u. The user then formulates a sequence of words w_u and articulates a speech signal x_u (see Levelt [22] for further details of the transition from intention to articulation performed by the user). The acoustic signal x_u is processed by both the ASR module and the AR module (the actual input of these modules is, $\tilde{x}_u$, a corrupted signal of x_u caused by the environment noise n_u). The output of the ASR module is a string of words $\tilde{w}_s$. This is then processed by the NLU module and the result is the observation of the user's action $\tilde{a}_u$. The output from the AR module ($\tilde{e}_u$) and NLU module ($\tilde{a}_u$) is sent to the DM. The DM selects a next system's action based on these inputs and the current system's belief b_s. The process is then repeated until either the system or the user terminates the interaction (e.g., the user hangs up the call in a telephone-based dialogue system or the system exits after providing a solution to the user).

The DM is a critical component in the dialogue system, and recent research has shown the advantages of modeling the DM as a POMDP (see Section 1). In the following, we will describe in detail the theory of POMDPs in the dialogue management context.

3 Theory of POMDPs

A POMDP is a generalization of a Markov Decision Process (MDP) which permits uncertainty regarding the state of the environment. Howard [19] described a transition in an MDP as a frog in a pond jumping from lily pad to lily pad. In a POMDP environment, the lily pond is covered by the mist, therefore the frog is no longer certain about the pad it is currently on [27]. Before jumping, the frog can observe information about its current location. This intuitive view is very appropriate to model the affective dialogue management problem as illustrated in Section 3.2.

In a dialogue management context, the agent is the dialogue manager, loosely called the system (Fig. 2). A part of the POMDP environment represents the user's state and user's action. Depending on the design for a particular dialogue application, the rest of the POMDP environment might be used to represent other modules such as the speech recognition and the emotion recognition [10, chap. 1]. Because the user's state cannot be directly observed, the agent uses a state estimator to compute its internal belief (called *belief state*) about the user's current state and an action selector where the policy, called π, is implemented to select actions. The state estimator takes as its input the previous belief state, the most recent system action and the most recent observation, and returns an updated belief state. The action selector takes as its input the agent's current belief state and returns an action that will be sent to the user.

In the following sections, we first describe a basic framework of POMDPs. Then, we present a simple empathic dialogue agent example for the tutoring domain. Two main tasks of the agent, *belief updating* and *finding an optimal policy*, are briefly described. Further details can be found in [10, chap. 2].

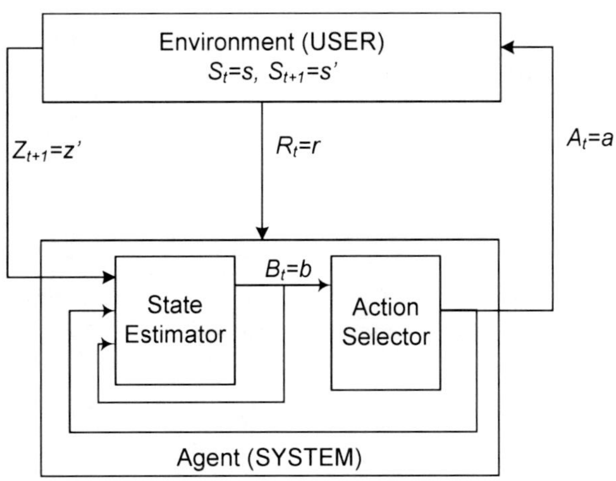

Fig. 2 Modularized view of the interaction between the dialogue manager and the user in a dialogue management context

3.1 Basic Framework

A POMDP [21] is defined as a tuple $\langle S,A,Z,T,O,R\rangle$, where S is a set of states of the environment (usually called *state space*), A is a set of the agent's actions, Z is a set of observations the agent can experience of its environment, T is a *transition function*, O is an *observation function*, and R is a *reward function*. We assume that S,A,Z are finite and that the interaction between the agent and environment follows a sequence of discrete time steps.

Let S_t,A_t,Z_{t+1}, and R_t be random variables taking their values from the sets S,A,Z, and $\mathbb{R}$ (the set of real numbers), respectively. At each time step t, the environment's state is S_t. The agent selects an action A_t and sends it to the environment. The environment's state changes to S_{t+1}. The agent receives an observation Z_{t+1} and a reward value R_t. Following this interaction description, the transition function T, observation function O, and reward function R are formally defined as follows.

- The transition function is defined as $T : S\times A\times S \rightarrow [0,1]$. Given any state s and action a, the probability of the next possible state s' is

$$\mathscr{P}^a_{ss'} = T(s,a,s') = P\{S_{t+1} = s'|S_t = s,A_t = a\}, \text{ for all } t. \tag{1}$$

  These quantities are called *transition probabilities*. Transition function T is time-invariant and the sum of transition probabilities over the state space $\sum_{s'\in S}\mathscr{P}^a_{ss'} = 1$, for all (s,a).
- The observation function is defined as $O : S\times A\times Z \rightarrow [0,1]$. Given any action a and next state s', the probability of the next observation z' is

$$\mathscr{P}^a_{s'z'} = O(s',a,z') = P\{Z_{t+1} = z'|A_t = a,S_{t+1} = s'\}, \text{ for all } t. \tag{2}$$

  These quantities are called *observation probabilities*. Observation function O is also time-invariant and the sum of observation probabilities over the observation space $\sum_{z'\in Z}\mathscr{P}^a_{s'z'} = 1$, for all (a,s').
- The reward function[2] is defined as $R : S\times A \rightarrow \mathbb{R}$. Given any current state s and action a, the *expected immediate reward* that the agent receives from the environment is

$$R^a_s = R(s,a) \tag{3}$$

Given the POMDP model, we want to design a framework in which the agent's goal is to maximize the *expected cumulative reward* over time

$$V = E\left(\sum_{t=0}^{\mathscr{T}-1}\gamma^t R_t\right) \tag{4}$$

[2] We can also define the reward function as $(R : S \rightarrow \mathbb{R})$ or $(R : S\times A\times S \rightarrow \mathbb{R})$ or $(R : S\times A\times S\times Z \rightarrow \mathbb{R})$. However, the first definition is sufficient and does not change the fundamental properties of the framework.

where $E(.)$ is the mathematical expectation, $\mathcal{T}$ is the *planning horizon* ($\mathcal{T} \geq 1$), γ is a *discount factor* ($0 \leq \gamma \leq 1$). The closer γ to 1, the more effect future rewards have on current agent action selection. This framework is called *finite-horizon* optimality. When $\mathcal{T} \to \infty$ and $\gamma < 1$, the framework is called *infinite-horizon* optimality. It is necessary to set the value of discount factor γ smaller than one in the infinite-horizon case to guarantee that the expected cumulative reward is bounded.

The state space might also contain some special *absorbing state* that only transitions to itself and the reward gained when the agent takes any action is 0. The absorbing state is useful for modeling finite-horizon POMDPs where the number of horizons cannot be determined in advance. Suppose s is an absorbing state, the transition function from s is as follows:

$$T(s,a,s') = \begin{cases} 1 \text{ if } s' = s \\ 0 \text{ otherwise} \end{cases} \text{ and } R(s,a) = 0, \text{ for all } a.$$

Let R_{min} and R_{max} be the lower bound and upper bound of the reward function, that is to say

$$R_{min} < R(s,a) < R_{max}, \text{ for all } (s,a). \tag{5}$$

From (4) and (5), we have for $\gamma < 1$ that:

$$E\left(\sum_{t=0}^{\infty} \gamma^t R_{min}\right) < V < E\left(\sum_{t=0}^{\infty} \gamma^t R_{max}\right) \Rightarrow \frac{R_{min}}{1-\gamma} < V < \frac{R_{max}}{1-\gamma}. \tag{6}$$

3.2 Empathic Dialogue Agent Example

To illustrate the main principle of a POMDP-based dialogue management, we use a simple empathic dialogue agent example for the tutoring domain, which is described as follows. A student ("the user") is interacting with the agent to learn a subject matter. The agent tries to infer the user's affective state to give an appropriate empathic feedback which aims to enhance the student's learning. Suppose that the user's affective state is either $s_1 = pos$ (positive) or $s_2 = neg$ (negative). The agent's goal is to select the most appropriate action from the following three actions: $a_1 = comfort, a_2 = check$, and $a_3 = encourage$. The *comfort* action is expressed by saying "I am sorry that you feel bad about the last question". The *check* action is the agent action to infer the user's affective state from outcome of an affect recognition module assumed to be available. The *encourage* action is expressed in the verbal form such as "Very good!" or "Well done!". If the agent knows exactly the user's true affective state, the action selection problem is trivial. The agent just selects the *encourage* action when the user's affective state is positive and the *comfort* action otherwise. Unfortunately, the user's affective state cannot be directly observed. Therefore, the agent must sometimes execute the *check* action to infer the user's affective state.

A POMDP model for this problem is represented by: (i) $S = \{s_1, s_2\} = \{pos, neg\}$, (ii) $A = \{a_1, a_2, a_3\} = \{comfort, check, encourage\}$, and (iii) $Z = \{z_1, z_2\} = \{\tilde{pos}, \tilde{neg}\}$. The transition function, observation function, and reward function are shown in Table 1. All transition and observation probabilities are handcrafted based on the common sense knowledge from the tutoring domain and affect recognition literature.

Table 1 Transition function, observation function, and reward function for the empathic dialogue agent

	$P(s'\|a,s)$			$P(z'\|a,s')$			$R(s,a)$	
a	s	$s' = pos$	$s' = neg$	s'	$z' = \tilde{pos}$	$z' = \tilde{neg}$	s	r
comfort	*pos*	0.80	0.20	*pos*	0.5	0.5	*pos*	-10
	neg	0.30	0.70	*neg*	0.5	0.5	*neg*	10
check	*pos*	0.90	0.10	*pos*	0.9	0.1	*pos*	-1
	neg	0.10	0.90	*neg*	0.1	0.9	*neg*	-1
encourage	*pos*	0.95	0.05	*pos*	0.5	0.5	*pos*	5
	neg	0.05	0.95	*neg*	0.5	0.5	*neg*	-10

The transition probability distribution in Table 1 is based on the following intuition. The user's affective state is *dynamic* and might change even without direct intervention from the agent. Therefore, when the agent selects the *check* action, which does not directly influence the user, the user's affective state might change from positive to negative or vice versa with probability 0.1. If the user is in a *pos* state and the agent selects the *encourage* action which is the right one, the user, therefore, remains in a positive state with a high probability (0.95). Vice versa, if the user is in a *neg* state, the *encourage* is not appropriate and therefore the chance that the negative state remains is high (0.95). Similarly, if the user's affective state is negative, the *comfort* action is appropriate and therefore the probability of changing to the positive state is higher than the *check* action.

When the agent selects the *check* action, it can infer the user's affective state with a 90% correctness (Table 1). The correctness rate here is interpreted as the classification accuracy of the affect recognition module. When the agent selects *encourage* or *comfort* action, the observation probabilities are equally distributed.

Reward values are specified by the designer. They represent *what* the designer wants the agent to achieve, not *how* the designer wants it achieved [42, pg. 57]. In this example, we want the agent to select an appropriate action given uncertainty about the user's affective state. When the user's affective state is positive, the *encourage* action is appropriate and therefore receives a positive reward. When the user's state is negative, the *comfort* action is appropriate and a positive reward is assigned for this action. When the agent selects *check* action, it incurs a small negative reward and in return the agent is more certain about the user's current affective state.

3.3 Computing Belief States

The state of the user cannot be directly observed. Therefore, in order to select good actions, the agent needs to maintain a complete trace of all the observations and actions that have happened so far. This trace is known as a *history*[3]. It is formally defined as:

$$H_{t+1} := \{A_0, Z_1, ..., Z_t, A_t, Z_{t+1}\}, \tag{7}$$

Astrom [3] showed that history H_{t+1} can be summarized via a *belief distribution*. A belief distribution is exactly the belief state of the agent.

$$B_{t+1}(s') = P\{S_{t+1} = s' | B_0, H_{t+1}\}, \tag{8}$$

Assuming the Markov property and using Bayes' rule, Equation 8 is transformed to the following equation (see the proof in [39, Appendix A]):

$$B_{t+1}(s') = P\{S_{t+1} = s' | Z_{t+1} = z', Z_t = a, B_t = b\} \tag{9}$$

Formally, let the belief space B be an infinite set of belief states. A belief state $b \in B$ is encoded as a $|S|$-dimensional column vector $(b_1, b_2, ..., b_{|S|})^T$, where each element $b_i = b(s_i)$ is the probability that the current state of the environment is s_i. Geometrically, a belief state is a point (called *belief point*) in a $(|S|-1)$-dimensional belief simplex.

Concretely, the agent starts with an initial belief state $B_0 = b_0$. At time t, the agent's belief is $B_t = b$, it selects action $A_t = a$ and sends this to the environment. The state changes to $S_{t+1} = s'$. State S_{t+1} cannot be directly observed and the agent only gets observation $Z_{t+1} = z$. The agent also receives a reward $R_t = r$, the value of which depends on the actual values of state s and agent's action a. At this moment, the agent needs to update its belief state $B_{t+1} = b'$ given known values for b,a,z. Starting from Equation 9, $b'(s')$ is computed using the basic laws from the probability theory as follows (see [10, chap. 2] for further details):

$$b'(s') = P(s' | z, a, b) = \eta \mathscr{P}^a_{s'z} T^a_{s'} b, \tag{10}$$

where $T^a_{s'}$ is a $|S|$-dimensional row vector:

$$T^a_{s'} = \left(\mathscr{P}^a_{s_1 s'}, ..., \mathscr{P}^a_{s_{|S|} s'}\right), |S| \text{ is the number of elements of } S.$$

$\eta = 1/P(z|a,b)$ is a normalizing constant, independent of state s'.

The belief state b' is represented as

$$b' = \eta W^a_z b \tag{11}$$

where W^a_z is a $|S| \times |S|$ matrix,

[3] In a dialogue management context, this trace is the dialogue history.

$$W_z^a = \begin{bmatrix} \mathcal{P}^a_{s_1 z}\mathcal{P}^a_{s_1 s_1} & \mathcal{P}^a_{s_1 z}\mathcal{P}^a_{s_2 s_1} & \dots & \mathcal{P}^a_{s_1 z}\mathcal{P}^a_{s_{|S|} s_1} \\ \mathcal{P}^a_{s_2 z}\mathcal{P}^a_{s_1 s_2} & \mathcal{P}^a_{s_2 z}\mathcal{P}^a_{s_2 s_2} & \dots & \mathcal{P}^a_{s_2 z}\mathcal{P}^a_{s_{|S|} s_2} \\ \dots & \dots & \dots & \dots \\ \mathcal{P}^a_{s_{|S|} z}\mathcal{P}^a_{s_1 s_{|S|}} & \mathcal{P}^a_{s_{|S|} z}\mathcal{P}^a_{s_2 s_{|S|}} & \dots & \mathcal{P}^a_{s_{|S|} z}\mathcal{P}^a_{s_{|S|} s_{|S|}} \end{bmatrix} \tag{12}$$

Given the current belief state the agent has to execute the optimal action, i.e. determining a policy that optimizes the rewards received. How an optimal policy is computed will be described in the next section.

3.4 Finding an Optimal Policy

A policy is a function:

$$\pi(b) \longrightarrow a, \tag{13}$$

where b is a belief state and a is the action chosen by the policy π.

An optimal policy π^* is a policy that maximizes the expected cumulative reward:

$$\pi^* = argmax_{\pi} E\left[\sum_{t=0}^{\infty} \gamma^t R_t\right], \tag{14}$$

where R_t is the reward when the agent follows policy π.

We define value functions $V_i : B \longrightarrow \mathbb{R}$. $V_n(b)$ is the maximum expected cumulative reward when the agent has n remaining steps to go. Its associated policy is denoted by π_n. When the agent has only one remaining step to go (i.e. $n = 1$), all it can do is to select an action and send it to the environment, we have:

$$\begin{aligned} V_1(b) &= \max_{a \in A} \sum_{s \in S} R(s,a)b(s) \\ &= \max_{a \in A} r_a b, \end{aligned} \tag{15}$$

where r_a is a row vector, $r_a = \left(R^a_{s_1},\dots,R^a_{s_{|S|}}\right)$.

When the agent has n remaining steps to go ($n > 1$), the value function V_n is defined inductively as [39]:

$$V_n(b) = \max_{a \in A}\left[r_a b + \gamma \sum_{z \in Z} P(z|a,b) V_{n-1}(b_a^z)\right] \tag{16}$$

where b_a^z is the belief state of the agent after selecting action a, and the observation of the environment changes to z.

When $n \longrightarrow \infty$, the optimal value function for the infinite-horizon case is denoted by V^*. Puterman [32, Theorem 6.9] proved that V_n converges to V^* when n goes to infinity. Therefore, from Equation 16 we have:

$$V^*(b) = \max_{a \in A} \left[r_a b + \gamma \sum_{z \in Z} P(z|a,b) V^*(b_a^z) \right] \tag{17}$$

For any positive number ε, the policy π_n is ε*-optimal* if

$$V^*(b) - V_n(b) \leq \varepsilon \text{ for all } b \in B. \tag{18}$$

Equation 16 is used to develop an important type of algorithm called Value Iteration (VI), which is an algorithm for finding ε*-optimal* policies. The approximation terminates when:

$$\sup_b |V_n(b) - V_{n-1}(b)| \leq \frac{\varepsilon(1-\gamma)}{2\gamma}, \tag{19}$$

where $\sup|X|$ stands for supremum norm of set X [32]. The left part of Equation 19 is called the *Bellman residual.*

Because there are an infinite number of belief states, we cannot compute V_{n-1} directly for each belief state b. Sondik [40] proved that V_{n-1} can be represented through a finite set of α-vectors $\Gamma_{n-1} = \{\alpha_1,...,\alpha_{|\Gamma_{n-1}|}\}$, where each vector $\alpha \in \Gamma_{n-1}$ is a $|S|$-dimensional row vector (also called a *hyperplane*, hereafter it is called an α*-vector*), and

$$V_{n-1}(b) = \max_{\alpha \in \Gamma_{n-1}} \alpha b \tag{20}$$

Therefore, from Equations 11 and 20 we can rewrite Equation 16 as

$$\begin{aligned}
V_n(b) &= \max_{a \in A} \left[r_a b + \gamma \sum_{z \in Z} P(z|a,b) \max_{\alpha \in \Gamma_{n-1}} \alpha b_a^z \right] \\
&= \max_{a \in A} \left[r_a b + \gamma \sum_{z \in Z} P(z|a,b) \max_{\alpha \in \Gamma_{n-1}} \alpha \frac{W_z^a b}{P(z|a,b)} \right] \\
&= \max_{a \in A} \left[r_a b + \gamma \sum_{z \in Z} \max_{\alpha \in \Gamma_{n-1}} \alpha W_z^a b \right] \\
&= \max_{a \in A} \left[r_a b + \gamma (\max_{l_1} \alpha_{l_1} . W_{z_1}^a b + ... + \max_{l_{|Z|}} \alpha_{l_{|Z|}} W_{z_{|Z|}}^a b) \right] \\
&= \max_{a \in A} \left[r_a b + \gamma \max_{l_1} ... \max_{l_{|Z|}} (\alpha_{l_1} W_{z_1}^a b + ... + \alpha_{l_{|Z|}} W_{z_{|Z|}}^a b) \right] \\
&= \max_{a \in A} \left[r_a b + \gamma \max_{l_1} ... \max_{l_{|Z|}} \sum_{k=1}^{|Z|} \alpha_{l_k} W_{z_k}^a b \right] \\
&= \max_{a \in A} \max_{l_1} ... \max_{l_{|Z|}} \left[r_a + \gamma \sum_{k=1}^{|Z|} \alpha_{l_k} W_{z_k}^a \right] b,
\end{aligned} \tag{21}$$

where $l_1, l_2, ..., l_{|Z|} \in [1, |\Gamma_{n-1}|]$.

The set Γ_n can now be generated from set Γ_{n-1} by the following update:

$$\Gamma_n \leftarrow \alpha' = r_a + \gamma \sum_{k=1}^{|Z|} \alpha_{l_k}.W_k^a, \forall a \in A, \alpha_{l_k} \in \Gamma_{n-1}, \tag{22}$$

where $n \geq 1$ and $\Gamma_1 = \{r_{a_1}, r_{a_2}, \ldots, r_{a_{|A|}}\}$.

Finding the optimal policy (for planning horizon $\mathscr{T} = n$) is now considered as solving a set of $|A||\Gamma_{n-1}|^{|Z|}$ linear constraints derived from Equation 21. To gain the computational tractability, it is necessary to keep only the vectors that contribute to the optimal value function because the number of α-vectors generated from Equation 22 is very large. We distinguish two types of α-vectors: *useful* vectors and *extraneous* vectors [54]. A vector $\alpha \in \Gamma_n$ is useful[4] if:

$$\exists b \in B : \alpha b > \alpha' b, \text{for all } \alpha' \in \Gamma_n - \alpha \tag{23}$$

A vector $\alpha' \in \Gamma_n$ that does not satisfy Equation 23 is an extraneous vector. A set Γ_n that is composed of useful vectors is called a *parsimonious* set [53]. From Equation 20 it is obvious that we can safely remove all the extraneous vectors from the set Γ_n. Monahan [27] proposed a procedure to remove extraneous vectors by solving the following linear program for each $\alpha \in \Gamma_n$:

$$\begin{aligned} &\textit{variables: } x, b_i, \forall i \in [1, |S|] \\ &\textit{maximize } x \\ &\textit{subject to constraints: } b(\alpha - \alpha') \geq x; \forall \alpha' \in \Gamma_n \ \& \ \sum_{i=1}^{|S|} b_i = 1 \end{aligned} \tag{24}$$

If $x < 0$, remove α from Γ_n.

When the set of useful α-vectors Γ_n is found. The agent's action $\hat{a}$ is determined as $\hat{a} \leftarrow \hat{\alpha}^5$, where

$$\hat{\alpha} = \arg\max_{\alpha \in \Gamma_n} \alpha b \tag{25}$$

4 Review of the POMDP-Based Dialogue Management

In this section, we focus on describing the POMDP-based dialogue management approaches. Reviews of other dialogue management approaches for spoken and multimodal dialogue systems are reviewed in McTear [26] and Bui [9], respectively.

Section 3 explained the basic activity of a POMDP-based dialogue management system. Young et al. [51] have argued that nearly all existing dialogue management systems, especially those based upon the information state approach [43], can be considered as direct implementations of the POMDP-based model with a *deterministic* (i.e., handcrafted) dialogue policy. These systems have a number of "severe

[4] Assume that the identical vectors in Γ_n are merged.

[5] Because each α-vector is associated with only one action, see Equation 22.

weaknesses" such as using unreliable confidence measures, having difficulty coping with the dynamic changing of the user's goal and intention. Moreover, tuning the dialogue policy is labor extensive, based on off-line analysis of the system logs [51].

The first work that applies the POMDP for the dialogue management problem was proposed by Roy et al. [34] for building a nursing home robot application. In this application, a flat POMDP model is used where the states represent the *user's intentions*; the observations are the *user's speech utterances*; and the actions are the *system responses*. They showed that the POMDP-based DM copes well with noisy speech utterances, for example their POMDP-based DM makes fewer mistakes than an MDP-based DM and it automatically adjusts the dialogue policy when the quality of the speech recognition degrades. Zhang et al. [52] extended the Roy model in several dimensions: (1) a factored POMDP [7] is deployed for the state and observation sets, (2) the states are composed of the *user's intentions* and *"hidden system states"*, (3) the observations are the *user's utterances* and other observations being inferred from *lower-level information of the speech recognizer, robust parser, and from other input modalities*. Williams et al. [47, 48] further extended the Zhang model by adding the state of the dialogue from the perspective of the user which is hidden from the system's view (called *user's dialogue state*) to the state set and adding the confidence score into the observation set. All these approaches have shown that POMDP-based dialogue strategies outperform MDP counterparts (e.g., Pietquin [31]). Furthermore, these strategies cope well with different types of errors in a Spoken Dialogue System (SDS), especially with ASR errors. Table 2 describes characteristics of these POMDP-based DMs.

Table 2 Characteristics of some POMDP-based dialogue managers (n is the number of slots)

Application	$n,\|S\|,\|A\|,\|Z\|$	Algorithm	Reward function
Nursing home robot [34]	4, 13, 20, 16	AMDP [35]	If the system action is labeled as *correct*: 100, *ok*: -1, *wrong*: -100.
Tour guide [52]	3, 40, 18, 25	QMDP [24], FIB [16]	If the answer matches user's request, the reward is positive. Otherwise, the reward is negative.
Travel booking [45]	2, 36, 5, 5	Perseus [41]	If the system action & dialogue state is *ask & not stated*: -1, *ask & stated*: -2, *ask & confirmed*: -3, *confirmed & not stated*: -3, *confirmed & stated*: -1, *confirm & confirmed*: -2. If the user's goal is determined correctly: 50, incorrectly: -50.

5 The Factored POMDP Approach

Extending the Williams model [47], we represent our affective dialogue model as a factored POMDP [7]. Factored representation of the state set and observation set allows for a natural and intuitive way to encode the information state of the dialogue domain. For example, a goal-oriented dialogue system needs to encode at least variables such as the user's actions and the user's goals. In our model, the state set is composed of the user's goal (Gu), the user's affective state (Eu), the user's action (Au), and the user's grounding state (Du) (similar to the user's dialogue state described in [47]). The observation set is composed of the observations of the user's action (OAu) and the observations of the user's affective state (OEu). Depending on the complexity of the application's domain, these features can be represented by more specific features. For example, the user's affective state can be encoded by continuous variables such as *valence* and *arousal* and can be represented using a continuous-state POMDP [8]. The observation of the user's affective state might be represented by a set of observable features such as response speech, speech pitch, speech volume, posture, and gesture [4]. Similarly, a continuous-observation POMDP [44] can be used to incorporate the continuous-valued features into the observation space.

Figure 3b shows the structure of our affective dialogue model. The features of the state set, action set, observation set, and their dependencies form a two time-slice Dynamic Decision Network (2TDN). Technically, a factored POMDP is equivalent to a 2TDN. Implicitly, some assumptions are made in this model: the user's goal only depends on the user's goal in the previous slice and the system action from the previous slice only influences the user's emotion, the user's action, and the grounding state. We can easily modify this model for representing other dependencies, for example the dependency between the user's emotion and the observation of the user's action. Parameters $p_{gc}, p_{ae}, p_{ec}, p_e, p_{oa}$, and p_{oe} are used to produce handcrafted transition and observation models in case no real data is available (e.g.,

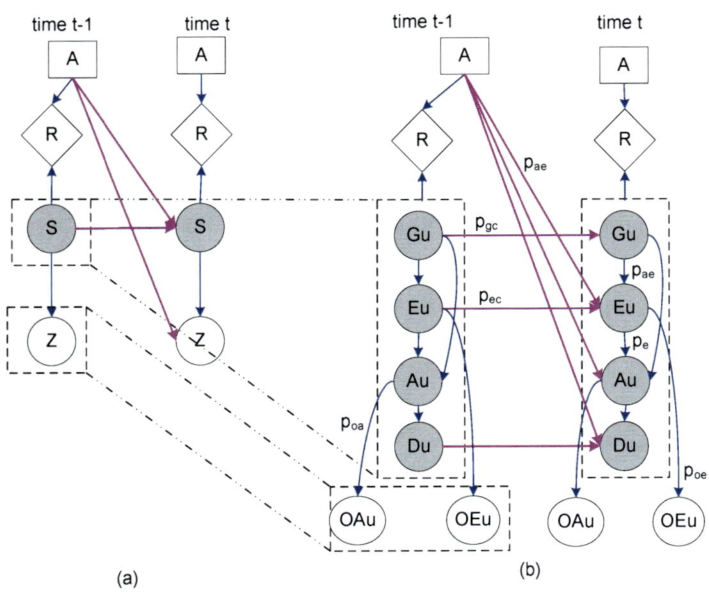

Fig. 3 (a) Standard POMDP and (b) Two time-slice of factored POMDP for the dialogue manager

at the initial phase of the system development), where p_{gc} is the probability of the user goal change; p_{ae} is the probability of the user's emotional change because of the influence of the system action such as when the system confirms an incorrect user's goal (represented by two causal links from the system action and user's goal to the user's affective state); p_{ec} is the probability that the user emotion change is due to the emotion decay and other causes; p_e is the probability of an error in the user's action being induced by emotion; p_{oa} and p_{oe} are the probabilities of the observation error of the user's action and the observation error of the user's affective state, respectively. The reward function is in principle different for each particular application. Therefore, it is not specified in our general affective dialogue model.

Suppose the set of user's goals has m values which are represented by $Gu = \{v_1, v_2, ..., v_m\}$. The features of S and Z and the action set A are formulated as follows:

- $Eu = \{neutral, stress, frustration, anger, happiness, ...\}$.
  Note that we can extend the representation of the user's emotion by adding more relevant features into the state space. For example, if the user's emotion is described by two dimensions *valence* and *arousal*. E_u then becomes a subnetwork with two continuous variables.
- $Au = \{answer(v), yes, no, ...\}$, where $v \in Gu$.
  The abstract format of Au is $userSpeechAct(v)$, where *userSpeechAct* is an element of the set of the user's speech acts.
- $Du = \{notstated, stated, confirmed, ...\}$.
- $OAu = \{answer(v), yes, no, ...\}$, where $v \in Gu$.
  The value $y \in OAu$ depends on the level of abstraction of the observation of the user's action. For example, if the observation of the user's action is sent by the ASR module, y is the word-graph or N-best hypotheses of the user's utterance. In our model, we assume a high level of abstraction for the observation of the user's action such as the output from a dialogue act recognition module or the intention level in the simulated user model [15]. In the latter case, OAu has the same set of values as Au.
- $OEu = \{neutral, stress, frustration, anger, happiness, ...\}$.
  Similar to the observation of the user's action, the observation of the user's affective state can be represented by a set of observable effects such as *response speed, speech pitch, speech volume, posture*, and *gesture* features [4]. In our current model, we assume that the observations of the user's affective states are the output of an AR module and therefore OEu has the same set of values as Eu.
- $A = \{ask, confirm(v), ...\}$, where $v \in Gu$.
  The abstract format of A is $systemSpeechAct(v)$, where *systemSpeechAct* is an element of the set of the system speech acts.

For a random variable X, we denote x and x' as the values of X at time $t-1$ and t, respectively. Based on the network structure shown in Figure 3b, the transition function is represented compactly as follows:

$$\mathscr{P}^a_{ss'} = P(g'_u|g_u)P(e'_u|a,e_u,g'_u)P(a'_u|a,g'_u,e'_u)P(d'_u|a,d_u,a'_u). \qquad (26)$$

$P(g'_u|g_u)$ is called the *user's goal model*, $P(e'_u|a,e_u,g'_u)$ is called the *user's emotion model*, $P(a'_u|a,g'_u,e'_u)$ is called the *user's action model*, and $P(d'_u|a,d_u,a'_u)$ is called the *user's grounding state model*. The observation function is as follows:

$$\mathscr{P}^{a}_{s'z'} = P(\tilde{a}'_u|a'_u)P(\tilde{e}'_u|e'_u), \tag{27}$$

where $\tilde{a}'_u \in OAu$ and $\tilde{e}'_u \in OEu$. $P(\tilde{a}'_u|a'_u)$ is called the *observation model of the user's actions* and $P(\tilde{e}'_u|e'_u)$ is called the *observation model of the user's emotions*.

6 User Simulation

The DM that we have described is a statistical DM. Dialogue corpora are usually used to train this type of DMs. However, the (PO)MDP-based DM has a huge number of states, therefore it is almost impossible to learn an optimal policy directly from a fixed corpus, regardless of its size [37]. To solve this problem, user simulation techniques have been used [15, 23, 31, 36, 38]. The main idea is a two-phase

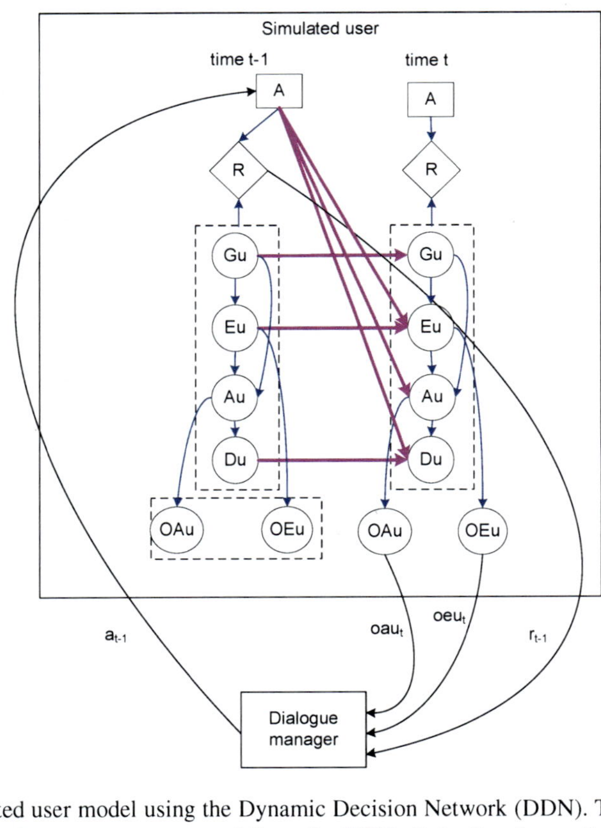

Fig. 4 Simulated user model using the Dynamic Decision Network (DDN). The user's state, action at each time-step are generated from the DDN. Only the observation of the user's action, affective state, and the reward are sent to the dialogue manager.

approach. A simulated user is first trained on a small human–computer dialogue corpus to learn responses of a real user given the dialogue context. The learning DM then interacts with this simulated user in a trial and error manner to learn an optimal dialogue strategy. Experimental results show that a competitive dialogue strategy can be learnt even with handcrafted user model parameters [51]. Recent work also demonstrated that user simulation can be used for testing dialogue systems in early phases of the iterative development cycle [1].

Our simulated user model, constructed based on the POMDP environment, is shown in Figure 4. The structure of this model is similar to the structure of the POMDP model (Fig. 3b), except that the state feature nodes (i.e., Gu, Eu, Au, and Du) in the simulated user model are observable from the user's perspective. If a corpus is available, we can use it to train the model structure and the parameters.

The process to generate observations of the user's actions and of the user's affective states is as follows: First, the value a_{t-1} from the DM is updated on node A of the time-slice $t-1$, the reward r_{t-1} is identified from node A of time-slice $t-1$. Second, the user's goal, affective state, action, and dialogue state are randomly generated based on the probability distributions of the nodes Gu, Eu, Au, and Du (of time-slice t), respectively. Third, the network is updated and the observation of the user's action oau_t and affective state oeu_t are randomly selected based on the probability distribution of nodes OAu and OEu. The tuple $< r_{t-1}, oau_t, oeu_t >$ is sent back to the DM.

7 Example: Single-Slot Route Navigation Example

We illustrate our affective dialogue model described in Section 5 by a simulated toy route navigation example: "A rescue worker (denoted by "the user") needs to get a route description to evacuate victims from an unsafe tunnel. To achieve this goal, he communicates his current location (one of m locations) to the system. The system can infer the user's stressed state and uses this information to adapt its dialogue policy."

In this simple example, the system can *ask* the user about their current location, *confirm* a location provided by the user, show the route description (*ok*) of a given location, and stop the dialogue (i.e., execute the *fail* action) by connecting the user to a human operator. The factored POMDP model for this example is represented by:

- $S = \langle Gu \times Au \times Eu \times Du \rangle \cup end$, where *end* is an absorbing state.

  1. $Gu = \{v_1, v_2, ..., v_m\}$
  2. $Au = \{answer(v_1), answer(v_2), ..., answer(v_m), yes, no\}$
  3. $Eu = \{e_1, e_2\} = \{nostress, stress\}$
  4. $Du = \{d_1, d_2\} = \{notstated, stated\}$

- $A = \{ask, confirm(v_1), ..., confirm(v_m), ok(v_1), ok(v_2), ..., ok(v_m), fail\}$,
- $Z = \langle OAu \times OEu \rangle$.

1. $OAu = \{answer(v_1), answer(v_2), ..., answer(v_m), yes, no\}$
2. $OEu = \{nostress, stress\}$

The full flat-POMDP model is composed of $4m^2 + 8m + 1$ states, $2m + 2$ actions, and $2m + 4$ observations, where m is the number of locations in the tunnel.

The reward model is specified in such a way that an (near) optimal policy helps the user obtain the correct route description as soon as possible and maintains the dialogue appropriateness [47]. Concretely, if the system *confirms* when the user's grounding state is *notstated*, the reward is -2, the reward is -3 for action *fail*, the reward is 10 when the system gives a correct solution (i.e., the system action is $ok(x)$ where x is the user's goal), otherwise the reward is -10. The reward for any action taken in the absorbing *end* state is 0. The reward for any other actions is -1. Designing a reward model that leads to a good dialogue policy is a challenging task. It requires both expert knowledge and practical debugging [13].

The probability distributions of the transition function and observation function are generated using the parameters $p_{gc}, p_{ac}, p_{ec}, p_e, p_{oa}, p_{oe}$ defined in Section 5 and two other parameters K_{ask} and $K_{confirm}$, where K_{ask} and $K_{confirm}$ are the coefficients associated with the *ask* and *confirm* actions. We assume that when the user is stressful, he will make more errors in response to the system *ask* action than the system *confirm* action because in our current model, the number of possible user actions in response to *ask* (m possible actions: $answer(v_1), answer(v_2), ..., answer(v_m)$) is greater than to *confirm* (2 actions: yes, no).

Concretely, the handcrafted models of the transition, observation and reward function are described as follows. The user's goal model is represented by:

$$P(g'_u|g_u) = \begin{cases} 1 - p_{gc} & \text{if } g'_u = g_u, \\ \frac{p_{gc}}{|G_u|-1} & \text{otherwise,} \end{cases} \quad (28)$$

where g_u is the user's goal at time $t-1$, g'_u is the user's goal at time t, $|G_u|$ is the number of the user's goals. This model assumes that the user does not change their goal at the next time step with the probability $1 - p_{gc}$.

The user's stress model:

$$P(e'_u|a, e_u, g'_u) = \begin{cases} 1 - p_{ec} & \text{if } e'_u = e_u \text{ and } a \in X, \\ p_{ec} & \text{if } e'_u \neq e_u \text{ and } a \in X, \\ 1 - p_{ec} - p_{ae} & \text{if } e_u = e'_u = e_1 \text{ and } a \notin X, \\ p_{ec} + p_{ae} & \text{if } e_u = e_1 \text{ and } e'_u = e_2 \text{ and } a \notin X, \\ p_{ec} - p_{ae} & \text{if } e_u = e_2 \text{ and } e'_u = e_1 \text{ and } a \notin X, \\ 1 - p_{ec} + p_{ae} & \text{if } e_u = e'_u = e_2 \text{ and } a \notin X, \end{cases} \quad (29)$$

where $p_{ec} \geq p_{ae} \geq 0$, $(p_{ec} + p_{ae}) \leq 1$ and $X = \{ask, confirm(g'_u), ok(g'_u)\}$. This model assumes that system mistakes (such as confirming the wrong item) would elevate the user's stress level.

The user's action model:

$$P(a'_u|a,g'_u,e'_u) = \begin{cases} 1 & \text{if } a = a_1,\, e'_u = e_1, \text{ and } a'_u = a_2, \\ 1 - p_1 & \text{if } a = a_1,\, e'_u = e_2, \text{ and } a'_u = a_2, \\ \frac{p_1}{|A_u|-1} & \text{if } a = a_1,\, e'_u = e_2, \text{ and } a'_u \neq a_2, \\ 1 & \text{if } a = a_3,\, e'_u = e_1, \text{ and } a'_u = a_4, \\ 1 - p_2 & \text{if } a = a_3,\, e'_u = e_2, \text{ and } a'_u = a_4, \\ \frac{p_2}{|A_u|-1} & \text{if } a = a_3,\, e'_u = e_2, \text{ and } a'_u \neq a_4, \\ 1 & \text{if } a = a_5,\, e'_u = e_1, \text{ and } a'_u = a_6, \\ 1 - p_2 & \text{if } a = a_5,\, e'_u = e_2, \text{ and } a'_u = a_6, \\ \frac{p_2}{|A_u|-1} & \text{if } a = a_5,\, e'_u = e_2, \text{ and } a'_u \neq a_6, \\ \frac{1}{|A_u|} & \text{if } a = ok(y) \text{ or } a = fail, \\ 0 & \text{otherwise,} \end{cases} \quad (30)$$

where $a_1 = ask$, $a_2 = answer(g'_u)$, $a_3 = confirm(g'_u)$, $a_4 = yes$, $a_5 = confirm(x)$, $a_6 = no$, $p_1 = p_e/K_{ask}$, $p_2 = p_e/K_{confirm}$, x & $y \in Gu$, and $x \neq g'_u$.

The main idea behind this handcrafted user's action model is explained as follows. When the user is not stressed, no communicative errors are made. When the user is under stress, they might make errors. The probability that the user makes no communicative errors when the system asks is $1 - p_1$ and when the system confirms is $1 - p_2$.

The user's grounding state model is handcrafted. It is represented by

$$P(d'_u|a,d_u,a'_u) = \begin{cases} 1 & \text{if } a = ask,\, d_u = d_1,\, a'_u = answer(x), \text{ and } d'_u = d_2, \\ 1 & \text{if } a = ask,\, d_u = d_1,\, a'_u \in \{yes,no\}, \text{ and } d'_u = d_1, \\ 1 & \text{if } a = confirm(x),\, d_u = d_1,\, a'_u = no, \text{ and } d'_u = d_1, \\ 1 & \text{if } a = confirm(x),\, d_u = d_1,\, a'_u \neq no, \text{ and } d'_u = d_2, \\ 1 & \text{if } a \in \{ask,confirm(x)\},\, d_u = d_2 \text{ and } d'_u = d_2, \\ 1 & \text{if } a \in \{ok(x),fail\} \text{ and } d'_u = d_1, \\ 0 & \text{otherwise.} \end{cases} \quad (31)$$

The observation model of the user's actions:

$$P(\tilde{a}_u|a_u) = \begin{cases} 1 - p_{oa} & \text{if } \tilde{a}'_u = a_u, \\ \frac{p_{oa}}{|A_u|-1} & \text{otherwise,} \end{cases} \quad (32)$$

where a_u and $\tilde{a}'_u$ are the user's action and the observation of the user's action, respectively. $|A_u|$ is the number of the user's actions. Parameter p_{oa} can be considered as the recognition error rate of the NLU module (see Section 2).

The observation model of the user's stress:

$$P(\tilde{e}_u|e_u) = \begin{cases} 1 - p_{oe} & \text{if } \tilde{e}_u = e_u, \\ \frac{p_{oe}}{|E_u|-1} & \text{otherwise,} \end{cases} \quad (33)$$

where e_u and $\tilde{e}_u$ are the user's stress state and the observation of the user's stress state, respectively. $|E_u|$ is the number of the user's stress states. Parameter p_{oe} can be considered as the recognition error rate of an affect recognition module used for the user's stress detection.

The reward function:

$$R(s,a) = \begin{cases} 0 & \text{if } s = end, \\ -2 & \text{if } a = ask \text{ and } g_u = stated, \\ -2 & \text{if } a = confirm(x) \text{ and } g_u = notstated, \\ 10 & \text{if } a = ok(x) \text{ and } g_u = x, \\ -10 & \text{if } a = ok(x) \text{ and } g_u \neq x, \\ -3 & \text{if } a = fail, \\ -1 & \text{otherwise.} \end{cases} \tag{34}$$

8 Evaluation

To compute a near-optimal policy for the route navigation example presented in Section 7, we use the Perseus solver[6] which is one of the state-of-the-art approximate POMDP solvers[7]. All experiments were conducted on a Linux server using a 3 GHz Intel Xeon CPU and a 24 GB RAM.

The performance of the computed policy is then evaluated using the simulated user presented in Section 6. The simulated user is implemented using the SMILE library[8]. The discount factor is only used for the planning phase. In the evaluation phase, the total reward for each dialogue session is the sum of all the rewards the system receives at each turn during the system–user interaction process (i.e., $\gamma = 1$). There are two reasons for this decision. First, in the dialogue domain, the number of turns in a dialogue session between the system and the user is finite. Second, the intuitive meaning of the discount factor ($\gamma < 1$) is reasonable for positive reward values but is not appropriate for negative reward values. For example, it is less likely that the second *confirm* action bears a smaller cost than the first one on a given piece of information provided by the user.

Note that when the discount factor γ is not used in the evaluation, it would be rational to set $\gamma = 1$ during the planning phase (our problem is guaranteed to terminate in a finite number of steps). However, Perseus requires that γ is smaller than 1. The planning time also increases when γ approaches 1. We argue that it does make sense to hand-tune the discount factor because it is not just a part of the problem description. As γ approaches 1, the agent becomes more farsighted [42].

---

[6] http://staff.science.uva.nl/ mtjspaan/pomdp/[accessed 2009-06-22]

[7] Although Perseus is made for flat POMDPs, this solver does exploit the factor representation of the state and observation space for computing near optimal policies. Alternatively, we can use the Symbolic Perseus solver. The advantage of this solver is mentioned in Section 8.4.

[8] http://genie.sis.pitt.edu[accessed 2009-06-24]

A dialogue session between the system and the user is defined as an *episode*. Each episode in the route navigation application starts with the system's action. Following the turn-taking mechanism, the system and the user exchange turns[9] until the system selects an *ok* or *fail* action (Table 3). The episode is then terminated[10].

The interaction between the DM and the simulated user is as follows. Both the dialogue manager (that uses the POMDP policy) and the simulated user exchange text messages through the iROS middleware[11]. Every time the dialogue manager performs *ok* or *fail* action, the current episode is finished and a new episode is started. The interaction ends when it reaches the number of episodes predefined by the experimenter.

Formally, the reward of each episode is calculated from the user side as follows:

$$R_e = \sum_{t=0}^{n} R_t(S_t, A_t), \tag{35}$$

where n is the number of turns of the episode, the reward at turn t $R_t(S_t,A_t)$ is equal to R_s^a (see Section 3) if the user's state and action at turn t is $S_t = s$ and the system action at turn t is a. Note that each turn is composed of a pair of system and user's actions except the last turn. Table 3 shows an example of a 3-turns episode of the single-slot route navigation application.

Table 3 An episode of the interaction between the system and the user

Turn	Utterance	Action	Reward
1	S_1 : Please provide the location of the victims?	*ask*	-1
	U_1 : Building 1.	$answer(v_1)$	
2	S_2 : The victims are in the Building 1. Is that correct?	$confirm(v_1)$	-1
	U_2 : Yes.	*yes*	
3	S_3 : Ok, the route description is shown on your PDA.	$ok(v_1)$	10
		Reward of the episode:	8

The average return of N episodes is defined as follows:

$$R_N = \frac{1}{N} \sum_{e=1}^{N} R_e \tag{36}$$

In the following sections, we first present the experiments for tuning three important parameters: the discount factor, the number of belief points (i.e., belief states) and the planning time. Second, we show the influence of the stress to the performance of computed policies. Third, we compare the performance of the approximate POMDP policies versus three handcrafted policies and the greedy action selection policy. We then conduct experiments to address tractable issues.

[9] Each time step in a formal POMDP definition is now considered as a turn.

[10] A telephone-based dialogue example in [13] shows that the episode ends when the user hangs up.

[11] http://sourceforge.net/projects/iros/[accessed 2009-06-24]

8.1 Parameter Tuning

Figures 5, 6, and 7 show the average returns of the simplest instance of the route navigation problem (three locations) where the near-optimal policy is computed based on different values of the discount factor, the number of belief points, and the run-time for the planning phase.

Based on the result shown in Figure 5, the discount factor value $\gamma = 0.99$ is selected for subsequent experiments that will be presented in this chapter. Interestingly, it turns out that the *de facto* discount factor value ($\gamma = 0.95$) that is usually used for POMDP-based dialogue problems from the literature (e.g., Williams et al. [49]) is not an optimal solution, at least for this example. It is worth noting that the off-line planning time to get an ε-optimal policy increases monotonically with the value of the discount factor especially when the convergence threshold ε is small. For example, the time to get a similar policy for γ is equal to 0.9, 0.95, and 0.99 is 7, 17, and 98 seconds, respectively. However, the time-bounded solution (Fig. 5) performs well when we conduct the test with our simulated user.

Figure 6 shows that a reasonable solution is achieved with the number of belief points starting from 200. Given a fixed threshold of the planning time, the number of iterations decreases when the number of belief points increases. That is why the average return of the case of 10000 belief points is low when the planning

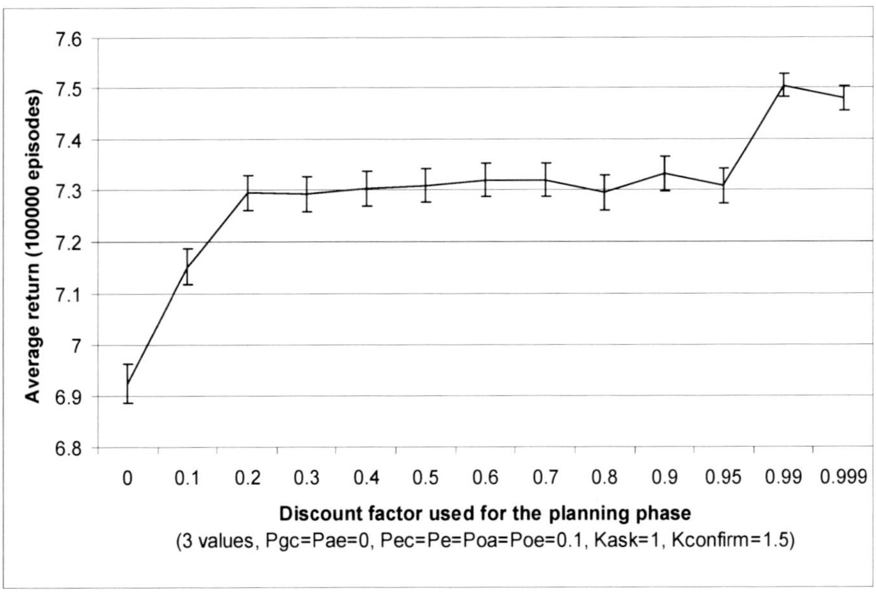

Fig. 5 Average return vs. the discount factor used for the planning phase. Error bars show the 95% confidence level. The threshold of the planning time is 60 seconds. Policies with $\gamma \leq 0.95$ converge ($\varepsilon = 0.001$) before this threshold.

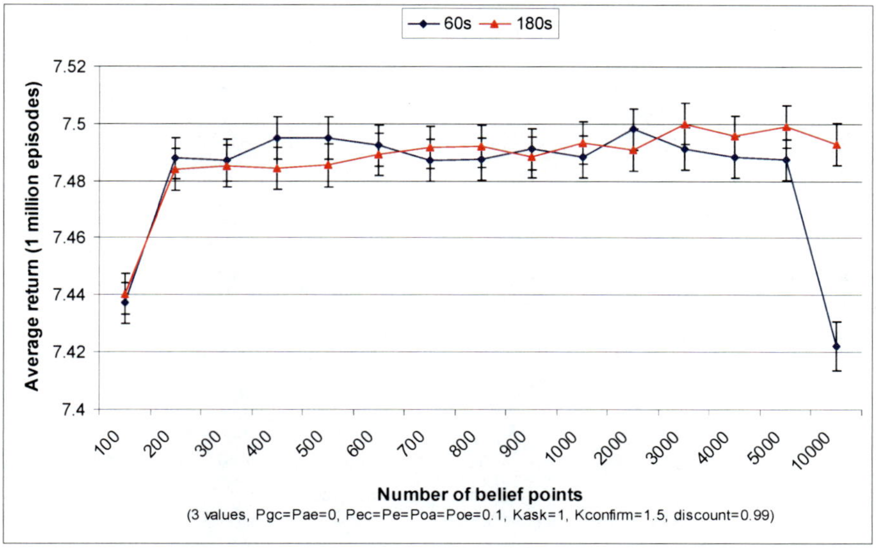

Fig. 6 Average return vs. number of belief points. Error bars show the 95% confidence level.

time is limited to 60 seconds. A default number of belief points for all subsequent experiments is set to 1000 (this value is a reasonable choice for the case that the number of locations is greater than 3).

Figure 7 shows that a stable solution is achieved when the planning time threshold is 60 seconds. For all the subsequent experiments for the 3-locations case, the default threshold for the planning time is set to 60 seconds. When the number of locations increases, the solver need a longer time to get a good solution. For example, the minimum time for the 10 locations case is 30 minutes.

8.2 *Influence of Stress to the Performance*

Figure 8 shows the influence of stress to the performance of two distinctive policies: the non-affective policy (SDS-POMDP) and the affective policy (ADS-POMDP). The SDS-POMDP does not incorporate the stress variable in the state and the observations set[12] (similar to the SDS-POMDP policy described in [49]). The ADS-POMDP is our affective dialogue model described in Section 5. The probability of the user's action error being induced by stress p_e changes from 0 (stress has no influence to the user's action selection) to 0.8 (the user is highly stressed and acts almost randomly). The average returns of both policies decreases when p_e increases. When stress has no influence on the user's action error, the average returns of the

[12] The SDS-POMDP policy is equivalent to the ADS-POMDP policy computed for the case $p_e = 0$.

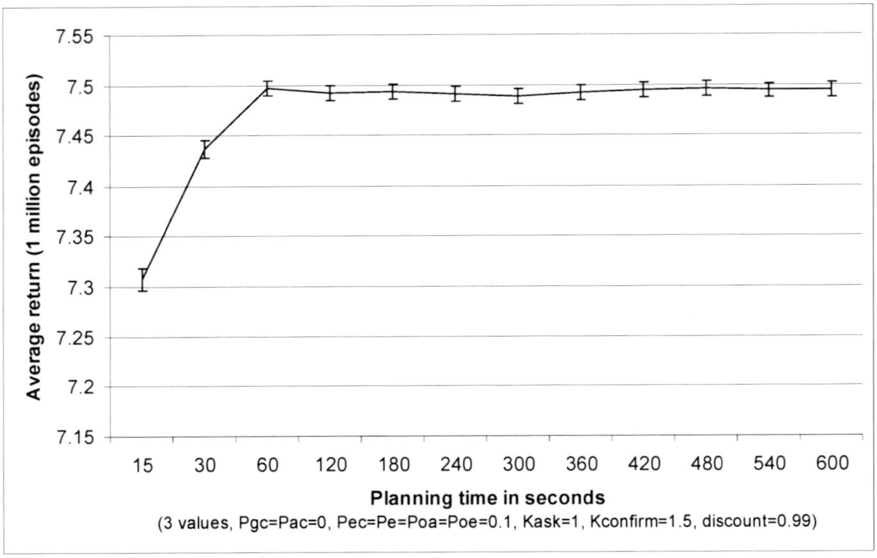

Fig. 7 Average return vs. planning time in seconds. Error bars show the 95% confidence level.

two policies are equal. When $p_e \geq 0.1$, the ADS-POMDP policy outperforms the SDS-POMDP counterpart[13].

Note that the affective POMDP policy performs better than its non-affective counterpart because it exploits the user's stress model. In reality, it is very challenging to obtain a reliable model of the user's stress. Therefore, it would be interesting to compare these two policies in an experiment with real users.

8.3 Comparison with Other Techniques

In this section, we evaluate the performance of the POMDP DM by comparing the performance of the approximate POMDP policy (ADS-POMDP) and four other dialogue strategies: HC1, HC2, HC3 (Fig. 9), and the greedy action selection strategy. HC1 is the optimal dialogue policy when $p_{gc} = p_e = p_{oa} = 0$ (the user's goal does not change; stress has no influence on the user's action and there is no error in observing the user's action, i.e., the speech recognition and spoken language understanding errors are equal to 0). HC1 and HC2 are considered as the non-affective dialogue strategies since they ignore the user's stress state. HC3 uses commonsense rules to generate the system behavior. The greedy policy is a special case of the POMDP-based dialogue with the discount factor $\gamma = 0$ (this strategy is similar to the one used in two real-world dialogue systems [29, 51]).

[13] We then assume that the ADS-POMDP policy is better than the non-affective MDP policy since Williams [46] demonstrated that the SDS-POMDP policy outperforms its MDP counterpart.

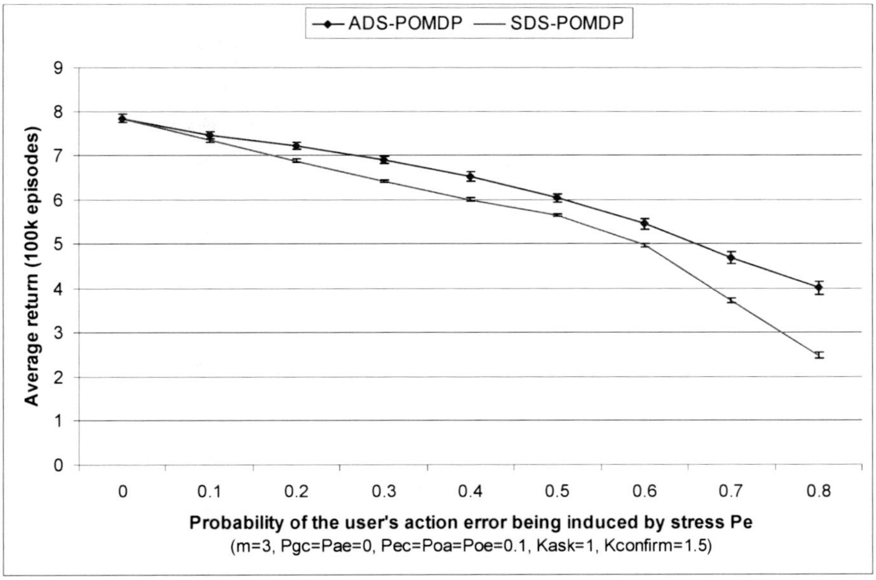

Fig. 8 Average returns of the affective policy and non-affective policy vs. the probability of the user's action error induced by stress p_e

As expected, the ADS-POMDP policy outperforms all other strategies (Fig. 10). HC3 outperforms its handcrafted counterparts.

8.4 *Tractability*

The ε-optimal policy presented in Sections 8.1, 8.2, and 8.3 is computed for the simplest instance of the single-slot route navigation example which is composed of only three locations ($m = 3$). In reality, even with a single-slot dialogue problem, the number of slot values m is usually large. Section 7 presents a connection between m and the size of the POMDP problem (state, action, and observation sets). For the single-slot route navigation example, the number of states is a quadric function of m. The number of actions and observations is also a linear function of m. In this section, we address the POMDP tractable issues by increasing the number of slot values gradually and trying to compute the ε-optimal policy for each case.

Perseus can only handle problems with $m \leq 15$. This is because although the approximate PBVI algorithms such as Perseus are able to handle the curse of history problem, the curse of dimensionality (i.e., the dimensionality of α-vectors grows exponentially with the number of states) remains (see [10, chap. 2] for further discussions about the POMDP tractability). Another practical issue is that the size of the optimized POMDP parameter file also increases exponentially in the dimension of m. A recently implemented POMDP solver, Symbolic Perseus, allows for a compact representation of the POMDP parameter files. Symbolic Perseus can help to

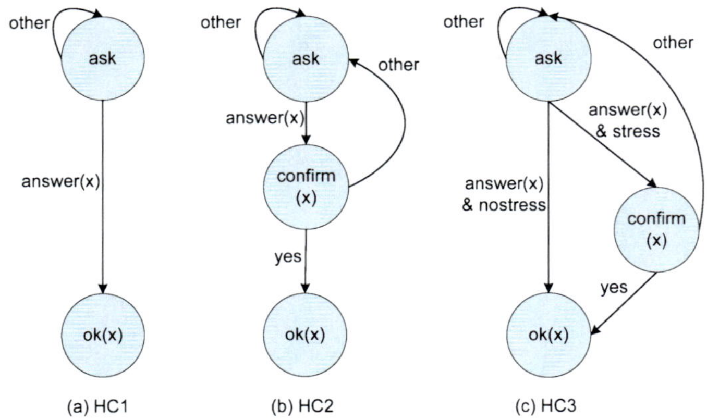

Fig. 9 Three handcrafted dialogue strategies for the single-slot route navigation problem (x is the observed location): (a) first *ask* and then select *ok* action if the observation of the user's action $\tilde{a}_u$ is *answer* (otherwise *ask*), (b) first *ask*, then *confirm* if $\tilde{a}_u = answer$ (otherwise *ask*) and then select *ok* action if $\tilde{a}_u = yes$ (otherwise *ask*), (c) first *ask*, then *confirm* if $\tilde{a}_u = answer$ & $\tilde{e}_u = stress$ and select *ok* action if $\tilde{a}_u = yes$

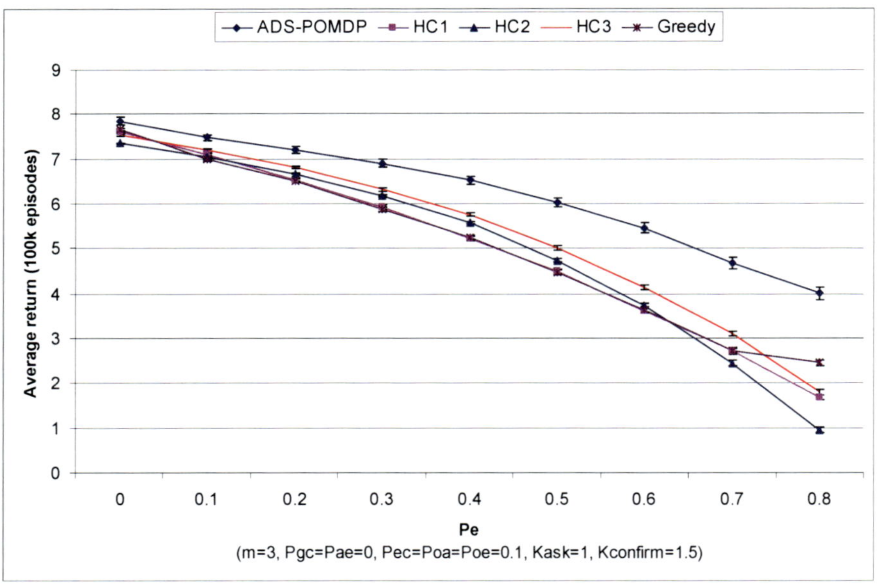

Fig. 10 Average return of the POMDP policy vs. other policies

scale the range of solvable problems to two orders of magnitude compared with the Perseus solver. As described in [18], it was demonstrated to solve a hand-washing problem with the size of 50 million states, 20 actions, and 12 observations. Although this is one of the most complex problems in the POMDP research community, it is only a toy problem compared with real-world dialogue problems. For example, the affective dialogue model for the RestInfo problem presented in [10, chap. 1] is composed of more than 600 million states and its spoken counterpart is composed of more than 300 million states. In [11], we proposed a solution to handle these complex problems by decomposing the dialogue model into two levels: global dialogue manger level and slot level dialogue manager level. The first is modeled using a set of simple rules. The second is first modeled as a factored POMDP similar to the one presented in this paper and then approximated as a set of Dynamic Decision Networks. More detailed information about this solution is presented in [11].

9 Conclusions

This chapter argues that POMDPs are appropriate for affective dialogue management. To support this argument, we have described the interaction process of a general affective dialogue system and illustrated how to specify the POMDP model for a simple empathic dialogue agent.

Extending from the previous POMDP-based dialogue management work, we have presented a factored POMDP approach for affective dialogue model design with a potential use for a wide range of applications. The 2TBN representation allows integration of the features of states, actions, and observations in a flexible way. The approach is illustrated through a route navigation example as a proof of concept. The experimental results showed that the affective POMDP policy outperformed its spoken counterpart and the handcrafted and greedy action selection strategies given the user's stress influences their behavior. We also conducted experiments to determine a best set of parameters (discount factor, number of belief points, and planning time). The results could be a good indicator for computing POMDP policy for other dialogue problems.

A key limitation of the current POMDP-based dialogue models is tractability. Given the mathematical soundness of the POMDP framework and recent efforts to resolve this limitation [11, 51], it would be worth to follow this direction for building robust dialogue systems.

References

1. Ai, H., Weng, F.: User simulation as testing for spoken dialog systems. In: Schlangen, D., Hockey, B.A. (eds.) Proceedings of the 9th SIGDial Workshop on Discourse and Dialogue (SIGdial 2008), Columbus, Ohio, USA, pp. 164–171 (2008)
2. André, E., Dybkjær, L., Minker, W., Heisterkamp, P. (eds.): ADS 2004. LNCS (LNAI), vol. 3068. Springer, Heidelberg (2004)
3. Astrom, K.J.: Optimal control of Markov processes with incomplete state information. Journal of Mathematical Analysis and Applications 10, 174–205 (1965)
4. Ball, E.: A Bayesian heart: Computer recognition and simulation of emotion. In: Robert Trappl, P.P., Payr, S. (eds.) Emotions in Humans and Artifacts, vol. 11, pp. 303–332. The MIT Press, Cambridge (2003)

5. Batliner, A., Fischer, K., Huber, R., Spilker, J., Nöth, E.: How to find trouble in communication. Speech Communication 40(1-2), 117–143 (2003)
6. Bhatt, K., Argamon, S., Evens, M.: Hedged responses and expressions of affect in human/human and human/computer tutorial interactions. In: Forbus, K., Gentner, D., Regier, T. (eds.) Proceedings of the 26th Annual Conference of the Cognitive Science Society (CogSci 2004), Chicago, Illinois, USA, pp. 114–119 (2004)
7. Boutilier, C., Poole, D.: Computing optimal policies for partially observable decision processes using compact representations. In: Proceedings of the 13th National Conference on Artificial Intelligence (AAAI 1996), Portland, Oregon, USA, vol. 2, pp. 1168–1175 (1996)
8. Brooks, A., Makarenkoa, A., Williamsa, S., Durrant-Whytea, H.: Parametric POMDPs for planning in continuous state spaces. Robotics and Autonomous Systems 54(11), 887–897 (2006)
9. Bui, T.H.: Multimodal dialogue management - State of the art. Tech. rep., University of Twente (2006)
10. Bui, T.H.: Toward affective dialogue management using partially observable markov decision processes. Ph.D. thesis, University of Twente (2008)
11. Bui, T.H., Poel, M., Nijholt, A., Zwiers, J.: A tractable hybrid DDN-POMDP approach to affective dialogue modeling for probabilistic frame-based dialogue systems. Natural Language Engineering 15(2), 273–307 (2009)
12. Bui, T.H., Rajman, M., Melichar, M.: Rapid dialogue prototyping methodology. In: Sojka, P., Kopeček, I., Pala, K. (eds.) TSD 2004. LNCS (LNAI), vol. 3206, pp. 579–586. Springer, Heidelberg (2004)
13. Bui, T.H., van Schooten, B., Hofs, D.: Practical dialogue manager development using POMDPs. In: Keizer, S., Bunt, H., Paek, T. (eds.) Proceedings of the 8th SIGdial Workshop on Discourse and Dialogue (SIGdial 2007), Antwerp, Belgium, pp. 215–218 (2007)
14. Bui, T.H., Zwiers, J., Nijholt, A., Poel, M.: Generic dialogue modeling for multi-application dialogue systems. In: Renals, S., Bengio, S. (eds.) MLMI 2005. LNCS, vol. 3869, pp. 174–186. Springer, Heidelberg (2006)
15. Eckert, W., Levin, E., Pieraccini, R.: User modelling for spoken dialogue system evaluation. In: Prococeedings of the IEEE Workshop on Automatic Speech Recognition and Understanding (ASRU 1997), pp. 80–87. IEEE, Santa Barbara (1997)
16. Hauskrecht, M.: Value-function approximations for partially observable Markov decision processes. Journal of Artificial Intelligence Research (JAIR) 13, 33–94 (2000)
17. Heylen, D., Nijholt, A., op den Akker, R.: Affect in tutoring dialogues. Applied Artificial Intelligence 19, 287–311 (2005)
18. Hoey, J., von Bertoldi, A., Poupart, P., Mihailidis, A.: Assisting persons with dementia during handwashing using a partially observable markov decision process. In: Proceedings of the 5th International Conference on Vision Systems (ICVS 2007), Bielefeld, Germany (2007)
19. Howard, R.A.: Dynamic Programming and Markov Process. The MIT Press, Cambridge (1960)
20. Jurafsky, D., Martin, J.: Speech and Language Processing: An Introduction to Natural Language Processing. Prentice Hall, Englewood Cliffs (2000)
21. Kaelbling, L.P., Littman, M.L., Cassandra, A.R.: Planning and acting in partially observable stochastic domains. Artificial Intelligence 101(1-2), 99–134 (1998)
22. Levelt, W.J.: Speaking: From Intention to Articulation. The MIT Press, Cambridge (1989)

23. Levin, E., Pieraccini, R., Eckert, W.: A stochastic model of human-machine interaction for learning dialogue strategies. IEEE Transactions on Speech and Audio Processing 8(1), 11–23 (2000)
24. Littman, M.L., Cassandra, A.R., Kaelbling, L.P.: Learning policies for partially observable environments: Scaling up. In: Prieditis, A., Russell, S.J. (eds.) Proceedings of the 12th International Conference on Machine Learning (ICML 1995), pp. 362–370. Morgan Kaufmann, Tahoe City (1995)
25. Martinovsky, B., Traum, D.R.: The error is the clue: Breakdown in human-machine interaction. In: Proceedings of the ISCA Tutorial and Research Workshop on Error handling in Spoken Dialogue Systems (EHSD 2003), Château d'Oex, Vaud, Switzerland, pp. 11–16 (2003)
26. McTear, M.: Spoken dialogue technology: Enabling the conversational user interface. ACM Computing Survey 34(1) (2002)
27. Monahan, G.E.: A survey of partially observable markov decision processes: Theory, models, and algorithms. Management Science 28-1, 1–16 (1982)
28. Ortony, A., Clore, G.L., Collins, A.: The Cognitive Structure of Emotions. Cambridge University Press, Cambridge (1988)
29. Paek, T., Horvitz, E.: Conversation as action under uncertainty. In: Proceedings of the 16th Conference on Uncertainty in Artificial Intelligence (UAI 2000), pp. 455–464. Morgan Kaufmann, San Francisco (2000)
30. Picard, R.W.: Affective Computing. The MIT Press, Cambridge (1997)
31. Pietquin, O.: A framework for unsupervised learning of dialogue strategies. Ph.D. thesis, Universitaires de Louvain (2004)
32. Puterman, M.L.: Markov decision processes. In: Heyman, D., Sobel, M. (eds.) Handbook in Operations Research and Management Science, vol. 2, pp. 331–434. Elsevier, Amsterdam (1990)
33. de Rosis, F., Novielli, N., Carofiglio, V., Cavalluzzi, A., Carolis, B.D.: User modeling and adaptation in health promotion dialogs with an animated character. Journal of Biomedical Informatics 39(5), 514–531 (2006)
34. Roy, N., Pineau, J., Thrun, S.: Spoken dialogue management using probabilistic reasoning. In: Proceedings of the 38th Annual Meeting of the Association for Computational Linguistics (ACL 2000), pp. 93–100. ACL, Hong Kong (2000)
35. Roy, N., Thrun, S.: Coastal navigation with mobile robots. In: Solla, S.A., Leen, T.K., Müller, K.R. (eds.) Advances in Neural Information Processing Systems 12, [NIPS Conference, Denver, Colorado, USA, November 29 - December 4, 1999], pp. 1043–1049. The MIT Press, Denver (2000)
36. Schatzmann, J., Thomson, B., Weilhammer, K., Ye, H., Young, S.: Agenda-based user simulation for bootstrapping a POMDP dialogue system. In: Human Language Technologies 2007: The Conference of the North American Chapter of the Association for Computational Linguistics (NAACL-HLT 2007), pp. 149–152. ACL, Rochester (2007)
37. Schatzmann, J., Weilhammer, K., Stuttle, M., Young, S.: A survey of statistical user simulation techniques for reinforcement-learning of dialogue management strategies. Knowledge Engineering Review 21(2), 97–126 (2006)
38. Scheffler, K., Young, S.J.: Automatic learning of dialogue strategy using dialogue simulation and reinforcement learning. In: Marcus, M. (ed.) Proceedings of the 2nd International Conference on Human Language Technology Research (HLT 2002), pp. 12–18. Morgan Kaufmann, San Francisco (2002)
39. Smallwood, R.D., Sondik, E.J.: The optimal control of partially observable Markov processes over a finite horizon. Operations Research 21-5, 1071–1088 (1973)

40. Sondik, E.J.: The optimal control of partially observable markov decision processes. Ph.D. thesis, Stanford University (1971)
41. Spaan, M.T.J., Vlassis, N.: Perseus: Randomized point-based value iteration for POMDPs. Journal of Artificial Intelligence Research (JAIR) 24, 195–220 (2005)
42. Sutton, R.S., Barto, A.G.: Reinforcement Learning: An Introduction. The MIT Press, Cambridge (1998)
43. Traum, D., Larsson, S.: The information state approach to dialogue management. In: van Kuppevelt, J., Smith, R.W. (eds.) Current and New Directions in Discourse and Dialogue, ch. 15, pp. 325–353. Kluwer Academic Publishers, Dordrecht (2003)
44. Williams, J., Poupart, P., Young, S.: Partially Observable Markov Decision Processes with Continuous Observations for Dialog Management. In: Williams, J., Poupart, P., Young, S. (eds.) Recent Trends in Discourse and Dialogue, chap, pp. 191–217. Springer, Heidelberg (2008)
45. Williams, J., Young, S.: Scaling up POMDPs for dialogue management: the summary POMDP method. In: Proceedings of the IEEE workshop on Automatic Speech Recognition and Understanding (ASRU 2005), Cancún, Mexico, pp. 250–255 (2005)
46. Williams, J.D.: Partially observable Markov decision processes for dialog management. Ph.D. thesis, Cambridge University (2006)
47. Williams, J.D., Poupart, P., Young, S.: Factored partially observable Markov decision processes for dialogue management. In: Zukerman, I., Alexandersson, J., Jönsson, A. (eds.) Proceedings of the 4th Workshop on Knowledge and Reasoning in Practical Dialog Systems (KRPD 2005), Edinburgh, Scotland, pp. 76–82 (2005)
48. Williams, J.D., Poupart, P., Young, S.: Partially observable Markov decision processes with continuous observations for dialogue management. In: Proceedings of the 6th SigDial Workshop on Discourse and Dialogue, SIGdial 2005 (2005)
49. Williams, J.D., Young, S.: Partially observable markov decision processes for spoken dialog systems. Computer Speech and Language 21(2), 393–422 (2007)
50. Young, S.: Talking to machines (statistically speaking). In: Proceedings of the 7th International Conference on Spoken Language Processing (ICSLP 2002), Denver, Colorado, USA, pp. 9–16 (2002)
51. Young, S., Gasić, M., Keizer, S., Mairesse, F., Schatzmann, J., Thomson, B., Yu, K.: The hidden information state model: A practical framework for pomdp-based spoken dialogue management. Computer Speech and Language (2009)
52. Zhang, B., Cai, Q., Mao, J., Guo, B.: Spoken dialog management as planning and acting under uncertainty. In: Proceedings of the 7th European Conference on Speech Communication and Technology (EUROSPEECH 2001), Aalborg, Denmark, pp. 2169–2172 (2001)
53. Zhang, N.L.: Efficient planning in stochastic domains through exploiting problem characteristics. Tech. Rep. HKUST-CS95-40, Hong Kong University of Science and Technology (1995)
54. Zhang, N.L., Zhang, W.: Speeding up the convergence of value iteration in partially observable markov decision processes. Journal of Artificial Intelligence Research (JAIR) 14, 29–51 (2001)

Context-Aware Multimodal Human–Computer Interaction

Siska Fitrianie, Zhenke Yang, Dragoş Datcu,
Alin G. Chiţu, and Léon J.M. Rothkrantz

1 Introduction

Crisis response and management involve the collaboration of many people. To perform and coordinate their activities, they must rely on detailed and accurate information about the crisis, the environment, and many more factors. To ensure collaboration of emergency services and high-quality care for victims, the ability to supply dynamic and contextually correlated information is necessary. However, current approaches to construct globally consistent views of crises suffer from problems identified in [60]: (a) the setting of events is constantly changing, (b) the information is distributed across geographically distant locations, and (c) the complexity of the

Siska Fitrianie
Man Machine Interaction, Delft University of Technology,
Mekelweg 4 2628CD The Netherlands
e-mail: s.fitrianie@tudelft.nl

Zhenke Yang
Man Machine Interaction, Delft University of Technology,
Mekelweg 4 2628CD The Netherlands
e-mail: z.yang@tudelft.nl

Dragoş Datcu
Man Machine Interaction, Delft University of Technology,
Mekelweg 4 2628CD The Netherlands
e-mail: d.datcu@tudelft.nl

Alin G. Chiţu
Man Machine Interaction, Delft University of Technology,
Mekelweg 4 2628CD The Netherlands
e-mail: a.g.chitu@tudelft.nl

Léon J.M. Rothkrantz
Man Machine Interaction, Delft University of Technology,
Mekelweg 4 2628CD The Netherlands and Netherlands Defence Academy,
Faculty of Technical Sciences, Den Helder, The Netherlands
e-mail: l.j.m.rothkrantz@tudelft.nl

R. Babuška & F.C.A. Groen (Eds.): Interactive Collaborative Information Systems, SCI 281, pp. 237–272.
springerlink.com

crisis management organization makes it difficult and time consuming to collaborate and verify obtained information.

The lack of consistent overviews is not the only limiting factor in adequate and timely response to crises. Acquisition of detailed and accurate information about the crisis situation is of key importance. Such information can be collected from a variety of sources, e.g., bystanders, rescue workers, and sensors. Analysis of past disasters (9/11, hurricanes Katrina and Rita by [51]) points to communication as a limiting factor. Current approaches to coordinate rescue and response activities suffer from the problem that information is neither current nor accurate. Unreliable communication networks, chaotic environments, and stressful conditions can make communication during crises difficult. The intense nature of crisis situations is believed to result in short-term memory loss, confusion, difficulties in setting priorities, and making decisions [26]. These result in fragmented and badly structured communication [64]. How well the user's situation is understood can impact the quality of service of the system.

Recent developments in technology offer citizens possibilities for diverse mobile communication devices, e.g., laptop/tablet PC, PDA, and smart-phone/mobile phone. Police departments show interest in utilizing mobile handheld devices, such as PDAs and smart phones, to allow officers to exchange information with control rooms and headquarters quickly and efficiently [36]. A demand for mobile devices and applications is also predicted for other service professionals, such as firefighters and paramedics [50]. As mobility becomes ubiquitous, multimodality becomes the inherent basis of the interaction paradigm. Multimodal user interfaces can offer various ways for users to interact in a more natural fashion [55]. By integrating various modalities, multimodal systems allow users to switch among modes or use one mode to enhance and complement the other. Multimedia services can be generated in a coordinated fashion on a combination of multiple modality channels [47]. This extension of interaction can provide users with more choices to find their own optimal experiences.

The integration of multimodal interaction possibilities into human–computer interaction (HCI) provides challenges on several levels, such as: (1) how to accommodate the flexible switching between communication modes and devices taking into account the different capabilities of the devices and of the available I/O modalities, (2) how to adapt the application and present information to the user taking into account context variables (e.g. profile, emotion, location) which may change over time due to mobility and dynamic environment, and (3) how to offer a unified view of services under the technology constraints and dynamic changes. These challenges yield serious interface and interaction design issues.

The introduction of novel information and communication technology in crisis management can help to provide more detailed and accurate situation overviews that can be shared among all management levels [51]. Toward this goal, the research reported here focuses on investigating methodologies to improve information acquisition and exchange during crisis response. This includes investigations on a HCI system that is able to adapt and improve its behavior based on the user's situation. We believe that our research can lead to better decision, planning, and reasoning in crisis situations. We developed a framework for building HCI systems for crisis management, which allows us to change, add or remove devices, modalities, roles, and functionalities in a convenient way.

In this chapter, we discuss the framework from a technological point of view. We present the proposed architecture and the different modules that have been developed. We focus on the communication between users via different devices. Each of these modules regards a different modality such as text, speech, lip movement, visual language, and face recognition. This allows us to determine which of the natural human communication channels are most appropriate for a given situation. The framework provides mechanisms to fuse these modalities into a context-dependent interpretation of the current situation and generate the appropriate multimodal information responses. Here, we present our approaches in (1) recognizing inputs, (2) interpreting and merging incoming information, (3) managing user interaction and (4) specifying and producing context-sensitive (and user-tailored) information—by the employment of scripts and ontology from and to multi-user, multi-device, and multimodal systems. The latest includes allocating and coordinating information across media, i.e. typed or spoken language, visual language, and graphics.

2 Related Work

2.1 Multimodal Systems

The use of human natural communication, such as speech and facial expressions, is often considered more intuitive and therefore expected to be more efficient in HCI. To date, a number of multimodal systems have been developed, e.g., SmartKom [71], COMIC [53], Match [37], MACK [13], and AdApt [33]. Some systems use an ontological language for encoding features and information presentation, for instance, M3L in SmartKom [71] and RDF-OWL in COMIC [32], others use XML (e.g. in Match [37] and MACK [13]). SmartKom utilizes an overlay operation [2] to integrate user inputs, while COMIC translates them into logical forms containing all possible dialogue moves. Most multimodal systems are designed for single-user services.

Effective HCI in multimodal systems requires an interaction manager that models the user's goal, the state of the dialogue, and the system's response at each turn of the dialogue. Many approaches have been proposed [48]. Dialogue strategies defined in tables [77], frames [14], or task hierarchies [58] have been used to select a specific action in a dialogue flow. Usually, a set of rules and a state machine are used to analyze the structure of the dialogue and to select a specific strategy. Others have used statistical techniques, e.g., Bayesian Networks [41]. Emotion has been included to help disambiguate user's dialogue acts [9]. Van Vark et al. analyzed over 5000 recorded dialogues of public transport information and classified them into a set of predefined dialogue acts [69]. The acts were interconnected by directed links that represent transitions from one act to another. To each link, a probabilistic value was assigned derived from the corpora analysis [61]. The chosen dialogue strategy selects the system action with the highest probability over the course of a dialogue. Based on the POMDP technique, Bui et al. model the user's affective state, intentions, and hidden states in a computation to select the next action [10]. The findings of ECA [12] address the impact of emotion on the dialogue strategy and service performance.

As part of the interaction manager, information presentation to the user plays a significant role. Information presentation often is framed into: deciding what to communicate and how to communicate it. A presentation planner retrieves the dialogue information and generates specifications for language and display presentation. Dialogue information contains information about user, system beliefs, dialogue history, and communication device constraints that are considered in all processes. The language generation typically uses some linguistics approaches, such as TAG [38] in SmartKom, n-grams and CCG [67] in COMIC, and template-based in Match and AdApt. Techniques like stack in Match, schema-based in COMIC and XIMS, frame-based in AdApt, and rule-based in MACK are applied for coordinating virtual agents' behaviors, graphical displays, and speech.

2.2 *Visual Languages for Human Observation Reporting*

After September 11, 2001, efforts to develop technology in crisis management emphasized the development of more sophisticated ICT-based emergency support systems. For example, a web-based reporting interface in CAMAS [49] and VCMC [54] allows users to send reports using natural language messages.

Entering text messages during (a time critical) crisis situation can be a difficult and tedious process. Visual languages provide an alternative to text and writing as a mode of comprehension and expression [65], for example, to communicate about topics that are difficult to speak about, or to support people not sharing a common language to communicate [15, 43], or for the speech impaired [1, 4]. In a visual language, a message can be composed using an arrangement of icons or using combinations of symbols to compose new symbols with new meanings [8, 34]. The symbols can be arranged in a rigid sequence order [5, 28, 43] or freely placed in a two-dimensional plane [29, 34]. A visual language-based communication interface for reporting observations in crisis situations has been developed in [28, 29]. A dedicated linguistic grammar was used to interpret the visual messages. Another visual language-based interface dedicated for sharing and merging icon-based topological maps in damaged buildings has been developed by [68]. The maps can be used for reasoning about the state of the building and providing guidance to given locations. A review of the use of existing icons for hazard and emergency maps, including those of US military and NATO, has been performed by [24]. Standard map icons are promoted for emergency response applications by the U.S. Government [35].

3 Human Computer Interaction Framework

We propose a framework that allows for the rapid construction and evaluation of multimodal HCI systems. The framework aims at module integration that is independent of the availability of modalities. Using the framework, a HCI system for reporting observations is developed, which is able to collect observations, interpret them automatically, and construct a global view of the reported events.

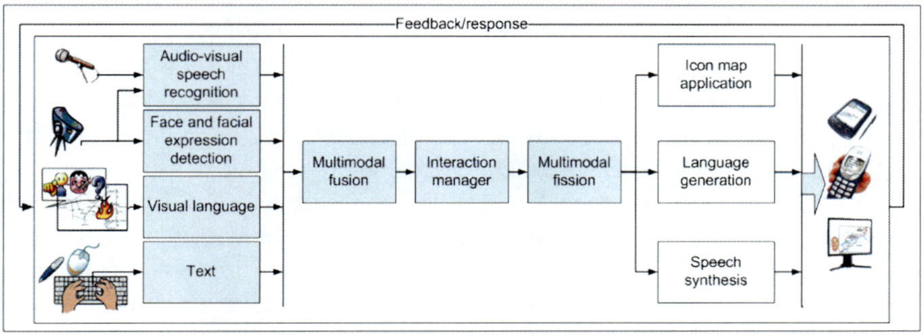

Fig. 1 The framework architecture for a single user

The framework includes input recognition modules of different modalities: text, speech, lip movements, visual language, and facial expression (Fig. 1). The output combines text, synthesized speech, visual language, and control of the underlying user interface. The multimodal input is combined and interpreted in the fusion component. This process includes interpretation of the user's affective state from facial expressions. The fusion here is local (each user). The Integration Manager-IM generates appropriate responses, and the fission component displays these using synchronized modalities.

The framework is designed to support various roles within the crisis management, including civilians and professionals in the field and control room. A centralized Fusion Manager processes and integrates every newly reported situation from all users,

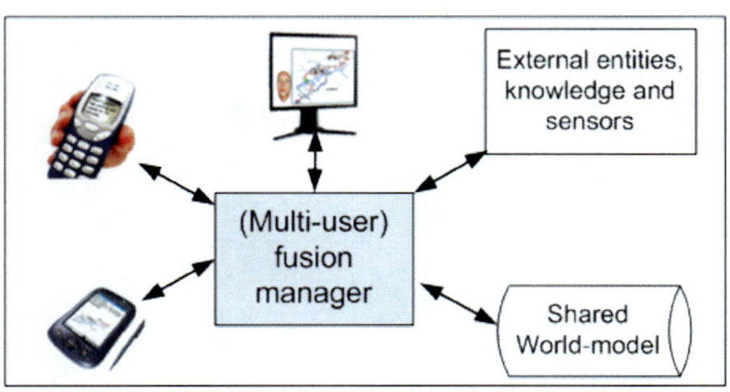

Fig. 2 A schematic overview of the communication system: multimodal, multi-device, and multi-user

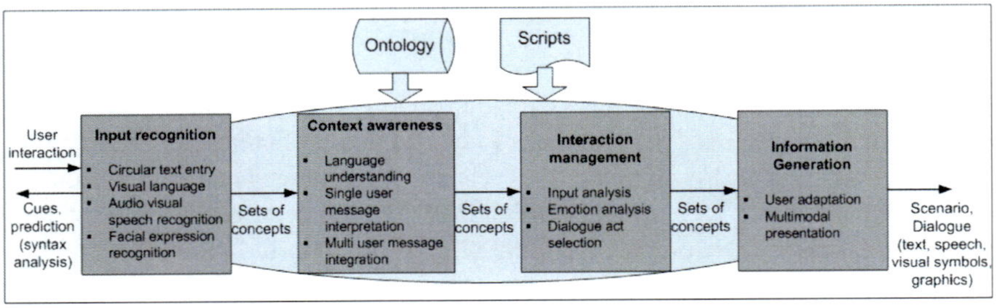

Fig. 3 A schematic overview of the research coverage

adapts the world view accordingly and then sends it back to the network (Fig. 2). All modules are integrated in an ad-hoc fashion using the iROS middleware [37].

The research reported here spans four fields (Fig. 3): the field of user input recognition, of message interpretation (context-aware computation), of user interaction management, and that of information generation. The knowledge representation used in the framework is introduced in the next section. After this, the different components in our developed framework architecture will be explained. Next, we present the results of some experiments with the framework. Finally, we conclude the chapter with a discussion on future work.

4 Corpus and Expert-Based Knowledge Representation

During crisis situations, responders in the field have to communicate with overseers in a crisis center to achieve effective crisis response. Figure 4 shows a schematic view of a crisis situation. A crisis responder (agent) is an autonomous participant that can report observations, e.g., what they see, hear, smell, and even feel or experience. These data are transformed into reports for others. At any time, each agent

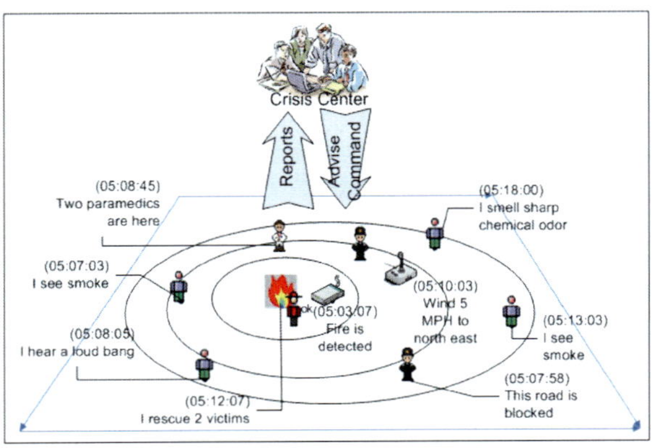

Fig. 4 Disaster dispersion (ellipses) occurs on the world; agents in the field report their observation

is associated with a location in the geographical space. The crisis center collects the reports and forms a global view about the crisis (e.g., its location and magnitude). An intelligent system helps to calculate the most probable causes, predict the impact of the crisis, and give advice.

In this research, world knowledge formation based on user reports was investigated. To understand how people report during crisis situations, three different data sources have been analyzed: (1) over 100 dialogues from 9-1-1 recording transcriptions [63, 74], (2) news articles with six different crisis topics (fire, bombing, flooding, tsunami, earthquake, and terrorist attack), and (3) interview results with experts from the Southern Rotterdam Fire Department and the crisis center of the Rijnmond environmental agency (DCMR) [6]. In addition, a workshop was held to acquire knowledge on formulating icon-based messages from text messages [27]. A large number of icon-based message corpora were analyzed. From these studies, we found that related icons are placed close to each other, grouped, or emphasized by links. These studies serve a number of purposes:

- developing a list of possible crisis scenarios,
- developing a list of concepts that play an important part in crisis situations,
- analyzing relations between concepts, and
- designing and implementing an interface for reporting observations.

The results show that the information extracted from the user reports is used to generate a hypothesis concerning the crisis. A report usually contains certain keywords and specific terms called features. A feature is described as the result of human perception, such as sound, vision, and odor. The hypothesis is based on these features and considerations of a set of typical crisis scenarios, such as fire, explosion, and chemical contamination. Each scenario has associated features, some of which can be important factors to distinguish the scenarios.

The information from different observers has to be combined to have a better view about the reported crisis. With the information from the aforementioned data sources, a set of predefined crisis scenarios and their associated features (concepts) were formulated. Next, the scenarios were tested to make sure that each scenario was

Table 1 Dialogue frames and their slots

Frame	Slots
user	{name, address(street, no, city, postcode), phone-no, role["VICTIM", "WITNESS", "PARAMEDIC", "FIREMEN", ...], emotion-type}
problem	{type["FIRE","TRAP",'SICK", ...], status, desc}
time	{hour, minute, second, day, month, year}
location	{street, no, city, postcode}
involved_party	{type["VICTIM", "SUSPECT", "PARAMEDIC", ...], status, address(street, no, city, postcode), name, phone-no}
reason	{type["HIT", "EXPLOSION","GAS-LEAKING", ...]}
weapon	{type["GUN", "KNIFE","BOMB", ...]}
urgency	{type[HIGH, MEDIUM, LOW]}

uniquely identifiable by its set of concepts. Finally, the concepts were developed into the system's ontology, while the scenarios were developed into scripts. The ontology defines and formalizes general semantic relations between concepts, while a script represents the chain of events that identify a possible scenario [62]. Both ontology and scripts are employed in the world knowledge construction.

Based on the findings, the dialogue model of the framework was designed. It is constructed using the principle to fill frames and their slots in an efficient manner (Table 1). Since we have only a limited number of recorded corpora, we cannot make a good assessment of the probability of success at every dialogue turn (as done in [61]). Instead, the implementation uses a heuristics rule-based approach. The choice of the next action is based on the evaluation of how much a dialogue step will reduce the distance to the goal state (see section 7).

4.1 Using Ontology to Represent the World, the User and the Task

To be able to form awareness of the current situation and the state of the user and to support consistent communication, knowledge about the world, the user, and the task is necessary. In our framework, the knowledge representation language is RDF-OWL. Classes in the ontology are specified according to how they stand in relation to other classes. Binary relations between classes are defined as "properties." We designed these relationships using thematic roles. The concept `Escort`, for example (Fig. 6(b)), has properties `agent`, `patient` (of type `Actor`), `source`, and `destination` (of type `Location`). Thus, an instance of `escort` describes that the agent escorts the patient from source to destination.

The world model is a representation of two geo-referenced contexts: (1) a sequence of time-specified events and a group of dynamic objects in action at a

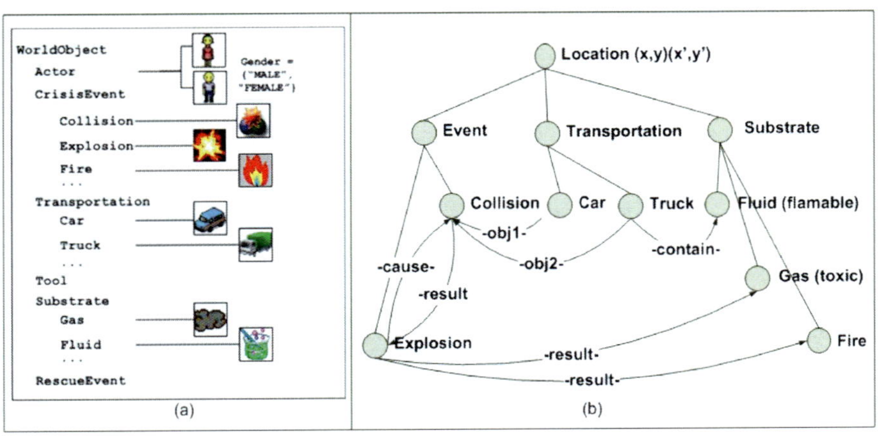

Fig. 5 (a) The WorldObject class Taxonomy; the icons are instances of a class, e.g., the icon depicting smoke is an instance of "Gas" and (b) a graph-based representation of "a Collision of a Car and a Truck result in an Explosion; the explosion causes toxic Gas and Fire"

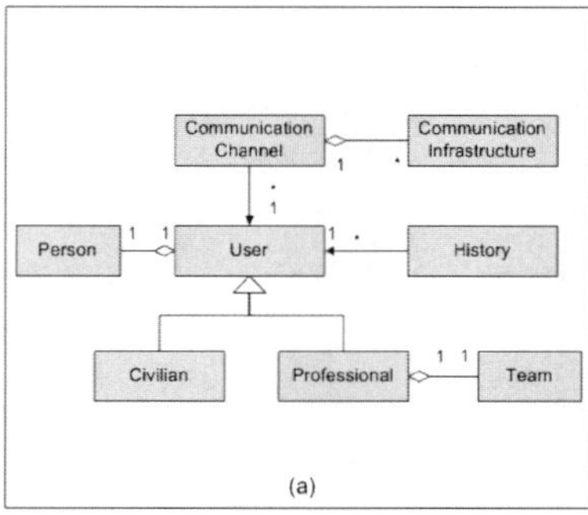


```
...
<dialogueActType>statement</dialogueActType>
<priority><value>1</value></priority>
<communicationchannel>
  <communicationinfrastructure>
    <mode>short-message</mode><mediatype>text</mediatype>
    ...
  <communicationInfrastructure>
  ...
</communicationchannel>
<concept>
  <name>Escort</name>
  <property><name>agent</name>
    <value><instance id="WM01Police01"type="Police"/></value>
  </property>
  <property> <name>patient</name>
    <value><instance id="WM01Paramedic01" type="Paramedic"/></value>
  </property>
  <property><name>source</name>...</property>
  <property><name>destination</name>...</property>
</concept>
...
```
(b)

Fig. 6 (a) User model class diagram and (b) a dialogue act: "informing a user that a policeman is going to escort the paramedic from a source location to a destination"; the police and the paramedic are referred by an id in the world model

certain location and (2) geographical information concerning fixed objects in the crisis location, such as buildings, streets, and parcels. The first context can be considered as the dynamic part, and the second as the static part of the world model. Both parts have direct links conceptually to the icons on the user interface (Fig. 5(a)). In the map interface in Fig. 9, concepts in the dynamic context are linked to icons overlayed on the map (static context) of the crisis area.

The geospatial model of crisis situations is described using graphs. The nodes in the graph represent objects, actions, or events in the world. They contain information including current status (e.g., living condition, spatial state), temporal information (e.g., frequency, timestamp), and spatial information (e.g., source, destination, path). The arcs represent the hierarchy of individuals, and the arrows show the relations among nodes (e.g. cause, contain). At the root, a node describes the perspective location of the event (see illustration in Fig. 5(b)).

A user model is a representation of a registered user (Fig. 6(a)). It contains the identity, emotional states, interaction time, communication media and modalities, and location in the world. Two types of users (civilian and professional) can be distinguished. `Person` stores most static information about a user, while the dynamic information is in `User` and archived in the `History`. One or more Communication Channels can be registered by a user; a channel can have one or more communication infrastructures, which contain information about available media and modalities.

The HCI flow is controlled by an IM, which defines dialogue acts and sends them to the Fission module. Ten possible dialogue acts have been implemented: (1) statement: to notify about an event, (2) request: to command into a specific action, (3) ask: to request specific information, (4) confirmation: to ask a clarification, (5) acknowledge: to display incoming information, (6) highlight: to highlight a certain information, (7) removal: to remove specific information from displays, (8) affirmation or (9) negation: to notify agreement to the user action, and (10) scenario: to display the crisis scenario. The dialogue acts are defined in the ontology as subclasses of a class "Process", which contains priority level, set of destination

```xml
<script name="accident" type="acknowledgement">
      <frame name="number" logic="and" weight="60">
            <condition operator="biggerThan" function="number">
                  <concept name="Transportation" propertyName="number"/>
                  <constant value="1" type="int"/>
            </condition>
            <condition operator="biggerEqualThan" function="number">
                  <concept name="Victim" propertyName="number"/>
                  <constant value="1" type="int"/>
            </condition>
      </frame>
      <frame name="location" logic="and" weight="1">
            <condition operator="lessEqualThan" unit="meter">
                  <condition function="distant">
                        <concept name="Transportation" propertyName="location"/>
                        <concept name="Road" propertyName="Topology"/>
                  </condition>
                  <constant value="1" type="double"/>
            </condition>
      </frame>
      <frame name="number" logic="and" weight="100">
            <condition operator="biggerThanEqual" function="number">
                  <concept name="Collision" propertyName="number"/>
                  <constant value="1" type="int"/>
            </condition>
      </frame>
      ...
</script>
```

Fig. 7 An Example of Scripts (in XML)

communication channels, timing, link to other processes, and reference to the world model. Figure 6(b) shows an example of a dialogue act message.

4.2 Scripts

In this research, we assume that knowledge about crisis situations is represented in the human mind as many scripts. Triggered by certain key events being perceived by the human senses, e.g., sight, hearing, smell, a script can be recognized even without having perceived all the events that constitutes the script [75]. A script consists of some frames of events, which are distinguished by some conditions (Fig. 7). A weight value is assigned on each frame to show how important the frame is. A frame is activated if all the conditions in the frame are satisfied. We sum the importance values of activated frames to measure how attractive the script is to be selected.

As the system gets a notion of the events occurring by interpreting the concepts that occur in the user messages, certain scripts become more plausible than others. As more concepts are perceived, the evidence in favor of the correct scenario increases. This in turn increases the plausibility of the corresponding of script.

Using only concepts that occur in the user message, only scripts that contain concepts entered by the user have a chance of being selected. Using knowledge in ontology, certain unobserved concepts can be inferred from the existence of other observed concepts even though not specifically entered by the user, e.g., if there is a collision there is an accident. As concepts in ontology have direct links to certain scripts, derived concepts can influence the importance value of the related scripts. If there is more than one script, the system will select one with the highest importance value.

5 Input Recognition

The variety of implemented input modules (Fig. 8) allows us to apply the framework to support various roles within the crisis management, including rescue workers, civilians, and control room operators. Since we consider that most of the roles are mobile users, the design takes the limitation of mobile devices, such as (1) small screen size and (2) limited options for user interaction, into account. The speech, text, and visual language input modules offer less error and are less tedious by adapting their predictive ability according to user's personal language usage, input context, and syntax rules. The learning component updates the personal dictionary from the user's inputs during interaction and personal document storages (offline).

A map interface for supporting people with geospatial information was designed (Fig. 9). Visual language, text, and speech are provided on the top of this map interface. A coherent and context-dependent interpretation and textual crisis scenario of the inputs are constructed as feedback to the user input. Figure 10 shows the implemented user interactions on the map interface. As messages are constructed, the system will continuously create or adapt its internal representation of the state of the world and try to find the scenario that matches the messages.

5.1 Audio–Visual Speech Input

Speech is the most important way of communication among human beings. Over the years much work has been done in the domain of automatic speech recognition

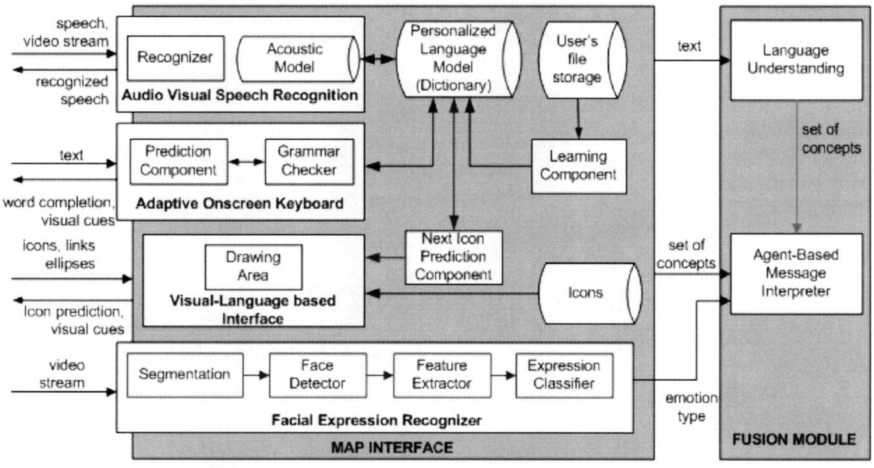

Fig. 8 A schematic view of the architecture of the input modules

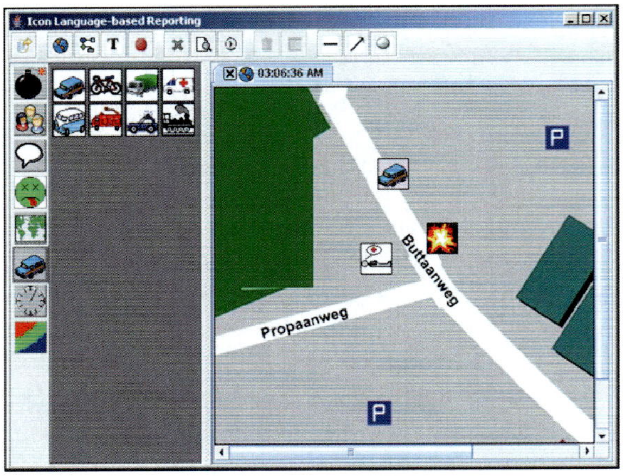

Fig. 9 Map interface

(ASR). The progress made is significant for small, medium, and even large vocabulary systems. However, the level of accuracy of the current ASR systems suffers greatly when the background noise increases. To cope with significant background noise in crisis situations (explosions, police and fire brigade cars), in recent years we developed an audio–visual speech recognition module (Fig. 11(a)) [17]. In our approach, features are extracted from the visual and audio modalities and combined into a common feature vector which is then fed into a set of HMMs. As the sampling rate of audio is much higher than video, the number of audio features (phonemes) is also much higher than the number of video features (visemes) per time interval. As a result, we have to interpolate the video features to match the audio stream.

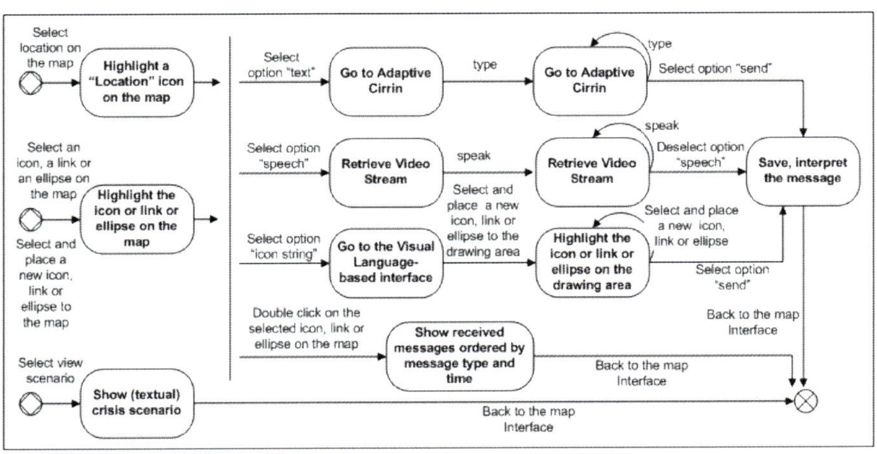

Fig. 10 The transition diagram of the current implemented user interaction on a map interface

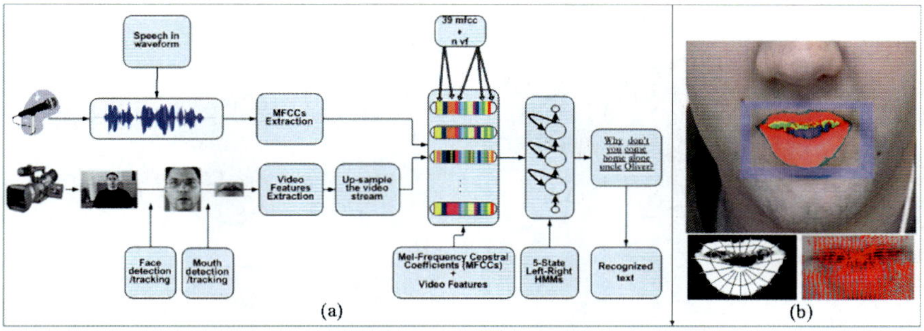

Fig. 11 (a) Audio–visual speech recognition module's architecture and (b) visual features extraction (left below: mouth shape estimation, right below: optical flow analysis)

We use the Mel-Frequency Cepstrum Coefficients (MFCC) [22] to parameterize the speech data. Our module uses two approaches for estimating the visual features: (1) estimating the shape of the mouth (lip contours and thickness) in time and (2) capturing the actual movement of the mouth using an optical flow-based algorithm [44]. On the source frame (Fig. 11(b)), we superimposed the lip area in red, the cavity of the mouth not obscured by tongue and teeth in green, and the teeth area in blue (the blue margins are highlighting the mouth region). Our current research is aimed at assessing the performance of our system both for clean audio–visual recordings and for situations where there is significant noise in both the audio and visual channels. A special case which we are about to investigate is the performance of our system on emotional data.

5.2 *Circular Text Entry*

The QWERTY layout is not optimal for pen-based text entry because the distance between common adjacent characters is too far [45]. Word prediction and completion can improve entry performance but searching through a (long) word list is considered as tedious and disruptive [3] or visually demanding due to the dynamic changing layout of characters. Additionally, research shows that visually guided tapping is easier for novice users, and gesture-based input is preferred by experts [76]. Based on this, an adaptive on-screen keyboard for both single-handed and zero-handed users is developed [30]. It combines tapping-based and motion-based text input with language-based acceleration techniques, including personalized and adaptive task-based dictionary, frequent character prompting, word completion, and grammar checker with suffix completion. We adopt the Cirrin device's [46] method to display all characters in a circular way (Fig. 12(a)).

The middle of the ring is an input area, where selected characters of a single word are displayed. The visual cue on a key gives information about the likelihood of the next character selection. In the fisheye mode (Fig. 12(b)), some keys are expanded based on their probability and distance to the cursor. The characters are arranged based on a scoring function to calculate the most used adjacent characters with two

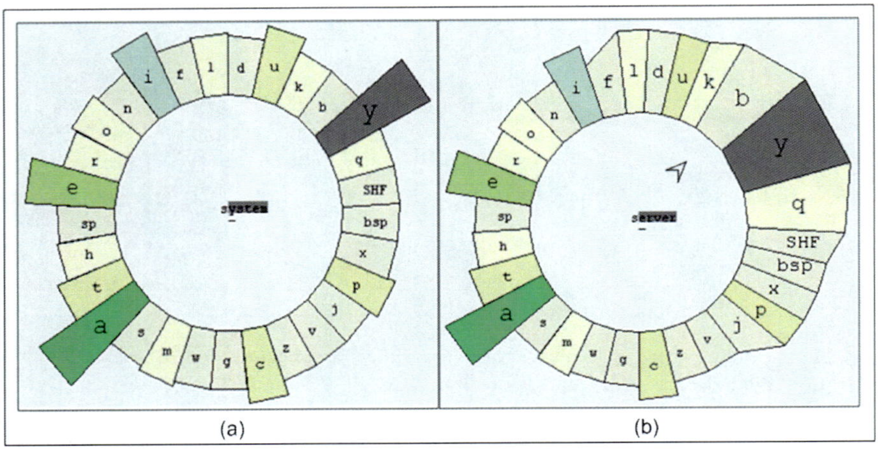

Fig. 12 Adaptive Cirrin with (a) visual cues and word completion, and (b) fisheye style

additional characters: space and backspace. An additional 6 x 5 matrix is placed on the right side of the circle for numbers, shift, return, control, period, punctuations, and comma. The user can select a word completion with a single tap (in tapping mode) or a left-to-right line motion (in motion mode) on the input area. The system will flush and append this word to the user input. The development of the prediction aims at improving the input speed by personalizing the prediction and the quality of syntax by suggesting only syntactically plausible words.

5.3 Visual Language-Based Message

Due to the small size of mobile devices, direct icon manipulation is one of the fastest interaction alternatives to date [42]. In addition, the use of icons to represent concepts makes user interaction in a language-independent context possible [56].

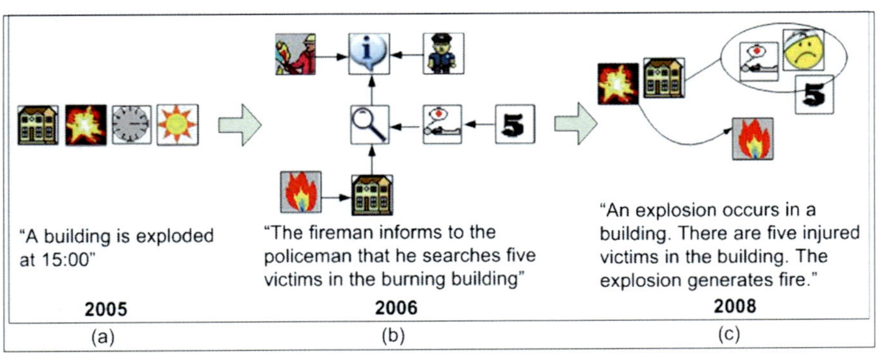

Fig. 13 Examples of visual language-based messages: in a sequential order, (b) in a 2D configuration and (c) using lines, arrows, and ellipses in sentence constructions

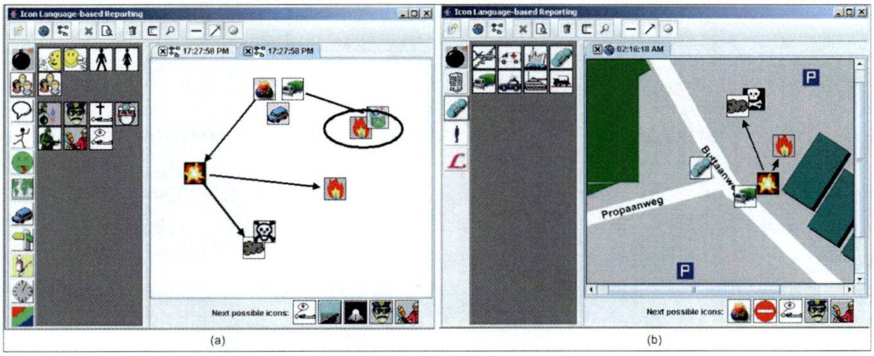

Fig. 14 Visual language-based messages: on (a) a drawing area and (b) a map

The icons represent objects and events in the world (e.g., fire, car, smoke). The descriptive meaning of icon-based messages can show a more direct link to multimodal concepts and a more direct mapping with "real-world" phenomenon [52]. In previous study, we have developed a translation module that uses a dedicated grammar to interpret and convert visual messages into natural language text and speech by adopting approaches of Natural Language Processing (Fig. 13(a) [28], Fig. 13(b) [29]). Hints, navigation and visual cues, next icon prediction, and search engine are provided to help users with message creation. However, the grammar still forces the users to follow linguistic lines behind sentence constructions. This makes the previous solutions somewhat rigid and requires the users to learn a "new" grammar. In contrast to the linear ordering of words in spoken language, a visual language has a multi-dimensional structure with meaning in temporal as well as in the spatial configuration, e.g., the sign language syntax for deaf people [7], comic illustrations [18], and diagrams [16].

Based on these properties of visual language, we explored a method to interpret the way in which people link and group certain concepts to describe events in [31]. This resulted in a communication interface that allows for a free and natural way of creating a spatial arrangement of graphics symbols (i.e. icons, lines, arrows, and shapes) to represent concepts or ideas (Fig. 13(c)). Figure 14(a) shows the developed interface, which provides a drawing area where users can attach and re-arrange icons. The users can group related icons by drawing an ellipse around them or placing them relatively near to each other but far from other icons. Icons (or groups of icons) can be linked using a line to represent a direct link or an arrow to represent a causality link (an icon generates/results another icon). The interface supports the users with mechanisms to draw and delete ellipses and links. An icon can be selected by three options: (a) the icon menu in which the icons are grouped based on their concepts, (b) the next icon prediction results, which are calculated by adapting an n-gram word prediction technique, and (c) the keyword-based search engine. Figure 14(b) shows a variation of the observation interface, where the icon language can be formed on a map. The map gives additional information such as the world coordinate position of an icon and the spatial relation of the icon to static objects, e.g., buildings, streets, and parcels. Currently, we are designing and developing a set of icons for each crisis scenario.

5.4 Face Detection and Facial Expression Recognition

The system assesses user's emotional state by using cues from nonverbal communication channels during HCI. We extract parametric information with high discrimination power from the face space and use it in a multi-layer data-driven classification environment [20, 21]. The dynamics of each facial expression are captured by considering the significant changes in feature domain that occur during the transition from neutral emotional state to the apex of each emotion. A fixed time window is set as support frame for performing temporal appreciation on the presence of the dominant emotion along the window video frames (see Fig. 15). Throughout the runtime, the results of distinct temporal analysis are combined to create smooth transitions between the outcomes from successive video segments. The facial expression recognition model is based on an algorithm that uses action units (AUs) that are part of the facial action coding system (FACS) [25], in combination with support vector machines (SVM) as a classifier for distinguishing the correct emotion. The AUs depict the correlation between the contractions of each facial muscle taken as single or in combinations with other muscles and the appearance of the face. Active appearance model [19] extracts information related to the shape and texture of each detected face and collects them as a set of facial characteristic points (FCPs). The

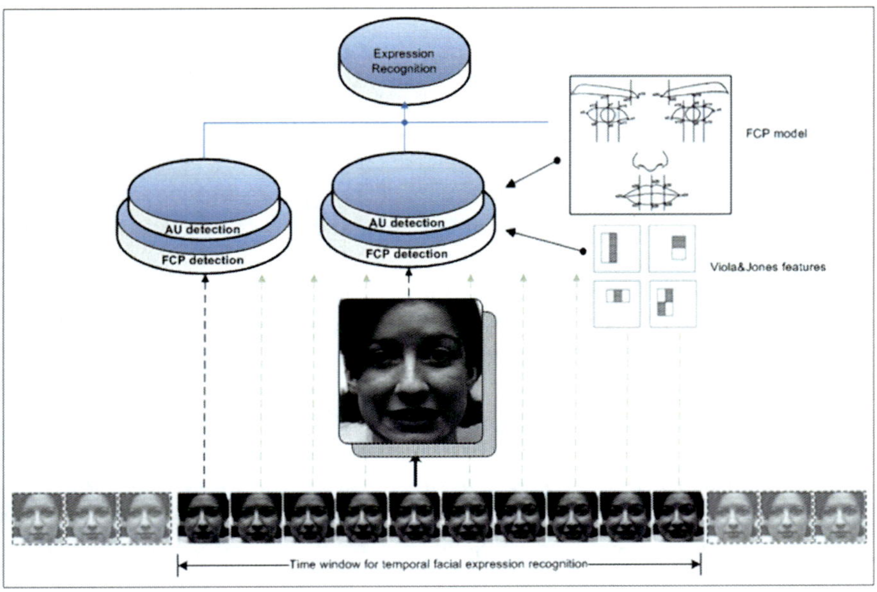

Fig. 15 Time window-oriented multi-layer facial expression recognition model for video sequences

presence of AUs is determined through a classification routine at a second layer of data processing. At this layer, the inputs of the classifier are parameters representing distances and angles from certain FCPs located on the face as detailed in the specifications of the Kobayashi and Hara model. Subsequently, the location of each FCP is determined during the third step again through a classification process. In this case, the inputs of the classifier are Viola&Jones features [70] computed directly from the frame data.

During the training stage, the classifier aims at identifying the set of the most relevant features that best represent the characteristics of each FCP. The data set we used for the research consisted of a reduced collection of emotional video samples selected from the Cohn-Kanade AU-Coded Facial Expression Database [39]. The facial expression recognition component operates in cascade immediately after the completion of a face detection module. The FCP localization module utilizes the proper set of Viola&Jones features associated with the appropriate degree of rotation of the face area in the video frame as provided by the face detection module. The true positive rates of six facial expressions recognition using an SVM classifier are above 80%.

6 Ontology-Based Computed Context Awareness

Our framework is designed to allow actors to send messages using a mobile communication device. Human actors can create messages using available modalities. The fusion module in the framework builds a structure of concepts locally (see Fig. 16). The concepts are supplied by all the available input recognition modules. The resulting structure is a contextually and temporally linked network of concepts. This

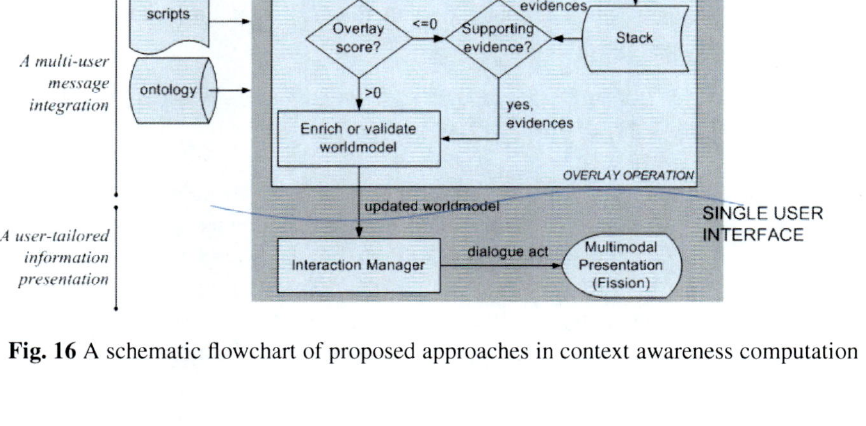

Fig. 16 A schematic flowchart of proposed approaches in context awareness computation

network is handed over to the IM to assist it in forming feedback to the user. The IM sends the single user's processed message to the (multi-user) Fusion Manager. The Fusion Manager attempts to relate the existing structure of the current system's world model with the new message and then broadcasts the latest world knowledge. Together with a selected dialogue act from the IM, the fission module of a single-user interface adapts the current world model and produces a user-tailored information presentation. All users receive the most up-to-date information about the current situations on their communication interface. Ontology and scripts are utilized in all processes.

6.1 Language Understanding

The language understanding component uses the Proxem Antelope [59] to extract syntactical and semantical aspects from the user text input. The Proxem splits text into sentences and sentences into words. It returns the grammatical relations between words in the sentence, like the part-of-speech of each word (e.g., noun, verb, and adjective) in a sentence and syntax dependency between words/phrases (e.g., subject, direct-object, preposition-object). The Proxem also returns the semantical aspects of each sentence, like the object and subject co-references (anaphora analysis) and thematic roles of the words/phrases (e.g., agent, experiencer, cause, location). Although the current implementation of the semantic parser is still experimental, it does give good performance on simple sentences. This language component retrieves a set of concepts from each sentence in the following algorithm:

- Replace all co-reference words (e.g., it, she, this) into the actual object if possible.
- Collect all words/phrases (mostly nouns and verbs) and find corresponding concepts in the ontology.
- Instantiate these concepts.
- Using the thematic roles of the words in the sentence, apply the relation between words (or their corresponding concepts) to fill in the properties of the selected concepts. This process can be done since the properties of the classes in the ontology were also designed using thematic roles.

All resulting instances of concepts found in the user text input are sent to the agent-based message interpretation.

6.2 Single-User Message Interpretation

In local message interpretation, each concept represents a puzzle piece (Fig. 17(b)). Each piece can have properties to be filled in by other concepts. This creates relations between them. A relation or some relations may create a new concept. By relating all or some important pieces, we can slowly see the picture emerging, i.e., the observer's view. This observer's view is used to activate a script. The selection of the right script is the goal of the interpretation process. Inspired by the work of

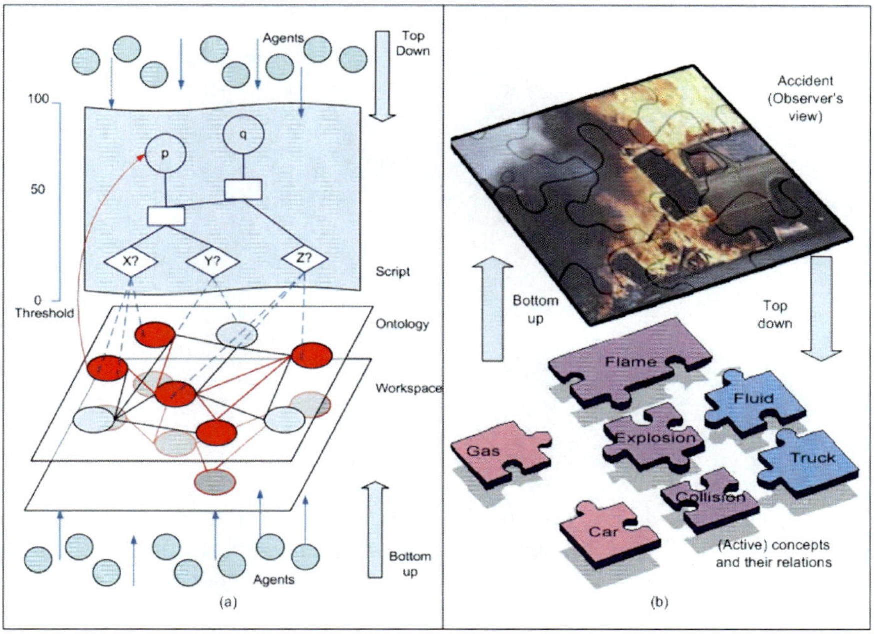

Fig. 17 (a) Agent-based approach for icon-based messages interpretation and (b) concept-based puzzle arrangement metaphor

Dor [23], we approach this puzzle arrangement problem by a hybrid agent-based architecture (Fig. 17(b)).

The interpretation of a user message is derived from purposive behavior emerging from interaction between multiple concepts on a user workspace (see Fig. 17(a)). *Workspace* is where perceptual structures are built on user input containing all instances of concepts retrieved from the user message. *Ontology* holds predefined concepts in the form of a graph of concepts and properties. As *agents* find matching instances in the workspace, they activate their corresponding concept property in the ontology and dynamically build relationships among these concepts and message structures. This is called the bottom-up pressure. In addition, when a concept in the ontology is activated, it will try to assign as many properties as possible by launching more agents to search for related values to strengthen the concept. As a result, an active concept will spread some activation to other relevant concepts. Furthermore, an active concept can activate one or more *scripts*. This is the top-down pressure shown in Fig. 17(a).

An active script can launch more agents to evaluate its conditions over time. The competing scripts with the most active concepts become more plausible. As a result, certain scripts become impossible and are removed from competition. This process continues until there is only one script left. At this point, it is assumed that a sufficient and coherence structure of the communicated message interpretation is produced.

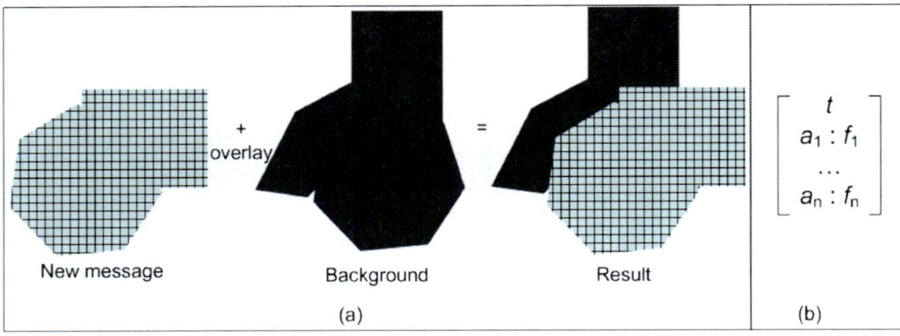

Fig. 18 (a) Schematic view of overlay operation and (b) graphical representation for denoting a TFS

6.3 Multi-user Message Integration

To form an aggregated world view based on observations of multiple users, the overlay operation is utilized [2]. Overlay is a formal operation based on unification of typed feature structures—TFS [11] (Fig. 18(b)). It can be seen as placing two typed feature structures—covering and background—on top of each other (Fig. 18(a)). The background can be viewed as the old state of the world, while the covering is an incoming message.

We assume a finite set of types $\{t, t_1, ...\}$ a finite set of typed features related within a inheritance hierarchy $\{a, a_1, ...\}$, where a_i denotes a feature whose value is restricted to the type t_i. A TFS is denoted as $\{t, [a_1 : f_1, \ldots, a_n : f_n]\} \in TFS$ where t is the type, and $[a_1 : f_1, \ldots, a_n : f_n]$ denotes the set of feature–value pairs of the TFS. With $[a_1 \ldots a_n]$ the features and $[f_1 \ldots f_n]$ the values which can be either atomic values or another TFS (allowing for recursion).

All inputs are aggregated in the following way:

1. Map the RDF-OWL document of a new message into TFS, as follows: (1) collect all active concepts, (2) fill all properties with actual values (from the instances), and (3) map all concepts into a corresponding TFS_i with properties $[a_{1i} \ldots a_{ni}]$ and values $[f_{1i} \ldots f_{ni}]$ (see example in Fig. 19).
2. Assimilate the background to the covering. The assimilation of a covering TFS (c) into the background TFS (b) (denoted by $c\|b$) transforms c such that $c\|b := (\{t_b, [a_{c1} : f_{c1}, \ldots, a_{cn} : f_{cn}]\}, b)$ if t_c is a super type of t_b. Otherwise (if t_b is a supertype of t_c), b is transformed: $c\|b := (c, \{t_c, [g_i : w_i, \ldots, g_i : w_i]\})$ with $g_i = lub(t_c, t_b)$. The function $lub(t_a, t_b)$ computes the most specific common super-type a and b.
3. Overlay the background b to the covering a where $a = \{t_a, [a_1 : f_1, \ldots, a_n : f_n]\}$ and $b = \{t_a, [b_1 : g_1, \ldots, b_m : g_m]\}$ and $t'_b = lub(a, b)$:

$$overlay(a, b) = overlay'(a, b\|a) \quad (1)$$

The $overlay'(a,b)$ function is defined as four cases:

a. If recursion:

$$overlay'(a,b) = \{t_a, [c_i : h_i | c_i = a_j = b_k, h_i = overlay(f_j, g_k)]\} \quad (2)$$

where $f_j, g_k \in TFS$

b. If the covering and the background have (atomic) values:

$$overlay'(a,b) = \{t_a, [c_i : h_i | c_i = a_j = b_k, h_i = f_i]\} \quad (3)$$

where $f_j \in A$

c. If a feature is absent in the background:

$$overlay'(a,b) = \{t_a, [c_i : h_i | c_i = a_j, h_i = f_i, c_i \neq b_k, 1 \leq k \leq m]\} \quad (4)$$

d. If a feature is absent or has no value in the covering:

$$overlay'(a,b) = \{t_a, [c_i : h_i | c_i = b_k, h_i = g_k]\}. \quad (5)$$

4. Calculate overlay score. This score [57] is defined to reflect how well the covering fits the background in terms of non-conflicting features:

$$score(co, bg, tc, cv) = \frac{co + bg - (tc + cv)}{co + bg + tc + cv} \in [-1 \ldots 1] \quad (6)$$

where co is the total number of non-conflicting or absent features for cases Eqs. 2, 3 and 4 and values in the covering for cases Eqs. 3 and 4; bg is the total number of non-conflicting or absent features in the background for cases Eqs. 2, 3 and 5 and values for cases Eqs. 3 and 5; tc is the total number of not-identical types of both the covering and the background; and cv is the total number of conflicting values (case Eq. 3, if the value of a feature of the background is overwritten). The $score(co, bg, tc, cv) = 1$ indicates the feature structures are unifiable, and $score(co, bg, tc, cv) = -1$ indicates all information from the background has been overwritten.

5. Enrich or validate the state of the world (see example in Fig. 19). The current implementation processes all messages that have the overlay score > 0. Messages with overlay score ≤ 0 are stored in a stack until some evidences support them (validated by a script). In particular, an activation level of active concepts in the ontology is defined. The level will grow if new messages include these concepts and decay at predefined intervals, otherwise. By this mechanism, only up-to-date (observed) concepts are active in the aggregated world model. We expect that the system's knowledge of the world is built based on reliable messages; while those unreliable messages by the mechanism eventually will be discarded.

15:20:10 An observer reports: **“A truck is on fire on Buttanweg”** **⇔ in XML:** ... **<fire>** **<causal>** **<location>** ... **</location>** **<number> 1 </number>** **</causal>** **<location>** ... **</location>** **<beginTime>** ... **</beginTime>** **</fire>** ...	**Initial world model ⇔** *TFS:* [*FIRE* *causal* : [*TRUCK* *location* : [*ADDRESS* *streetName* : *Buttanweg*] *number* : 1] *location* : ... *beginTime* : ...]
15:21:02 A policeman reports: **“A truck is exploded on Buttanweg 321 creating fire”** ***co*** **= 14;** ***bg*** **= 12;** ***tc*** **=1;** ***cv*** **= 7;** ***score*(*co, bg, tc, cv*) = 0.529** **The message is validated by a script: the fire is caused due to explosion. It enriches and validates the world model.**	**updated world model ⇔** *TFS:* [*FIRE* *causal* : [*EXPLOSION* *causal* : [*TRUCK* *location* : [*ADDRESS* *streetName* : *Buttanweg* *number* : 321] *number* : 1] *location* : ... *beginTime* : ...] *location* : ... *beginTime* : ...]
... a conflicting message coming: **15:22:07 An observer reports:** **“Awful smell in the air near Propaanweg”** ***co*** **= 0;** ***bg*** **= 0;** ***tc*** **= 1;** ***cv*** **= 0;** ***score*(*co, bg, tc, cv*) = -1**	**a new message (stored in the stack) ⇔** *TFS:* [*GAS* *hazardouslevel* : *unknown* *location* : ... *beginTime* : ...]
... after sometime, a new evidence is coming .. **15:29:55 A fireman reports: “The truck (on Buttanweg 321) contains toxic fluid”** ***co*** **= 3;** ***bg*** **= 21;** ***tc*** **= 0;** ***cv*** **= 0;** ***score*(*co, bg, tc, cv*) = 1** **This evidence supports the previous message. It is validated by a script: the toxic fluid in fire may release toxic gas. Both messages enrich the world model.**	**updated world model ⇔** *TFS:* [*GAS* *causal* : [*FIRE* *causal* : [*EXPLOSION* *causal* : [*TRUCK* *location* : [*ADDRESS* *streetName* : *Buttanweg* *number* : 321] *contain* : [*FLUID* *hazardousLevel* : *HIGH* ....] *number* : 1] *location* : ... *beginTime* : ...] *location* : *beginTime* : ...] *hazardousLevel* : *HIGH* ... *beginTime* : ... *location* : ... ...]
15:29:57 A fireman reports: “Toxic smoke is discovered ... ”	**A Smoke is a subclass of a Gas. This message will validate the world model.**

Fig. 19 Applying overlay on observation reports

The entire interpretation process and results do not include any private information of users, except their location. All observers are treated as (anonymous) actors doing a certain action in a certain location. The information about these actors, combined with information about specific objects and events appeared in the world, builds up a scenario of a certain crisis event, which is validated by the scripts.

7 Interaction Manager

The interaction manager module works based on predefined dialogue strategies (Fig. 21). The `<frame>` tag marks a dialogue position, and the `<concern>` tag marks user's emotion state. The dialogue strategy provides some dialogue acts that can be selected depending on the value of both tags. Each component (Fig. 20) in the dialogue manager is explained below.

When the (local) world model is updated based on the user message, the *input analyzer* receives the user world model and selected script. It sends it to the (multi-user) Fusion Manager, then the analyzer compares both selected scripts (from the fusion module and the (multi-user) Fusion Manager) and checks the overlay score. If they are the same and the overlay score > 0, it fills slots' value of each known frame. Otherwise, this component triggers the *emotion analyzer* component to check the user's emotion. The analyzer decides the value's type whether it is extracted from the input (`filled`), assumed by the dispatcher (`assume`), or uncertain (`uncertain`). For example, when the system receives `A truck is on fire`, the component sets the `value` attribute of `frame.problem.type` to `"FIRE"` and `type` to `"filled"`. Figure 22 shows that the problem frame becomes semi-filled.

At every state of user-system dialogue, there are different ways to continue a dialogue. One way is based on user's current emotional state, which is indicated by the `<concern>` tag. The current implementation of the IM retrieves this information from the user model. Another way to select the next action is based on an evaluation of whether by selecting an action a maximal slot-filling is achieved. For this purpose, the *dialogue act selector* uses a heuristics rule-based approach. First,

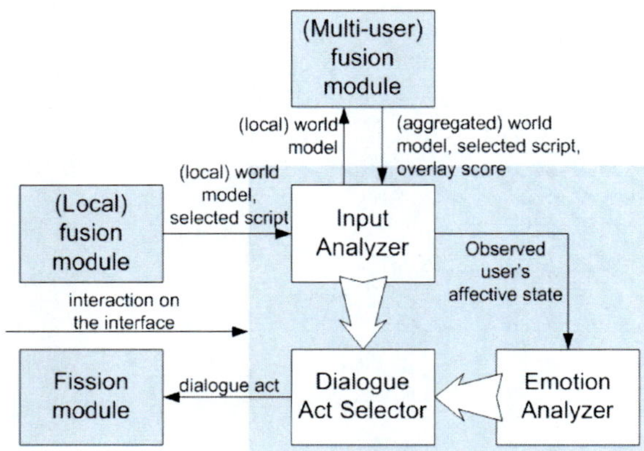

Fig. 20 The architecture of the interaction manager module

optional frames and slots are indicated based on the nature of the emergency problem, e.g., in the case of the `problem.type` is `FIRE` some optional slots such as `involved_party.name` and `problem.desc` may not be filled in. Finally, the heuristics rules have been designed, such as:

- if there is a (non-empty) subset of open frames and a (non-empty) subset of filled frames, ask for information covering as much as the open frames as possible;
- if one frame in the current state is not completely filled, immediately ask for the missing information to solve the ambiguity concerning this frame;
- if the subset of not completely filled frames contains more that one slot, handle the individual slots one after the other; and
- as long as new information can be provided assumptions of the system are not verified.

Figure 22 shows the differences in the system's memory in every dialogue state. Given a set of possible (open) frames and slots to choose, the interaction manager

```
...
<dialoguestrategies>
  <dialoguestrategy>
    <slot name="frame.problem.type" type="uncertain" value="FIRE"/>
    <slot name="frame.problem.status" type="assume" value="DANGER"/>
    <slot name="frame.reason.type" type="uncertain" value="EXPLOSION"/>
    <slot name="frame.urgency.status" type="assume" value="HIGH"/>
    <slot name="frame.user.type" type="filled" value="WITNESS"/>
    <slot name="frame.time.*" type="assume" value="system.getDateTime()"/>
    <slot name="scenario" value="FIRE"/>

    <frame name="reason.*">
      <concern type="anger">
        <dialogueActType>clarify</dialogueActType>
        <concept>
          <name>Fire</name>
          <property>
            <name>cause</name><value><concept>Explosion</concept></value>
          </property>
        </concept>
      </concern>
      <concern type="neutral">
        <dialogueActType>acknowledge</dialogueActType>
        ... // semantic content
      </concern>
      ... // other concern
    </frame>
    <frame name="involved_party.status">
      <concern type="neutral"> ... </concern>
      ... // other concern
    </frame>
    ... // other frame
  </dialoguestrategy>
...
</dialoguestrategies>
```

Fig. 21 A dialogue strategy in XML

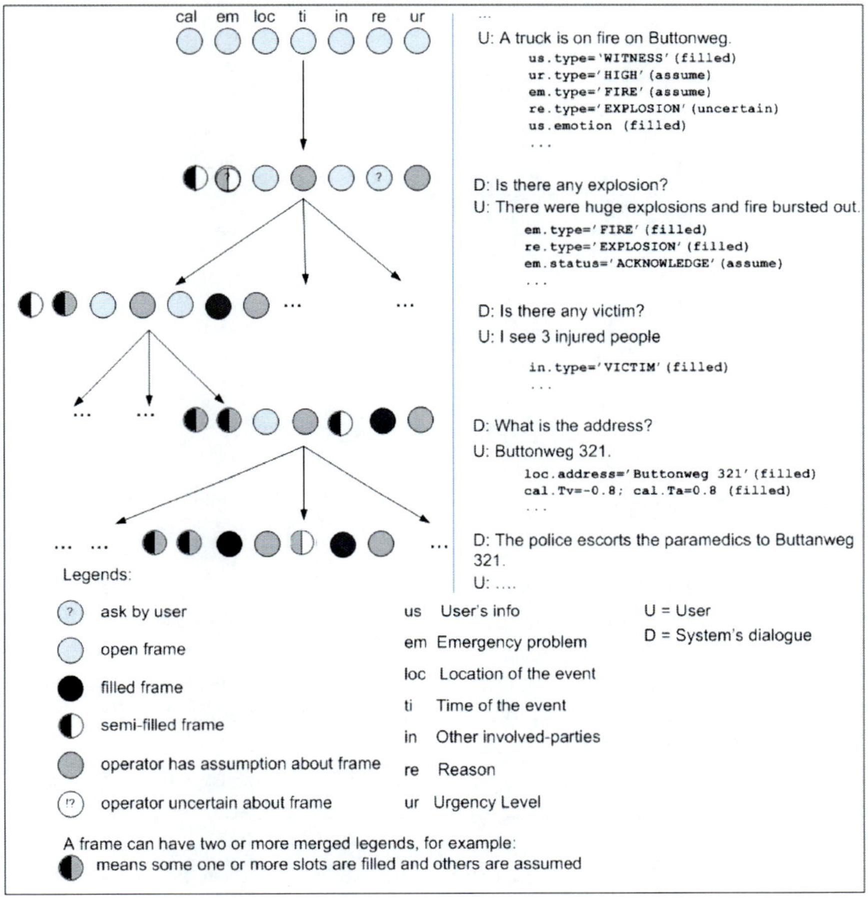

Fig. 22 Dialogue model from the system perspective (circles represents frames)

selects an appropriate frame's name and slot's name from the sets of dialogue strategies. The most specific frame's and slot's name is matched first before any frames or slots. Within the selected frame tag, given the user emotional state, the module selects the most appropriate response based on concern tag. Then, the selected dialogue act and its semantic content are sent to the fission module.

8 Information Generation

The fission module receives a dialogue act for a specific user from the IM and the current world model (Fig. 23). To support the rapid changes of a crisis situation, the presentation responds to triggers by changes in the world model.

With direct access to and the availability of real-time context and technical information of three knowledge sources: the user, the task, and the world, the *user*

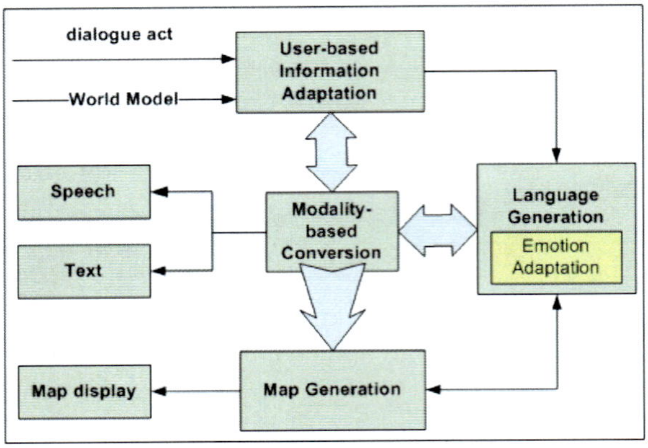

Fig. 23 The architecture of the fission module

adaptation component plans presentation contents based on the following user information.

- *User role*. Filtering messages using predefined flags, i.e., public (for all), protected (for rescue workers and crisis room operators), and private (for a certain group or operators).
- *Contextual user location*. Using a set of policies based on the current position of the user and locations of physical objects and phenomena including the dispersion of the crisis and its impact.
- *Available modalities*. Selecting output modalities based on the modalities selected by the IM, the input modalities used by the user, and the output modalities available.
- *Interaction time*. Presenting the most up-to-date information based on the user's dialogue history.
- *User emotion*. Output emotion that is suitable considering the input emotion and the current situation/world state.

The *Modality-based Conversion* component allocates the user-adapted information across the underlying modality-specific output generation components, e.g., language generation and the map display generation component. The Modality-based Conversion component divides and distributes all active concepts to the generation components as processing segments. Upon receiving a processing segment, the generation component estimates the time it needs to process the segment. The Modality-based Conversion component uses this timing information to create a full schedule for the processing of all segments.

Concepts are linked and synchronized as they are processed. Taking the example of the language generation and the map display generation component, language segments corresponding to concepts are generated sequentially and synchronized with the display (of active or highlighted concepts). The component also informs the generation components about concepts that are not longer active.

The *Language Generation* component works by a simple recognition of the dialogue act and the concept name and a substitution of the property names by their values controlled by a modified-AIML format. AIML [72] is extended-XML specifications for language generation in dialogue systems. A category (a knowledge unit in a dialogue) is defined within a certain topic. It provides templates (of the output messages) that can be selected depending on the <concern> tag, which refers to the current user's emotional state. The same concept and dialogue act can have many categories, but a different set of properties. Inside the template, <get> tags can be substituted by a value (a) from a property, for example: <get name = "source" type = "Location"/> and (b) of a function, for example:

```
<get function = "getStaticObjectInfo (upperLeft,
bottomUp)" type = "String")/>.
```

If the type is a concept name, the component will search the category with the corresponding concept recursively (Fig. 24).

```
<aiml>
 <topic name="CAR ACCIDENT">
  <category>
   <dialogAct>statement</dialogAct>
   <concept name="Escort"/>
   <properties>
     <property name="patient"/><property name="source"/><property name="patient"/>
   </properties>
   <template type="short-message">
    <concern name="neutral" value=""/><random>
      <li>A <get name="agent" type="Class.name"/> will escort the
          <get name="patient" type="Class.name"/> from <get name="source" type="Location"/> to
          <get name="destination" type="Location"/>.
     </li>
     <li>A <get name="agent" type="Class.name"/> and a <get name="patient" type="Class.name"/>
         will come to <get name="destination" type="Location"/>
         from <get name="source" type="Location"/>
     </li>
    ...
    </random></concern>
   ...
   </template>
  </category>
  <category>
   <dialogAct>*</dialogAct>
   <concept name="Location"/>
   <properties>
     <property name="upperLeft"/> <property name="bottomUp"/>
   </properties>
   <template type="*">
      <get function="getStaticObjectInfo(upperLeft, bottomUp)" type="String"/>
   </template>
  </category>
 ...
 </topic>
...
</aiml>
```

Example of resulted text:
"A police will escort a paramedic from A4 West BeneluxTunnel km 30 to Buttanweg no. 321 "

Fig. 24 An example of the text generation from two AIML units (an asterisk "*" means any)

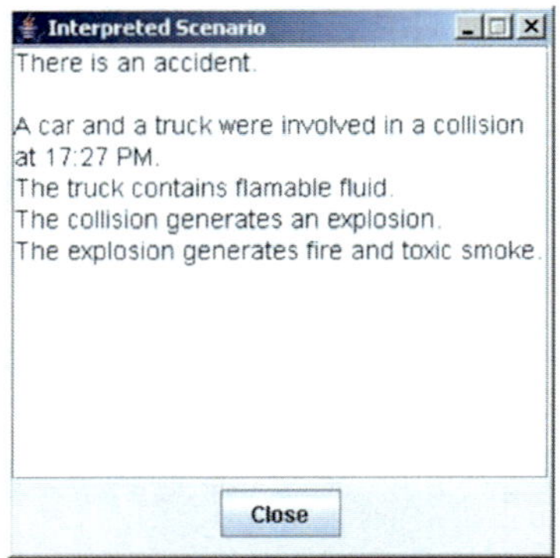

Fig. 25 An example of the resulted crisis scenario

The dialogue act `scenario` is triggered by the user selection on "view scenario" option. The (natural language) crisis scenario is generated based on the current world model as feedback on the user messages. Figure 25 shows the example of the resulted scenario of Fig. 14(a).

The *Map Generation* component generates a representation of a map interface as follows:

- *Adding icons.* All icons corresponding to active concepts are collected and displayed while ensuring none of the instances of the concepts are displayed more than once. A function is used to map world coordinates to screen coordinates.
- *Adding links between icons.* An arrow shows a causality (the property "cause"). A line shows relations between icons: (1) if one is the value of the other's property and (2) if they are the property values of an active concept that does not have a correspond icon.

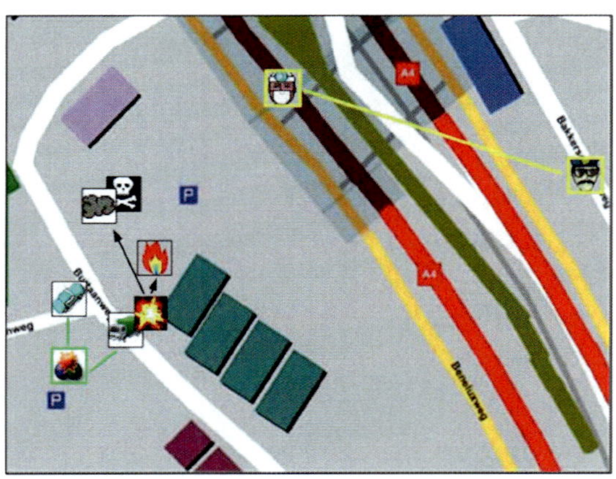

Fig. 26 A display shown while informing the user that a policeman and a doctor will come to the crisis scene; the concept "Collision" (green bordered) is added to the user's workspace by the system; the information is retrieved from the world model

- *Update icons and links*. The component updates all correspondence icons (and their links) except those that are explicitly specified by the user (only the user can remove them). This action can be forced by the dialog act "remove."
- *Highlight concepts or certain locations on the map*. This action is triggered by a dialog act "highlight". It also supports visually the language component presentation, e.g., the highlighted icons of "Paramedic" and "Policeman" in Fig. 26 are displayed, while the system synthesizes the text resulted in Fig. 24. The highlight is removed after some predefined time or forced by the dialogue act "remove".

To distinguish the user interaction and the system action on the graphical objects, three different colors are used: (1) black borders for the concepts that are specified by the user, (2) green borders for the concepts that are specified by the system, and (3) thick yellow borders for highlighting concepts or locations.

9 Experiments

It is difficult to create a controlled experiment of a disaster to test the demonstrator system. Therefore, some experiments have been done in simulated crisis situations.

Two laboratory tests have been conducted. The first experiment was performed to assess whether users were able to express their ideas using the provided icons and to address the usability of the visual language interface. Eight people took part in the test and played the role of human observers, while the crisis center was simulated by software. The participants were selected in the age range of 25–50 years. The simulation provides a methodology for modeling the dynamic nature of disasters. Thereby, we expected to be able to capture the dynamic creations of messages. First a crisis scenario was run on the simulator. Triggered by the events in the scenario, the simulator sends images (photos) of the crisis situation to the participants based on their location in the simulated world. The participants were asked to report what they saw using the visual language interface on their communication device. Figure 27(a) shows some examples of generated events in a scenario. Figure 27(b) shows examples of images that were sent to the participants (professionals in action). All activities were recorded and logged to be analyzed afterward. After the experiments, we interviewed the participants' satisfaction with the interface.

As our visual language interface is made up of interlinked icons, the method for measuring the sense of being lost in hypermedia [66] was adopted to measure disorientation in the interface. The experiment results show that our target users were capable of expressing the concepts and ideas that were in their minds using the interface. The users accomplished their tasks with relevant icon messages. However, some users had problems finding the right icon for certain concepts in mind. This was indicated by the high number of the lostness rating of five sequence messages (Fig. 27(c)). This was in some cases due to problems of unrecognized icons and limited number of provided icons. It appeared that the test users tended to keep searching for the icon they intended. This slowed down the interaction. Despite these problems, we saw decreasing lostness rating in terms of improvement of user

Generated event	Possible action
05:03 Fire at area (x,y)(x',y') A fire detector detects fire and activates alarm.	*Human observers*: see fire and report it to the crisis center. *Crisis center*: receive reports and send a call to Fireman for checking.
05:07 Fire and smoke develop A thermometer measures the current temperature. Firemen are on their way to (x,y)(x',y').	*Human observers*: report to the crisis center about developed smoke and the present of firemen. *Crisis center*: receive reports, send policemen to (x,y)(x',y').
05:08 Explosion Policemen block roads in the danger area.	*Human observers*: report to the crisis center about damage and casualties, a loud bang and the present of policemen. *Crisis center*: order extra units firemen and paramedics to (x,y)(x',y').
05:10 Professionals in action A wind meter measures the wind speed and direction. Paramedics arrive at (x,y)(x',y'). Firemen start extinguishing fire and rescuing victims, paramedics help victims and policemen guard the area.	*Human observers*: report to the crisis center about the professionals' activities. *Crisis center*: send advise to the professionals and civilians.
...	...

(a)

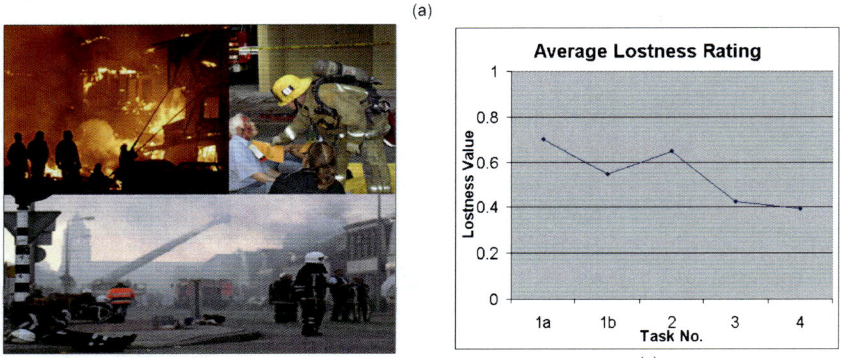

(b) (c)

Fig. 27 (a) First minutes-scenario based on Fig. 4, (b) example photographs of crisis situations and (c) average lostness rating for five sequence messages during experiments

performances. Our test users had taken benefits provided by the interface, for example, visual cues and the next icon prediction tool, in creating iconic messages. We also concluded that our test users only needed a small period of time to adapt to the developed visual language interface.

After redesigning the unrecognized icons and providing more icons (analyzing more corpora), we performed the second experiment to test the HCI system built based on the proposed HCI framework. It focused on the assessment of the performance of the individual modules. The design of the second test was similar to the first one with a different group of (eight) people playing the role of an observer. In this experiment, the participants had a communication device that allowed them to send text, speech, or visual language-based messages. Preliminary results show that all participants accomplished their tasks with relevant observation messages. The system is able to integrate the observations and generate appropriate multimodal responses based on available modalities and user contexts. The users often checked the resulted scenarios and rearranged their messages if necessary. Some issues due to the simulated experiment setup: out of domain reports, unobserved key events, and undetected user emotions, occurred and still need to be resolved.

At the time this chapter was being written, another experiment was performed. Learning from the previous experiments, we have redesigned our laboratory experiments. In this test, the participants are still asked to play the role of a civilian or a crisis worker. The difference is that this time they are actually doing a certain task depending on the role they have been assigned, for example, a paramedic helps a victim and a firefighter trying to put out a fire. While they are busy with the given task, the crisis simulator generates events based on a scenario. Instead using static images, we use video clips. It does not only contain animated images but also sounds when an event occurs, such as people screaming, smashing objects, and explosion. Similar to the previous setting, these videos are sent to the participants based on their location in the world.

The participants are asked to report their observations using their communication device. The evaluation of such framework, with respect to this design, allows us to determine how people use the systems developed. We expect that the result could give some indications whether they improve the efficiency of a given task.

10 Conclusions

In this chapter, we proposed a framework that consists of a set of modules that can be combined in an ad-hoc fashion to form a multimodal system. We have discussed how to apply it within a current crisis management environment by an example of an observation report system. Such a system, we believe, can aid research into crisis management.

The research reported here focused on investigating HCI technology for coping with the dynamic nature of crisis environments together with a context-dependent interpretation of input, relevant response selection, and appropriate response generation. The input and fusion modules are designed to cope with the intense nature of the crisis situations and noisy measurements. In the dynamic setting of information retrieval during a crisis, the crisis response and management teams must deal with several sources of uncertainty, for example, spontaneous speech input under stress. Communicated messages in such setting may turn out to be ambiguous or even irrelevant. Although this may be true for each modality on its own, we believe that utilizing different modalities to form a coherent picture of the situation at hand is the right way to go to reduce both ambiguity and incompleteness. From such a view point, rather than adding complexity to the scene, the added (and possibly redundant) information from any modality is seen as enhancing and complementing data from other modalities.

Our approach in computing context awareness involves the mechanism of building coherent semantic structures from dynamic concepts. Although a collection of such concepts may be ambiguous and fragmented, it is, by its very nature, contextually and temporally correlated. These characteristics together with the advantageous multimodal redundancy may be used for coming up with coherent, context-dependent interpretation of the communicated situation. With the availability of the knowledge of user, of the world and task, such a system can generate integrated and dynamic multimodal responses to the HCI. It is possible to realize a coherent and synchronized multimodal presentation with constraints defined by communication technology and context information.

The current approaches do not address the full complexity of any crisis management setting. In our view such controls should be still in human hands. This could be done since obtained data still can be verified and updated by multiple observers and intelligent processes. Our proposal includes direct feedback to user inputs, allowing for verifying and altering information, and ways for collaboration information. Moreover, the use of ontology, scripts, and AIML make inferences easier to verify afterward. All knowledge may be executed in a different order, but they can be designed and specified sequentially in order of the basic physical events of a crisis scenario. As a drawback, for the system to be fully expressive, all possible scenarios have to be converted into sets of concepts, scripts, and AIML in advance since reports about scenarios that have not been specified are not possible. A topic for further research will be deeper evaluations of the integrated system performance in real crisis settings. This allows us to see how people that experience stress use the system and how the application of the system aids or interferes with the (traditional) crisis management process.

References

1. Albacete, P.L., Chang, S., Polese, G., Baker, B.: Iconic language design for people with significant speech and multiple impairments. In: Proc. of the ACM Assets 1994, CA, USA, pp. 23–30. ACM, New York (1994)
2. Alexandersson, J., Becker, T.: The Formal Foundations Underlying Overlay. In: IWCS-5, The Netherlands, pp. 22–36 (2003)
3. Anson, D.K., Moist, P., Przywara, M., Wells, H., Saylor, H., Maxime, H.: The Effects of Word Completion and Word Prediction on Typing Rates using On-Screen Keyboards. In: Proc. of RESNA. RESNA Press, Arlington (2005)
4. Basu, A., Sankar, S., Chakraborty, K., Bhattarcharya, S., Choudhury, M., Patel, R.: Vernacula Educational Communication Tool for the People with Multiple Disabilities. In: Development by Design Conference, Bangalore (2002)
5. Beardon, C.: CD-Icon, an iconic language-based on conceptual dependency. Intelligent Tutoring Media 3(4) (1992)
6. Benjamins, T.: MACSIM: Multi-Agent Crisis Simulator, Interpreter and Monitor. Master thesis, TU Delft (2006)
7. Bevelier, D., Corina, D.P., Neville, H.J.: Brain and Language: a Perspective from Sign Language. Neuron 21, 275–278 (1998)
8. Bliss, C.K.: The Blissymbols Picture Book. Semantography Press, Sidney (1984)
9. Bosma, W., André, E.: Exploiting emotions to disambiguate dialogue acts. In: Proc. of IUI, Portugal, pp. 85–92 (2004)
10. Bui, T.H., Poel, M., Nijholt, A., Zwiers, J.: A Tractable DDN-POMDP Approach to Affective Dialogue Modeling for General Probabilistic Frame-based Dialogue Systems. In: IJCAI 2007, India (2007)

11. Carpenter, B.: The Logic of Typed Feature Structures. Cambridge University Press, England (1992)
12. Cassell, J.: Embodied conversational interface agents. Communication ACM 43(4), 70–78 (2000)
13. Cassell, J., Stocky, T., Bickmore, T., Gao, Y., Nakano, Y., Ryokai, K., Tversky, D., Vaucelle, C., Vilhjálmsson, H.: MACK: Media lab Autonomous Conversational Kiosk. In: Proc. of Imagina 2002, Monte Carlo (2002)
14. Catizone, R., Setzer, A., Wilks, Y.: Multimodal dialogue management in the Comic project. In: Proc. of EACL, Hungary (2003)
15. Champoux, B., Fujisawa, K., Inoue, T., Iwadate, Y.: Transmitting Visual Information: Icons become Words. IEEE - Information Visualization, 244–249 (2000)
16. Chin Jr., G., Stephan, E.G., Gracio, D.K., Kuchar, O.A., Whitney, P.D., Schuchardt, K.L.: Developing Concept-Based User Interfaces for Scientific Computing. Computer 39(9), 26–34 (2006)
17. Chiţu, A.G., Rothkrantz, L.J.M., Wiggers, P.: Comparison between different feature extraction techniques for audio-visual. Journal on Multimodal User Interfaces 1(1), 7–20 (2007)
18. Cohn, N.: Visual Syntactic Structures, Towards a Generative Grammar of Visual Language (2003), http://www.emaki.net/essays/
19. Cootes, T.F., Edwards, G.J., Taylor, C.J.: Active Appearance Models. In: European Conf. Computer Vision, pp. 484–498 (1998)
20. Datcu, D., Rothkrantz, L.J.M.: Multimodal workbench for human emotion recognition. In: Software Demo at IEEE CVPR 2007, USA (2007)
21. Datcu, D., Rothkrantz, L.J.M.: The use of Active Appearance Model for facial expression recognition in crisis environments. In: Proc. of ISCRAM 2007, USA, pp. 515–524 (2007)
22. Davis, S., Mermelstein, P.: Comparison of Parametric Representations for Monosyllabic Word Recognition. IEEE ASSP 28, 357–366 (1980)
23. Dor, R.: The Ear's Mind, a Computer Model of the Fundamental Mechanisms of the Perception of Sound, Master thesis, TU Delft (2005)
24. Dymon, U.J.: An Analysis of Emergency Map Symbology. Int. Journal of Emergency Management 1(3), 227–237 (2003)
25. Ekman, P., Friesen, W.F.: Unmasking the Face. Prentice-Hall, Inc., Englewood Cliffs (1975)
26. Farberow, N.L., Frederick, C.J.: Training Manual for Human Service Workers in Major Disasters. Rockville, Maryland - National Institute of Mental Health (1978)
27. Fitrianie, S.: An Icon-based Communication Tool on a PDA, Postgraduate Thesis. TU Eindhoven (2004)
28. Fitrianie, S., Rothkrantz, L.J.M.: Communication in Crisis Situations using Icon Language. In: Proc. of IEEE ICME 2005, The Netherlands, pp. 1370–1373 (2005)
29. Fitrianie, S., Datcu, D., Rothkrantz, L.J.M.: Constructing Knowledge of the World in Crisis Situations using Visual Language. In: Proc of. IEEE SMC 2006, Taiwan, pp. 121–126 (2006)
30. Fitrianie, S., Rothkrantz, L.J.M.: An Adaptive Keyboard with Personalized Language-based Features. In: Matoušek, V., Mautner, P. (eds.) TSD 2007. LNCS (LNAI), vol. 4629, pp. 131–138. Springer, Heidelberg (2007)

31. Fitrianie, S., Yang, Z., Rothkrantz, L.J.M.: Developing concept-based user interface using icons for reporting observations. In: ISCRAM 2008, USA (2008)
32. Foster, M.E., White, M., Setzer, A., Catizone, R.: Multimodal generation in the COMIC dialogue. In: ACL 2005, USA, pp. 45–48 (2004)
33. Gustafson, J., Bell, L., Beskow, J., Boye, J., Carlson, R., Edlund, J., Granström, B., House, D., Wirén, M.: Adapt - a Multimodal Conversational Dialogue System in an Apartment Domain. In: ICSLP 2000 (2), pp. 134–137 (2000)
34. Housz, T.I.: The elephant's memory (1994–1996), `http://www.khm.de/timot`
35. Homeland Security Working Group: Symbology Reference (2003), `http://www.fgdc.gov/HSWG/index.html`
36. IBM e-business Case Studies: Bullhead City Police Department, `http://www.ibm.com/e-business/doc/content/casestudy/47405.html`
37. Johanson, B., Fox, A., Winograd, T.: The Interactive Workspaces Project: Experiences with Ubiquitous Computing Rooms. IEEE Pervasive Computing 1(2), 67–74 (2002)
38. Joshi, A.K., Vijay-Shanker, K.: Compositional Semantics for Lexicalized Tree-Adjoining Grammars. In: Proc. of Computational Semantics, The Netherlands (1999)
39. Kanade, T., Cohn, J.F., Tian, Y.: Comprehensive database for facial expression analysis. In: Proc. of IEEE Automatic Face and Gesture Recognition, France (2000)
40. Karlson, A., Bederson, B., Contreras-Vidal, J.: Understanding Single Handed Use of Handheld Devices. In: Jo, L. (ed.) Handbook of Research on User Interface Design and Evaluation for Mobile Technology (2006) (in press)
41. Keizer, S.: Reasoning under uncertainty in Natural Language Dialogue using Bayesian Networks, Doctoral Dissertation. Twente University (2003)
42. Kjeldskov, J., Kolbe, N.: Interaction Design for Handheld Computers. In: Proc. of APCHI 2002. Science Press, China (2002)
43. Leemans, N.E.M.P.: VIL: A Visual Inter Lingua. Doctoral Dissertation, Worcester Polytechnic Institute, USA (2001)
44. Lucas, B.D., Kanade, T.: An Iterative Image Registration Technique with an Application to Stereo Vision. In: IJCAI 1981, pp. 674–679 (1981)
45. MacKenzie, I.S., Zhang, S.X., Soukoreff, R.W.: Text Entry using Soft Keyboards. Behaviour and Information Technology 18, 235–244 (1999)
46. Mankoff, J., Abowd, G.D.: Cirrin - A Word-Level Unistroke Keyboard for Pen Input. In: ACM UIST 1998, pp. 213–214 (1998)
47. Maybury, M.: Intelligent User Interfaces: An Introduction. In: Proc. of IUI 1999, pp. 3–4. ACM, New York (1999)
48. McTear, M.F.: Spoken dialogue technology: enabling the conversational user interface. ACM Computing Surveys 34(1), 90–169 (2002)
49. Mehrotra, S., Butts, C., Kalashnikov, D., Venkatasubramanian, N., Altintas, K., Hariharan, P., Lee, H., Ma, Y., Myers, A., Wickramasuriya, J., Eguchi, R., Huyck, C.: Camas - a Citizen Awareness System for Crisis Mitigation. In: Proc. of ACM SIGMOD 2004, New York, USA, pp. 955–956 (2004)
50. MESA (Mobile Broadband for Emergency and Safety Applications) project, `http://www.projectmesa.org`
51. Moore, L.K.: CRS Report for Congress: Public Safety Communication Policy, Confessional Research Service, the Library of Congress (2006)
52. Norman, D.: Things That Make Us Smart. Addison-Wesley Publishing Co., Reading (1993)
53. den Os, E., Boves, L.: Towards Ambient Intelligence: Multimodal Computers that Understand Our Intentions. In: Proc. of eChallenges e-2003 (2003)

54. Otten, J., van Heijningen, B., Lafortune, J.F.: The Virtual Crisis Management center - An ICT implementation to canalize information! In: ISCRAM 2004, Brussels (2004)
55. Oviatt, S., Coulston, R., Lunsford, R.: When Do We Interact Multimodally?: Cognitive Load and Multimodal Communication Patterns. In: Proc. of ICMI 2004, pp. 129–136. ACM, New York (2004)
56. Perlovsky, L.I.: Emotions, Learning and Control. In: Proc. of International Symposium: Intelligent Control, Intelligent Systems and Semiotics, pp. 131–137 (1999)
57. Pfleger, N., Alexandersson, J., Becker, T.: Scoring Functions for Overlay and their Application in Discourse Processing. In: KONVENS 2002, Germany, pp. 139–146 (2002)
58. Potamiano, A., Ammicht, E., Kuo, H.-K.J.: Dialogue Management the Bell Labs Communicator System. In: Proc. of ICSLP, Beijing, China, vol. 2, pp. 603–606 (2000)
59. Proxem, Proxem Antilope (2008), http://www.proxem.com
60. Ramaswamy, S., Rogers, M., Crockett, A.D., Feaker, D., Carter, M.: WHISPER – Service Integrated Incident Management System. Int. Journal of Intelligent Control and Systems 11(2), 114–123 (2006)
61. Rothkrantz, L.J.M., van Vark, R.J., Peters, A., Andeweg, N.A.: Dialogue control the Alparon system. In: Sojka, P., Kopeček, I., Pala, K. (eds.) TSD 2000. LNCS (LNAI), vol. 1902, pp. 333–338. Springer, Heidelberg (2000)
62. Schank, R., Abelson, R.: Scripts, Plans, Goals and Understanding. Erlbaum, Hillsdale (1977)
63. Museum, S.F.: The Virtual Museum of the City of San Francisco: San Francisco 9-1-1 Dispatch Tapes, October 17 (1989), http://www.sfmuseum.org/1989/sf911.html, http://www.sfmuseum.net/1989/sc911.html
64. Sharma, R., Yeasin, M., Krahnstoever, N., Rauschert, I., Cai, G., Brewer, I., MacEachren, A.M., Sengupta, K.: IEEE, Speech-Gesture Driven Multimodal Interfaces for Crisis Management. 91(9), 1327–1354 (2003)
65. Singhal, A., Rattine-Flaherty, E.: Pencils and Photos as Tools of Communicative Research and Praxis. International Communication Gazette 68(4), 313–330 (2006)
66. Smith, P.: Toward a Practical Measure of Hypertext Usability. Interacting with Computers 8(4), 365–381 (1996)
67. Steedman, M., Baldridge, J.: Combinatory Categorial Grammar. In: Borsley, R., Borjars, K. (eds.) Non-Transformational Syntax. Blackwell, Malden (2005)
68. Tatomir, B., Rothkrantz, L.J.M.: Intelligent System for Exploring Dynamic Crisis Environments. In: Van de Walle, B., Turoff, M. (eds.) ISCRAM 2006, Newark, NJ, SA (2006)
69. Vark, R.J.V., de Vreught, J.P.M., Rothkrantz, L.J.M.: Classification of public transport information dialogues using an information-based coding scheme. In: Proc. of ECAI, pp. 55–69 (1996)
70. Viola, P., Jones, M.: Robust Real-time Object Detection. In: 2nd Int. Workshop on Statistical and Computational Theories of Vision-Modeling, Learning, Computing, and Sampling (2001)
71. Wahlster, W.: Dialogue Systems Go Multimodal: The Smartkom experience. In: SmartKom: Foundations of Multimodal Dialogue Systems. Springer, Heidelberg (2006)
72. Wallace, R.S.: The Elements of AIML Style (2003), http://alicebot.org
73. Ward, D.A., Blackwell, A., MacKay, D.: Dasher - a Data Entry Interface Using Continuous Gesture and Language Models. In: ACM UIST, pp. 129–136. ACM, NY (2000)

74. WNBC: Exclusive - 911 tapes tell horror of 9/11 (part 1 and 2): Tapes released for first time (2002), http://www.wnbc.com/news/1315646/detail.html
75. Yang, Z., Rothkrantz, L.J.M.: Dynamic Scripting in Crisis Environments. In: HCI International 2007, China, pp. 554–563 (2007)
76. Zhai, S., Kristensson, P.-O.: Shorthand Writing on Stylus Keyboard. In: ACM CHI, pp. 97–104 (2003)
77. Zue, V., Seneff, S., Glass, J., Polifroni, J., Pao, C., Hazen, T., Hetherington, L.: Jupiter - A telephone-based conversational interface for weather information. IEEE Trans. on Speech and Audio Processing 8(1) (2000)

Design Issues for Pen-Centric Interactive Maps

Louis Vuurpijl, Don Willems, Ralph Niels, and Marcel van Gerven

Abstract. Recent advances in interactive pen-aware systems, pattern recognition technologies, and human–computer interaction have provided new opportunities for pen-based communication between human users and intelligent computer systems. Using interactive maps, users can annotate pictorial or cartographic information by means of pen gestures and handwriting. Interactive maps may provide an efficient means of communication, in particular in the envisaged contexts of crisis management scenarios, which require robust and effective exchange of information. This information contains, e.g., the location of objects, the kind of incidents, or the indication of route alternatives. When considering human interactions in these contexts, various pen input modes are involved, like handwriting, drawing, and sketching. How to design the required technology for grasping the intentions of the user based on these pen inputs remains an elusive challenge, which is discussed in this chapter. Aspects like the design of a suitable set of pen gestures, data collection in the context of the envisaged scenarios, and the development of distinguishing features and

Louis Vuurpijl
Radboud University Nijmegen, Donders Institute for Brain, Cognition and Behavior, Montessorilaan 3, 6525 HR, Nijmegen, The Netherlands
e-mail: l.vuurpijl@donders.ru.nl

Ralph Niels
Radboud University Nijmegen, Donders Institute for Brain, Cognition and Behavior, Montessorilaan 3, 6525 HR, Nijmegen, The Netherlands
e-mail: r.niels@donders.ru.nl

Don Willems
Radboud University Nijmegen, Donders Institute for Brain, Cognition and Behavior, Montessorilaan 3, 6525 HR, Nijmegen, The Netherlands
e-mail: d.willems@donders.ru.nl

Marcel van Gerven
Radboud University Nijmegen, Institute for Computing and Information Sciences, Heyendaalseweg 135, 6525 AJ, The Netherlands
e-mail: marcelge@cs.ru.nl

R. Babuška & F.C.A. Groen (Eds.): Interactive Collaborative Information Systems, SCI 281, pp. 273–297.
springerlink.com

pattern recognition technologies for robustly recognizing pen input from varying modes are described. These aspects are illustrated by presenting our recent results on the development of interactive maps within the framework of the ICIS project on crisis management systems.

1 Introduction

The use of information systems for supporting various services in crisis management is rapidly increasing. In the ICIS project (interactive collaborative information systems) [58], novel interactive collaborative information systems are being designed for such applications. The ICIS/CHIM (computational human interaction modeling) research cluster has pursued the development of multimodal interfaces for crisis management tasks and incident response systems. The technologies developed in ICIS/CHIM are mainly targeted at professional users operating in crisis control rooms, mobile communication centers at the scene of an incident, and first responders carrying light-weighted communication equipment. The motivation for the research in ICIS/CHIM is based on the fact that for conveying information in such applications, robust and efficient human to computer and computer to human interaction is vital. It is known that humans interact best with computer systems when they are allowed to do so in a natural way, just as when they interact with other humans [13, 14, 47, 57, 74]. Since for this purpose a varied set of modalities is available for communication (like speech, gesture, facial expressions, and drawing or writing), research on multimodal systems has become increasingly important in the last decade.

1.1 Multimodal Interaction

The research in ICIS/CHIM has explored different input modalities [22], including hand/arm gestures [53], speech [17, 52], facial expressions [17], or pen gestures and handwriting [70] to communicate with the system. Recognition and interpretation of these various forms of user input is performed by specialized modules, using pattern recognition techniques to transform raw data inputs (like speech signals, pen input trajectories, or hand movements) to a ranked list of potential classification results.

The resulting interpretations provided by each single module are combined by the so-called fusion module [9, 16]. The process of "fusing" information entails the integration of semantically related topics provided by different modules [21]. For example, descriptive speech utterances like "This road is blocked" may be combined with deictic pen gestures indicating the location of the road. The integrated information is subsequently processed by the dialog action manager, which decides on the response of the system. Information provided to the user is rendered in, e.g., synthetic speech or computer graphics as controlled by the fission module [25]. The design and implementation of multimodal systems like the one pursued in ICIS/CHIM is a complex task. It requires specialized skills and knowledge from a multi-disciplinary research team, involving significant collaborative software engineering efforts to

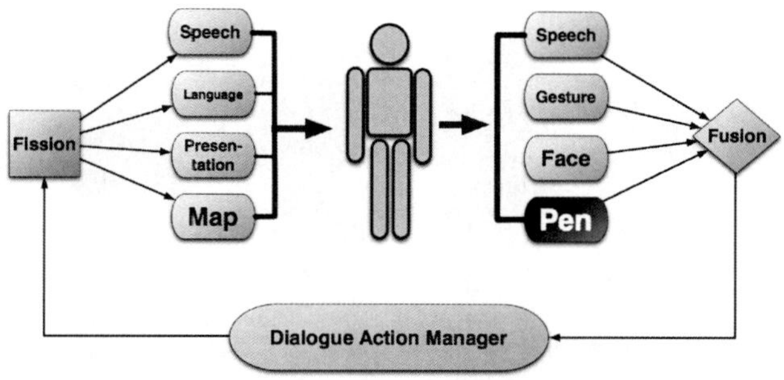

Fig. 1 Different modalities involved in multimodal interaction in the ICIS project [22, 70]. Input modalities comprise pen, hand gestures, speech, and facial expressions. System responses are rendered on the display, depicting a map of an incident. Outputs may comprise symbolic information on the map, or messages using synthetic speech. The "pen" module is marked bold, since this chapter focuses on pen-centric systems.

ensure the correct mutual operation of the distinct processing modules involved. For a detailed treatment of design issues for multimodal systems, the reader is referred to, e.g., [4, 19, 47, 57, 65].

1.2 Pen-Centric Interactive Maps

Over the past decades, increasingly more usable pen-centric systems have been developed, profiting from technical advances in computer hardware, human–computer interaction, and pattern recognition techniques. Using pen-centric systems, human pen input or hand/finger gestures can be captured directly from the display of interactive devices. In particular with the advent of the tabletPC and other pen-aware devices like pocketPCs, PDAs, or cellular phones, the number of application domains has evolved to basically anything that can be done with a pen and paper. Categories of pen input comprise, for example, handwriting [51, 55], music scores [42], mathematical or chemical expressions [75], command gestures [2, 10, 54], iconic gestures [40, 67], and drawing or sketching [3, 8, 27, 29, 34].

Our research has focused on *interactive maps*, a challenging application of pen-aware systems in which many of these categories join [8, 69]. Using interactive maps, users can annotate displayed photographic or cartographic map information using the pen (see Fig. 2). Typically, annotations contain [74]: (i) deictic gestures for marking particular locations or objects on the display, (ii) handwritten texts for adding textual notes, explanations, or other descriptive content, (iii) iconic gestures for indicating events or objects from a particular gesture repertoire, and (iv) free-hand drawing or sketching. The use of interactive maps is believed to provide an important tool to enhance interaction between different actors, in particular where spatial information is concerned [14, 43]. For example, Cohen et al. [13, 14] have shown that a pen interface improves the efficiency of communications in military applications. There are only a few studies that explore pen interactions in crisis management situations. In [57], a thorough discussion of multimodal interfaces in crisis

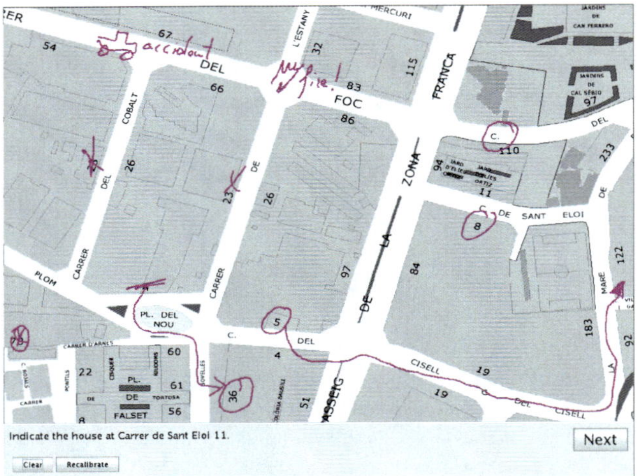

Fig. 2 Pen gestures in interactive maps. The user replied by pen to a sequence of questions, like indicating objects and events. Replies contained deictic gestures such as cross marks, arrows, encirclements, and route specifications; iconic gestures (car and fire); and handwritten texts.

management has been conducted, focusing on the fusion between pen, speech, and gesture modalities. Another multimodal interactive system for crisis management (i2Map) is described in [35] in which users use hand gestures and speech to interact with the system. Finally, in a recent survey on interactive maps for geoinformation and cartography applications, Doyle et al. describe their COMPASS system, which uses speech, pen gestures, and handwriting for mobile mapping services [19].

The task of robustly understanding the intention of the user in interactive map applications, where a myriad of unknown pen input has to be processed, is huge. As recently underlined by Cheriet et al., in particular in situations where mixed pen input modes have to be distinguished and recognized and where inputs may be fully unconstrained in terms of layout and gesture repertoires, pattern recognition remains an elusive goal [12].

In most interactive systems, the (only) solution to this problem is to constrain the user. Two broad ways to achieve this are to either minimize the variability in pen gestures by means of a limited gesture lexicon or to limit the user by following a human–computer dialog in which the system requests for specific information in a step-by-step manner. Indeed, a wide range of pen gesture recognition systems follow this approach, like Quickset [15], Rubine's GRANDMA system [54], and SILK [37]. For a recent review, the reader is referred to [59]. The majority of these systems target the recognition of a limited set of command gestures [2, 10, 20, 39] (e.g., arrow up/down/left/right for scrolling, or gestures for performing select/delete/undo actions). We have pursued the recognition of pen gestures in the context of multimodal interaction for bathroom design [8, 18]. As an example of the complexity of unconstrained user input, consider Fig. 3. Only using

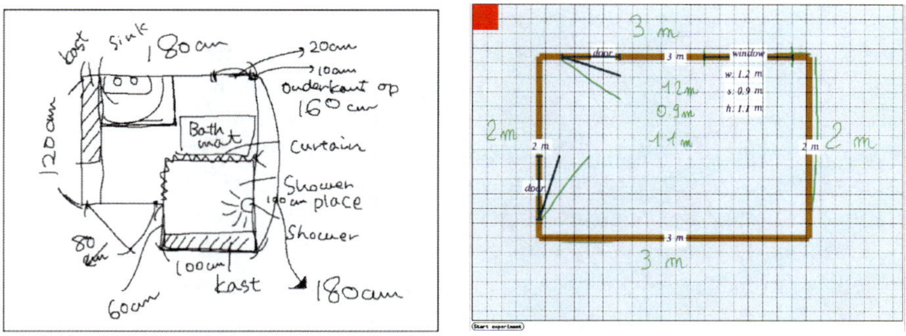

Fig. 3 Unconstrained gestures, sketches, and handwriting in bathroom design (left picture) versus step-by-step entering information following a strict dialog protocol (right picture). The red square indicates that the system is processing information and that user input should be delayed until the next information is requested by the system.

a strict turn-taking protocol, in which the user can only enter information requested by the system, acceptable recognition results were achieved [8, 18].

1.3 A Brief Primer on Pattern Recognition for Pen-Centric Systems

There is a wide body of literature on machine learning techniques and pattern recognition methods, which provides a variety of solutions for recognizing pen input [33, 51]. Pattern recognition systems require the availability of sufficient data to train and fine tune a model for a certain domain. To assess the generalization performance of a system after training, it should be evaluated on data collections which differ from the data observed during training and fine tuning. A generally accepted method employs k-fold cross validation to perform *k* sessions during which the system is trained on all but $1/k$th randomly selected data samples, which are used for evaluating the trained system. Subsequently, average recognition performance of the *k* sessions is reported. Alternatively, the available data may be split into three separate subsets: (i) the training set, (ii) the test set for testing a trained system, and (iii) the evaluation set which should be kept apart for reporting the final recognition accuracies. To assess the so-called "walk-up" recognition performance, this constraint is tightened such that the final evaluation data set may only contain data acquired from users the system has never seen before. In case of pen-centric systems, *writer-dependent* recognition rates are reported in the case that data from the same writer may be used during training, fine tuning, and evaluation. For *writer-independent* recognition rates, it is prohibit that the system is trained and evaluated with data from the same writer. As a consequence, systems trained in a writer-independent fashion (or trained on data that somehow differ from the eventual real-life conditions) perform worse. Note that there are many ways to compute the recognition accuracy of a pattern recognition system. A simple and easily computable measure is average recognition performance, which counts the number of cases in which the classifier outcome was correct and divides this number by the total number of data samples. A pattern recognition system operates in four phases [33]. In the first phase, "raw" data acquired through some recording device (like a digitizer tablet or microphone) is segmented in meaningful fragments, like spoken words or handwritten characters. In many cases, pre-segmented databases are used, for example, containing isolated handwritten words, gestures, or characters. Before and/or after segmentation, the data may be preprocessed, e.g., for filtering noise and jitter caused by the recording equipment, or by using normalization procedures to reduce variations in size of characters or writing slant. Third, distinguishing features are computed, like the width, height, average curvature, or area of handwritten segments in the case of pen-centric systems. The fourth phase entails classifier design and learning. Popular classifiers used in pattern recognition are neural networks [7], support vector machines [61], or template matching [33]. The latter method computes the similarity between two data samples. In our work, we have used Dynamic Time Warping [46] for this purpose (see Fig. 4).

For pen-centric systems, the raw input data contains a sequence of coordinates recorded at the pen tip of a digitizing pen, using a digital writing tablet sampled at a certain temporal (100–512 Hz) and spatial resolution (100–2540 dpi). The resulting signal contains the $(x(t),y(t))$ position sampled at time t at a certain tablet position. Other recorded values can be the tilt and azimuth orientation of the pen and $p(t)$, which represents pen pressure. Pen tip movements slightly above the tablet surface

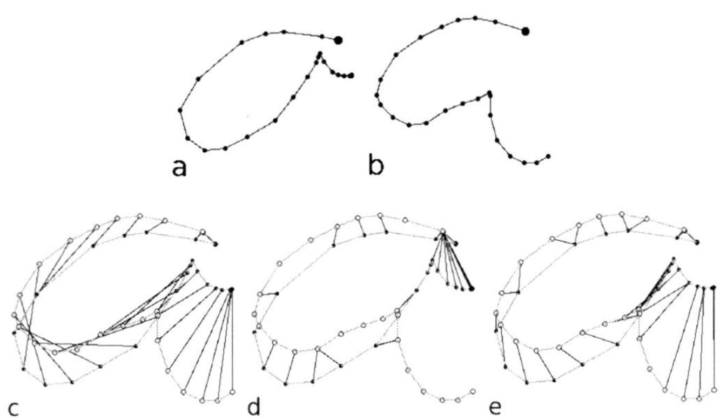

Fig. 4 Examples of template matching techniques, which perform a point-by-point comparison of two trajectories. Samples (a) and (b) are matched using (c) linear matching (every point i of trajectory 1 matches with point i of trajectory 2), (d) complete matching (every point of trajectory 1 matches with the nearest point of trajectory 2), and (e) DTW matching. DTW uses the production order of the coordinates, and is able to match the coordinates that are placed in the same position in the two curves. Note that "strange" matches like between the points at the bottom of (a) and the left of (b) (as occur in (c)) and between the points like those at the end of (a) and the beginning of (b) (as occur in (d)) do not occur in the DTW match. Furthermore, DTW does not require spatial resampling (because it can match trajectories of different length), whereas linear matching does.

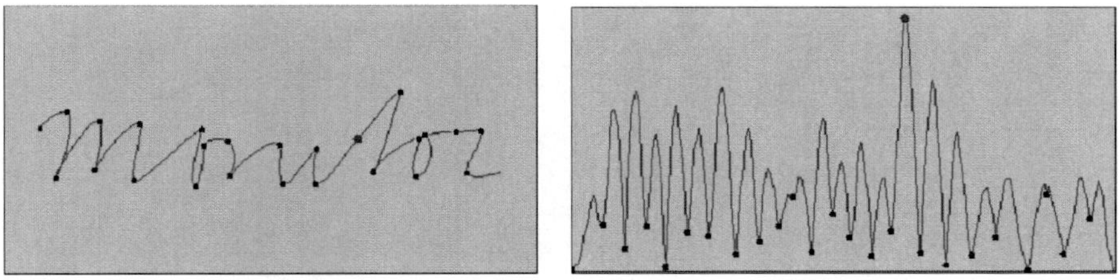

Fig. 5 The handwritten word "monitor" and corresponding velocity profile. Dark green dots represent minima in the velocity signal, which correspond to extrema in vertical position. Note the red dot, indicating the point where maximum velocity is achieved (Picture from http://www.ai.rug.nl/~lambert/recog/hwr-tutor/velocity.html).

can be recorded (up to 10 mm), where $p(t) < \theta$ indicates so-called *penup* movements. For *pendown* movements, $p(t) \geq \theta$ (for some threshold θ) holds. Temporal or spatial filtering may be employed for preprocessing. Character segmentation is mostly performed based on the velocity profile, where points of low velocity indicate stable segmentation points, co-occurring with extrema in the Y position [56] (see Figure 5). Other segmentation clues that are particularly useful for the segmentation of pen gestures in interactive maps employ pressure (since users lift the pen during transition between pen inputs), temporal (there is a brief pause between meaningful segments), and spatial layout characteristics [74].

1.4 Data Collections for Developing Pen-Centric Systems

Databases for training and testing pen-centric recognition systems have become readily available for the handwriting recognition community. The first large publically available datasets were UNIPEN [28], consisting of online isolated characters and handwritten words from Western scripts, and the CEDAR database [30], containing offline scans of address information like city names, state names, and zip codes. Note here the distinction between online and offline data, the former referring to pen input acquired through a digital writing tablet and/or digitizing pen. Other well-known databases containing Western handwriting are IRONOFF [62] and the IAM database [41]. In the past decade, many other scripts like Japanese [32, 44], Tamil [5, 6], and Arabic [48] have become available. As a result, many researchers in handwriting recognition have been able to develop and assess their recognition technologies on generally accepted benchmarks. Besides handwriting, other categories of pen input are musical scores, mathematical expressions, command gestures, and sketches. Unlike the availability of handwriting databases, collections of these other categories are scarcely available. Unfortunately, the same holds for dedicated applications like the interactive maps technologies discussed in this chapter. Therefore, preceding the phases in the design of pen-centric systems, a proper data acquisition process has to be performed. To illustrate the data acquisition process, feature extraction and classifier design phases for developing pen-centric systems, this chapter discusses the processing of so-called *iconic gestures* [67].

1.5 Organization of the Remainder of This Chapter

In the remainder of this chapter, the reader is provided with a worked-out case study of the design of a specific component of the pen gesture recognition module from the ICIS/CHIM multimodal system [22]. Our approach tries to bridge the gap between allowing unconstrained user inputs and limiting the user through constrained gesture repertoires. As we will explain in the next section, Section 2, the design of our iconic gesture database for interactive maps is constrained by graphical symbolic representations used in cartographic applications and connected to the gesture repertoires, which were collected via our exploratory studies reported in [68]. The database is called NicIcon and was introduced in [45]. In this chapter, we briefly describe the data collection process in Section 2. Based on the acquired data, three subsets were selected for a proper distinction into train, test, and evaluation sets.

Sections 3 and 4 of this chapter use the NicIcon database for the design and evaluation of discriminative feature sets and pattern recognition systems. For this purpose, a number of common features in gesture- and handwriting recognition will be discussed and a new set of powerful features selected through the Best Individual N technique [66] is presented. Furthermore, the design of a multiple classifier system comprising a number of well-known classifiers (multi-layered perceptrons, support vector machines, and dynamic time warping) is presented. In Section 3, we will elaborate on the feature extraction and selection process. In particular, we will present new features that are based on the observation that 90% of the iconic gestures contained in the NicIcon dataset are drawn in multiple strokes. For each gesture, features are computed along the complete gesture shape as well as along each individual stroke. As we will show through feature selection and an evaluation of recognition performances, adding mean and standard deviations of the individual stroke features has a dramatic impact, which may also be of value for other applications in pattern recognition.

In our concluding chapter, we will recapitulate the main issues of designing pen-centric systems as discussed in this chapter.

2 The Design of the NicIcon Database of Iconic Pen Gestures

We previously reported about a large study on the typical pen interactions that emerge in crisis management scenarios using interactive maps [68]. In this study, we asked users to annotate pictorial or cartographic image contents similar to the setup depicted in Fig. 2. The categorization of the obtained pen gestures reported in [68] showed that next to route descriptions and markings of locations, the iconic sketchings of, e.g., cars, fires, casualties, accidents, or persons occurred quite frequently. Unfortunately, the processing of these unconstrained iconic gestures (with a large variability within and between users), resulted in a recognition performance that was unacceptable for the envisaged target domain. Based on these findings, we concluded that natural and unconstrained interactions in pen-centric interactive

Fig. 6 Examples of graphical symbols from the homeland security working group [1]

maps are indeed goals which remain unattainable, as recently emphasized by [12]. However, iconic gestures have a meaningful shape and are therefore easier to learn and remember by users of pen-centric systems than abstract gestures which have no obvious relation between shape and semantics [24, 59]. Furthermore, since these gestures are used frequently during unconstrained pen interactions, we have designed and collected a suitable set of iconic gestures for specifying objects and events.

As described in Fitriani and Rothkrantz [24], information about the use of symbols (or icons) in the field of crisis management is not readily available. The research described in [24] considered a large set of 209 graphical symbol classes used by the governments of the United States, Australia, and New Zealand [1]. This collection contains a distinction in the categories incidents (50 symbols), natural events (30), operations (46), and infrastructures (83).

The IconMap application described by Fitriani and Rothkrantz was originally developed with the goal to provide language-independent communication for mobile use [23]. Using IconMap, users can convey complex messages by clicking on a sequence of symbols. We cooperated with the developers of IconMap in the context of the ICIS project. Inspired by their work, we selected a subset of 14 icon shapes representing "atomic concepts" [24], like fire, ambulance, police, car, and victim. These concepts relate to the gesture categories obtained in our previous experiments in which we collected unconstrained gestures [68]. Since these gestures described in [68] have been acquired in a setting where users were not constrained by visual examples, they are considered as natural sketches that users find appropriate. The latter is important for enabling proper learning and recall [59]. Finally, our gesture shapes were designed such that the recognition systems are able to disambiguate the individual gesture classes (see Fig. 7 below). Note that iconic pen gestures comprise a limited number of pen strokes, rather than the stylized graphical symbols from [1], depicted in Fig. 6.

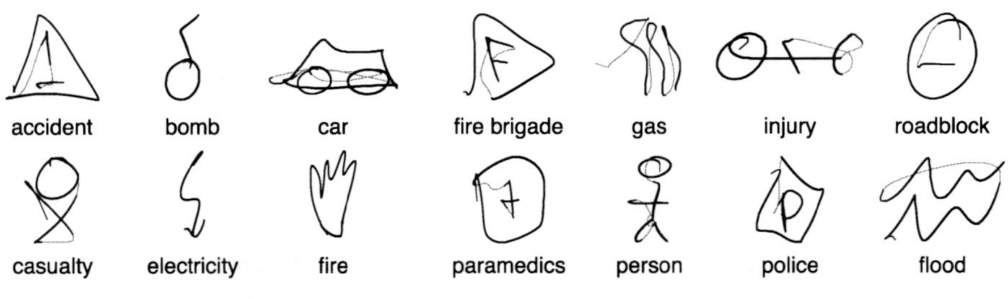

Fig. 7 Examples of 14 different iconic gestures, produced by the same participant contributing to the NicIcon data collection. Penup strokes are depicted in light gray.

Fig. 8 A piece of paper was clamped to a writing tablet. The participants used an inking stylus to draw the icons. The resulting data contains online trajectories and offline scans of the paper.

Although the selected 14 classes of iconic gestures comprise a small subset of the symbol classes used in Refs. [1, 24], they form an interesting data collection for our research goals in which we explore distinguishing features and pattern recognition techniques for interactive maps.

2.1 The Data Collection Setup

The data collection process of the NicIcon database is described in detail in [45]. Both online data (dynamic pentip traces acquired through a digitizing tablet) and offline data (ink traces scanned from paper) were simultaneously acquired. The work described in this chapter focuses on the online data, collected from 34 writers. The offline data provide opportunities for other research purposes, like assessing the performance of offline icon recognition, or validating the outcomes of automatic trajectory extraction techniques [46]. Each writer was requested to fill in 22 paper sheets, each containing seven rows of five columns, resulting in 35 drawing areas and 770 iconic gestures per participant (see Figure 8). For further details on the data collection setup and recording equipment used, the reader is referred to [45].

2.2 Data Segmentation and Statistics

Because of the convenient organization of the data entry forms, the combination of spatial layout characteristics and temporal information could successfully be used by a novel segmentation algorithm [67], yielding in total 26,163 iconic gestures (participants skipped 17 gestures). Table 1 shows the distribution of gesture samples distinguished in the 14 gesture classes. The table also shows the average number of

Table 1 Distribution of the 26,163 gestures over the 14 gesture classes, indicating that for some classes in total 17 gestures were not successfully acquired. For each class, the average number of strokes μ and standard deviation σ is given as well, counting both pendown and penup strokes.

Description	Icon	n	μ	σ	Description	Icon	n	μ	σ
Accident		1857	5.5	1.5	Gas		1870	5.1	0.8
Bomb		1869	2.8	0.8	Injury	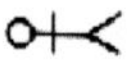	1870	7.3	1.6
Car	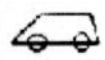	1870	5.8	1.9	Paramedics		1870	5.6	2.1
Casualty		1870	4.8	0.8	Person		1870	7.5	1.8
Electricity		1870	2.7	1.3	Police		1870	4.3	2.1
Fire		1867	1.3	0.8	Roadblock		1870	3.1	1.2
Fire brigade		1870	7.2	1.9	Flood	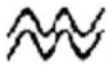	1870	3.1	0.7

strokes that users employ to draw an iconic gesture class. On average, 4.7 strokes are used for each iconic gesture.

2.3 Data Subsets for Training, Testing, and Evaluation

A common practice in pattern recognition is to device three datasets for the design of a classifier (a training set and test set for optimizing the classifier and an evaluation set which is kept hidden for final assessment). Accordingly, the data were divided into three subsets, the *trainset* (containing 36% of the icons), the *testset* (24%), and the *evalset* (40%). This distinction into subsets was performed twice: once for writer-dependent (WD) experiments, where gesture samples produced by one writer occur in all three subsets, and one for writer-independent (WI) settings, where no writer occurs in more than one subset. As explained in Section 1.3, the distinction between WD and WI settings can be used to compare how well a system performs on data acquired from known versus unseen users. For both the WI and WD settings, the subsets were selected through stratified random sampling, such that each subset contains the same relative number of icons per class.

3 Design of Pattern Recognition Systems for Iconic Gestures

In the previous section, it was described that for the development of interactive technology for the recognition of pen gestures, a suitable collection of training and testing data is required. This section highlights a number of issues that are important in the design of distinguishing features and robust pattern recognition techniques.

3.1 Feature Extraction and Selection

For our review on features for gesture recognition, we have considered applications in, e.g., handwriting recognition, (mathematical) symbol recognition, sketching, and

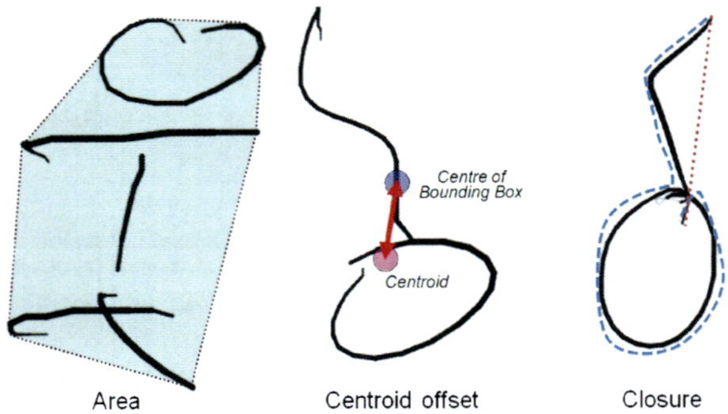

Fig. 9 Examples of the global g-28 features: area, centroid, and closure

design. For example, the studies presented in [31, 34, 38, 49, 54, 67, 71, 72, 76] provide detailed descriptions of features that have proven their suitability for recognizing pen input in applications similar to interactive maps. Based on these works, four feature sets were computed from the NicIcon dataset, referred to as the *g-28*, the *t-30*, the *t-60*, and the *m-fs* features. The g-28 features represent global information over a complete gesture trajectory, like its length, average curvature, or area. The t-30 and t-60 features are computed at spatially equidistant points along the trajectory and thus represent local trajectory information, like (x,y) coordinates and running angular information. The m-fs features contain both global information computed from the total trajectory and local features computed along the trajectory. In the following, each of these feature sets is briefly described.

For each feature set, we normalized each gesture with respect to rotation, orientation, and scale. These normalization steps are quite common in handwriting recognition [51, 63], where different writers produce gestures and handwriting in different sizes (requiring scale normalization) and slants (requiring normalization on rotation in case of isolated characters or gesture segments). After normalization on scale and orientation, each data sample is sized within a standardized bounding box and positioned in, e.g., the top-left corner or center of the bounding box. Rotation was determined by calculating the principal components over the coordinate points in the gesture; the orientation of the first principal axis then denotes the required rotation. Scale and orientation were normalized by computing the bounding box for each gesture, translating all coordinates such that the center of the bounding box is at origin and dividing all coordinates by MAX(height,width) of the gesture. After feature extraction, all features were normalized through mean shifting to a common scale with an average of zero and standard deviation of one [26].

Global features: the g-28 feature set

The g-28 features are described in [72, 73]. They have been developed for a pen gesture recognition system used for interactive map applications. The g-28 features are known to distinguish classes of pen input modes such as deictic gestures (gestures used for marking objects, like arrows, encirclements, and crosses), handwriting, and geometric shapes (squares, triangles, ovals, lines, etc.). Each feature is computed on the complete gesture trajectory (see Figure 9).

The g-28 set contains the following 28 features (for details, the reader is referred to [71–73]): (1) length of pen stream, (2) area of convex hull, (3) compactness, (4) eccentricity, (5) ratio of the coordinate axes, (6) closure, (7) circular variance, (8) curvature, (9) average curvature, (10) perpendicularity, (11) average perpendicularity, (12) centroid offset along major axes, (13) length along first principle axis, (14) rectangularity, (15) initial horizontal offset, (16) final horizontal offset, (17) initial vertical offset, (18) final vertical offset, (19) straight line count, (20) largest straight line ratio, (21) straight line ratio, (22) largest straight line, (23) macro perpendicularity, (24) average macro perpendicularity, and (25) ratio of the principal axes. Also included are three features based on the pressure data, which is available in online data: (26) average pressure, (27) pendown count, and (28) penup/down ratio.

Local trajectory information: the t-30 and t-60 features

The t-30 and t-60 features are described in [64]. These trajectory-based features are computed from spatially resampled (to n=30 or n=60 coordinates) gestures. The t-30 features have been used extensively for character recognition. For each out of n points, the (x,y,z) coordinates, the running angles, and angular velocities are computed, resulting in $3 \cdot n+2 \cdot (n-1)+2 \cdot (n-2)$ features. As explained in [64], a typical resampling of Western characters requires $n=30$ (204 features). Given that many of the collected iconic gestures have a more complex shape than the average Western character, we also explored resampling to $n=60$ (414 features), resulting in a better coverage of the original trajectory with resampled points.

The mixed features m-fs containing both global and local features

The m-fs feature set is a large new feature set which was recently developed in our department [71]. The set contains global features computed from the full trajectory and local features along the running trajectory. Many of these features were inspired by published works from the literature on pen input recognition; the interested reader may consider [31, 38, 49, 54, 72, 76]. For each of these "plain" features, eight extra features are computed which appear to dramatically boost recognition performances, as we will discuss in Section 5. These features are based on the observation that the majority of iconic gestures contain multiple strokes. For most plain features computed for a gesture trajectory containing n strokes, we computed the same feature over each of the corresponding n subtrajectories. Subsequently, we included the mean and standard deviation of the n computed feature values as extra features. Furthermore, this same computation was performed for penup strokes only and pendown strokes only. So, as depicted in Figure 10, for each plain feature, we added the feature computed for penup only, pendown only, μ and σ of all strokes, penup strokes only, and pendown strokes only. Note that for some features, such

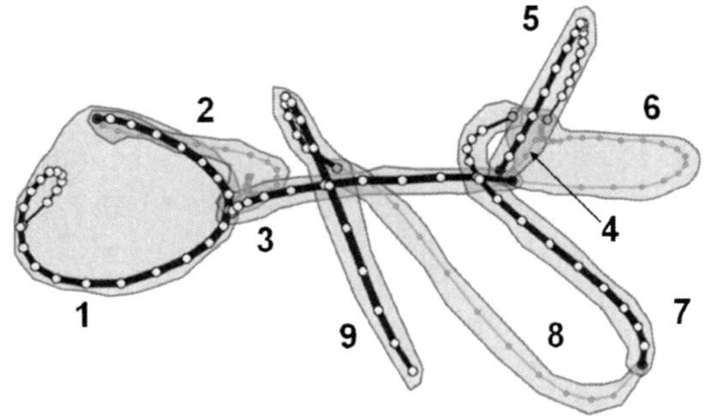

Fig. 10 Most features in the m-fs feature set were not only calculated over the complete gesture but also over each of the strokes (segmented by transitions between pendown and penup) in the gesture. The average value (μ) and the standard deviation (σ) of a feature over the strokes were used as feature values. This was repeated for (i) all strokes, (ii) only the pendown strokes, and (iii) only the penup strokes. For the gesture depicted here, μ and σ were therefore calculated over all strokes, over the odd numbered strokes (pendown) and the even numbered strokes (penup). The penup stroke "4" is marked with an arrow to distinguish it for stroke "6".

as the penup/pendown ratio, it is not relevant to differentiate between penup and pendown coordinates.

The m-fs feature set contains a total of 1185 features [71]. Note that this set is not completely independent of the other three sets, since it subsumes the g-28 set. Furthermore, from the t-30 and t-60 sets, the running angle (represented as chain code features) is also included.

3.1.1 Feature Selection from the m-fs Set

The Best Individual N (BIN) feature selection algorithm [66] was used to select a suitable subset of features from the m-fs set. BIN tests the discriminative power of each of the features individually. The features are ranked according to the recognition performance for each individual feature using an SVM classifier [61]. The SVM was trained on each feature, and the recognition performance was tested on the testset. Training the SVM was performed using the default parameter settings. Each SVM was trained with cost parameter $C = 1$, using a nonlinear radial basis function (RBF) kernel. For other default settings ruling the performance of the implementation of support vector machines used in our experiments, the reader is referred to [11].

Like most other feature selection methods, the BIN method cannot guarantee to yield the best features [33]. Furthermore, BIN does not explore linear or nonlinear combinations of selected features, which in certain cases will lead to suboptimal solutions [50]. Yet, we decided to use this method because it is relatively straightforward and can be used to discard a relevant subset of suboptimal features [33].

For the WI test, the best performance on the testset (97.6%) was achieved with 545 selected features. For the WD tests, the best performance (99.5%) was achieved

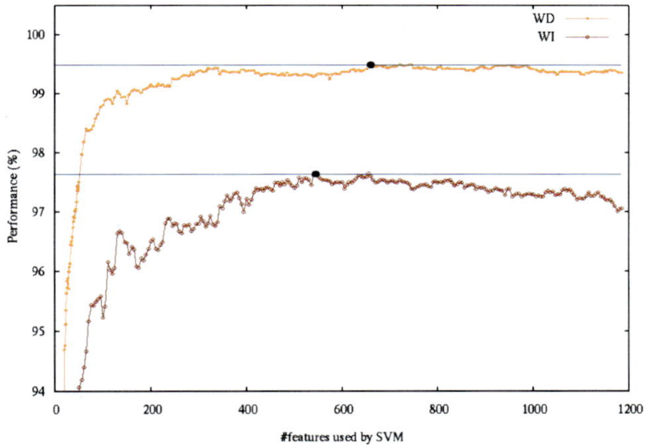

Fig. 11 Recognition scores of SVMs trained with varying number of features ranked according to BIN. The maximum scores are indicated with boldface dots and correspond to the number of features selected by BIN.

with 660 features. To examine: (i) whether certain strong features were incorrectly discarded by this method and (ii) whether a smaller subset of the selected features achieved a better performance, we computed the recognition scores of a number of SVMs with an increasing number of features (ranked according to BIN). As described above, each SVM was trained using the default settings. The results are depicted in Figure 11 below. As can be observed, BIN was able to find a suitable subset of m-fs features that achieve maximum performance.

From the 1185 features contained in the m-fs set, two subsets of features were selected, the m-fsD and m-fsI set. Most (19) of the features from the g-28 feature set were part of both subsets. However, length, area, average perpendicularity, length of the first principal axis, straight line count, and average macro-perpendicularity were not selected in either subset. Furthermore, eccentricity and ratio of the coordinate axis were not selected in the WI set, and average pressure was not selected in the WD set. Four chaincode features from the m-fs set (encoding the angular information from the t-30 and t-60 sets) were contained in the m-fsD subset and two in the m-fsI set. Based on this information, it can be deduced that the m-fs subsets have a very high overlap with the g-28 features and only limited overlap with the local trajectory-based information encoded in t-30 and t-60.

Both the m-fsD and m-fsI sets contain penup features. This implies that, apparently, recognition benefits from information contained in the penup strokes. Furthermore, a striking number of the new mean and standard deviation features computed from constituent pen strokes was contained in both subsets. In Table 2, the distribution of the different feature groups is depicted, for penup (PU) strokes, pendown (PD) strokes, and strokes from both types. Mean and standard deviation features comprise about two-third of all selected features by the BIN method.

In this section, we described how we selected two new feature sets from the m-fs set. It is notable that a significant part of the selected features contains the new

Table 2 Distribution of different feature groups in m-fsI and m-fsD

	m-fsD				m-fsI			
Stroke type	Plain	μ	σ	Total	Plain	μ	σ	Total
PU	.13	.11	.08	.31	.14	.12	.06	.31
PD	.10	.12	.13	.34	.09	.12	.13	.34
PU+PD	.11	.11	.13	.35	.10	.11	.14	.35
Total	.33	.33	.34	1.00	.32	.34	.33	1.00

mean and standard deviation features. In the sections to follow, we will assess the recognition performance of a number of classifiers using the m-fs, g-28, t-30, and t-60 feature sets. Furthermore, the recognition performance of a template matching method based on dynamic time warping (DTW) is compared.

3.2 *Classifier Design and Learning*

In this section, we elaborate on the design of various classifiers for the recognition of iconic gestures. For each of the three classifiers involved (MLP, SVM, and DTW), a similar design approach was followed. Each classifier used the *trainset* with a specific set of parameters for training. Subsequently, classification performance of the resulting classifier was assessed on the *testset*. This was repeated for a number of parameter combinations. The combination that yielded the best classification performance on the *testset* was selected for the final evaluation. Finally, the classification performance on the *evalset* was used as the evaluation measure. This approach was followed for each of the four feature sets described above.

3.2.1 Feature-Based Classification Using MLP and SVM

We used two common feature-based classifiers: the multi-layered perceptron (MLP) and the support vector machine (SVM) [61]. For both techniques, a number of parameters can be set, which results in a different architecture and performance of the classifier for a given problem. Since each combination of a classifier architecture and the feature set used for training can be considered as a distinct classifier, eventually eight classifiers were optimized. For the process of fine tuning classifiers, elaborate experimentation is required in which various settings for these parameters are explored. Our MLP neural network implementation uses the generalized delta rule with momentum. The parameters which were varied for the MLP were learning rate (between 0.001 and 0.05), momentum (between 0.5 and 0.8), and number of hidden units (between 32 and 128 with steps of 8). Training each MLP was performed until performance on the testset (cross-validation) reached a maximum performance. The best results were achieved with a momentum of 0.7 and small learning rates of 0.001. The optimal number of hidden units varied between the feature sets.

For fine tuning the SVMs, we used LIBSVM [11], public domain software implementing a multi-class recognition algorithm. The SVM type we used was C-SVC. Besides the traditional linear kernel, we also used nonlinear kernels to achieve the highest correct classification performance. We tested a polynomial kernel, a radial basis function kernel, a sigmoid kernel, and different cost parameters *C*. Each of these kernels have their own parameters which we varied: *gamma* for all nonlinear kernels and *degree* and *coef0* for the polynomial and the sigmoid kernel [11].

3.2.2 Template Matching Using DTW

Our DTW algorithm [46] computes the DTW distance between two data samples *A* and *B* as the sum of the Euclidean distances between their consecutive matching points A_i and B_j. Three conditions are examined: (i) the continuity condition, which is satisfied when index *i* is on the same relative position on *A* as index *j* is on *B* (the amount in which the relative positions are allowed to differ is controlled by a parameter *c*), (ii) the boundary condition, which is satisfied if *i* and *j* are both at the first, or both at the last position of their trajectory, and (iii) the penup/pendown condition, which is satisfied when both *i* and *j* are produced with the pen on the tablet, or when they are both produced with the pen above the tablet. A_i and B_j match if either the boundary condition or both other conditions are satisfied. Classification of a test sample is performed through nearest neighbor matching with the DTW distance function. Each DTW classifier was optimized by varying parameter *c*, which controls the strictness of the continuity condition.

3.3 Multiple Classifier System

As described above, nine classifiers were used to classify the data: SVMs and MLPs (each using the four described feature sets), and the DTW classifier. The classifiers were organized in a multiple-classifier system (MCS) [36], which employed plurality voting [60] for combining classifier outputs. In case of a tie, the output of the classifier with the highest individual performance on the testset was used to rule the outcome. Optimization of the MCS system was done by exhaustively computing all results on the testset for the different combinations of classifiers (2^9-1 combinations). The best combination of classifiers in both WD and WI setting was used to compute the final recognition performances of the evalset.

4 Results

Each of the developed classifiers was assessed on the final evaluation set. The resulting classifications were combined in a simple plurality voting MCS. Table 3 shows the correct classification performance for each classifier (where a classifier is the combination between the used classification method and the used features) and the overall performance of the MCS, for both WD and the WI setting.

Table 3 Correct classification performances of the individual classifiers using the different feature sets and the multi-classifier system on the WD and WI data. From the m-fs features, the m-fsD set was used for assessing WD performance and the m-fsI set for assessing WI performance. Underlined numbers indicate that this particular classifier was used for the two MCS systems.

		New results		Previous results	
Classifier	Feature set	WD perf.	WI perf.	WD perf.	WI perf.
SVM	g-28	84.80%	73.04%	84.39%	79.99%
	t-30	98.13%	90.30%	98.57%	92.63%
	t-60	98.01%	88.62%	98.51%	92.16%
	m-fs	99.25%	96.43%	-	-
MLP	g-28	84.03%	78.52%	94.00%	90.72%
	t-30	94.52%	85.24%	96.63%	92.40%
	t-60	94.44%	85.10%	97.79%	92.92%
	m-fs	98.72%	96.31%	-	-
DTW	-	98.41%	93.60%	98.06%	94.70 %
MCS	-	**99.51%**	**97.83%**	**99.32%**	**96.49%**

In our previous experiments with data from the NicIcon set [45], we trained similar classifiers as used in the work presented here. For a comparison between both experiments, we add the previous recognition performances in Table 3. Please note that the results presented in [45] are based on significantly more training data (60% versus 36% in the current work), which makes the comparison of individual classifiers unfair. Still, the performances on the new m-fs features outperform any of the classifiers used in our previous work.

Table 3 shows that the trajectory-based feature sets t-30 and t-60 result in a better performance than the global features g-28. To our surprise, the t-60 features (with more coordinate points along the gesture trace) are slightly worse than the t-30 features. This is reflected in the selection of classifiers for the MCS systems (marked by underlined cells in Table 3). The new m-fs features introduced in this work outperform all other feature sets, which is also reflected in the selection of the corresponding classifiers for the MCS. The DTW algorithm achieves high performance in the WD setting, which can be explained by the fact that a relatively large number of templates (the trainset) are used for classification. The MCS system significantly outperforms each single classifier.

When considering the error rates for the individual classifiers, the impact of the new m-fs features become apparent. The lowest error rates for the SVM using the other three feature sets, are respectively, 9.70% and 1.87% for the WI and WD settings (both using the t-30 feature set). Using the m-fs features, error rates drop to, respectively, 3.57% and 0.75%. This involves an improvement of 63.3% and 59.9%. Similarly, for MLP, improvements of 75.0% and 76.6% are achieved for the WI and WD settings, respectively.

Fig. 12 Examples of the three main conflicting cases: (i) retraces, (ii) sloppy drawings, and (iii) wrong box shapes. These cases were selected as follows. During evaluation, the evaluation samples E_i were selected for which all three classifiers involved in the MCS system yielded a wrong result. Subsequently, for each of these selected E_i, the corresponding classification output from the DTW classifier was considered. Note that DTW uses a number of training prototypes which are matched to E_i. Consequently, the best matching training sample T_k can be considered as a conflicting case. For the examples depicted here, we selected a number of characteristic T_ks.

To explore conflicting classes between the different icon shapes, we computed the confusion matrices for the WD and WI settings on the evaluation sets. Since the achieved error rates are very low, most entries in the confusion matrix are not very informative (near zero confusion between classes). What is noted is that especially the boxed icons ⚠, ⊖, F▷, and ◇P are harder for the system to distinguish. As illustrated in Figure 12, the main reason for this confusion is that users in some occasions mix up the shape and orientation of the boxes. Two other categories of typical misclassifications observed from the confusion matrices are cases where users retrace the pen trajectories and cases where the pen input is sloppy.

5 Discussion and Future Perspectives

This chapter has highlighted the use of pen-based interactions in the broad context of interactive collaborative information systems. Various issues in the design of the required recognition technology were discussed: (i) the design and acquisition of a benchmark database relevant to the domain of crisis management; (ii) the design and assessment of different subsets of distinguishing features; and (iii) the development of (combinations of) classifier systems for the robust recognition of pen gestures on the basis of these features.

As an illustrative example, we provided an overview of our recent work on iconic gesture recognition. The NicIcon benchmark database of iconic gestures was briefly introduced in Section 2. Three subsets were selected from a total of 26,163 iconic gestures collected from 34 writers. Training sets, test sets, and evaluation sets were selected by means of stratified random sampling. Different sets were extracted for WI and WD settings. These subsets may serve as future reference benchmark databases for fellow researchers pursuing iconic gesture recognition.

The process of selecting distinguishing features has been discussed in Section 3. Three existing feature sets were compared to our new m-fs feature set. We have demonstrated high recognition performances using these feature sets, using optimization strategies on different multiple classifier systems. The m-fs feature set contains a large number of features described in the literature. Additionally, we have included two new strong features: the mean and standard deviation of features computed from the constituting penup and pendown strokes of iconic gestures. The BIN feature selection method appeared to be very successful in selecting subsets of features for both WI and WD settings. A particularly interesting conclusion to draw is that a major part of the selected features (about 2/3) comprise our new features. The impact on recognition performance is significant: error rates have dropped between 59.9% and 76.6%.

We are currently further developing and assessing our pen input recognition technologies in more elaborate experiments, involving pen input data acquired in more realistic situations. Furthermore, in a recent paper [67], we demonstrated that the developed technologies are generalizable to other domains from pen input, i.e., handwriting recognition and the recognition of special symbols. However, since the envisaged context is emergency services personnel working in stressful circumstances, we are considering experimental conditions including drawing from memory, drawing under pressure of time, or drawing in multi-task settings.

The NicIcon database of iconic gestures was collected for the design of distinguishing features and accurate pattern recognition techniques. It should be noted, however, that for real usage in crisis management scenarios, most probably additional gesture classes are required. Furthermore, our studies on pen-centric interactive maps have shown that other modes of pen input should be handled as well, requiring parallel recognizers for the classification of handwriting and deictic gestures. The results from the current study provide a well-described recipe for the design of the corresponding additional recognition technologies.

Acknowledgments. This research was supported by the Netherlands Organization for Scientific Research (NWO), project nr: 634.000.434.

References

1. Homeland Security Working Group, Symbology Reference, version 2.20 (released September 14, 2005)
2. Agarawala, A., Balakrishnan, R.: Keepin' it real: Pushing the desktop metaphor with physics, piles and the pen. In: CHI 2006 - the ACM Conf. on Human Factors in Computing Systems, pp. 1283–1292 (2006)
3. Alvarado, C., Davis, R.: SketchREAD: A multi-domain sketch recognition engine. In: Proc. of the 17th annual ACM symposium on user interface software and technology, Santa Fe, New Mexico, USA, pp. 23–32 (2004)
4. Benoit, C., Martin, J.-C., Pelachaud, C., Schomaker, L., Suhm, B.: Audio-visual and multimodal speech-based systems. In: Gibbon, D., Mertins, I., Moore, R. (eds.) Handbook of multimodal and spoken dialogue systems: Resources, terminology and product evaluation, pp. 102–203. Kluwer Academic Publishers, Dordrecht (2000)
5. Bhaskarabhatla, A.S., Madhvanath, S.: Experiences in collection of handwriting data for online handwriting recognition in Indic scripts. In: Proc. of the 4th Int. Conf. Linguistic Resources and Evaluation (LREC), CDROM (2004)
6. Bhattacharya, U.: Handwritten character databases of Indic scripts (2004), `http://www.isical.ac.in/~ujjwal/download/database.html`
7. Bishop, C.M.: Neural Networks for Pattern Recognition. Oxford University Press, Oxford (1995)
8. Boves, L., Neumann, A., Vuurpijl, L., ten Bosch, L., Rossignol, S., Engel, R., Pfleger, N.: Multimodal interaction in architectural design applications. In: 8th ERCIM Workshop on User Interfaces for AI, Vienna, Austria (2004)
9. Bui, T.: Toward affective dialogue management using partially observable Markov decision processes. PhD thesis, University of Twente, Enschede (October 2008)
10. Buxton, W.: Chunking and phrasing and the design of human-computer dialogues. In: Proc. of the IFIP World Computer Congress, Dublin, Ireland, pp. 475–480 (1986)
11. Chang, C.-C., Lin, C.-J.: LIBSVM: a library for support vctor machines (2001), `http://www.csie.ntu.edu.tw/~cjlin/libsvm`
12. Cheriet, M., El Yacoubi, M., Fujisawa, H., Lopresti, D., Lorette, G.: Handwriting recognition research: Twenty years of achievement ... and beyond. Pattern Recognition 42(12), 3131–3135 (2009)
13. Cohen, P.R., Johnston, M., McGee, D., Smith, I., Oviatt, S., Pittman, J., Chen, L., Clow, J.: Quickset: Multimodal interaction for simulation set-up and control. In: Proc. of the Fifth Applied Natural Language Processing meeting, Association for Computational Linguistics (1997)
14. Cohen, P.R., Johnston, M., McGee, D., Oviatt, S., Pittman, J., Smith, I., Chen, L., Clow, J.: Quickset: Multimodal interaction for distributed applications. In: Proc. of the Fifth ACM Int. Multimedia Conf., pp. 31–40 (1997)
15. Cohen, P.R., Johnston, M., McGee, D., Oviatt, S., Pittman, J., Smith, I., Chen, L., Clow, J.: Quickset: multimodal interaction for distributed applications. In: MULTIMEDIA 1997: Proc. of the fifth ACM international conference on Multimedia, pp. 31–40. ACM, New York (1997)
16. Datcu, D., Rothkrantz, L.: A dialog action manager for automatic crisis management. In: Proc. of the 6th Int. Conf. on Information Systems for Crisis Response and Management (ISCRAM2008), pp. 384–393 (2008)
17. Datcu, D., Rothkrantz, L.: Semantic audio-visual data fusion for automatic emotion recognition. In: Euromedia 2008, April 2008, pp. 1–6. Eurosis, Ghent (2008)

18. den Os, E., Boves, L., Rossignol, S., ten Bosch, L., Vuurpijl, L.: Conversational agent or direct manipulation in human system interaction. Speech Communication 47, 194–207 (2005)
19. Doyle, J., Bertolotto, M., Wilson, D.: A survey of multimodal interfaces for mobile mapping applications. In: Meng, L., Zipf, A., Winter, S. (eds.) Map-based Mobile Services. Lecture Notes in Geoinformation and Cartography, pp. 146–167. Springer, Heidelberg (2008)
20. Egger, M.: Find new meaning in your ink with tablet PC APIs in Windows Vista. MSDN Magazine, Microsoft Corporation (May 2006)
21. Engel, R., Pfleger, N.: Modality Fusion. Cognitive Technologies, pp. 223–235. Springer, Heidelberg (2006)
22. Fitrianie, S., Poppe, R., Bui, T.H., Chitu, A.G., Datcu, D., Dor, R., Hofs, D.H.W., Wiggers, P., Willems, D.J.M., Poel, M., Rothkrantz, L.J.M., Vuurpijl, L.G., Zwiers, J.: Multimodal human-computer interaction in crisis management environments. In: Burghardt, P., Van de Walle, B., Nieuwenhuis, C. (eds.) Proc. of the 4th Int. Conf. on Information Systems for Crisis Response and Management ISCRAM 2007, pp. 149–158 (2007)
23. Fitrianie, S., Rothkrantz, L.: Language-independent communication using icons on a PDA. In: Matoušek, V., Mautner, P., Pavelka, T. (eds.) TSD 2005. LNCS (LNAI), vol. 3658, pp. 404–411. Springer, Heidelberg (2005)
24. Fitrianie, S., Rothkrantz, L.: A visual communication language for crisis management. Int. Journal of Intelligent Control and Systems (Special Issue of Distributed Intelligent Systems) 12(2), 208–216 (2007)
25. Fitrianie, S., Tatomir, I., Rothkranz, L.: A context aware and user tailored multimodal information generation in a multimodal HCI framework. In: Euromedia 2008. Eurosis, Ghent (2008)
26. Fukunaga, K., Hostetler, L.: The estimation of the gradient of a density function, with applications in pattern recognition. IEEE Transactions on Information Theory 21, 32–40 (1975)
27. Gennari, L., Kara, L.B., Stahovich, T.F., Shimada, K.: Combining geometry and domain knowledge to interpret hand-drawn diagrams. Computers and Graphics 29, 547–562 (2005)
28. Guyon, I., Schomaker, L., Plamondon, R., Liberman, M., Janet, S.: UNIPEN project of on-line data exchange and recognizer benchmarks. In: Proc. ICPR 1994, October 1994, pp. 29–33 (1994)
29. Hammond, T., Davis, R.: Tahuti: A geometrical sketch recognition system for UML class diagrams. Papers from the 2002 AAAI Spring Symposium on Sketch Understanding, Stanford, California, USA, pp. 59–68. AAAI Press, Menlo Park (2002)
30. Hull, J.: A database for handwritten text recognition research. IEEE Transactions on Pattern Analysis and Machine Intelligence 16(5), 550–554 (1994)
31. Iivarinen, J., Peura, M., Särelä, J., Visa, A.: Comparison of combined shape descriptors for irregular objects. In: Clark, A.F. (ed.) 8th British Machine Vision Conf., BMVC 1997, Essex, UK, pp. 430–439 (1997)
32. Jaeger, S., Nakagawa, M.: Two on-line Japanese character databases in Unipen format. In: Proc. Intl. Conf. Document Analysis and Recognition, ICDAR 2001, pp. 566–570. IEEE Computer Society, Los Alamitos (2001)
33. Jain, A., Duin, R., Mao, J.: Statistical pattern recognition: A review. IEEE Trans. PAMI 22(1), 4–37 (2000)
34. Kara, L.B., Stahovich, T.F.: An image-based trainable symbol recognizer for hand-drawn sketches. Computers and Graphics 29, 501–517 (2005)

35. Kettebekov, S., Krahnstöver, N., Leas, M., Polat, E., Raju, H., Schapira, E., Sharma, R.: i2map: Crisis management using a multimodal interface. In: ARL Federate Laboratory 4th Annual Symposium, College Park, MD (March 2000)
36. Kittler, J., Hatef, M., Duin, R., Matas, J.: On combining classifiers. IEEE Trans. PAMI 20(3) (1998)
37. Landay, J.A., Dannenberg, R.B.: Interactive sketching for the early stages of user interface design. In: CHI 1995 Computer Human Interaction, pp. 43–50 (1995)
38. LaViola Jr., J., Zeleznik, R.C.: A practical approach for writer-dependent symbol recognition using a writer-independent symbol recognizer. IEEE Transactions on pattern analysis and machine intelligence 29(11), 1917–1926 (2007)
39. Lipscomb, J.: A trainable gesture recognizer. Pattern Recognition 24(9), 895–907 (1991)
40. Long, C., Landay, J., Rowe, L., Michiels, J.: Visual similarity of pen gestures. In: CHI 2000: Proc. of the SIGCHI conference on Human factors in computing systems, pp. 360–367 (2000)
41. Marti, U., Bunke, H.: A full English sentence database for off-line handwriting recognition. In: Proc. of the 5th Int. Conf. on Document Analysis and Recognition (ICDAR 1999), Bangalore, India, pp. 705–708 (1999)
42. Miyao, H., Maruyama, M.: An online handwritten music score recognition system. In: 17th Int. Conf. on Pattern Recognition (ICPR 2004), vol. 1, pp. 461–464 (2004)
43. Montello, D.R.: Spatial cognition. In: Smelser, N.J., Baltes, P.B. (eds.) Int. Encyclopedia of the Social & Behavioral Sciences, pp. 14771–14775. Oxford Pergamon Press, Oxford (2001)
44. Nakagawa, M., Matsumoto, K.: Collection of on-line handwritten Japanese character pattern databases and their analysis. Int. Journal on Document Analysis and Recognition 7(1), 69–81 (2004)
45. Niels, R., Willems, D., Vuurpijl, L.: The NicIcon collection of handwritten icons. In: ICFHR8, the 11th Int. Conf. on Frontiers of Handwriting Recognition, Montreal, Canada, August 2008, pp. 296–301 (2008)
46. Niels, R., Vuurpijl, L., Schomaker, L.: Automatic allograph matching in forensic writer identification. Int. Journal of Pattern Recognition and Artificial Intelligence 21(1), 61–81 (2007)
47. Oviatt, S.L.: Multimodal interfaces. In: Jacko, J., Sears, A. (eds.) The Human-Computer Interaction Handbook: Fundamentals, Evolving Technologies and Emerging Applications, ch. 14, pp. 286–304. Lawrence Erlbaum Assoc., Mahwah (2003)
48. Pechwitz, M., Maddouri, S., Märgner, V., Ellouze, N., Amiri, H.: Ifn/enit-database of handwritten Arabic words. In: 7th Colloque Int. Francophone sur l'Ecrit et le Document (CIFED 2002), Hammamet, Tunis, October 21-23, pp. 1–8 (2002)
49. Peura, M., Iivarinen, J.: Efficiency of simple shape descriptors. In: Arcelli, L.P., Cordella, C., Sanniti di Baja, G. (eds.) Advances in Visual Form Analysis, pp. 443–451. World Scientific, Singapore (1997)
50. Piramuthu, S.: Evaluating feature selection methods for learning in data mining applications. European Journal of Operational Research 156, 483–494 (2004)
51. Plamondon, R., Srihari, S.: On-line and off-line handwriting recognition: A comprehensive survey. IEEE PAMI 22(1), 63–84 (2000)
52. Poel, M., Poppe, R.W., Nijholt, A.: Meeting behavior detection in smart environments: Nonverbal cues that help to obtain natural interaction. In: Cohn, J., Huang, T.S., Pantic, M., Sebe, N. (eds.) Proc. of the IEEE Int. Conf. on Automatic face and Gesture Recognition (FGR 2008), Amsterdam, The Netherlands, September 2008, pp. 1–6. IEEE Computer Society Press, Los Alamitos (2008)

53. Poppe, R.W.: Discriminative Vision-Based Recovery and Recognition of Human Motion. PhD thesis, University of Twente (April 2009)
54. Rubine, D.: Specifying gestures by example. Computer Graphics 25(4), 329–337 (1991)
55. Schomaker, L.R.B.: From handwriting analysis to pen-computer applications. IEE Electronics Communication Engineering Journal 10(3), 93–102 (1998)
56. Schomaker, L.R.B., Teulings, H.-L.: A handwriting recognition system based on properties of the human motor system. In: Proc. of the First Int. Workshop on Frontiers in Handwriting Recognition (IWFHR), Montreal, Canada, pp. 195–211 (1990)
57. Sharma, R., Yeasin, M., Krahnstöver, N., Rauschert, I., Cai, G., Brewer, I., Maceachren, A.M., Sengupta, K.: Speech-gesture driven multimodal interfaces for crisis management. In: Proc. of the IEEE, pp. 1327–1354 (2003)
58. The ICIS project. Interactive collaborative information systems ICIS (2004), `http://www.icis.decis.nl/`
59. Tian, F., Cheng, T., Wang, H., Dai, G.: Research on User-Centered Design and Recognition of Pen Gestures. In: Nishita, T., Peng, Q., Seidel, H.-P. (eds.) CGI 2006. LNCS, vol. 4035, pp. 312–323. Springer, Heidelberg (2006)
60. van Erp, M., Vuurpijl, L., Schomaker, L.: An overview and comparison of voting methods for pattern recognition. In: Proc. of IWFHR8, Niagara-on-the-Lake, Canada, August 2002, pp. 195–200 (2002)
61. Vapnik, V.: The nature of statistical learning theory. Springer, Berlin (1995)
62. Viard-Gaudin, C., Lallican, P.M., Binter, P., Knerr, S.: The IRESTE On/Off (IRONOFF) dual handwriting database. In: Proc. Intl Conf. Document Analysis and Recognition, ICDAR 1999, Bangalore, India, pp. 455–458 (1999)
63. Vinciarelli, A.: A survey on off-line cursive script recognition. Pattern Recognition 35(7), 1433–1446 (2002)
64. Vuurpijl, L., Schomaker, L.: Finding structure in diversity: A hierarchical clustering method for the categorization of allographs in handwriting. In: Proc. ICDAR4, pp. 387–393. IEEE Computer Society, Los Alamitos (1997)
65. Wahlster, W. (ed.): SmartKom: Foundations of Multimodal Dialogue Systems. Springer, Berlin (2006)
66. Webb, A.R.: Feature selection and extraction. In: Statistical Pattern Recognition, 2nd edn., ch. 9, pp. 305–359. John Wiley & Sons Ltd., Chichester (2002)
67. Willems, D.M., Niels, R., van Gerven, M., Vuurpijl, L.: Iconic and multi-stroke gesture recognition. Pattern Recognition 42(12), 3303–3312 (2009)
68. Willems, D., Vuurpijl, L.: Pen gestures in online map and photograph annotation tasks. In: Proc.of the tenth Int. Workshop on Frontiers in Handwriting Recognition (IWFHR 2006), La Baule, France, October 2006, pp. 297–402 (2006)
69. Willems, D., Vuurpijl, L.: Designing interactive maps for crisis management. In: Proc. of the 4th Int. Conf. on Information Systems for Crisis Response and Management (ISCRAM 2007), pp. 159–166 (2007)
70. Willems, D.J.M.: Interactive Maps: using the pen in human-computer interaction. PhD thesis, Donders Institute for Brain, Cognition and Behavior, Radboud University Nijmegen, The Netherlands (in press)
71. Willems, D.J.M., Niels, R.: Definitions for features used in online pen gesture recognition. Technical report, NICI, Radboud University Nijmegen (2008)
72. Willems, D.J.M., Rossignol, S.Y.P., Vuurpijl, L.: Mode detection in online pen drawing and handwriting recognition. In: Proc. of the 8th Int. Conf. on Document Analysis and Recognition (ICDAR 2005), Seoul, Korea, pp. 31–35 (2005)

73. Willems, D.J.M., Vuurpijl, L.: A Bayesian network approach to mode detection for interactive maps. In: Proc. of the 9th Int. Conf. on Document Analysis and Recognition (ICDAR 2007), Curitiba, Brazil, pp. 869–873 (2007)
74. Willems, D.J.M., Vuurpijl, L.G.: Pen gestures in online map and photograph annotation tasks. In: Proc. of the Tenth Int. Workshop on Frontiers in Handwriting Recognition, pp. 397–402 (2006)
75. Zeleznik, R., van Dam, A., Li, C., Tenneson, D., Maloney, C., LaViola Jr., J.: Applications and issues in pen-centric computing. IEEE Multimedia 14(4), 14–21 (2008)
76. Zhang, L., Sun, Z.: An experimental comparison of machine learning for adaptive sketch recognition. Applied Mathematics and Computation 185(2), 1138–1148 (2007)

Interacting with Adaptive Systems*

Vanessa Evers, Henriette Cramer, Maarten van Someren, and Bob Wielinga

Abstract. This chapter concerns user responses toward adaptive and autonomous system behavior. The work consists of a review of the relevant literature off-set with the findings from three studies that investigated the way people respond to adaptive and autonomous agents. The systems evaluated in this chapter make decisions on behalf of the user, behave autonomously and need user feedback or compliance.

Vanessa Evers
Human-Computer Studies, University of Amsterdam, Science Park 107, Amsterdam, The Netherlands
e-mail: v.evers@uva.nl

Henriette Cramer
Human-Computer Studies, University of Amsterdam, Science Park 107, Amsterdam, The Netherlands
e-mail: hcramer@science.uva.nl

Maarten van Someren
Human-Computer Studies, University of Amsterdam, Science Park 107, Amsterdam, The Netherlands
e-mail: M.W.vanSomeren@uva.nl

Bob Wielinga
Human-Computer Studies, University of Amsterdam, Science Park 107, Amsterdam, The Netherlands
e-mail: B.J.Wielinga@uva.nl

* This chapter re-uses material from the following previous publications:
– Henriette Cramer, Vanessa Evers, Maarten van Someren, Bob Wielinga (2009) "Awareness, Training and Trust in Interaction with Adaptive Spam Filters", Proceedings of the 27th International Conference on Human Factors in Computing Systems (CHI'09), Boston, USA, pp. 909-912. http://doi.acm.org/10.1145/1518701.1518839. ©2009 Association for Computing Machinery, Inc. Reprinted by permission.
– Henriette Cramer, Vanessa Evers, Maarten van Someren, Satyan Ramlal, Lloyd Rutledge, Natalia Stash, Lora Aroyo, Bob Wielinga (2008) The effects of transparency on trust and acceptance in interaction with a content-based art recommender, User Modeling and User-Adapted Interaction, 18, 5, pp. 455-496. DOI 10.1007/s11257-008-9051-3. ©2008 Springer. With kind permission of Springer Science and Business Media.
– Henriette Cramer, Vanessa Evers, Maarten van Someren, Bob Wielinga, Sam Besselink, Lloyd Rutledge, Natalia Stash, Lora Aroyo (2007). User Interaction with user-adaptive information filters. Proceedings of the Second International Conference on Usability and

R. Babuška & F.C.A. Groen (Eds.): Interactive Collaborative Information Systems, SCI 281, pp. 299–325.
springerlink.com

Apart from the need for systems to competently perform their tasks, people will see adaptive and autonomous system behavior as social actions. Factors from humans' social interaction will therefore also play a role in the experience users will have when using the systems. The user needs to trust, understand and control the system's autonomous actions. In some cases the users need to invest effort in training the system so that it can learn. This indicates a complex relationship between the user, the adaptive and autonomous system and the specific context in which the system is used. This chapter specifically evaluates the way people trust and understand a system as well as the effects of system transparency and autonomy.

1 Introduction

In technological solutions, it is becoming more commonplace to implement adaptive and autonomous system capabilities. Adaptivity and autonomy are used to deal with an overload of information or to take over repetitive tasks from the user. For instance, in crisis response settings where service personnel is pressed for time and where people work under high risk and high-stress circumstances, technology can take away some of the burden of decision making and information filtering. Examples of technologies that make decisions or independently carry out tasks for the users are adaptive information filters, distributed sensor networks, and autonomous mobile agents such as robots.

Autonomous and adaptive systems adapt to the users' needs and are able to improve over time with implicit or explicit feedback from the user [45] to build a user profile or user model. For instance, for an adaptive spam filter, the user needs to indicate which messages are spam in order for the filter to improve. Since performance relies heavily on implicit or explicit feedback from the user, optimizing the design of the interaction is very important. Users need to trust an adaptive information filter before they choose to depend on the system. In the case of a spam filter, users risk losing important information when the spam filter deletes messages autonomously. Trust is even more complex when the adaptive system is based on machine learning technology, as the system improves over time. Hopefully, the system learns and improves from user feedback, but performance could also decrease. Performance can decrease when the interaction is not well designed. For instance, a user may not be aware which icon is to delete a message and which icon is to mark a message as spam. It is not trivial to program a system that can determine if a message is relevant

Internationalization, UI-HCII 2007, Held as Part of HCI International 2007, Beijing, China, Lecture Notes in Computer Science, Usability and Internationalization. Global and Local User Interfaces. pp. 324-333. DOI 10.1109/WIIAT.2008.326 ©2008 Springer. With kind permission of Springer Science and Business Media.
– Henriette Cramer, Vanessa Evers, Nicander Kemper, Bob Wielinga (2008) "Effects of Autonomy, Traffic Conditions and Driver Personality Traits on Attitudes and Trust towards In-Vehicle Agents", Workshop on Human Aspects in Ambient Intelligence (HAI'08), Proceedings of the International Conference on Web Intelligence and Intelligent Agent Technology (WI-IAT'08), Sydney, Australia, pp. 477-482, DOI 10.1109/WIIAT.2008.326. ©2008 IEEE. Reprinted by permission.

for a user in a certain context. Therefore, such systems are trained using data from exercises and earlier events. Although systems that are programmed by hand are not free of errors, constructing or tuning a system by training can result in errors that may be unexpected for a user. Using the system thus requires the user to understand the system. In order for the user to understand the way the system works, it needs to be transparent. By evaluating previous research we cannot gain full understanding of how users respond to system autonomy and adaptivity and what factors influence user attitudes, acceptance, and trust. Studies generally focus on technical aspects, such as the accuracy of different machine learning techniques. A few studies are available on user interaction with recommender systems [71] [1]. Research about trust in and interaction with user-adaptive information filters has focused mostly on issues for collaborative recommenders (recommender systems that use the opinions of a community of users to help individuals in that community more effectively identify content of interest), for example [67][59][56]. Rather than trust in the decisions of other users, in this chapter we are interested in the ways in which users trust the decisions of the adaptive or autonomous system and what influences user attitudes toward these systems. In the next section, we will first review the relevant theory and formulate our research questions. The literature focuses on trust in and acceptance of autonomous system and social aspects of interaction with autonomous systems. In the section that follows, we will discuss the results from three studies that have been carried out as part of the ICIS-CHIM project on user responses to autonomous and adaptive system behavior. In the final section, we will explain how the results of these studies contribute to a deeper understanding of human interaction with adaptive systems and offer specific guidelines for the development and design of such systems.

2 Related Work

2.1 User Interaction with User-Adaptive Systems

User-adaptive systems adapt to the interests and context of a specific user and adapt when the interests or situation of a user changes [3] [7]. As such, they make autonomous or semi-autonomous decisions on what is relevant in a specific context. Authors such as Keeble and Macredie [38] and Benyon [3] give a number of reasons to make a system adaptive to its user. First, users change over time and a system needs to adjust to the user's changed interests or preferences. Second, for applications in situations that cannot be predicted beforehand, it may not be possible to develop an effective system without adaptivity. For information filters for example, it can be event dependent which information will interest the user. Apart from the advantages, user adaptivity also introduces new challenges for human–system interaction. Alpert et al. [1] state that adaptivity is only wanted when the system is perceived to be capable and is trusted; otherwise, the system should not be adaptive at all. Adaptivity inherently means a system will not always behave the same way and will not always present the same information; it will change with user behavior.

This means that a system might not behave as expected (illustrated by, e.g., [20]). This might cause interaction problems. Billsus [5], for example, identifies common interface problems for recommender systems that proactively recommend information items that might be relevant to the user on the basis of earlier interaction. First, the messages can either be too obtrusive or too subtle to notice. Task-focused users might not be willing to interrupt their current tasks by examining new recommendations. Not all recommendations might be relevant either, which might cause users to ignore recommendations altogether. Not sufficiently explaining why particular recommendations are made might contribute to this problem. Third, a system that is based on machine learning technology will undoubtedly make mistakes. If it is a system that can be trained then it can improve by learning from its mistakes. It is, however, important to understand whether and under what circumstances users are willing to accept a certain level of mistakes. An additional challenge is to get feedback from the user so that a user-adaptive system can learn from and adapt to that specific user. Appropriate and convenient ways to enable the user to provide feedback to the system need to be devised. Feedback can be given implicitly (e.g., by measuring reading time of an item or from sensor data), explicitly (which requires user action, e.g., marking information items as relevant or irrelevant, or by direct editing of a user's profile), or by a mixed approach. User feedback to adaptive systems can be offered by direct manipulation of profiles and user models [71][1]. Waern [71], however, reports that even when users appear to appreciate this direct form of control, it is difficult for them to improve their own user profiles and recognize high-quality profiles, because they do not fully understand their profiles or how the system uses them. Jameson [32] provides an overview of challenges related to user adaptivity, such as controllability, privacy, obtrusiveness, breadth of the user's experience, predictability, and transparency. User-adaptive systems take decisions on behalf of the user; this influences controllability of a system and might conflict with the user's need to control a system's actions or be "in the loop." User-adaptive systems use data about the user. A concern might be that this information is used in undesirable ways, affecting the user's privacy. Obtrusiveness might be an issue when the adaptive system itself requires attention from the user, taking away attention from the user's primary task. User-adaptive systems that take over tasks from the user can also affect the breadth of the user's experience. For example, a user might not learn as much about a domain when information is filtered for him or her and the user may inappropriately rely on the information provided by the system alone. Höök [28] also notes that user adaptivity might conflict with usability principles such as making a system transparent, understandable, and predictable. Adaptive systems might change and adapt in unpredictable ways without the user being able to understand why the system behaves in a certain way. User understanding of the system can affect both the user's attitude toward the system and the user's feedback. When the system behaves in ways the user does not expect (e.g., recommending something he or she does not like or need), the user may find the system unpredictable and unreliable [28]. If the user has an inaccurate mental model of the way the system learns, he or she might also change his or her feedback toward the system in such a way that performance decreases (e.g., [71]). Different contexts and types of users might ask for different types of adaptivity and user control. Alpert et al. [1] note that a personalization technique may be considered desirable by users in one context but not in other. Alpert and colleagues studied user interaction with a user adaptive e-commerce web site for computer equipment sales and support. In the study of Alpert et al., participants had mixed reactions to the collaborative filtering feature. Users felt positive about the system that presented new products and services related to the users' current search activities and their recent past activity on the website. However, Alpert and colleagues found that participants reacted positively to the theoretical possibility of such a feature, but that they did not expect the system to actually be able to predict their needs in practice. They noted that in other e-commerce settings such as movie or book recommendations, users appear to have no objections toward collaborative techniques. Not only the application context but also characteristics of users themselves might influence acceptance of adaptive systems. Specifically in the domain of user interaction with personalized museum guides, Goren-Bar et al. [24] found that personality traits relating to the notion of control have a selective effect on the acceptance of the adaptivity dimensions. One of their findings was that users who feel less in control of their own beliefs and actions prefer to maintain control over a system and prefer non-adaptivity. Jameson and Schwarzkopf [33] investigated the level of controllability of a user-adaptive system, and performed a study in which users got to determine the nature and timing of adaptations. Jameson and Schwarzkopf concluded that the preferable form of control and automation for a task depend on factors ranging from individual differences and system properties that are relatively stable to unpredictable situational characteristics; there is no one level that fits all situations.

2.2 User Attitudes toward Systems' Autonomous Behavior

Balancing user control and system autonomy is a vital issue in achieving trust. Autonomous system behavior can lead to a sense of loss of user control. (Perceived) Control over adaptivity has been shown to affect user attitudes toward adaptive systems [33] and user trust [9]. Control is also crucial in maintaining combined human–system performance. Desmond et al. [17] found in a car simulator study that it is hard for drivers to recover performance when an automatic driving system makes a mistake. Participants who were in full control of driving performed better than participants who used an automatic driving system. Perceived control over a situation or outcome additionally has been shown to be related to the level of stress experienced [43]. This is unsurprising as stress involves perceptions that situational demands exceed a person's abilities and/or resources. An instructive system that makes decisions for the user or instructs the user to take action can diminish the sense of control, possibly decreasing trust and positive attitudes toward such a system. Instead of taking over the driving task from the driver, a system could offer information about stressful conditions ahead. Providing preparatory information can decrease stress reactions and increase perceived control, self-efficacy, and performance [31]. Perceived risk might differ between circumstances, as users might expect the system's performance (and their own) to vary across situations as well. In time-critical situations where the user cannot spend much attention to system information or when a user is stressed, instructions might be more preferable to information the user has to process him or herself. However, stressful (high-risk) situations might also make the user feel less secure about depending on a system. Thus, context is expected to affect user responses to autonomous system behavior as well.

Which system adaptations and level of user control are suitable have to be determined for a particular context, application, and the specific user. This needs to be reflected in the user–system dialogue. This dialogue between the user-adaptive system and user appears crucial in achieving both excellent performance and user satisfaction. In the context of user-adaptive recommender systems, this dialogue should provide the user with high-quality recommendations and information. At the same time, the dialogue needs to provide the system with information to improve its recommendations. The user should be persuaded to invest time and effort to provide the system with enough useful training data to improve its recommendations. As Benyon [3] points out, user preferences for information items can change over time. This means that a user's profile that might have been suitable at one time might not always be suitable at other occasions. This might necessitate continuous adaptation to the user's feedback while taking into account the issues described above, such as controllability, obtrusiveness, and user understanding. User-adaptive systems raise the expectation to "know what the user wants" and to make (semi-)autonomous decisions on what is relevant to the user. The system's user profiling has to be adequate to process the user's feedback, and the criteria used for recommending have to be suitable for the user's task. The user on the other hand has to be willing to release control, accept the system, and delegate tasks. The dialogue between system and user should overcome challenges inherent to user adaptivity and instill trust in the user. This so that they can indeed let a system make decisions for them in their specific situation.

2.3 Trust in Adaptive and Autonomous Systems

Reeves and Nass [63] have shown that people tend to interact with computer systems as if they were social actors. Users' affective and social reactions to automated tools affect acceptance of and interaction with such systems. A possible implication is that social rules and principles of human-to-human interaction can also be applied to human–computer interaction. This includes the concept of trust. Trust is thought to be one of the important affective elements that play a role in human interaction with technology [41][19]. Making a system reliable is not enough to achieve appropriate delegation to that system (as observed by, e.g., [57]). Trust also depends on the interaction with the user. Users need to adapt their levels of trust to optimize human–machine performance and benefit fully from the cooperation [60][61][62]. Users may have a level of trust in a device, but also have a level of trust in the success of cooperation with the device [52][53]. Even if an appropriate level of automation can be determined, there is no guarantee that users will trust a system to an appropriate degree and will actually decide to use it. In the context of adaptive cruise control assisting drivers, Rajaonah et al. [61] found that trust in the cooperation with a device determines the quality of interaction between the user and system, as well as appropriate use. In an experiment using a flight simulator and an engine monitoring system that provided advice on engine fault diagnosis, Parasuraman and Miller [57] found that user performance and trust were lowered by poor automation etiquette, regardless of reliability. In their study, a non-interruptive and patient communication style, instead of using interruptive and impatient worded system messages, increased trust and performance. Even though a highly reliable, good etiquette system version yielded the best user performance and the highest trust, they found that good automation etiquette can compensate to a certain extent for low automation reliability. It seems that performance alone does not completely determine whether users take full advantage of a system. In this chapter we will adopt the definition of trust by Jøsang and Lo Presti [36]:

> the extent to which one party is willing to depend on somebody or something, in a given situation with a feeling of relative security, even though negative consequences are possible.

A similar definition is offered by Lee and See [41], who define trust as

> the attitude that an agent will help achieve an individual's goals in a situation characterized by uncertainty and vulnerability.

Parasuraman and Miller [57] state that trust is a user's willingness to believe information from a system or make use of its capabilities. This definition is related to an alternative concept proposed by Fogg and Tseng [22], who introduced the concept of credibility. Credibility (synonymous to believability) is proposed to consist of trustworthiness (the perception that the source has good intentions) and expertise (or competence) of a system. In this chapter, the concept of trust consists of trust in the intentions of a system (goal alignment) and trust in the competence of the system. Competence is seen as the perceived skill of the system: the extent to which it is able to offer the right recommendations. The perception of the alignment of goals of the system and the user's goals, coupled with a belief that a system will perform its task competently, form the basis of trust. In the context of this chapter, trust refers to the user's willingness to depend on a system and its recommendations/decisions in the specific context of the user and his or her task(s), even though the system might make mistakes. A number of factors might play a role in the perception users have of the goals of the system and the intentions of its developers and other users. Briggs and colleagues [6] note that personalization itself plays a role in trust formation. In the context of e-commerce websites, in an experiment using a (mock) website offering travel insurance, Briggs found that participants felt more positive about the site when it offered a personalized quote based on the participants' circumstances. They felt, for example, that if they had been offered more choice, they would have found the site more predictable and that it was easier to use. Such perceived benefits of personalization might, however, be dependent on whether the intentions and capabilities of a personalization system instill trust in the user. In some situations, trust may be a prerequisite to get the personal information needed to personalize a system, invoking privacy concerns [40]. In recommender systems, trust in the information source can also play a role in whether a user trusts the system. Authors such as Wang and Benbasat [72] and Xiao and Benbasat [75] studied trust in online e-commerce recommender settings. In such settings, trust in the online vendor's goals in providing recommendations might also play a role. In collaborative-based recommenders, users need to assess whether they trust the intentions and actions of other users [70]. In addition, in contexts where a system might provide information about a recommended item not immediately verifiable by users, trust in the correctness and credibility of the information might play a role as well. For example, when a recommender system provides recommendations based on how soft a certain product will feel, users cannot know whether this is correct before they actually buy and feel the product [22]. Whether a personalized system is found competent can be influenced by issues inherent to adaptivity. User-adaptive systems may not offer a high level of performance from the start. If performance is low in the beginning, users' trust levels could start out low as well, and gradually increase as training progresses and performance increases. However, trust rapidly decreases when users notice errors and only slowly increases as a system performs without errors [19][52][53]. This makes achieving trust difficult in adaptive systems that have not been pre-trained. Lee and See [41] use the terms calibration, resolution, and specificity of trust to describe such (mis)matches between trust and the capabilities of automation. Calibration refers to the correspondence between a person's trust in a system and the system's capabilities (see also [42][52]). Resolution of trust refers to what extent such changes in system capability affect trust. Trust also has specificity; trust can be more or less specific to part of a system or to changes in a system's capability. Getting these aspects of trust right is especially challenging in a personalized system, as the nature of these systems implies that the system's capabilities change; the specificity and resolution of trust in the user might not match these changes.

Reliance on a system is guided by trust, but not completely determined by it; trust does not necessarily imply reliance (see also [11]). Castelfranchi and Falcone [9], for example, note that a user might trust a system to work well but will not rely on the system as the user may find the risk associated with a possible mistake unacceptable. Users have to calibrate their level of trust and their level of reliance on the system to its capabilities [42][52][53]. To ensure effective performance, an appropriate level of automation needs to be determined while preserving an appropriate level of human control [51]. Understanding the reasoning of a system and the reasons for results could potentially help users calibrate their level of trust. System transparency has been reported to increase users' trust, which also has been shown in the context of information filtering and recommender systems (e.g., [67][27][59]).

2.4 Transparency of Adaptive Systems

Maes (in a discussion between Shneiderman and Maes [65]) states that one of the great interaction design challenges is how to achieve understanding and control in user–agent interaction. If the user understands how a system works and can predict system actions and outcomes, the user can focus on his or her task instead of trying to figure out the system. In the case of user-adaptive systems, it may be even more imperative for the users to understand how the system decides on a recommendation or action. In order for the system to reach an optimal level of performance, it needs to learn from the users' implicit or explicit input, for example by analyzing user's search behavior or by receiving explicit feedback. When a user is not aware of or does not have an accurate understanding of how the system makes decisions, the user will not "train" the system properly and it may not be possible for the system to improve. Better understanding of the system could help improve such aspects of interaction. Transparency aims to increase understanding and entails offering the user insight in to how a system works, for example by offering explanations for system behavior.

System transparency may not always improve user interaction. It is not always possible for the user to construct an accurate mental model in a reasonable timeframe. Fogg and Tseng [21] note that in the context of credibility, the impact of any interface element on users' attitudes depends on to what extent it is noticed (prominence) and what value users assign to the element (interpretation). Parasuraman and Miller [57] note that as automation gets more complex, users will be less willing and able to learn about the mechanisms that produce the system's behaviors. Höök [28] notes that it is not necessarily desirable to have a system explain how it works in full detail because the full details might be alienating to a layman user. It might not be necessary to explain all details to achieve adequate understanding of the system's adaptivity. Waern [71] shows that users cannot always recognize what is a high-quality user profile and cannot always improve them when given the opportunity. Apart from the effort the user needs to invest to process transparency information, other adverse effects have to be considered as well. Herlocker et al. [27] found that poorly designed explanations can lower the acceptance of individual recommendations. Dzindolet and colleagues [19] found that explanations on why errors occur increased trust in automated tools, even when this trust was not justified. Therefore, transparency and task performance can influence each other negatively (see also [10]). Bilgic and Mooney [4] note that explanations should not just promote ("sell") a system's recommendations, but that explanations should enable the user to accurately assess the quality of a system's recommendations or decisions. Well-designed transparency needs to overcome these challenges.

In spite of these potential problems, previous authors have emphasized the importance of transparency [32][30]. Sinha and Swearingen [67], for instance, evaluated transparency and recommendations for five music recommender systems. People liked recommendations more and had more confidence in those that were more understandable to participants. Other prior studies also suggest that making adaptive systems more transparent to the user could lead to a more positive user attitude toward using a system and increases in system performance [37][30][29][8][25]. Lee and See [41] state that trust and reliability depend on how well the capabilities of a system are conveyed to the user. McAllister[47], quoting [66], notes that

> the amount of knowledge necessary for trust is somewhere between total knowledge and total ignorance. Given total knowledge, there is no need to trust, and given total ignorance, there is no basis upon which to rationally trust.

Lee and See argue that promoting appropriate trust may depend on presenting information about a system in a manner compatible with analytic, analogical, and affective processes that influence trust. They identify three types of goal-oriented information that appear to contribute to the development of trust: information on current and past performance (system expertise), information on the system's process (to assess appropriateness of the automation's algorithms for the user's situation and goals), and purpose (to match the designer's intent for the system to the user's goals). Increasing transparency of a system can help users decide whether they can trust the system.

A number of authors describe systems with various features that aim to support trust by making a system's reasoning process understandable and providing insight in to system competence. In the context of the Semantic Web, McGuinness and Pinheiro da Silva [48] describe an explanation system that explains answers from Semantic Web agents that use multiple sources to devise an answer to user questions. Their explanations include information on the origin of answers or how they were derived. Other examples of transparency features can be found in the context of e-commerce product recommendations. McSherry [50] discusses an approach to find a fitting product recommendation by guiding the user through the features of the available products, which can, for example, help understand the trade-offs and interdependencies between product features. Shimazu [64] discusses a system that mimics the interaction between a customer and a store clerk to come to a final understandable product recommendation. Pu and Chen [59] have focused on trust in the recommender based on the user's perception of its competence and its ability to explain the recommended results. Their organization interface, which organized recommender results by trade-off properties, was perceived as more capable and efficient in assisting user decisions. Cortellessa et al. [12] also evaluated the importance of explanations in interactive problem-solving systems and found that explanations are needed more when the system's problem solving fails.

Various studies have investigated the effects of different types of transparency features. McNee and colleagues [49] found that adding a confidence metric to a movie recommender, indicating which recommendations were "more risky," increased user satisfaction and influenced user behavior. However, they also found that more experienced users of the movie recommender were less satisfied with the system after being instructed about the confidence rating. Wang and Benbasat [73] compared the effects of three types of explanations on user trust in an e-commerce recommender. The recommender system used in their study advised users what digital camera to buy based on their answers to questions about their product needs (not on the basis of the user's ratings of other products). Wang and Benbasat compared the effects of, in their terminology, "why," "how", and "trade-off" explanations. "Why" explanations were designed to show that the recommender is designed to fulfil the user's needs and interests. These "why" explanations justified the importance and purpose of questions posed by the recommender and justified final recommendations. "How" explanations revealed the line of reasoning based on consumer needs and product attributes preferences in reaching recommendations. "Trade-off" explanations offered decisional guidance, helping users to make proper trade-offs among product features and costs of these features (e.g., a digital camera offering more zoom capabilities will also cost more). Wang and Benbasat found that "why" explanations increased benevolence beliefs; the perception that a recommender acts in the consumer's interests. They found that "how" explanations increased participants' belief in competence and benevolence of the recommender. "Trade-off" explanations increased integrity beliefs. Herlocker et al. [27] compared the effects of 21 different explanations for a movie recommendation on participants' acceptance of the recommendation. The participants had to indicate how likely it was that they would go and see the movie that was recommended by the system. Participants reported that the explanations were important. Histograms of other users' ratings of an item were found to be the most convincing way to explain why a recommendation had been made. Indications of a system's past performance, likeness to similar previously rated items, and domain-specific content features (such as favorite actress) were most compelling in justifying a recommendation. Some other types of explanations, such as correlations between information items, were difficult to understand and actually decreased acceptance of recommendations. Tintarev and Masthoff [68] provide a number of guidelines for recommender system explanations. They advise that explanations needs to be tailored to the user and context, as different users might find other criteria important in selecting recommendations.

2.5 The Role of Trust in Acceptance of Adaptive Systems

The preceding section described how trust guides reliability or dependence on a system. Reliability is related to acceptance and use of a system. The Technology Acceptance Model [16] describes a number of concepts that influence acceptance of a system. According to the model (updated in [69]), performance expectancy, effort expectancy (how much effort it will take to learn and use a system), social influence, and facilitating conditions (whether the user has enough time and resources to use the technology) are expected to influence intent to use a system and actual usage behavior. Several authors have studied possible extensions to the Technology Acceptance Model (see, e.g., [54][44][39][26]). Van der Heijden [26], for example, showed that the purpose of a system plays an important role in determining the predictive importance of respective elements of the technology acceptance model.

Perceived usefulness is important in achieving user acceptance when interacting with work- and task-oriented systems. If users interact with "hedonic" systems that are aimed at fun and entertaining the user instead, perceived enjoyment and perceived ease of use are more important than perceived usefulness. Van der Heijden suggests that focusing on the type of system could increase understanding of technology acceptance in addition to extending acceptance models with more factors that influence system acceptance.

Trust has been shown to affect system acceptance and has been combined with the technology acceptance model in, for example, [74][73][58] and [23]. These authors did so mostly in the context of e-commerce and on trust in services. They primarily focused on trust in online transactions and the relationship between a buyer and an online vendor. Integration of acceptance models with other trust-related aspects, such as trust in a system itself, in other contexts is still scarce. In a non-commercial context, such a consumer–buyer relationship is arguably absent. The trust-building process might therefore be more focused on whether the system goals match the users' goals and whether it would be competent in giving recommendations. This chapter focuses more on the trust a user feels toward a system itself and delegation of tasks to that system. It can be argued that trust is the attitude that arises from both the (affective) evaluation of the intentions of a system (or rather its developers) and whether they match the user' goals for using that system, plus an (affective) evaluation of the competence of a system (will the system actually work well and not make too many errors).

2.6 Conclusions

The review of the literature in the preceding section highlights the important role of trust and transparency for the acceptance of adaptive systems. Adaptive systems make decisions for the user or exhibit other autonomous behaviors based on reasons that are not always transparent to the user. The authors therefore expect that transparency is an important feature in the design of adaptive systems.

Moreover, as people tend to interact with a technology as if it were a social actor, social aspects of interaction are important to consider. Technology that adapts itself to the user or the situation and behaves autonomously may be perceived as even more anthropomorphic, and social schemas of interaction may even be more strongly invoked in the human users of the systems.

The level of autonomy a system has and the control that users have over the system or perceive to have over the system is likely to impact the users' attitude toward the system. An intelligent system that is highly autonomous may need to "prove" that it is competent before the user is willing to rely on it. However, for less critical tasks, a user may be less inclined to control the system.

We therefore feel there are a number of important questions to consider:

- Q1: To what extent does trust in an adaptive system influence acceptance of the system?

- Q2: What is the role of transparency in increasing the trust a user has in an intelligent system?
- Q3: Does the context of the task (critical or less critical) and environment (high versus low risk) influence the users' opinion of the appropriate level of system autonomy and user control?

In the following section, we will discuss a series of studies that have been carried out as part of the ICIS-CHIM project that aim to answer these questions.

3 Results from Studies That Investigate User Interaction with Intelligent Systems

3.1 Trust in Interaction with Adaptive Spam Filters

This study[2] (also reported in [15]) aimed to investigate user attitudes and behaviors when using trainable and non-trainable spam filters. We were especially interested in understanding the role of trust and the factors that motivate users to spend time and effort training a spam filter system. Data was collected through 30-min. to 1.5-hour sessions combining observation, in-depth interviews, and a survey. Participants were observed while using their regular email clients at their own workplace. Participants were asked to refrain from using email on the day of the session so that their usual routine of checking email and dealing with spam messages could be observed. The number of email messages, number of spam messages, and the filter settings were recorded. Afterwards, a semi-structured interview was conducted. Twelve open-ended questions concerned users' experiences in interacting with (adaptive) spam filter, their attitudes toward spam, and spam filter use. In addition, we asked participants to explain the way they thought their email filter worked and to explain why they did (not) train their spam filters. A survey that measured acceptance of and trust in information filters and in the training of information filters concluded each session. The questionnaire consisted of 7 items relating to the participants' background and 22 items based on Jian et al. [34] and Venkatesh et al. [69] concerning perceived filter usefulness, perceived ease of use, attitude toward the spam filter, and

[2] This section re-uses material from:
-Henriette Cramer, Vanessa Evers, Maarten van Someren, Bob Wielinga (2009) "Awareness, Training and Trust in Interaction with Adaptive Spam Filters", Proceedings of the 27th International Conference on Human Factors in Computing Systems (CHI'09), Boston, USA, pp. 909-912. http://doi.acm.org/10.1145/1518701.1518839. ©2009 Association for Computing Machinery, Inc. Reprinted by permission.
-Henriette Cramer, Vanessa Evers, Maarten van Someren, Bob Wielinga, Sam Besselink, Lloyd Rutledge, Natalia Stash, Lora Aroyo (2007). User Interaction with user-adaptive information filters. Proceedings of the Second International Conference on Usability and Internationalization, UI-HCII 2007, Held as Part of HCI International 2007, Beijing, China, Lecture Notes in Computer Science, Usability and Internationalization. Global and Local User Interfaces. pp. 324-333. DOI 10.1109/WIIAT.2008.326 ©2008 Springer. With kind permission of Springer Science and Business Media.

Table 1 Example items final scales, participant mean scores, and standard deviations. Cronbach's α as reliability measure. All 7-point (0-6) Likert-type scales. Copyright Â© 2009 ACM, Inc. Reprinted from [15].

Perceived usefulness of the filter α=.86, M=5.3, SD=1.1 2 items, e.g., Use of a spam filter enables me to deal with my email more quickly.
Perceived ease of use filter α=.82, M=4.5, SD=1.4 2 items, e.g., I find the spam filter easy to use.
Attitude toward filter M=5.7, SD=.7 Using a spamfilter is a good idea.
Trust in filter α=.77, M=3.2, SD=1.3 4 items, e.g., I think the spam filter correctly assesses email as spam or non-spam.
Perceived usefulness of training α=.74, M=4.9, SD=1.3 2 items, e.g., I find it useful to train the spam filter.
Perceived ease of use of training α=.86, M=4.5, SD=1.5 3 items, e.g., Training the spam filter is clear and understandable to me.
Attitude toward training M=5.1, SD=1.4 Training the spam filter is a good idea.
Trust in training α=.85, M=4.3, SD=1.3 3 items, e.g., I trust the outcome of the training process of the spam filter.

dependability of the filter (See Table 1). The questionnaire addressed both using and training the spam filter using two separate sections.

Forty-three participants took part in the study at their place of employment at two research organizations. Thirty were male. The mean age was 38 (range: 24–59, SD=11.7). A total of 28 worked in an area related to computer science, the others did not. The average number of legitimate emails per day was 18 (range: 2–58, SD=15.5), and the average of spam emails was a relatively modest 26 (range: 0–270, SD =31.8).

Results indicated that significant differences were found between the three types of filters for usefulness (H(2)=13.896, p=.001), ease of use (F=3.655, p=.036), and attitude toward the filter (H(2)=11.844, p=.003). Jonckheere's test revealed a significant trend in the data: participants who would allow for more autonomy of the system (from labeling, to moving, to deleting spam) were more positive on system usefulness, ease of use of the filter, and attitude toward it. Trust, however, was not found to be related to the choices for delegation.

A significant difference was found for trust in the system's training process (U=118.5, p(1-tailed)=.036). Scores for participants who trained their filters were significantly higher (Mdn=5) than for participants who did not do so (Mdn=3.5). These results indicate that a user's decision to train or correct an adaptive spam filter has an important relation with trust.

The observation and interviews yielded interesting insights into the importance of awareness. In this study, all of the participants who had a server-side filter were aware of the filter's activities, while a considerable portion (29%) of the participants who had an active Thunderbird filter were not. Even though the Thunderbird filter showed spam icons, a "mark as spam" button and a warning message in the content of mails that are labeled as spam, not all users noticed and recognized these.

Such lack of awareness of participants of both filter activity and interface items led to less-than-optimal training behavior in a number of ways. For instance, knowing that the filter was active did not guarantee correct usage of interface items related to training. Sometimes, the spam button and icons were mistaken for delete buttons. This led to inadvertent training of the filter when users used the button to delete no-longer needed, but legitimate email. Also, users occasionally consciously decided to not mark specific spam messages as spam. This decision concerned spam messages that in their opinion were very similar to non-spam messages to "not confuse the filter". Ironically, these messages would be the most informative for the filter to improve.

The findings of this study show system designers need to pay special attention to ensuring awareness about system activity and adaptivity. Trust in the effectiveness of training was found to play an important role in the user's willingness to invest effort in a system that can be taught to improve over time. Systems that depend on explicit user feedback need to be designed in such a way that trust is optimized. An overview of filtering activities should be available to the user. How the system learns and can be taught should be made clear, but only on a level necessary to use the right interface items [71]. Interface items specifically dedicated to training of the system should be recognizable as such. Training a system on "borderline cases" has to be encouraged. Risks may be decreased by providing an opportunity for users to tell the system a case is special, e.g., by explicitly showing the system which similar messages they are worried about might be inadvertently labeled as spam on the basis of their feedback.

The findings reported in this chapter indicate that a more positive attitude toward a system and a more positive assessment of a system's ease of use and usefulness increases the likelihood that a user delegates higher risk tasks to a system; in this case, automatic deletion of messages marked as spam. Users' choices to actively offer feedback to (train) an adaptive system related to trust in the trainability of the filter. Interestingly, while many of our participants invested extensive effort in training their filters, training did not appear to increase reliance on a system. This raises the question whether this investment has positive effects on user attitudes at all. Finally, qualitative findings indicate that facilitating awareness about system activity and adaptivity is extremely important in ensuring trust and useful training behavior.

3.2 The Effects of Transparency on Trust in and Acceptance of a Content-Based Art Recommender

This study[3] (reported in [14]) aimed to discover whether transparency increases trust in and acceptance of adaptive systems (in this case a content-based art recommender). It also aimed to find out whether different types of system transparencies influenced the user attitudes differently.

The experiment had a between-subject design with independent variable transparency (with a transparent condition, a non-transparent condition, and a condition that showed information on how certain the system was of its recommendation, but did not offer more transparency of reasons behind recommendations). Dependent variables included perceived competence, understanding, trust in, and acceptance of the content-based recommender system.

There were three between-subject conditions to manipulate transparency:

- A non-transparent condition ("non"): no transparency feature was offered (Figure 1).
- A transparent ("why") condition: below each thumbnail of a recommended artwork a "why?" link was shown (Figure 2). This link opened a pop-up window listing those properties the recommended artwork had in common with artworks the user had previously rated positively (Figure 3).
- Confidence rating ("sure") condition: below each recommended artwork, a percentage was shown which indicated the confidence the system had that the recommendation was interesting to the user (Figure 4).

Each participant (N=60) participated in one of the three conditions and took part in individual, task-oriented sessions that lasted 45 minutes to 3 hours. Each individual session with a participant consisted of three parts: (1) observation of the participant carrying out a task using the CHIP system (described in more detail below) (2) a questionnaire evaluating participants' acceptance and trust in the system, and (3) a structured interview to further explore participants' attitudes toward the system. The study was conducted by two researchers; each observed 50% of the participants for each of the three conditions. Each of these parts of the study is explained in the following paragraph.

[3] This section re-uses material from:
-Henriette Cramer, Vanessa Evers, Maarten van Someren, Satyan Ramlal, Lloyd Rutledge, Natalia Stash, Lora Aroyo, Bob Wielinga (2008) The effects of transparency on trust and acceptance in interaction with a content-based art recommender, User Modeling and User-Adapted Interaction, 18, 5, pp. 455-496. DOI 10.1007/s11257-008-9051-3. ©2008 Springer. With kind permission of Springer Science and Business Media.
-Henriette Cramer, Vanessa Evers, Maarten van Someren, Bob Wielinga, Sam Besselink, Lloyd Rutledge, Natalia Stash, Lora Aroyo (2007). User Interaction with user-adaptive information filters. Proceedings of the Second International Conference on Usability and Internationalization, UI-HCII 2007, Held as Part of HCI International 2007, Beijing, China, Lecture Notes in Computer Science, Usability and Internationalization. Global and Local User Interfaces. pp. 324-333. DOI 10.1109/WIIAT.2008.326 ©2008 Springer. With kind permission of Springer Science and Business Media.

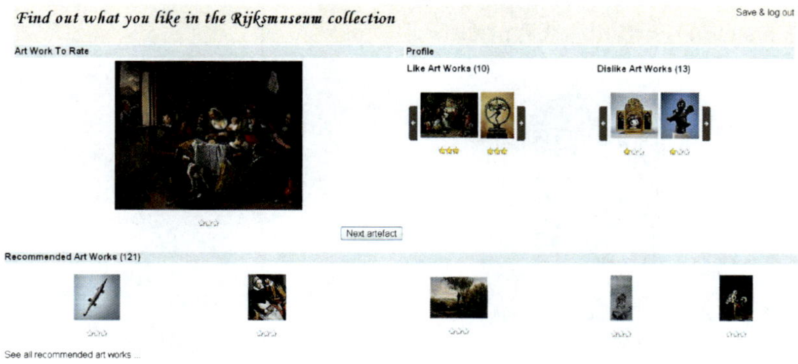

Fig. 1 Non-transparent condition interface. Reprinted from [14]. With kind permission of Springer Science and Business Media.

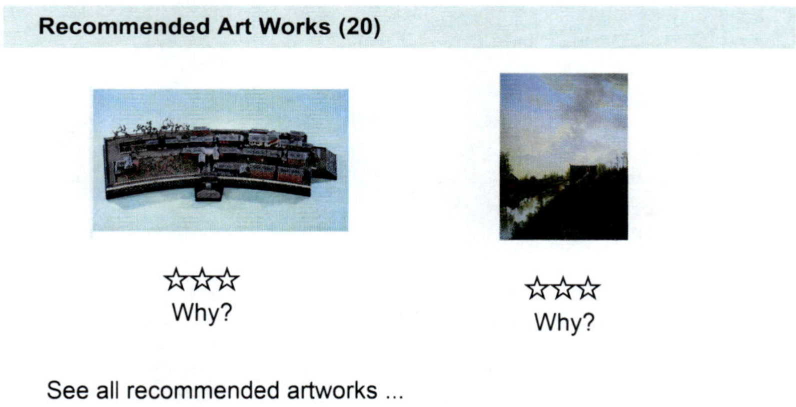

Fig. 2 Detail of the transparent ("why") condition interface, with "why" link (replicated for printing purposes). Reprinted from [14]. With kind permission of Springer Science and Business Media.

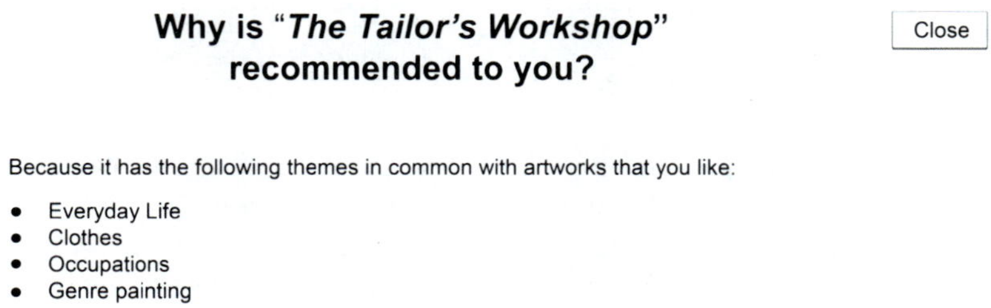

Fig. 3 The "why" feature, pop-up window that is shown when "why" link is clicked by the user (replicated for printing purposes). Reprinted from [14]. With kind permission of Springer Science and Business Media.

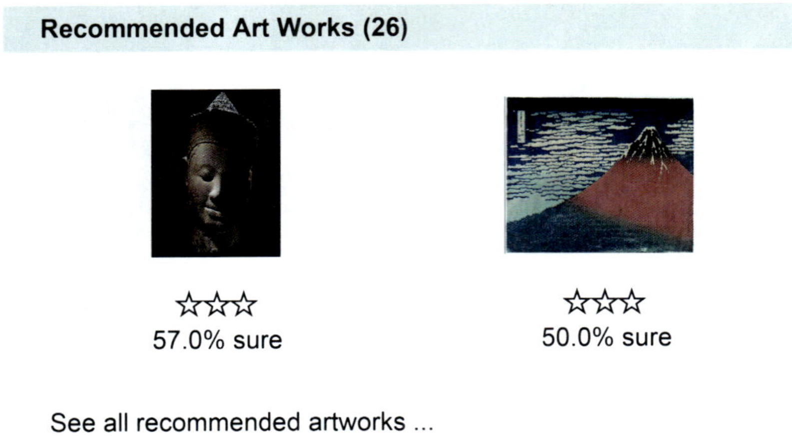

Fig. 4 Detail of the confidence rating ("sure") condition. Reprinted from [14]. With kind permission of Springer Science and Business Media.

Table 2 Participant characteristics. Reprinted from [14]. With kind permission of Springer Science and Business Media.

Age	Mean = 34.4 years SD = 13.63 range 18–68
Gender	Male N = 31 Female N = 29
Max level Education	1.67% (N = 1) primary education 13.3% (N = 8) secondary education 21.7% (N = 13) professional tertiary education 63.4% (N = 38) academic tertiary education
Previous experience with recommenders	31.7% (N = 19) previous experience
Computer experience	Mean = 5.4, SD = 1.2, 1–7 scale
Knowledge of art	Mean = 3.6, SD = 1.7, 1–7 scale 13.3% high level
Interest in art	Art in general: Mean = 5.1, SD = 1.6, 1–7 scale Art in Rijksmuseum: Mean = 4.5, SD = 1.4, 1–7 scale

Table 2 includes the participant's characteristics, and Table 3 includes the final scales and reliability for the measures used in this study. Questionnaire items were all 7-point Likert-type scale questions.

As expected, participants did find the transparent ("why") condition more transparent than the non-transparent condition (Mann–Whitney U=136.500, p(1-tailed)= .025, N"non"=22, N"why"=19, Mean"non"=4.4, St.dev"non"=1.1, Mdn"non"=5.0, Mean"why"=4.9, St.dev"why"=1.4, Mdn"why"=5.0). This finding shows that offering explanations for the systems recommendations indeed causes participants to feel they understand the system better, while offering a confidence percentage does

Table 3 Final scales and variables included in overall analysis, including Cronbach's alpha for the final scale. Questionnaire items were seven-point Likert-type scale questions, with scale ranging from 1, very strongly disagree, to 7,very strongly agree. Reprinted from [14]. With kind permission of Springer Science and Business Media.

Perceived transparency of the system α=.74, M=4.5, SD=1.2, range: 2.0–6.5 2 items, e.g., I understand why the system recommended the artworks it did.
Perceived competence α=.91, M=4.1, SD=1.15, range: 1.6–6.5 8 items, e.g.: I think that the system's criteria in choosing recommendations for me are similar to my own criteria.
Intent to use the system α=.91, M=4.4, SD=1.4, range: 1.0–6.7 3 items, e.g., I would like to use the system again for similar tasks.
Acceptance of Recommendations M=2.0, SD=1.9, range: 0.0–6.0 Number of recommendations that the participant included in final selection of 6 artworks.
Acceptance of the System Participants choosing system selection= 32, Participants choosing manual selection=28 Scenario measuring participant's willingness to delegate task to system.
Trust α=.90, M=4.2, SD=1.1, range: 1.8–6.2 10 items, e.g., The system is reliable.

not. It is interesting that for the participants whose understanding was analyzed, there was no significant correlation between how well participants thought they understood the system and their actual understanding (Spearman's rho=.005, p(1-tailed)=.491, N=21). This illustrates the necessity of measuring participants' actual understanding as well as participants' perceived understanding; not all participants appear capable of assessing their own level of understanding of a system.

Participants in the confidence rating ("sure") condition did not think they understood the system better than those in the non-transparent condition. The confidence rating ("sure") condition was included to gage the effect of different types of information offered about a system's workings. The "sure" feature's numerical rating, however, did not offer participants insight into the criteria the system uses to choose recommendations. The "sure" feature did not also increase participants' trust, or acceptance of the system and its recommendations. Offering type of additional information about a system's decision process does not automatically increase acceptance of a system.

The final selection of artworks for participants in the transparent ("why") condition, however, contained significantly more recommendations than in the non-transparent condition (Mann–Whitney U=125.500, p(1-tailed)=.013, N"non"=22, N"why"=19, Mdn"non"=1.0, Mdn"why"=2.0). Acceptance of the recommendations was indeed higher in the transparent condition. The data of this study indicate that transparency does indeed influence acceptance of the system's recommendations, but that the transparency feature offered here does not influence acceptance or adoption of the system itself. No significant differences were found either on acceptance of the system or its recommendations between the non-transparent and "sure" condition. Contrary to our hypothesis, participants in the transparent condition did not trust the system more. No significant difference was found on the scores on trust in the system between the conditions. Trust is correlated with intent to use (rho=.805, p(1-tailed)=.000, N=60), the decision whether to use the system or catalogue (rho=453, p(1-tailed)=000, N=60), and to the acceptance of recommendations (rho=.363, p(1-tailed)=.002, N=60). We conclude that trust is related to the decision whether to use the system.

To conclude this experiment, transparency increased the acceptance of the system's recommendations. Even though users' acceptance of and trust in the system itself were not affected by transparency, transparency can be considered as an important aspect of interaction with user-adaptive systems. Findings show that the transparent version was perceived as more understandable, and perceived understanding, correlated with perceived competence, trust, and acceptance of the system. Future research is necessary to evaluate the effects of transparency on trust in and acceptance of user-adaptive systems. A distinction has to be made in future research between acceptance of the system in users' context and acceptance of its results.

3.3 Effects of Autonomy, Traffic Conditions, and Driver Personality Traits on Attitudes and Trust toward In-Vehicle Agents

The study, previously reported in [13], investigates the effects that level of system autonomy and criticalness of the situation influences user's trust in and acceptance of an in-vehicle intelligent agent that uses voice interaction.[4]

The study is an online survey addressing participants' attitudes toward an in-vehicle agent shown in a short movie clip. Our 2x2 survey-based experiment varied agent autonomy (agent providing information on traffic in the merging lane vs. agent instructing the driver the right moment to merge) and complexity of the situation (heavy traffic or light traffic), resulting in four between-subject conditions:

- Heavy traffic, instructive system
- Heavy traffic, informational system
- Light traffic, instructive system
- Light traffic, informational system

[4] This section re-uses material from:
-Henriette Cramer, Vanessa Evers, Nicander Kemper, Bob Wielinga (2008) "Effects of Autonomy, Traffic Conditions and Driver Personality Traits on Attitudes and Trust towards In-Vehicle Agents", Workshop on Human Aspects in Ambient Intelligence (CHI'09), Proceedings of the International Conference on Web Intelligence and Intelligent Agent Technology (WI-IAT'08), Sydney, Australia, pp. 477-482, DOI 10.1109/WIIAT.2008.326.

Table 4 Final scales (Cronbach's alphas, means, and ranges) included in the analysis. Copyright ©2008 IEEE. Reprinted with permission from [13].

Personality traits: Locus of control α =.702 M=5.3, SD=.98, r: 1.0–7.0 3 items, e.g., I like jobs where I can make decisions and be responsible for my own work. Adapted from [18]
Personality traits: Driver characteristics Adapted from [46] **Thrill seeking** α =.61 M=3.1, SD=1.5, r: 1.0–6.5 2 items, e.g., I like to raise my adrenaline levels while driving. **Hazard monitoring** M=4.5, SD=1.5, r:1.0–7.0 I try very hard to look out for hazards even when it's not strictly necessary. **Aggression** M=5.0, SD=1.5, r:2.0–7.0 I really dislike other drivers who cause me problems. **Anxiety/dislike of driving** M=3.6, SD=1.5, r:1.0–7.0 I find myself worrying about my mistakes when driving.
Perceived competence car agent α =.72 M=3.9 SD=1.0, r:1.0–6.7 3 items, e.g., The car agent was very capable at performing its job.
Perceived usefulness car agent α=.77, M=3.7, SD=1.2, r: 1.0–6.3 3 items, e.g., I think that the car agent is useful. Adapted from [69]
Attitude towards agent alpha=.77, M=3.2, SD=1.1, r:1.0–5.7 3 items, e.g., Using the car agent is a good idea. Adapted from [69]
Trust α=.81, M=4.2, SD=.6, r:2.4–5.8 16 items: 2 scale items, e.g. I trust the car agent. 14 bipolar semantic differential pairs, e.g. careful – not careful, reliable – unreliable. Taken from [35]
Intent to use α=.89, M=2.6, SD=1.4, r:1.0–5.8 4 items, e.g., I would like to have the car agent. Adapted from [69]
Intent to follow-up on decision α=.80, M=3.6, SD=1.6, r:1.0–7.0 2 items, e.g., I would follow the car agent's directions

Participants were randomly assigned to one of the four conditions. They watched a 15-second movie in which a driver is assisted by an in-vehicle agent to merge with highway traffic and then answered survey items addressing their perceptions of and attitudes toward the agent. In the instructive condition the driver was informed about the distances between cars and instructed to merge when sufficient space to merge occurred. In the informative condition the driver was merely informed about the distances between cars and made his own decision to merge.

Table 4 lists the final 7-point Likert-type scales used in the analysis.

A result of 100 participants were included in the study. The majority were Dutch (70.7%), with a mean age of 33. Twenty-six participants were female. All participants had a driver's license (M=15 years). No differences were found between conditions on age, education level, locus of control, and driving behavior traits.

Results show interaction effects for perceived usefulness (F(1,95)=4.023, p=.048), attitude toward the agent (F(1,95)=6.520, p=.012), trust (F(1,85)=4.100, p=.046), and intent to use (RT F(1,95)=7.303, p=.008). No interaction effect was found for perceived agent competence. Simple effects analysis showed that in heavy-traffic conditions, scores for perceived usefulness, attitude, trust, and intent to use did not significantly differ between the instructive and informational agent. However, in light-traffic scores for the instructive agent on perceived usefulness, attitude, trust, and intent to use were all significantly higher (p<.05) than for the informational agent. Participants preferred the instructive agent, but this preference only holds in light traffic conditions. A significant main effect of type of agent was found for perceived competence. The instructive agent was perceived as more competent (M=4.0) than the informational agent (M=3.7). A main effect of traffic conditions was found for participants' intent to follow-up on the system's instructions. Participants were less willing to let the driver follow up on agent decisions in heavy traffic (M=3.3) than in light traffic (M=3.9), F(1,54)=1.693, p(1-tailed)<.001. Locus of control affected compliance. Participants who scored high on locus of control tended to report the driver should follow the agent's instructions more (Mdn=4.0) than those scoring low on locus of control (Mdn=3.0) (U=255.500, p(1-tailed)=0.048). No other effects of locus of control were found. High-anxiety participants thought the situation in the video was more time-critical (U=576.500, p(2-tailed) =.022, Mdn=6.0) than those participants scoring low in dislike of driving/anxiety (Mdn=5.0). Participants who scored high on aggression also thought the situation was more time-critical (U=475.000, p(2-tailed)=.004, Mdn(H)=5.0, Mdn(L)=4.0). Anxiety and aggressive driving tendencies affect the perceptions of traffic conditions.

This study shows that perceptions of, trust in, and attitudes toward in-vehicle agents are affected by both traffic context and autonomy of an agent in terms of the level of information or instructions offered. Only in the context of light traffic participants preferred the instructive agent, in the heavy traffic, attitudes were at a similar, slightly negative level, for both types of agents. An in-vehicle agent thus might not be accepted in all contexts. The study's results showed that personality traits can affect attitudes toward agents, with locus of control affecting the willingness to follow an agent's instructions, supporting the findings of Goren-Bar et al. [24]. We could not confirm that traits specific to driving affect attitudes toward in-vehicle agents. These traits did, however, affect perceptions of the interaction scenario.

4 Discussion and Conclusion

After careful examination of the literature in interaction with adaptive systems, we identified three key research questions. After carrying out the studies described in the preceding section we are able to start answering these questions.

Question 1 was "To what extent does trust in an adaptive system influence acceptance of the system?" The study discussed in section 3.1 indicates that for trainable spam filters, trust in the effectiveness of training was found to play an important role in the user's willingness to invest effort in a system that can be taught to improve over time. While the reported level of trust in the spam filter was not always related to the choice for delegation, for content-based recommenders, we found that trust was related to the decision to use a system. People with a more positive assessment of the systems' ease of use and usefulness tend to delegate higher risk tasks to the system. Please note that we distinguished between system competence (e.g., how many mistakes the system makes or how fast it is) and trustworthiness of the system which can be understood as the "intentions" of the system. In future research it will be important to compare different error rates for autonomous and adaptive systems and the effects of this on trust and acceptance.

Question 2 stated: "What is the role of transparency in increasing the trust a user has in a system?" The second study shows that transparency does indeed influence acceptance of the system's recommendations, but that the transparency feature offered (why a recommendation was made) does not influence acceptance or adoption of the system itself. The transparent version of a recommender system was perceived as more understandable, and perceived understanding correlated with perceived competence, trust, and acceptance of the system. Explaining why a certain recommendation was made increased the user's acceptance of the recommendations.

Question 3 was "Does the context of the task (critical or less critical) and environment (high versus low risk) influence the users' opinion of the appropriate level of system autonomy and user control?". In the third study, participants reported to prefer higher autonomy systems (e.g., instructive rather than informative agent) in low-risk settings. Participants seemed averse to autonomy in higher risk settings.

Specific guidelines for developers of adaptive and autonomous systems that will aid the design of such technology are the following:

- It is important to make the users aware of the fact that the system adapts itself and what it adapts to. If a system makes decisions or takes over tasks autonomously, the user should be offered a clear and concise explanation of the reason why the decision was made.
- When the system operates in a high-risk setting, people are less inclined to delegate tasks to an intelligent system.
- When the system operates in a low-risk setting, people are more inclined to delegate tasks to an intelligent system.

This chapter offered a review of relevant literature concerning interaction with adaptive systems. It discussed four studies that were carried out to gain insight into the effects of transparency, trust, and social interaction types on user acceptance of and attitudes toward intelligent systems. The conclusions of this chapter indicate that it is important to make the system's decision-making process transparent to the user. Also, acceptance of adaptive systems can be optimized by carefully considering the level of autonomy it should have and the modality of interaction with the user.

References

1. Alpert, S., Karat, J., Karat, C., Brodie, C., Vergo, J.: User attitudes regarding a user-adaptive e-commerce web site. User Modeling and User-Adapted Interaction 13(4), 373–396 (2003)
2. Avesani, P., Massa, P., Tiella, R.: A trust-enhanced recommender system application: Moleskiing. In: ACM symposium on Applied computing, pp. 1589–1593 (2005)
3. Benyon, D.: Adaptive systems: A solution to usability problems. User Modeling and User-Adapted Interaction 3(1), 65–87 (1993)
4. Bilgic, M., Mooney, R.: Explaining recommendations: Satisfaction vs. promotion. In: Beyond Personalization Workshop, The International Conference on Intelligent User Interfaces, San Diego, California, USA (2005)
5. Billsus, D., Hilbert, D., Maynes-Aminzade, D.: Improving proactive information systems. In: IUI 2005, pp. 159–166. ACM Press, New York (2005)
6. Briggs, P., Simpson, B., De Angeli, A.: Trust and Personalisation: A Reciprocal Relationship? In: Designing Personalised User Experiences for eCommerce, pp. 39–55. Kluwer, Dordrecht (2004)
7. Brusilovsky, P.: Methods and techniques of adaptive hypermedia. User Modeling and User-Adapted Interaction 6(2-3), 87–129 (1996)
8. Carmichael, D., Kay, J., Kummerfeld, B.: W., N.: Why did you show/tell/hide that?: The need for scrutability in ubiquitous personalisation. In: ECHISE Workshop Exploiting Context Histories in Smart Environments at UbiComp, Irvine, CA, USA (2006)
9. Castelfranchi, C., Falcone, R.: Trust and control: a dialectic link. Applied Artificial Intelligence 14(8), 799–823 (2000)
10. Cheverst, K., Byun, H., Fitton, D., Sas, C., Kray, C., Villar, N.: Exploring issues of user model transparency and proactive behaviour in an office environment control system. User Modeling and User-Adapted Interaction 15(3-4), 235–273 (2005)
11. Corritore, C., Kracher, B., Wiedenbeck, S.: On-line trust: concepts, evolving themes, a model. International Journal of Human-Computer Studies 58(6), 737–758 (2003)
12. Cortellessa, G., Giuliani, M., Scopelliti, M., Cesta, A.: Issues in interactive problem solving: An empirical investigation on users attitude. In: Interact, Rome, Italy, pp. 657–670 (2005)
13. Cramer, H., Evers, V., Kemper, N., Wielinga, B.: Effects of autonomy, traffic conditions and driver personality traits on attitudes and trust towards in-vehicle agents. In: Workshop on Human Aspects in Ambient Intelligence at WI-IAT 2008, Sydney, Australia, pp. 477–482 (2008)
14. Cramer, H., Evers, V., Van Someren, M., Ramlal, S., Rutledge, L., Stash, N., Aroyo, L., Wielinga, B.: The effects of transparency on trust and acceptance in interaction with a content-based art recommender. User Modeling and User-Adapted Interaction 18(5), 455–496 (2008)
15. Cramer, H., Evers, V., Van Someren, M., Wielinga, B.: Awareness, training and trust in interaction with adaptive spam filters. In: CHI, pp. 909–912 (2009)
16. Davis, F.: Perceived usefulness, perceived ease of use and user acceptance of information technology. MIS Quarterly 13(2), 318–340 (1989)
17. Desmond, P., Hancock, P., Monette, J.: Fatigue and automation-induced impairments in simulated driving performance. Transportation Research Record 1628, 8–14 (1998)
18. Duttweiler, P.: The internal control index: a newly developed measure of locus of control. Educational and Psychological Measurement, 61–74 (1984)

19. Dzindolet, M., Peterson, S., Pomranky, R., Pierce, L., Beck, H.: The role of trust in automation reliance. International Journal of Human-Computer Studies 58(6), 697–718 (2003)
20. Erickson, T.: Designing Agents as if People Mattered. In: Software Agents. MIT Press, Cambridge (1997)
21. Fogg, B.: Prominence-interpretation theory: Explaining how people assess credibility online. In: Extended Abstracts CHI 2003, pp. 722–723 (2003)
22. Fogg, B., Tseng, H.: The elements of computer credibility. In: CHI 1999, pp. 80–87. ACM Press, New York (1999)
23. Gefen, D., Karahanna, E., Straub, D.: Trust and tam in online shopping: An integrated model. MIS Quarterly 27(1), 51–90 (2003)
24. Goren-Bar, D., Graziola, I., Pianesi, F., Zancanaro, M.: The influence of personality factors on visitor attitudes towards adaptivity dimensions for mobile museum guides. User Modeling and User-Adapted Interaction 16(1), 31–62 (2006)
25. Gregor, S., Benbasat, I.: Explanations from intelligent systems: Theoretical foundations and implications for practice. MIS Quarterly 23(4), 497–530 (1999)
26. van der Heijden, H.: Acceptance of hedonic information systems. MIS Quarterly 28, 695–704 (2004)
27. Herlocker, J., Konstan, J., Riedl, J.: Explaining collaborative filtering recommendations. In: CSCW 2000, Philadelphia, Pennsylvania, USA, pp. 241–250 (2000)
28. Höök, K.: Evaluating the utility and usability of an adaptive hypermedia system. In: IUI 1997, pp. 179–186 (1997)
29. Höök, K.: Steps to take before intelligent interfaces become real. Interacting with Computers 12(4), 409–426 (2000)
30. Höök, K., Karlgren, J., Waern, A., Dahlbck, N., Jansson, C., Karlgren, K., Lemaire, B.: A glass box approach to adaptive hypermedia. User Modeling and User-Adapted Interaction 6(2-3), 157–184 (1996)
31. Inzana, C., Driskell, J., Salas, E., Johnston, J.: Effects of preparatory information on enhancing performance under stress. Applied Psychology 81(4), 429–435 (1996)
32. Jameson, A.: Adaptive Interfaces and Agents. In: The human-computer interaction handbook: Fundamentals, evolving technologies and emerging applications, pp. 305–330. Erlbaum, Mahwah (2003)
33. Jameson, A., Schwarzkopf, E.: Pros and cons of controllability: An empirical study. In: De Bra, P., Brusilovsky, P., Conejo, R. (eds.) AH 2002. LNCS, vol. 2347, pp. 193–202. Springer, Heidelberg (2002)
34. Jian, J., Bisantz, A., Drury, C.: Foundations for an empirically determined scale of trust in automated systems. International Journal Cognitive Ergonomics 4(1), 53–71 (2000)
35. Jonnson, I.-M., Harris, H., Nass, C.: How accurate must in-car information systems be? Consequences of accurate and inaccurate information in cars. In: CHI 2008, Florence, Italy, pp. 1665–1674 (2008)
36. Jøsang, A., Lo Presti, S.: Analysing the relationship between risk and trust. In: Trust Management (2004)
37. Kay, J.: Scrutable adaptation: Because we can and must. In: Adaptive Hypermedia and Adaptive Web-Based Systems, Dublin, Ireland, pp. 11–19 (2006)
38. Keeble, R., Macredie, R.: Assistant agents for the world wide web intelligent interface design challenges. Interacting with Computers 12(4), 357–381 (2000)
39. Klopping, I.M., McKinney, E.: Extending the technology acceptance model and the task-technology fit model to consumer e-commerce. Information Technology, Learning, and Performance Journal 22(1), 35–48 (2004)

40. Langheinrich, M.: Privacy by design - principles for privacy-aware ubiquitous systems. In: International Symposium on Ubiquitous Computing, Atlanta, GA, USA (2001)
41. Lee, J., See, K.: Trust in automation: designing for appropriate reliance. Human Factors 42(1), 50–80 (2004)
42. Lee, J.D., Moray, N.: Trust, self-confidence, and operator's adaptation to automation. International Journal of Human-Computer Studies 40, 153–184 (1994)
43. Lu, L., Wu, H., Cooper, C.: Perceived work stress and locus of control: a combined quantitative and qualitative approach. Research & Practice in Human Resources Management 6, 51–64 (1999)
44. Ma, Q., Liu, L.: The technology acceptance model: A meta-analysis of empirical findings. Journal of Organizational and End User Computing 16(1), 59–72 (2004)
45. Maes, P.: Agents that reduce work and information overload. Communications of the ACM 37(7), 30–40 (1994)
46. Matthews, G.: Desmond, P.A., Joyner, L., Carcary, B., Gilliland, K.: Validation of the driver stress inventory and driver coping questionnaire. In: International Conference on Traffic and Transport Psychology, Valencia, Spain (1996)
47. McAllister, D.: Affect and cognition-based trust as foundations for interpersonal cooperation in organizations. Academy of Management Journal 38(1), 24–59 (1995)
48. McGuinness, D., Pinheiro da Silva, P.: Explaining answers from the semantic web: The inference web approach. Web Semantics: Science, Services and Agents on the World Wide Web 1(4), 397–413 (2004)
49. McNee, S., Lam, S., Guetzlaff, C., Konstan, J., Riedl, J.: Confidence metrics and displays in recommender systems. In: Interact, Zurich, Switzerland, pp. 176–183 (2003)
50. McSherry, D.: Explanation in recommender systems. Artificial Intelligence Review 24, 179–197 (2005)
51. Miller, C., Parasuraman, R.: Designing for flexible interaction between humans and automation: Delegation interfaces for supervisory control. Human Factors 49, 57–75 (2007)
52. Muir, B.: Trust in automation: Part i. theoretical issues in the study of trust and human intervention in automated systems. Ergonomics 37, 1905–1922 (1994)
53. Muir, B., Moray, N.: Trust in automation. part ii. Ergonomics 39, 429–460 (1996)
54. Ndubisi, N.O., Gupta, O., Ndubisi, G.: The moguls' model of computing: Integrating the moderating impact of users' persona into the technology acceptance model. Journal of Global Information Technology Management 8(1), 27–47 (2005)
55. Netten, C., Bruinsma, G., van Someren, M., de Hoog, R.: Task-adaptive information distribution for dynamic collaborative emergency response. International Journal of Intelligent Control Systems 11(4), 238–247 (2006)
56. O'Donovan, J., Smyth, B.: Trust in recommender systems. In: IUI, pp. 167–174. ACM Press, New York (2005)
57. Parasuraman, R., Miller, C.: Trust and etiquette in high-criticality automated systems. Communications of the ACM 47(4), 51–55 (2004)
58. Pavlou, P.: Consumer acceptance of electronic commerce: Integrating trust and risk with the technology acceptance model. International Journal of Electronic Commerce 7(3), 101–134 (2003)
59. Pu, P., Chen, L.: Trust-inspiring explanation interfaces for recommender systems. Knowledge-Based Systems Journal 20, 542–556 (2007)
60. Rajaonah, B., Anceaux, F., Vienne, F.: Study of driver trust during cooperation with adaptive cruise control. Le Travail Humaine 69, 99–127 (2006)

61. Rajaonah, B., Anceaux, F., Vienne, F.: Trust and the use of adaptive cruise control: a study of a cut-in situation. Cognition, Technology & Work 8(2), 146–155 (2006)
62. Rajaonah, B., Tricot, N., Anceaux, F., Millot, P.: Role of intervening variables in driver-acc cooperation. International Journal of Human Computer Studies 66(3), 185–197 (2008)
63. Reeves, B., Nass, C.: The media equation: how people treat computers, television, and new media like real people and places. Cambridge University Press, New York (1996)
64. Shimazu, H.: Expertclerk: A conversational case-based reasoning tool for developing salesclerk agents in e-commerce webshops. Artificial Intelligence Review 18, 223–244 (2002)
65. Shneiderman, B., Maes, P.: Direct manipulation vs. interface agents. Interactions 4(62), 42–61 (2007)
66. Simmel, G.: The Sociology of George Simmel. Free Press, New York (1964)
67. Sinha, R., Swearingen, K.: The role of transparency in recommender systems. In: Proceedings CHI 2002, pp. 830–831. ACM Press, New York (2002)
68. Tintarev, N., Masthoff, J.: Effective explanations of recommendations: User-centered design. In: ACM Conference on Recommender systems, pp. 153–156 (2007)
69. Venkatesh, V., Morris, M., Davis, G., Davis, F.: User acceptance of information technology. MIS Quarterly 27(3), 425–478 (2003)
70. Victor, P., Cornelis, C., De Cock, M., da Silva, P.: Gradual trust and distrust in recommender system. Fuzzy Sets and Systems 160(10), 1367–1382 (2009)
71. Waern, A.: User involvement in automatic filtering: an experimental study. User Modeling and User-Adapted Interaction 14(2-3), 201–237 (2004)
72. Wang, W., Benbasat, I.: Trust in and adoption of online recommendation agents. Journal of the Association for Information Systems 6(3), 72–101 (2005)
73. Wang, W., Benbasat, I.: Recommendation agents for electronic commerce: Effects of explanation facilities on trusting beliefs. Journal of Management Information Systems 23(4), 217–246 (2007)
74. Wu, I., Chen, J.: An extension of trust and tam model with tpb in the initial adoption of on-line tax: an empirical study. International Journal of Human-Computer Studies 62(6), 784–808 (2005)
75. Xiao, S., Benbasat, I.: The formation of trust and distrust in recommendation agents in repeated interactions: a process-tracing analysis. In: 5th international conference on Electronic commerce, pp. 287–293 (2003)

Example-Based Human Pose Recovery under Predicted Partial Occlusions

Ronald Poppe

Abstract. For human pose recovery, the presence of occlusions due to objects or other persons in the scene remains a difficult problem to cope with. However, recent advances in the area of human detection allow for simultaneous segmentation of humans and the prediction of occluded regions. In this chapter, we present an example-based pose recovery approach where this information is used. We effectively used the grid-based nature of histograms of oriented gradients descriptors to ignore part of the image observation space. This allowed us to recover poses directly, even in the presence of significant occlusions. We evaluated our approach on the HumanEva-I dataset, where we simulated different occlusion conditions. Without occlusion, we obtained relative 3D errors of approximately 69 mm. Our results showed approximately 10% increase in error when 20% of the observation is occluded. When 33% of the observation is occluded, the error is on average 15% higher compared to the observations without occlusions. These results showed that poses can be recovered from partially occluded observations, with a moderate increase in error. To the best of our knowledge, our approach is the first to investigate the effect of partial occlusions in a direct matching approach. Future work is aimed at combining our work with human detection.

1 Introduction

Human body pose recovery, or pose estimation, is the process of estimating the configuration of body parts from sensor input. When poses are estimated over time, the term human motion analysis is used. Traditionally, motion capture systems require that markers are attached to the body. These systems have some major drawbacks as they are obtrusive, expensive, and impractical in applications in which the observed humans are not necessarily cooperative. As such, many applications, especially in

Ronald Poppe
Human Media Interaction Group, University of Twente, Enschede, The Netherlands
e-mail: poppe@ewi.utwente.nl

R. Babuška & F.C.A. Groen (Eds.): Interactive Collaborative Information Systems, SCI 281, pp. 327–354.
springerlink.com

surveillance and human–computer interaction (HCI), would benefit from a solution that is markerless. Vision-based motion capture systems attempt to provide such a solution using cameras as sensors. In this chapter, we focus on applications that require real-time, robust pose estimates.

In general, there are two main approaches in vision-based human motion analysis: model-based (or generative) and model-free (or discriminative) [57]. The former approach uses a parameterized human body model which, for a given model instantiation, is matched to the visual observation. Guided by the projection error, the pose estimate is refined until a (local) optimum is found. This method is flexible as many parameters such as limb lengths, viewpoint, and appearance can be modeled. The drawbacks are the computational costly steps of model projection and projection-to-observation matching. As the pose space is generally large (20–60 dimensions), global optimization is not very practical. Therefore, local optimization is often used, which presents the risk of getting stuck at a local optimum. When multiple observations in time are available, this issue can be solved by taking a sampling-based approach (e.g., Condensation [30]), where many particles are used to estimate the cost surface. Due to the high dimensionality, a sufficiently large number of particles are needed to obtain a good estimate. Each particle brings an additional computational cost of model projection and projection-to-observation matching. In practice, model-based approach are therefore not suitable for real-time applications. Moreover, initialization remains a difficult problem with these methods.

Discriminative approaches do not use a human body model, but approximate the mapping from image to pose directly. They can be applied in real time and automatically provide initialization, but are less flexible in terms of encoding of parameters compared to generative approaches. Variations such as visual appearance and viewpoint should be encoded either explicitly in the parameter space or implicitly in the image representation. In this research, we take a discriminative approach where we explicitly encode viewpoint, as changes in viewpoint have a large impact on the image representation. We require our image representation to implicitly encode lighting variations and variations in body dimensions and clothing.

The focus of this chapter is on robust invariant image representations. We use a variant of histograms of oriented gradients (HOG, [11]), a holistic image representation. Discriminative pose recovery approaches are either example-based or regression-based. Regression-based approaches allow for faster evaluation but their precise implementation and parameter setting (number of experts, regression function, learning of the regression, and gating functions) influences the performance. This is undesirable, because it is less intuitive to attribute the performance to image representation or regression approach. Therefore, we use an example-based approach instead. The n visually most similar examples (nearest neighbors) are selected and the final pose is estimated to be the weighted interpolation of the n corresponding poses. We discuss regression-based and example-based approaches in Section 2.3.

In realistic scenarios, partial occlusion of the human figure in the image due to other persons or objects in the environment is common. We present an approach to handle these partial occlusions, if these can be predicted from a foreground segmentation process. We assume that the location and scale of a human figure can be detected from an image or video. When occlusions occur, we also require that these areas are labeled. Due to the importance of this preliminary step, we discuss the process of human detection in slightly more detail in Section 2.1.

The remainder of this chapter is structured as follows. Section 2 discusses literature in the field of human motion analysis, with an emphasis on discriminative work. In Section 3, we introduce our adaptation of HOG and the occlusion-sensitive example-based pose recovery approach. Results on the publicly available HumanEva benchmark dataset are presented in Section 4, and we conclude in Section 5.

2 Related Work on Human Motion Analysis

Human motion analysis from images or video comprises many topics. Here, we focus on human pose recovery in Section 2.2. We discuss discriminative approaches in Section 2.3 as these can be applied in real time. Since our approach assumes that a human subject is segmented in an image, we briefly discuss the state-of-the-art in human detection first in Section 2.1.

2.1 Human Detection

Human detection and pose recovery can be seen as complementary tasks. In the detection task, the aim is to generalize over different poses, whereas in the recovery task, one wants to discriminate between them. We advocate a separation of these tasks. This eliminates the need to perform human detection and pose recovery simultaneously, as this would require large amounts of pose-annotated training pairs, which are costly to obtain.

The detection of human subjects from images is an important first step in the analysis of human pose or action. Recently, there has been an increased interest in this topic (see [33] for an overview). In general, human detection methods are either holistic or part-based. A *holistic* approach considers the human body as a whole [11, 21]. In many cases, a retinoscopic representation is used, where the human is assumed to be centered within a defined region of interest (ROI). Human detection is performed by sliding the window over the image and performing binary classification at each location. Dalal and Triggs [11] train a support vector machine on positive and negative examples, encoded as HOGs. All training examples are retained in [21], where Chamfer matching between a large set of pedestrian examples is performed hierarchically. Dong et al. [12] explicitly take into account inter-human occlusion. They extract foreground blobs of a single person, or a group of people. An example-based approach, assuming that for each blob the corresponding number of persons with their exact locations are annotated, is used to segment each person individually. Earlier work by Elgammal and Davis [14] used known color distributions of each person to segment persons under occlusion. The advantage of holistic approaches is that they generate relatively few false positives since much information about the human body can be incorporated. The main drawback of a holistic approach is that occlusions cannot be dealt with without closer inspection of the scene.

In contrast, *part-based* approaches divide the human body into several parts, each of which can be modeled individually. Human detection is performed by looking at assemblies of these individual parts (e.g. [43, 44, 80]). Mohan et al. [44] find the head, legs, and the separate arms in a window by applying component-based classifiers. Then, an SVM over the individual part detectors is used to classify the entire window as human or non-human. The work of Mikolajczyk et al. [43] is similar in nature, but the focus is on the face and shoulders, which are encoded for frontal and side views separately. Felzenszwalb et al. [17] describe a person with a deformable part model, where each part is discriminatively learned from HOG descriptors. Niebles et al. [47] use a similar approach, but reduce the search space by applying a holistic detector first and relying on temporal continuity. Recent work by Wu and Nevatia [80] takes into account occlusions between persons in the scene. Such an approach is not only able to detect persons but can also determine which part of the observation is occluded. In recent work [81], they extend their approach to output pixel-level segmentations. Lin et al. [37] address the problem of finding suitable assemblies by introducing a tree of parts. Using re-evaluation, their method is also able to segment occluded persons from an image. They extend their approach in [36] to better handle variation in pose.

In general, part-based approaches generate many false positives for individual body segments. This can be explained since a body segment alone is often less discriminative compared to a full body. However, part-based approaches have a number of advantages over holistic methods. First, geometric constraints can be encoded efficiently. Second, by learning the detectors for parts individually, the combinatorial problem is effectively decomposed. Therefore, fewer training data of body-part templates is needed. Third, given a sensible assembly algorithm, humans can still be detected and segmented from the image even if parts are missing. This allows part-based methods to cope with partial occlusions from the environment or other persons.

Summarized, recent work has greatly advanced the quality of human detection. Especially the successful combination of detection and segmentation leaves us to believe that it is realistic to assume that a separation into foreground, background, and occluded area can be made. In the remainder of this chapter, we assume that such a segmentation is available.

2.2 Human Pose Recovery

The aim of human pose recovery is to find a numerical solution for a parameter estimation problem. The parameters that need to be recovered include not only those that describe the human body configuration but also viewpoint, body dimensions, body appearance, and clothing description can be part of the parameter space. Usually, we are only interested in the body configuration, and want to generalize over body dimensions, appearance, and clothing and environmental factors such as illumination. Still, these have an effect on the observation. Dealing with those variations is one of the challenges of human pose recovery. Usually, an image representation is used that is partly invariant to illumination and clothing, for example, by considering silhouettes. Another challenge is to deal with the high dimensionality of the pose (or rather parameter) space, especially when real-time performance in needed. A global search for the best parameter combination is computationally infeasible.

As mentioned before, human pose recovery approaches are either generative (model-based) or discriminative (model-free). We argue that the high computational cost of the projection-to-observation matching of generative approaches makes them unsuitable for real-time applications. In this chapter, we therefore take a discriminative approach. In the next section, we describe literature that reports on discriminative work.

2.3 Discriminative Approaches to Pose Recovery

If no explicit human body model is available, a direct relation between image observation and pose must be established. In practice, this means that the image representation must generalize over variations in body dimensions, appearance, and clothing. In general, a far-off view is assumed, where perspective effects are negligible. When multiple cameras are employed, calibration is assumed. Discriminative algorithms automatically perform (re)initialization and can be used to initialize model-based approaches (e.g. [25, 65, 68]). Several works first use a discriminative approach to find certain key poses in time, and use a model-based algorithm to recover the poses in the intermediary frames (e.g. [18, 61]).

The training data must account for those parameters that we wish to recover, usually the pose representation and the viewpoint. Not all kinematically possible poses are also likely, and the training data implicitly forms a manifold in pose space. Due to the high nonlinearity of this manifold, the pose space should be covered densely to obtain faithful mappings. Dimensionality reduction can be used in pose or image space to facilitate the learning of the mapping as in [13, 15].

Two main classes of pose estimation approach can be identified: example-based (Section 2.3.1) and learning-based (Section 2.3.2). Example-based approaches retain all image–pose training examples. For a given input image, a similarity search is performed, and candidate poses are interpolated to obtain the pose estimate. Learning-based approaches avoid having to store a large amount of examples and approximate the mapping from image to pose space functionally by training on image–pose pairs.

2.3.1 Example-Based

Example-based approaches use a database of examples that describe poses in both image space and pose space. While no mapping from image to pose space has to be learned, the drawback of example-based approaches is the large amount of space that is needed to store the database. Moreover, matching can be computationally costly, depending on the search scheme that is used.

In its simplest form, example-based approaches encode both the image part of the database and the observation into image representations and perform a linear search to obtain the closest matches, the nearest neighbors. The associated poses of these matches can be interpolated to allow for a more continuous range of pose estimates. Poppe [56] uses HOG representations, which encode edges while allowing for small variations in spatial arrangement. Results are presented from monocular and multi-view settings. In the multi-view case, the camera arrangement in training and test conditions is required to be the same. Silhouettes described using turning angle and Chamfer distance are considered by Howe [26]. In later work [27], optical flow information is used in addition. Fathi and Mori [16] only use motion information, which is invariant to illumination and texture.

When multiple synchronized cameras are available, a visual hull can be constructed. Van den Bergh et al. [3] approximate this hull using 3D haarlets, an extension to 3D of the haarlets proposed in [79]. They focus on pose recognition and learn a discriminative set of haarlets to maximize recognition performance.

Instead of using a direct distance measure, Sullivan and Carlsson [72] use deformation cost between examples and an input image. To improve the robustness of the point transferral, the spatial relationship of the body points and color information is exploited. Mori and Malik [45] employ shape contexts to encode edges. In an estimation step, the stored example are deformed to match the image observation. In this deformation, the location of the hand-labeled 2D locations of joints also changes. The most likely 2D joint estimate is found by enforcing 2D image distance consistency between body parts.

Temporal information can be used to overcome ambiguities from the image to some extent. Toyama and Blake [76] incorporate examples in a probabilistic temporal framework. By employing an HMM, they approximate a low-dimensional manifold by linear segments. Similar in concept is the work by Ong et al. [52], who cluster the examples and determine flow vectors for each cluster. A particle filter framework is used where the particles are guided by the flow vectors. Particle likelihoods are based on the matching distance to the closest example in the cluster.

The computational complexity of a naive nearest neighbor search is linear in the number of examples. For recovering more unconstrained movements or high number of DOF, the number of required examples grows substantially. Therefore, Shakhnarovich et al. [67] introduce parameter sensitive hashing (PSH) to rapidly estimate the pose given a new image. Because of the ambiguity in the use of silhouettes alone, they use edge direction histograms within a contour. An alternative approach to reduce the computational complexity of the matching is by storing the examples in a tree (e.g. [21]). Given an input image, a top-down matching procedure is used. Starting from the highest level node, a matching is performed for each of the child nodes. Only those subtrees that satisfy a certain criterium (e.g., threshold or best match) are further evaluated. This significantly reduces the computation time needed to select similar examples. This approach has also been taken by Yang and Lee [83], who construct the pose estimate as a linear combination of the selected examples from the bottom level of the tree. Rogez et al. [63] use a collection of trees. The nodes in each tree are trained to be discriminative and take into account a single dimension from a HOG representation. By using a collection of trees, many features can be used, and the resulting algorithm is more robust to noise.

2.3.2 Learning-Based

Learning-based approaches approximate the mapping from image to pose space functionally. The advantage of these regression methods is that inference can be performed efficiently and training data can be discarded after training. The drawback is learning the mapping. Especially for large amounts of training examples, computation requirements might be prohibitively large.

Xu and Hogg [82] present one of the earliest uses of regression in human pose recovery. A neural network is employed to map silhouette representations to pose representations. Agarwal and Triggs [2] use nonlinear relevance vector machine (RVM) regression over both linear and kernel bases to model the relation between histograms of shape contexts and 3D poses. Ambiguities are resolved using dynamics. Agarwal and Triggs [1] use direct regression to recover upper-body poses. Non-negative matrix factorization (NMF) on grid-based edge histograms is used to obtain a set of basis vectors that correspond to local features on the human body and ignore the presence of clutter. This enables them to recover poses without relying on background segmentation. The work by Onishi et al. [53] is similar in spirit, and extends the work of Poppe [56] with a noise-reduction step. Instead of applying NMF, they perform PCA in each block of cells in the grid to reduce the influence of backgrounds.

Instead of using a single view, information from multiple synchronized views can be combined into a voxel model. Ambiguities caused by using a single view are thus avoided. Also, a 3D voxel representation is independent of the camera setup. The approach is most suitable for controlled settings, where clean silhouettes can be obtained. Sun et al. [73] use an adaptations of the RVM to recover the pose from 3D shape context descriptors. When rotation normalization can be performed, such an approach can be used to learn view-independent regressors. Gond et al. [22] fit the voxel model in a 3D circular grid. This descriptor is rotation-normalized after recovering the orientation of the torso. The normalized feature representation is then used as an input for a sparse regressor. The approach has the advantage that significantly less training data is required, at the cost of an additional normalization step.

The space of common human poses is much smaller than the space of kinematically possible poses and these poses usually occupy a well-defined area in this high-dimensional space. This has led to the introduction of dimensionality reduction techniques. These techniques are also well suited for learning-based approaches as they can simplify the regression functions. For example, Grauman et al. [23] describe a distribution over both multi-view silhouettes and 3D joint locations with a mixture of probabilistic PCA. A pose estimate is obtained from the Bayesian reconstruction given the image representation. Similar in concept is the work of Bowden et al. [8], who fit a nonlinear point distribution model (PDM) to 2D position of head and hands, the 2D body contour, and the 3D pose representation. The feature space is projected on a lower dimensional space and allows for reconstruction of the pose given an input image. Ong and Gong [51] include views from multiple cameras in the PDM and recover a pose from multi-view images. Rogez et al. [62] use single view and learn separate models. Temporal and spatial constraints are further used to solve pose ambiguities. This concept is similar to Brand's [9], who models a manifold of pose and velocity configurations with an HMM. Temporal ambiguities are resolved by recovering poses over an entire sequence by applying the Viterbi algorithm.

Due to depth ambiguities in image space, the mapping from image to pose space is multi-valued and cannot be determined with a single regressor. Therefore, mixtures of regressors have been introduced. These divide the image space into clusters, where a regressor is learned for each cluster. Rosales and Sclaroff [64] cluster the 2D pose space and learn specialized functions for each cluster from image descriptors to pose space. A neural network is used as mapping function. In [66], the work is extended to allow input from multiple cameras. The pose is estimated for each camera individually and in a subsequent step, the hypotheses are combined into a set of self-consistent 3D pose hypotheses. Thayananthan et al. [74] use a mixture of regressors but validate the pose estimate for each by matching it against the input image to select the most likely pose. A similar approach is used by Sminchisescu et al. [70], who jointly learn mappings between image and pose space. The processes are guaranteed to converge to equilibrium. During inference, the results of the mapping from image to pose is validated using the mapping back.

Sminchisescu et al. [71] take a probabilistic approach and model the multi-valued nature of the mapping with Bayesian mixture of experts (BME). Each expert has an associated gating function, which gives the conditional probability that the regressor should be used given an input image. Guo and Qian [24] adapt the initialization using k-means and use stereo observations to reduce the multi-modality of the mapping. Ning et al. [48] initialize the experts on a partitioned subset of the image space. In the BME framework, experts and gating functions are learned simultaneously. This requires a double-loop optimization approach, which is computationally costly. Therefore, Bo et al. [6] train both models sequentially, which results in a decrease of both memory and computation requirements. Their algorithm thus can handle much larger numbers of examples.

Usually, not the whole image representation is useful for learning the regression. Redundancy and noise in the training data can thus affect the learning and performance of the regressors. This can be avoided by selecting only the relevant features. Additionally, this lowers the dimensionality of the image space and thus the complexity of the regressor. Ning et al. [49] jointly learn the BME regressors and the selection of visual words in a supervised manner. A similar approach is taken by Kanaujia et al. [31], who focus on hierarchical image representations and semi-supervised learning. Okada and Soatto [50] discriminatively select those orientations within HOG cells that are meaningful for predefined class of poses. An input image is first classified to a pose class, before recovering the pose. Bissacco et al. [4] use boosting to select a limited set of discriminative binary edge features and to learn the mapping directly.

Instead of learning the regression function offline, Urtasun and Darrell [78] learn it online, given an input image. With an example-based approach, the closest examples are selected. A local regression is then learned from these matches. Their approach can handle large numbers of examples, but is computationally more costly due to the selection of the nearest neighbors.

Learning these mappings depends largely on the availability of pose–observation pairs, which are difficult to obtain, especially in more unconstrained scenarios. Several authors have used synthetic observations generated by character modeling software (e.g. [2, 70]). Instead, Navaratnam et al. [46] use unlabeled examples to improve the regression functions. These examples can be easily obtained by using images of humans and by considering motion capture data.

3 Pose Recovery Using HOG

To describe an image, we can either use a holistic descriptor or a local (or patch-based) descriptor. The former encodes the image observation as a whole. Local deformations in the image will affect the entire descriptor. In contrast, local descriptors describe the image observation as a collection of local regions. Usually, these regions are extracted at interest points (local features), which are expected to be invariant to changes in viewpoint and illumination [41, 77]. Currently, a popular local descriptor is the scale invariant feature transform (SIFT [39]) and extensions (SIFT-PCA [32] and GLOH [42]). Local descriptors have the advantage that they can cope with variations in illumination, pose, and viewpoint to some extent. However, they strongly rely on robust extraction of interest points, which might be difficult due to differences in subject and background appearance. Moreover, extraction of local descriptors is more time-consuming due to the localization of interest points and the calculation of the local descriptor. Also, matching of bags of local descriptors is less straightforward. Therefore, we use a holistic descriptor in our work.

In this section, we present an example-based approach to human pose recovery. We use HOG as image representation. This holistic representation has been introduced by Dalal and Triggs [11]. Their HOG descriptor is inspired by work on orientation histograms [19], but uses dense sampling instead. The key idea is to calculate HOG (edges) within each cell of a regular spatial grid. This grid has a fixed number of cells which cover an area that is determined by a rectangular ROI. The HOG descriptor is a concatenation of all cell histograms. Several alternatives to HOGs have been proposed in literature. Levi and Weiss [35] use edge orientation histograms that contain ratios between orientation responses, dominant orientation, and symmetry features, calculated exhaustively over all rectangular subwindows of an image. Adaboost is used to select the relevant features. In contrast, HOGs use a fixed spatial structure, which allows for direct matching. The pyramid of histograms of oriented gradients (PHOG) proposed by Bosch et al. [7] is a generalization of the HOG where the notion of a block of cells is extended to multiple levels. At the lowest level, the ROI is described as a single edge orientation histogram. For each higher pyramid level, a division into 2×2 cells is made. The PHOG approach is suitable when there is variation in the localization of the ROI but restricts the number of rows and columns in the grid to be equal and to be powers of 2. As the height of a human figure in the image is larger than its width, we use the original HOG concept.

HOGs have been used for several human motion analysis tasks. Dalal and Triggs initially used HOGs for pedestrian detection, a binary classification task. Variations in clothing, lighting, and body dimensions, but also viewpoint and pose, were implicitly encoded. Such an approach is reasonable since there are clear cues such as head and shoulder lines, which remain present also when seen from different viewpoints. Gandhi and Trivedi [20] use HOG descriptors to classify the orientation of pedestrians, thus explicitly encoding the (relative) viewpoint. Thurau [75] used HOGs to model human shape for human action recognition. Both Liu et al. [38] and Chakraborty et al. [10] use body part classifiers based on HOG descriptors. Such an approach is suitable for 2D location of limbs. However, without strong pose priors, lifting these to 3D will lead to ambiguities as there is no verification step where the observation is used in a holistic manner.

While HOGs have been shown to be robust descriptors for the aforementioned tasks, we believe that HOG descriptors are even sufficiently rich for recovery of human poses, including the viewpoint. This task is, however, more demanding as we do not have to distinguish between a small number of classes, but instead aim at *regression* of 60-dimensional poses. Moreover, the HOG descriptors still have to be invariant to lighting, clothing, and body dimensions.

Since we need to recover more information from the HOG descriptors, we also require more precise HOG extraction. The above-mentioned works extract the HOG descriptors directly from the image, which has two drawbacks. First, the ROI needs to be determined, which is computationally expensive. The ROI can vary in position and scale (we do not regard rotation, upright recordings are assumed), and many possible ROI candidates within an image have to be validated. Zhu et al. [84] introduce an efficient approach based on the integral image [60], but real-time performance still cannot achieved. Moreover, there will be false positives in the neighborhood of the actual ROI, which makes determination of the actual location and scale difficult.

Second, there is the problem of background clutter. Edges within the ROI that do not belong to the person, but to the background, will affect the HOG descriptor. Therefore, a number of works have explored ways to focus on edges that belong to the foreground. Poppe [56] uses background subtraction and only uses those edge responses that fall within the foreground region. Agarwal and Triggs [1] use NMF to suppress background edges. They demonstrate their work on recovery of frontal poses. Both Sminchisescu et al. [70] and Okada and Soatto [50] implicitly determine a set of discriminative features by learning regression functions from the HOG space to the pose space. Due to the use of multiple regressors, this selection is pose-dependent. Rogez et al. [63] use randomized trees, each of which is trained discriminatively. However, none of these works has explicitly addressed partial occlusions.

Given the common presence of partial occlusions due to other persons or the environment, there has been surprisingly little work that explicitly addresses this issue. Poppe and Poel [59] detect humans and recover their poses in single images. They use body-part templates and, for a match, vote over all joints in the human body. Such an approach can deal with severe occlusions, but is restricted to a limited class of motions (e.g., walking). Peursum et al. [55] use factored-state hierarchical HMM to model the motion of one given action. Occlusion of the feet can be detected using [54], and the likelihood function is adapted by ignoring the occluded area. The learned dynamical model will ensure stable tracking but also poses restrictions on the movement that can be recovered.

For example-based approaches, occlusions are variations that are not explicitly modeled in the example set. To be able to handle these variations, there must be a way to take into account the missing (or ambiguous) information in the matching. To the best of our knowledge, only one paper takes into account occlusion in an example-based approach. Howe [28] uses boundary fragment matching to match partial shapes. Boundary fragments are small parameterized outlines of an extracted silhouette. Background and occlusion areas need to be labeled, so the matching algorithm knows which boundary fragments belong to the actual shape.

In many cases, silhouettes can be obtained relatively reliably using background substraction. Similar to [56], we assume that such a segmentation into foreground and background can be made. Employing this segmentation has two main advantages. First, determination of the ROI is straightforward. This relieves the burden on the detection task, as only a single detection window has to be processed. This will significantly aid in achieving real-time performance. Second, by considering only foreground edges, we effectively ignore background clutter. The resulting HOG descriptor is therefore not dependent on the background, which increases generalization. In addition, we assume that it is known which parts of the human subject in the image are occluded, similar to [28]. We effectively use the grid-based nature of the HOG descriptor to use only those dimensions in the matching procedure that correspond to non-occluded cells. Our work is an adaptation of [56] where descriptor normalization and matching procedure are adapted. In this section, we explain the steps of our approach and show that poses can be recovered accurately, even when foreground segmentation is noisy. We further demonstrate that our approach can recover poses under partial occlusion. To the best of our knowledge, our approach is the first to address partial occlusions in a direct matching approach. We believe that this is a key characteristic of any human pose recovery approach that is to perform in real time in more realistic environments.

In our contribution, we neither regard the temporal aspect nor do we apply any measures to reduce the computational complexity. This allows us to focus on the performance of the HOG descriptors. In Section 3.1, we discuss our HOG variant, and how we obtain the descriptor from an image. The nearest neighbor pose recovery approach is explained in Section 3.2. Our experiments on the HumanEva datasets are presented in Section 4.

3.1 *Histogram of Oriented Gradients*

Our descriptor differs from the HOG descriptor as described by Dalal and Triggs [11]. First, we only take into account the edges within a foreground mask. This requires background segmentation, but allows us to focus only on those edges that are meaningful. Second, we do not use the notion of (overlapping) blocks, which results in a significantly reduced descriptor size. Third, we do not apply color normalization, which further reduces the computational costs of calculating the descriptor. Fourth, we use a different grid size. In our experiments (Section 4), we divide the ROI into a grid with 6 rows and 5 columns. This is an arbitrary choice, but the height of each cell roughly corresponds with the height of the head in a standing position. Similarly, in a relaxed standing pose, the body covers approximately 3 columns horizontally.

Within each cell in the grid, we calculate the orientation and magnitude of each pixel that appears in the foreground mask. We apply a $[-1\ 0\ 1]$ gradient filter to each pixel in horizontal and vertical direction independently. The orientation of the edge is given by the angle between these two derivatives. The magnitude is given by the square root of the sum of the two squared derivatives. We divide the absolute orientations over 9 equally sized bins in the $0°$–$180°$ range. Each pixel contributes the magnitude of its orientation to the according histogram bin, which results in a 9-bin histogram per cell. Note that this binning is slightly different than proposed in [11], where votes are interpolated bi-linearly between the neighboring cells and orientation bins. The total length of the descriptor is 270. Poppe [56] normalized the entire descriptor to unit length to overcome differences in scale. To further reduce the size of the HOG descriptor and suppress background noise, PCA is applied by Lu and Little [40] and Onishi et al. [53] in the context of human motion analysis. However, both a global normalization and the use of PCA make the descriptor holistic, as local variations affect the entire descriptor. However, partial occlusions cause some of the observation to be uninformative, or even misleading. Normalization of the entire descriptor depends on all individual cells. In case of occlusion, some of the cells will have different edge responses. By normalizing descriptor d to unit length, these cells will affect all others. Therefore, we normalize each cell $h_{i,j}$ (i and j are a row and column index, respectively) individually to be of unit length. This has the advantage that we can still deal with variations in scale, as each cell is approximately equal in size. On the other hand, we discard the global character, and each cell contributes equally to the final descriptor. Specifically, individual cells that have a relatively low edge response have a similar summed weight as cells that have a high response. Alternatively, we could have normalized each cell by its area size. This would make the descriptor invariant to scale, but would not take into account variations due to different lighting settings and clothing.

3.1.1 Determination of ROI and Foreground Mask

We calculate HOGs within an image's ROI, in our case, the bounding box around the subject. While HOGs can be used to determine this region, as in [11, 75, 84], we

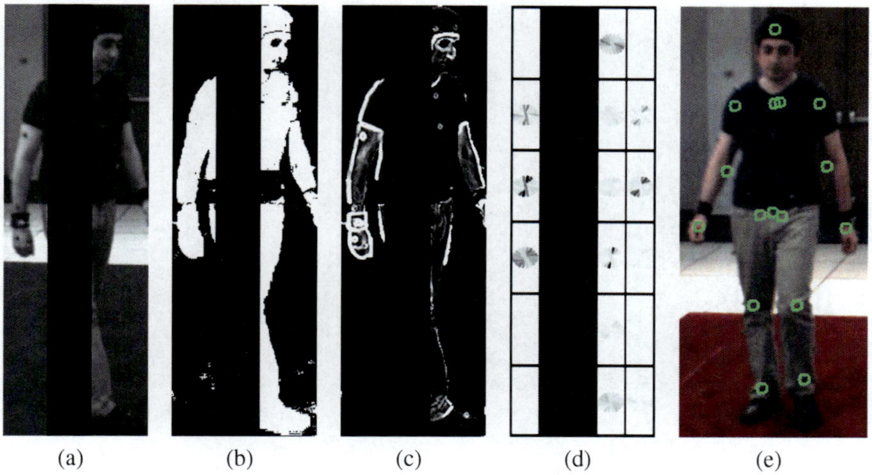

Fig. 1 Different steps in the calculation of a HOG descriptor. (a) the input image, with simulated occlusion, (b) the foreground mask. For clarity, the occlusion is not shown. (c) the edge magnitudes and (d) the HOG descriptor. The cells that are (partly) occluded are shown black. (e) Locations of the 20 joints.

rely on background subtraction. As discussed previously, this significantly speeds up the process, and we suppress background edges at the same time. We describe the process here in detail to allow for replication. First, we apply the background subtraction with the suggested risk values, as included in the HumanEva source code [69]. The background is modeled as a mixture model with 3 Gaussians per color channel. The minimum enclosing box of all foreground areas larger than 600 pixels is obtained. After conversion to HSV color space, we apply shadow removal in the lower 20% of the ROI. Pixels that have a saturation that is between 0 and 25 higher than the saturation of any of the means in the background mixture model are removed from the foreground mask. We again obtain the minimum enclosing box, which is our final ROI. Figure 1 shows an example of background subtraction and displays the HOG descriptor.

It may seem that our approach is highly sensitive to good background subtraction, but the shadow removal is only needed to ensure that the ROI fits the subject reasonably. For certain cases, we slightly adjusted the parameters. For camera 1 in HumanEva-I, only for subject 2, we multiplied the risk with factor 10^{12} to remove artifacts from the foreground. For cameras 2 and 3 in HumanEva-I, we lowered the shadow threshold to 10. We did not use the additional four grayscale cameras in HumanEva-I. For HumanEva-II, we reduced the background risk with factor 10^{50}, only for camera 3. Still, errors in the background segmentation frequently result in incorrect determination of the ROI and inaccurate foreground masks (see also Figure 2).

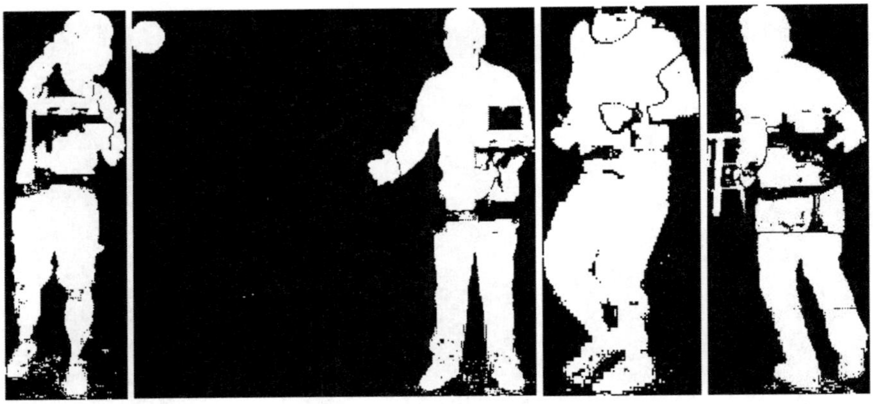

(a) (b) (c) (d)

Fig. 2 Background subtraction errors that result in inaccurate foreground masks and incorrect placement of the ROI

3.2 Pose Recovery Using Nearest Neighbor Interpolation

In an example-based approach, each image observation is encoded and matched against an example set of encoded observations. We use the previously described HOGs as encodings. To match a HOG with those in the example set, we need to define a distance measure between the two descriptors. To deal with partial occlusions, we match descriptors at the cell level. We introduce weights $\phi_{i,j}$ for each cell. These weights scale the feature space and therefore affect the distance functions that we define. While these weights can take any value (e.g., in the $[0,1]$ range), we use here binary values. That is, $\phi_{i,j} = 0$ in the case of occlusion within the cell, and $\phi_{i,j} = 1$, otherwise. This weighting effectively determines a lower-dimensional subspace, in which we can perform matching. The distance between descriptor d with cell-normalized histograms $\hat{h}_{i,j}$ and descriptor d' with cell-normalized histograms $\hat{h'}_{i,j}$ is calculated as

$$D(d,d') = \frac{\sum_{i=1}^{n_r} \sum_{j=1}^{n_c} \phi_{i,j} \delta(\hat{h}_{i,j}, \hat{h}'_{i,j})}{\sum_{i=1}^{n_r} \sum_{j=1}^{n_c} \phi_{i,j}} \tag{1}$$

Here, n_r and n_c are the number of rows and columns, respectively. In our experiments, $n_r = 5$ and $n_c = 6$. δ is the distance measure between two histograms, for which we use Manhattan distance. Note that, since we predict which cells are partly occluded, we could ignore these cells and normalize the descriptor to unit length. This approach would maintain the advantages of a global normalization but has the drawback that normalization of all n examples in the database needs to be performed at run time. This will severely affect the computational performance.

Matching a HOG with the entire example set results in a distance value for each of the m examples. We could choose the example with the lowest distance, as this is the example that best matches the image observation. However, in practice, taking the n best matches (nearest neighbors) results in more accurate pose recovery. It should be noted that n is the only parameter in our approach. Of course, n will depend on the number of examples in the example set that are close to the presented frame. Here, we use $n = 25$, in accordance with [58]. To determine the final pose estimate, we use the poses that correspond to the n best examples. We determine the final pose estimate $\boldsymbol{p}$, the normalized weighted interpolation of these poses, as $\boldsymbol{p} = \sum_{i=1..n} w^i \boldsymbol{p^i} / (\sum_{j=1..n} w^j + \delta)$ and $w^i = \frac{1}{d^i+\delta}$ $(1 \leq i \leq n)$. Here, δ is a small number to avoid division by zero in the rare case which, retrieved examples, exactly that match. $\boldsymbol{p^i}$, d^i, and w^i correspond to the pose vector, HOG distance value, and weight of the ith best matching example, respectively. This implies that close HOG matches contribute more to the final estimate than HOG matches at a larger distance. One word of caution is in its place here. Since we interpolate poses, the final joint estimates are likely to lie closer to the mean distance for this joint, so closer to the body. This effect is especially visible for examples that have similar image observations but are distant in pose space.

Since we do not determine the correspondences between our localized subject in the image and the estimated pose, we are not able to estimate the global position of each joint. Instead, we report the distances of each joint relative to the pelvis (*torsoDistal*) joint, as suggested in [69] (see Figure 1(e)).

4 Experiment Results

In our experiments, we use the HumanEva-I dataset [69], which we describe in Section 4.1. We introduce the example set in Section 4.2. To the best of our knowledge, there is no dataset that contains both ground truth motion capture data and video data with occlusions. Therefore, we use the HumanEva dataset and simulate different types of occlusion. The test sets are presented in Section 4.3. Results on both non-occluded and occluded observations are presented in Section 4.4 and discussed in Section 4.5.

4.1 HumanEva Dataset

The HumanEva-I dataset [69] contains several sequences, divided into training, validation, and test sets.

The training and validation sequences in HumanEva-I contain synchronized video and motion capture (mocap) data. There are 4 subjects that perform 5 actions (Walking, Jog, Box, Gesture, and Throw/Catch). In addition, there is one sequence for each subject–action pair that contains only mocap data. In the test set of HumanEva-I, all these subject–action pairs also appear. Also, for each subject, there is an additional Combo sequence that contains walking, jogging movements, and some additional balancing movements that do not appear in any training or validation trial. Example frames are shown in Figure 3.

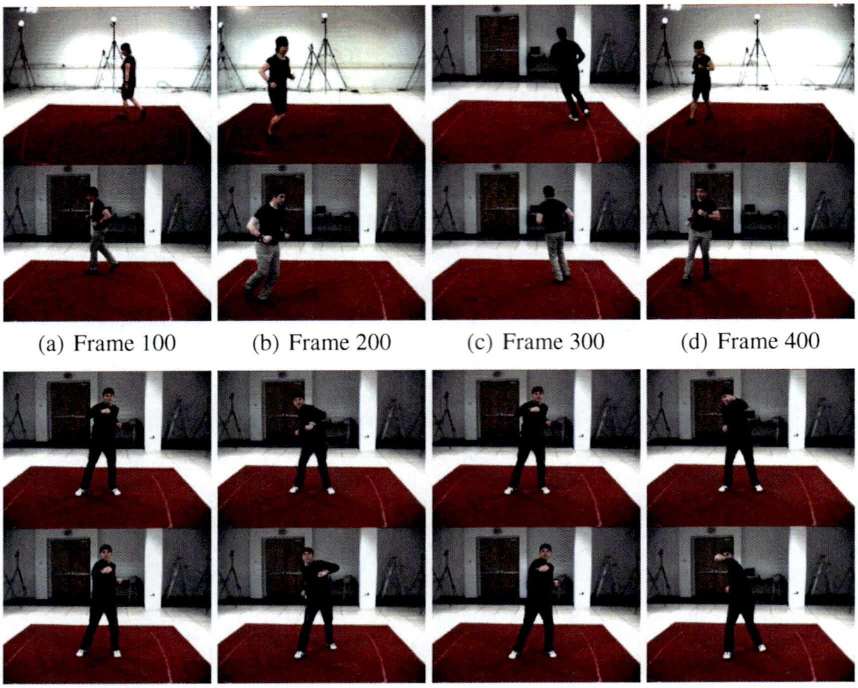

(a) Frame 100 (b) Frame 200 (c) Frame 300 (d) Frame 400

(e) Frame 100 (f) Frame 200 (g) Frame 300 (h) Frame 400

Fig. 3 Single best estimate (top row) and original frame (bottom row) for the HumanEva-I Jog sequence performed by subject 2 (a–d) and HumanEva-I Throw/Catch sequence performed by subject 3 (e–h), both evaluated using a single camera (C1)

The walking and jogging actions are performed by moving in a circle, counter-clockwise. Boxing, gesturing, and throwing and catching a ball are performed while facing camera C1. There is some variation in these sequences, though. Especially, the body orientation while catching the ball is heavily dependent on were the ball appears.

In HumanEva-I, each sequence has been recorded with 3 color cameras and 4 grayscale cameras. These have all been synchronized and calibrated. The frame rate of the video sequences is 60 frames per second. The dataset comes with source code to temporally align different video streams with the motion capture data. Also, background subtraction is included, which describes the background with a Gaussian mixture model. We have used these provided algorithms where possible. In our experiments, we use only the color cameras. For the monocular case, we only regard camera 1. Sequences of subject 4 were not evaluated due to difficulties with the background subtraction.

For the test sets, ground truth pose information is held back. An online validation system, as described in [69], is used to validate the pose recovery results. This system ensures that results of different parties can be compared, and frustrates

Table 1 Number of valid examples per action and subject in HumanEva-I training and validation sequences

Action	Subject 1	Subject 2	Subject 3	Total
Walking	1176	876	895	2947
Jog	439	795	831	2065
Throw/Catch	217	806	0	1023
Gestures	801	681	214	1696
Box	502	464	933	1899
Combo	0	0	0	0
Total	3135	3622	2873	9630

parameter tuning. Specifically, ground truth information in our case are the 3D positions of 20 key joints, relative to the pelvis (*torsoDistal*) joint (see Figure 1(e)). This is the full set of joints, and we report the root mean squared error (RMSE) in mm, averaged over all joints, as described in [69].

4.2 Example Sets

We describe our example set for monocular pose recovery. We associate the HOGs for an individual view with their corresponding poses. Only the examples that contain valid mocap data are included in the example set.

When given a new image observation, together with the knowledge from which camera the observation is obtained, we can estimate the relative pose. We observe that the elevation (rotation in vertical direction) and roll (rotation around line of sight) of all cameras are approximately the same. In other words, the orientation of all cameras is almost equal except for the orientation around a vertical axis. If we would rotate the subject in the scene around a vertical axis, we would theoretically be able to generate very similar observations for all cameras. In practice, view-specific parameters such as backgrounds and lighting conditions are likely to result in observations that are somewhat different. However, we want our approach to be robust against these image deformations and therefore we perform this rotation virtually. This has the additional advantage that the number of examples is effectively tripled, resulting in a total of 28,890. Table 1 summarizes the origins of the examples.

We transform the mocap data in such a way that we obtain the joint positions as if we were looking through another camera. With an observation from camera i, and the projection onto camera j, our pose vector $\boldsymbol{p_i} = (x_i, y_i, z_i, 1)^T$ is transformed into $\boldsymbol{p_j}$ as follows: $\mathbf{p_j} = M_j M_i^{-1} \boldsymbol{p_i}$, where M_i and M_j are the rotation matrices of cameras i and j, respectively.

4.3 Test Sets

We use the HumanEva-I test sequences of subjects 1–3 for testing. Evaluation of the Combo and Throw/Catch test sequences of subject 1 failed repeatedly.

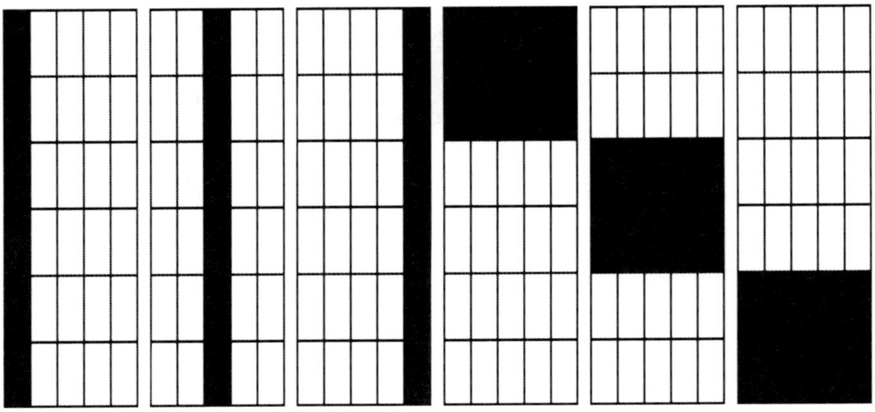

Fig. 4 Occlusion settings (left to right) *v_left*, *v_center*, *v_right*, *h_top*, *h_center*, and *h_bottom*

Consequently, we cannot report our results on these trials. Note that part of the Combo sequences contains movements that are not in the example set, which our algorithm cannot handle.

The original frames are used to evaluate the pose recovery accuracy without occlusion. In addition, we simulate different occlusion settings. We define six different occlusion settings. Each of these settings is a fixed combination of weights $\phi_{i,j}$. Effectively, we simulate occlusion on a fixed set of cells, not regarding the location of the subject in the image. As such, we can test the influence of occlusion on a large number of poses. The six settings that we define are *v_left*, *v_center*, *v_right*, *h_top*, *h_center*, and *h_bottom*, see also Figure 4. In the vertical and horizontal conditions, 20% and 33% of the image observation is occluded, respectively.

In addition to the fixed occlusion settings, we also created a *pole* sequence. This is the walking sequence of subject 1, but with a fixed occluded area in the image, not relative to the ROI. The area is a vertical pole with a width of 40 pixels. For comparison, the subject in the image is on average approximately 115 pixels in width. Due to the scale and pose, this can vary between 70 and 170 pixels. The *pole* sequence is a more realistic scenario and can show how the estimation error changes as the subject moves through an occlusion.

4.4 Results

Table 2 summarizes the average 3D errors over all joints, relative to the pelvis. The last column shows the results without occlusion. The average 3D relative error per joint is approximately 69 mm over all actions. For walking and jogging, the error is significantly lower at 45.05 mm and 45.98 mm, respectively. The larger number of available examples and lower variation in movement are the most important causes for the lower errors.

Table 2 Mean relative 3D error in *mm* per joint for HumanEva-I test sequences, evaluated with camera C1. For each fixed occlusion setting, the increase over the non-occluded observations and the amount of occlusion are given.

Subject	Action	Vertical			Horizontal			None
		Left	Center	Right	Top	Center	Bottom	
S1	Walking	39.80	43.28	42.10	47.52	43.83	45.83	39.03
S1	Jog	58.24	57.16	54.09	59.34	52.27	63.42	48.75
S1	Throw/Catch	N/A	N/A	N/A	N/A	N/A	N/A	N/A
S1	Gestures	30.72	28.93	30.75	30.58	30.44	31.79	30.11
S1	Box	84.40	93.13	84.80	97.81	90.30	79.41	81.38
S1	Combo	N/A	N/A	N/A	N/A	N/A	N/A	N/A
S2	Walking	37.60	40.18	39.22	42.23	39.53	41.24	37.27
S2	Jog	43.11	41.66	43.34	49.00	45.16	50.79	41.37
S2	Throw/Catch	73.53	72.13	71.34	76.81	74.06	76.33	72.18
S2	Gestures	95.30	95.65	86.14	84.90	85.25	93.16	82.81
S2	Box	109.17	120.03	107.43	107.14	112.11	115.45	105.08
S2	Combo	71.20	72.72	72.67	78.30	74.14	76.05	68.73
S3	Walking	58.38	64.73	64.23	61.31	61.22	65.79	58.86
S3	Jog	52.57	54.55	50.77	53.26	52.55	53.63	47.83
S3	Throw/Catch	120.06	112.22	94.38	110.81	103.15	122.22	101.40
S3	Gestures	80.09	76.85	97.63	72.49	91.20	114.00	82.63
S3	Box	105.64	110.48	98.69	110.15	105.65	106.99	98.44
S3	Combo	117.75	116.97	111.11	119.18	112.84	116.02	106.85
Average		73.60	75.04	71.79	74.43	73.36	78.26	68.92
Increase (%)		6.79	8.88	4.16	7.99	6.44	13.56	0.00
Occluded (%)		20.00	20.00	20.00	33.33	33.33	33.33	0.00

For each occlusion condition, the table shows the increase in error over all evaluated actions and subjects. For the vertical conditions, each of which occludes 20% of the image, the average error is approximately 7% higher. The horizontal settings result in a 9% increase, while occlusion covers one third of the observation. There are, however, large differences in performance for different trials. We will discuss these in the next section.

For the *pole* sequence, we obtained a 3D error of 43.54 mm averaged over all joints and relative to the pelvis joint. We also discuss these results in the next section.

4.5 Discussion

Several sequences show slightly lower errors when occlusion is added. This is most likely caused by bad foreground segmentation. In the occluded conditions, these distracting regions are ignored.

In general, we observe differences in accuracy between occlusion settings. The horizontal settings have a slightly higher error (9%) compared to the vertical occlusion settings (7%). This effect can be partly explained by the higher percentage

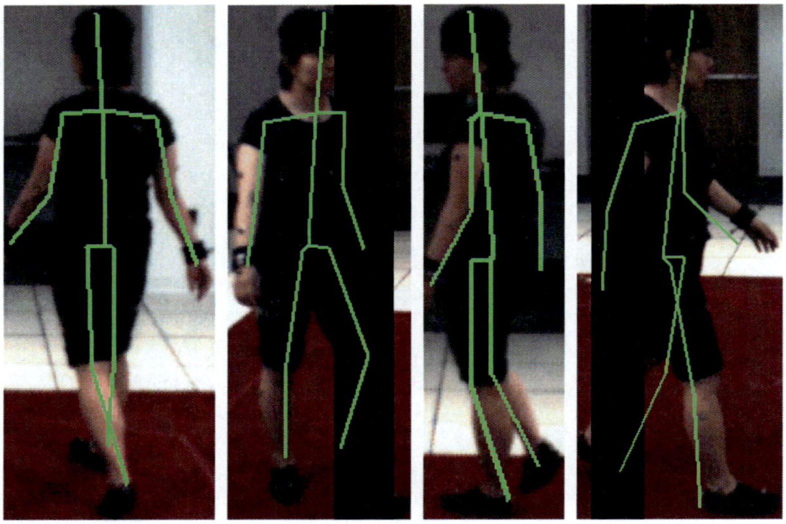

(a) (b) (c) (d)

Fig. 5 Recovered poses of frames 250, 500, 750, and 1000. In frames 500 and 1000, the pole is shown, but a larger region is not taken into account (60 and 40%, respectively). The 3D pelvis location was set manually, as we only estimate relative joint locations.

of occlusion (33%, compared to 20%). However, there are differences between settings. Most of these differences are caused by several individual trials. We will discuss these sequences that largely contribute to the increased error.

A significant part of the increase in recovery error is due to the boxing action. Especially for subject 3, almost all occlusion settings result in significantly higher errors. Analysis of these results reveal that many examples from the gesture action are selected. For the gesture action, a similar trend is visible, especially in the *h_bottom* condition. We expect that this is mainly due to the fact that subject 3 wears dark clothes, which results in relatively few edge responses between arms and body. Therefore, examples from beckoning gestures and box punches are selected interchangeably. The same effect is visible for subject 2, but to a much lesser extent.

For the jog action, recovery accuracy is significantly lower in the *h_bottom* occlusion setting. As the hands do not move a lot while jogging, the legs are most informative in this case as well.

The throwing and catching trials of subject 2 have a relatively low increase in error, whereas the *h_bottom* condition for subject 3 shows much higher errors. Similar to earlier observations, the relatively low number of edge responses for subject 3 is probably the reason that the missing edges for the feet result in decreased accuracy. Also, the *v_left* occlusion setting shows higher error values, which can be explained since the subjects both throws and catches the ball right to him. In the image, this corresponds to the left side of the observation.

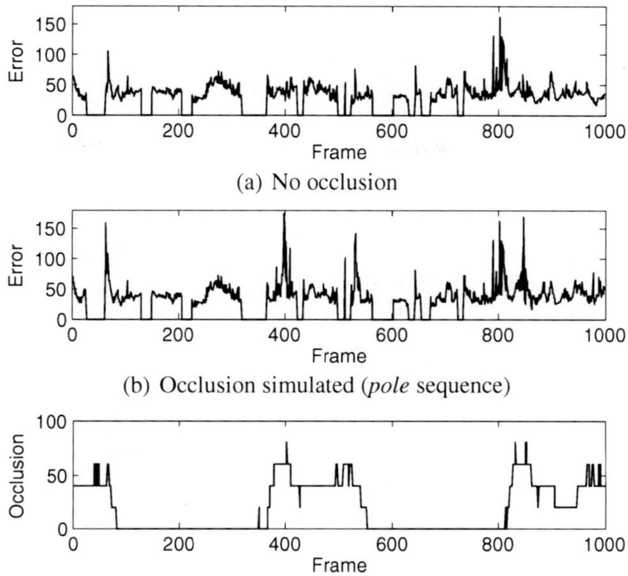

(a) No occlusion

(b) Occlusion simulated (*pole* sequence)

(c) Percentage of occlusion

Fig. 6 Mean relative 3D error (in mm) plots for HumanEva-I Walking, performed by subject 1 and viewed with a single camera (C1), (a) without occlusion, (b) for *pole* condition, and (c) percentage of occlusion for *pole* condition. Instances that have a zero error contain invalid mocap.

The error plots for the *pole* sequence are shown in Figure 6. The amount of occlusion for each frame is in the range of 0–80%. On average, 19.06% of the observation is occluded. The graphs clearly show that the error increases under occlusion. The average increase in error is 11.56%. It should be noted that this is mainly due to those frames where more than half of the observation is occluded. Here, the average error is 57.00 mm, compared to 39.17 mm for the non-occluded sequence. When occlusion is in the range of 20–40%, the mean error is 43.63 mm compared to 37.49 mm for the original. Differences between the upper two graphs for those frames where no occlusion is present are due to differences in normalization. The pose estimates for several frames are shown in Figure 5.

4.5.1 Comparison with Related Discriminative Work

The HumanEva dataset has been used by others to evaluate their work. In Table 3, we compare previous reports on the HumanEva-I dataset, obtained using discriminative approaches. Only non-occluded observations are used. The errors in Table 3 are indicative, as there are various differences between methods. First, Elgammal and Lee [15, 34] normalize the errors for rotation. Second, some works evaluate their approach on the validation sets, which allows for parameter tuning since the online evaluation system is not required. Also, the observations might be closer to

Table 3 Comparison of results of discriminative approaches, reported on HumanEva-I. Dynamics (*dyn.*) indicates whether a dynamical model (possibly activity-specific) is employed. All errors are relative to the *torsoDistal* joint. Errors in 3D are in *mm*, for 2D in pixels. Direct comparison is hindered due to differences between evaluations (different subjects, validation set instead of test set, only part of the sequence). [15, 34] use a rotation-normalized error measure.

	Image representation	Dyn.	Walking	Box	2D/3D
Bo et al. [6]	Histogram of SC	N	25.66	30.40	3D
Bo and Sminchisescu [5]	HOG descriptor	N	37.07	89.47	3D
Elgammal and Lee [15]	Silhouette	Y	31.36		3D
Lee and Elgammal [34]	Silhouette	N	76.56		3D
Okada and Soatto [50]	HOG descriptor	N	37.98		3D
Poppe (this work)	HOG descriptor	N	45.05	94.97	3D
Poppe [56]	HOG descriptor	N	45.36	94.48	3D
Urtasun et al. [78]	Hierarchical features	N	32.70	38.50	3D
Howe [29]	Silhouette	Y	15.00		2D
Urtasun et al. [78]	Hierarchical features	N	5.18	6.68	2D

the training set as training and validation sets are obtained from the same movement sequence. Third, different sets are used for training. Several authors have used person-specific training sets (e.g. [78]). Such an approach does not give any information about the ability to generalize over subjects.

Howe [29] presents an example-based approach and uses a batch approach to recover the most likely poses over a sequence of observations. This allows to avoid forward–backward ambiguities, which are common when using silhouettes. Our approach is closely related to Poppe's [56], who uses a global normalization and matching distance. Differences in accuracy between a global and cell-level normalization are small. Moreover, our approach can deal with partial occlusions.

Regression-based approaches (e.g. [6, 50, 78]) in general show lower errors. However, this comes at the cost of a computationally costly training phase, which has proved difficult to generalize to larger numbers of training samples. Also, it remains to be investigated how these approaches can cope with partial occlusions.

5 Conclusion

We presented an approach to recover human poses directly from image observations, even under significant occlusions. Our method requires segmentation of the human subject in the image and prediction of the occluded areas. Recent work in the domain of human detection can produce this information. We take an example-based approach, where we encode the image observation using HOG. For each grid cell, we assign a weight that indicates whether the cell is occluded. As such, we can use the same example database, regardless of the type of occlusion. Experiments were performed on the HumanEva-I dataset, which was also used to construct a database of image–pose pairs. Since the dataset does not contain occlusions, we additionally simulated occlusion for six different conditions. Poses are estimated by interpolating the poses corresponding to the 25 closest matches from a database. In this matching, we treat the occluded cells as missing observations. This allows us to recover poses even if a significant percentage of the image observation is occluded.

For non-occluded observations, we report 3D errors of approximately 69 mm, relative to the pelvis and averaged over a set of 20 joint locations. For walking and jogging, this error is approximately 45 mm. Our results show approximately 10% increase in error when 20% of the observation is occluded. When 33% of the observation is occluded, the error is on average 15% higher compared to the observations without occlusions. We plan to combine our work with human detection. One avenue of future research is to analyze how stereo information can aid in the human detection and occlusion prediction tasks. Also, we are looking at ways to reduce the linear complexity in the number of examples, either using hashing or a regression-based approach.

Acknowledgment. We wish to thank the authors of [69] for making their database available.

References

1. Agarwal, A., Triggs, B.: A local basis representation for estimating human pose from cluttered images. In: Narayanan, P.J., Nayar, S.K., Shum, H.-Y. (eds.) ACCV 2006. LNCS, vol. 3851, pp. 50–59. Springer, Heidelberg (2006)
2. Agarwal, A., Triggs, B.: Recovering 3D human pose from monocular images. IEEE Transactions on Pattern Analysis and Machine Intelligence (PAMI) 28(1), 44–58 (2006)
3. van den Bergh, M., Koller-Meier, E., van Gool, L.J.: Real-time body pose recognition using 2D or 3D haarlets. International Journal of Computer Vision (IJCV) 83(1), 72–84 (2009)
4. Bissacco, A., Yang, M.-H., Soatto, S.: Fast human pose estimation using appearance and motion via multi-dimensional boosting regression. In: Proceedings of the Conference on Computer Vision and Pattern Recognition (CVPR 2007), Minneapolis, MN, June 2007, pp. 1–8 (2007)
5. Bo, L., Sminchisescu, C.: Twin Gaussian processes for structured prediction. International Journal of Computer Vision (IJCV) 87(1-2), 28–52 (2010)
6. Bo, L., Sminchisescu, C., Kanaujia, A., Metaxas, D.: Fast algorithms for large scale conditional 3D prediction. In: Proceedings of the Conference on Computer Vision and Pattern Recognition (CVPR 2008), Anchorage, AK, June 2008, pp. 1–8 (2008)
7. Bosch, A., Zisserman, A., Muñoz, X.: Representing shape with a spatial pyramid kernel. In: Proceedings of the International Conference on Image and Video Retrieval (CIVR 2007), Amsterdam, The Netherlands, July 2007, pp. 401–408 (2007)
8. Bowden, R., Mitchell, T.A., Sarhadi, M.: Non-linear statistical models for the 3D reconstruction of human pose and motion from monocular image sequences. Image and Vision Computing 18(9), 729–737 (2000)
9. Brand, M.: Shadow puppetry. In: Proceedings of the International Conference on Computer Vision (ICCV 1999), Kerkyra, Greece, September 1999, vol. 2, pp. 1237–1244 (1999)

10. Chakraborty, B., Rudovic, O., Gonzàlez, J.: View-invariant human-body detection with extension to human action recognition using component-wise HMM of body parts. In: Proceedings of the International Conference on Automatic Face and Gesture Recognition (FGR 2008), Amsterdam, The Netherlands, September 2008, pp. 1–6 (2008)
11. Dalal, N., Triggs, B.: Histograms of oriented gradients for human detection. In: Proceedings of the Conference on Computer Vision and Pattern Recognition (CVPR 2005), San Diego, CA, June 2005, vol. 1, pp. 886–893 (2005)
12. Dong, L., Parameswaran, V., Ramesh, V., Zoghlami, I.: Fast crowd segmentation using shape indexing. In: Proceedings of the International Conference On Computer Vision (ICCV 2007), Rio de Janeiro, Brazil, October 2007, pp. 1–8 (2007)
13. Ek, C.H., Rihan, J., Torr, P.H.S., Rogez, G., Lawrence, N.D.: Ambiguity modeling in latent spaces. In: Popescu-Belis, A., Stiefelhagen, R. (eds.) MLMI 2008. LNCS, vol. 5237, pp. 62–73. Springer, Heidelberg (2008)
14. Elgammal, A.M., Davis, L.S.: Probabilistic framework for segmenting people under occlusion. In: Proceedings of the International Conference On Computer Vision (ICCV 2001), Vancouver, Canada, July 2001, vol. 2, pp. 145–152 (2001)
15. Elgammal, A.M., Lee, C.-S.: Trackingpeople on a torus. IEEE Transactions on Pattern Analysis and Machine Intelligence (PAMI) 31(3), 520–538 (2009)
16. Fathi, A., Mori, G.: Human pose estimation using motion exemplars. In: Proceedings of the International Conference On Computer Vision (ICCV 2007), Rio de Janeiro, Brazil, Ocotber 2007, pp. 1–8 (2007)
17. Pedro, F.: Felzenszwalb, David McAllester, and Deva Ramanan. Discriminatively trained multiscale deformable part models. In: Proceedings of the Conference on Computer Vision and Pattern Recognition (CVPR 2008), Anchorage, AK, June 2008, pp. 1–8 (2008)
18. Fossati, A., Arnaud, E., Horaud, R., Fua, P.: Tracking articulated bodies using generalized expectation maximization. In: Proceedings of the Conference on Computer Vision and Pattern Recognition Workshops (CVPRW 2008), Washington, DC, June 2008, pp. 1–6 (2008)
19. William, T.: Freeman and Michal Roth. Orientation histograms for hand gesture recognition. In: Proceedings of the Workshop on Automatic Face and Gesture Recognition, Zurich, Switzerland, June 1995, pp. 296–301 (1995)
20. Gandhi, T., Trivedi, M.M.: Image based estimation of pedestrian orientation for improving path prediction. In: Proceedings of the Intelligent Vehicles Symposium (IV 2008), Eindhoven, The Netherlands, June 2008, pp. 506–511 (2008)
21. Gavrila, D.M.: A Bayesian, exemplar-based approach to hierarchical shape matching. IEEE Transactions on Pattern Analysis and Machine Intelligence (PAMI) 29(8), 1408–1421 (2007)
22. Gond, L., Sayd, P., Chateau, T., Dhome, M.: A 3D shape descriptor for human pose recovery. In: Perales, F.J., Fisher, R.B. (eds.) AMDO 2008. LNCS, vol. 5098, pp. 370–379. Springer, Heidelberg (2008)
23. Grauman, K., Shakhnarovich, G., Darrell, T.: Inferring 3D structure with a statistical image-based shape model. In: Proceedings of the International Conference on Computer Vision (ICCV 2003), Nice, France, October 2003, vol. 1, pp. 641–647 (2003)
24. Guo, F., Qian, G.: Human pose inference from stereo cameras. In: Proceedings of the Workshop on Applications of Computer Vision (WACV 2007), Austin, TX, February 2007, p. 37 (2007)

25. Gupta, A., Chen, T., Chen, F., Kimber, D., Davis, L.S.: Context and observation driven latent variable model for human pose estimation. In: Proceedings of the Conference on Computer Vision and Pattern Recognition (CVPR 2008), Anchorage, AK, June 2008, pp. 1–8 (2008)
26. Howe, N.R.: Silhouette lookup for automatic pose tracking. In: Proceedings of the Conference on Computer Vision and Pattern Recognition Workshops (CVPRW 2004), Los Alamitos, CA, June 2004, p. 15 (2004)
27. Howe, N.R.: Flow lookup and biological motion perception. In: Proceedings of the Internation Conference on Image Processing (ICIP 2005), Genova, Italy, September 2005, vol. 3, pp. 1168–1171 (2005)
28. Howe, N.R.: Boundary fragment matching and articulated pose under occlusion. In: Perales, F.J., Fisher, R.B. (eds.) AMDO 2006. LNCS, vol. 4069, pp. 271–280. Springer, Heidelberg (2006)
29. Nicholas, R.: Howe. Recognition-based motion capture and the HumanEva II test data. In: Proceedings of the Workshop on Evaluation of Articulated Human Motion and Pose Estimation, Minneapolis, MN (June 2007)
30. Isard, M., Blake, A.: CONDENSATION - conditional density propagation for visual tracking. International Journal of Computer Vision 29(1), 5–28 (1998)
31. Kanaujia, A., Sminchisescu, C., Metaxas, D.: Semi-supervised hierarchical models for 3D human pose reconstruction. In: Proceedings of the Conference on Computer Vision and Pattern Recognition (CVPR 2007), Minneapolis, MN, June 2007, pp. 1–8 (2007)
32. Ke, Y., Sukthankar, R.: PCA-SIFT: A more distinctive representation for local image descriptors. In: Proceedings of the Conference on Computer Vision and Pattern Recognition (CVPR 2004), Washington, DC, June 2004, vol. 2, pp. 506–513 (2004)
33. Stephen, J.: Krotosky and Mohan M. Trivedi. On color-, infrared-, and multimodal-stereo approaches to pedestrian detection. IEEE Transactions on Intelligent Transportation Systems 8(4), 619–629 (2007)
34. Lee, C.-S., Elgammal, A.M.: Simultaneous inference of view and body pose using torus manifolds. In: Proceedings of the International Conference on Pattern Recognition (ICPR 2006), Kowloon Tong, Hong Kong, August 2006, vol. 3, pp. 489–494 (2006)
35. Levi, K., Weiss, Y.: Learning object detection from a small number of examples: the importance of good features. In: Proceedings of the Conference on Computer Vision and Pattern Recognition (CVPR 2004), Washington, DC, June 2004, vol. 2, pp. 53–60 (2004)
36. Lin, Z., Davis, L.S.: A pose-invariant descriptor for human detection and segmentation. In: Forsyth, D., Torr, P., Zisserman, A. (eds.) ECCV 2008, Part IV. LNCS, vol. 5305, pp. 423–436. Springer, Heidelberg (2008)
37. Lin, Z., Davis, L.S., Doermann, D., DeMenthon, D.: Hierarchical part-template matching for human detection and segmentation. In: Proceedings of the International Conference on Computer Vision (ICCV 2007), Rio de Janeiro, Brazil, October 2007, pp. 1–8 (2007)
38. Liu, X., Yu, T., Sebastian, T., Tu, P.: Boosted deformable model for human body alignment. In: Proceedings of the Conference on Computer Vision and Pattern Recognition (CVPR 2008), Anchorage, AK, June 2008, pp. 1–8 (2008)
39. Lowe, D.G.: Distinctive image features from scale-invariant keypoints. International Journal of Computer Vision (IJCV) 60(2), 91–110 (2004)
40. Lu, W.-L., Little, J.J.: Simultaneous tracking and action recognition using the PCA-HOG descriptor. In: Proceedings of the Canadian Conference on Computer and Robot Vision (CRV 2006), Quebec City, Canada, June 2006, p. 6 (2006)

41. Mikolajczyk, K., Schmid, C.: Scale and affine invariant interest point detectors. International Journal of Computer Vision (IJCV) 60(1), 63–86 (2004)
42. Mikolajczyk, K., Schmid, C.: A performance evaluation of local descriptors. IEEE Transactions on Pattern Analysis and Machine Intelligence (PAMI) 27(10), 1615–1630 (2005)
43. Mikolajczyk, K., Schmid, C., Zisserman, A.: Human detection based on a probabilistic assembly of robust part detectors. In: Pajdla, T., Matas, J(G.) (eds.) ECCV 2004. LNCS, vol. 3021, pp. 69–82. Springer, Heidelberg (2004)
44. Mohan, A., Papageorgiou, C., Poggio, T.: Example-based object detection in images by components. IEEE Transactions on Pattern Analysis and Machine Intelligence (PAMI) 23(4), 349–361 (2001)
45. Mori, G., Malik, J.: Recovering 3D human body configurations using shape contexts. IEEE Transactions on Pattern Analysis and Machine Intelligence (PAMI) 28(7), 1052–1062 (2006)
46. Navaratnam, R., Fitzgibbon, A.W., Cipolla, R.: Semi-supervised learning of joint density models for human pose estimation. In: Proceedings of the British Machine Vision Conference (BMVC 2006), Edinburgh, United Kingdom, September 2006, vol. 2, pp. 679–688 (2006)
47. Niebles, J.C., Han, B., Ferencz, A., Fei-Fei, L.: Extracting moving people from internet videos. In: Forsyth, D., Torr, P., Zisserman, A. (eds.) ECCV 2008, Part IV. LNCS, vol. 5305, pp. 527–540. Springer, Heidelberg (2008)
48. Ning, H., Hu, Y., Huang, T.S.: Efficient initialization of mixtures of experts for human pose estimation. In: Proceedings of the International Conference on Image Processing (ICIP 2008), San Diego, CA, October 2008, pp. 2164–2167 (2008)
49. Ning, H., Xu, W., Gong, Y., Huang, T.S.: Discriminative learning of visual words for 3D human pose estimation. In: Proceedings of the Conference on Computer Vision and Pattern Recognition (CVPR 2008), Anchorage, AK, June 2008, pp. 1–8 (2008)
50. Okada, R., Soatto, S.: Relevant feature selection for human pose estimation and localization in cluttered images. In: Forsyth, D., Torr, P., Zisserman, A. (eds.) ECCV 2008, Part II. LNCS, vol. 5303, pp. 434–445. Springer, Heidelberg (2008)
51. Ong, E.-J., Gong, S.: A dynamic 3D human model using hybrid 2D-3D representations in hierarchical pca space. In: Proceedings of the British Machine Vision Conference (BMVC 1999), Nottingham, United Kingdom, September 1999, pp. 33–42 (1999)
52. Ong, E.-J., Micilotta, A.S., Bowden, R., Hilton, A.: Viewpoint invariant exemplar-based 3D human tracking. Computer Vision and Image Understanding (CVIU) 104(2-3), 178–189 (2006)
53. Onishi, K., Takiguchi, T., Ariki, Y.: 3D human posture estimation using the HOG features from monocular image. In: Proceedings of the International Conference on Pattern Recognition (ICPR 2008), Tampa, FL, December 2008, pp. 1–4 (2008)
54. Peursum, P., Venkatesh, S., West, G.: Observation-switching linear dynamic systems for tracking humans through unexpected partial occlusions by scene objects. In: Proceedings of the International Conference on Pattern Recognition (ICPR 2006), Kowloon Tong, Hong Kong, August 2006, vol. 4, pp. 929–934 (2006)
55. Peursum, P., Venkatesh, S., West, G.: Tracking-as-recognition for articulated full-body human motion analysis. In: Proceedings of the Conference on Computer Vision and Pattern Recognition (CVPR 2007), Minneapolis, MN, June 2007, pp. 1–8 (2007)
56. Poppe, R.: Evaluating example-based pose estimation: Experiments on the HumanEva sets. In: Proceedings of the Workshop on Evaluation of Articulated Human Motion and Pose Estimation (CVPR-EHuM), Minneapolis, MN (June 2007)

57. Poppe, R.: Vision-based human motion analysis: An overview. Computer Vision and Image Understanding (CVIU) 108(1-2), 4–18 (2007)
58. Poppe, R., Poel, M.: Comparison of silhouette shape descriptors for example-based human pose recovery. In: Proceedings of the International Conference on Automatic Face and Gesture Recognition (FGR 2006), Southampton, United Kingdom, April 2006, pp. 541–546 (2006)
59. Poppe, R., Poel, M.: Body-part templates for recovery of 2D human poses under occlusion. In: Perales, F.J., Fisher, R.B. (eds.) AMDO 2008. LNCS, vol. 5098, pp. 289–298. Springer, Heidelberg (2008)
60. Porikli, F.: Integral histogram: A fast way to extract histograms in cartesian spaces. In: Proceedings of the Conference on Computer Vision and Pattern Recognition (CVPR 2005), San Diego, CA, June 2005, vol. 1, pp. 829–836 (2005)
61. Ramanan, D., Forsyth, D.A., Zisserman, A.: Tracking people by learning their appearance. IEEE Transactions on Pattern Analysis and Machine Intelligence (PAMI) 29(1), 65–81 (2007)
62. Rogez, G., Orrite-Uruñuela, C., del Rincón, J.M.: A spatio-temporal 2D-models framework for human pose recovery in monocular sequences. Pattern Recognition 41(9), 2926–2944 (2008)
63. Rogez, G., Rihan, J., Ramalingam, S., Orrite-Uruñuela, C., Torr, P.H.S.: Randomized trees for human pose detection. In: Proceedings of the Conference on Computer Vision and Pattern Recognition (CVPR 2008), Anchorage, AK, June 2008, pp. 1–8 (2008)
64. Rosales, R.E., Sclaroff, S.: Learning body pose via specialized maps. In: Advances in Neural Information Processing Systems (NIPS), Vancouver, Canada, December 2001, vol. 14, pp. 1263–1270 (2001)
65. Rosales, R.E., Sclaroff, S.: Combining generative and discriminative models in a framework for articulated pose estimation. International Journal of Computer Vision (IJCV) 67(3), 251–276 (2006)
66. Rosales, R.E., Siddiqui, M., Alon, J., Sclaroff, S.: Estimating 3D body pose using uncalibrated cameras. In: Proceedings of the Conference on Computer Vision and Pattern Recognition (CVPR 2001), Kauai, HI, December 2001, vol. 1, pp. 821–827 (2001)
67. Shakhnarovich, G., Viola, P.A., Darrell, T.: Fast pose estimation with parameter-sensitive hashing. In: Proceedings of the International Conference on Computer Vision (ICCV 2003), Nice, France, October 2003, vol. 2, pp. 750–759 (2003)
68. Sigal, L., Bălan, A.O., Black, M.J.: Combined discriminative and generative articulated pose and non-rigid shape estimation. In: Advances in Neural Information Processing Systems (NIPS), Vancouver, Canada, December 2008, vol. 20, pp. 1337–1344 (2008)
69. Sigal, L., Black, M.J.: HumanEva: Synchronized video and motion capture dataset for evaluation of articulated human motion. Technical Report CS-06-08, Brown University, Department of Computer Science, Providence, RI (September 2006)
70. Sminchisescu, C., Kanaujia, A., Metaxas, D.: Learning joint top-down and bottom-up processes for 3D visual inference. In: Proceedings of the Conference on Computer Vision and Pattern Recognition (CVPR 2006), New York, NY, June 2006, vol. 2, pp. 1743–1752 (2006)
71. Sminchisescu, C., Kanaujia, A., Metaxas, D.N.: BM^3E: Discriminative density propagation for visual tracking. IEEE Transactions on Pattern Analysis and Machine Intelligence (PAMI) 29(11), 2030–2044 (2007)
72. Sullivan, J., Carlsson, S.: Recognizing and tracking human action. In: Heyden, A., Sparr, G., Nielsen, M., Johansen, P. (eds.) ECCV 2002. LNCS, vol. 2350, pp. 629–644. Springer, Heidelberg (2002)

73. Sun, Y., Bray, M., Thayananthan, A., Yuan, B., Torr, P.H.S.: Regression-based human motion capture from voxel data. In: Proceedings of the British Machine Vision Conference (BMVC 2006), Edinburgh, United Kingdom, September 2006, vol. 1, pp. 277–286 (2006)
74. Thayananthan, A., Navaratnam, R., Stenger, B., Torr, P.H.S., Cipolla, R.: Pose estimation and tracking using multivariate regression. Pattern Recognition Letters 29(9), 1302–1310 (2003)
75. Thurau, C.: Behavior histograms for action recognition and human detection. In: Elgammal, A., Rosenhahn, B., Klette, R. (eds.) Human Motion 2007. LNCS, vol. 4814, pp. 271–284. Springer, Heidelberg (2007)
76. Toyama, K., Blake, A.: Probabilistic tracking with exemplars in a metric space. International Journal of Computer Vision 48(1), 9–19 (2002)
77. Tuytelaars, T., Mikolajczyk, K.: Local invariant feature detectors: A survey. Foundations and Trends in Computer Graphics and Vision 3(3), 177–280 (2008)
78. Urtasun, R., Darrell, T.: Sparse probabilistic regression for activity-independent human pose inference. In: Proceedings of the Conference on Computer Vision and Pattern Recognition (CVPR 2008), Anchorage, AK, June 2008, pp. 1–8 (2008)
79. Viola, P.A., Jones, M.J.: Rapid object detection using a boosted cascade of simple features. In: Proceedings of the Conference on Computer Vision and Pattern Recognition (CVPR 2001), Kauai, HI, December 2001, vol. 1, pp. 511–518 (2001)
80. Wu, B., Nevatia, R.: Detection and tracking of multiple, partially occluded humans by Bayesian combination of edgelet based part detectors. International Journal of Computer Vision (IJCV) 75(2), 247–266 (2007)
81. Wu, B., Nevatia, R.: Detection and segmentation of multiple, partially occluded objects by grouping, merging, assigning part detection responses. International Journal of Computer Vision (IJCV) 82(2), 185–204 (2009)
82. Xu, L.-Q., Hogg, D.C.: Neural networks in human motion tracking - an experimental study. Image and Vision Computing 15(8), 607–615 (1997)
83. Yang, H.-D., Lee, S.-W.: Reconstruction of 3D human body pose from stereo image sequences based on top-down learning. Pattern Recognition 40(11), 3120–3131 (2007)
84. Zhu, Q., Avidan, S., Yeh, M.-C., Cheng, K.-T.: Fast human detection using a cascade of histograms of oriented gradients. In: Proceedings of the Conference on Computer Vision and Pattern Recognition (CVPR 2006), New York, NY, June 2006, vol. 2, pp. 1491–1498 (2006)

Part IV
Architectures for Distributed Agent-Actor Communities

Agility and Adaptive Autonomy in Networked Organizations

Martijn Neef and Bob van der Vecht

Abstract. In any multi-actor environment, there is an inevitable trade-off between achieving global coordination of activities and respecting the autonomy of the actors involved. Agile and resilient behavior demands dynamic coordination capabilities, but task and resource allocation quickly becomes challenging because of individual constraints and demands. In this study, we present research on adaptive autonomy in multi-agent organizations. We have studied the relationship between autonomy and coordination, and developed an agent reasoning model not only that enables collaborative task coordination, but also guarantees individual autonomy—the capability to self-manage behavior. We define autonomy as the amount of influence other agents have on one's decision-making process. We have given the agent options to adapt its openness to external influences, so it can change its own level of autonomy. This allows agents to select the level of autonomy that best fits the circumstances, given a certain tasking, individual policies and organizational structure. We have incorporated this concept in a practical model and added heuristics for environmental events, information relevance and organizational rules. Our approach addresses fundamental collaborative challenges in dynamic environments, and may bring about new perspectives on autonomy in collaborative environments.

1 Introduction

Many organizations are rethinking their coordination strategies to cope with the complexity of modern challenges. In both civil and military settings, organizations need to adopt new ways of working to achieve higher levels of agility and resilience.

Martijn Neef
TNO Defence, Security and Safety
e-mail: martijn.neef@tno.nl

Bob van der Vecht
TNO Defence, Security and Safety
e-mail: bob.vandervecht@tno.nl

R. Babuška & F.C.A. Groen (Eds.): Interactive Collaborative Information Systems, SCI 281, pp. 357–373.
springerlink.com

Traditional centralized coordination strategies do not lead to the degree of agility and adaptivity that modern organizations seek. At the same time, more parties become involved in the coordination process, all with individual constraints and demands. For instance, in most current military engagement, there are multi-national coalitions involved instead of single nations. The adaptive coordination of work and the shared use of resources in such settings pose complex planning challenges that traditional command and control strategies cannot easily uphold [1]. In modern safety and security arenas, the same observations can be made. Crisis response organizations are made up of many loosely connected parties (firefighters, police, medical workers and other services) which usually uphold their autonomy and limit close collaboration to management levels. Recent large-scale incidents have shown that this way of working can lead to problems in reacting to changing conditions, with severe consequences to performance. A possible solution is to work toward a more distributed way of working where more collaboration takes place on lower organizational levels. Efforts to implement such solutions are usually complicated for parties that value their autonomy highly. Therefore, we need new ways of thinking about coordination in networked organizations that facilitate multi-party collaboration and respect individual autonomy at the same time.

Our work on adaptive autonomy in agent systems might provide some interesting perspectives on agility and resilience in dynamic environments. We have been researching the topic of autonomy in multi-agent systems and in particular the relationship between autonomy and coordination. In any multi-actor environment, there is an inevitable trade-off between achieving global coordination of activities and respecting the autonomy of the actors involved. If decision-making processes and operational tasks are distributed over many parties, respect of autonomy becomes an increasingly important issue. In this research, we work towards collaborative decision-making models that respect the agent's own autonomy, but at the same time take organizational roles and operational conditions into account.

In this study, we make the connection with the network-enabled capabilities (NEC) paradigm [1]. NEC is a popular future perspective on command-and-control principles in military organizations and crisis management organizations. Every actor in the organization is considered to be a node in a network. The philosophy behind the networked structure is that organizations become flexible in assigning roles and tasks to actors, and that makes it possible to share resources across the boundaries of organizations. We believe that our experiences from multi-agent research on autonomy and coordination are relevant for the NEC community, as it deals with joint and combined missions where there might be conflicts of individual interests. It brings about new perspectives on autonomy and agility in collaborative, networked environments.

In sections 2 and 3, we present research on adaptive autonomy in multi-agent organizations. We explore the relationship between autonomy and coordination in agent organizations and present a decision-making model for agents. In sections 4 and 5, we demonstrate how our approach can be used to model organizational behaviors and discuss its relevance for NEC organizations. Section 6 discusses our approach in relation to related work, and section 7 concludes this chapter.

2 Autonomy and Agility in Agent Systems

Academic research on artificial agents has had limited impact on practical applications for complex environments so far. However, there are opportunities, because research into agent organizations revolves around many of the same issues that complicate the development of NEC organizations, such as collaborative decision making, decentralized coordination strategies and dynamic organizational structures. Actor autonomy is another important trait that both agent organizations and NEC organizations share. In this section, we discuss agent autonomy, and explain why autonomy is important for agile coordination in NEC organizations.

2.1 The Role of Autonomy in Agent Systems

Autonomy is an important aspect of artificial agents. It is usually regarded as one of the defining features of an agent [2], [3]. Agents have, by definition, control over both their internal state (e.g. beliefs, desires, intentions) and their behavior. We consider agents that act goal-oriented and exert their independence and problem-solving capacities to reach their objectives. In multi-agent systems, there is communication and coordination of activities between agents.

Coordination implies that agents enter into a work agreement and agree to a certain interdependence and interaction. This implies that agents will influence each other and possibly make demands that may affect each other's degree of autonomy. For instance, a supervising agent may demand that one of his subordinate agents performs a certain task. This demand challenges the autonomy of the subordinate agent.

If the relationship between the supervisor and the subordinate agent is clear and agreed upon by both parties, then the subordinate agent will not object to the request, since it will also advance its own goals (successfully follow orders and help to fulfill the organizational goals). Moreover, you could say that agreements on relationships actually define the level of autonomy that either party may exhibit. Figure 1 illustrates the importance of such agreements from the perspective of an single individual agent. From a wide range of ideally possible actions, agents may only be able to perform a certain subset of actions and may only be allowed to perform another subset [4]. Agreements between agents set the 'Permitted Actions' and 'Obligated Actions' boundary, and thus confine their autonomy.

On the other hand, if the demand directly opposes the agent's beliefs or own objectives, then the agent will find itself challenged in autonomy. Such a situation might be a cause to question the demand itself, or the interaction agreement. Should the agent follow orders and perform an act that is detrimental to its own goals or exert autonomy and let his own motivations prevail over demands from others? Such situations are reminiscent of the difficult autonomy challenges that Asimov's Laws of Robotics brought about [5], and many ethical questions in warfare. The relationship between ethics, policies and autonomy is an important but precarious one, and not easily captured in a model. We will not diverge into a discussion on the role of human ethics in decision making. However, for coordination purposes in multi-actor

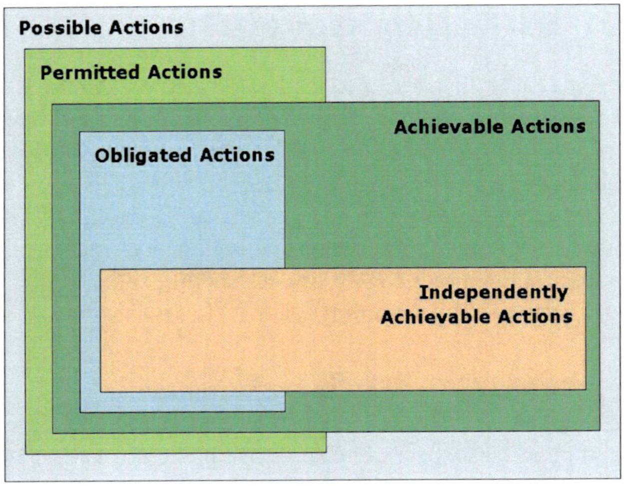

Fig. 1 Basic dimensions of autonomy, after [4]

environments, we do need a way to model and deal with autonomy in multi-agent coordination. We believe that it is essential to give agent autonomy a central role in coordination activities. Most conventional coordination tactics do not explicitly give the affected actors a say in the process, and consequently negate the essence of autonomy: giving actors the freedom to object or accept a proposal, based on their own interest.

2.2 Autonomy and Coordination Mechanisms

Since, by definition, agents need to be in control of their own internal state and behavior, the question rises how agent decision making is actually affected by external influences. How can an agent maintain control over his own behavior, but at the same time cooperate with other agents to achieve coordination? The traditional way to achieve coordination is by developing a top-down coordination mechanism. The designer of a multi-agent system specifies the tasks and interaction mechanisms that the agents will follow, e.g. the Gaia methodology [6]. The rules of the coordination mechanism are embedded in the decision-making process of the agent. This allows the agents to jointly find a correct division of labor. One can argue that such agents are not truly autonomous, since they can only behave in line with commands that are set forth at design time. The agent has no means to enforce its autonomy, and cannot pro-actively deviate from plans. Of course, in many systems, this is a desirable feature: we want the system to do what it was designed for. In some cases, however, it might be favorable to grant agents a degree of sovereignty. One could think of situations where standard procedures fail, and agents are left to their own devices to create alternative plans. For instance, if the chain of command collapses because of communication breakdown, deployed agents need to be able to take matters into their own hands. In other words, they need to adapt their autonomy from being commanded to self-ruling. In order to fully benefit from the autonomy of agents, we need to define the coordination mechanism separately from the internals of the agent. This approach to system design is also taken by methodologies for agent organizations. Agent organizations are motivated by descriptions of human organizations. One of the first was the Agent Group Role model [7], known as the AGR model, which introduced the concepts of roles, groups and interaction between roles. A more expressive model is provided by the OperA approach [8]. It adds the notions of organizational norms, which are general behavioral rules an agent is supposed to follow. The role descriptions and the norms are translated to contracts. Actors fulfilling a role have to sign the contract.

Both approaches focus on defining the structural aspects of the organization. The organizational model is constructed disregarding the agents who will take up the roles. It does not matter how an agent achieves its goals as long as it follows the organizational rules. In fact, a role can be fulfilled by a human as well as by an artificial agent.

2.3 *Adaptive Autonomy and Agile Organizations*

Many systems exist of autonomous actors. In our perspective, autonomy is about the level of independence of decision making. The degree of autonomy of decision making can be defined as the degree of intervention by other agents on the decision-making process of one agent [9]. An agent that is heavily influenced by other agents in its decision making is displaying obedient behavior. An agent that does not allow any external influence in its decision making is being ultimately independent. By altering the amount of external influence on its decision making, an agent can adapt its own level autonomy. In this fashion, agents can actively select the level of autonomy that best fits the circumstances. In effect, an agent that changes its level of autonomy in response to changing conditions shows adaptive autonomy.

Adaptive autonomy of individual agents facilitates the agility and resilience of an organization. Agility refers to the ability of an entity to quickly and gracefully respond to a changing environment. Resilience refers to the ability of an entity to withstand disturbance and to readily recover. Both are very desirable features for organizations, and, understandably, very much in the spotlight of networked organizations and distributed systems research. An agile and resilient organization must be able to cope with changes in circumstances (e.g. unexpected events, new objectives) and changes in organizational structure (e.g. failing elements, modified chain of command), and respond promptly with a new course of action or a new organizational layout (e.g. new coordination scheme). A traditional top-down approach to coordination will not yield agile behavior in a modern network-centric organization, because of its transient nature. There are simply too many actors and systems involved to create a course of action that fits all individual capabilities and norms. We believe that agility in a dynamic network-centric organization must find its roots in individual adaptive autonomy.

3 A Model for Adaptive Autonomy

We have developed a reasoning model for artificial agents that implements the notion of adaptive autonomy as described above. We give the agent means to manage incoming external events (observations, messages) through an influence control process. This process controls which events affect the internal state of the agent and thus influence the decision making and behavior. With the process of influence control, we refer to a conscious process, where the agent is aware of the external events and uses its knowledge to determine the effect on the mental state. This makes perception an active process, similar to the psychological Perceptual Control Theory [10], [11], instead of a passive process.

The agent itself is in command of the influence-control process, and can configure the process to fit his own objectives. This approach guarantees the autonomy of the agent [12], because it makes the agent ultimately responsible for its own behavior. Figure 2 shows the position of the influence-control process in relation to the decision-making process of an agent. The internal state holds the agent's current beliefs, goals and plans. The decision-making process deliberates on the elements of the internal state to derive decisions and courses of action.

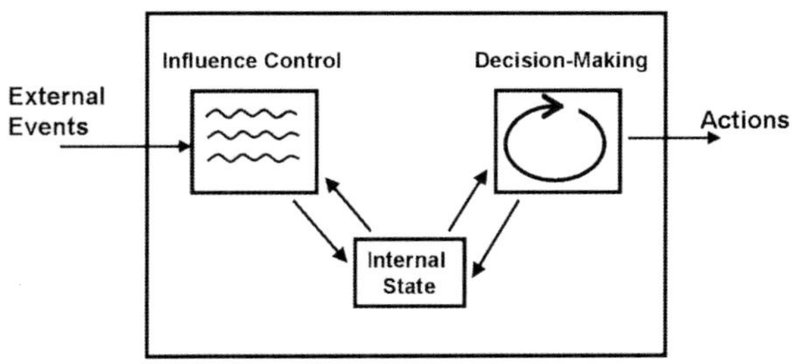

Fig. 2 Schematic reasoning model of an adaptive autonomous agent

3.1 Event Processing

The influence-control process uses event-processing rules to decide on adoption or rejection of certain influences. We propose to use rules of the form:

```
<EVENT> <- <CONSTRAINTS> | <EFFECT>
```

The event is the trigger of the rule; it should match the incoming event. There are three types of events: observations, inform messages and request messages. The first two contain information, whereas the latter contains a task or a goal.

The constraints describe situational constraints for the rule. They should match the beliefs and goals of the agent to make the rule applicable.

The effect of the rule specifies an internal action of the agent that holds the effect on the mental state. The possible effects depend on how the mental state of the agent is constructed. Receiving new information generally leads to belief updates.

Concerning tasks, an agent can add a goal, as a consequence of accepting a request. Another option is to specify rules that result in ignoring the external event, for example, when information is irrelevant, or when the agent is busy. We propose to use three internal actions and the corresponding effect on the mental state:

- Update Beliefs
- Add Goal
- Ignore Event

As we stated, this set does not contain all possible results of event-processing rules. For example, another internal action can be to drop a goal, as response to a prohibition. Furthermore, a belief update is not necessarily a straight-forward process, as adopting a new belief can lead to inconsistent beliefs. However, in the examples presented in this work, the presented three are sufficient to demonstrate control over external influences.

3.2 *Basic Attitudes*

The event-processing rules specify the attitude of how agents perceive the world. The attitudes are directly related to levels of autonomy of the decision-making process. For all events the agent can choose to adopt or reject them. When combining the three event types and the two processing options (adopt or ignore an event), we can construct general attitudes, as summarized in Table 1.

Table 1 Basic Attitudes

Event	Effect	Basic Attitude
Observation	Update Beliefs	Self-reliant
	Ignore Event	Non-self-reliant
Inform message	Update Beliefs	Trusting
	Ignore Event	Non-trusting
Request message	Add Goal	Cooperative
	Ignore Event	Non-cooperative

The three types of basic attitudes (self-reliant/non-self-reliant, trusting/non-trusting, cooperative/non-cooperative) can be combined to form a total of eight attitudes. As an illustration, we present three attitudes and show how they are constructed from the event-processing rules.

- Non-social (self-reliant/non-trusting/non-cooperative): An agent with a non-social attitude does not interact with other agents. It adopts its own observations, but messages from others are ignored. Therefore, the goal/task determination is free from external influences. The agent creates and selects its own goals and plans. Influence via environmental modification is still possible; it is possible to influence the agent's behavior by manipulating the environment. The following event-processing rules for messages and observations are active:

```
observation(X) <- TRUE | UpdateBeliefs(X)
message(Sender, inform, X) <- TRUE | IgnoreEvent()
message(Sender, request, X) <- TRUE | IgnoreEvent()
```

- Self-reliant/trusting/non-cooperative: A self-reliant, trusting agent will process messages from others and believe the content. Its beliefs are influenced by others, but it will not accept requests from others. The agent adopts its own observations, and determines its goals and tasks by itself, without taking any orders. We have implemented the *self-reliant, trusting* attitude by the following event-processing rules:

```
observation(X) <- TRUE | UpdateBeliefs(X)
message(Sender, inform, X) <- TRUE | UpdateBeliefs(X)
message(Sender, request, X) <- TRUE | IgnoreEvent()
```

- Self-reliant/non-trusting/cooperative: If agent A is cooperative with respect to agent B, it will do what agent B asks for, without considering other options. Its tasks and goals are determined by agent B. Agent B can send a request message to agent A. A processes the message by adding the request to its goal base. The agent adopts own observations in order to create an own world perspective, and will not adopt (trust) any observations from others. The following rules specify the self-reliant, cooperative attitude:

```
observation(X) <- TRUE | UpdateBeliefs(X)
message(S, inform, X) <- TRUE | IgnoreEvent()
message(S, request, X) <- TRUE | AddGoal(X)
```

These are just three examples of the eight possible attitudes. Of course, it is possible to construct more complex profiles by specifying the event-processing rules differently, for example, by specifying the situational constraints.

These constraints use local knowledge from the agent's internal state to permit or bar external events from influencing the agent's beliefs. Therefore, sensible knowledge should be used. These reasoning rules are effectively heuristics.

3.3 Meta-knowledge for Influence Control

What knowledge should be taken into account by the reasoning rules? We have explored several heuristics that seem appropriate to control external influences. One of the heuristics is relevance of information. If an agent can determine the relevance of information with respect to a certain goal, it can focus itself on a specific type of information, or prevent itself from information overload by filtering incoming information on relevance. Information relevance is important for influence control. Another related heuristic is the proverbial 'state of mind' of an agent. An agent might react differently when busy than free, or when its objectives are at stake. We cluster such heuristics as *self-knowledge*. Examples of event-processing rules using this type of knowledge are

```
observation(X) <- relevant(X) | UpdateBeliefs(X)
message(Sender, Performative, X) <- busy() | IgnoreEvent()
```

Self-knowledge creates heuristics for an agent to determine how it is influenced. The agent should also be able to control by whom it is influenced. The reasoning rules for event control can use knowledge about the existing organization or about the agent's social context. Can the sender of a message be trusted? Does a request originate from a superior or from an unfamiliar source? Agents can achieve coordination by allowing influence on the internal state based on *social and organizational knowledge*, for example:

```
message(S, inform, X) <- trusted(S) | UpdateBeliefs(X)
message(S, request, X) <- superior(S) | AddGoal(X)
```

One can think of several other reasons to allow or disallow influence on the internal state in a certain context. A specific coordination type puts requirements on the environment, such as availability of communication or information resources. Critical changes in the environment can be used to determine the proper level of autonomy and thus provide another important heuristic for influence control: *environmental knowledge*.

Table 2 summarizes the above types of knowledge that can be used in reasoning rules for influence control. This is obviously not an exhaustive list, but these three types of knowledge seem to capture the factors that affect agent autonomy most.

Table 2 Examples of meta-knowledge for influence control

Type of knowledge	Examples
Self knowledge	Is this information relevant for my objectives? Does my state of mind permit new requests?
Organizational/Social knowledge	What is my relation with information source? Can the source be trusted?
Environmental knowledge	Availability of communication Availability of information sources

When we use the meta-knowledge in the event-processing rules, we create agents with adaptive autonomy. It depends on the situation whether an agent allows influences on decision making or not and thus makes the agent locally responsible for its own behavior. Furthermore, it is a powerful feature to design dynamic coordination mechanisms, as we will see in the next section.

4 Using Adaptive Autonomy for Coordination

When designing a coordination mechanism, we use rules to specify desired collaborative patterns for all members and instruct them how they should exchange information and delegate tasks. This allows us to create different types of coordination strategies. For example, Dekker [13] presents a taxonomy of coordination mechanisms in networked organizations. He distinguishes two dimensions: homogeneity/heterogeneity of the members and value-symmetry/non-value-symmetry of the roles they perform. From these dimensions, he creates eight different coordination models by taking the extremes and a midpoint of both dimensions. The three main types are: *centralized control* (hetergeneous/non-value-symmetric), *request-based coordination* (heterogeneous/value-symmetric) and *emergent coordination* (homogeneous/value-symmetric). In a centralized control setting, the coordination of activities is done in a hierarchical manner with a central commander on top. In request-based coordination, there is no central command, but rather a service-oriented type of coordination strategy, where there is joint coordination among members. In a emergent settings, there is deliberate coordination among members, and the perception of coordinated behavior stems from the interaction of members. These types of coordination and all variations thereof can be implemented by using our reasoning rules, as presented earlier.

In agent research many work has been done on specifying organizational models, for example, OperA [8]. The organizational model describes the roles and relations between actors and specifies behavioral rules in terms of norms. It is constructed based on functional requirements of the organization. The behavioral guidelines for the roles are described in contracts. Taking up a role in an organization means that an agent is expected to act following the contracts.

OperA describes a coordination mechanism based on organizational structures at an abstract level. The organizational model is separated from the reasoning process of the agents that will fulfill the roles. Earlier work has shown that contracts specified in OperA can be translated to event-processing rules for the agent [14]. However, the specified behavioral rules in the organizational model are static, which has drawbacks for agility.

4.1 Agile Organizations

Static coordination mechanisms are usually best suited for a particular type of environment [15], [16]. Centralized strategies, for example, work best in familiar, well-structured situations, but fall short in situations where an organization needs to rely on improvisation and ad-hoc collaborations. On the other hand, decentralized approaches work best in unknown, dynamic situations, but take a lot more time than centralized strategies in familiar environments. In highly dynamic situations, it is impossible to select a single coordination strategy that fares well over the entire duration of an operation. The answer lies in the use of adaptive coordination strategies that change with the circumstances and thus make it possible for an organization to act in an agile manner.

There are two ways to achieve agility in an organization:

- *Top-down*: a new organizational model is defined, and the agents change their contracts with the organization. As a consequence, they adopt different reasoning rules for influence control, which will change the coordination arrangement.

- *Bottom-up*: the agents change the (priority of) reasoning rules for influence control by themselves if they notice that the organizational model fails. They adjust their autonomy to repair the organizational failure.

In the top-down manner, there is an appointed actor or procedure in place that produces a demand for structural change. Upon receiving this command, agents will commence to renegotiate their contracts to accommodate the new structure (i.e. a new line of command, or new organizational rules). In practice, this leads to a new set of reasoning rules for all agents concerned. In the bottom-up case, the agents themselves initiate the change. They have their own rules to estimate the situation and organizational performance and have the autonomy to self-adjust their behavior as to adopt a better-suited rules. In practice, they may do so by giving certain reasoning rules a higher priority over others. For instance, they may choose to give rules that imply self-government preference over rules that imply following commands from others.

In both situations, organizational adaptation implies changes in the reasoning process of individual agents. The model of adaptive autonomy presented here enables to achieve the required adaptivity in the reasoning process.

4.2 Example Application Scenario

We use a simple maritime NEC scenario to demonstrate the approach. A combined task force is in charge of ensuring a safe transit for a high value unit (HVU). The coalition exists four vessels from three nations. Each vessel has specific capabilities. Each nation has specific operational policies.

There are different ways to organize tasks, as shown in the C2 taxonomy of Dekker [13]. Additionally, there are other circumstances that require dynamic reorganization, such as a necessary switch between decentralized and centralized coordination, upsizing or downscaling of the organization and a change in the rules of engagement. Figure 3 illustrates some situations in which it is necessary to reorganize. The leftmost situation on the bottom row shows a change in coordination strategy, from a centralized to a decentralized approach. The middle figure shows the situation in which a new vessel is added to the organization. The right figure show the situation in which communication with the commanding vessel is lost and the remaining vessels are left to their own devices. We will discuss these cases in the following sections, from a bottom-up and a top-down perspective.

Bottom-up dynamics. First, we show bottom-up dynamics in the organization. We achieve this by changing the autonomy level of the decision-making process of the agents. We start with a hierarchical organization with one commander and three followers. We have implemented the followers with the following rules for event processing:

- IF request from commander THEN follow request
- IF I am in danger THEN ignore requests and follow own goals
- IF no communication THEN follow own goals

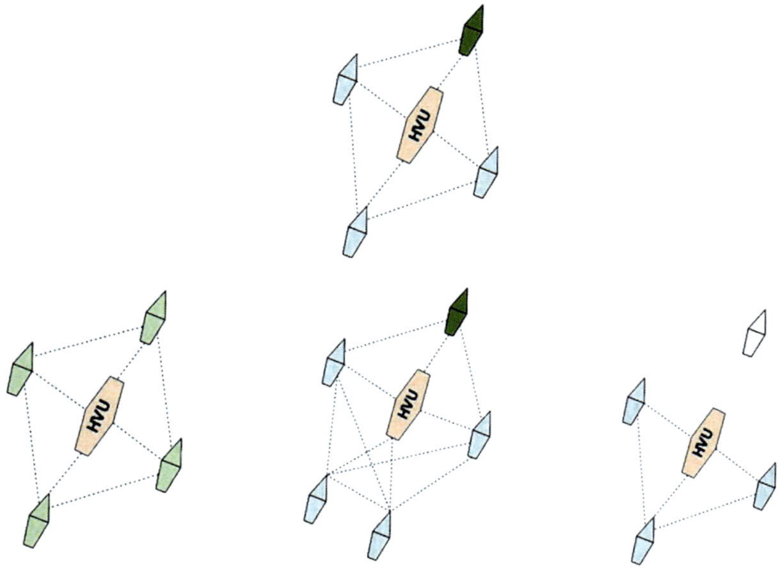

Fig. 3 Example situations in which reorganization is necessary. The top figure shows the default mode, with a single command ship (dark), three additional vessels (light) and a high value unit (HVU). The bottom row shows (a) decentralized command, (b) addition of a vessel and (c) loss of command. Lines denote communication paths.

The rules ensure that the agents follow the commands, but if communication fails, they will pursue their goal using local observations. Also, in case of danger the agent will take care of its own safety.

Under these rules, fleet members might be disobedient to the requests of the commander at certain times. When they feel that they are not in danger (based on their own local beliefs), and are in clear communication with the commander, they will follow the orders. In this manner, the organization effectively switches between different coordination mechanisms. Although the dynamics of the coordination mechanism have been designed in advance, the responsibility of the choice for coordination type is in hands of the actors themselves.

Top-down dynamics. Top-down dynamics can be achieved by carrying out structural changes. In our coordination model, it means that the contracts between the agents and the organization need to be changed. The modular approach in the agent's reasoning model provides a mechanism to adopt organizational rules into the decision-making process. This can be done dynamically by changing contracts at runtime [14]. For instance, the commander may need to put a new set of rules of engagement into effect. He can implement these rules by issuing new social contracts to all his fleet members. These contracts put the new rules of engagement into practice.

Let us describe some simple scenarios that show how the adaptive autonomy model facilitates agile behavior. We start with the hierarchical organization of the fleet. The vessels have adopted the corresponding organizational rules as event-processing rules. The rules specify a coordination mechanism in which the agents are obliged to inform the commander about a status change, and that they have to answer an extinguish request from the commander with an accept or reject message. The commander is the highest in the hierarchy. We assume that he can take the initiative to change the coordination process.

If, for some reason, the current coordination mechanism causes too much communication over the network, the commander can take the initiative to change the interaction protocol. The agents can decide on a new protocol that leaves the reply of the vessels out. The vessels translate the new interaction contract to event-processing rules and ends up with the same set of rules, but without the rule saying to explicitly answer the request. The new interaction protocol has of course an effect on the information circulation within the organization, and therewith possibly on the performance.

A more rigid example of reorganization can lead to a new role division. Imagine that the commander agent gets overloaded by too many tasks. It can assign the new commander role to a vessel agent to take over the coordination in a certain area. Furthermore, he informs other vessels that they get a new superior. The vessel agent becoming coordinator adopts the behavior rules belonging to the new role. The vessels who get a new superior update their beliefs according to the new situation.

These scenarios describe top-down dynamics in the organization. Coordination in the fleet is guided by contracts, and agile behavior results from adaptation of the contracts. The modular approach in the agent's reasoning model makes this possible, and still leaves room for the individual agent to implement its preferences.

5 Practical Application in NEC Environments

How does the above model translate to operational environments? The most evident use is in supporting coordination activities in multi-party environments. Agile and resilient behavior demands dynamic coordination capabilities. Task and resource allocation quickly becomes a demanding challenge in joint and combined NEC environments because of individual constraints and demands [1]. Artificial agents could support this process by acting as proxies: mediating representatives for all parties involved.

Such proxy agents could support mission planning and resource sharing, and make it easier to respect individual constraints and policies. Each force member can instruct its proxy agent by tuning the various attitudes to influences. For example, a unit may not be allowed to engage in offensive measures because of local rules of engagement, but may allow its resources to be used in defensive actions. It could instruct its representative agent accordingly by configuring its openness to external influences. The agent would decide to filter out requests for offensive capabilities, but join the coordination process when dealing with defensive goals. It would use its delegated autonomy to actively accept or refuse requests. In a hectic conflict, there may not be enough time to deal with such individual constraints or resolve

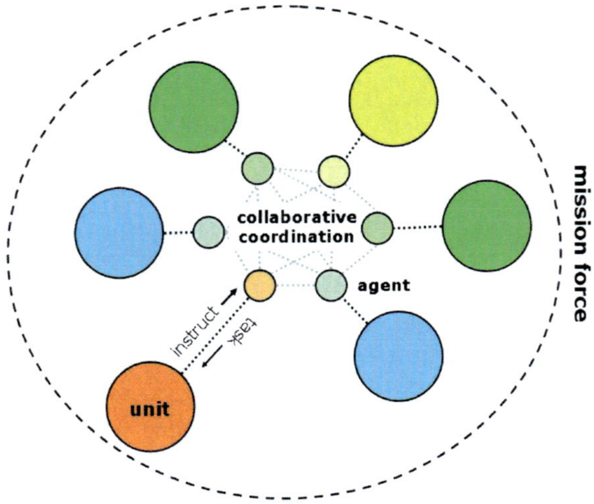

Fig. 4 Using mediating agents for collaborative coordination

potential conflicts through ordinary communication. Artificial agents may help to cope with the dynamics of multi-party collaborations and improve agility. When the organization changes structurally because of leadership changes or the arrival of extra units, the embedded organizational knowledge will need to adopt the new structure. Artificial agents can facilitate such processes and lessen the costs usually associated with updating organizational awareness [17].

Such autonomy-related configurations could be further facilitated by using meta-reasoning models. In a meta-reasoning model, the agent does not just reason about particular external events, but also about the relationship between various goals and information. This approach gives us a way to use prioritization of decisions and in effect construct 'attitudes'. For example, given a certain operational event, the agent's reasoning model may conclude that, based on internal knowledge and the agent's attitude, it prioritizes to force protection over self-defense. This means that the agent has received information that it is relevant for the success of two goals (self-defense and force protection), but that it actively chooses to let only the force-protection rule succeed. It blocks the information for the self-defense rule that would lead to retreat. Because of the modularity of the autonomy model, it is relatively easy to develop and implement such meta-reasoning models and their corresponding attitudes. In practice, such attitudes could serve as a way to implement doctrines.

6 Discussion

We have argued that an autonomous agent should control external influences on its decision-making process. Coordination between autonomous actors, then, can be achieved by explicitly allowing influences on the decision-making process. By making influence control a dynamic process, the agent can adjust its autonomy relative to others. Therewith, individual actors can play an important role to the agility of an organization. This way of relating autonomy and coordination is new in agent research. We believe it is a powerful basis to design autonomous actors for dynamic organizations and to model organizational performance.

Currently, organizational simulation in the NEC community is mainly done at an abstract level, for example, based on social simulations. Dekker [18] analyzes the network structure via the information flow within the network. However, no tasks are actually executed. We believe that simulations can be made more powerful when the individual behavior of the actors is implemented as well. Then, the organizational performance is dependent on both individual behavior and organizational structure. The taxonomy of coordination types [13] can perfectly be specified using OperA [8]. We have shown how autonomous agents can adopt the organizational rules. We include individual behavior in the organizational simulations.

Another approach to achieve dynamic coordination in group decisions is presented by Martin and Barber [9]. They propose several decision-making styles for groups ranging from command driven to locally autonomous. In between the two extremes there are several ways to achieve consensus about a decision, for example, via voting or a market-based approach. The best decision-making style depends on the situations. The researchers present a model where the agents switch from one decision-making style to the other. Their approach matches well to ours. The decision-making styles can be described in contracts of behavioral rules. When a decision-making style is chosen, the individual agents will follow the corresponding contract. Our addition is that we separate the group decision-making styles from the internals of the agent. Therewith, we can easily develop and implement new decision-making styles at runtime.

In section 5, we described an application of our work using a network of proxy agents that take up a part of the coordination within the network. Scerri et al. [19] use a similar approach in their coordination framework. In their work, the proxy network has the ability to transfer the control over a decision from one agent to the other. The agents have pre-planned transfer-of-control strategies using coordination constraints. For example, an agent should take a decision before a certain deadline, otherwise the decision-making control is passed to another agent. This way, the proxy network plays a part in the coordination of tasks. With respect to their work, our approach has a more broad perspective on coordination. Coordination in our context consists of influencing each other not only by passing on tasks but also by exchanging information. We believe that information sharing plays an important role to coordinate activities. In the organizational model that we use interaction between agents is part of the coordination mechanism, and belief influence is part of the reasoning model of the agents.

7 Conclusions

In this study, we briefly introduced our work on adaptive autonomy in agent systems. We have developed a decision-making model for artificial agents in which coordination can be defined in terms of organizational norms and rules, but that also guarantees autonomy. We have given the agent capabilities to self-adapt its openness to external influences, so it can change its own level of autonomy. We have distinguished several types of external influences that may impact decision making, such as environmental events, information relevance and organizational rules, and have given practical means to define attitudes and preferences.

We believe that our approach addresses some fundamental challenges in the progress toward higher NEC maturity levels. Agility and self-synchronization can only be achieved when participants in a NEC organization have adopted practical methods to manage their autonomy. We recognize the essence of having local autonomy, but we also recognize the necessity of coordinated activities. Our model shows that it is possible to relate autonomy and global coordination and define simple mechanisms that enable adaptive behavior. The underlying concept is relevant for understanding and facilitating task coordination in NEC environments. For instance, networked parties might interact through the use of mediating agents, that represent a party, and guards its autonomy. There will be many issues in implementing such a model, but it may inspire NEC developments and bring about new perspectives on autonomy in collaborative environments.

References

1. Alberts, D.S., Hayes, R.E.: Power to the Edge. CCRP Publication Series (2003)
2. Jennings, N.R.: On agent-based software engineering. Artificial Intelligence 117(2), 277–296 (2000)
3. Castelfranchi, C.: Guarantees for autonomy in cognitive agent architecture. In: Wooldridge, M.J., Jennings, N.R. (eds.) ECAI 1994 and ATAL 1994. LNCS (LNAI), vol. 890, pp. 56–70. Springer, Heidelberg (1995)
4. Bradshaw, J.M.: Dimensions of adjustable autonomy and mixed-initiative interaction. In: Nickles, M., Rovatsos, M., Weiss, G. (eds.) AUTONOMY 2003. LNCS (LNAI), vol. 2969, pp. 17–39. Springer, Heidelberg (2004)
5. Asimov, I.: Robot Visions. Visual Publications, New York (1990)
6. Zambonelli, F., Jennings, N., Wooldridge, M.: Developing multi-agent systems: the gaia methodolody. ACM Transactions on Software Engineering and Methodology (TOSEM) 12(3), 317–370 (2003)
7. Ferber, J., Gutknecht, O.: A meta-model for the analysis and design of organizations in multi-agent systems. In: Proceedings of the 3rd International Conference on Multi Agent Systems, p. 128. IEEE Computer Society, Los Alamitos (1998)
8. Dignum, V.: A Model for Organizational Interaction: Based on Agents, Founded in Logic. PhD Thesis, Universiteit Utrecht, The Netherlands (2004)
9. Martin, C.E., Barber, K.S.: Adaptive decision-making frameworks for dynamic multi-agent organizational change. Autonomous Agents and Multi-Agent Systems 13(3), 391–428 (2006)
10. Powers, W.T., Clark, R.K., McFarland, R.L.: A general feedback theory of human behavior: Part 1. Perceptual and Motor Skills (11), 71–88 (1960)

11. Farrell, P.S.E., Hollands, J.G., Taylor, M.M., Gamble, H.D.: Perceptual control and layered protocols in interface design: fundamental concepts. International Journal of Human-Computer Studies (50), 489–520 (1999)
12. van der Vecht, B., Meyer, J.-J.C., Dignum, F., Neef, M.: A dynamic coordination mechanism using adjustable autonomy. In: Sichman, J.S., Padget, J., Ossowski, S., Noriega, P. (eds.) COIN 2007. LNCS (LNAI), vol. 4870, pp. 83–96. Springer, Heidelberg (2008)
13. Dekker, A.: A taxonomy of network centric warfare architectures. In: SETE 2005 Systems Engineering, Test and Evaluation Conference, Brisbane, Australia (2005)
14. van der Vecht, B., Dignum, F., Meyer, J.J.C.: Autonomous agents adopting organizational rules. In: Dignum, V. (ed.) Multi-Agent Systems: Semantics and Dynamics of Organizational Models, ch. 13, pp. 314–333. IGI Global Press (2008)
15. Ghijsen, M., Jansweijer, W., Wielinga, B.: Towards a framework for agent coordination and reorganization, agentcore. In: Sichman, J.S., Padget, J., Ossowski, S., Noriega, P. (eds.) COIN 2007. LNCS (LNAI), vol. 4870, pp. 1–14. Springer, Heidelberg (2008)
16. Neef, M.: A taxonomy of human - agent team collaborations. In: Proceedings of the 18th BeNeLux Conference on Artificial Intelligence (BNAIC 2006), Namur, Belgium, pp. 245–250 (2006)
17. Oomes, A., Neef, R.: Scaling-up support for emergency response organizations. In: van der Walle, B., Carle, B. (eds.) Proceedings of the Second International Conference on Information Systems for Crisis Response and Management (ISCRAM 2005), pp. 29–41 (2005)
18. Dekker, A.: Agility in networked military systems: A simulation experiment. In: 11th International Command and Control Research and Technology Symposium (ICCRTS), Cambridge, UK (2006)
19. Tambe, M., Scerri, P., Pynadath, D.: Adjustable autonomy for the real world. Journal of Artificial Intelligence Research (17), 171–228 (2002)

Adaptive Hierarchical Multi-agent Organizations

Mattijs Ghijsen, Wouter N.H. Jansweijer, and Bob J. Wielinga

Abstract. In this chapter, we discuss the design of adaptive hierarchical organizations for multi-agent systems (MAS). Hierarchical organizations have a number of advantages such as their ability to handle complex problems and their scalability to large organizations. By introducing adaptivity in the structure of hierarchical MAS organizations, we enable agents to balance resources in their organization. We will first provide a number of generic principles for the design of hierarchical MAS organizations. We show how these principles are used to design three different hierarchical organizations for a search and rescue task in the RoboCupRescue simulation environment. The first two of these organizations are static, and the third is able to adapt its structure. An empirical study on the performance of these three organizations shows that the dynamic organization performs better than the two static organizations.

1 Introduction

For distributed intelligent systems, such as multi-agent systems (MAS), collaboration is essential for the system to achieve its goals. Collaboration between agents

Mattijs Ghijsen
Human Computer Studies Group, Informatics Institute, Universtiteit van Amsterdam, Science Park 107, 1098 XG Amsterdam, The Netherlands
e-mail: m.ghijsen@uva.nl

Wouter N.H. Jansweijer
Human Computer Studies Group, Informatics Institute, Universtiteit van Amsterdam, Science Park 107, 1098 XG Amsterdam, The Netherlands
e-mail: w.n.h.jansweijer@uva.nl

Bob J. Wielinga
Human Computer Studies Group, Informatics Institute, Universtiteit van Amsterdam, Science Park 107, 1098 XG Amsterdam, The Netherlands
e-mail: b.j.wielinga@uva.nl

R. Babuška & F.C.A. Groen (Eds.): Interactive Collaborative Information Systems, SCI 281, pp. 375–400.
springerlink.com

can be enforced in many different ways by imposing some form of organization on a MAS [12]. How a MAS organization can best be designed depends on the task the organization has to perform and the environment in which the organization is operating [26, 27]. When a MAS is operating in a dynamic environment, changes in the environment can cause the initial organization structure to become ineffective or inefficient. In our research, we study how MAS organizations can adapt themselves to changes in their environment.

Organization theory provides a number of concepts that can be used to design organizations for MAS [6]. The behavior of individual agents can be described by roles [29], and the behavior of a multi-agent system can be seen as the result of the behavior of agents enacting these roles and the interaction patterns between the agents. Paradigms from organizational theory as described by Mintzberg [22] have already been applied to the design of MAS [1, 16].

For the design of MAS, a range of design methodologies are available. Ferber and Gutknecht [5] proposed a model which links agents to roles and defines groups of agents that enact certain roles. The organization is defined by one or more organizational groups which can each have their own structure consisting of a set of roles, interaction patterns, and an interaction language. A more formal approach is the Opera methodology [3], which provides an extensive formal framework for designing organizations. In this framework, agents adopt a role by creating a contract which specifies the organizational rules for that role. The Opera approach ensures that an organization structure is made explicit for the agents and enables the agents to reason about organizational concepts. In the work by van Aart et al. [1], a number of building blocks are described that can be used to design distributed intelligent systems. Their framework consists of four components: an organizational model, a coordination model, an operational model, and a task model. The organizational model is composed of a number of units that contain agents or other units. Operator agents are responsible for performing operational tasks, and manager agents are responsible for coordinating activities between operator agents.

As stated by Levchuk et al. [18], several approaches can be adopted to enable a MAS to operate in dynamic environments. On one end of the spectrum, one can design an organization by making it robust, such that the organization is capable of handling all foreseeable scenarios. At the other end of the spectrum, one can design an organization capable of adapting its structure when faced with a scenario for which its current structure is not appropriate. The latter approach has been taken by Carley [2], who has shown that adaptive organizations tend to perform better than static organizations.

Glaser and Morignot [9] show how organizations can adapt themselves when agents join or leave the organizations. By means of a utility function the organization can decide whether to accept new agents and an agent can decide whether to join an organization. In [13], reorganization takes place by adapting the MOISE+ organization model. An organization manager may adapt the organization when an organizational event occurs. These events range from the entering or leaving of agents in the organization, agents achieving, abandoning or committing to goals, agents forming new groups to agents adopting or abandoning roles. The organization manager ensures that the behavior of the agents does not conflict with existing organizational constraints.

For some environments, it is not feasible to foresee every possible situation or the number of possible situations is so large that it becomes impossible to adopt the robustness approach by designing an organization capable of handling all these situations. Therefore, in our research, we adopt the approach of creating dynamic organizations. As an organization performs its task, it will gather knowledge about the state of its environment. We will use this knowledge to adapt the organizational structure when the current structure is no longer appropriate.

In this chapter, we focus on collaboration that is achieved by designing adaptive hierarchical MAS organizations. Hierarchical organizations have a number of advantages such as their ability to handle complex tasks. This ability of handling complex tasks lies in the fact that every agent in a hierarchy can decompose a task in a number of smaller tasks and assign these tasks to agents that are lower in the hierarchy. Using this mechanism of repeated decomposition, large and complex tasks can ultimately be decomposed into many small tasks that can be performed by individual agents. Hierarchical organizations also scale well to large organizations. Another advantage is that authority and responsibility of agents can be clearly defined in hierarchical organizations.

In this chapter, we show how we can use adaptivity in a hierarchical organization to balance workload and reduce the amount of information that is exchanged between agents. The driving force behind the adaptivity is an unbalance in workload among the agents. To resolve this unbalance, the organization structure is adapted such that the decision-making authority for a task is transferred to the agent that has the best knowledge for resolving the detected unbalance in workload.

The organization of this chapter is as follows. First, we will discuss how we design adaptive hierarchical multi-agent organizations. In Section 3, we describe three organizations we have developed for a search and rescue task in the RoboCupRescue simulation environment [15]. Section 4 describes an experiment in which we show how these organizations are capable of dealing with heterogeneous distribution of workload and limited communication. Related work is discussed in Section 5, and our conclusions are presented in Section 6. Directions for future research are discussed in Section 7.

2 Hierarchical MAS Organizations

For defining a MAS organization, one often distinguishes between the structure of the organization and the behavior of the organization. In our approach, we define the structure of a MAS organization using organizational roles enacted by agents and relations between those roles. The behavior of a MAS organization is defined by tasks performed by agents. Figure 1 gives an overview of the basic concepts of a MAS organization. An agent may know about other agents, is enacting one or more roles, is capable of performing one or more tasks and may be assigned to

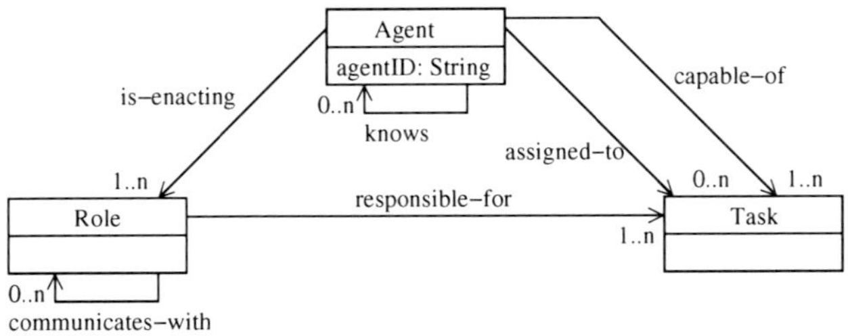

Fig. 1 Basic organizational concepts

perform one or more tasks. When an agent is capable of performing a task, it has the necessary skills and knowledge to perform that task.

An agent that is enacting a role is responsible for one or more tasks. Furthermore, a role may require an agent to communicate with other agents. An agent can only communicate with another agent if it knows that agent.

2.1 Organization Structure

On top of the basic concepts shown in Figure 1, we define a hierarchical organization by distinguishing between operator and manager roles (see Figure 2). This distinction between operator and manager roles is similar to van Aart [1], who distinguishes between technical and management activities in agent organizations. Between these two roles, authority relations exist. An agent with a manager role is the boss of one or more other agents (regardless of their role). Operator agents have one boss with a manager role. An agent with a manager role also has one boss, except when this is the agent in the top of the hierarchy. Using the authority relations, a tree-structured organization can be formed. As long as their capabilities permit and it is allowed by organizational rules, agents can enact both a manager role and operational role simultaneously.

When an authority relation between two agents exists, the superior agent has full control over its subordinate agent and the subordinate agent will obey all orders from its superior agent. A manager agent can use its subordinate agents to assign tasks, monitor task progress, allocate resources, etc.

Although authority relations are intuitively transitive, we do not allow for transitive authority relations. If we would allow this, agents in the top of the hierarchy would have direct control over all agents below them in the organization. Thus, agents would have multiple superior agents, which can result in conflicts when an agent receives orders from different superior agents. This approach of authority results in a limited scope of authority where an agent only has to obey orders from its direct superior and an agent only has full control over its direct subordinates.

Furthermore, in our definition of a hierarchical organization, an authority relation acts as a constraint on the `communicates-with` relation. In a hierarchical organization, we only allow for a `communicates-with` relation to exist between two roles if an authority relation (`subordinate-of` and `boss-of`)

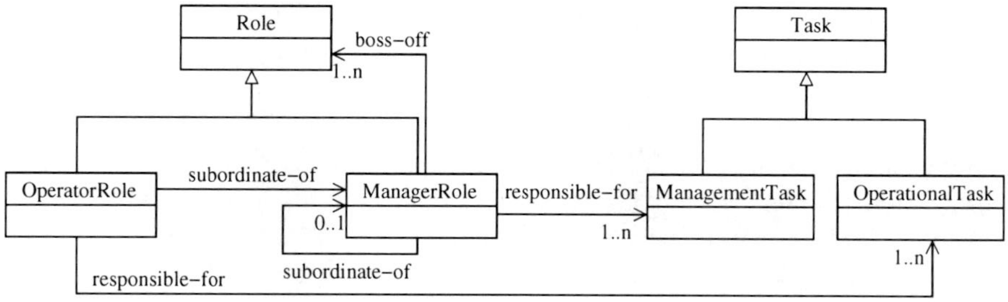

Fig. 2 Basic hierarchical concepts

exists between those two roles. By allowing agents only to communicate with each other if an authority relation exists, we enforce communication to follow the lines of the organizational structure. If one would allow for direct interaction between agents that do not have any authority relation, additional coordination effort is required to manage such interactions. Furthermore, allowing additional communication links could cause problems when scaling to larger organizations. Although we recognize that other less strict views on communication in hierarchical organizations are possible, we maintain this strict view in the remainder of this chapter.

2.2 Organization Behavior

We distinguish between operational and management tasks in hierarchical organizations. An agent with an operator role is responsible for performing operational tasks. An agent with a manager role is responsible for management tasks concerned with coordination of work within the organization.

Coordination in a hierarchical organization follows the principle of master-driven control. Orders to perform tasks or requests for information flow from the top of the hierarchy down to the agents that have to perform these tasks or gather the information. This information flows back to the top of the hierarchy. This information can then be used to update planning and to generate new task orders. The flow of command and information from the perspective of a single agent is illustrated in Figure 3.

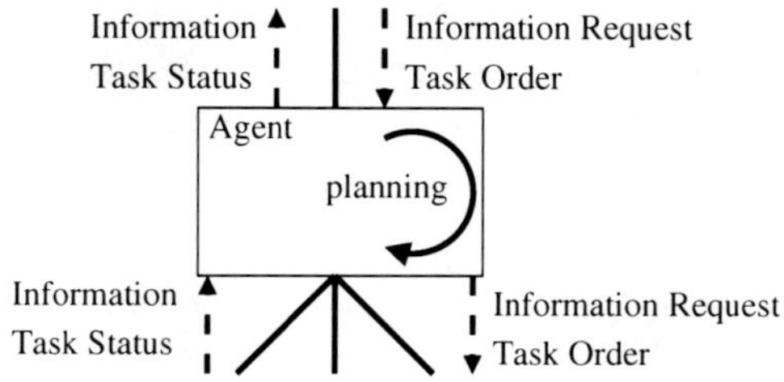

Fig. 3 Command and information flow

Two important types of tasks of a manager agent are task decomposition and task assignment. To perform these tasks well, a manager agent needs to have organizational awareness. For each of its subordinate agents, a manager agent needs to know the size of the organizational branch that each of its subordinates is responsible for and the capabilities that are present in these different branches of the organization. Furthermore, a manager agent needs to have an overview of the workload of its subordinates. Such an overview can be achieved by requesting status updates on each of the assigned tasks on a regular basis.

By decomposing a task into smaller and less complex subtasks and assigning subordinate agents to these subtasks, a hierarchical organization is able to perform large and complex tasks. Not only the tasks themselves are decomposed but also the decision-making authority on how to perform tasks is distributed over the organization. In our research, we adopt a simple approach to task decomposition and task assignment. More advanced techniques such as the GPGP/TÆMS approach [17] are outside the scope of the research in this chapter.

When an agent receives an order to perform a certain task, implicitly that agent is given the authority to make the decisions on how to perform that task. When the agent decides it cannot perform this task itself, it will try to find a subordinate agent that is capable of performing this task. If this fails, the agent will try to decompose the task into subtasks, such that one or more of its subordinates is able to perform these subtasks. If it cannot find any suitable decomposition, the task cannot be executed and the agent will send this information back to the agent that originally gave the order for that task.

When information is sent up into an organizational hierarchy, the amount of information that has to be processed by the agents increases as the agents are higher in the hierarchy. To avoid information overload, agents can use aggregation and abstraction of information.

Aggregation of information takes place when an agent receives information from multiple sources. The purpose of aggregation is to fuse data from multiple sources into a consistent view by removing duplicate information and resolving conflicting information [10, 19]. Data and information fusion is outside the scope of our research.

Abstraction of information can take place when an agent sends information to another agent. In a hierarchical organization, abstraction can be used to prevent a superior agent to become overloaded with information, especially when the superior agent receives information from multiple subordinates. In cases where communication is a problem, abstraction of information can also be used to reduce the load on the communication network. In such cases, a risk exists that essential information is not communicated.

2.3 Dynamic Hierarchies

As mentioned in [4], adaptivity in an agent organization can relate to the change of behavior or the change of the structure of the organization. In this chapter, we focus on the adaptation of the structure of a hierarchical organization and more specifically, the adaptation of roles and authority relations. Similar to the task decomposition and task assignment in a hierarchical organization, adaptation of the organization is also authority driven. This means that an agent can adapt the roles of its subordinate agents and the authority relations to and from its subordinate agents.

A possible task of an agent with a manager role in an adaptive organization is the adaptation of the part of the organization it is responsible for (not all managers should necessarily be allowed to adapt the organization). In general, the hierarchy can be adapted by

1. a manager (`m`) sending a message to a subordinate agent (the target agent `t`) which contains an order to adopt or abandon a role (`r`) (see Figure 4).

sender -> receiver(s)	**message content**
`m -> t`	`AdoptRole(t,r)`
`m -> t`	`AbandonRole(t,r)`

Fig. 4 Communication for changing roles

2. a manager agent (`m`) assigning one or more subordinate agents (the subject agents `[s]`) to work for another subordinate agent (the target agent `t`). First, the authority relations between the manager agent `m` and the subject agents `[s]` have to be removed. Then, authority relations between the target agent and the subject agents have to be created. In this type of adaptation, agents are moved toward the bottom of the hierarchy and additional layers can be added to the hierarchy (see Figure 5).

sender -> receiver(s)	**message content**
`m -> ([s],t)`	`Assign([s],t)`

Fig. 5 Communication for moving agents toward the bottom of the hierarchy

3. a manager agent (`m`) (re)claiming one or more agents (the subject agents `[s]`) from one of its direct subordinate agents (the target agent `t`). This is done in two steps. First, the manager agent sends a claim message to target agent `t` for the subject agents `[s]`. In the second step, the target agent `t` has to assign the subject agents `[s]` to the manager agent `m`. Using this type of adaptation, agents can be moved toward the top of the hierarchy and layers can be removed from the hierarchy (see Figure 6).

sender -> receiver(s)	**message content**
`m -> t`	`Claim([s],t)`
`t -> ([s],m)`	`Assign([s],m)`

Fig. 6 Communication for moving agents toward the top of the hierarchy

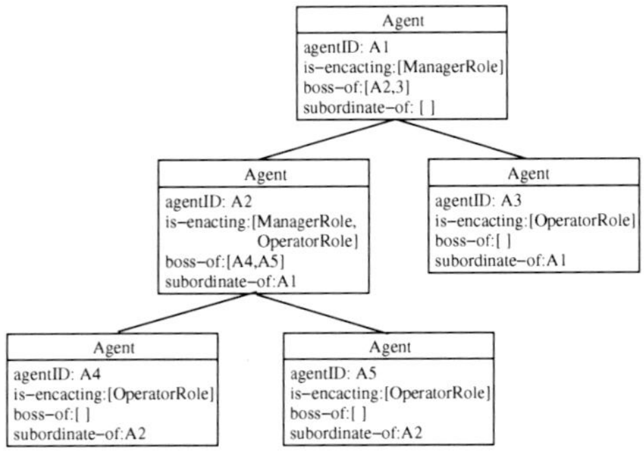

(a) Initial Hierarchical Organization

time	**sender -> receiver(s)**	**message content**
t	A1 -> A2	Claim([A4,A5])
t+1	A2 -> ([A4,A5],A1)	Assign([A4,A5],A1)
t+2	A1 -> A2	AbandonRole(A2,ManagerRole)

(b) Communication to transform the organization to the intermediate stage

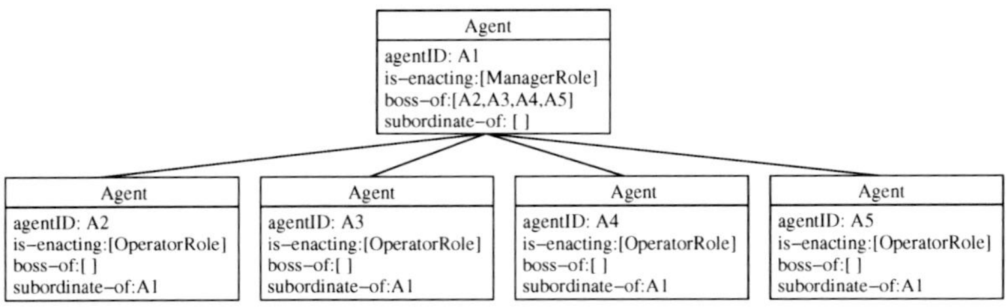

(c) Intermediate Hierarchical Organization

time	**sender -> receiver(s)**	**message content**
t+3	A1 -> A3	AdoptRole(A3,ManagerRole)
t+4	A1 -> ([A4,A5],A3)	Assign([A4,A5],A3)

(d) Communication to transform the organization to the target stage

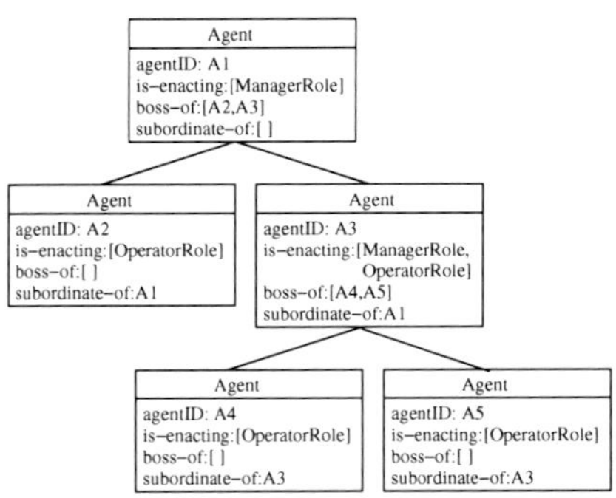

(e) Target Hierarchical Organization

Fig. 7 Example of an adaptive hierarchy

In a claim message, a manager agent can provide specific agent identifiers for [s] but instead it can also indicate the amount of agents it requires or a number of capabilities that are required.

Using these three forms of adaptation, agents can change the structure of the organization while the organization retains its hierarchical form. These changes only require communication between agents that have an authority relation. Furthermore, the organizational adaptation described above assumes that agents have accurate organizational awareness. Depending on the domain in which the organization operates, this may or may not be a realistic assumption. Thus, when errors in organizational awareness are likely to occur, additional measures for failure in reorganization will have to be taken. This issue will not be addressed in this chapter.

To illustrate the process of organizational change, consider the following example. Figure 7(a) shows an initial hierarchical organization structure. The lines between the agents depict authority relations between the agents. The agent in the top of the hierarchy (A1) has a manager role and has two subordinate agents A2 and A3. Both subordinate agents have an operator role and agent A2 also has a manager role and it has two subordinate agents (A4 and A5).

Suppose at some moment in time *t* the organizational structure is in its initial state as shown in Figure 7(a). The manager agent A1 wants to assign a new task to agent A3 but in order for agent A3 to be successful it will require the help of two more agents. At the same time, agent A2 is working on a less complex task which it can perform by itself. As a result, agent A1 decides to reorganize by claiming agents A4 and A5 from agent A2 and assigning these agents to work for agent A3. This also means that agent A2 will have to abandon its manager role. The organizational change will consist of two steps. In the first step, we move from the initial organization structure in Figure 7(a) to the intermediate organization structure shown in Figure 7(c). In the second step, we move from the intermediate structure to the target organizational structure shown in Figure 7(e).

To transform the organization to the intermediate stage, communication will take place as shown in Figure 7(b). When these actions have been completed successfully, the organizational structure will be as shown in Figure 7(c). This is an intermediate structure in which agent A1 has gained authority over agents A4 and A5. This intermediate structure is a necessary step in the reorganization process because agent A1 needs to have direct control over A4 and A5 and be able to communicate with them and to assign A4 and A5 to work for agent A3.

To move from the intermediate structure to the target structure, agent A1 will first tell agent A3 to adopt a manager role and then assign agents A4 and A5 to agent A3 (see Figure 7(d). This will result in the following communication and the organizational structure shown in Figure 7(e).

3 Hierarchical Search and Rescue Organizations

In this section, we discuss the design of three multi-agent organizations that perform a search and rescue task. The aim of this section is to illustrate how the theory presented in Section 2 can be applied to design hierarchical agent organizations. In Section 4, these three organizations are used in a number of experiments in which we show how these organizations are affected by an uneven distribution of workload and/or limited communication.

3.1 RoboCupRescue

In our experiments, we use the RoboCupRescue[1] (RCR) simulator [15]. In this simulation environment, agents are deployed that jointly perform a rescue operation. When a simulation starts, buildings collapse, civilians get injured and buried under the debris, buildings catch fire, and fires spread to neighboring buildings. Debris of the collapsed buildings falls on the roads causing roads to be blocked. For this rescue operation, three main tasks can be distinguished and for each of these tasks, a type of agent with appropriate capabilities is available. Fires are extinguished by fire brigade teams, blocked roads are cleared by police agents, and injured civilians are rescued by ambulance teams.

In this study, we only use AmbulanceTeam agents and an AmbulanceCenter agent and focus on the search and rescue task. An AmbulanceTeam agent is able to move around the city and walk into buildings to see if there are any injured civilians. AmbulanceTeam agents have limited observability. When an AmbulanceTeam agent is on the same position as an injured civilian, it can perform dig actions to remove the debris. The amount of digging effort that is required depends on the burriedness of the civilian. As long as a civilian is covered with debris its health state will decrease. The rate of health decrease is different for each civilian and it correlates to the burriedness of the civilian. When all debris has been removed, a civilian can be put into the ambulance and transported to a hospital. An AmbulanceTeam agent can communicate directly with another AmbulanceTeam agent when they are within each others hearing distance. An AmbulanceTeam agent can also send a message to the AmbulanceCenter agent.

An AmbulanceCenter agent cannot move or observe its environment. It only acts as a relay station for communication. It can receive messages from each of the AmbulanceTeam agents and it can send a broadcast message which will be received by all the AmbulanceTeam agents. As shown in the previous section, for our approach to reorganization, agents need to be able to send directed messages to each other instead of broadcasts. Because the RCR environment does not support direct communication between agents, we have enabled our agents to communicate via a communication service outside the RCR simulator, which supports unicast and multicast messages.

3.2 Organization Design

The organizations discussed in this section consist of one AmbulanceCenter agent and nine AmbulanceTeam agents in which the AmbulanceCenter enacts a manager

[1] In our work, we have used RoboCupRescue version 0.48.

role and the AmbulanceTeam agents enact an operator role. The coordination styles that are used in these organizations are inspired by the five organization types described by Mintzberg [22]. The first organization we describe is based on small organizations which consist of a few number of people. In such organizations, a manager can use *direct supervision* to coordinate work between the operator agents. This coordination style is characterized by a manager who delegates small tasks and keeps a detailed overview of its subordinates and their environment. For search and rescue, a manager agent can use this style of coordination to tell other agents where to search and where and when to rescue a civilian.

As organizations grow larger and tasks get more complex, a manager using direct supervision will become overloaded. For larger organizations, Mintzberg describes a machine bureaucracy (e.g., a car factory) and a professional bureaucracy (e.g., a hospital or university). In the machine bureaucracy, *standardization of work processes* is used to coordinate work. This type of coordination is suitable for routine work in large volumes in predictable and static environments. The search and rescue tasks in RoboCupRescue contain many uncertain factors such as the location and health state of injured civilians. As such, this organization style is not suited for a search and rescue task.

The professional bureaucracy, however, is more suited for search and rescue, and we have used this organization style for the design of both our second and third search and rescue organizations. In a professional bureaucracy, *standardization of skills* is used as a coordination style. In this coordination style, a manager relies on the high-level of skills of its operator agents. The operator agents have a large amount of autonomy in their decision making on how to perform their tasks. For the search and rescue task, the area to be searched can be divided by a manager agent into smaller areas and each operator agent can use its skills to search and rescue civilians in these areas.

A division organization is used for very large organizations that perform a wide variety of tasks and is, therefore, not suited for the design of a single-purpose organization such as a search and rescue organization. The adhocracy uses *mutual adjustment* to coordinate work and is capable of handling complex tasks in dynamic environments. Therefore, it is highly suited to perform a search and rescue task. However, mutual adjustment is not based on hierarchical structures and, therefore, we leave this type of coordination style outside the scope of the research in this chapter. We do, however, recognize the potential of mutual adjustment for future research.

3.2.1 Static: Direct Supervision

In this organization, the AmbulanceCenter agent adopts a direct supervision coordination style. We first define the main task for the organization which is the task of the search and rescue of all civilians on a map in the RCR environment. The manager then decomposes this task into a large number of small search tasks for housing blocks (SearchBlock). For each civilian that is found, a RescueCivilian task is generated. Figure 8 shows an example of a task decomposition for direct supervision.

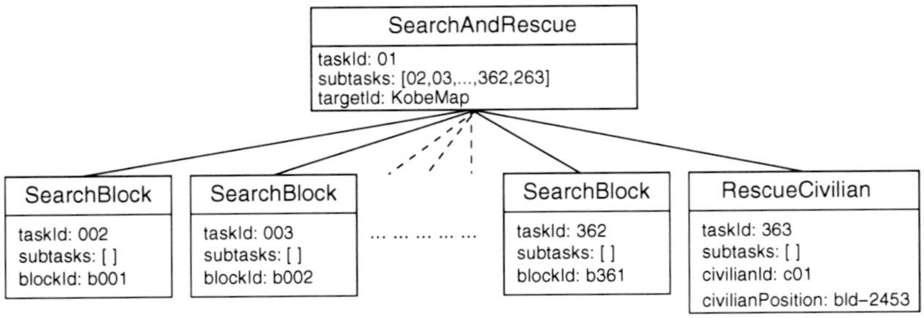

Fig. 8 Direct Supervision Task Decomposition

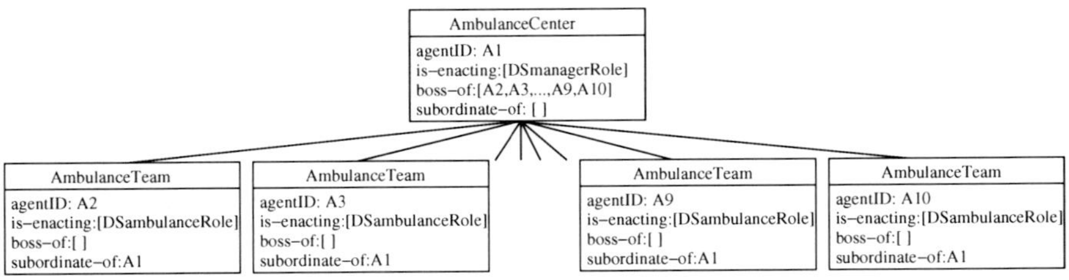

Fig. 9 Direct Supervision Organization

In this example, the SearchAndRescue task for the city of Kobe is decomposed in SearchBlock tasks for all 361 housing blocks and one RescueCivilian task for a civilian c01 which is located in building b2453. For this organization and the other organizations presented in this section, we assume that agents will always perform a task and will never cease task execution.

Besides the operational tasks (SearchAndRescue, SearchBlock, and Rescue-Civilian), we define the following management tasks for this organization: a DS-decomposeSRtask[2] task and an AssignTasks task. An agent that performs the DSdecomposeSRtask task will decompose a SearchAndRescue task into subtasks as explained in the example above. An agent that performs the AssignTasks task will first prioritize all the subtasks and then assign the task with the highest priority to an agent that has yet no task assigned. The priority of RescueCivilian tasks is based on a civilian's health status. As long as the injury of the civilian is not critical, the priority of the RescueCivilian task is lower than the SearchBlock tasks. As a civilian becomes more injured, the priority of the RescueCivilian task becomes larger than the SearchBlock tasks. By adjusting the priority of RescueCivlian tasks, the agents search the map as fast as possible but prevent civilians from dying. When multiple tasks have equal priority, the task that is closest to an agent will be assigned to that agent.

Using these tasks described above, we can now define the roles for the agents in this organization as shown in Figure 10.

---

[2] DS stands for Direct Supervision and SR stands for Search and Rescue.

```
DSmanagerRole:    [DSdecomposeSRtask,
                   AssignTasks]
DSambulanceRole: [SearchBlock,RescueCivilian]
```

Fig. 10 Responsible Tasks for Direct Supervision Roles

These roles are used to define the organization structure as shown in Figure 9. The ambulance center agent enacts the DSmanagerRole while all the ambulance agents enact the DSambulanceRole and are subordinate to the ambulance center agent.

The agents in this organization as well as the organizations in the following sections have been implemented using the AgentCoRe framework [8]. This framework provides decision-making processes that enable agents to decompose tasks, assign tasks, and reorganize.

3.2.2 Static: Standardization of Skills

In this second organization, work is coordinated by standardization of skills. The main search and rescue task is decomposed by the manager agent into large subtasks as shown in Figure 11. For this decomposition, the map is divided into nine equally large sectors, and a search and rescue task is created for each of these sectors. The manager will rely on the skills of the ambulance agents to search and rescue all civilians in their sector as fast as possible. The strategy that is used by the AmbulanceTeam agents is similar to the strategy used by the manager agent in the first organization: try to search as fast as possible and only start to rescue a civilian when its health reaches a critical state. The assignment procedure assumes equal priority to each sector and assigns each sector to the closest ambulance agent.

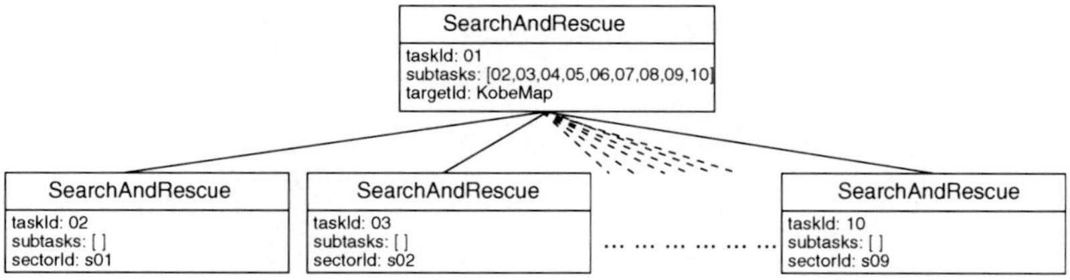

Fig. 11 Standardization of Skills Task Decomposition

For the task decomposition shown in Figure 11, we define the STdecomposeSRtask[3] task. We can now define the two roles as shown in Figure 12.

These roles are used to define the organizational structure as shown in Figure 13. The ambulance center agent enacts the STmanagerRole, and all ambulance agents enact the STambulanceRole.

[3] ST stands for STandardization.

```
STmanagerRole:    [STdecomposeSRtask,AssignTasks]
STambulanceRole: [SearchAndRescue]
```

Fig. 12 Responsible Tasks for Static Standardization of Skills Roles

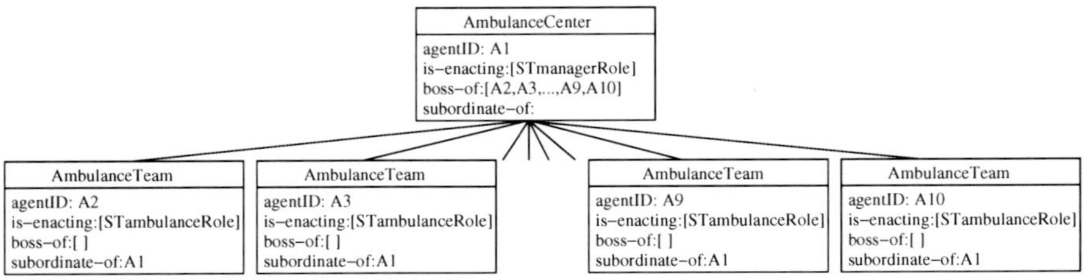

Fig. 13 Static Standardization of Skills Organization

3.2.3 Dynamic: Standardization of Skills

The third organization uses the same approach to task decomposition and task assignment as the static standardization of skills organization. However, in this organization, we added dynamics in the hierarchy by creating an additional AdaptOrganization task. Using this task, we can define the roles for the dynamic organization as shown in Figure 14 below.[4] The initial organization structure is the same as shown for the static organization in Figure 13; only the AmbulanceCenter agent will now enact a DSTmanagerRole and the AmbulanceTeam agents will enact the DSTambulanceRole. We start with an initial 2-layered organization because the agents have no prior knowledge about the distribution of workload.

```
DSTmanagerRole:    [STdecomposeSRtask,AssignTasks,
                    AdaptOrganization]
DSTambulanceRole: [SearchAndRescue]
```

Fig. 14 Responsible Tasks for Dynamic Standardization of Skills Roles

When an agent with a DSTmanagerRole is performing the AdaptOrganization task, organizational change is triggered when an ambulance agent reports that it has successfully finished its SearchAndRescue task while there are other ambulance agents still working on their SearchAndRescue tasks. Based on the workload of the other subordinate agents, the agent that is now idle will be assigned to work for the agent with the highest workload. Workload of an agent is determined by the number of civilians that have to be rescued in the sector the agent is working in and the total number of ambulances working in that same sector. The agent with the highest workload is first ordered to adopt a DSTmanagerRole (only if it does not already have this role). Then, it will receive the ambulance agent that was idle and

[4] DST stands for Dynamic Standardization of Skills.

this agent will become its subordinate. We will refer to an ambulance agent with an operational and a manager role as a local manager.

The decision by the manager to give an ambulance agent the responsibility to coordinate work on the search and rescue task for a specific sector is based on the following heuristic. The agent that has been working on a task for the longest time will be the agent that has acquired the most knowledge that is relevant for performing that task. Because this agent has the best knowledge for that specific task, it is also the most suitable agent to coordinate work on that task. By the transfer of decision-making authority of a part of the overall search and rescue task, the manager agent prevents the exchange of large amounts of information that it would have needed to do the decision making by itself.

Because a local manager is also allowed to change the organization of its subordinates, the same process of organizational change will continue to occur at the lower levels of the hierarchy. Search and rescue tasks with a large workload will be cut into smaller search and rescue tasks and divided among subordinate agents. Because a local manager also has an operational role, it will include itself in the task assignment process. Over time, this mechanism will cause the organization structure to form large organizational branches under the agents with a large workload.

When a local manager transfers part of its search and rescue task to another ambulance agent, it will include detailed information that is relevant for the other ambulance agent. This information includes buildings that have already been searched and locations and health status of civilians that have already been found but have yet to be rescued. This prevents buildings to be searched twice and ensures that civilians will be rescued in time. Although a large amount of information may be exchanged, it is unlikely that this will cause information overload at the receiving agent. This is because the information exchange will occur only once and the direction of information flow is toward the bottom of the organization. Information overload is more likely to occur when a continuous stream of information flows to the top of the organization.

4 Experiment

The goal of this experiment is to study the performance of the three search and rescue organizations presented in the previous section. These organizations are faced with two different challenges in their environment, heterogeneous distribution of workload and limited communication. Our experiment consists of three parts. First, we will run a number of simulations that show how performance of these organizations is influenced by the distribution of workload. A second series of simulations shows the effects of limiting the amount of information that can be exchanged between agents. The third series of simulations shows how the organizations perform when faced with both problems simultaneously.

4.1 Distribution of Workload

Depending on the environment and the type of coordination mechanism used in an organization, the assignment of tasks of different size can cause an unbalance in the distribution of workload among the agents. An unbalanced workload among agents will result in some agents working on their tasks while others finish their tasks early and remain idle. In the RoboCupRescue environment, the distribution of civilians on the map can cause agents to be assigned with tasks of varying workload. We will show the effect of this distribution of civilians on the three organizations discussed in the previous section. Performance of these organizations is measured under two conditions. In the first condition, civilians are evenly distributed over the search area. In the second condition, civilians are clustered in three groups.

On the Kobe map, we created 5 different homogeneous distributions of civilians and 5 different heterogeneous distribution. Each distribution contains 20 civilians that have to be rescued. The initial positions of the ambulance agents are the same for each distribution. Figure 15 shows examples of both distributions.

We compare organizational performance to a baseline performance for each organization. This baseline was determined by running one simulation using that organization on each homogeneous map. Data for each organization in the heterogeneous condition was gathered by running one simulation on each heterogeneous map. Performance was measured by taking the average number of civilians that were rescued successfully when the simulations finished after 300 time steps. The results in Figure 16 show the baseline performance and the performance on the heterogeneous maps.

The baseline shows that, on average, direct supervision performs worse than standardization and the adaptive hierarchy. This can be explained by the interaction pattern used in direct supervision. Every time an ambulance agent indicates it has completed its task, it will take some time before the manager assigns a new task. In the time that ambulance agents are waiting for new tasks, they remain idle. In the

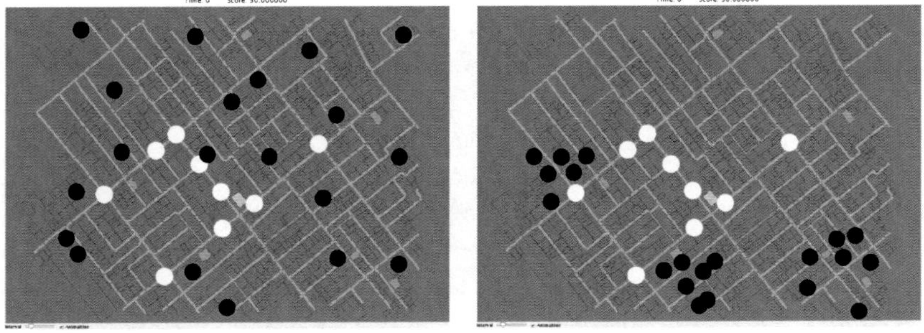

(a) Homogeneous Distribution (b) Heterogeneous Distribution

Fig. 15 Distribution of civilians on the Kobe map. White dots are AmbulanceTeam agents, and black dots are injured civilians.

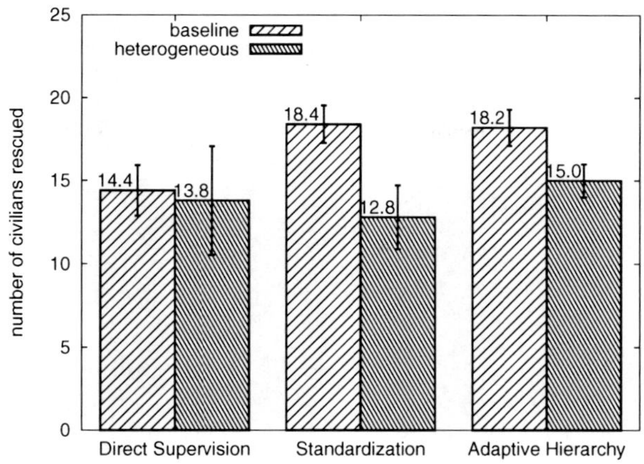

Fig. 16 Effects of heterogeneous task distribution compared to baseline performance

other two organizations, ambulance agents have larger tasks and, therefore, have to spend less time waiting for new task assignments.

A comparison between the baseline condition and the heterogeneous condition shows that a heterogeneous distribution of civilians has a minor effect on direct supervision (4% decrease compared to the baseline performance), while the decrease compared to the baseline for standardization and the adaptive hierarchy is considerably larger (30% and 18% decrease compared to the baseline, respectively). Statistical analysis[5] indicates that the decrease in performance for direct supervision is not significant. The decrease in performance for standardization and the adaptive hierarchy was found to be significant (both with $p < 0.01$). Direct supervision benefits from its task decomposition into fine-grained subtasks by the manager agent. Because these subtasks are small, differences in workload also become small and, therefore, it is easier for the manager to balance workload among the ambulance agents.

When using standardization, task decomposition results in large subtasks that are assigned to the ambulance agents. Because these tasks are larger in size, the differences in workload can potentially also be larger. When civilians are clustered, this will result in a large workload for the sectors on the map that contain these clusters while other sectors contain no civilians at all. Standardization has no mechanism to handle these differences in workload, hence the large decrease in performance.

In the case of the adaptive hierarchy, task decomposition also results in large tasks which causes problems in the case of heterogeneous distribution of civilians. However, because of its flexibility, the adaptive hierarchy is able to mitigate some of the negative effect caused by this heterogeneity. During a simulation on a heterogeneous map, this will result in the assignment of ambulance agents with a low

[5] For each organization we used a t-test to compare the baseline with the heterogeneous maps. The reported p values are two-tailed.

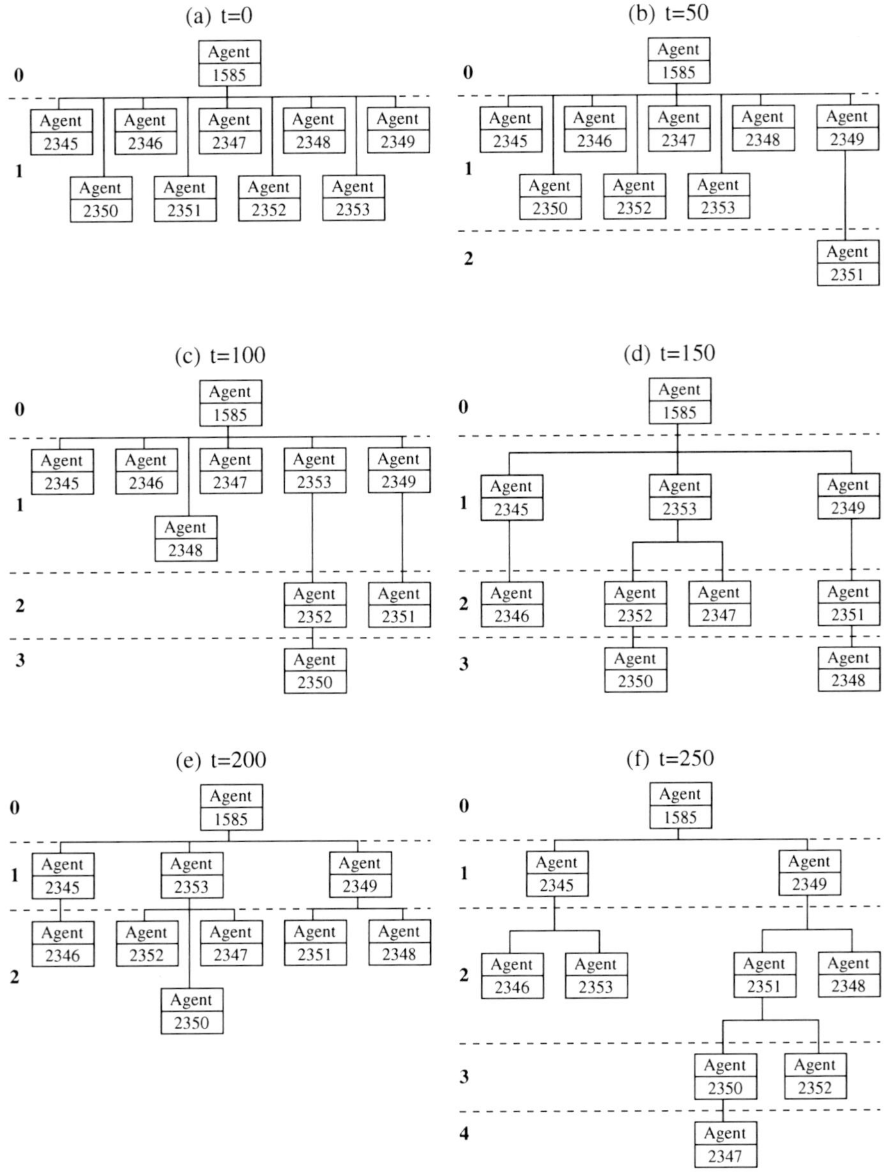

Fig. 17 Evolving organizational structure over time

workload to ambulance agents with a high workload. Figure 17 shows an example of an evolving organizational structure during a simulation on a heterogeneous map. What we see is that the structure of the organization adapts itself to the workload and over time, the hierarchy forms deeper organizational branches under agents with a high workload.

4.2 Limited Communication

The amount of information that can be exchanged between agents in an organization can have a large influence on the performance of an organization. In Section 4.1, we saw that direct supervision is well capable of balancing workload among its agents. However, to accomplish this, it requires a lot of information to be sent from and to the manager agent.

In this part of the experiment, we show the effect of restricting the amount of communication by setting a threshold to the amount of information an agent can receive over time. We run this series of experiments using two conditions. The first condition is the baseline performance in which no threshold is set on the amount of information that can be exchanged. In the second condition, agents have to keep their communication below the threshold. This means that agents that would send large amounts of information to their superior agent in the baseline condition will omit some information to prevent this superior to become overloaded by the information received from its subordinates.

In the case of direct supervision, we omit the health status of civilians from their status reports. This is shown in Figures 18 and 19, which show examples of a full health report and an reduced health report that are sent by an ambulance to the manager agent. The full health report contains the id of a civilian, the id of the building it is located in, its current health state (indicated by "hp"), the decrease of its health state (indicated by "damage"), the amount of effort required to extract the civilian from the debris (indicated by "burriedness"), and whether the civilian has already been transported to a hospital (indicated by "isrescued"). In the reduced form, all information is omitted that is not essential for the manager to generate rescue tasks. Thus, a reduced health report only indicates that there is some injured civilian in the indicated building. Using this information, the manager agent can send an ambulance agent to that building for the rescue operation. Although the manager is still able to generate rescue tasks for injured civilians, it will no longer be able to prioritize the rescue tasks because the necessary information is unknown to the manager.

In the cases of standardization and the adaptive hierarchy, agents keep their communication well below the threshold and thus they will not have to adapt their communication to stay below this threshold.

The baseline data is the same as in the previous experiment. For the limited communication condition, we ran one simulation on each homogeneous map for each organization. The results in Fig. 20 show a large decrease in performance (31% compared to the baseline) when agents use direct supervision and can no longer include health reports in their status updates to the manager agent. This decrease in performance was found to be significant with $p < 0.01$. The manager agent who decides on the priority of each of the civilians can no longer make good decisions about when to rescue civilians if their health state is unknown to the manager agent. Thus, when receiving incomplete information, the quality of the decision making decreases and less civilians are rescued.

```
<rdf:Description rdf:about="http://www.uva.nl/rcrc#c2330">
    <rdf:type rdf:resource="http://www.uva.nl/rcrc#Civilian"/>
    <rcr:id>2330</rcr:id>
    <rcr:position>1849</rcr:position>
    <rcr:hp>5992</rcr:hp>
    <rcr:damage>83</rcr:damage>
    <rcr:burriedness>60</rcr:burriedness>
    <rcr:isrescued>false</rcr:isrescued>
  </rdf:Description>
```

Fig. 18 Full health report

```
<rdf:Description rdf:about="http://www.uva.nl/rcrc#c2330">
    <rdf:type rdf:resource="http://www.uva.nl/rcrc#Civilian"/>
    <rcr:position>1849</rcr:position>
  </rdf:Description>
```

Fig. 19 Reduced health report

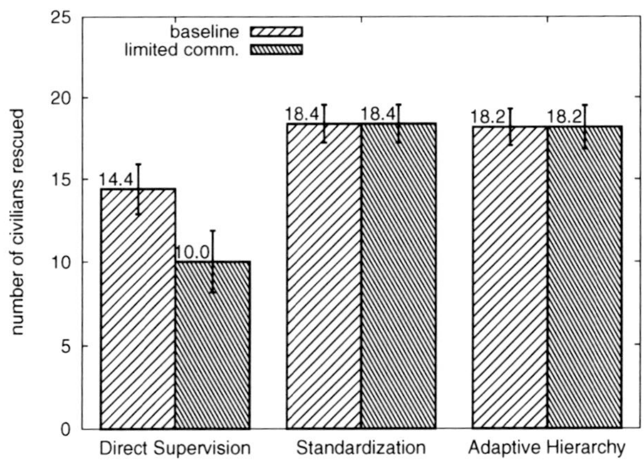

Fig. 20 Effect of limited communication

Standardization and the adaptive hierarchy are not affected by this threshold because of the small amount of information communicated by the agents. This is due to the standardization of skills mechanism that is used to coordinate work in these two organizations. Using this coordination mechanism, a manager agent requires only the number of houses that have to be searched in a sector and the number of civilians that have to be rescued in that sector.

These results show that a direct supervision approach is vulnerable when the amount of communication between agents is limited. This is because not all information is communicated, which reduces the quality of decision making by a manager agent. Thus, in cases where communication is limited, coordination by

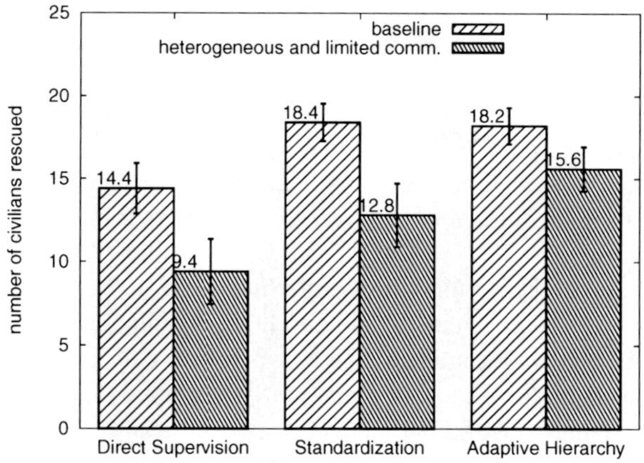

Fig. 21 Effect of heterogeneous workload and limited communication

standardization of skills would be more suitable because it requires less detailed information to be sent to a manager agent.

4.3 Heterogeneous Workload and Limited Communication

In this series of simulations, we look at the performance of the organizations when faced with both heterogeneous distribution of workload and limited communication. For each organization, we ran one simulation on a homogeneous map with unlimited communication and one simulation on a heterogeneous map with limited communication.

The results shown in Figure 21 show that the performance of the adaptive hierarchy is the least affected by these two problems. The decrease of performance of the adaptive hierarchy compared to the homogeneous map using unlimited communication is 13%. This is considerably less than direct supervision (31% decrease) and standardization (30% decrease). All differences in performance are significant with $p < 0.01$. Further statistical analysis (two way independent ANOVA) shows no significant interaction effects between heterogeneity of task distribution and limited communication. This is understandable because communication is only influenced by the amount of civilians, not by how they are distributed.

5 Related Work

So and Durfee [26, 27] present a performance model for hierarchical organizations. In this model, they show how span of control (i.e., the amount of agents controlled by an agent) influences the response time of a hierarchical organization. They show for which level of task granularity what the most appropriate span of control in the hierarchy should be and vice-versa. Their performance model can be used by agents to decide whether and how to reorganize based on the granularity of their task decomposition and the span of control in the organization. The work by So and Durfee does not show the actual adaptation of the organizational structure. Another difference with our work is that they assume a complete overview of how a task is decomposed in an organization and full knowledge about the structure of an organization. In our approach, a manager agent changes the organization based on the local state of the organization and how workload is divided in that specific part of the organization.

In the area of distributed problem solving, Shehory et al. [25] use agent cloning to balance workload among agents in a hierarchical organization. When an agent is overloaded, it will clone itself and hand over some of the task load to the new agent. This same approach to reorganization has recently been used and extended by Kamboj et al. [14] to balance workload and adapt task allocation, also for distributed problem solving. Agent cloning is a useful method of adaptation in the area of distributed problem solving; but for domains where agents cannot be duplicated (such as the search and rescue domain), this approach is not feasible and other methods of adaptation have to be considered.

Mathieu et al. [21] first describe strictly hierarchical organizations in which all communication flow follows the structure of the hierarchy, and then adapt this organization by creating additional relations between agents. By creating these relations, shortcuts for communication are created that enable faster communication and improve performance. However, by circumventing the standard communication paths in a hierarchy, additional coordination is required to manage these dependencies that fall outside the organizational hierarchy. Furthermore, this requires agents to have knowledge about agents and their capabilities in other parts of the organization. We believe that for large hierarchies, maintaining an overview of the organization and managing additional communication paths is difficult. Instead, by adapting the hierarchical structure itself using only local knowledge, we believe our approach will scale better to larger organization structures.

In van der Vecht et al. [28], the concept of adjustable autonomy [23, 24] is used to adapt agent behavior in a hierarchical organization. In this approach, hierarchical relations in the organization are not changed, only the behavior of the individual agents is changed by adapting the level of autonomy the agents have. The difference with our approach is that in [28], the subordinate agents decide for themselves to increase their autonomy and they will only do this to handle exceptional events that occur in the environment. When such an event has been dealt with, the agents will resume their original level of autonomy. Adjustable autonomy is also used by Martin and Barber [20] to adapt the organization. In this approach, not only the autonomy is adapted but also the relations between agents. Different types of coordination mechanisms are considered such as master driven coordination for hierarchical organizations and consensus based coordination for decentralized organizations. Adaptation of the organization in [20] means that the complete organizational structure is changed. This can be seen as a rather abrupt change in organizational structure while in our approach, changes are more subtle and gradual over time, as shown in our evolving organizational structure in figure 17.

The approach taken by Martin and Barber seems especially suitable for adapting to large changes in the environment such as communication failures (as shown in [7]), while our approach is better suited for responding to relative small changes in the environment.

Hoogendoorn et al. [11] present a formal model of organizations and organizational change. This model is based on how human organizations tend to adapt their organization structure and it describes several phases in organizational change such as unfreezing, movement, and refreezing of the organization. In the unfreezing phase, it is decided between the agents whether to start changing the organization. The change of the organization takes place during the movement phase, and in the refreezing phase, the organization checks if the movement was completed successfully and if the behavior of the agents is as desired. When the correct structure and behavior has been observed for a sufficient amount of time, the organizational change is said to be completed. This model of organizational change is suitable for analyzing and predicting whether organizations can and will change. However, the reasoning engine behind this model assumes full knowledge of each agent's internal state. Thus, it cannot be used to build adaptive multi-agent organizations where the agents internal state is unknown or partially known.

6 Conclusions

In general, the principle of standardization of skills is beneficial because it reduces the need for detailed information in the top of the hierarchy. Although the large amounts of detailed information in direct supervision enables the manager to reduce the negative effect of heterogeneous task distribution, the amount of delay that is caused between a subordinate agent sending information and receiving a new order for a task has a large negative effect on the performance. This is clearly visible in the baseline performance of the three organizations where the standardization and the adaptive hierarchical organizations outperform direct supervision.

By adapting the hierarchical structure of an organization that is based on standardization of skills, the organization is capable of reducing the negative effects caused by an unbalanced workload among the agents. Our approach for the design of adaptive hierarchical organizations is based on the following principles:

1. Agents in a hierarchical organization have a limited scope of authority. They have full control over their direct subordinates but not over agents that are further below in the organizational hierarchy. Thus, the amount of effort that has to be spent on maintaining organizational awareness is reduced. Especially when building large hierarchical organizations, this principle of limited scope of authority will improve scalability of these organizations.
2. Agents can only communicate when an authority relation exists between them. This causes interaction between agents to follow the hierarchical structure. This reduces complexity and ensures the scalability to larger organizations.

3. The mechanism of adaptation ensures that the hierarchical structure remains intact. Furthermore, the adaptation mechanism itself respects the first two principles for the design of hierarchical organizations.
4. Decision-making authority for a task is transferred to the agent with the best knowledge for that task. Instead of sending large amounts of information between agents, we transfer decision-making authority to the agents that have the best knowledge for making these decisions. We assume that the agent that has been working for the longest time on a task will be the most suitable to become a manager for that task when additional coordination is required.

7 Future Work

In our experiments, we have used the RoboCupRescue simulation environment. Although this environment provides a detailed representation of the search and rescue domain, this simulator lacks control over the environment and manipulation of the environment. For example, in our experiment, we manipulated the distribution of civilians by manually creating these distributions by placing victims in buildings. Because each building has different properties (construction materials, size, and distance to the fault line of the earthquake), each building sustains a different degree of damage. The health and health decrease of civilians is related to the damage of the building they are located in and thus the location of a civilian will influence the performance of the search and rescue organization. This influence is the most likely cause for the small difference in performance between homogeneous and heterogeneous workload for direct supervision while we did not expect to find a statistically significant difference in performance. A possible explanation for this difference is that civilians in the heterogeneous maps sustained more damage after the earthquake than the civilians in the homogeneous maps. However, this remains uncertain due to the lack of control over the environment. Therefore, we see the need for a simulation environment which allows for a more systematic variation of task-environment parameters.

For our experiments, we have compared two static hierarchies (direct supervision and standardization of skills) with a dynamic hierarchy. For the RoboCupRescue scenario, a 2-layered hierarchy was chosen as the initial organizational structure. For other case scenarios, however, other (multi-level) organizational structures might be more appropriate as initial organizational structure. Furthermore, for our experiments, we have only considered hierarchical organizations in which communication follows only authority relations. Despite the disadvantages of communication outside the organizational hierarchy, for some scenarios communication relations between agents outside the organizational hierarchy may be beneficial. An in-depth comparison between strict hierarchies and hierarchies in which other communication relations are allowed is recommended for future research. As mentioned in Section 3.2, other organizations are possible besides hierarchies (e.g., an adhocracy). As the basic concepts of roles, agents, and task are also suited to design such organizations, a comparison between hierarchical organizations and decentralized organizations is also recommended for future research.

For an agent to operate as a manager in a dynamic hierarchy, we have identified a number of generic tasks such an agent has to be capable of. These generic tasks are task decomposition, task assignment, and organizational adaptation. As such tasks will occur in any adaptive organization, future work should aim at a generic approach to enable agents to perform these tasks.

Acknowledgments. The authors wish to thank the reviewers for their helpful comments and suggestions for improvements of our work reported in this chapter.

References

1. van Aart, C., Wielinga, B., Schreiber, G.: Organizational Building Blocks for Design of Distributed Intelligent System. International Journal of Human-Computer Studies 61, 567–599 (2004)
2. Carley, K.M.: Organizational adaptation. Annals of Operations Research 75, 25–47 (1997)
3. Dignum, V.: A model for organizational interaction: Based on agents, founded in logic. Ph.D. thesis, Universiteit Utrecht (2003)
4. Dignum, V., Dignum, F., Sonenberg, L.: Towards dynamic reorganization of agent societies. In: Proceedings of CEAS: Workshop on Coordination in Emergent Agent Societies at ECAI 2004, pp. 22–27 (2004)
5. Ferber, J., Gutknecht, O.: A meta-model for the analysis and design of organizations in multi-agent systems. In: Proceedings of the 3rd International Conference on Multi Agent Systems, p. 128 (1998)
6. Fox, M.S.: An organizational view of distributed systems. IEEE Trans. on System, Man, and Cybernetics SMC-11(1), 70–80 (1981)
7. Ghijsen, M., Jansweijer, W., Wielinga, B.: Agent decision making for dynamic selection of coordination mechanisms. In: Proceedings of the 2008 IEEE/WIC/ACM International Conference on Web Intelligence and Intelligent Agent Technology, pp. 87–91 (2008)
8. Ghijsen, M., Jansweijer, W., Wielinga, B.J.: Towards a framework for agent coordination and reorganization, AgentCoRe. In: Sichman, J.S., Padget, J., Ossowski, S., Noriega, P. (eds.) COIN 2007. LNCS (LNAI), vol. 4870, pp. 1–14. Springer, Heidelberg (2008)
9. Glaser, N., Morignot, P.: The reorganization of societies of autonomous agents. In: Boman, M., Van de Velde, W. (eds.) MAAMAW 1997. LNCS, vol. 1237, pp. 98–111. Springer, Heidelberg (1997)
10. Hall, D., Llinas, J.: An introduction to multisensor data fusion. Proceedings of the IEEE 85(1), 6–23 (1997)
11. Hoogendoorn, M., Jonker, C., Schut, M., Treur, J.: Modeling centralized organization of organizational change. Computational and Mathematical Organization Theory, 147–184 (2006)
12. Horling, B., Lesser, V.: A Survey of Multi-Agent Organizational Paradigms. The Knowledge Engineering Review 19(4), 281–316 (2005)
13. Hübner, J.F., Sichman, J.S., Boissier, O.: S-moise$^+$: A middleware for developing organised multi-agent systems. In: Boissier, O., Padget, J., Dignum, V., Lindemann, G., Matson, E., Ossowski, S., Sichman, J.S., Vázquez-Salceda, J. (eds.) ANIREM 2005 and OOOP 2005. LNCS (LNAI), vol. 3913, pp. 64–78. Springer, Heidelberg (2006)

14. Kamboj, S., Decker, K.: Organizational self-design in semi-dynamic environments. In: AAMAS 2007: Proceedings of the Sixth International Joint Conference on Autonomous Agents and Multi-Agent Systems, pp. 1220–1227 (2007)
15. Kitano, H., Tadokoro, S., Noda, I., Matsubara, H., Takahashi, T., Shinjou, A., Shimada, S.: Robocup-rescue: Search and rescue for large scale disasters as a domain for multi-agent research. In: Proceedings of IEEE Conference on Man, Systems, and Cybernetics (SMC 1999), vol. 6, pp. 739–743 (1999)
16. Kolp, M., Giorgini, P., Mylopoulos, J.: Multi-agent architectures as organizational structures. Autonomous Agents and Multi-Agent Systems 13(1), 3–25 (2006)
17. Lesser, V., Decker, K., Wagner, T., Carver, N., Garvey, A., Horling, B., Neiman, D., Podorozhny, R., Prasad, M.N., Raja, A., Vincent, R., Xuan, P., Zhang, X.Q.: Evolution of the GPGP/TÆMS domain-independent coordination framework. Autonomous Agents and Multi-Agent Systems 9, 87–143 (2004)
18. Levchuk, G.M., Meirina, C., Levchuk, Y.N., Pattipati, K.R., Kleinma, D.L.: Design and analysis of robust and adaptive organizations. In: 2001 Command and Control Research and Technology Symposium, A2C2 session (2001)
19. Little, E.G., Rogova, G.L.: Designing ontologies for higher level fusion. Information Fusion 10, 70–82 (2009)
20. Martin, C., Barber, K.S.: Adaptive decision-making frameworks for dynamic multi-agent organizational change. Autonomous Agents and Multi-Agent Systems 13(3), 391–428 (2006)
21. Mathieu, P., Routier, J., Secq, Y.: Dynamic organization of multi-agent systems. In: AAMAS 2002: Proceedings of the First International Joint Conference on Autonomous Agents and Multi-Agent Systems, pp. 451–452 (2002)
22. Mintzberg, H.: Structures in fives: designing effective organizations. Prentice Hall, Englewood Cliffs (1993)
23. Scerri, P., Pynadath, D.V., Tambe, M.: Towards adjustable autonomy for the real world. Journal of Artificial Intelligence Research 17, 171–228 (2002)
24. Schurr, N., Marecki, J., Kasinadhuni, N., Tambe, M., Lewis, J.P., Scerri, P.: The defacto system for human omnipresence to coordinate agent teams: the future of disaster response. In: AAMAS 2005: Proceedings of the Fourth International Joint Conference on Autonomous Agents and Multi-Agent Systems, pp. 1229–1230 (2005)
25. Shehory, O., Sycara, K., Chalasani, P., Jha, S.: Agent cloning: An approach to agent mobility and resource allocation. IEEE Communications 36(7), 58–67 (1998)
26. So, Y., Durfee, E.: Designing tree-structured organizations for computational agents. Computational and Mathematical Organization Theory 2(3), 219–246 (1996)
27. So, Y., Durfee, E.: Designing organizations for computational agents. In: Prietula, M., Carley, K., Gasser, L. (eds.) Simulating Organizations, pp. 47–64. AAAI Press/ MIT Press (1998)
28. van der Vecht, B., Dignum, F., Meyer, J.J.C., Neef, M.: A dynamic coordination mechanism using adjustable autonomy. In: Coordination, Organizations, Institutions, and Norms in Agent Systems III, pp. 83–96 (2008)
29. Zambonelli, F., Jennings, N.R., Wooldridge, M.: Organisational abstractions for the analysis and design of multi-agent systems. In: 1st International Workshop on Agent-Oriented Software Engineering at ICSE 2000 (2000)

Method for Designing Networking Adaptive Interactive Hybrid Systems

Leon Kester

Abstract. Advances in network technologies enable distributed systems, operating in complex physical environments, to co-ordinate their activities over larger areas within shorter time intervals. Some envisioned application domains for such systems are defence, crisis management, traffic management and public safety. In these systems, humans and intelligent machines will, in close interaction, be able to adapt their behavior under changing conditions and situations to reach their common goals. Various architecture models are proposed for such Networking Adaptive Interactive Hybrid Systems (NAIHS) from different research areas such as sensor web technology, sensor fusion, command and control, cognitive science, ergonomics, agent technology, multi-agent systems and robotics. However, most of these models only cover part of the system are too much focussed on specific research areas and use different design philosophies for intelligent systems. In this article a top-down design methodology is proposed that may combine the merits of the various approaches and pave the way to efficient design and effective operation of such systems.

1 Introduction

Technological advances in many different research areas have since centuries inspired people to envision a future society in which intelligent machines would coexist with humans. However, only recently the steady exponential nature of these technological advances is convincing many that such a society could become reality before the end of the century. As always there are those who embrace this

Leon Kester
TNO Defence, Security and Safety, Oude Waalsdorperweg 63, 2597 AK The Hague, The Netherlands
University of Amsterdam, Science Park 107, 1098 XG Amsterdam, The Netherlands
e-mail: leon.kester@tno.nl

R. Babuška & F.C.A. Groen (Eds.): Interactive Collaborative Information Systems, SCI 281, pp. 401–421.
springerlink.com

development with optimism [18] while others have a more cautious mind [19]. The main technological developments that feed this belief are rather arbitrarily listed:

- *Nanotechnology*
  Since the famous speech of Feynman, *there's plenty of room at the bottom* [11], much progress is made in exploiting that room. However, plenty of room is still left resulting in even smaller sensors, actuators, computers or complete robots. Other interesting developments are in the area energy where thin film technologies may decimate the price of solar cells and nanotechnology is used to develop cheap rechargeable batteries with high energy density.
- *Computation*
  Moore's law is already reigning for decades and is expected to continue to do so for at least the coming two decades. The computational power comparable to the human brain will become available at affordable prices. Even if the methodologies used for artificial intelligence will not improve much, which is not to be expected, a million-fold increased in computational power will enhance machine capabilities considerably.
- *Communication*
  Our society becomes more and more connected by wired and wireless communication facilities and although advances in this area are more constraint by physical processes communication capabilities will increase substantially using smarter (computational intensive) and more efficient methods.
- *Bioengineering*
  Ongoing discoveries about he mechanisms of life including those of the human brain will narrow the gap between artificial and real life and will open up new forms of human machine interaction. Furthermore, the potential power, as already demonstrated by nature, of learning and evolutionary processes applied to many forms of artificial and real life will be substantial.

These technological developments will enable future systems to communicate more easily and co-ordinates their activities over larger areas within shorter time intervals. Also these systems of humans and intelligent machines will, in close interaction with each other and their complex physical environment, be able to adapt their behavior under changing conditions and situations to reach their common goals. However, there is surprisingly little consensus on how to integrate these technologies into systems that are able to meet future challenges e.g. in the field of military defence, crisis management, traffic management and public safety. Several reasons can be given for this situation:

- The field of enabling technologies is very broad which makes it difficult to have a good and unbiased view on their capabilities. The field of artificial intelligence already covers a lot of different technologies where the state of the art is developed by specialists. Their focus is usually on exploring and advancing their technology and less on exploiting it in combination with complementary technologies.
- System architects usually have a good overview of enabling technologies. Their mind is usually focussed on what technology to use to generate a certain

performance but have difficulty with integrating artificial intelligence into their methodologies.
- Researchers, system architects, operators and costumers usually have different (philosophical) views on human and artificial intelligence and the interaction between them which makes it difficult to agree on high level system design issues.

In this chapter, we propose a design methodology that tries to unite the various views on humans and intelligent machines, taking a system designer approach with a focus on how to generate system functionality, trying to have an unbiased view on artificial intelligence and dealing with the difficulties that arise in the systems engineering methodology from the incorporation of artificial intelligence techniques.

In the following sections, we use the system engineering process SIMILAR [6] as recommended by INCOSE [1] and use the various tasks in this process to structure the sections of this chapter. In Section 2, the design methodology is discussed. In Section 3, the problem is stated in more detail. In Section 4, different functional and physical decomposition strategies are considered and compared to models in the domains of sensor fusion, robotics and cognitive science. In Section 5, the integration of the functional and physical components is treated using interoperability and interaction models. In Section 6, the stage of launching the system and producing output is presented. In Section 7, performance assessment is covered. In Section 8, some examples are discussed and, finally, Section 9 concludes this paper.

2 Design Process

To structure the design methodology the system engineering process SIMILAR is used. This process comprises the following tasks:

1. *State the problem*
   Stating the problem is the most important systems' engineering task. It entails identifying customers, understanding customer needs, establishing the need for change, discovering requirements and defining system functions.
2. *Investigate Alternatives*
   Alternatives are investigated and evaluated based on performance, cost and risk.
3. *Model the System*
   Using models clarifies requirements, reveals bottlenecks and fragmented activities, reduces cost and exposes duplication of efforts.
4. *Integrate*
   Integration means designing interfaces and bringing system elements together so they work as a whole. This requires extensive communication and coordination.
5. *Launch the system*
   Launching the system means running the system and producing outputs – making the system do what it was intended to do.

6. *Assess performance*
   Performance is assessed using evaluation criteria, technical performance measures and measures of effectiveness—measurement is the key. If you cannot measure it, you cannot improve it.
7. *Re-evaluation*
   Re-evaluation should be a continual and iterative process with many parallel loops.

A graphical representation of this process is given in Figure 1.

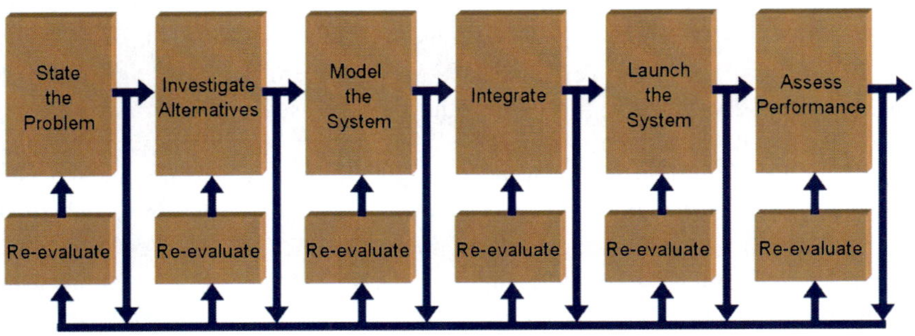

Fig. 1 The SIMILAR process

An important aspect of this process is that for each task there is a re-evaluation sub-process. This can be used for further optimization as well as engineering the system in more and more detail similar to the approach taken by the Vee model [3]. In this article mainly the higher levels of this iterative process will be discussed since the models used are still generic and most problems arise from implicit choices in sub-system models at the early stages in the design that are not well aligned with each other.

3 Problem Statement

Before we start with the statement of the problem, we must first have a clear view on what the system comprises. According to the definition of INCOSE, a system is:

Definition 1. A construct or collection of different elements that together produce results not obtainable by the elements alone. The elements, or parts, can include people, hardware, software, facilities, policies, and documents; that is, all things required to produce system-level results. The results include system-level qualities, properties, characteristics, functions, behavior and performance. The value added by the system as a whole, beyond that contributed independently by the parts, is primarily created by the relationship among the parts; that is, how they are interconnected [25].

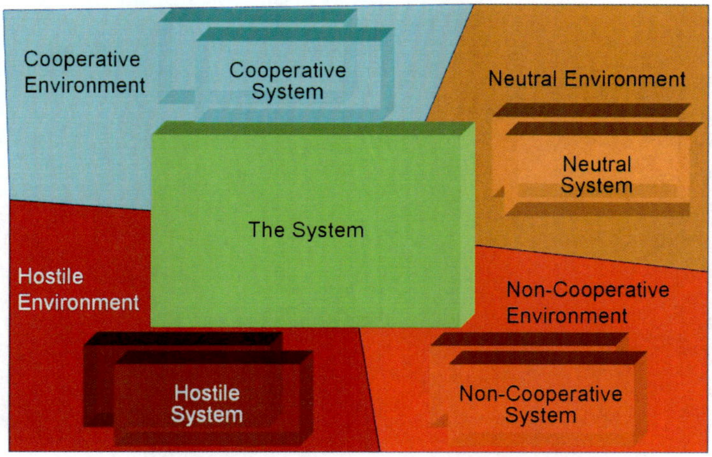

Fig. 2 The system in its environment

The system under consideration will be embedded in an environment with which it interacts to reach its goals. As depicted in Figure 2, other systems may be present in the environment. These systems may have different intentions ranging from cooperative, neutral, non-cooperative to hostile. The cooperative systems, e.g. the internet with computers connected to it or humans that respond to SMS messages can according to the system definition of INCOSE be considered as part of the operational system but from a designer's point of view, however, these cooperative systems are part of the environment since they are not part of the system that needs to be designed. The main characteristics of the system suitable for military, crisis management, traffic management and public safety are that it is Networking, Adaptive to changing and unexpected situations, Interactive within the system and with the environment and consisting of humans and machines (Hybrid). In the remaining part of this chapter, we will refer to this type of system as NAIHS.

4 System Modeling

The next task in the SIMILAR process is system modeling. In this task, we will start at a high level modeling and progress to more and more detail. There is already some earlier work published on the modeling of NAIHS [16]. In this section, we build on this earlier work which is referred to as the original NAIHS model.

4.1 High-Level Model

As the expression *hybrid* already suggests, one would be tempted to decompose the system in a human and machine part. Although different in nature, both can perform similar tasks, such as sense the outside world, recognize, act on the outside world and reason how to reach certain goals. Therefore, the approach taken in modeling

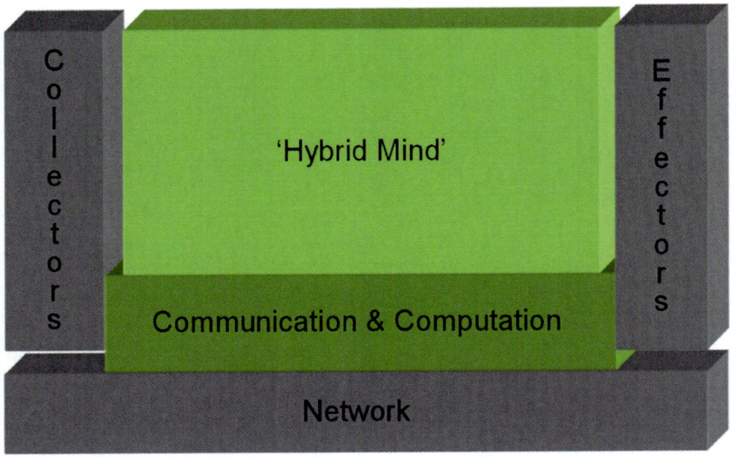

Fig. 3 High-level system model

NAIHS is to focus on modeling the system according to specific functional components, postponing the decision if they should be performed by humans or machines. In Figure 3 a schematic picture is given for such a system. One can characterize the system as a distributed set of hybrid functional components (or hybrid mind) that interacts with collectors (sensors, observers or other information sources), effectors (actuators or actors) and a (distributed) service that takes care of communication and computation between the distributed functional components.

A common approach is to decompose the system in the chain from collectors to effectors. In this 'dimension' a well-established model is in use, the Observe Orient Decide Act (OODA) loop [8]. According to the OODA cycle, functional components can be identified, which are engaged in creating situation awareness and components that are engaged in deciding which effects to generate. This view is depicted in Figure 4.

The goal of such a system is to grasp that part of the situation, defined as the system in its environment, based on which the most efficient and effective actions can be taken.

An important aspect that has to be addressed now is what strategies and principles should be used to further decompose *create situation awareness* and *decide on action* into more detailed functional components.

4.2 Decomposition Strategies

Three principles that are distinguished in modeling NAIHS are Information abstraction, time horizon of its goals and physical structure of the system in its environment (the situation). In the next paragraphs, these principles are discussed.

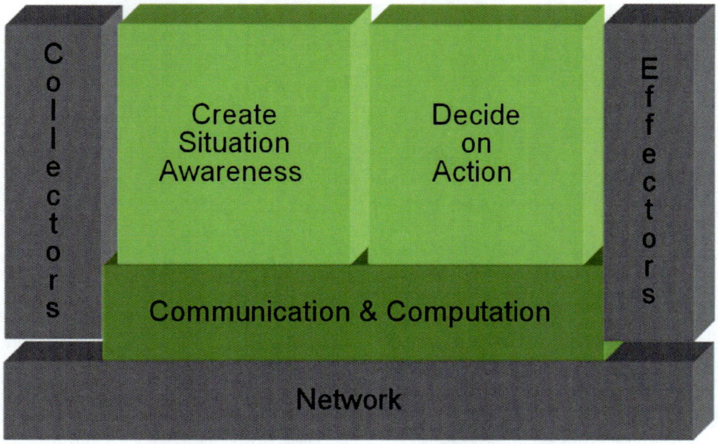

Fig. 4 Separation of the hybrid mind in *creating situation awareness* and *decision making*

4.2.1 Information Abstraction

A well-known model of information abstraction is the data, information, knowledge, wisdom or DIKW model [4, 29]. Similar to this abstraction the JDL model [2] uses:

- *Level 0* Estimation of States of Sub-Object Entities (e.g. signals, features, data)
- *Level 1* Estimation of States of Discrete Physical Entities (e.g. vehicles, buildings)
- *Level 2* Estimation of Relationships Among Entities (e.g. aggregates, cueing, intent, acting on)
- *Level 3* Estimation of Impacts (e.g. consequences of threat activities on one's own assets and goals)

We can relate these models easily with the high-level model of Figure 4 to *create situation awareness*.

A model also very similar to that but more focussed on the human perspective is the model of Endsley [10] schematically depicted in Figure 5. While the JDL model is confined to creating situation awareness the model of Endsley includes decision making and performing actions. In the original NAIHS model [16], the 'process cycle' distinguishes:

- *N1* Signal (Pre)-Processing - generating the feature space from the collector raw data
- *N2* Filtering in Feature Space - selecting phenomena from the feature space likely to originate from objects (detection)
- *N3* Filtering in Time - associating detected phenomena in time and estimate state, tracking
- *N4* Recognition - classification and identification
- *N5* Situation Assessment - Relationships among entities, similar to level 2 of the JDL model.

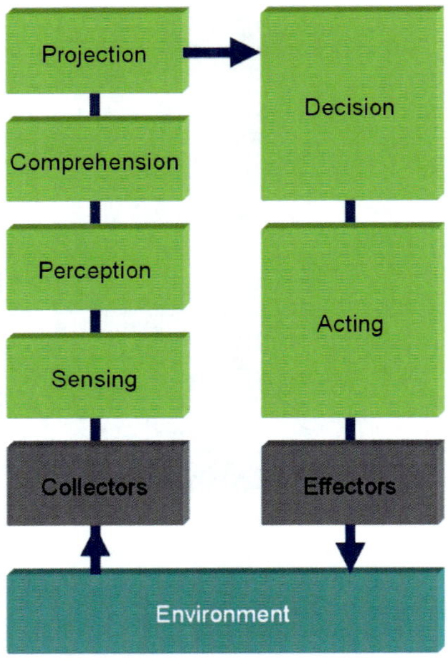

Fig. 5 Endsley's situation awareness model

- *N6* Relevance Assessment - threat evaluation, risk assessment, similar to level 3 of the JDL model.
- *N7* Action Assessment - decide on what actions to take
- *N8* Execution - execute the actions

The main difference with the other models is that JDL level 1 is further decomposed into three components. If we turn our view to the cognitive domain an influential model is that proposed by Fuster [13] represented in Figure 6.

The model describes the perception process situated in back part of the brain as well as the execution process situated in the front part of the brain. These two processes can be associated with the *generate situation awareness* and *decision making* processes in the high-level model of Figure 4. The information abstraction levels that are distinguished at the side of perception are:

- *Phyletic sensory* This is the level at which specific signal processing of the individual human sensors takes place.
- *Polysensory* This is the level at which information on the various human sensors is integrated.
- *Episodic* This level can be related to situation assessment.
- *Semantic* At this level semantic interaction between humans (language) takes place.
- *Conceptual* This is the level at which conceptualization of complex problems and reasoning on intentions of intelligent entities takes places.

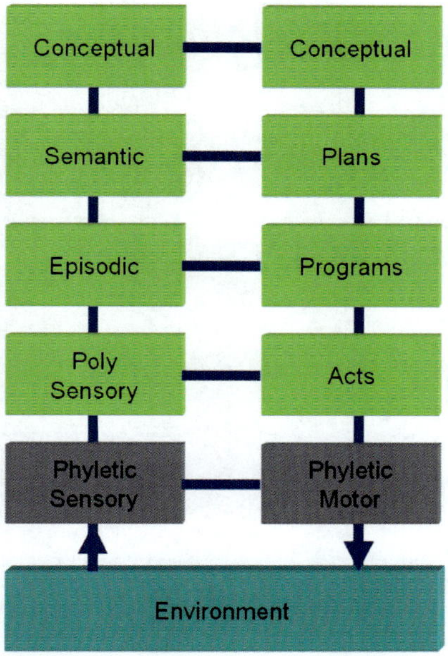

Fig. 6 Fuster's model of the mind

The levels in this *perception hierarchy* coincide fairly well with the DIKW and JDL hierarchy with the exception of the semantic level. This level is primarily concerned with transforming thoughts to semantic presentations in order to be able to communicate with other humans. Further evidence for compatibility of Fuster's model with the DIKW and JDL hierarchy was presented at [21]. Fuster links the perception hierarchy on all levels to an *execution hierarchy* consisting of:

- *Phyletic motor*
- *Acts*
- *Programs*
- *Plans*
- *Conceptual*

In a similar way Steinberg [27] proposes to link situation awareness on each level of the JDL model to a management process that uses this information to base its decisions on. Comparing the various models on information abstraction, a suitable NAIHS model can be made by adopting the abstraction levels of the DIKW hierarchy and the JDL model as the second level (the decomposition of the system in section 4.1 being the first level) of decomposition. The decomposition of the object assessment function, N2, N3 and N4 in the original NAIHS model can then be considered to be a deeper level of decomposition. To each level a 'management' function is coupled that, based on the generated situation awareness, at each level generates intentions. The combined intentions finally lead to signals that are sent to the effectors. The resulting functional model of NAIHS is depicted in Figure 7.

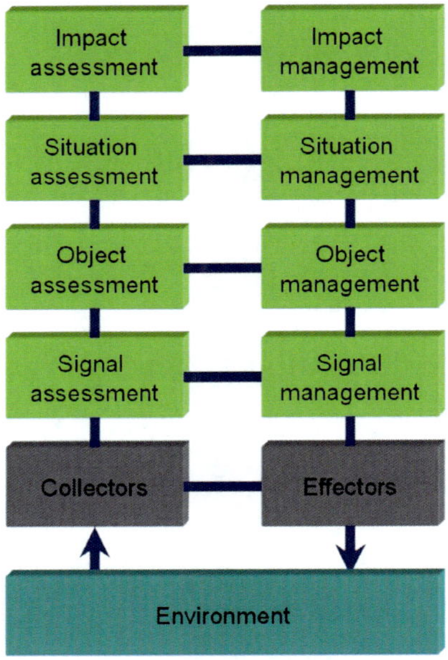

Fig. 7 Functional model of NAIHS

4.2.2 Time Abstraction

Every decision making or management process reasons about possible outcomes of the future taking into account possible actions. The time horizon of these processes may have a wide range. Therefore, a decomposition of the process cycle in this dimension has been adopted in various application domains. In the military field a strategic, operational and tactical level is in use. For (business) planning the same levels are in use, however, contrary to the military case the operational level acts on a shorter time scale than the tactical level. In the AI domain Brooks [9], who discards the decomposition in information abstraction, proposes a hierarchical composition of process cycles based on reaction or cycle time. The need for decomposition into temporal abstractions is also acknowledged in the case of decision making processes in complex situations [7]. The time scale may differ widely for various applications. A suitable decomposition therefore depends on the application. Most models use only one type of abstraction, there are some that distinguish these different principles but integrate them in one abstraction hierarchy [5, 27]. Although it is a tempting thought to do so in the model for NAIHS we would like to consider the two principles on which decomposition can be based separately, resulting in a two-dimensional decomposition of the 'hybrid mind'. The reason for this is that it is well imaginable that information processing at a high information abstraction level is very fast and relevant for short-term decision making or that information processing from one sensor at one platform needs to be decomposed in many components at the higher information levels. Furthermore, results presented in [24] indicate that both abstraction principles play a role in hierarchical behavior. In the functional model for NAIHS, therefore, the freedom is maintained of accommodating both abstractions.

4.2.3 Decomposition due to Physical Structure

A third principle on which decomposition can be based is physical structure. In systems considered here the physical structure is due to the network of platforms and the various collectors and effectors, in other words the situated system. Each 'assessment-management' component in the two-dimensional cognitive hierarchy could therefore have multiple instantiations in the system resulting in a three-dimensional structure of 'assessment-management' components. Besides that also the assessment and management components could be on different platforms. This distributed system model is symbolized in Figure 8. Note that each component in this figure may be hybrid and consisting of multiple levels due to time abstraction.

4.3 Topology of the Functional Components

Now that we have analyzed the three dimensions of possible decompositions of the 'hybrid mind' one may wonder if there is a preferred structure or topology for

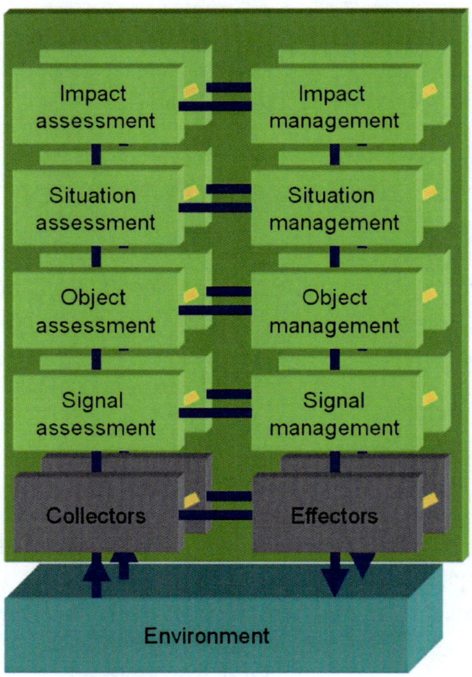

Fig. 8 Distributed cognitive system model

distributed cognition. We first assess the relation between information abstraction and time abstraction. Then we assess the relation with the physical structure.

Information abstraction and time abstraction: The lower levels of the information abstraction hierarchy will usually only have valid information for deciding on effects on a short-time horizon. The higher we go up in the information abstraction hierarchy the wider the range in time may be. Therefore, one expects that going up the information abstraction hierarchy the range of time horizons will be more extended and the need for time abstraction is larger.

Information abstraction and physical structure: The lower levels of information abstraction tend to be more decentralized than the higher levels of information abstraction. There are several reasons for that:

- The lower levels of information abstraction tend to be more heterogeneous due to heterogeneous sensors and effectors.
- The range of the collectors and effectors may be limited. In order to cover the whole situation a large number distributed sensors in the environment is needed.
- At the lower information abstraction levels the amount of data is larger and the results are more time critical. Therefore the cost of utile communication at these levels is higher due to bandwidth and latency requirements.
- Because of their closer connection to the environment the lower information abstraction levels are more vulnerable to malfunctioning due to physical interaction with the environment or hostile action. Therefore redundancy in these lower levels may help to make systems more robust.

4.4 Hybrid Components

As mentioned earlier, each component may be artificial, human or hybrid. Besides a decomposition strategy, also a strategy for dividing the system into a human and an artificial part is needed.

A significant difference between the human mind and the artificial mind is that the human mind is very good at dealing with unstructured and imprecise information but is rather slow at reaching results and is not able to handle large amounts of (dissimilar) information simultaneously. In contrast, the artificial mind usually lacks *a priori* knowledge to deal with imprecise information while large amounts of well-modeled data can be processed with relative ease.

This encourages the notion that systems may support humans in their decisions but that humans should always be in control. This must, however, not be exaggerated. Being faithful to the underlying principles artificial (aided) decision making is sometimes preferred, e.g. in the case of mathematically well-defined problems where a large amount of (correlated) possible actions needs to be considered or where synchronized action is needed on a very short-time scale. In the latter case also the fact that a high level of interoperability between artificial minds is usually easier to establish than between human minds increases the preference for an artificial implementation. This was further analyzed, using the NAIHS functional model, in a recent study on organizing smart networks and human teams [20].

Limitations of humans, with respect to artificial minds like decomposition possibilities, communication capabilities, interoperability and adaptivity will, in future systems, shift the interaction border between humans and artificial components upward.

5 Integration

Now that the dimensions of decomposition for NAIHS are analyzed; the next phase in the design process is to address how the functional components can be integrated in such a way that the system as a whole is able to reach its goals in the most effective way. First we reconsider the model in Figure 8. In this model it is assumed that each functional component only interacts with its direct neighbors, either in the information abstraction hierarchy, in the temporal hierarchy, social interaction or between the assessment and management components. This modular view is disputed, e.g. in the field of neuroscience, particularly at the higher levels of information abstraction [13]. Here we assume modularity keeping in mind that we might have to include additional interactions.

5.1 Interoperability

An important characterization related to distributed behavior is interoperability that characterizes how well the networked components 'understand' each other. A common model for interoperability is the LISI model [28] that distinguishes five levels:

- L0: no interaction or unintelligible information
- L1: unstructured representation of information
- L2: a common representation of information (syntactic level)
- L3: a common understanding of information (semantic level)
- L4: common understanding of how information is interpreted (pragmatic level)

Obviously, the perspectives for coherent behavior increase with a higher level of interoperability. Note that the denomination 'semantic level' here originates from the field of semiotics to characterize that a common understanding of the 'language' is assumed. In the cognitive model of Fuster (Figure 6) 'semantic' and 'episodic' characterize the type of memory involved in different mental processes.

5.2 Interaction

Another important concept is that of service orientation in which the components that are in need of a service (information, communication, etc.) express their need to service providers who try to provide the most utile service. In [16] it is described in detail how with a service-oriented approach and interoperability level 4, adaptivity to changing needs in information can be accommodated in a distributed setting.

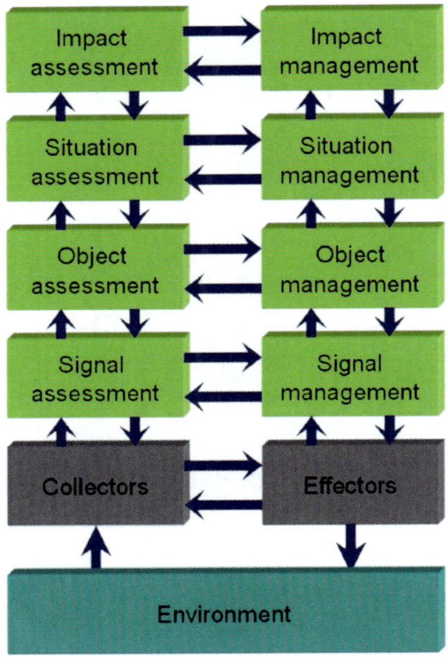

Fig. 9 Model of NAIHS with service-oriented interaction between the components

Since we have adapted the 'process cycle' of the original NAIHS model to the information abstraction hierarchy depicted in Figure 7, service-oriented interactions have to be added between the assessment and management modules at the lower levels. This adds extra service requests from the management modules to the assessment modules but does not influence the coordination mechanism of distributed modules. This is schematically shown in Figure 9. In the perspective of distributed behavior it is now interesting to look at a roadmap depicted by NATO on Networked Enabled Capabilities (NEC) in Figure 10. In this roadmap three system layers are distinguished, namely, communication, information integration and a functional area. Apart from the confusing use of terminology this corresponds well to the three layers in the high level system model of Figure 3. Each layer delivers services in support of the desired effects. Different from the functional model for NAIHS, however, the human capabilities are considered to be separated from the functional area services.

Furthermore, four phases of 'co-evolution' or NEC maturity of the three service layers are identified:

- Deconflict: the services are not conflicting but act separated.
- Coordinate: information and applications are available to the system.
- Integrate: capabilities adjust their services to changing situations.
- Coherent (also known as Agile): the whole system reorganizes itself to changing situations

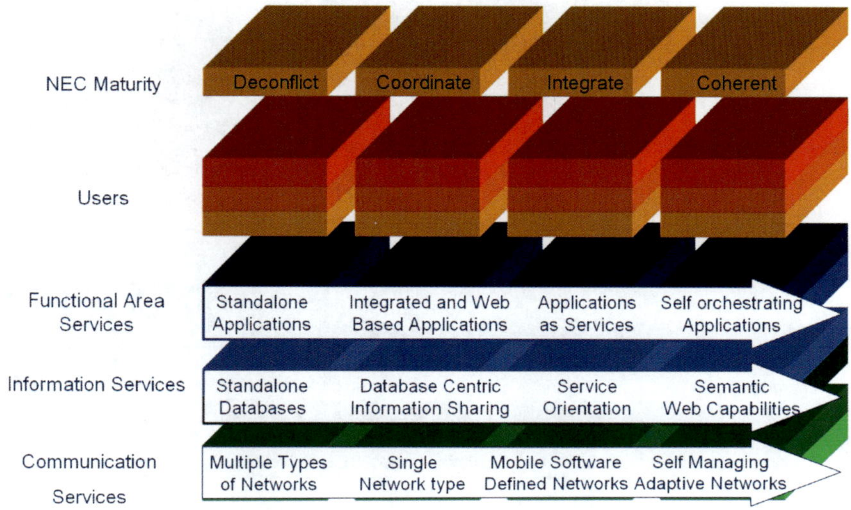

Fig. 10 NATO NEC maturity roadmap model

When compared to this characterization the NAIHS design methodology goes further than the integrated/collaborative phase since the services in the functional area's act as agents that try to adjust their behavior to generate the most utile service. It is furthermore important to note that the NAIHS model considers distributed behavior throughout the whole information abstraction and time hierarchy while the NEC roadmap usually only refers to the higher levels in the hierarchies.

The term Integrate is ambiguous in the context of the integration phase of the SIMILAR process. Moreover, the term collaborative is more common for this type of interaction. Taken all these issues into account the NEC maturity roadmap model is adjusted to be used as a high-level NAIHS evolutionary interaction model (Figure 11).

An important question to ask here is what is needed to characterize NAIHS as a coherent system? An interesting research domain to study in this context is that of distributed cognition, e.g. [14]. An analysis shows three subjects that need to be addressed further: how does the system learn, how does it determine its (high level) goals in a changing situation and how does the system generate self-awareness? Since the system consists of agents, how the system learns can be rephrased to: how do the individual agents learn? The behavior of the agent is determined by its place in the three-dimensional functional topology. For these agents the NAIHS method distinguishes six different interactions:

- with agents higher in the information abstraction hierarchy
- with agents lower in the information abstraction hierarchy
- with agents similar in functionality/behavior
- with agents higher in time hierarchy
- with agents lower in the time hierarchy
- with its counterpart in the assessment/management pair

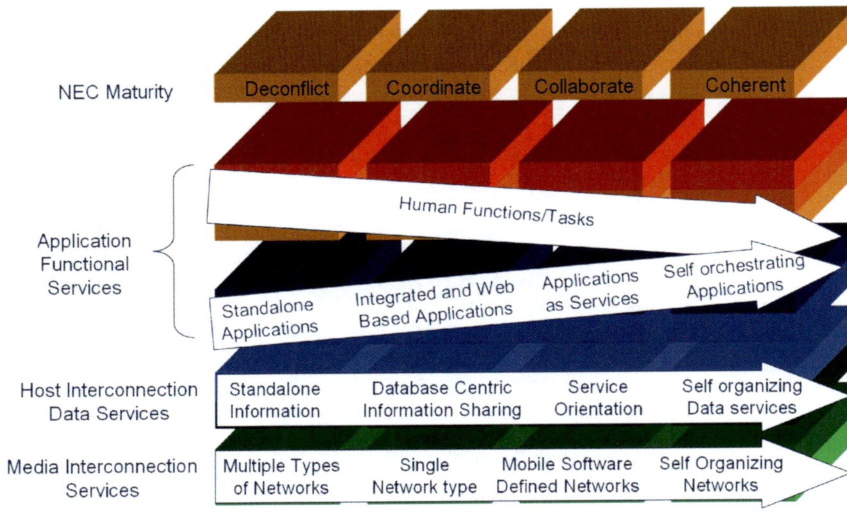

Fig. 11 NAIHS evolutionary interaction model

The behavior of the agent can therefore be influenced by these interactions. On a higher abstraction level there are, however, only two basic interactions, namely, requests and deliveries. The agent memory can be split in declarative and procedural memory. It can learn by testing how well different procedures are able to meet the requested deliveries. For improving its declarative memory the agent is dependent on the other agents with which he interacts. Its learning capability is therefore dependent on its ability to identify a lack of information and formulate requests to the other agents, usually but not necessarily lower in the information abstraction hierarchy. Each agent has a specific place in the time abstraction hierarchy. Information in its declarative memory is therefore only useful if it is valid for at least the duration of the time horizon. It should therefore also have a deterioration model of its declarative memory. A step further is that the agent also learns how to communicate, i.e. improve its interoperability and interaction mechanism, with techniques from the semantic web domain.

The second problem is very difficult to address in a generic sense. If the goals of the system are complex, it is even more difficult to generate new goals under changing situations. The research area of Complex Adaptive Systems may come up with mechanisms to deal with this. For the more simple goals adaptive (hierarchical) reinforcement methods are promising, in particular because they appear to be suitable to emulate human decision making processes [21]. The third problem of system or self-organization has many aspects. Within the framework of service and agent-oriented architectures a lot of these aspects can be covered by the NAIHS method. Two issues need to be examined further, though; swarming and reorganization of the functional structure in information abstraction and time abstraction.

Most swarming applications are dealing with entities with identical behavior, functional equivalent behavior or a common interaction mechanism. Swarming techniques are therefore interesting candidates for efficiently coordinating the behavior of agents at the same information abstraction and/or time abstraction level. They might be interesting for coordinating behavior between agents in adjacent levels. In dynamic situations it may be useful to change dynamically the number of levels in the time hierarchy or change the granularity in information abstraction. A simple example of this is the change from an object assessment agent to separate agents for filtering in feature space, filtering in time and recognition as suggested by the original NAIHS model. At this moment it is difficult to imagine how dynamic reorganization of the functional structure itself could be established, because first the utility infrastructure of a service-oriented system is needed before one can reason about the utility of the structure itself. One may conclude, however, that the design methodology of NAIHS seems an appropriate framework to develop further these capabilities.

6 Launching the System

The systems that are considered here are usually very complex. Just building them and assess their performance is therefore too costly or infeasible. Therefore a stepwise approach, also known as model based design, is used where only part of the system is implemented and part of the system is modeled. The situation that is created for experimenting with the system is therefore partly real and partly virtual, also known as mixed reality. If we map this idea on the abstract representation of the system in its environment as depicted in Figure 2 we can see that the setup of the mixed reality experiment is in itself a systems engineering problem but now the goal of this system is to assess the validity and performance of the system under design at the proper level of abstraction of the design process. So, following the system engineering process we need to state the problem; that is what is the purpose of the experiment, which part of our system in its environment do we want to implement and which part do we want to simulate. In addition, we have to consider to which level of detail we would like to simulate things to get valid results from our mixed reality experimentation. Note here that since we may deal with hybrid systems, choices have to be made on putting either humans in the loop as part of an implementation or that we simulate (relevant part of) the human functionalities. It is further interesting to note that the development of NAIHS type of systems can also be used to simulate intelligent (human) behavior of the environment. An important problem that arises in these mixed reality environments if we go to more and more intelligent systems is that the types of interactions between the functional components are alien to most simulation environments. Therefore, they need to be simulated explicitly which will make the mixed reality environment run many times slower than the real system would do with the same amount of computational power. This leads to the conclusion that for high demand real-time applications modeling that part of the system is not an option and experiments need to be performed on specific implementations.

7 Performance Assessment

An important issue in system design is how the performance of such systems can be assessed. Not only is this needed to demonstrate its value but also to be able to re-evaluate the system. It is interesting to remark that in the more traditional system engineering approaches the human was seen as the user of the system and not as integral part of the system. Measuring the performance is then measured separate from the human performance. It was also believed that these systems could be tested such that they would work flawlessly. From that perspective one was very sceptical of introducing artificial intelligence techniques because these techniques could not be guaranteed to work in a flawless manner. After some major disappointments [22] this perspective changed and it was realized that any system that interacts with the real world cannot grasp all relevant elements of that world, just like humans, and therefore a flawless performance cannot be guaranteed. Performance measurement under these conditions is, similar to *the logic of scientific discoveries* [23], in this case probably the best method to measure the performance with a certain probability where artificial intelligence techniques are trying optimize this performance. Realizing this, going to more intelligent systems changes the perspective of performance measurement quite dramatically. This can be illustrated best with an example. Suppose that we have a system designed at networking maturity level 2 and we would like to measure the performance of the object assessment component. We can assess the effectiveness of the whole system in an experimentation environment, however, measuring the performance of the object assessment component is more difficult since it is unclear what is most important, e.g. accuracy, object identity, track ambiguity, etc. Moreover, what is important may change over time. If, however, the system operates at network maturity level 3, the performance of the object assessment component is determined by how well it can adapt to its changing agent environment, in this case for example the different needs that where expressed by the situation assessment agent. More specific the utility function mentioned in section 5, which is necessary to let the system work in a service-oriented approach can be used to measure the performance of the individual agents. In order to do this only the local environment, i.e. the local set of agents it interacts with, needs to be implemented. The implementation of a certain agent that is able to generate the most utile information on request is the agent that performs best. First evaluations on single agents can therefore easily be performed by comparing the different scores on utility it is able to generate. A direct consequence of having a formalized procedure to measure performance is the possibility of implementing this in the system and let the system itself identify badly functioning agents or identify agents with suspicious behavior. This is particularly the case if local evaluation results are exchanged. In this way the system may also protect itself against agent or platform hijacking on the functional level, next to security measures on the communication level. Furthermore, the system may decide to replace the malfunctioning agents with ones performing better. These types of functionalities are necessary ingredients for coherent systems.

8 Applicability of the NAIHS Design Method

Firstly this method is used to co-ordinate the research on ‘enhanced situation awareness’ in a large research project on Interactive Collaborative Information Systems (ICIS). This research ranges from distributed tracking, modular Bayesian networks, semantic networks, swarming, risk assessment and hierarchical reinforcement learning. Other research clusters in this project are Collaborative Decision Making, Computer Human Interaction and ICIS Architecture. The original NAIHS model was considered to be a good dynamical model for the whole ICIS project and is used to link various research topics. Furthermore, we have applied this method to the case of multi-platform engagement capability. In this case there are multiple ships with multiple dissimilar sensors that have to cooperate to counter common threats like incoming missiles and hostile fighters. This case is primarily focussed on the tactical level. For this application, the time hierarchy, physical structure and information abstraction are organized in the same way. For this application, we have focused on interacting tracking algorithms at the object assessment level. These tracking algorithms are capable of fusing information from radar and electro-optic sensors. This was tested in a demonstration environment where radar and electro-optic sensors were simulated but the real tracking and information management functions were implemented [12, 15]. A third case in which the NAIHS method was used is for a system for indoor security and safety. This system consists of a network of camera’s, electro-magnetic imaging gateways, security personnel and aid workers, a positioning system for these people, smart phones and a control room with personnel. Furthermore, many algorithms for information processing and decision support are used. The NAIHS method was used to design the three modes of operation; surveillance, alert and threat countering, with specific attention to human machine interaction [20]. A fourth case is an application for sensor management in *cognitive* radar networks [26]. Finally the NAIHS method is used to integrate concepts of hierarchical networks, social networks, information networks and communication networks in an NEC context [17]. From this exercise a coherent vision on NEC will have to emerge, which will help to define the research activities of TNO, Defence, Security and Safety in the field of NEC.

9 Conclusions

In this article we have considered the design process of Networking Adaptive Interactive Hybrid Systems (NAIHS). For such systems the different tasks of the SIMILAR systems engineering process have been analyzed, i.e. problem statement, system modeling, integration, launching the system and performance assessment. For the system modeling task the principles of decomposition of the system were reviewed by comparing it to other models in the field of fusion, robotics and cognitive systems. The structure is however revised by coupling each component involved with creating situation awareness to a decision making component similar to human cognition models. From this analysis a generic NAIHS system modeling method was proposed. For NAIHS system integration, interoperability and interaction were analyzed. The principles of interaction were related to the phases identified in the NATO Networked Enabled Capabilities roadmap and the developments in the field of distributed cognition. From this analysis a high level NAIHS evolutionary integration model was proposed. Finally the use of the NAIHS design process was discussed for various applications. From these exercises it can be concluded that the proposed design method helps to design efficiently effective systems and that its models help to accommodate artificial intelligence technology more easily.

References

1. INternational Council On Systems Engineering (INCOSE), `http://www.incose.org`
2. Data fusion lexicon, US DoD, subpanel of the joint directors of laboratories (1991)
3. Vee model of systems engineering design and integration, INCOSE G2SEBOK 3.30 (2003)
4. Ackoff, R.: From data to wisdom. Journal of Applied Systems Analysis 16 (1989)
5. Albus, J.: Outline for a theory of intelligence. IEEE Transactions on Systems, Man and Cybernetics 21(3) (1991)
6. Bahill, A.T., Gissing, B.: Re-evaluating systems engineering concepts using systems thinking. IEEE Transactions on Systems, Man and Cybernetics, Part C: Applications and Reviews 28(4), 516–527 (1998)
7. Bakker, B., Kester, L., Vlassis, N., de Jong, E.: Workshop on hierarchical autonomous agents and multi-agent systems (H-AAMAS), AAMAS 2006 (2006)
8. Boyd, J.: A discourse on winning and losing. Maxwell AFB Lecture (1987)
9. Brooks, R.: How to build complete creatures rather than isolated cognitive simulators. Lawrence Erlbaum Associates, Hillsdale (1991)
10. Endsley, M.: Design and evaluation for situation awareness enhancement. In: Proceedings of the Human Factors Society 32nd annual meeting, Santa Monica, CA, pp. 97–101 (1988)
11. Feynman, R.: There's plenty of room at the bottom. In: Proceedings of the annual meeting of the American Physical Society (1959)
12. van Foeken, E., Kester, L.: Shared belief in communication constrained networks. In: Proceedings of IASTED International Conference on Sensor Networks SN 2008 (2008)
13. Fuster, J.: Upper processing stages of the perception-action cycle. Trends in Cognitive Sciences 8, 143–145 (2004)
14. Heylighen, F., Heath, M., van Overwalle, F.: Issue of cognitive systems research the emergence of distributed cognition: a conceptual framework (2008)
15. van Iersel, M., et al.: Creating shared situation awareness in a multi-platform sensor network. In: Proceedings of the 11th International Conference on Information Fusion, FUSION (2008)
16. Kester, L.J.: Model for networked adaptive interactive hybrid systems. In: Proceedings of COGnitive sytems and Interactive Sensors COGIS 2006 (2006)
17. Kester, L.J.: Rethinking the NAIHS model for NEC. In: Proceedings of SCI symposium on Agility and Resilience in NEC (2008)
18. Kurzweil, R.: The Singularity is Near. Viking, New York (2005)
19. Martin, J.: The meaning of the 21st century. Penguin Books, London (2006)

20. Neef, M., et al.: Organizing smart networks and humans into augmented teams. In: Proceedings of HCI 2009 (2009)
21. Niv, Y., Botvinick, M., Barto, A.: Workshop on hierarchical organization of behavior: Computational, psychological and neural perspectives (2007), http://www.princeton.edu/~yael/NIPS_workshop
22. Parnas, D.L.: Professional responsibility to blow the whistle on SDI. Abacus 4(2), 46–52 (1987)
23. Popper, K.: Logik der Forschung. Springer, Vienna (1934)
24. Rasmussen, J.: The role of hierarchical knowledge representation in decision making and system management. IEEE Transactions on Systems, Man and Cybernetics 15, 234–243 (1985)
25. Rechtin, E.: The Art of Systems Architecting. CRC Press, New York (2000)
26. Smits, F., Huizing, A., Rossum, W.V., Hiemstra, P.: A cognitive radar network: Architecture and application to multiplatform radar management. In: Proceedings of the 5th European Radar Conference (2008)
27. Steinberg, A., Bowman, C.: Rethinking the JDL data fusion levels. In: Proceedings of National Symposium on Sensor Data Fusion JHUAPL (2004)
28. US Department of Defense C4ISR Architecture Working Group: Levels of Information Systems Interoperability (LISI) (1998)
29. Zeleny, M.: Management support systems: Towards integrated knowledge management. Human Systems Management 7(1) (1987)

Part V
Case Studies and Applications

A Call for Sensemaking Support Systems in Crisis Management

Willem J. Muhren and Bartel Van de Walle

Abstract. In this chapter, we explore four information processing challenges commonly experienced in crisis situations, which form the basis of the design of information systems that should support actors in these situations. When we explore the difference between Sensemaking and decision making, two activities that are undertaken to cope with information processing challenges, we can understand the two types of information systems support that are needed. The first type—decision support systems—supports actors in dealing with information-related problems of uncertainty and complexity, and is the traditional focus of information systems design. The second type—sensemaking support systems—should support actors in dealing with problems of frames of reference, ambiguity, and equivocality, but is not commonplace yet. We conducted three case studies in different crisis situations to explore these information processing challenges: A case study of the sudden crisis of an airplane crash in the Barents Rescue Exercise, a case study of the yearly recurring forest fires crises in Portugal, and a case study of the post-conflict European Union Police Mission in Bosnia and Herzegovina. We discuss design premises for crisis management information systems and compare these to our findings, and observe that systems designed accordingly will provide for the necessary Sensemaking support.

1 Introduction

The term "crisis" derives from the ancient Greek word $\kappa\rho\iota\sigma\iota\varsigma$ (krisis), meaning moment of decision, judgment, or choice. In Greek tragedies, for example, $\kappa\rho\iota\sigma\varepsilon\iota\varsigma$ (kriseis) were turning points where human choice could make a fundamental

Willem J. Muhren
Department of Information Systems and Management, Tilburg University,
PO Box 90153, 5000 LE Tilburg, The Netherlands
e-mail: w.j.muhren@uvt.nl

Bartel Van de Walle
Department of Information Systems and Management, Tilburg University,
PO Box 90153, 5000 LE Tilburg, The Netherlands
e-mail: bartel@uvt.nl

R. Babuška & F.C.A. Groen (Eds.): Interactive Collaborative Information Systems, SCI 281, pp. 425–452.
springerlink.com

difference to the future [25]. Nowadays, we use the term crisis for either "a crucial stage or turning point in the course of anything," which reflects the original meaning of the word, or "a time of extreme trouble or danger" [9]. In this chapter, we use the term crisis to describe the latter type of events, when people are struck by disastrous circumstances. As a consequence of such events, however, people will inherently find themselves in a crucial stage or turning point in which they do not only need to make decisions on the course of action they will pursue, but also need to make judgments on what is happening around them and on what the decision context is. We use the theory of Sensemaking to study exactly this: how people make sense of their environment, and how they give meaning to what is happening. Sensemaking is a crucial process in crises, as the manner and thereby the success of how one deals with crucial events is determined by the grasp one has of a situation.

Crisis environments are characterized by various types of information problems that complicate the response, such as inaccurate, late, superficial, irrelevant, unreliable, and conflicting information [30, 32]. This poses difficulties for actors to make sense of what is going on and to take appropriate action. Such issues of information processing are a major challenge for the field of crisis management, both conceptually and empirically [19]. Commonly, research is aimed at how often, when and why process members use a certain technology. However, such research does not address the more fundamental and underlying question of what kind of processes take place when people make use of the information they retrieve. This involves less explicit information processing mechanisms that are considered a "black box" [23]. To design appropriate supporting systems, we need to know more about what is being supported [40], and that is what we want to investigate in this contribution.

We focus on how people cope with information processing challenges in crisis situations to better understand what type of Information Systems (IS) support people need in crisis management. More specifically, we examine how information processing challenges of ambiguity, uncertainty, equivocality, and complexity are related to Sensemaking and/or decision making, and observe through case study research how actors deal with these information processing challenges in three specific crisis situations: A case study of an aviation crash in the Barents Rescue Exercise, a case study on forest fires in Portugal, and a case study in Bosnia–Herzegovina of the European Union Police Mission (EUPM).

This chapter is outlined as follows: First, we discuss Sensemaking and the constructs that characterize Sensemaking. We then discuss several information processing challenges, how they relate to Sensemaking and decision making, and how IS can support them. Thereafter, we discuss the case studies we conducted, the methodology we used, the findings related to how actors deal with information processing challenges, and analyze the findings. Finally, we discuss implications for the design of crisis management IS and present our conclusions.

2 Sensemaking

Sensemaking literally means making sense of things, making things sensible [37, p. 16]. Organizational and management scholars have defined and used the concept of Sensemaking in different ways. March and Olsen related Sensemaking to experiential learning [4, p. 77], as "individuals and organizations make sense of their experience and modify behavior in terms of their interpretations" [15, p. 56]. Huber and Daft talked about Sensemaking as the construction of sensible and sensable events [11, p. 154]. From Starbuck and Milliken's perspective, "Sensemaking has many distinct aspects: comprehending, understanding, explaining, attributing, extrapolating, and predicting, at least. (...) What is common to these processes is that they involve placing stimuli into frameworks (or schemata) that make sense of the stimuli" [27, p. 51]. This broader notion of Sensemaking is also acknowledged by Thomas et al., who view information seeking, processing, creating, and using to be central activities of Sensemaking [28]. This means that Sensemaking is not a noun, but a verb; that it is a process, with sense as its product [19].

We heavily rely on Weick's extensive work on Sensemaking [34–39]. He matured the concept of Sensemaking in organizations, among others by defining its underlying constructs. In the following section, we expound on these properties.

2.1 Sensemaking Constructs

Weick [37] distinguishes between seven properties of Sensemaking. Although they might not be fully exhaustive nor exclusive in the scientific sense, they still are a grand attempt to render the way people deal with interruptions more tangible [19]. The seven different properties of Sensemaking can be captured by the acronym SIR COPE: Social context, Identity construction, Retrospection, Cue extraction, Ongoing projects, Plausibility, and Enactment [17–21, 37–39].

Social context

"People learn about events when they compare what they see with what someone else sees and then negotiate some mutually acceptable version of what really happened" [34]. Cognitive and social aspects of Sensemaking are inextricably linked. People need social anchors and a form of social reality [38], because what we say or think or do is contingent on what others say and think and do. Sensemaking requires talking, interaction, conversation, argument, and dialogue with others [18].

Identity construction

Depending on who the Sensemaker is, the definition of what is happening will also change. What the situation means is defined by who one becomes while dealing with it or what and who one represents. "The Sensemaker is himself or herself an ongoing puzzle undergoing continual redefinition, coincident with presenting some self to others and trying to decide which self is appropriate" [37, p. 20]. An organization seeks to discover what it "thinks" and "knows" about itself and its environment, and this construction of identity is the basis for imparting meaning to information within the organization and, eventually, determining what problems must be solved.

Retrospection

"Sensemaking is influenced by what people notice in elapsed events, how far back they look, and how well they remember what they were doing" [38]. Weick, Sutcliffe, and Obstfeld [41] point out that answers to the question, "what's the story?" emerge from retrospect, connections with past experience, and dialogue among people who act on behalf of larger social units. Answers to the question, "now what?" emerge from presumptions about the future, articulation concurrent with action, and projects that become increasingly clear as they unfold.

Cue extraction

Sensemaking is influenced by both individual preferences for certain cues as well as environmental conditions that make certain cues figural and salient [38]. We notice some things and not others. We pay attention and extract a particular cue and then link it with some other idea that clarifies the meaning of the cue, which then alters the more general idea to which we linked the cue, and on and on. Extracted cues enable us to act, which increases our confidence and confirms our faith in earlier cues [18].

Ongoing projects

Sensemaking has neither a beginning nor a formal end. Instead, it "takes place in a continuing and dynamic fashion as events unfold and we continually seek to understand what events mean in relationship to our organizations" [21]. Most of us at any given time find ourselves "in the middle of something." As we move from one situation to another, we make and revise assumptions and beliefs along the way. Once you cannot keep pace with the action, you lose context, information, situated cognition, and tools made meaningful by actual use [39].

Plausibility

"Sensemaking is about coherence, how events hang together, certainty that is sufficient for present purposes, and credibility" [38]. Looking for what is plausible is often of more practical help than finding accuracy [21]. Plausibility helps us explore what we see and energizes us to act; the search for accuracy can de-energize us as the search drags on and on.

Enactment

People often do not know what the "appropriate action" is until they take some action, guided by preconceptions, and see what happens. "Action is a means to gain some sense of what one is up against, as when one asks questions, tries a negotiating gambit, builds a prototype to evoke reactions, makes a declaration to see what response it pulls, or probes something to see how it reacts" [38]. Action determines the situation, as it creates an orderly, material, social construction that is subject to multiple interpretations [35]. The basic premise is that there is no objective environment out there separate from one's interpretation of it. Thus, the organization creates or enacts parts of its environment through selective attention and interpretation.

Weick et al. [41, p. 419] formulate a gripping conclusion on what the seven Sensemaking properties are all about: "Taken together these properties suggest that increased skill at Sensemaking should occur when people are socialized to make do, be resilient, treat constraints as self-imposed, strive for plausibility, keep showing up, use retrospect to get a sense of direction, and articulate descriptions that energize. These are micro-level actions. They are small actions. But they are small actions with large consequences."

3 Information Processing Challenges and Support

Information-related problems cause people to have difficulties in processing information in crisis situations. Very often the terms uncertainty, complexity, ambiguity, and equivocality are used in an attempt to stress these "difficult circumstances" people have to cope with. However, these terms are mostly used interchangeably, without exactly describing what is meant. Zack [42] distinguished these four terms according to two dimensions: the nature of what is being processed and the constitution of the processing problem.

The nature of what is being processed is either information or frames of reference. With information, we mean "observations that have been cognitively processed and punctuated into coherent messages" [42]. Frames of reference [4, p. 108], on the other hand, are the interpretative frames which provide the context for creating and understanding information. There can be situations in which there is a lack of information or a frame of reference, or too much information or too many frames of reference to process.

Table 1 Information processing challenges (adapted from [42])

-	**Information**	**Frame(s) of reference**
Lack of...	Uncertainty	Ambiguity
Variety/diversity of...	Complexity	Equivocality

As shown in Table 1, this breakdown into two dimensions leads to four different types of information processing challenges [42]: uncertainty, complexity, ambiguity, and equivocality.

Uncertainty is a situation in which there is not enough information already possessed by the organization to perform the task [5, 8]. *Complexity* is the second information-based challenge, and arises when there is more information than one can easily process [42]. Although information-related problems are not the only type of problems that lead to complexity, this narrow definition suffices for our present focus on information-related processing challenges. When there is a situation in which one does not have a framework for interpreting information, there is *ambiguity* [42]. Finally, *equivocality*—or confusion—is a situation in which one has several competing or contradictory frameworks [6]. Ambiguity and equivocality may at first sight seem to be synonymous terms, but they are used throughout literature to distinguish between unclear meaning (ambiguity) and the confusion created by two or more meanings as in a pun or equivoque (equivocality) [37, p. 92].

The four information processing challenges are not mutually exclusive, but often exhibit a natural hierarchy of difficulty in practice. Ambiguity is the most difficult challenge to overcome, since it involves developing a frame of reference when none is available. When people negotiate their interpretations and share their understandings, a situation of equivocality can arise as there are multiple conflicting frames of reference. There is a balance needed for, on one side, creating new frames of references and reducing the frames on the other side. Once an appropriate frame is constructed, the situation may reveal itself to be uncertain, complex, or both. This will determine whether a strategy of information seeking or information reduction should be adapted.

In the following section, we will highlight the difference between Sensemaking and decision making, and discuss which information processing challenges they both are aimed at.

3.1 Sensemaking versus Decision Making

Decision making is traditionally viewed as a sequential process of problem classification and definition, alternative generation, alternative evaluation, and selection of the best course of action [26]. This process is about strategic rationality, aimed at reducing uncertainty [6, 36]. Uncertainty can be reduced through objective analysis because it consists of clear questions for which answers exist [5, 40]. Complexity can also be reduced by objective analysis, as it requires restricting or reducing factual information and associated linkages [42].

On the contrary, Sensemaking is about contextual rationality, built out of vague questions, muddy answers, and negotiated agreements that attempt to reduce ambiguity and equivocality. The genesis of Sensemaking is a lack of fit between what we expect and what we encounter [40]. With Sensemaking, one does not look at the question of "which course of action should we choose?", but instead at an earlier point in time where users are unsure whether there is even a decision to be made, with questions such as "what is going on here, and should I even be asking this question just now?" [40]. This shows that Sensemaking is used to overcome situations of ambiguity. When there are too many interpretations of an event, people engage in Sensemaking too, to reduce equivocality.

Sensemaking is concerned with making things that have already happened meaningful [3] and is more than problem definition, as Weick and Meader [40] explain: "to label a small portion of the stream of experiences as a 'problem' is only one of many options. The stream could also be labeled a predicament, an enigma, a dilemma, or an opportunity. Each of these labels has a different implication for action. If it is a problem, then solve it; but if it is a predicament, then accept it; if it is an enigma then ignore it; if it is a dilemma then define it anyway; and if it is an opportunity then exploit it. To call something a problem is the outcome of Sensemaking."

People usually enhance their Sensemaking efforts after a sudden loss of meaning, when they experience a diffuse sense of unease that perhaps something needs to be done, although no one can say for sure. People then "must make sense of an uncerta in situation that initially makes no sense" [24, p. 40], and try to shape and give definition to the decision context by processes of Sensemaking [40]. Sensemaking differs from interpretation as Sensemaking "is about the ways people generate what they interpret" [37, p. 13].

Just as the information processing challenges from Table 1 are not mutually exclusive, Sensemaking and decision making cannot be separated, but instead operate simultaneously. Meaning must be established and then sufficiently negotiated prior to acting on information [42]: Sensemaking shapes events into decisions, and decision making clarifies what is happening [40].

The previous discussion does not imply that dealing with information challenges is not important for Sensemaking. It is not possible to separate the two activities of coping with information challenges and interpretation challenges. However, the main activity of Sensemaking is ascribing meaning to what is really happening and not gathering information on a situation. More information does not automatically lead to better Sensemaking [12]. The central problem requiring Sensemaking is mostly that there are too many potential meanings, and so acquiring information can sometimes help but often is not needed. Instead, triangulating information [34], socializing and exchanging different points of view [20], and thinking back of previous experiences to place the current situation into context, as the retrospection property showed us, are a few strategies that are likely to be more successful for Sensemaking.

The previous discussion enables us to make a clear distinction between decision making and Sensemaking: Decision making is about coping with information processing challenges of uncertainty and complexity by dealing with information, whereas Sensemaking is about coping with information processing challenges of ambiguity and equivocality by dealing with frames of reference. This information processing distinction between decision making and Sensemaking has not been made previously in literature. We will apply this dichotomy in the discussion of the case studies in Section 4.

3.2 Information Systems

Some interesting work has been done on the intersection between IS and crisis management, for example [10, 29], but recent research and practice have shown that current crisis management IS are long overdue [30]. Interesting and substantial research exists on Sensemaking and crises situations such as [36], but relatively few studies use Sensemaking as an analytical lens for the design of information technology [40], and there is scarce research on how IS can support information processing challenges—specifically related to Sensemaking—in crisis management [14]. In the hectic circumstances of crises, people could benefit from IS support in making sense of what is going on. There are, however, many challenges to accomplish this, since data in IS "contain only what can be collected and processed through machines. That excludes sensory information, feelings, intuitions, and context — all of which are necessary for an accurate perception of what is happening. (…) To withhold these incompatible data is to handicap the observer. And therein lies the problem" [34].

The problems of managing information and managing frames of reference are "tightly linked in a mutually interacting loop" and require "managing information and the systems that provide it" [42]. IS have been generally designed to overcome the information problems from Table 1. Most IS are aimed at either storing and retrieving information to reduce uncertainty, such as database management systems and document repositories, or at analyzing and processing large amounts of information to reduce complexity, such as decision support systems [31]. However, as we have previously discussed, information related strategies are not always helpful in coping with a variety of potential meanings.

Problems of interpretation and the creation and management of frames of reference, which aids Sensemaking, have generally not been taken into account when designing IS. Most IS currently seem to intend the opposite because they aim at replacing or suppressing the possibility to make sense of situations. For example, heavily inspired by Herbert Simon's work, IS research and practice use structured data as a substitute for information [2]. However, information is not a commodity; it is a skilled human accomplishment. Information is meaning resulting from a person's engagement with data [7]. IS should thus be designed to take dialogue, interpretation, and an individual's search for meaning as sacred [2]. We use the term "Sensemaking Support Systems" [18, 37, 40] to denote systems that should be designed in the future to support Sensemaking. "We need to understand more about Sensemaking Support Systems as well as Decision Support Systems, which means we need to know more about what is being supported" [37, p. 179]. In the following section we discuss three case studies we conducted in different crisis situations, in which we examined how people handle and process information in crises to understand how supporting IS should be designed.

4 Case Studies

In this chapter, we report on three case studies we conducted [16]: the sudden crisis of an airplane crash in the Barents Rescue Exercise, the yearly recurring crises of forest fires in Portugal, and the post-conflict state building EUPM in Bosnia and Herzegovina. We selected these case studies on grounds of their differing crisis characteristics, as we wanted to investigate how people handle and process information and make sense in a broad spectrum of crisis management situations. In Table 2,

Table 2 Taxonomy of crisis management case studies

-	Case study 1: Barents Rescue Exercise	Case study 2: Forest fires in Portugal	Case study 3: EUPM in Bosnia and Herzegovina
Crisis type	Accident	Natural disaster	Conflict
Timing and type of operations	In casu; dealing with the plane crash	Ex ante and in casu; preventing and dealing with forest fires	Ex post; state building
Level of management	Operational	Strategic	Operational
Time span	Short	Ongoing, long term	Medium term
Predictability	Sudden	Expected	Expected

we show how these case studies differ from each other on varying aspects of crisis management.

We first describe the common methodology used. After this, we describe the three case studies: an introduction on each case study, how we conducted it, the findings from the case study, and a discussion of the findings. The findings are for each case study organized according to the nature of the processing of information: the dealing with information to reduce uncertainty and complexity, and the dealing with frames of reference to reduce ambiguity and equivocality. Literal quotations from interviewees are indicated with quotation marks.

4.1 Methodology

For these three case studies, we used an interpretive approach and conducted in total 20 interviews with a common approach and research aim. Interpretive research attempts to understand phenomena through the meanings that people assign to them [22]. Interpretive methods of IS research take into account the context in which the information system is used with that particularity that it also acknowledges the mutual interaction between the system and its context. To succeed in the opening up of these mutual interactions, the researcher has to interact with the research participants. Klein and Myers [13] state that the "data are not just sitting there waiting to be gathered, like rocks on the seashore." Data are produced in a social interaction of the researchers with the participants.

For our research design, we drew on Walsham [33] and Klein and Myers [13], who provide comprehensive guidelines on how to conduct interpretive case study research in the IS domain. On the practical level this shows itself throughout our research by means of a colorful interviewing style with which we stimulated our respondents to answer us difficult questions related to the Sensemaking constructs [19], among others by using statements, dichotomies, metaphors and dilemmas, relying heavily on examples and anecdotes, and calling upon their imagination to find out the bottom-line. In our interpretive case studies, we adjusted our style to the respondent, such as to his/her language, world view, professional experience, and personality.

The interviews were semi-structured, in the sense that we knew which topics to touch upon and had a list of the general points we wanted to find out, related to the seven Sensemaking constructs, but adjusted the questions to how the interview was evolving. Permission for tape recording was granted for 19 interviews; only notes were taken at the interview in which the respondent was not comfortable with the use of a tape recorder. Confidentiality was guaranteed in all interviews.

We supplemented data and understanding of the case studies with other types of field data, such as reports, background stories on various websites, press articles, brochures, and informal interaction at the sites of the case studies.

4.2 Case Study 1: Barents Rescue Exercise

Barents Rescue is a series of field training exercises which are organized and conducted by the countries within the Barents Euro-Arctic Council (BEAC): Norway, Sweden, Finland, and Russia. The Barents Rescue 2007 Exercise was held in October 2007 in Saariselkä and Ivalo, located in the Northern part of Finland. The exercise aimed to facilitate communication, coordination, and cooperation between countries and civil–military services that may be involved in crises relevant to the Barents region. The project consisted of a series of planning conferences, training events, and exercises, of which the Barents Rescue Exercise in October 2007 was the final and main event.

There are several challenges to crisis response operations in the Barents region: The distances between cities are big, there is limited infrastructure, because of the scarce population there are limited resources for rescue operations, and the climate conditions are severe in winter. For this reason, it is important that the countries in the region plan on how to join forces when responding to a crisis. The Barents Rescue Exercise was aimed at training such cooperation and improving crisis preparedness.

The scenario for the exercise was an aviation accident. A British aircraft executed an emergency landing in the uninhabited areas of the Inari municipality. The more than 200 passengers were mainly tourists from the United Kingdom. The reason for the crash was not immediately clear, but it was very likely that many passengers were injured or deceased. The scenario involved different stakeholders from the BEAC countries, such as alarm centers, national rescue services, hospitals, the military, private companies, and voluntary organizations.

The exercise included three phases with different approaches to crisis management. The first phase, the alarm exercise, was aimed at exercising the alarming and gathering of possible resources in the Barents region in case of a major crisis. The second phase, the table top exercise, was aimed at exercising the practical response in the crisis area, consisting of a command post exercise, table top exercise, and exercise with utilizing virtual tools. The third and final phase, the field training exercise, was aimed at training the capabilities of organizations and agencies involved in the direct response to a crisis, both on the operational level and on the strategic level.

4.2.1 Case Study Implementation

In October 2007, we traveled to the Northern part of Finland for the Barents Rescue Exercise. We conducted four interviews with key people involved in the crisis response operation: One person working at the On Site Operations Coordination Center (OSOCC), one person working at the Local Emergency Management Authority (LEMA), one person working at the On Site Command Center (OSC), and one person in charge of leading the medical team.

Besides these interviews, we got a good overview of how the actors handle and process information in the exercise through observations [1]. As the table top exercise and the field training exercise were each organized on one location, we could observe how all the actors dealt with the crisis. These observations were used for the interviews, as we could ask specific questions on the actions of the observed people and what had happened in the exercise. The Barents Rescue Exercise provided a good opportunity for us to observe a crisis on the operational level, as it is difficult to observe a crisis in a real-life situation.

4.2.2 Case Study Findings on Dealing with Information

We observed several information problems in the exercise. There were problems of conflicting information when air traffic control and the rescue services labeled the accident site with two different geographical coordinates. Information often arrived late in the alarm exercise, as the primary communication technology that was used was the fax, the most commonly used tool at the time the procedures were put in place. Other unnecessary delays were caused in the field because it was not clear which organization was in charge of the rescue operation. Moreover, not all actors were informed about the Emergency Rescue Center in Tromsø as a contact point for the response operation. Another factor delaying the response was the information provision to the medical team. After the initial alarm, it took 90 minutes before the medical team received the first information about the victims. The medical team was only able to determine the kind of assistance that was needed after this information was received.

Actors dealt with these information problems in different ways. People indicated to have preferences concerning the level of detail of information that is useful to have. Although most interviewees clearly stated that more detail was better for them, there were also cases in which actors argued that anything more than essential information was not needed and in fact distracted them from their job.

People also had different ways for making sure they were communicating as effectively as possible. For example, one actor started handling the crisis by listing his key contact persons and phone numbers to prepare himself for information exchanges in the upcoming hectic circumstances.

What happened in practice when people communicated in the crisis was that messages were double-checked to make sure that they arrived properly, for example, in radio communications by the recipient repeating the main words.

The interviewee who was working in the OSC said that "you are forced to trust the people with whom you are exchanging information, no matter who is on the other side" when responding to a crisis. He felt that there is not much time to think about whether the information received from others is right or wrong, and one therefore has to act on the information that is available.

For the people in the OSOCC, it was important to receive information on what is happening on a continuous basis: "We need to get the key cards, to keep being informed". The OSOCC and LEMA were operating in two separate but adjacent rooms. A liaison officer of LEMA was appointed to regularly brief the people in the OSOCC on what was happening. This briefing was, however, not frequent enough, especially since LEMA was very busy. Moreover, the people in the OSOCC did not use any system to receive information and communicate with the LEMA.

4.2.3 Case Study Findings on Dealing with Frames of Reference

Some people identified experience to be a helpful resource for crisis response while others felt it to be essential: "From your experience you cannot remember everything, but many things stay in the back of your head and become a routine."

For one key actor it was important to create time to think about the situation: "I read the documents and think about what's next. And if there is too much noise, I go perhaps out, take a cup of coffee, and smoke a cigarette. Because when it's so hectic, you have to clear your mind and think about what's going on."

Most of the interviewees indicated that they value effectiveness over efficiency when responding to a crisis. For example, they would request more resources than deemed absolutely necessary, just to be sure to have enough.

In the alarm exercise, actors had different expectations of each other. For example, regarding the procedure to follow, Finland expected a very fast response from other countries, while Norwegian actors took their time to find out how many resources were needed and available.

The comparison of frames of reference on the strategic level by exchanging points of view and understanding of the crisis was hampered because the OSOCC and LEMA worked more or less independently.

Language was an issue of concern in this international exercise. As responders cannot use their native language when cooperating with responders from other countries, it was difficult for them to choose the right words to use. Culture is also important, as an interviewee emphasized: "We [Finnish people], together with the Swedes and Norwegians, think in the same way and have the same kind of picture in mind all the time. We have the same kind of systems [structures] in place, and we understand each other. But with the Russians it's a bit harder."

4.2.4 Discussion of the Findings

The initial response to an airplane crash should be a rapid response, as surviving passengers need immediate medical aid. Problems of uncertainty—a lack of information—delayed the response: the alarm center could not inform actors swiftly due to an outdated alarm procedure by fax, the contact point and focal point for the response operation was unknown to many actors, and the medical team received information on the victims far too late. Moreover, the information exchange between LEMA and OSOCC was not good. In the future this can be avoided by having LEMA and OSOCC in the same location, or by using shared IS. Then, the people in the OSOCC are not dependent on briefings but can actively follow what is going on. Problems of complexity showed up, for example, when determining the accident site, as the IS that were used by various actors were not interoperable.

People try to cope with uncertainty and complexity in different ways. Interviewees indicated to have different preferences concerning the level of detail of information, and, consequently, either seek more detailed information or refrain from this level of detail. Actors take precautionary measures against complexity by organizing information in certain ways, double checking information they receive to avoid having conflicting information, and acting on uncertain information as it is the only information they have.

Situations of ambiguity and equivocality are more difficult to observe, but revealed themselves in the interviews. Interviewees indicated that when they have no accurate frame of reference, they rely on their experience, take time to think about the situation, and act according to the (inaccurate) frame of reference they have. It was difficult for the actors to cope with equivocality. Communication and information exchange between actors was not sufficient, due to physical separation and lack of appropriate systems to support this. Also, the international context of response hampered the discussion and exchange of frames of reference, as actors had different cultural backgrounds and were speaking different languages.

4.3 Case Study 2: Forest Fires in Portugal

Forest fires are a great concern for Portugal, as more than a quarter of the country is covered by forests, and droughts in summertime increase the likelihood of such natural disasters. In recent years, Portugal has been facing extremely hot and dry summers, with highest peaks in 2003 and 2005. The fire risk in Portugal has been increased by changes in land use practices. Rural exodus has left a large area of land uncultivated, where combustible materials can now unnoted trigger big fires when droughts occur.

The National Authority for Civil Protection (Portuguese acronym ANPC) has the primary role in planning, coordinating, and implementing the Civil Protection policy. The ANPC is a central operational service under the direct administration of the Ministry of Interior. The ANPC maintains its own operational structure, the National Command for Relief Operations that ensures the operational command in terms of relief operations and the integrated operational command of all the fire brigades in accordance with the legal system. The Integrated System for Relief and Protection Operations is a set of structures, norms, and procedures which ensures that all civil protection agents act under a sole command. This integrated system aims at responding to (imminent) crisis situations.

The Portuguese State does not have its own fire brigades. The great majority of the fire brigades are volunteer fire fighter associations; others belong to the city councils and private companies. The Portuguese Forest Services play an important role in the management of the forests in Portugal, especially in mapping the risks of hazards and in educating people on how to prevent forest fires.

In 2003, 2004, and 2005, Portugal could not cope with the forest fires themselves and, therefore, requested outside assistance through the European Commission (EC)'s Community Civil Protection Mechanism. The Community Civil Protection Mechanism was established in October 2001 and is an operational instrument designed to enhance preparedness and to mobilize immediate civil protection assistance in the event of disasters. It can be activated in case of natural and man-made disasters by any country in the world, after which one or more of the 30 participating states—the European Union (EU) member states as well as Liechtenstein, Norway, and Iceland—will try to offer their assistance. The mechanism is coordinated by the Monitoring and Information Center (MIC) of the EC in Brussels.

4.3.1 Case Study Implementation

For the Portuguese forest fires case study, we conducted interviews using the previously discussed interpretive approach, interviewing six people who are both involved in managing and preventing forest fires in Portugal, and who are involved in coordinating the international response to the Portuguese forest fires through the Community Civil Protection Mechanism. The interviews were aimed at finding out how these actors handle and process information related to the Portuguese forest fires.

In December 2007, the first author traveled to Lisbon to interview two people working at the ANPC and two people at the Portuguese Forest Services, and got a chance to observe the operations center of the ANPC. Two follow-up interviews were conducted in February 2008 in Brussels: One interview with a person working at the EC, Directorate-General for the Environment, who also demonstrated the MIC, and an interview with a person working at the Civil Protection Unit of the EU's Council Secretariat.

4.3.2 Case Study Findings on Dealing with Information

The Portuguese Forest Services are mainly focused on preventing forest fires. They produce two kinds of fire hazard maps and share these with the district level and the ANPC: One structural, not very detailed map with a large pixel size, and one map with a smaller pixel size. The latter is better suited for teams on the ground and indicates the hazards in summer; in winter, it is used to find areas where the authorities can conduct "prescribed firing," which are techniques to manage fuel.

Historical information on where the forest fires took place is important information for producing these maps. Depending on the tree species, it takes on average 5 years for a forest to become a potential "problem situation" when forest fires are a substantial risk, but it takes many more years for a forest to return to their pre-fire state. The Portuguese Forest Services experience problems of acquiring up-to-date information for producing the fire hazard maps, as these often have to be bought and land-use data is very expensive.

Both detection of forest fires and first response to forest fires are very important, as the difficulty of fighting such fires increases nearly exponentially in time. Forest fires can be detected in three ways: by somebody from the population calling 117 (the dedicated forest fire emergency phone number) or 112 (the normal emergency phone number), by a surveillance post, or by a surveillance brigade. The ANPC is responsible for informing the population, the people who work for civil protection, the media, and the national and district level command structure. The district command for relief operations is responsible for the deployment of means, both terrestrial and aerial, and conducting the first intervention. If they do not succeed, or if the fire crosses the district level, the ANPC is responsible for the enlarged combat situation. The ANPC then sends more resources to the field, which can be requested from other districts.

When the crisis is overwhelming and national resources are insufficient, the minister—advised by the national commander—can activate the Community Civil Protection Mechanism by sending a message to the MIC in Brussels. The ANPC is permanently connected to the Common Emergency Communication and Information System (CECIS) of the MIC. CECIS allows for sharing of current information on the situation as well as identification of what is needed. Countries can also indicate whether they can provide assistance or not.

Daily briefings take place at the ANPC in summer on the general hazard situation of the country with representatives from the major players. The National Guard and the armed forces have permanent liaison officers stationed at the ANPC. In the case of a severe crisis, liaison officers from other civil protection agencies are present at the ANPC, such as officers from the forest department, the maritime authority, the police, the medical services, and the meteorological institute: "In the daily briefings we bring information from all the agents who share responsibility in terms of civil protection." At the end of these briefings, the ANPC provides all actors with a summary of these briefings in writing. The national commander takes decisions according to the information shared and the analysis conducted in the meeting, for example, on whether to increase the readiness level or to pre-mobilize resources.

The ANPC mainly uses the media to communicate with the population and other external parties. When a threatening situation arises, the ANPC sends out an alert and the key players from the media come to the office where they are briefed. The media outlets then disseminate the message to the citizens, including any measures that should be taken. The ANPC also organizes a press conference every week on what is happening and what the expectations are for the following week. In severe crisis situations there is a daily press conference.

The ANPC also displays information for the citizens on their website, such as information on where current fires are located and how many fire fighters are working in the area. The website also advices people on what to do for each level of forest fire risk. The people at the ANPC realize that it is important to provide accurate information: "We cannot always tell people that tomorrow will be the worst day."

There is a lot of redundancy in the sharing of information. When civil protection actors send a message, they make sure it arrives by using as many means as possible. At the ANPC, people are updated on the situation by text messages, and senders usually immediately call to check whether the message has been received and read. When the Community Mechanism for Civil Protection is activated, fax messages are sent and phone calls are made to other countries besides the formal request through CECIS: "In a crisis situation this is quite normal as you want to make sure the message gets to the person."

4.3.3 Case Study Findings on Dealing with Frames of Reference

Forest fires in Portugal do not only happen during the summer, they also occur in wintertime. Winter fires are good if they are controlled, since what burns at that time cannot burn in summer when the fires are mostly not controllable. However, the concept of "good fires" is new and generally unknown by the population. It therefore is important to inform the people that fires during winter are not bad if they are managed appropriately. The Portuguese Forest Services promote this idea on their website and have launched a campaign on television, radio, and in newspapers. Another campaign highlighted that people should not light fires near forests during summer, and that they should clear all combustible materials. Besides these nation-wide campaigns, the Portuguese Forest Services have engaged in direct contact with local shepherds, farmers, and forest owners to try to change their behavior and spread the word that not all fires are bad.

The ANPC's main target is currently to educate people. They do this together with big companies such as supermarket chains. One initiative included printing the phrases "Portugal without fires depends on everyone" and "You should not use fire on a hot day" on the supermarket's plastic bags. The ANPC has also been advertising in football stadiums and on football shirts, and has broadcasted their campaigns on television and advertised in newspapers. "You have to give good and correct information to your population, otherwise you can have situations as in 2003, 2004 and 2005", one interviewee said. It has been successful until now: the number of ignitions has been reduced since the start of these campaigns.

There have been instances in which the forest fire situation was under control, but people started to panic as images of the fire were aired on television. One interviewee stated that the best way to prevent these kinds of situations is to actively cooperate with the media.

To aid interpreting new information on the hazard situation, the ANPC visualizes information about the major incidents that are occurring in the country on a geographical IS-based map. There is a special screen on which the forest fire situation is projected.

After each big crisis in Portugal, the ANPC invites all the actors that were involved in the response to discuss the process, what was done and what worked or did not work. These lessons learned are then incorporated into their procedures.

The MIC has the overall European perspective on civil protection. The mechanism can facilitate and give an "educated hint" as to which country to help, but in the end the countries themselves decide on where they want to provide or accept assistance. Countries do not always accept assistance, because the help that is offered may not be exactly what they wanted, or it comes with a price tag they are not willing to pay.

Notably, the use of language in CECIS is not standardized. This sometimes causes problems in interpreting what is meant by a request or an offer, and has resulted in countries bringing resources that were not of use in that crisis situation.

"Nothing helps more in emergency situations than people knowing each other," many interviewees mentioned. The Community Mechanism for Civil Protection and the MIC bring together people from all 30 countries for events like training sessions, workshops, common exercises, and meetings. Through these types of interaction people become more familiar with the realities of the other countries: "The more you exchange information, the more you know where and how to target your request for assistance. And the better you make your request, the better you get answered."

4.3.4 Discussion of the Findings

In summer, the ANPC needs to get accurate and timely information on the actual situation of the forest fires, and needs to provide the actors on the ground and the citizens with information concerning the forest fires. As forest fires pose a continuous threat to the country, there are good systems in place for obtaining and providing this information, such as the different ways a forest fire can be reported, daily briefings with all key response actors, and good cooperation with the media.

The importance of timely and accurate information is evident in the behavior of actors when communicating through the Community Mechanism for Civil Protection, as redundant information exchanges take place to make sure that recipients have read a message and understood it correctly. This, however, might lead to a situation of complexity, as people receive many notifications on different communication media and might lose track of which message is new for them and which message is part of a "reminder."

With such yearly recurring crises, the actors have chosen for a strategy of prevention rather than only response. The Portuguese Forest Services play a leading role in this by mapping the risk of forest fires in the different areas. They, however, experience problems of uncertainty as they have difficulties in obtaining all relevant information for such maps.

Informing citizens is important in Portugal's crisis situation. The Portuguese Forest Services cooperate with the ANPC for prevention campaigns, the ANPC provides information on forest fires on their website, and there are press conferences in summertime. All these activities are aimed at removing uncertainty that citizens face.

Some campaigns are also aimed at removing ambiguity by providing a frame of reference for citizens to make sense of what is happening. For example, the concept of "good fires" was unknown to the majority of people, but after the campaign citizens would better understand the fires that occur in wintertime. Such educational campaigns could also prevent interpretation problems leading to panic, such as the broadcast of fires on television at a time when fires are under control.

The main strategy to cope with equivocality is to socially mix and engage in discussion. After a crisis, the key actors discuss their understanding of what happened and what went wrong, and try to create a common frame of reference which is then incorporated into their procedures and used to deal with the next crisis. Moreover, one interviewee mentioned the importance for actors to socially mix between different country representatives, as it helps civil protection actors to understand the other countries' frames of reference: the situation in their respective country, how they respond, their concerns, etc. This is very useful for future collaboration and assistance, as the other actors' frames of reference are then better understood.

4.4 Case Study 3: European Union Police Mission in Bosnia and Herzegovina

Since 1991, the EC has set aside more than 2.5 billion euros to support Bosnia and Herzegovina with public administration reform, justice and home affairs-related issues, and improvement of the investment climate as key target areas. Police reform has been one of the main obstacles to Bosnia and Herzegovina's integration in the EU. On July 1, 2005, the EC concluded that police reform was the single remaining obstacle to the Stabilization and Association Agreement.

International efforts to reform the Bosnian police force started immediately after the peace agreement when the United Nations led International Police Task Force (IPTF) was deployed. The IPTF comprised more than 2,000 international police officers from 43 countries.

On January 1, 2003, the EU launched the EUPM in Bosnia and Herzegovina. It was the first mission initiated under the European Security and Defense Policy (ESDP), initially intended to cover a three-year period and included around 500 police officers from more than 30 countries. Following an invitation by the Bosnian authorities, the EU decided to establish a continued and refocused EUPM of 200 international staff members and a mandate of 2 years, until the end of 2008. The EUPM supports the police reform process and continues to develop and consolidate local capacity and regional cooperation in the fight against organized crime.

In addition to the EUPM, the EU has an European Union Special Representative (EUSR) and a military crisis management mission, European Union Force (EUFOR) Althea, in Bosnia and Herzegovina. The EUSR is in charge of assuring the coherence of the ESDP activities. The EUSR can offer political advice to the EUFOR among others regarding organized crime, and facilitates coordination between Brussels and Sarajevo. The EUFOR also has a paramilitary police force under its command. The EUPM, however, is the lead in the coordination of policing aspects of the ESDP efforts in the fight against organized crime.

4.4.1 Case Study Implementation

In April 2008, we traveled to Sarajevo and conducted a total of 12 interviews using the interpretive approach. We conducted interviews with EUPM people from the general management, EU Coordination Office, Security Department, and Press and Public Information Department. Moreover, we conducted interviews with people who are directly or indirectly involved in the EUPM: representatives of the EC Department Police Projects, EUFOR, EUSR, Canton Sarajevo Police Department, Organization for Security and Co-operation in Europe, and the Populari think-tank.

4.4.2 Case Study Findings on Dealing with Information

To support the local police in fighting organized crime, it is important for EUPM to collect criminal intelligence from everywhere in Bosnia and Herzegovina and analyze it to create a good situational overview. The main sources of information for EUPM's headquarters are the local police, the daily reports from EUPM's field offices, and the security awareness working group. EUPM follows the general statistics of everyday crime in the country to notice crime trends that could be part of a bigger picture. Information that relates to police crime and police corruption is also important. The main source of such information is not the local police, but the local community. EUPM has informants who have high positions in society.

EUPM tries to obtain local information on the overall security situation together with EUFOR through EUFOR's Liaison Observation Team (LOT) house concept. These are small groups of military people living in normal houses and liaising with the local population. Because people from LOTs are military peacekeepers, they are sometimes not very good at collecting criminal intelligence for EUPM.

The local police are an important source of information for EUPM, but EUPM often needs to "press out" information from them. This goes back to the time of the former Yugoslavia when having information implied having power and information would not be shared unless ordered. This culture of not sharing information with internationals was partially enforced during the IPTF time.

Traditionally, as effective tools were missing, the writing of many reports was the main means of analysis for the police after information was gathered, and these reports were stored in nationwide data systems. After the war there was no money to maintain these systems, and they were cut into smaller elements. Nowadays, the police have to be aware of the fact that informants from different political parties and criminal groups could reside in their units. That is why the police do not document much anymore, but rather "store" important information in the heads of some trusted members of the police agency.

EUPM does not have a central database to retrieve important information from. All information is stored on local hard drives. There is a need for some kind of web-based application, but nobody thought of this when EUPM was deployed and now it is too late to implement something like this. EUPM does have an archive of aggregated reports that are sent on a weekly basis to the headquarters in Brussels, but these only indicate trends and no detailed information, as they are intended for member states representatives and other people in Brussels who "are not interested in details, but that is exactly what we need here."

EUPM has experienced difficulties in distributing its message to the citizens: "With normal press work you fail to communicate with the public." The local media in the country are biased, and citizens tend to watch and read their own particular news and media. Whenever EUPM issues a press release it is reported in three different ways: from Serb, Croat, and Bosnian angles. This means that it is not easy to properly reach out to the public. Bosnia and Herzegovina's television networks are still underdeveloped and underfunded but constitute the most unbiased media in the country. However, the ratings are not very high. It is also difficult for EUPM to trust local journalists. This is why EUPM has set up their own relatively large media department, used to distribute their message to the citizens. EUPM launches their own television programs, does public information campaigns, organizes round-table discussions, and produces their own radio programs. An example of a newly set-up television program is the Bosnian version of "America's most wanted." This is a tool for EUPM to distribute police-related or EUPM-related topics. In addition, EUPM is producing a print supplement, called "Kronika 112" (112 is the Bosnian emergency phone number).

EUPM has introduced public complaint bureaus throughout the country, where citizens can report bribe demands or any other suspicious police activities. Once a year EUPM conducts comprehensive public opinion surveys in which they test the messages that have been sent out. EUPM also trained press officers of local police forces and police chiefs in media management.

Within the European community in Bosnia and Herzegovina, there is a system of liaison officers to communicate and share information between the different EU organizations. There is a liaison officer dedicated to EUPM in each EU organization. There are bi-weekly meetings to share security information between all the international organizations in the so-called "security awareness working group," which is one of the most relevant tools or systems to share information regarding the security and safety situation.

Some interviewees complained that very few people understand that information sharing and coordination takes a lot of time: "Sometimes you have some things which cannot be e-mailed, or which have to be encrypted. All these kind of things take time and have to be organized."

4.4.3 Case Study Findings on Dealing with Frames of Reference

In a complex country such as Bosnia and Herzegovina, EUPM needs information to be able to put police information and crime security information in the right context, interpret it in the right way, and identify the ways to proceed based on those interpretations. That is why EUPM collects a lot of political information. And not only is the local and national Bosnian political information important but so too is international political information such as that related to the situation in Kosovo.

The Bosnia and Herzegovina authorities carried out an organized crime overview and analysis, but this did not result in an accurate picture of the situation. There are many reliability problems regarding the information that is gathered and put together by the public administration because of different legislations, different practices, and often an unwillingness to give out information. This is not necessarily because of corruption, but mostly caused by political preferences. The kind of information the political decision makers are giving out depends on which view of the security situation they want to show. Every now and then there are some situations that are politically fueled, and people are manipulated.

Other doubts were raised as to whether the international community in Bosnia and Herzegovina is basing its analysis on the right information. One interviewee was very critical of the international community, arguing that they do not conduct enough field-based research. The failure of the international community to get information from local people in multi-ethnic areas has led to "a gap between the Bosnian reality and the way it is presented by the media and policy makers."

There is a problem within EUPM concerning their institutional memory because of the high turnover of staff. EUPM does not have information sharing tools within the mission, causing internal information management to be based on daily, weekly, and monthly incident reports and the institutional memory to be situated in the heads of "veterans."

4.4.4 Discussion of the Findings

To cope with uncertainty, EUPM engages in different information gathering activities. First, they communicate with the local police, but as they do not have any system in place to store information and there are cultural differences in sharing information, it is difficult to obtain information from them. Second, EUPM is very active in media outreach to the citizens, for which they created a big media department, and has installed public complaint bureaus. Finally, EUPM has informants and liaising teams in place in the local community.

It is difficult for EUPM to deal with complexity as their institutional memory is not sufficient. EUPM also lacks an IS with a repository function to compensate for this, which should store all information they receive. This not only leads to loss of details but also to problems in interpreting what is going on in Bosnia and Herzegovina. EUPM tries to gather all kinds of contextual information, such as political information, to create a frame of reference and thereby to cope with ambiguity. But the question is how effective they can be at doing this, if there is no appropriate system in place.

The doubts that were raised about whether the analysis of the situation is based on the right information is a question concerning equivocality: there are several competing or contradictory frames of reference, and there is confusion on the appropriate frame for this situation.

5 Design of Crisis Management Information Systems

The three case studies gave us insight into how actors cope with information processing challenges in various crisis situations. But what can we learn from this for the design of IS for crisis management? To answer this question, we examined IS design guidelines for crisis management as developed for a Dynamic Emergency Response Management Information System (DERMIS) to support the response to crises, and compared these guidelines to our findings. In the work on DERMIS [29], which is based on extensive practical experiences in the field of crisis management and academic literature study, a set of nine design premises is introduced to guide the future design of crisis management IS. Because we focus on the information, information processing, decision making, and Sensemaking aspects of crisis management, we use the six design premises shown in Table 3 on page 447 for this discussion. We do not include the other three DERMIS design premises in our analysis, as they are more about the organizational aspects of crisis management, such as coordination, roles, and training.

DERMIS' first design premise clearly focuses on dealing with information, especially on dealing with complexity. In the Barents Rescue Exercise, some actors argued that they need a lot of detailed information, while others just wanted the most relevant information. The premise on information focus is about filtering out information so actors work with the level of information that is most suitable and/or preferable to them. The people working for EUPM experienced the problem that there is a lot of information "out there," but that there are no systems in place to make use of that information and analyze it. Design premise 1 also focuses on the importance of creating a good picture of what is going on by giving actors access to all contextual information. This overview will aid in creating a suitable frame of reference, as we saw at EUPM where police information, criminal intelligence, and political information is collected from all around the country and is used as contextual information to construct good frames of reference for their operations.

The design premise of "crisis memory" only supports dealing with frames of reference: By having access to historical information on the situation, actors will have the foundation for establishing a suitable frame of reference. In the Barents Rescue Exercise, respondents have explicitly mentioned the importance of experience in crisis situations. EUPM also acknowledges that experience is important, but faces problems of crisis memory due to high staff turnover. The ANPC in Portugal makes sure that the lessons learned from a crisis are incorporated into their procedures, to improve their response to following crises.

Design premise 3 also only deals with managing frames of reference, as it refers to the fact that there cannot be one frame of reference for a crisis situation. You cannot know in advance what will happen in a crisis and how it will evolve over time. Instead, flexible systems are needed that support changing frames of reference. Planning for a crisis is therefore difficult, and expectations of other actors—perhaps guided by previous experiences, agreements, or "common sense" as perceived by one side—are not always met, as Norwegian and Finnish actors experienced in the Barents Rescue Exercise.

Table 3 DERMIS design premises [29]

DERMIS design premise	Explanation of the design premise
Design premise 1: Information focus	During a crisis, those who are dealing with the emergency risk are flooded by information. Therefore, the support system should carefully filter information that is directed toward actors. However, they must still be able to access all (contextual) information related to the crisis as information elements that are filtered out by the system may still be of vital importance.
Design premise 2: Crisis memory	It is important that the system is able to log the chain of events during a crisis without imposing an extra workload on those involved in the crisis response. This information can be used to improve the system for use in future crises, but it can also be used to analyze the crisis itself.
Design premise 3: Exceptions as norms	Due to the uniqueness of most crises, usually a planned response to the crisis cannot be followed in detail. Most actions are exceptions to the earlier defined norms. This implies that the support system must be flexible enough to allow reconfiguring and reallocation of resources during a crisis response.
Design premise 4: Scope and nature of crisis	Depending on the scope and nature of the crisis, several response teams may have to be assembled with members providing the necessary knowledge and experience for the teams' tasks. Special care should also be given to the fact that teams may operate only for a limited amount of time and then transfer their tasks to other teams or actors. The same goes for individual team members who may, for example, become exhausted after an amount of time.
Design premise 5: Information validity and timeliness	As actions undertaken during crises are always based on incomplete information, it is of paramount importance that the emergency response system makes an effort to store all available information in a centralized database. Thus, those involved in the crisis response can rely on a broad base of information, helping them making decision that are more effective and efficient in handling the crisis.
Design premise 6: Free exchange of information	During crisis response, it is important that a great amount of information can be exchanged among stakeholders so that they can delegate authority and conduct oversight. This, however, induces a risk of information overload, which, in turn, can be a risk to the crisis response effort. The response system should somehow protect participants from information overload.

The fourth design premise is overarching for our case studies, and is reflected by the different types of management of the crises we saw in the case studies and the taxonomy shown in Table 2. Crises differ in their nature and their scope, and to counteract this, the associated management of the crisis—either it is prevention, response or recovery—must be adapted to that situation.

At first sight, the design premise on "information validity and timeliness" only seems to be about supporting the coping with uncertainty, as it describes the "usual" situation that there is incomplete information. This point was made by the people in the OSOCC at the Barents Rescue Exercise, as they stressed the importance of continuous up-to-date information on the situation. In the Portuguese case study, we saw that the Forest Services were continuously gathering information on fire risks and were making it available to everybody by means of fire hazard maps. The importance of information validity revealed itself in the observation of redundant information exchange behavior at the Community Mechanism for Civil Protection. However, design premise 5 is also about organizing all information available, enabling actors to construct the best possible frame of reference. In Portugal, we saw the example of the use of media to provide citizens with timely and good information that they should be careful with combustible materials and the use of fire, and the new concept that in winter fires can be good. In Bosnia and Herzegovina, actors have to be careful for manipulation of their frames of reference through other parties' partial or biased information sharing, mostly caused by political preferences.

The last design premise, "free exchange of information," is also intended to support dealing with both information and frames of reference. Design premise 6 is about the importance of the social context that is used to gather and exchange information. In the Barents Rescue Exercise, there was no free exchange of information between the important actors, such as between the OSOCC and LEMA. In Portugal, the Forest Services experienced problems of acquiring up-to-date information for their fire hazard maps. Cultural differences, fear of political and criminal infiltration, and ethnical barriers made free exchange of information nearly impossible in the post-conflict setting of EUPM. But we also saw good examples of free exchange of information, such as CECIS which allows actors to share information on the crisis situation and ANPC's use of the media to quickly reach out to the citizens. The social context is also used to compare frames of reference. Free exchange of information for this purpose was hampered in the Barents Rescue Exercise by physical separation of key actors without sufficient IS support to overcome this, and different languages and cultures. In the Portuguese case study, we saw interpretation problems arising from a lack of standardized language in CECIS. There was, however, an example of a good infrastructure for exchanging information in the Community Mechanism for Civil Protection that led to better familiarity of the other countries' problems, resources, and capabilities.

6 Conclusion

In this contribution, we have explored four common information processing challenges in three different crisis situations. Although all information processing challenges were present in all case studies, some challenges were more common in specific situations. At the airplane crash in the Barents Rescue Exercise, there were at first many uncertainty-related information problems. As this was a sudden crisis, actors obviously did not know what was happening in the first stages of the crisis. For an ongoing crisis situation as the forest fires in Portugal, we saw the importance of providing citizens with frames of reference. By supporting them in their situation of ambiguity, more severe forest fires can be prevented as people learn how to handle and prevent them, and people do not start to panic immediately when they see a fire. Finally, in the EUPM in Bosnia and Herzegovina, we saw the main problem of dealing with complexity, as a lot of information is gathered but cannot be analyzed and processed adequately due to a lack of systems and procedures.

We observed that the common trend in these three crisis situations was that actors at first focus on information-related problems, especially uncertainty problems of acquiring information, and then shift to strategies of coping with the complexity of too much information. This is an interesting finding and needs to be validated in future research. IS traditionally play a significant role in these areas of information-related problems, as they can support people in storing, retrieving, and analyzing huge amounts of data. This can be considered to be the realm of Decision Support Systems (DSS).

The challenges of ambiguity and equivocality, on the other hand, were mentioned less often and were not the specific aim of the actors, as they deal with them more implicitly. For these challenges, it is less important to search for more information. Rather people try to manage their frames of reference for interpreting the information by activities of Sensemaking. When we compared our findings to the DERMIS design premises, we found that all premises in some way contribute to the support of Sensemaking. If crisis management systems are designed to support access to all contextual information (design premise 1) and storage of historical information and incorporation of lessons learned (design premise 2), actors are supported to construct good frames of reference; if these systems are designed to be flexible during the response (design premise 3) and adaptable to the nature and scope of the crisis (design premise 4), actors are supported to update and change their frames of reference. Finally, as general requirements, if systems facilitate interaction and collaboration by supporting timely and valid information exchange (design premise 5) without any impediments (design premise 6), actors are supported to create, compare, update, and change their frames of reference. If crisis management IS are designed accordingly, they will become true Sensemaking Support Systems (SSS).

Acknowledgments. The case studies described in this article were part of a research project [16] initiated by the Crisis Management Initiative (CMI). The authors are grateful to CMI's Kristiina Rintakoski and Meeri-Maria Jaarva for enabling their active participation in this case study. Special thanks to Gerd Van Den Eede (Mechelen University College), Ilkka Demander (Elisa Corporation) and Damir Durbic (Tilburg University) for their help in conducting the interviews reported in this article and for sharing with us their insights and experience.

References

1. Angrosino, M.V.: Recontextualizing observation: Ethnography, pedagagy, and the prospects for a progressive political agenda. In: Denzin, N.K., Lincoln, Y.S. (eds.) Handbook of qualitative research, 3rd edn. Sage Publications, Thousand Oaks (2005)
2. Boland Jr., R.J.: The information of information systems. In: Boland Jr., R.J., Hirschheim, R.A. (eds.) Critical issues in information systems research. John Wiley & Sons, New York (1987)
3. Boland Jr., R.J.: Decision making and sensemaking. In: Burstein, F., Holsapple, C.W. (eds.) Handbook on decision support systems, vol. 1. Springer, Heidelberg (2008)
4. Choo, C.W.: The knowing organization: How organizations use information to construct meaning, create knowledge, and make decisions. Oxford University Press, New York (2006)
5. Daft, R.L., Lengel, R.H.: Organizational information requirements, media richness and structural design. Management Science 32(5), 554–571 (1986)
6. Daft, R.L., MacIntosh, N.B.: A tentative exploration into the amount and equivocality of information processing in organizational work units. Administrative Science Quarterly 26(2), 207–224 (1981)
7. Davenport, T.H., Prusak, L.: Working knowledge: How organizations manage what they know. Harvard Business School Press, Boston (1998)
8. Galbraith, J.: Organizational design. Addison-Wesley, Reading (1977)
9. Gilmour, L. (ed.): Collins concise dictionary, 3rd edn. HarperCollins Publishers, Glasgow (2003)
10. Housel, T.J., El Sawy, O.A., Donovan, P.F.: Information systems for crisis management: Lessons from southern california edison. MIS Quarterly 10(4), 389–400 (1986)
11. Huber, G.P., Daft, R.L.: The information environments of organizations. In: Jablin, F.M., Putnam, L.L., Roberts, K.H., Porter, L.W. (eds.) Handbook of organizational communication. Sage Publications, Newbury Park (1987)
12. Klein, G., Phillips, J.K., Rall, E.L., Peluso, D.A.: A data-frame theory of sensemaking. In: Hoffman, R.R. (ed.) Expertise out of context: Proceedings of the sixth international conference on naturalistic decision making. Lawrence Erlbaum Associates, New York (2007)
13. Klein, H.K., Myers, M.D.: A set of principles for conducting and evaluating interpretive field studies in information systems. MIS Quarterly 23(1), 67–93 (1999)
14. Landgren, J.: Supporting fire crew sensemaking enroute to incidents. International Journal of Emergency Management 2(3), 176–188 (2005)
15. March, J.G., Olsen, J.P.: Ambiguity and choice in organizations. Universitetsforlaget, Bergen, Norway (1976)
16. Muhren, W.J., Jaarva, M.-M., Rintakoski, K., Sundqvist, J.: Information sharing models and interoperability in national, cross-border and international crisis management. Crisis Management Initiative, Helsinki (2008)
17. Muhren, W.J., Van de Walle, B.: Sensemaking and information management in humanitarian disaster response: Observations from the triplex exercise. In: Landgren, J., Nulden, U., Van de Walle, B. (eds.) Proceedings of the 6th International Conference on Information Systems for Crisis Response and Management (ISCRAM 2009), Gothenburg (2009)
18. Muhren, W.J., Van Den Eede, G., Van de Walle, B.: Sensemaking and implications for information systems design: Findings from the democratic republic of congo's ongoing crisis. Information Technology for Development 14(3), 197–212 (2008)

19. Muhren, W.J., Van Den Eede, G., Van de Walle, B.: Sensemaking as a methodology for iscram research: Information processing in an ongoing crisis. In: Fiedrich, F., Van de Walle, B. (eds.) Proceedings of the 5th International Conference on Information Systems for Crisis Response and Management (ISCRAM), Washington DC (2008)
20. Muhren, W.J., Van Den Eede, G., Van de Walle, B.: Making sense of media synchronicity in humanitarian crises. IEEE Transactions on Professional Communication (2009) (forthcoming)
21. Nathan, M.L.: How past becomes prologue: A sensemaking interpretation of the hindsight-foresight relationship given the circumstances of crisis. Futures 36(2), 181–199 (2004)
22. Orlikowski, W.J., Baroudi, J.J.: Studying information technology in organizations: Research approaches and assumptions. Information Systems Research 2(1), 1–28 (1991)
23. Savolainen, R.: Information use as gap-bridging: The viewpoint of sense-making methodology. Journal of the American Society for Information Science and Technology 57(8), 1116–1125 (2006)
24. Schon, D.A.: The reflective practitioner: How professionals think in action. Basic, New York (1983)
25. Shrivastava, P.: Crisis theory/practice: Towards a sustainable future. Organization & Environment 7(1), 23–42 (1993)
26. Simon, H.: Administrative behavior, 3rd edn. The Free Press, New York (1976)
27. Starbuck, W.H., Milliken, F.J.: Executives' perceptual filters: What they notice and how they make sense. In: Hambrick, D. (ed.) The executive effect: Concepts and methods for studying top managers. JAI Press, Greenwich (1988)
28. Thomas, J.B., Clark, S.M., Gioia, D.A.: Strategic sensemaking and organizational performance: Linkages among scanning, interpretation, action, and outcomes. Academy of Management Journal 36(2), 239–270 (1993)
29. Turoff, M., Chumer, M., Van de Walle, B., Yao, X.: The design of a dynamic emergency response management information system. Journal of Information Technology Theory and Application (JITTA) 5(4), 1–36 (2004)
30. Van de Walle, B., Turoff, M.: Emergency response information systems: Emerging trends and technologies. Communications of the ACM 50(3), 29–31 (2007)
31. Van de Walle, B., Turoff, M.: Decision support for emergency situations. In: Burstein, F., Holsapple, C.W. (eds.) Handbook on decision support systems, vol. 2. Springer, Heidelberg (2008)
32. Van de Walle, B., Van Den Eede, G., Muhren, W.J.: Humanitarian information management and systems. In: Löffler, J., Klann, M. (eds.) Mobile Response. LNCS, vol. 5424, pp. 12–21. Springer, Heidelberg (2009)
33. Walsham, G.: Doing interpretive research. European Journal of Information Systems 15(3), 320–330 (2006)
34. Weick, K.E.: Cosmos vs. Chaos: Sense and nonsense in electronic contexts. Organizational Dynamics 14(2), 51–64 (1985)
35. Weick, K.E.: Enacted sensemaking in crisis situations. Journal of Management Studies 25(4), 305–317 (1988)
36. Weick, K.E.: The collapse of sensemaking in organizations: The mann gulch disaster. Administrative Science Quarterly 38(4), 628–652 (1993)
37. Weick, K.E.: Sensemaking in organizations. Sage Publications, Thousand Oaks (1995)

38. Weick, K.E.: Sensemaking as an organizational dimension of global change. In: Making sense of the organization. Blackwell Publishing, Malden (2001); Originally published in Dutton, J., Cooperrider, D. (eds) The Human Dimensions of Global Change. Sage Publications, Thousand Oaks (1999)
39. Weick, K.E.: Managing the unexpected: Complexity as distributed sensemaking. In: Mc Daniel, R.R., Driebe, D.J. (eds.) Uncertainty and surprise in complex systems: Questions on working with the unexpected. Springer, Heidelberg (2005)
40. Weick, K.E., Meader, D.K.: Sensemaking and group support systems. In: Jessup, L., Valacich, J. (eds.) Group support systems: New perspectives. MacMillan, New York (1993)
41. Weick, K.E., Sutcliffe, K.M., Obstfeld, D.: Organizing and the process of sensemaking. Organization Science 16(4), 409–421 (2005)
42. Zack, M.H.: The role of decision support systems in an indeterminate world. Decision Support Systems 43(4), 1664–1674 (2007)

A Distributed Approach to Gas Detection and Source Localization Using Heterogeneous Information

Gregor Pavlin, Frans Groen, Patrick de Oude, and Michiel Kamermans

Abstract. This chapter introduces a system for early detection of gaseous substances and coarse source localization by using heterogeneous sensor measurements and human reports. The system is based on Distributed Perception Networks, a Multi-agent system framework implementing distributed Bayesian reasoning. Causal probabilistic models are exploited in several complementary ways. They support uniform and efficient integration of very heterogeneous information sources, such as different static and mobile sensors as well as human reports. In principle, modular Bayesian networks allow creation of complex probabilistic observation models which adapt to changing constellations of information sources at runtime. On the other hand, Bayesian networks are used also for coarse modeling of transitions in the gas propagation processes. By combining dynamic models of gas propagation processes with the observation models, we obtain adaptive Bayesian systems which correspond to Hidden Markov Models. The resulting systems facilitate seamless combination of prior domain knowledge and heterogeneous observations.

Gregor Pavlin
Thales Research & Technology Netherlands, Delftechpark 24, 2628 XH, Delft, The Netherlands
e-mail: `Gregor.Pavlin@icis.decis.nl`

Frans Groen
Intelligent Autonomous Systems Group, Informatics Institute, Faculty of Science, University of Amsterdam, Science Park 107 1098 XG Amsterdam, The Netherlands
e-mail: `F.C.A.Groen@uva.nl`

Patrick de Oude
Intelligent Autonomous Systems Group, Informatics Institute, Faculty of Science, University of Amsterdam, Science Park 107 1098 XG Amsterdam, The Netherlands
e-mail: `P.deOude@uva.nl`

Michiel Kamermans
Thales Research & Technology Netherlands, Delftechpark 24, 2628 XH, Delft, The Netherlands
e-mail: `Michiel.Kamermans@icis.decis.nl`

R. Babuška & F.C.A. Groen (Eds.): Interactive Collaborative Information Systems, SCI 281, pp. 453–474.
springerlink.com

1 Introduction

This chapter introduces a distributed approach to automated detection of gaseous substances and coarse localization of leaks by using heterogeneous reports originating from sensors and humans. Such estimation is relevant for situation assessment in environmental monitoring applications. The presented system makes use of Bayesian networks, which support (i) efficient construction of complex domain models and (ii) robust inference. The focus of this chapter is discussion on why and how different, well-understood properties of Bayesian networks can be exploited in creation of flexible and robust gas detection and leak localization systems.

We illustrate the challenges and proposed solutions with the help of the case provided by the DCMR Milieudienst Rijnmond. The DCMR is an environmental protection agency that operates in Rijnmond and the larger Port of Rotterdam area, a densely populated industrial region with high concentration of chemical industrial facilities. The risk of exposure to gaseous pollutants released during production or through accidents is in such an environment relatively high. Clearly, environmental protection organizations, such as DCMR, play a critical role in minimizing the health risks through (i) inspection of industrial facilities and (ii) the capability to quickly detect harmful gaseous substances and locate the leak. In this chapter, we focus on the latter task.

In such settings, unusual gas concentrations must quickly be inferred through interpretation of large quantities of uncertain and very heterogeneous information, i.e., information fusion based on adequate inference algorithms and modeling techniques. For example, operators at the DCMR Monitoring and Control Center process messages from people reporting observations about unusual situations that range from "strange smells" to obligatory reports from local companies. The latter could be about detected leakages or changes in production processes. Besides information from callers, valuable information can be obtained from field inspectors (experts who can recognize gases by the way they smell), readings from a few stationary gas sensors, and meteorological sensors. In general, the control center operators make conclusions about the presence of a particular gas type in a certain area and the pollution sources by interpreting different clues.

An important tool is complaint plot, which displays spatial distribution of complaints collected during a certain period of time. Often the complaint plot is used for the detection of anomalies, i.e., based on reports from complainers and simple sensors, we know that unusual gas mixtures are present at a certain location. By combining such a plot with the information on the wind direction, the operators form hypotheses about possible sources. Based on this information, further investigations concentrate on a smaller set of locations/companies, and additional information on the source of the pollution is obtained directly by contacting the companies. In this way, information on the type of the gas and the released quantities can be obtained.

In addition, the operators are sometimes able to estimate the type of gases by interpreting different types of complaints. The operators know what kind of observable effects different gases might cause. With the help of such a causal knowledge and by taking into account information about the effects, the operators reason about the existence of gases. Since the reporting persons are often not able to directly identify the chemicals, the operators use additional clues for determining the type of the gas. For this, they use models to systematically obtain the information they need to pinpoint the type of the gas. That is, they guide the information acquisition process by posing simple questions to the caller such as "does it smell like a hospital" to obtain more clues. Sometimes, gas classification can be based on calibrated sensors.

Since the information obtained from sensors and humans is inherently noisy and often ambiguous, the gas detection process should not rely on a single sensory or a human report. To achieve good assessment quality, the inference should take into account as many observations as possible. On the other hand, state-of-the-art systems for the detection and tracking of annoying or dangerous gases exploit only a small fraction of the huge body of the relevant information. Namely, as noted by experts, detection and tracking of gases by manually gathering and interpreting relevant information is often associated with communication and processing bottlenecks. Often several persons call the control center at the same time, while the interrogation procedure takes significant time and the number of operators is limited. Hence, many callers are put on hold. In other words, either most callers cannot be attended and valuable information is ignored or a conclusion is made late.

Obviously, an automated system is needed which can query people, process data from arbitrary chemical sensors, and interpret the obtained information. This interpretation is typically based on correlation of very heterogeneous information of various quality. Such a correlation must be based on adequate domain models which capture relations between observations which are spatially and temporally distributed. In the targeted application, a lot of useful information is discrete (e.g., human reports) and it is difficult to obtain reliable concentration measurements. Therefore, we propose a solution based on the principles of discrete Bayesian filtering, smoothing, and prediction [14]. However, an adequate solution requires a wide spectrum of modeling and inference techniques which support (i) efficient construction of fusion systems and (ii) robust estimation despite noisy information sources and uncertain domain models.

In this chapter, we present a system which exploits causal Bayesian networks (BN) in different, complementary ways. We show how BNs (i) facilitate distribution of domain models, inference processes, and automated information acquisition processes in a decentralized monitoring system and (ii) support robust reasoning with respect to the modeling parameters. In particular, we make use of Distributed Perception Networks (DPN) [8, 11], a framework which support creation of self-configurable systems implementing exact Bayesian inference. In principle, the presented system can be viewed as an automated complaint plot which supports automated information acquisition as well as automated reasoning about the possible sources given heterogeneous information.

The main purpose of the chapter is the explanation of why and how Bayesian networks can be used for efficient real-world gas detection and leak localization.

2 Causal Probabilistic Models

Often monitoring processes can be viewed as causal stochastic processes, where hidden events cause observations according to certain probability distributions. Let us assume that the presence of a certain toxic gas[1] causes conditions under which a certain type of semiconductors features distinctive conductivity. Similarly, a particular conductivity could be observed in an ionized gas mixture if the toxic gas is present. In this chapter, we assume two types of sensors, evaluating conductivity in semiconductors and in an ionized gas mixture, respectively. Introduction of a sensor measuring a particular type of conductivity will spawn various processes in the sensor's electronic circuitry which in turn will result in a certain state of the sensor. Depend on the sensor state we will obtain a sequence of reports, either confirming or refuting the presence of the gas. Similarly, humans are likely to report about typical symptoms (e.g., smell).

Such a causal process can be described through the graph shown in Figure 1, where each node represents a binary variable, e.g., $G = true$ if the gas is present, otherwise $G = false$. The situation under which a semiconductor element and ionized gas mixture feature typical conductivity is represented by variables *CondC* and *Ion*, respectively. States of the binary variable *CondC* correspond to the situations where electrical current under ideal circumstances would either exceed some detection threshold (i.e., $CondC = true$) or remain below that threshold (i.e. $CondC = false$). The states of the ith sensor of type x are represented by S_i^x, while a sequence of sensory reports is denoted by binary variables $E_1^x, \ldots, E_n^x$; $E_k^x = true$ if a report confirms the presence of the gas. In this example, subgraphs containing nodes $S_1^C, C_1^C, E_1^C, \ldots, E_m^C$ and $S_2^C, C_2^C, E_{m+1}^C, \ldots, E_n^C$ describe processes in two sensors measuring the conductivity of local semiconductor elements. Subgraph consisting of nodes $S_3^I, C_3^{I1}, C_3^{I2}, C_3^{I3}, E_1^I, \ldots, E_o^I$, on the other hand, corresponds to the third sensor measuring conductivity of the ionized gas mixture. Variables S_1^C and S_2^C denote the measured conductivity on the semiconductor elements in the first two sensors, while S_3^I denotes the measured conductivity of the ionized gas mixture in the third sensor. Moreover, variables $C_1^c, C_2^c, C_3^{I1}, C_3^{I2}, C_3^{I3}$ represent the states of critical electronic components of the three sensors. We also assume that the causal process is influenced by the air humidity and temperature represented by variables M and T, respectively. Note that with each sensor an independent local causal process is introduced to the domain.[2] By introducing a new gas sensor, the presence of gas will initiate different processes in the sensor's circuitry which will eventually produce sensor reports represented. Since our estimation is based on models which describe how observations are produced, the used Bayesian networks must be updated by nodes and relations capturing critical states of the processes in a sensor by adding

[1] For the sake of simplicity we say that a toxic gas is present if its concentration exceeds some critical value. Otherwise we say that the gas is absent.

[2] In principle, the exposure to a gas the measurement elements in a sensor and the circuitry produce observations. In other words, the local processes of a sensor become part of the overall mechanisms in the domain which generate the observations.

the corresponding nodes. For example, nodes $S_3^I, C_3^{I1}, C_3^{I2}, C_3^{I3}, E_1^I, \ldots, E_o^I$ in Figure 1 capture causal relations between the observations and states of sensor components.

Besides the sensors, we assume that there are humans in the area, who have olfactory reactions to G and who submit reports of what they smell via a call service or a web interface. The states of binary variable $Smell$ represent situations in which people familiar with a typical smell of G either do or do not recognize the smell. Moreover, each individual report is represented by a node E_i^{Smell}. Similarly, first aid workers might be able to report about health symptoms which are typical results of exposure to G. A situation in which observable symptoms take place is denoted by variable $Sympt$, and the reports are denoted by variables E_i^{Sympt}.

Stochastic aspects in such a causal process can be described through conditional probabilities. Consequently, the presented process can efficiently be modeled with the help of a causal BN [13], which exploits conditional independence reflected in the topology of the directed acyclic graph (DAG) from Figure 1. In general, a BN encodes a joint probability distribution $P(V)$ over a set of variables V through a product of conditional probabilities $\prod_i P(X_i|\pi(X_i))$, where $X_i \in V$. The factorization of $P(V)$ corresponds to the topology of the DAG. The stochastic relations between discrete variables in BNs are captured by conditional probability tables (CPT). Moreover, there exist efficient approaches to inference [2] with Bayesian networks, which exploit the locality of relations for efficient computation of marginal distributions. Note that the model depicted in Figure 1 corresponds to a single time slice in which a set of observations was obtained in the time slice while the states of the non-leaf variables remained constant.

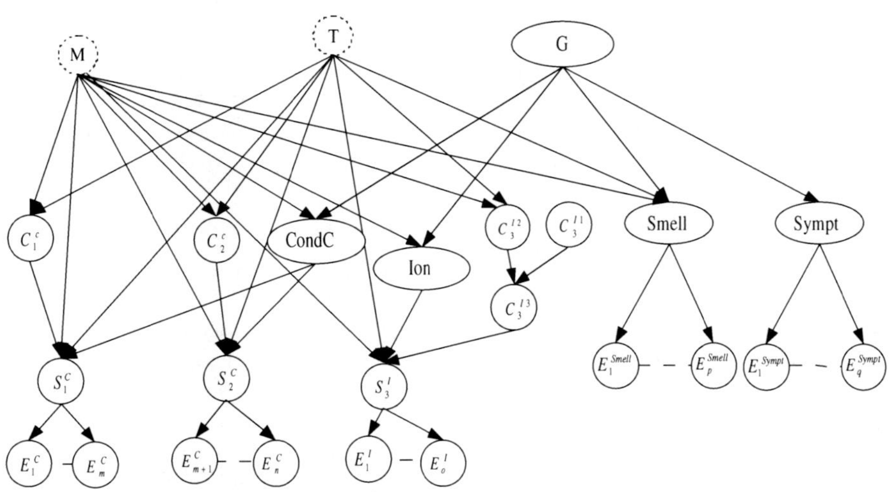

Fig. 1 A causal model of a monitoring process capturing relations between the possible states within a particular time slice, such as presence/absence of G and different types of observations represented by leaf nodes

However, each sensor and report must be captured by a node in a BN. In other words, each particular fusion process requires a specific domain model which maps observations from the current constellation of information sources to the hypotheses of interest. On the other hand, constellations of information sources are often not known prior to the operation and they change at runtime. In addition, heterogeneity and large numbers of information sources imply complex domain models and it is likely that huge quantities of noisy information should be processed. Given these challenges, a modular solution seems to be an obvious choice; fusion modules could assemble adequate domain models at runtime and the processing load could be distributed throughout a system of networked devices.

Note that causal models can be used for the classification of gases or just detection of anomalies (i.e., identification of abnormal mixtures of gases). Classification of gases requires calibrated chemical sensors and reports on very specific symptoms associated with different gases, which must be captured in the causal model. Detection of anomalies, on the other hand, can be achieved with simpler models, uncalibrated sensors, and simple reports on unusual chemical smells.

2.1 Modeling Dynamic Processes

Besides the momentary observations, also the observation sequences provide important clues about the hidden phenomena of interest and the source of the gas release. In this chapter, we assume plumes caused by gas leaks, which are active for a significant period of time.[3] Moreover, we assume that the plume life-cycle can be subdivided into three phases. In the first phase, the plume grows, i.e., the area in which the gas detectable by the sensors and people increases. In this phase, the concentrations at different locations in the area of influence steadily increase. In addition, we assume that the speed of gas dispersion v_g is approximately equal to the speed of wind v_w along the line which is parallel to the wind direction and intersects the leak location. This is a plausible assumption in case the wind speed exceeds a certain level.

During the second phase, gas concentrations at any point in the affected area remain approximately constant for a certain period of time, i.e., if the concentration at a particular location exceeded certain detection threshold during the first phase, then the concentration will remain above this threshold throughout the second phase. The third phase takes place after the source is eliminated and the gas concentrations are gradually reduced. In this study, we focus on the first two phases, which are critical for the detection of gases and coarse tracking to the sources.

In the presented approach, the evolution of plumes is captured by causal probabilistic models which correspond to discrete Hidden Markov Models (HMM). The causal probabilistic networks are coarse models of monitoring processes and gas propagation mechanisms (see Figure 2). The gas propagation from a source is viewed as a sequence of discrete state transitions. A state in the domain at time t is represented by a set of binary variables $G_1^t, G_2^t, \ldots, G_N^t$. States of each G_i^t correspond

[3] Often a pollution is caused by a leak that is active for a few hours.

Table 1 Conditional probability tables capturing relations $P(G_i^t|G_i^{t-1},G_{i-1}^{t-1})$

G_i^{t-1}	*true*		*false*	
G_{i-1}^{t-1}	*true*	*false*	*true*	*false*
$G_i^t = true$	1	1	p	$1-q$
$G_i^t = false$	0	0	$1-p$	q

to the situations in which the gas concentration at location l_i at time t was above or below some critical level, respectively.

The causal model in Figure 2 also specifies relations between each G_i^t and the observations collected at the corresponding location l_i. Each small network attached to some of the G_i^t nodes captures the causal stochastic processes producing observations at location l_i. These networks can be viewed as arbitrarily complex observation models which describe how different observation patterns at each location are generated, given the presence of gas.

Moreover, the BN shown in Figure 2 captures relations which can be described with the so-called *Left to Right HMM* [1]. Namely, the propagation is assumed to have a distinctive direction; the gas can propagate from location l_1 to l_2, from l_2 to l_3, and so on. Since we are dealing with plumes, the concentrations at some location l_i remain high for a significant period of time after the gas reached l_i. Thus, the states at such locations can be considered static for the duration of the estimation process. This is reflected by the deterministic transitions between the states of variables representing the presence of gas at the same location at different times, i.e., $P(G_i^t = true|G_i^{t-1} = true) = 1$. In addition, the state of G_i^t can change only through influence of a state from a neighboring location G_{i-1}^{t-1}, if the gas was not present at l_i at time $t-1$. In other words, we assume that pollution is irreversible for a significant period of time. This is reflected in the conditional probability tables of the non-root hidden state nodes G_i^t shown in Table 1. In this table, p and q represent transition probabilities, which are typically greater than 0.5.

The causal model shown in Figure 2 captures *correlations between heterogeneous, spatially and temporally distributed observations, and hidden states.* In conjunction with, well-known inference algorithms [2, 13], causal model can be used for the computation of probability distribution of arbitrary sets of variables represented in the model. In this chapter, we are interested in computation of any $P(G_i^t|\varepsilon)$, which is based on the entire evidence ε collected in a certain time interval[4] in a specific area. This probability is used as a score for ranking of the leak-source hypotheses introduced in the following section.

3 Estimating the Source

The leak localization is based on the creation of hypotheses. For each potential source s, a hypothesis $h(s)$ assumes that if s were indeed the source, the gas would propagate in a certain way, and typical observation sequences (i.e., patterns) in a

[4] A sequence of time slices.

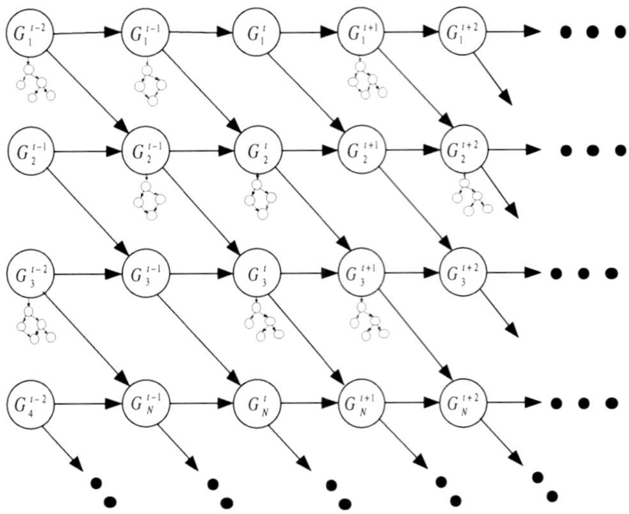

Fig. 2 A causal model (i) describing gas propagation through a discretized area and (ii) capturing relations between the hidden states and the observations. This model corresponds to a HMM. The subscripts and superscripts indicate location and time slice of an estimate, respectively.

discretized downwind area should be obtained. This requires assumption about the downwind area A_h in which the relevant observations for each hypothesis $h(s)$ must be collected. For the sake of clarity, we assume that the gas would spread from the source within an area with a simple form, whose symmetry axis intersects the location of s and is parallel to the wind direction (see example in Fig. 3). For the sake of clarity, in this chapter we assume that the wind has a constant direction and speed. In principle, we can use areas A_h of arbitrary shape which reflect complex airflows and dispersion models, if available. For example, A_h could be determined by using arbitrarily advanced gas dispersion models, such as, for example, RimPuff [4].

3.1 Dedicated Domain Models

For each hypothesis, a BN capturing a specific HMM is created, which is used for the computation of the hypothesis score. The area of each hypothesis is subdivided into several sections l_i (i.e., locations), each associated with a hidden binary variable G_i^t representing the presence/absence of gas. In principle, the HMM is used to compute the posterior probability of $P(h|\varepsilon_h)$ of hypothesis h being true given a set of observations ε_h collected in the down-wind area A_h relative to the location s. $P(h|\varepsilon_h) = P(G_1^{t-k}|\varepsilon_h)$, where G_1^{t-k} represents the states corresponding to the location of a hypothesized leak at the time of trigger minus the time the gas would need to travel from the source location to the trigger location. The presence of gas at the location of the triggering report is represented by G_i^t, where i corresponds to the distance between the potential source and the trigger.

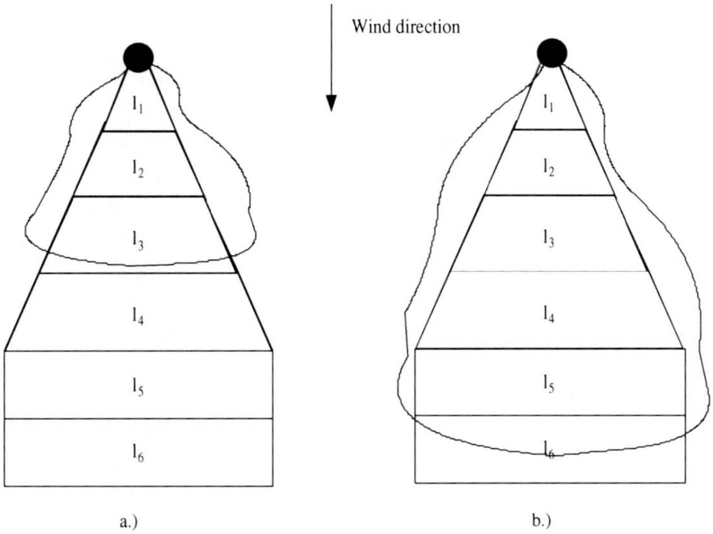

Fig. 3 A possible segmentation of the hypothetical down wind area affected by a leak

A hypothesis creation is initiated through anomaly detection based on data from a single chemical sensor or a report from a human. For each potential source, a hypothesis is spawned if the corresponding downwind area contains the triggering report.

3.2 Modeling Temporal Aspects

Gas propagation through time is modeled by controlling evidence collection according to some time function. Note that evidence is represented by leaf nodes which do not influence belief if they are not instantiated according to some evidence (see the concept of barren nodes [2]). The gas propagation is thus reflected in the following instantiation rules:

- All available evidence in sections of A_h upwind of the triggering report is immediately used for the instantiation of the observation nodes corresponding to the upwind locations.
- Collection of evidence in the sections downwind from the triggering sensor is delayed according to a simple function of wind speed and the distance. Evidence from a particular segment l_i downwind from the triggering sensor is collected and used for instantiation after the time has passed that the gas needed to travel to l_i.

Note that this temporal control is very coarse. In case advanced models for gas propagation are available, better activation rules can be deployed. However, it is important to note that irrespective of the propagation model, the timing is introduced through a sequence of variable instantiations.

4 Distributed Modeling and Inference

Gas detection and leak localization in the targeted domain requires complex causal models containing many variables and relations. In principle, the graph in Figure 2 corresponds to a factorization of a joint probability over a large number of variables corresponding to many interrelated observable and hidden phenomena. The models grow with the spatial and temporal resolution of the model as well as the heterogeneity and quantities of used observations. In addition, certain modeling fragments might appear in multiple HMMs of different hypothesis simultaneously.

To cope with such complexity and to be able to reuse modeling fragments, we apply a distributed approach to modeling and inference. In particular, we use the Distributed Perception Networks framework (DPN), a modular approach to Bayesian information fusion [8]. Modules in DPNs are implemented through agents which cooperate to form arbitrarily large and complex distributed fusion systems. DPN systems are, in essence, distributed self-organizing classifiers, which consist of agents with very heterogeneous domain expertise. Each agent processes information by using a local causal BN, which captures partial domain knowledge; each agent basically provides a fusion service. Local BNs in DPN agents are based on design and organization rules, which guarantee that the resulting distributed fusion system implements exact belief propagation without any centralized configuration and fusion control [7]. An example of a system of collaborating DPN agents is shown in Figure 4.

The models can be distributed in different ways using the DPN framework. However, in the given application, it makes sense to distinguish between (i) modules that model relations between G_i^t and observations, and thus implement observation

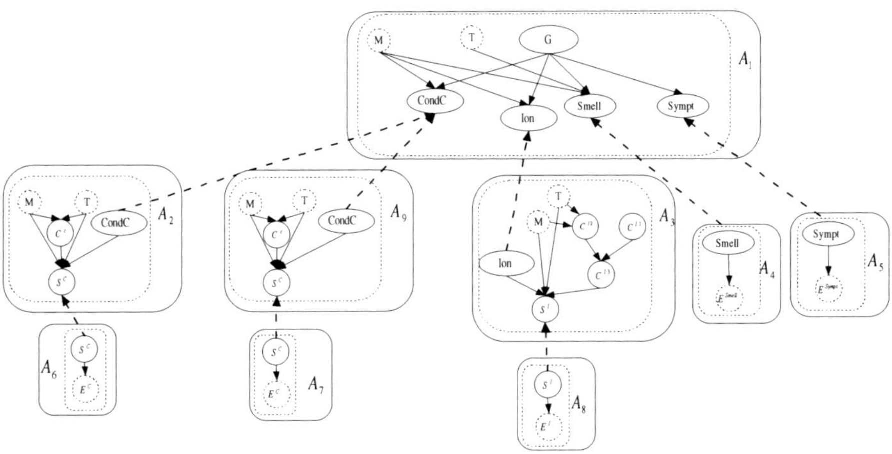

Fig. 4 An organization of fusion agents implementing a distributed causal model. Each solid rectangle corresponds to an agent with a modeling fragment (dashed rectangles). Dashed arrows show the flow of the inter-agent messages containing partial fusion results. The dashed circles correspond to observed variables.

models, and (ii) modules which can compute transitions between the hidden states of variables G_i^t (i.e., implement coarse models of gas propagation). In the following text, we discuss the design principles for the observation models and dynamic gas propagation models.

4.1 Distributed Observation Models

For each location, we introduce a set of DPN agents which implement an arbitrarily complex distributed observation model.[5] Such a model describes causal processes leading to the observations obtained at a particular location in a certain time interval as a function of the presence of gas at that location. An example of a set of modules implementing a complex observation model is shown in Figure 4. This system of fusion agents supports Bayesian inference which is equivalent to inference based on the monolithic observation model from Figure 1. Note that an observation model is created for each location l_i.

4.2 Dynamic Gas Propagation Models

DPN agents can implement dynamic gas propagation models in several ways. In the simplest case, a DPN agent uses a monolithic BN describing a HMM over a limited set of locations and time steps (see, e.g., Figure 5). Such a model contains interface variables ${}^M G_i^t$, which in principle represent the estimates of the presence of gas at location l_i based purely on the observations collected at l_i during a time interval $[t,t+1]$ (i.e., time slice). The CPT between ${}^M G_i^t$ and the G_i^t is deterministic[6]: $P({}^M G_i^t = true|G_i^t = true) = 1$ and $P({}^M G_i^t = false|G_i^t = false) = 1$.

Distributed *observation models*, such as the one shown in Figure 4, are used for the computation of soft[7] evidence $P({}^M G_i^t = true|\varepsilon_i^t)$ for variables ${}^M G_i^t$, where ε_i^t denotes all evidence obtained at location l_i within the time interval between t and $t+1$. Acquisition of new evidence in this interval triggers new computation of

[5] This model can be viewed as a complex probabilistic sensor model relating the presence of gas and complex observation patterns originating from multiple, often heterogeneous sources.

[6] This is a modeling trick which facilitates creation of modular systems without compromising correctness of distributed belief propagation. Variables ${}^M G_i^t$ represent the only link between the modules estimating the gas dispersion processes and the modules forming arbitrarily complex *observation models*. This approach simplifies construction and maintenance of models because the models capturing gas propagation processes remain simple; they contain only variables representing the presence of the gas at different locations and points in time and do not have to contain variables which are specific to particular sensor constellations at particular locations. The *observation models*, on the other hand, can easily accommodate new sensing modalities without changing the modules that reason about the gas propagation. In other words, a change in a sensor constellation at a certain location does not require any update of the gas propagation model and facilitates efficient propagation in dynamic BNs by using network fragments.

[7] See [2, 12] for the discussion on soft evidence.

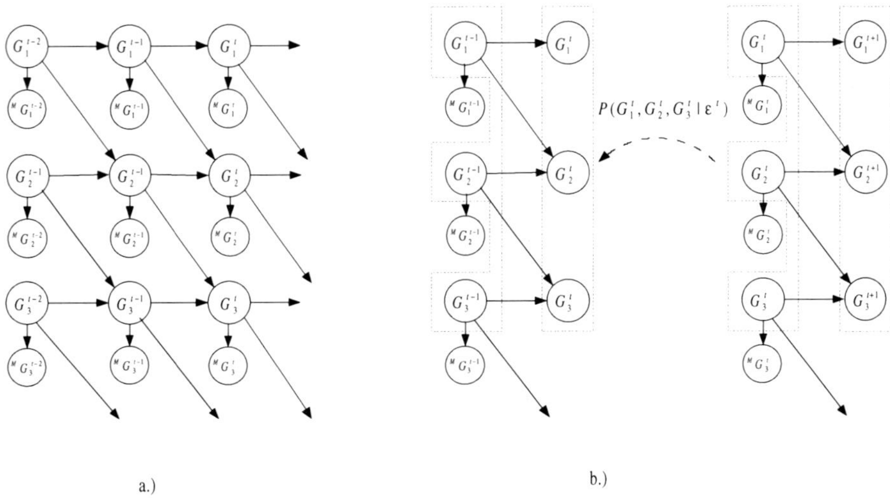

Fig. 5 Coarse causal models capturing dynamic processes. a.) a monolithic gas propagation model. b.) a single module of a gas propagation model is used in subsequent inference steps. Inference results from one step are used as soft evidence in another inference process corresponding to an earlier time slice.

$P({}^M G_i^t = true|\varepsilon_i^t)$, which in turn triggers update of the posterior $P(G_1^{t-k}|\varepsilon)$. Note that $P({}^M G_i^t = true|\varepsilon_i^t)$ can be simultaneously delivered to multiple HMM if they share the corresponding location. In other words, the processing resources as well as the information can be reused.

By combining $P({}^M G_i^t = true|\varepsilon_i^t)$ based on distributed *observation models* (see Figure 4) with the dynamic model from Figure 5a, the estimation results (i.e., $P(G_1^{t-k}|\varepsilon)$) are the same as if the monolithic model in Figure 2 were used.

The monolithic approach is not practical in case of many time slices, because it requires large dynamic BNs in which inference can become very inefficient.

This challenge can be tackled by using a single network fragment capturing relations between the variables in two subsequent time slices. By taking into account the concept of d-separation, we can derive from the monolithic BN a simpler BN fragment which relates variables from two consecutive time slices (see example in Figure 5b). Such a fragment captures a recurring pattern in the monolithic BN. In other words, the monolithic dynamic BN can be viewed as a chain of identical BN fragments. By using such a fragment in an iterative belief updating process, we can obtain results identical to the outcomes of exact belief propagation in the monolithic BN.

The agent reasoning about the gas distribution uses the same fragment in a series of subsequent inference steps, each corresponding to a time step. After a new time interval at time t is initiated, the agent uses the model fragment to estimate $P(G_1^t,\ldots,G_n^t|\varepsilon^t)$. Set of variables $\{G_1^t,\ldots,G_n^t\}$ can be viewed as a hyper-variable whose states capture all possible gas distributions over the locations in the downwind area A_h during the time interval between t and $t+1$. Set ε^t denotes a union of all observations from all locations in the same time interval. In principle, the multiply connected monolithic BN from Figure 5a can be transformed to a singly connected network in which the variables representing gas distribution in a particular time slice can be represented through hyper-variables. Consequently, the derived BN fragment can be represented as a singly connected BN module containing hyper-variables corresponding to sets $\{G_1^t,\ldots,G_n^t\}$ and $\{G_1^{t+1},\ldots,G_n^{t+1}\}$, respectively. The sets corresponding to the hyper variables are indicated by dashed frames in Figure 5b.

The computed $P(G_1^t,\ldots,G_n^t|\varepsilon^t)$ over the root nodes in a BN fragment is used in the next iteration as soft evidence for the hyper-variable which is a leaf node in the same fragment (see example in Figure 5b.). Thus, in the next iteration, $P(G_1^{t-1},\ldots,G_n^{t-1}|\varepsilon^{t-1},\varepsilon^t)$ is estimated by using the estimated distribution over the hyper-variable corresponding to set $\{G_1^t,\ldots,G_n^t\}$. $P(G_1^{t-1},\ldots,G_n^{t-1}|\varepsilon^{t-1},\varepsilon^t)$ is then used in the following iteration as soft evidence. This process continues until $P(G_1^{t-k},\ldots,G_n^{t-k}|\varepsilon^{t-k},\ldots\varepsilon^t)$ is obtained. By marginalizing over this distribution, we obtain the posterior probability $P(G_i^{t-k}|\varepsilon^{t-k},\ldots,\varepsilon^t)$ of a leak at location i at time t minus estimated propagation time k between the source and location i. Note that with the number of locations, the set $\{G_1^t,\ldots,G_n^t\}$ increases, which can result in hyper-variables with many states.

In this approach, $P(^M G_i^t = true|\varepsilon_i^t)$ obtained through the propagation in *distributed observation models* have to be stored for each time slice t. These distributions are reused as soft evidence each $P(G_1^{t-k},\ldots,G_n^{t-k}|\varepsilon^{t-k},\ldots,\varepsilon^t)$ is computed, which is triggered by newly obtained observations in the last, currently active time slice. Note that the modeling fragment contains variables from two time slices. It is obtained by using the design rules introduced in [10]; a set of nodes from a single time slice was augmented with the children.[8]

Moreover, the agent using gas propagation models also activates gas estimation processes providing $P(^M G_i^t = true|\varepsilon_i^t)$ at different segments in the downwind area at appropriate times according to the activation rules introduced in Section 3.2.

5 Construction and Maintenance of Adequate Domain Models

The proposed system is inherently complex. It implements complex causal domain models involving many variables and parameters. On the other hand, causal models used in the presented system have a critical impact on the fusion quality. In general, domain models must describe the processes of interest sufficiently well. In the case of causal BNs, this requires (i) adequate causal graphs and (ii) parameters captured by the CPTs. However, real-world domains are often so complex that not all relevant phenomena and relations can be observed, i.e., we are confronted with the acquisition intractability [16]. Thus, the models are inevitably erroneous to a certain

[8] This is a special case of the rule, which requires that all children and children's parents are added to a certain partition (see appendix in [10]). In this case, all parents of the children are already included in the original partition.

extent; important dependencies in causal models might be ignored and parameters are likely to deviate from the true distributions significantly.

In the following text, we discuss the modeling challenges and explain why the presented approach can support reliable detection and leak localization despite inability to obtain perfect domain models.

5.1 Determination of Causal Dependencies

Explicit representation of direct dependencies between phenomena in stochastic processes supports sound inference and facilitates distribution of models. In case of domains which can be represented by sparsely connected BNs, the inference can be efficient and determination of adequate parameters (i.e., conditional probabilities) is simpler. However, the question is whether such relations can be identified sufficiently well.

By explicitly considering causality, it turns out that in domain models capturing monitoring processes, the critical dependencies can easily be determined. In such domains, the causal processes are understood well, since they are to a great extent created through designers of monitoring systems and operators; such processes are created by (i) putting together various man-made components, such as sensors, communication systems and (ii) exploitation of well-known procedures and management structures. We can safely assume that sensor developers understand causal mechanisms in the sensing devices, while the professionals in an estimation process follow well-established procedures.

Usually, each type of sensors provides observations about a single phenomenon represented by a process variable. In addition, components of one sensor do not influence components of another sensor. Similarly, reports from one sensor do not influence reports from another sensor. In other words, we can represent a process generating observations of a certain type with a network fragment which is sparsely connected with other fragments (see, e.g., Figure 1). This means that correct inference can be achieved if belief propagation is distributed over different modules and by considering only few dependencies between the local inference processes in the modules. In addition, because of aforementioned sparse dependencies, different designers can build adequate local BNs by taking into account only a few dependencies, which makes the design process efficient.

5.2 Determination of Adequate Modeling Parameters

In the presented models of gas propagation and monitoring processes, we have to determine (i) the transition probabilities between the hidden variables representing the presence of gas and (ii) the conditional probabilities between the phenomena in the monitoring process. In either case, it is very likely that given the limited amounts of data and lack of detailed domain knowledge, the estimated probability distributions in the model (i.e., the parameters) significantly deviate from the true conditional probabilities. However, in the context of monitoring applications, ranking of hypothesis is very important and this can be achieved without having precise probabilities.

If the transition probabilities shown in Table 1 merely capture the correct tendencies, we might in many cases be able to obtain useful ranking. For example, if the true probabilities p and q are both greater than 0.5 and our modeling parameters capture the same greater-than/smaller-than relations, then the models still capture important tendencies, which have to be considered in the hypothesis scores. Namely, the presence of gas at time t at location l_i makes the presence of gas at location l_{i+1} at time $t+1$ more likely. Moreover, the impact of the evidence on the hypothesis score falls with the distance from the potential source, which correctly captures the fact that the relations between the source and observations become increasingly uncertain with the distance. Moreover, the model roughly captures the partial (temporal) order of observations throughout the sections of the downwind area. Temporal order of observations is an additional evidence about the likelihood of a particular source; the hypotheses associated with observation sequences which follow a temporal order expected according to the dispersion models obtain higher scores. While the dispersion processes are inherently complex, they can be captured sufficiently well by simple models taking into account the wind direction and speed.

Observation models support detection/classification of gases at different locations. Each observation model supports interpretation of observations acquired in a particular section l_i in the downwind area and provides inputs to the inference using the gas propagation models. An observation model dedicated to location l_i captures causal relations between heterogeneous phenomena in a process producing observations at l_i. The strength of such relations is captured by conditional probabilities, which are likely to deviate from the true probabilities significantly as well. However, by considering the theory of BNs, it was shown that for a significant class of networks, we can achieve very accurate detection despite imprecise parameters. The recently introduced theory of Inference Meta Models (IM) is based on very coarse assumptions and exposes properties of BNs that are relevant for the construction of inherently robust fusion systems [9]. With the help of IM, we can show that the expected detection (i.e., classification) performance improves with the growing number of fragments that are conditionally independent given the variable representing the hypotheses,[9] even if the modeling parameters (i.e., CPTs) deviate from the true distributions significantly. This is the case if the modeling parameters in each conditionally independent fragment correctly capture simple greater-than/smaller-than relations between the probabilities in the true distributions.[10] Often it is plausible to assume that such relations can easily be identified by experts or extracted from relatively small datasets with the help of machine learning techniques. This is

[9] A special case of causal models featuring tree topologies with great branching factors with respect to the hypothesis variable.

[10] For example, assume the true conditional probabilities over binary variables E and C: $P(e|c) = 0.7$, $P(\overline{e}|c) = 0.3$, $P(e|\overline{c}) = 0.4$, and $P(\overline{e}|\overline{c}) = 0.6$. We say that a conditional probability table correctly captures relations between these probabilities if its parameters (i.e., conditional probabilities $\hat{P}(E|C)$) satisfy very simple relations: $\hat{P}(e|c) > 0.5$ and $\hat{P}(e|\overline{c}) < 0.5$.

especially relevant for real-world applications where it is often very difficult or even impossible to obtain precise domain models. For example, it might be very difficult to obtain modeling parameters that precisely describe the true probability of obtaining certain types of reports from humans (e.g., $P(E_m^{Smell}|G)$). Similarly, it is very difficult to find parameters which precisely describe distributions over the combinations of the sensor states and states of the related phenomena, such as, for example, $P(S_1^C|CondC)$. However, if many sensors of different types are available, we can obtain BNs with great numbers of conditionally independent network fragments, which makes fusion very reliable even if we use CPT parameters that deviate from the true distributions significantly. Also, the IM theory provides a guidance for the integration of sensors, such that the fusion robustness is improved; we add sensors, whose corresponding causal models will increase the number of network fragments rooted in the hypothesis variable.

5.3 Determination of Downwind Areas

The downwind area A_h associated with each hypothesis h is a very crude approximation of the boundaries within critical gas concentrations are exceeded. This is a consequence of the lack of a dense network of calibrated sensors and detailed knowledge of the actual gas propagation mechanisms. Consequently, information sources which are not influenced by a plume originating from a hypothesized source might be considered as relevant and, conversely, outputs of sources that are relevant for a particular hypothesis might be ignored. In other words, the system will be fed by partially misleading observations.

This problem can be alleviated by increasing the density of sensors, such that sufficient reports are obtained from the area close to the line which passes the hypothetical source location and is parallel to the wind direction. Another possibility is to send mobile sensor units or concentrate querying of people to this area.

In addition, by using coarse Bayesian models, the impact of occasionally misleading sensor observations can be compensated, i.e., the system is inherently robust and correct hypothesis will still be associated with the greatest probability. Despite the fact that many misleading measurements were used, relevant observations were ignored and the models use conditional probability tables which deviate from the true distributions. If the computed probability is used merely as a score for ranking of hypotheses, then the presented fusion system can provide good decision support, despite sensor imperfection and coarse models.

6 System Design

The presented modeling techniques are implemented in a decentralized fusion system, which supports (i) automated creation of hypotheses upon receiving initial trigger, (ii) collection of the relevant information, and (iii) hypothesis scoring based on the interpretation of the observed patterns. Since the actual constellations of information sources cannot be anticipated, the system must be able to create adequate domain models correlating all information gathered at runtime.

6.1 DPN Layer

Central to the implementation of the proposed discrete Bayesian filters is the aforementioned DPN framework [11]. The DPN is a multi-agent systems approach, which supports efficient creation of theoretically rigorous distributed Bayesian fusion systems. Each agent is composed of predefined software components; all agents use identical components implementing communication and collaboration protocols and inference algorithms. However, the functionality of each agent is determined through a local BN. Such local BNs are defined in advance by experts or with the help of appropriate machine learning techniques. Thus, the configuration of the fusion systems does not require any programming but merely specification of local BNs used by DPN agents. Self-organization of a DPN fusion system is based on the local models. Contrary to other approaches to distributed inference [3, 16], the DPN framework does not rely on the computation of secondary inference structures such as junction trees, which span multiple modules. Compilation of such structures can be computationally expensive; it requires recursion through a system of inference modules (i.e., agents), which can be a serious obstacle if the sensor constellations change frequently at runtime. In DPNs, such compilation is avoided through systematic instantiation of variables [6].

In the presented approach, we assign to each source hypothesis a DPN agent labeled D_h, which fuses outcomes of several anomaly/gas detection systems from different locations in the downwind area (see example in Figure 6). Agent D_h estimates the likelihood of a leak by correlating spatially and temporally distributed detection results by using either the monolithic model shown in Figure 5a or the model shown in Figure 5b.

In addition, agent D_h determines the downwind area A_h and initiates self-configuration of anomaly/gas detection systems for each segment in A_h. In each segment l_i of the downwind area, a set of DPN agents dedicated to processing of information acquired at this location is created. The DPN agents associated with a location autonomously assemble distributed detection systems, each detecting a certain gas or an anomaly using particular types of observations. For example, in Figure 6 a set of DPN agents dedicated to each segment l_i forms distributed anomaly detectors of types A_i, B_i, and C_i; each of these detectors estimates anomaly based on different domain models and observations. Agents labeled A_i, B_i, and C_i interface each distributed anomaly detection system with agent D_h, which correlates detection/estimation outcomes from different locations. Each DPN-based anomaly/gas detector is an organization of collaborating agents implementing a distributed detector of a particular substance; see, for example, a DPN depicted in Figure 4. Such a DPN organization is formed by multiple agents at runtime through service discovery as the relevant information sources become available.

Note that at different locations detectors of the same type can be implemented through DPN systems with differing structure (see Figure 6). This is a consequence of variable availability of information sources; the DPN detectors are assembled at runtime as information sources become available.

In principle, by using the DPN framework we can implement distributed fusion systems which support collaborative, decentralized reasoning about very complex domains. In the DPN approach, simpler models are combined with the help of collaborating agents in such a way that the collaborative inference is equivalent to the belief propagation in complex monolithic domain models describing gas propagation as well as mechanisms generating observations (see Figure 2); in other words, with the help of the DPN, very complex processing is systematically reduced to a collection of smaller processes which combine their local outputs in such a way that the overall fusion is globally coherent.

In addition, DPN agents can wrap mobile sensors. These agents continuously check their location and switch between cells as they transverse the borders between them. Such agents make their service available only to the agents which are located in the current cell. Consequently, the DPN systems served by such mobile fusion agents frequently change, which corresponds to adding/removing nodes in graphs depicted in Figure 6.

Moreover, fusion systems consisting of DPN agents facilitate maintenance. Functionality of the fusion systems is defined primarily through local causal models; each DPN agent is using the same SW components, while its fusion functionality is determined by a local BN. If the expertise about a certain sub-domain changes, only the partial BNs implementing that expertise must be replaced without any further actions such as compilation of Junction trees, etc. In other words, fusion capabilities can be adapted through hot-plugging and swapping of fusion modules, without any compilation.

6.2 Incorporation of Arbitrary Information Sources

It turns out that BNs facilitate incorporation of very heterogeneous sensors of different quality in an efficient way. In this chapter, we assume that sensors are used as simple detectors; analog signals are preprocessed and the sensors return discrete information indicating the most likely states. In the presented setting, electronic noses can be used as simple anomaly detectors returning true and false corresponding to the presence and absence of an anomalous gas mixture, respectively. In other words, simple preprocessing algorithms transform raw sensor data into discrete information which can directly be incorporated in the introduced discrete Bayesian estimation system.

By knowing the performance of such simple detectors, each sensor can easily be described in the causal probabilistic model with the help of a sensor model corresponding to an arbitrarily complex subgraph. Such a sensor model captures conditional probability of different detection results given a particular phenomenon that a sensor is measuring.

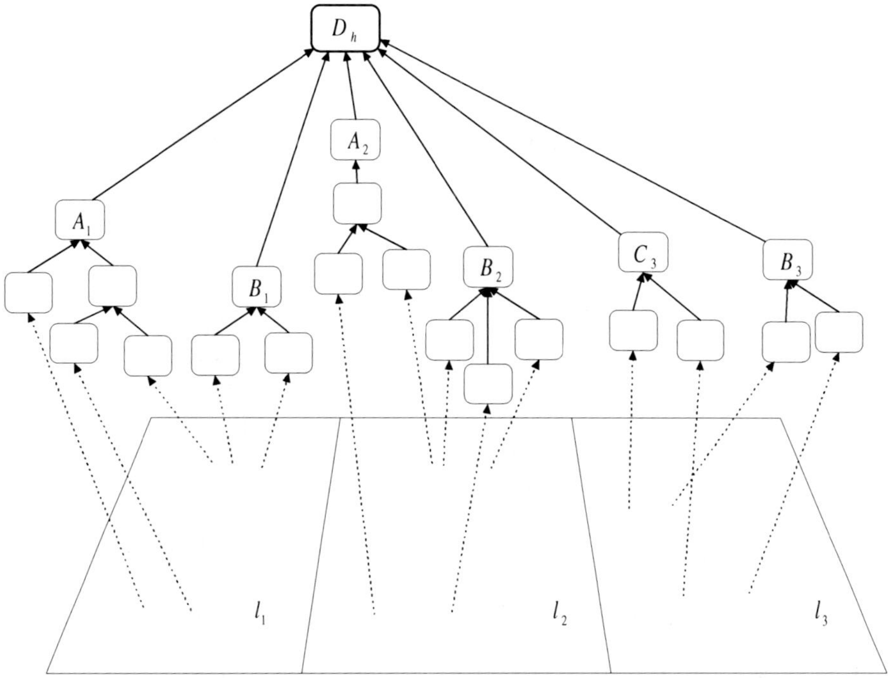

Fig. 6 A DPN fusion system implementing Bayesian filtering of observation sequences gathered at locations (segments) l_1, l_2, and l_3. The rounded rectangles represent DPN agents of various types, and the solid arrows show the flow of information within a particular system. Agent D_h (thick rectangle) uses coarse gas dispersion models to fuse sequences of detection outcomes from agents A_i, B_i, and C_i, each interpreting observations from location l_i. Note that each rounded rectangle corresponds to an arbitrarily complex BN (see Figure 4). Dashed arrows represent observations which are inputs to various agents "wrapping" sensors or humans.

In the same manner, we capture the conditional probability of a certain response to a particular query given some phenomenon the humans can observe. In other words, the models for human and sensor reports do not differ with respect to the representation and inference processes within the fusion system.

Clearly, it might be very difficult to obtain reliable estimates of conditional probability distributions for the human information sources. However, preliminary results from a recently concluded pilot study [15] imply that humans can be viewed as sensors which provide useful information. We analyzed a database with reports from complainers which has been compiled by the DCMR. Often such reports are associated with the information on the cause of the complaints, which allows estimation of relevant conditional probability distributions. In addition, performance of gas detectors using human observations can be analyzed. It turned out that for several types of gases/gas families gas detection systems could be built, which performed better than random [15]. In other words, the performance of human information sources can be described with statistical models sufficiently well to provide useful inputs to effective Bayesian detectors.

6.3 *Automated Querying*

Explicit causal models used by DPNs support also systematic acquisition of large amounts of information that is relevant for a particular fusion task. This is especially useful for the fusion of information obtained from humans. Causal models explicitly encode the relevant concepts through random variables which, in turn, can be mapped to meaningful queries. For example, a causal model can capture the conditional probability that people exposed to a particular gaseous substance will develop certain easily observable symptoms or be able to identify typical smells (e.g., rotten eggs). The observations can be captured by binary variables which can be associated with simple queries. Such queries can be answered by a simple yes or no. If a DPN agent contains variables corresponding to human observations, the queries about the states of such variables are delivered to human observers via SMS, Web interfaces, or voice. The replies in the form of yes or no are routed back to the querying agent and used for the instantiation of the corresponding variable representing observations. In other words, the DPN agents can generate relevant queries from the concepts encoded in their local BNs.

In addition, human information sources have limited capacities; a person can reply to a limited number of queries. Therefore, a DPN system should first generate queries that will have the greatest impact on its belief about the hidden events of interest. Fortunately, rigorous causal models support theoretically sound and efficient approach to the ranking of evidence types with respect to the relevance [5]. The same method can be used for the selection and allocation of scarce/expensive sensors.

7 Conclusions and Future Work

This chapter presents a discrete Bayesian filter for automated detection of gaseous substances and estimation of pollution sources. The approach supports detection and leak localization based on interpretation of complex patterns originating from heterogeneous information sources, such as different types of chemical sensors and humans. In the simplest case, leak localization is based on the detection of anomalous gas compositions. By combining sequences of results from spatially distributed anomaly detectors, the source of anomaly is estimated. Such correlation of spatially and temporally distributed detections is based on causal probabilistic models describing dynamic gas propagation processes. The presented approach in principle allows also classification of gases, if calibrated sensors or information about typical symptoms associated with different gases are available. This requires more complex *observation models*, but the modeling and inference principles remain the same.

The presented approach exploits well-known properties of Bayesian networks. In particular, BNs explicitly capture locality of causal relations, which facilitates theoretically sound and efficient distribution of domain models and inference processes over systems of networked devices. In this way, computation of joint probability distributions over large numbers of variables can be efficiently carried out in work flows of collaborating processing modules distributed over arbitrary networks of computing devices. These properties of BNs are exploited in the DPN, a framework that distributed Bayesian inference. The DPNs provide the processing infrastructure in the presented detection and leak localization approach.

In addition, the locality of causal relations is reflected in the factorization of the domain models, which influences the robustness of Bayesian classification/detection processes. We can show that the types of BNs relevant for the discussed detection problems are often inherently robust with respect to the deviations between the modeling parameters and the true distributions over the variables of interest. This is important since in the targeted domains it is often very difficult to obtain parameters which precisely describe the true distributions. Because of this property we can often use reports from humans and sensors in real-world settings. A recent pilot study carried out in cooperation with the University of Amsterdam and the DCMR shows that reports from humans support effective detection.

An initial prototype system implementing the presented detection/leak localization techniques has been created. Simple anomaly detector using signals from the electronic noses has been developed and incorporated into a distributed Bayesian gas detection/leak localization system. A preliminary validation of the system was carried out with a few selected sequences of real-world sensor data collected by the DCMR Milieudienst Rijnmond. A systematic evaluation of the system performance will be carried out in the near future. Real-world evaluation and adaptation of the presented system is being carried out in cooperation with the DCMR Milieudienst Rijnmond. The future research will focus on improved determination of relevant downwind areas for information gathering and improved models of gas propagation and observation processes.

Acknowledgment. This study was partially funded by the European Union under the Information and Communication Technologies (ICT) theme of the 7th Framework Program for R&D (DIADEM project, ref. no: 224318).

References

1. Bishop, C.: Pattern Recognition and Machine Learning. Springer, Heidelberg (2006)
2. Jensen, F.V.: Bayesian Networks and Decision Graphs. Springer, New York (2001)
3. Mark, P., Carlos, G., McFadden, J.: A robust architecture for inference in sensor networks. In: Fourth International Conference on Information Processing in Sensor Networks, Los Angelos, California, pp. 55–62 (2005)
4. Mikkelsen, T., Desiato, F.: Atmospheric dispersion models and pre-processing of meteorological data for real-time application. Journal of Radiation Protection Dosimetry 50(2-4), 205–218 (1993)
5. Nunnink, J., Pavlin, G.: A probabilistic approach to resource allocation in distributed fusion systems. In: Proc. Int. Conf. on Autonomous Agents and Multi-Agent Systems (AAMAS 2005), Utrecht, Netherlands, pp. 846–852 (2005)

6. de Oude, P., Pavlin, G.: Efficient distributed bayesian reasoning via targeted instantiation of variables. To appear in IEEE/WIC/ACM Joint Conference on Intelligent Agent Technology (IAT 2009), Milano, Italy, pp. 323–330 (2000)
7. de Oude, P., Pavlin, G., Hood, T.: A modular approach to adaptive bayesian information fusion. In: The 10th International Conference on Information Fusion, Quebec City, Quebec, Canada, pp. 1–8 (2007)
8. Pavlin, G., Maris, M., Nunnink, J.: An agent-based approach to distributed data and information fusion. In: IEEE/WIC/ACM Joint Conference on Intelligent Agent Technology (IAT 2004), pp. 466–470 (2004)
9. Pavlin, G., Nunnink, J.: Inference meta models: Towards robust information fusion with bayesian networks. In: Proceedings of the 9th IEEE/ISIF International Conference on Information Fusion, Florence, Italy, pp. 1–8 (2006)
10. Pavlin, G., de Oude, P., Maris, M., Nunnink, J., Hood, T.: A distributed approach to information fusion based on causal probabilistic models. Tech. rep. no. IAS-UVA-07-03 (2007)
11. Pavlin, G., de Oude, P., Maris, M., Nunnink, J., Hood, T.: A multi agent systems approach to distributed Bayesian information fusion. Information Fusion, 458 (2009), doi:10.1016/j.inffus.2009.09.007
12. Pearl, J.: Fusion, propagation, and structuring in belief networks. Artificial Intelligence 29(3), 241–288 (1986)
13. Pearl, J.: Probabilistic Reasoning in Intelligent Systems: Networks of Plausible Inference. Morgan Kaufmann, San Francisco (1988)
14. Russell, S., Norvig, P.: Artificial Intelligence - A Modern Approach, 2nd edn. Prentice Hall, Englewood Cliffs (2002)
15. van Schuppen, D.: An investigation of the DCMR domain for automated compound detection. Master Thesis in Artifcial Intelligence, Informatics Institute, University of Amsterdam (2008)
16. Xiang, Y.: Probabilistic Reasoning in Multiagent Systems: A Graphical Models Approach. Cambridge University Press, Cambridge (2002)

Traffic Light Control by Multiagent Reinforcement Learning Systems

Bram Bakker, Shimon Whiteson, Leon Kester, and Frans C.A. Groen

Abstract. Traffic light control is one of the main means of controlling road traffic. Improving traffic control is important because it can lead to higher traffic throughput and reduced traffic congestion. This chapter describes multiagent reinforcement learning techniques for automatic optimization of traffic light controllers. Such techniques are attractive because they can automatically discover efficient control strategies for complex tasks, such as traffic control, for which it is hard or impossible to compute optimal solutions directly and hard to develop hand-coded solutions. First, the general multi-agent reinforcement learning framework is described, which is used to control traffic lights in this work. In this framework, multiple local controllers (agents) are each responsible for the optimization of traffic lights around a single traffic junction, making use of locally perceived traffic state information (sensed cars on the road), a learned probabilistic model of car behavior, and a learned value function which indicates how traffic light decisions affect long-term utility, in terms of the average waiting time of cars. Next, three extensions are

Bram Bakker
Informatics Institute, University of Amsterdam, Science Park 107, 1098 XG Amsterdam, The Netherlands
e-mail: p.b.bakker@uva.nl

Shimon Whiteson
Informatics Institute, University of Amsterdam, Science Park 107, 1098 XG Amsterdam, The Netherlands
e-mail: s.a.whiteson@uva.nl

Leon Kester
TNO Defense, Security and Safety, Oude Waalsdorperweg 63, 2597 AK Den Haag, The Netherlands
e-mail: leon.kester@tno.nl

Frans C.A. Groen
Informatics Institute, University of Amsterdam, Science Park 107, 1098 XG Amsterdam, The Netherlands
e-mail: f.c.a.groen@uva.nl

R. Babuška & F.C.A. Groen (Eds.): Interactive Collaborative Information Systems, SCI 281, pp. 475–510.
springerlink.com

described which improve upon the basic framework in various ways: agents (traffic junction controllers) taking into account congestion information from neighboring agents; handling partial observability of traffic states; and coordinating the behavior of multiple agents by coordination graphs and the max-plus algorithm.

1 Introduction

Traffic light control is one of the main means of controlling urban road traffic. Improving traffic control is important because it can lead to higher traffic throughput, reduced congestion, and traffic networks that are capable of handling higher traffic loads. At the same time, improving traffic control is difficult because the traffic system, when it is modeled with some degree of realism, is a complex, nonlinear system with large state and action spaces, in which suboptimal control actions can easily lead to congestions that spread quickly and that are hard to dissolve.

In practice, most traffic lights use very simple protocols that merely alternate red and green lights for fixed intervals. The interval lengths may change during peak hours but are not otherwise optimized. Since such controllers are far from optimal, several researchers have investigated the application of artificial intelligence and machine learning techniques to develop more efficient controllers. The methods employed include fuzzy logic [7], neural networks [22] and evolutionary algorithms [9]. These methods perform well but can only handle networks with a relatively small number of controllers and roads.

Since traffic control is fundamentally a problem of sequential decision making, and at the same time, is a task that is too complex for straightforward computation of optimal solutions or effective hand-coded solutions, it is perhaps best suited to the framework of Markov decision processes (MDPs) and reinforcement learning (RL), in which an agent learns from trial and error via interaction with its environment. Each action results in immediate rewards and new observations about the state of the world. Over time, the agent learns a control policy that maximizes the expected long-term reward it receives.

One way to apply reinforcement learning to traffic control is to train a single agent to control the entire system, i.e. to determine how every traffic light in the network is set at each timestep. However, such centralized controllers scale very poorly, since the size of the agent's action set is exponential in the number of traffic lights.

An alternative approach is to view the problem as a multiagent system where each agent controls a single traffic light [3, 33]. Since each agent observes only its local environment and selects only among actions related to one traffic light, this approach can scale to large numbers of agents. This chapter describes multiagent reinforcement learning (MARL) techniques for automatic optimization of traffic light controllers. This means that *variations on* the standard MDP framework must be used to account for the multiagent aspects. This chapter does not contain a full review of MDPs; see [14, 26] for such reviews, and [10, 15, 31] for reviews of multi-agent variations of MDPs.

The controllers are trained and tested using a traffic simulator which provides a simplified model of traffic, but that nevertheless captures many of the complexities of real traffic. Next to the relevance for traffic control per se, this work provides a case study of applying such machine learning techniques to large control problems, and it provides suggestions of how such machine learning techniques can ultimately be applied to real-world problems.

This chapter first describes the general multiagent reinforcement learning framework used to control traffic lights in this work. In this framework, multiple local controllers (agents) are each responsible for the optimization of traffic lights around a single traffic junction. They make use of locally perceived traffic state information (sensed cars on the road), a learned probabilistic model of car behavior, and a learned value function which indicates how traffic light decisions affect long-term utility, in terms of average waiting time of cars.

Next, three extensions are described which improve upon the basic framework in various ways. In the first extension, the local (traffic) state information that is used by each local traffic light controller at one traffic junction is extended by incorporating additional information from neighboring junctions, reflecting the amount of traffic (congestion) at those junctions. Using the same reinforcement learning techniques, the controllers automatically learn to take this additional information into account, effectively learning to avoid to send too much traffic to already congested areas.

In the second extension, more realism is added to the model by having only partial observability of the traffic state. Methods to deal with this situation, based on belief state estimation and POMDP value function approximation, are presented and shown to be effective.

The third extension extends the basic approach to include explicit coordination between neighboring traffic lights. Coordination is achieved using the max-plus algorithm, which estimates the optimal joint action by sending locally optimized messages among connected agents. This extension presents the first application of max-plus to a large-scale problem and thus verifies its efficacy in realistic settings.

The remainder of this chapter is organized as follows. Section 2 introduces the traffic model used in our experiments. Section 3 describes the traffic control problem as a reinforcement learning task. Section 4 describes the work on better handling of congestion situations, section 5 describes the work on partial observability, and Section 6 describes the work on coordination with coordination graphs and the max-plus algorithm. Section 7 presents general conclusions.

2 Traffic Model

All experiments presented in this chapter were conducted using The green light district (GLD) traffic simulator[1] [3, 33]. GLD is a *microscopic* traffic model, i.e. it simulates each vehicle individually, instead of simply modeling aggregate properties of traffic flow. The state variables of the model represent microscopic

[1] Available at http://sourceforge.net/projects/stoplicht.

properties such as the position and velocity of each vehicle. Vehicles move through the network according to their physical characteristics (e.g. length, speed, etc.), fundamental rules of motion, and predefined rules of driver behavior. GLD's simulation is based on *cellular automata*, in which discrete, partially connected cells can occupy various states. For example, a road cell can be occupied by a vehicle or be empty. Local transition rules determine the dynamics of the system and even simple rules can lead to a highly dynamic system.

The GLD infrastructure consists of roads and nodes. A road connects two nodes, and can have several lanes in each direction. The length of each road is expressed in cells. A node is either an intersection where traffic lights are operational or an edge node. There are two types of agents that occupy such an infrastructure namely, vehicles and traffic lights (or intersections). All agents act autonomously and are updated every timestep. Vehicles enter the network at edge nodes and each edge node has a certain probability of generating a vehicle at each timestep (spawn rate). Each generated vehicle is assigned one of the other edge nodes as a destination. The distribution of destinations for each edge node can be adjusted.

There can be several types of vehicles, defined by their speed, length, and number of passengers. In this chapter, all vehicles have equal length and an equal number of passengers. The state of each vehicle is updated every timestep. It either moves the distance determined by its speed and the state around it (e.g. vehicles in front may limit how far it can travel) or remains in the same position (e.g. the next position is occupied or a traffic light prevents its lane from moving).

When a vehicle crosses an intersection, its driving policy determines which lane it goes to next. Once a lane is selected, the vehicle cannot switch to a different lane. For each intersection, there are multiple light configurations that are safe. At each timestep, the intersection must choose from among these configurations, given the current state.

Figure 1 shows an example GLD intersection. It has four roads, each consisting of four lanes (two in each direction). Vehicles occupy n cells of a lane, depending on their length. Traffic on a given lane can only travel in the directions allowed on that lane. This determines the possible safe light configurations. For example, the figure shows a lane where traffic is only allowed to travel straight or right.

The behavior of each vehicle depends on how it selects a path to its destination node and how it adjusts its speed over time. In our experiments, the vehicles always select the shortest path to their destination node.

3 Multiagent Reinforcement Learning for Urban Traffic Control

Several techniques for learning traffic controllers with model-free reinforcement learning methods like Sarsa [30] or Q-learning [1, 19] have previously been developed. However, they all suffer from the same problem: they do not scale to large networks since the size of the state space grows rapidly. Hence, they are either applied only to small networks or are used to train homogeneous controllers (by training

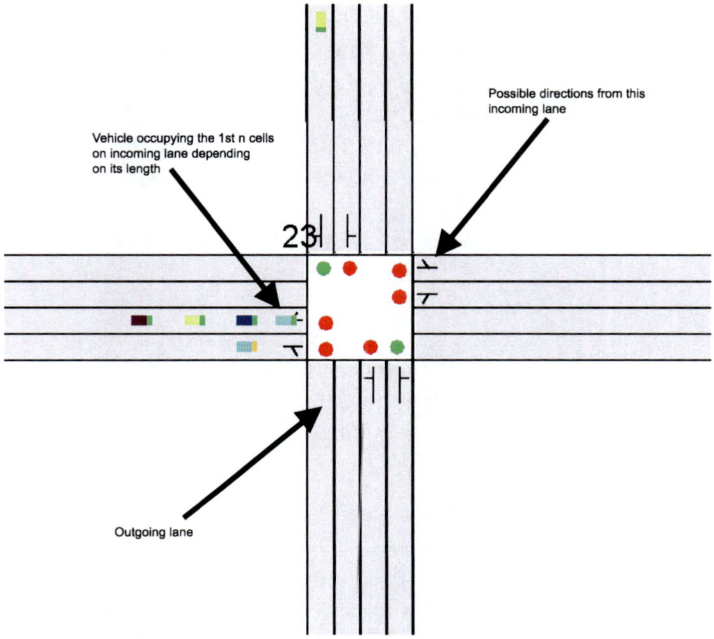

Fig. 1 An example GLD intersection

on a single isolated intersection and copying the result to each intersection in the network).

Even though there is little theoretical support, many authors (e.g. [2, 4, 27]) have found that often, in practice, a more tractable approach, in terms of sample complexity, is to use some variation of model-based reinforcement learning, in which the transition and reward functions are estimated from experience and afterward or simultaneously used to find a policy via planning methods like *dynamic programming* [5]. An intuitive reason for this is that such an approach can typically make more effective use of each interaction with the environment than model-free reinforcement learning. This is because the succession of states and rewards given actions contains much useful information about the dynamics of the environment which is ignored when one just learns a value function as in model-free learning, whereas it is fully used when one learns to predict the environment when learning the model. In practice, a learned model can typically be sufficiently accurate relatively quickly, and thus provides useful guidance for approximating the value function relatively quickly, compared to model-free learning.

In the case of our traffic system, a full transition function would have to map the location of every vehicle in the system at one timestep to the location of every vehicle at the next timestep. Doing so is clearly infeasible, but learning a model is nonetheless possible if a *vehicle-based* representation [33] is used. In this approach, the global state is decomposed into local states based on each individual vehicle. The transition function maps one vehicle's location at a given timestep to its location at the next timestep. As a result, the number of states grows linearly in the number of cells and can scale to much larger networks. Furthermore, the transition function can generalize from experience gathered in different locations, rather than having to learn separate mappings for each location.

To represent the model, we need only keep track of the number of times each transition (s,a,s') has occurred and each state-action pair (s,a) has been reached. The transition model can then be estimated via the maximum likelihood probability, as described below. Hence, each timestep produces new data which is used to update the model. Every time the model changes, the value function computed via dynamic programming must be updated too. However, rather than having to update each state, we can update only the states most likely to be affected by the new data, using an approach based on *prioritized sweeping* [2]. The remainder of this section describes the process of learning the model in more detail.

Given a vehicle-based representation, the traffic control problem consists of the following components:

- $s \in S$: the fully observable global state
- $i \in I$: an intersection (traffic junction) controller
- $a \in A$: an action, which consists of setting to green a subset of the traffic lights at the intersection; $A_i \subseteq A$ is the subset of actions that are safe at intersection i
- $l \in L$: a traffic lane; $L_i \subseteq L$ is the subset of incoming lanes for intersection i
- $p \in P$: a position; $P_l \subseteq P$ is the subset of positions for lane l

The global transition model is $P(s'|s,a)$ and the global state s decomposes into a vector of local states, $s = \langle s_{p_{l_i}} \rangle$, with one for each position in the network.

The transition model can be estimated using maximum likelihoods by counting state transitions and corresponding actions. The update is given by

$$P(s'_{p_{l_i}}|s_{p_{l_i}},a_i) := \frac{C(s_{p_{l_i}},a_i,s'_{p_{l_{ij}}})}{C(s_{p_{l_i}},a_i)} \tag{1}$$

where $C(\cdot)$ is a function that counts the number of times the event (state transition) occurs. To estimate the value of local states (discussed below), we also need to estimate the probability that a certain action will be taken given the state (and the policy), which is done using the following update:

$$P(a_i|s_{p_{l_i}}) := \frac{C(s_{p_{l_i}},a_i)}{C(s_{p_{l_i}})} \tag{2}$$

It is important to note that this does not correspond to the action selection process itself (cf. Eq. 9) or learning of the controller (cf. Eq. 7), but rather an estimate of the expected action, which is necessary in Equation 8. The global reward function decomposes as

$$r(s,s') = \sum_i \sum_{l_i} \sum_{p_{l_i}} r(s_{p_{l_i}},s'_{p_{l_i}}) \tag{3}$$

and

$$r(s_{p_{l_i}}, s'_{p_{l_i}}) = \begin{cases} 0 & s_{p_{l_i}} \neq s'_{p_{l_i}} \\ -1 & \text{otherwise} \end{cases} \tag{4}$$

Thus, if and only if a car stays in the same place the reward (cost) is -1. This means that values estimated by the value function learned using (multiagent) reinforcement learning will reflect total waiting time of cars, which must be minimized. Since the value function essentially represents the sum of the waiting times of individual cars at individual junctions, the action-value function decomposes as

$$Q(s,a) = \sum_i Q_i(s_i, a_i) \tag{5}$$

where s_i is the local state around intersection i and a_i is the local action (safe traffic light configuration) at this intersection, and

$$Q_i(s_i, a_i) = \sum_{l_i} \sum_{p_{l_i}} Q_{p_{l_i}}(s_{p_{l_i}}, a_i). \tag{6}$$

Given the current model, the optimal value function is estimated using dynamic programming, in this case value iteration [4, 5, 26], with a fixed number of iterations. We perform only one iteration per timestep [2, 4] and use ε-greedy exploration to ensure the estimated model obtains sufficiently diverse data. ε-greedy exploration usually takes the action that is currently estimated to be optimal given the current value function, but with probability ε takes a different random action. The vehicle-based update rule is then given by

$$Q_{p_{l_i}}(s_{p_{l_i}}, a_i) := \sum_{s'_{p_{l_i}} \in S'} P(s'_{p_{l_i}} | a_i, s_{p_{l_i}}) [r(s_{p_{l_i}}, s'_{p_{l_i}}) + \gamma V(s'_{p_{l_i}})] \tag{7}$$

where S' are all possible states that can be reached from $s_{p_{l_i}}$ given the current traffic situation and the vehicle's properties (e.g. its speed). $V(s_{p_{l_i}})$ estimates the expected waiting time at p_{l_i} and is given by

$$V(s_{p_{l_i}}) := \sum_{a_i} P(a_i | s_{p_{l_i}}) Q(s_{p_{l_i}}, a_i). \tag{8}$$

We use a discount parameter $\gamma = 0.9$. In reinforcement learning and dynamic programming, the discount parameter discounts future rewards, and depending on the value (between 0 and 1) lets the agent pay more attention to short-term reward as opposed to long-term reward (see [26]). Note also that similar to prioritized sweeping [2], at each timestep only local states that are directly affected (because there are cars driving over their corresponding cells) are updated.

At each time step, using the decomposition of the global state, the joint action (traffic light settings) currently estimated to be optimal (given the value function) is determined as follows:

$$\pi^*(s) = \arg\max_a Q(s,a) \tag{9}$$

where $\arg\max_a Q(s,a)$, which is the optimal global action vector a, in this model corresponds to the optimal local action for each intersection i, i.e. $\arg\max_{a_i} Q(s_i,a_i)$, which can in turn be decomposed as

$$\arg\max_{a_i} Q(s_i,a_i) = \arg\max_{a_i} \sum_{l_i} \sum_{p_{l_i}} O_{p_{l_i}} Q_{p_{l_i}}(s_{p_{l_i}},a_i) \tag{10}$$

where $O_{p_{l_i}}$ is a binary operator which indicates occupancy at p_{l_i}:

$$O_{p_{l_i}} = \begin{cases} 0 & \text{if } p_{l_i} \text{ not occupied} \\ 1 & \text{otherwise} \end{cases} \tag{11}$$

4 Representing and Handling Traffic Congestion

The method described above has already been shown to outperform alternative, non-learning traffic light controllers [33–35]. Nevertheless, there is still considerable room for improvement. In the basic method (called Traffic Controller-1 or TC-1 in the original papers [33–35], and for this reason we will stay with this naming convention), traffic light decisions are made based only on local state information around a single junction. Taking into account traffic situations at other junctions may improve overall performance, especially if these traffic situations are highly dynamic and may include both free flowing traffic and traffic jams (congestion). For example, there is no use in setting lights to green for cars that, further on in their route, will have to wait anyway because there traffic is congested completely.

The two extensions described in this section extend the method described above based on these considerations. The basic idea in both new methods is to keep the principle of local optimization of traffic light junctions, to keep the methods computationally feasible. However, traffic light junctions will now take into account, as extra information, the amount of traffic at neighboring junctions, a "congestion factor". This means that next to local state information, more distant, global (non-local) information is also used for traffic light optimization.

4.1 TC-SBC Method

The first extension proposed in this section takes into account traffic situations at other places in the traffic network by including *congestion* information in the state representation. The cost of including such congestion information is a larger state space and potentially slower learning.

The value function $Q_{p_{l_i}}(s_{p_{l_i}},a_i)$ is extended to $Q_{p_{l_i}}(s_{p_{l_i}},c_{p_{l_i}},a_i)$ where $c_{p_{l_i}} \in \{0,1\}$ is a single bit indicating the congestion level at the next lane for the vehicle currently at p_{l_i}. If the congestion at the next lane exceeds a threshold then $c_{p_{l_i}} = 1$ and otherwise it is set to 0. This extension allows the agents to learn different state transition probabilities and value functions when the outbound lanes are congested. We call this method the "State Bit for Congestion" (TC-SBC) method. Specifically, first a real-valued congestion factor $k_{\text{dest}_{p_{l_i}}}$ is computed for each car at p_{l_i}:

$$k_{\text{dest}_{p_{l_i}}} = \frac{w_{\text{dest}_{p_{l_i}}}}{D_{\text{dest}_{p_{l_i}}}} \tag{12}$$

where $w_{\text{dest}_{p_{l_i}}}$ is the number of cars on destination lane $\text{dest}_{p_{l_i}}$, and $D_{\text{dest}_{p_{l_i}}}$ is the number of available positions on that destination lane $\text{dest}_{p_{l_i}}$. The congestion bit $c_{p_{l_i}}$, which determines the table entry for the value function, is computed according to

$$c_{p_{l_i}} = \begin{cases} 1 & \text{if } k_{\text{dest}_{p_{l_i}}} > \theta \\ 0 & \text{otherwise} \end{cases} \tag{13}$$

where θ is a parameter acting as a threshold.

Like before, the transition model and value function are estimated online using maximum likelihood estimation and dynamic programming. But unlike before, now the system effectively learns different state transition probabilities and value functions when the neighboring roads are congested and when they are not congested (determined by the congestion bit). This makes sense, because the state transition probabilities and expected waiting times are likely to be widely different for these two cases. This allows the system to effectively learn different controllers for the cases of congestion and no congestion. An example of how this may lead to improved traffic flow is that a traffic light may learn that if (and only if) traffic at the next junction is congested, the expected waiting time for cars in that direction is almost identical when this traffic light is set to green compared to when it is set to red. Therefore, it will give precedence to cars going in another direction, where there is no congestion. At the same time, this will give the neighboring congested traffic light junction more time to resolve the congestion.

4.2 TC-GAC Method

A disadvantage of the TC-SBC method described above (cf. Section 4.1, Eq. 12 and 13) is that it increases the size of the state space. That is, in fact, the main reason for restricting the state expansion to only one bit indicating congestion for only the immediate neighbor. The second new method to deal with congestion investigated in this section does not expand the state space, but instead uses the congestion factor $k_{\text{dest}_{p_{l_i}}}$ (described above) in a different way.

Rather than quantizing the real-valued $k_{\text{dest}_{p_{l_i}}}$ to obtain an additional state bit, it is used in the computation of the estimated optimal traffic light configuration (cf. Eq. 10):

$$\arg\max_{a_i} Q(s_i,a_i) = \arg\max_{a_i} \sum_{l_i} \sum_{p_{l_i}} O_{p_{l_i}} (1 - k_{\text{dest}_{p_{l_i}}}) Q_{p_{l_i}}(s_{p_{l_i}},a_i) \tag{14}$$

Thus, the congestion factor $k_{\text{dest}_{p_{l_i}}}$ is subtracted from 1 such that the calculated value for an individual car occupying position p_{l_i} given a particular traffic light configuration (representing its estimated waiting time given that it will see a red or a green light) will be taken fully into account when its next lane is empty (it is then multiplied by 1), or will not be taken into account at all if the next lane is fully congested (it is then multiplied by 0).

We call this method the "Gain Adapted by Congestion" (TC-GAC) method, because it does not affect the value function estimation itself, but only the computation of the gain, in the decision theory sense, of setting traffic lights to green as opposed to red for cars that face or do not face congestion in their destination lanes. The advantage of this method is that it does not increase the size of the state space, while making full use of real-valued congestion information. The disadvantage is that unlike TC-SBC, TC-GAC never learns anything permanent about congestion, and the method is even more specific to this particular application domain and even less generalizable to other domains—at least the principle of extending the state space to represent specific relevant situations that may require different control, as used by TC-SBC, can be generalized to other domains. Note that GAC can in fact be combined with SBC, because it is, in a sense, an orthogonal extension. We call the combination of the two the TC-SBC+GAC method.

4.3 Experimental Results

4.3.1 Test Domains

We tested our algorithms using the GLD simulator (section 2). We used the same traffic network, named Jillesville, as the one used in [34, 35] (see Figure 2), and compared our test results to the results of the basic MARL algorithm (TC-1, section 3).

The original experiments [33–35] were all done with fixed spawning rates, i.e. fixed rates of cars entering the network at the edge nodes. We replicate some of those experiments, but also added experiments (using the same road network) with *dynamic* spawning rates. In these experiments, spawning rates change over time, simulating the more realistic situation of changes in the amount of traffic entering and leaving a city, due to rush hours etc.

In all simulations we measure the Average Trip Waiting Time (ATWT), which corresponds to the amount of time spent waiting rather than driving. It is possible that traffic becomes completely jammed due to the large numbers of cars entering the road network. In that case we set ATWT to a very large value (50).

We did preliminary experiments to find a good parameter setting of the congestion threshold parameter θ for the TC-SBC method. $\theta = 0.8$ gave the best results in all cases, also for the TC-SBC+GAC method, so we are using that value in the experiments described below. It is worth noting that even though TC-SBC adds another parameter, apparently one parameter value works well for different variations

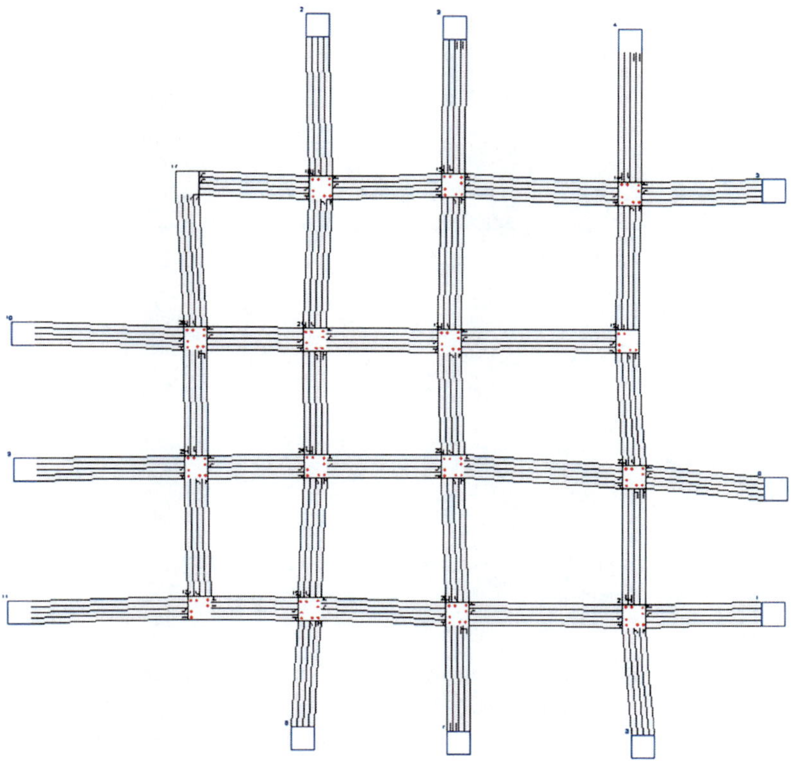

Fig. 2 The Jillesville infrastructure

of the problem and the algorithm, and no parameter tuning is required for every specific case.

4.3.2 Experiment 1: Fixed Spawning Rate

As described above, we performed simulations to compare our new congestion-based RL controllers (TC-SBC, TC-GAC, TC-SBC+GAC, section 3) to the original RL controller (TC-1, section 2.3) in one of the original test problems. This test problem corresponds to Wiering's experiment 1 [34, 35] and uses the road network depicted in Figure 2. The spawning rate is fixed and set to 0.4. Each run corresponds to 50,000 cycles. For each experimental condition (controller type) 5 runs were done and the results were averaged. We found that 5 runs sufficed to obtain sufficiently clear and distinct results for the different experimental conditions, but no statistical significance was determined.

Figure 3 shows the overall results of the experiment. The average trip waiting time (ATWT) is depicted over time. Each curve represents the average of 5 runs. The most important result is that all of our new methods lead, in the long run, to better performance, i.e. lower ATWT, than the original method TC-1. The best performance in this test problem is obtained by TC-SBC+GAC, the method that

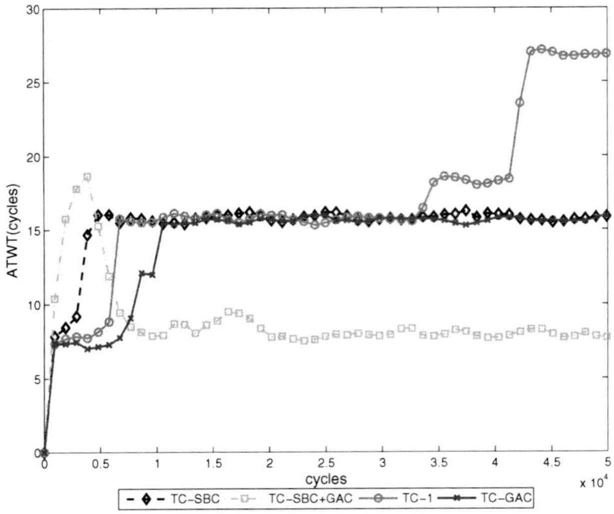

Fig. 3 Performance of our congestion-based controllers versus TC-1, in the fixed spawning rate simulation (experiment 1)

combines both ways of using congestion information. Interestingly, in the initial stages of learning this method has, for a while, the worst average performance, but then it apparently learns a very good value function and corresponding policy, and reduces ATWT to a very low level. This relatively late learning of the best policy may be caused by the learning system first learning a basic policy and later fine-tuning this based on the experienced effects of the GAC component.

Figure 4 shows a snapshot of part of the traffic network in the traffic simulator during one run, where the controller was TC-SBC+GAC. A darker traffic junction indicates a more congested traffic situation around that junction. The lower right junction is fairly congested, which results in the traffic lights leading to that junction to be set to red more often and for longer periods of time.

4.3.3 Experiment 2: Dynamic Spawning Rate, Rush Hours

Experiment 2 uses the same traffic network (Figure 2) as experiment 1, but uses dynamic spawning rates rather than fixed ones. These simulations similarly last for 50,000 cycles and start with spawning rate 0.4. Every run is divided into five blocks of 10,000 cycles, within each of which at cycle 5,000 "rush hour" starts and the spawning rate is set to 0.7. At cycle 5,500 the spawning rate is set to 0.2, and at cycle 8,000 the spawning rate is set back to 0.4 again. The idea is that especially in these dynamic situations traffic flow may benefit from our new methods, which dynamically take into account congestion.

Figure 5 shows the results. As in experiment 1, all of our new methods outperform the original, TC-1 method. As expected, the difference is actually more pronounced (note the scale of the graph): TC-SBC, TC-GAC, and TC-SBC+GAC are much

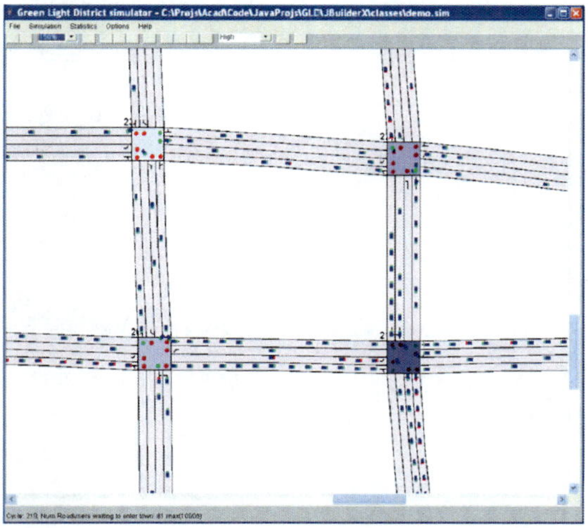

Fig. 4 Snapshot of part of the traffic network in the traffic simulator. Darker nodes indicate more congested traffic

better in dealing with the changes in traffic flow due to rush hours. Rush hours do lead to longer waiting times temporarily (the upward "bumps" in the graph). However, unlike the original, TC-1 method, our methods avoid getting complete traffic jams, and after rush hours they manage to reduce the average trip waiting time. The best method in this experiment is TC-GAC. It is not entirely clear why TC-GAC outperforms the version with SBC, but it may have to do with the fact that the increased state space associated with SBC makes it difficult to obtain a sufficient amount of experience for each state in this experiment where there is much variation in traffic flow.

5 Partial Observability

In the work described so far we assumed that the traffic light controllers have access, in principle, to all traffic state information, even if the system is designed such that agents make decisions based only on local state information. In other words, we made the assumption of complete observability of the MDP (see Figure 6). In the real world it is often not realistic to assume that all state information can be gathered or that the gathered information is perfectly accurate. Thus decisions typically have to be made based on incomplete or erroneous information. This implies dealing with *partial observability*.

In order to obtain full observability in a real-world implementation of the traffic control system, a similar grid like state discretization as is the case in our GLD traffic simulation would have to be made on roads, with a car sensor (e.g. an induction loop sensor) for each discrete state. On a real road the amount of sensors needed to realize this is immense; moreover, road users do not move from one discrete state to

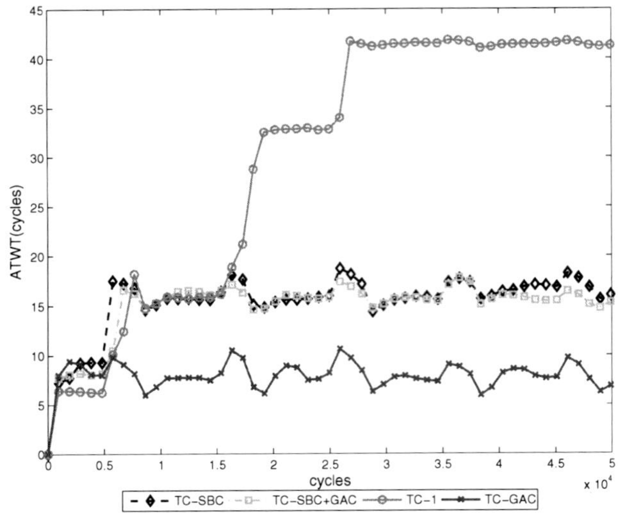

Fig. 5 Performance of our congestion-based controllers versus TC-1, in the dynamic spawning rate simulation (experiment 2, "rush hours")

another, but move continuously. Approximating the state space using fewer sensors converts the problem from completely to partially observable. The controller no longer has direct access to all state information. The quality of the sensors, i.e. the possibility of faulty sensor readings, also plays a role in partial observability.

In this section, we study partial observability within our multiagent reinforcement learning for traffic light control framework. To model partial observability in the—up to now fully observable—traffic simulator an observation layer is added. This allows algorithms to get their state information through this observation layer instead.

Multiagent systems with partial observability constitute a domain not heavily explored, but an important one to make a link to real-world application and realistic traffic light control. Most of the already developed algorithms for partial observability focus on single-agent problems. Another problem is that many are too computationally intensive to solve tasks more complex than toy problems. Adding the multiagent aspect increases the complexity considerably, because the size of the joint action space is exponential in the number of agents. Instead of having to compute a best action based on one agent, the combined (joint) best action needs to be chosen from all the combinations of best actions of single agents. In the traffic light controller work described in this chapter so far, this problem is addressed by decomposing into local actions, each of which is optimized individually, and this same approach is used here.

For the single-agent case, to diminish the computational complexity several heuristic approximations for dealing with partial observability have been proposed. Primary examples of these heuristic approximations are the most likely state (MLS)

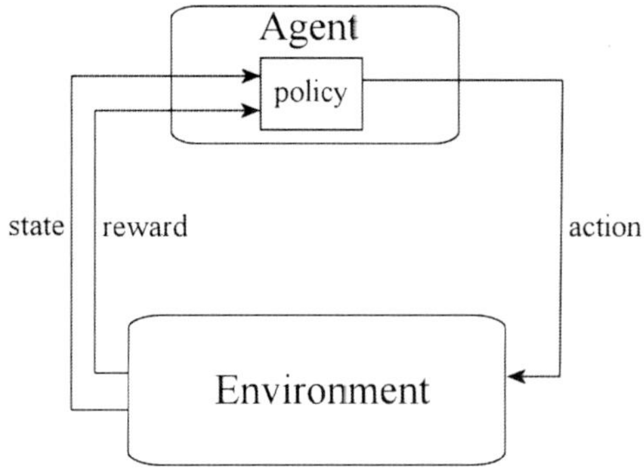

Fig. 6 The generic model of an agent in a (completely observable) MDP. At each time step, the agent performs actions in its environment, and in response receives information about the new state that it is in as well as an immediate reward (or cost).

approach and Q-MDP [6]. Our approach will be based on these approximations, and can be viewed as a multiagent extension.

5.1 POMDPs

Partially observable Markov decision processes, or POMDPs for short, are Markov Decision Processes (the basis for all work described above) in which the state of the environment, which itself has the Markov property, is only partially observable, such that the resulting observations do not have the Markov property (see [6, 13, 20] for introductions to POMDPs). For clarity we will, in this section, refer to MDPs, including the traffic light systems described above which are non-standard variations of MDPs, as completely observable MDPs (or COMDPs).

A POMDP is denoted by the tuple $\Xi = (S,A,Z,R,P,M)$, where S, A, P, and R are the same as in COMDPs. Z is the set of observations. After a state transition one of these observations o is perceived by the agent(s). M is the observation function, $M : S \times A \rightarrow \Pi(Z)$, mapping states-action pairs to a probability distribution over the observations the agent will see after doing that action from that state. As can be seen, the observation function is dependent on the state the agents is in and the action it executes. Importantly, in contrast to (standard) COMDPs, the agent no longer directly perceives the state (see Figure 6), but instead perceives an observation which may give ambiguous information regarding the underlying state (i.e. there is hidden state), and must do *state estimation* to obtain its best estimate of the current state (see Figure 7).

5.2 Belief States

A POMDP agent's best estimate of the current state is obtained when it remembers the entire history of the process [32][6]. However it is possible to "compress" the

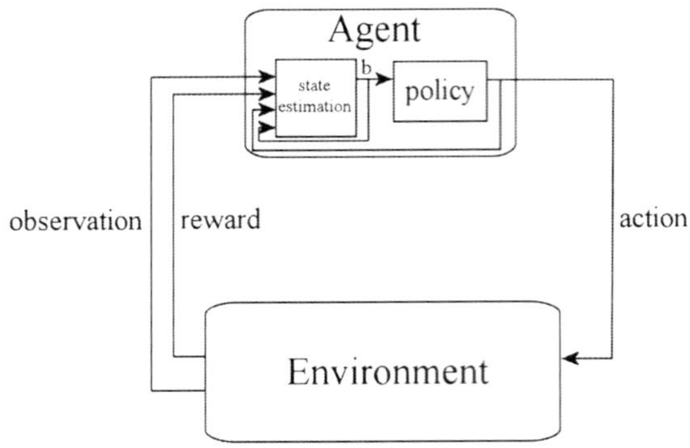

Fig. 7 The POMDP model: since states are no longer directly observable, they will have to be estimated from the sequence of observations following actions. This is done by the state estimation module, which outputs a belief state b to be used by the policy.

entire history into a summary state, a belief state, which is a sufficient statistic for the history. The belief state is a probability distribution over the state space: $B = \Pi(S)$, denoting for each state the probability that the agent is in that state, given the history of observations and actions. The policy can then be based on the entire history, $\pi : B \rightarrow A$. Unlike the complete history, however, the belief state B is of fixed size, while the history grows with each time step. In contrast to the POMDP observations, the POMDP belief state has the Markov property, subsequent belief states only depend on the current belief state and no additional amount of history about the observations and actions can improve the information contained in the belief state. The notation for the belief state probability will be $b(s)$, the probabilities that the agent is in each (global) state $s \in S$. The next belief state depends only on the previous belief state and the current action and observation, and is computed using Bayes' rule [6].

5.3 *Partial Observability in the Traffic System*

Figure 8 shows a single-traffic lane in our simulator, both for the completely (fully) observable case, as was used in the work described in the previous sections, and for the partially observable case. In the latter case, there are car sensors e which provide noisy information e(s) about the occupation of the network position where they are placed, and they are placed only at the very beginning and very end of each lane, which means that for the intermediate positions nontrivial state estimation must be done.

In order to get a good belief state of a drive lane b^d such as depicted in Figure 8, it is necessary first of all to create a belief state for individual road users, b_f. To create b_f, sensor information can be used in combination with assumptions that can be made for a road user.

Actions u_f are the actions of the road user f. Suppose, the sensors inform the system of road user being on the beginning of the road on time step $t = 0$, then given that cars may move at different speeds, the car may move 1, 2, or 3 steps

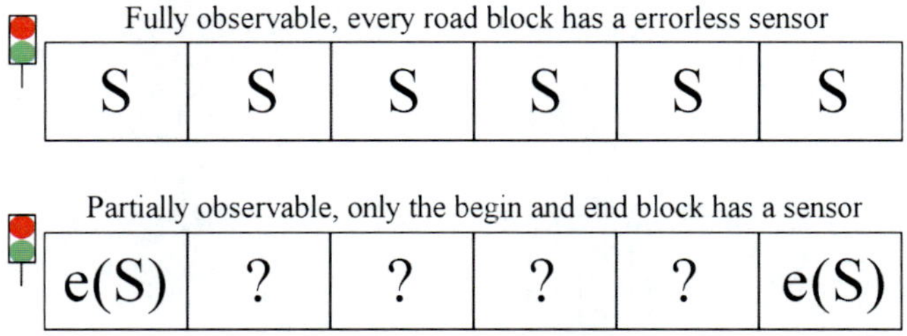

Fig. 8 Sensors on fully and partially observable roads

forward. This means that the road user is on position 1, position 2, or position 3 of the road on time step $t = 1$. By the same logic, the whole b_f of all cars can be computed for all time steps, using the Bayes update rule:

$$b_f(s_t) = p(z_t|s_t) \sum_{s_{t-1}} p(s_t|u_{f,t}, s_{t-1}) p(s_{t-1}|z_{t-1}, u_{t-1}) \tag{15}$$

Since there are many cars, calculating only belief states of individual road users and assuming that those individual road users are not dependent on each other's position is not correct. Therefore, to calculate the belief state for the whole driving lane, b^d, all the b_f for that driving lane must be combined, taking into account their respective positions. $b^d_{p_{l_i}}$ denotes the individual probability, according to the complete belief state b^d, of a car being at position p of lane l at intersection (junction) i. Three additional principles guide the algorithm that computes the belief state for the combination of all cars in the network. First, road users that arrive on a lane at a later t cannot pass road users that arrived at an earlier t. Second, road users cannot be on top of each other. Third, road users always move forward as far as they can, leaving room behind them when they can, such that road users arriving at a later time step have a spot on the road. The exact algorithm incorporating this can be found in [24]. Figure 9 shows an example belief state for the whole driving lane, b^d, for one lane in a traffic network.

5.4 Multiagent Most-Likely-State and Q-MDP

This section describes multiagent variations to two well-known heuristic approximations for POMDPs, the most likely state (MLS) approach and Q-MDP [6]. Both are based on estimating the value function as if the system is completely observable, which is not optimal but works well in many cases, and then combining this value function with the belief state.

The MLS approach simply assumes that the most likely state in the belief state is the actual state, and chooses actions accordingly. Two types of most likely state implementations are investigated here. In the first, we use the individual road user belief state (for all road users):

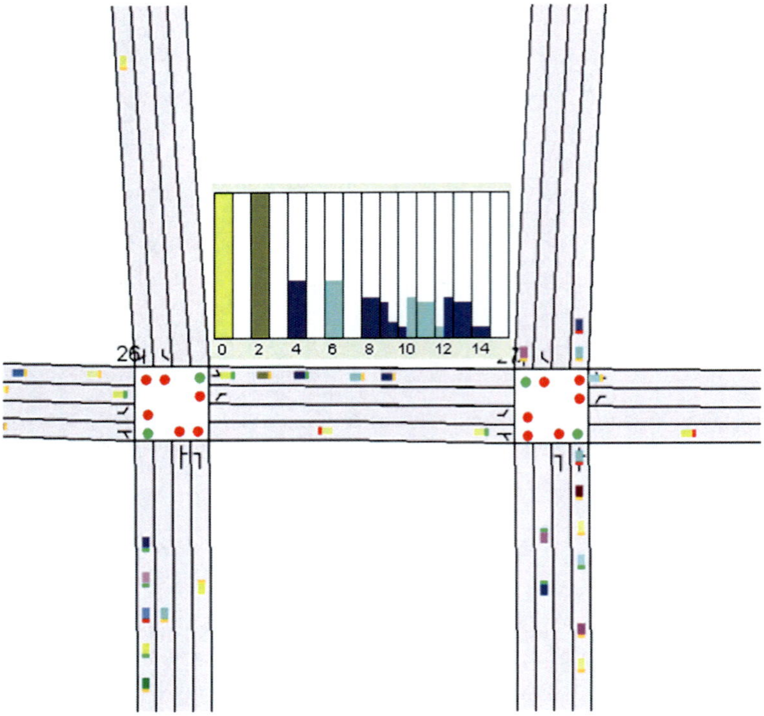

Fig. 9 The probability distribution of b^d is shown for a drive lane of the Jillesville infrastructure

$$s_f^{\mathrm{MLS}} = \arg\max_s b_f(s) \tag{16}$$

and, as before, a binary operator which indicates occupancy of a position in the road network by a car is defined:

$$O_{p_{l_i}}^{\mathrm{MLS}} = \begin{cases} 1 & \text{if } p_{l_i} \text{ corresponds to an } s_{p_{l_i}} \in s_f^{\mathrm{MLS}} \\ 0 & \text{otherwise.} \end{cases} \tag{17}$$

Thus, the difference with occupancy as defined in the standard model in Eq. 11 is that here occupancy is assumed if this is indicated by the most likely state estimation. Action selection is done as follows:

$$\arg\max_{a_i} Q(s_i,a_i) = \arg\max_{a_i} \sum_{l_i} \sum_{p_{l_i}} O_{p_{l_i}}^{\mathrm{MLS}} Q_{p_{l_i}}(s_{p_{l_i}},a_i) \tag{18}$$

The second type of MLS implementation uses b^d, the belief state defined for a drive lane, which is more sophisticated and takes into account mutual constraints between cars. With this belief state the state, can be seen as the possible queues of road users for that drive lane. The MLS approximation for this queue is called the most likely queue, or MLQ. The most like queue is

$$s_d^{\mathrm{MLQ}} = \arg\max_s b^d(s) \tag{19}$$

and

$$O_{p_{l_i}}^{\mathrm{MLQ}} = \begin{cases} 1 & \text{if } p_{l_i} \text{ corresponds to an } s_{p_{l_i}} \in s_d^{\mathrm{MLQ}} \\ 0 & \text{otherwise} \end{cases} \tag{20}$$

and action selection is done as follows:

$$\arg\max_{a_i} Q(s_i,a_i) = \arg\max_{a_i} \sum_{l_i} \sum_{p_{l_i}} O_{p_{l_i}}^{\mathrm{MLQ}} Q_{p_{l_i}}(s_{p_{l_i}},a_i). \tag{21}$$

Instead of selecting the most like state (or queue) from the belief state, the idea of Q-MDP is to use the entire belief state: the state-action value function is weighed by the probabilities defined by the belief state, allowing all possible occupied states to have a "vote" in the action selection process and giving greater weight to more likely states. Thus, action selection becomes

$$\arg\max_{a_i} Q(s_i,a_i) = \arg\max_{a_i} \sum_{l_i} \sum_{p_{l_i}} b_{p_{l_i}}^d Q_{p_{l_i}}(s_{p_{l_i}},a_i). \tag{22}$$

In single-agent POMDPs, Q-MDP usually leads to better results than MLS [6].

In order to provide a baseline method for comparison, a simple method was implemented as well: the All In Front, or AIF, method assumes that all the road users that are detected by the first sensor are immediately at the end of the lane, in front of the traffic light. The road users are not placed on top of each other. Then the decision for the configuration of the traffic lights is based on this simplistic assumption. Furthermore, we compare with the standard TC-1 RL controller, referred to as COMDP here, that has complete observability of the state.

5.5 *Learning the Model*

MLS/MLQ and Q-MDP provide approximate solutions for the partially observable traffic light control system, given that the state transition model and value function can still be learned as usual. However, in the partially observable case the model cannot be learned by counting state transitions, because the states cannot directly be accessed. Instead, all the transitions between the belief state probabilities of the current state and the probabilities of the next belief states should be counted to be able to learn the model. Recall, however, that belief states have a continuous state space even if the original state space is discrete, so there will in effect be an infinite number of possible state transitions to count at each time step. This is not possible. An approximation is therefore needed.

The following approximation is used: learning the model without using the full belief state space, but just using the belief state point that has the highest probability, as in MLS/MLQ. Using the combined road user belief state b^d, the most likely states in this belief state are s^{MLQ}. The local state (being occupied or not) at position p of lane l at intersection (junction) i, according to the most likely queue estimate, is denoted by $s_{p_{l_i}}^{\mathrm{MLQ}}$. Assuming that these most likely states are the correct states, the system can again estimate the state transition model by counting the transitions of road users as if the state were fully observable (cf. eq. 1):

$$P(s'_{p_{l_i}}|s_{p_{l_i}},a_i) := \frac{C(s_{p_{l_i}}^{\mathrm{MLQ}},a_i,s_{p_{l_{ij}}}^{\mathrm{MLQ}})}{C(s_{p_{l_i}}^{\mathrm{MLQ}},a_i)} \tag{23}$$

and likewise for the other update rules (cf. section 3). Although this approximation is not optimal, it provides a way to sample the model and the intuition is that with many samples it provides a fair approximation.

5.6 Test Domains

The new algorithms were tested on two different traffic networks, namely, the Jillesville traffic network used before (see Figure 2), and a newly designed simple network named LongestRoad.

Since Jillesville was used in previous work it is interesting to use it for partial observability as well, and to see if it can still be controlled well under partial observability. Jillesville is a traffic network with 16 junctions, 12 edge nodes or spawn nodes and 36 roads with 4 drive lanes each.

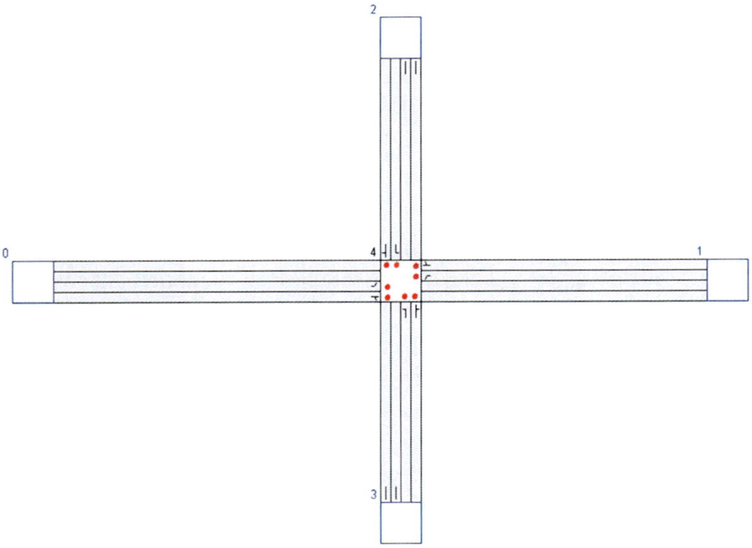

Fig. 10 The LongestRoad infrastructure

LongestRoad (Figure 10) is, in most respects, a much simpler traffic network than Jillesville because it only has 1 junction, 4 edge nodes and 4 roads with 4 drive lanes each. The biggest challenge, however, in this set-up is that the drive lanes themselves are much longer than in Jillesville, making the probabilities of being wrong with the belief states a lot higher, given sensors only at the beginnings and ends of lanes. This was used to test the robustness of the POMDP algorithms.

In the experiments described above (Section 4.3), the main performance measure was average traffic waiting time (ATWT). However, in the case of partial observability we found that many suboptimal controllers lead to complete congestion with no cars moving any more, in which case ATWT is not the most informative measure. Instead, here we use a measure called total arrived road users (TAR) which, as the name implies, is the number of total road users that actually reached their goal node. TAR measures effectively how many road users actually arrive at their goal and do not just stand still forever. Furthermore, we specifically measure TAR as a function of the spawning rate, i.e. the rate at which new cars enter the network at edge nodes. This allows us to gauge the level of traffic load that the system can handle correctly, and observe when it breaks down and leads to complete or almost complete catastrophic congestion.

5.7 *Experimental Results: COMDP vs. POMDP Algorithms*

Figure 11 shows the results, TAR as a function of spawning rate, for the different algorithms described above, when tested on the partially observable Jillesville traffic infrastructure. A straight monotically increasingly line indicates that all additional road users injected at higher spawning rates arrive at their destinations (because TAR scales linearly with the spawning rate, unless there are congestions). Here we do not yet have model learning under partial observability; the model is learned under full observability, and the focus in on how the controllers function under partial obsrvability.

The results in Figure 11 show that Q-MDP performs very well, and is comparable to COMDP, with MLQ slightly behind the two top algorithms. AIF and MLS perform much worse; this is not very surprising as they make assumptions which are overly simplistic. The highest throughput is thus reached by Q-MDP since that algorithm has the most total arrived road users. This test also shows that the robustness of Q-MDP is better than the alternative POMDP methods since the maximum spawn rates for Q-MDP before the simulation gets into a "deadlock" situation (the sudden drop in the curves when spawn rates are pushed beyond a certain level) are higher. Generally, the algorithms MLQ, Q-MDP, and COMDP perform at almost an equal level.

With the LongestRoad network, in which individual roads are substantially longer, a slightly different result can be observed (see Figure 12). In the Jillesville traffic network, MLS was underperforming, compared to the other methods. On the LongestRoad network on the other hand the performance is not bad. COMDP appears to perform slightly worse than the POMDP methods (except for AIF).

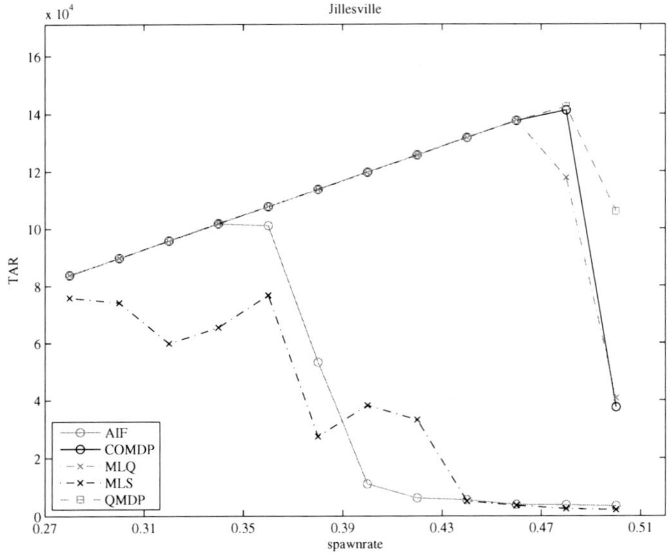

Fig. 11 Results on the partially observable Jillesville traffic infrastructure, comparing POMDP with COMDP algorithms

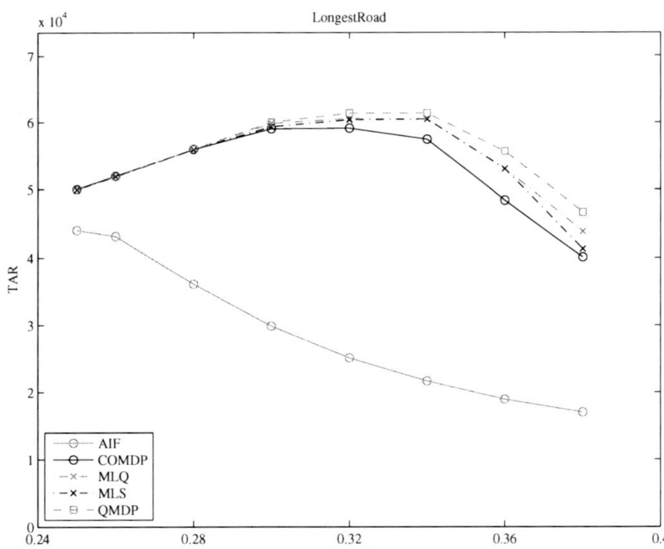

Fig. 12 Results on the partially observable LongestRoad traffic infrastructure, comparing POMDP with COMDP algorithms

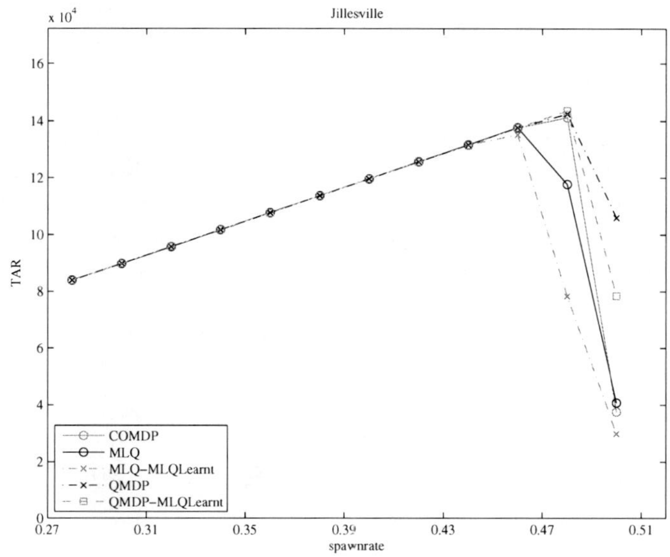

Fig. 13 Results on the partially observable Jillesville traffic infrastructure, comparing POMDP algorithms which learn the model under partial observability (MLQLearnt) with POMDP algorithms which do not have this additional difficulty

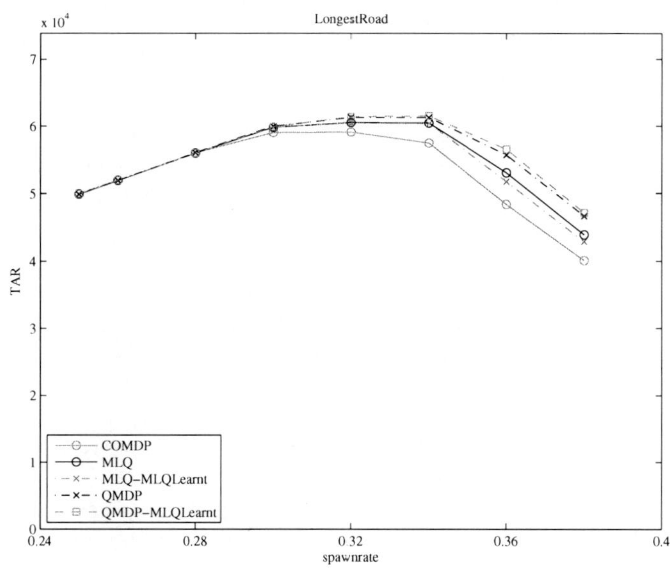

Fig. 14 Results on the partially observable LongestRoad traffic infrastructure, comparing POMDP algorithms which learn the model under partial observability (MLQLearnt) with POMDP algorithms which do not have this additional difficulty

In theory COMDP should be the upper bound, since its state space is directly accessible and the RL approach should then perform best. One possibility is that the COMDP algorithm has some difficulties converging on this infrastructure compared to other infrastructures because of the greater lengths of the roads. Longer roads provide more states and since a road user will generally not come across all states when it passes through (with higher speeds many cells are skipped), many road users are needed to get good model and value function approximations. The POMDP algorithms' state estimation mechanism is somewhat crude, but perhaps this effectively leads to fewer states being considered and less sensitivity to inaccuracies of value estimations for some states.

In any case, the performance of the most advanced POMDP and COMDP methods is comparable, and only the baseline AIF method and the simple MLS method perform much worse.

5.8 Experimental Results: Learning the Model under Partial Observability

The experiment described above showed good results of making decisions under partial observability. Next, we test the effectiveness of learning the model under partial observability. The methods doing that do so using the MLQ state (see above) and are indicated by MLQLearnt suffixes. We again compare with COMDP, and with the POMDP methods with complete observability model learning. We do not include AIF in this comparison, as it was already shown to be inferior above.

The results on Jillesville are shown in Figure 13. It can be seen that the MLQLearnt methods works very well. MLQ-MLQLearnt (MLQ action selection and model learning based on MLQ) does not perform as well as Q-MDP-MLQLearnt (Q-MDP action selection and model learning based on MLQ), but still has a good performance on Jillesville. It can be concluded that in this domain, when there is partial observability, both action selection and model learning can be done effectively, given that proper approximation techniques are used.

We can derive the same conclusion from a similar experiment with the LongestRoad traffic network (see Figure 14). Again learning the model under partial observability based on the MLQ state does not negatively affect behavior.

6 Multiagent Coordination of Traffic Light Controllers

In the third and final extension to the basic multiagent reinforcement learning approach to traffic light control, we return to the situation of complete observability of the state (as opposed to partial observability considered in the previous section), but focus on the issue of multiagent coordination. The primary limitation of the approaches described above is that the individual agents (controllers for individual traffic junctions) do not coordinate their behavior. Consequently, agents may select individual actions that are locally optimal but that together result in global inefficiencies. Coordinating actions, here and in general in multiagent systems, can be difficult since the size of the joint action space is exponential in the number of agents. However, in many cases, the best action for a given agent may depend on only a small subset of the other agents. If so, the global reward function can be decomposed into local functions involving only subsets of agents. The optimal joint action can then be estimated by finding the joint action that maximizes the sum of the local rewards.

A *coordination graph* [10], which can be used to describe the dependencies between agents, is an undirected graph $G = (V,E)$ in which each node $i \in V$ represents an agent and each edge $e(i,j) \in E$ between agents i and j indicates a dependency between them. The global coordination problem is then decomposed into a set of local coordination problems, each involving a subset of the agents. Since any arbitrary graph can be converted to one with only pairwise dependencies [36], the global action-value function can be decomposed into pairwise value functions given by

$$Q(s,a) = \sum_{i,j \in E} Q_{ij}(s,a_i,a_j) \tag{24}$$

where a_i and a_j are the corresponding actions of agents i and j, respectively. Using such a decomposition, the *variable elimination* [10] algorithm can compute the optimal joint action by iteratively eliminating agents and creating new conditional functions that compute the maximal value the agent can achieve given the actions of the other agents on which it depends. Although this algorithm always finds the optimal joint action, it is computationally expensive, as the execution time is exponential in the induced width of the graph [31]. Furthermore, the actions are known only when the entire computation completes, which can be a problem for systems that must perform under time constraints. In such cases, it is desirable to have an *anytime* algorithm that improves its solution gradually.

One such algorithm is *max-plus* [15, 16], which approximates the optimal joint action by iteratively sending locally optimized messages between connected nodes in the graph. While in state s, a message from agent i to neighboring agent j describes a local reward function for agent j and is defined by

$$\mu_{ij}(a_j) = \max_{a_i} \{Q_{ij}(s,a_i,a_j) + \sum_{k \in \Gamma(i) \setminus j} \mu_{ki}(a_i)\} + c_{ij} \tag{25}$$

where $\Gamma(i) \setminus j$ denotes all neighbors of i except for j and c_{ij} is either zero or can be used to normalize the messages. The message approximates the maximum value agent i can achieve for each action of agent j based on the function defined between them and incoming messages to agent i from other connected agents (except j). Once the algorithm converges or time runs out, each agent i can select the action

$$a_i^* = \arg\max_{a_i} \sum_{j \in \Gamma(i)} \mu_{ji}(a_i) \tag{26}$$

Max-plus has been proven to converge to the optimal action in finite iterations, but only for tree-structured graphs, not those with cycles. Nevertheless, the algorithm has been successfully applied to such graphs [8, 15, 36].

6.1 Max-Plus for Urban Traffic Control

Max-plus enables agents to coordinate their actions and learn cooperatively. Doing so can increase robustness, as the system can become unstable and inconsistent when agents do not coordinate. By exploiting coordination graphs, max-plus minimizes the expense of computing joint actions and allows them to be approximated within time constraints.

In this chapter, we combine max-plus with the model-based approach to traffic control described above. We use the vehicle-based representation defined in Section 3 but add dependence relationships between certain agents. If $i,j \in J$ are two intersections connected by a road, then they become neighbors in the coordination graph, i.e. $i \in \Gamma(j)$ and $j \in \Gamma(i)$. The local value functions are

$$Q_i(s_i,a_i,a_j) = \sum_{l_i} \sum_{p_{l_i}} Q_{p_{l_i}}(s_{p_{l_i}},a_i,a_j) \tag{27}$$

Using the above, we can define the pairwise value functions used by max-plus:

$$Q_{ij}(s,a_i,a_j) = \sum_{p_{l_i}} O_{p_{l_i}} Q_{p_{l_i}}(s_{p_{l_i}},a_i,a_j) + \sum_{p_{l_j}} O_{p_{l_j}} Q_{p_{l_j}}(s_{p_{l_j}},a_j,a_i) \tag{28}$$

where $O_{p_{l_i}}$ is the binary operator which indicates occupancy at p_{l_i} (eq. 11).

These local functions are plugged directly into Equation 25 to implement max-plus. Note that the functions are symmetric such that $Q_{ij}(s,a_i,a_j) = Q_{ji}(s,a_j,a_i)$. Thus, using Equation 28, the joint action can be estimated directly by the max-plus algorithm. Like before, we use one iteration of dynamic programming per timestep and ε-greedy exploration. We also limit max-plus to 3 iterations per timestep.

Note that there are two levels of value propagation among agents. On the lower level, the vehicle-based representation enables estimated values to be propagated between neighboring agents and eventually through the entire network, as before. On the higher level, agents use max-plus when computing joint actions to inform their neighbors of the best value they can achieve, given the current state and the values received from other agents.

Using this approach, agents can learn cooperative behavior, since they share value functions with their neighbors. Furthermore, they can do so efficiently, since the number of value functions is linear in the induced width of the graph. Stronger dependence relationships could also be modeled, i.e. between intersections not directly connected by a road, but we make the simplifying assumption that it is sufficient to model the dependencies between immediate neighbors in the traffic network.

6.2 Experimental Results

In this section, we compare the novel approach described in Section 6.1 to the TC-1 (Traffic Controller 1) described in section 3 and the TC-SBC (traffic controller with state bit for congestion) extension described in section 4 (we do not compare to TC-GAC as that is not really a learning method and is very specific to the particular application domain and not easily generalizable). We focus our experiments on comparisons between the coordination graph/max-plus method, TC-1, and TC-SBC in saturated traffic conditions (i.e. a lot of traffic, close to congestion), as these are conditions in which differences between traffic light controllers become apparent and coordination may be important.

These experiments are designed to test the hypothesis that, under highly saturated conditions, coordination is beneficial when the amount of *local traffic* is small. Local traffic consists of vehicles that cross a single intersection and then exit the network, thereby interacting with just one learning agent (traffic junction controller). If this hypothesis is correct, coordinated learning with max-plus should substantially outperform TC-1 and TC-SBC in particular when most vehicles pass through multiple intersections.

In particular, we consider three different scenarios. In the *baseline* scenario, the traffic network includes routes, i.e. paths from one edge node to another, that cross only a single intersection. Since each vehicle's destination is chosen from a uniform distribution, there is a substantial amount of local traffic. In the *nonuniform destinations* scenario, the same network is used but destinations are selected to ensure that each vehicle crosses two or more intersections, thereby eliminating local traffic. To ensure that any performance differences we observe are due to the absence of local traffic and not just to a lack of uniform destinations, we also consider the *long routes* scenario. In this case, destinations are selected uniformly but the network is altered such that all routes contain at least two intersections, again eliminating local traffic.

While a small amount of local traffic will occur in real-world scenarios, the vast majority is likely to be non-local. Thus, the baseline scenario is used, not for its realism, but to help isolate the effect of local traffic on each method's performance. The nonuniform destinations and long routes scenarios are more challenging and realistic, as they require the methods to cope with an abundance of non-local traffic.

We present initial proof-of-concept results in small networks and then study the same three scenarios in larger networks to show that the max-plus approach scales well and that the qualitative differences between the methods are the same in more realistic scenarios.

For each case, we consider again the metric of average trip waiting time (ATWT): the total waiting time of all vehicles that have reached their destination divided by the number of such vehicles. All results are averaged over 10 independent runs.

6.2.1 Small Networks

Figure 15 shows the small network used for the baseline and nonuniform destinations scenarios. Each intersection allows traffic to cross from only one direction at a time. All lanes have equal length and all edge nodes have equal spawning rates (vehicles are generated with probability 0.2 per timestep). The left side of Figure 16 shows results from the baseline scenario, which have uniform destinations. As a result, much of the traffic is local and hence there is no significant performance difference between TC-1 and max-plus. TC-SBC performs worse than the other methods,

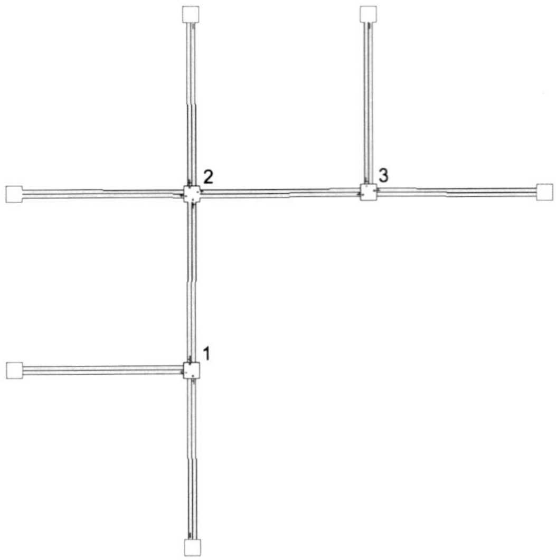

Fig. 15 The small network used in the baseline and nonuniform destinations scenarios

which is likely due to slower learning as a result of a larger state space, and a lack of serious congestion, which is the situation that TC-SBC was designed for.

The right side of Figure 16 shows results from the nonuniform destinations scenario. In this case, all traffic from intersections 1 and 3 is directed to intersection 2. Traffic from the top edge node of intersection 2 is directed to intersection 1 and traffic from the left edge node is directed to intersection 3. Consequently, there is no local traffic. This results in a dramatic performance difference between max-plus and the other two methods.

This result is not surprising since the lack of uniform destinations creates a clear incentive for the intersections to coordinate their actions. For example, the lane from intersection 1 to 2 is likely to become saturated, as all traffic from edge nodes connected to intersection 1 must travel through it. When such saturation occurs, it is important for the two intersections to coordinate, since allowing incoming traffic to cross intersection 1 is pointless unless intersection 2 allows that same traffic to cross in a "green wave".

To ensure that the performance difference between the baseline and nonuniform destinations scenarios is due to the removal of local traffic and not some other effect of nonuniform destinations, we also consider the long routes scenario. Destinations are kept uniform, but the network structure is altered such that all routes involve at least two intersections. Figure 17 shows the new network, which has a fourth intersection that makes local traffic impossible. Figure 18 shows the results from this scenario.

As before, max-plus substantially outperforms the other two methods, suggesting that its advantage is due to the absence of local traffic rather than other factors. TC-1 achieves a lower ATWT than TC-SBC but actually performs much worse. In fact, TC-1's joint actions are so poor that the outbound lanes of some edge nodes become

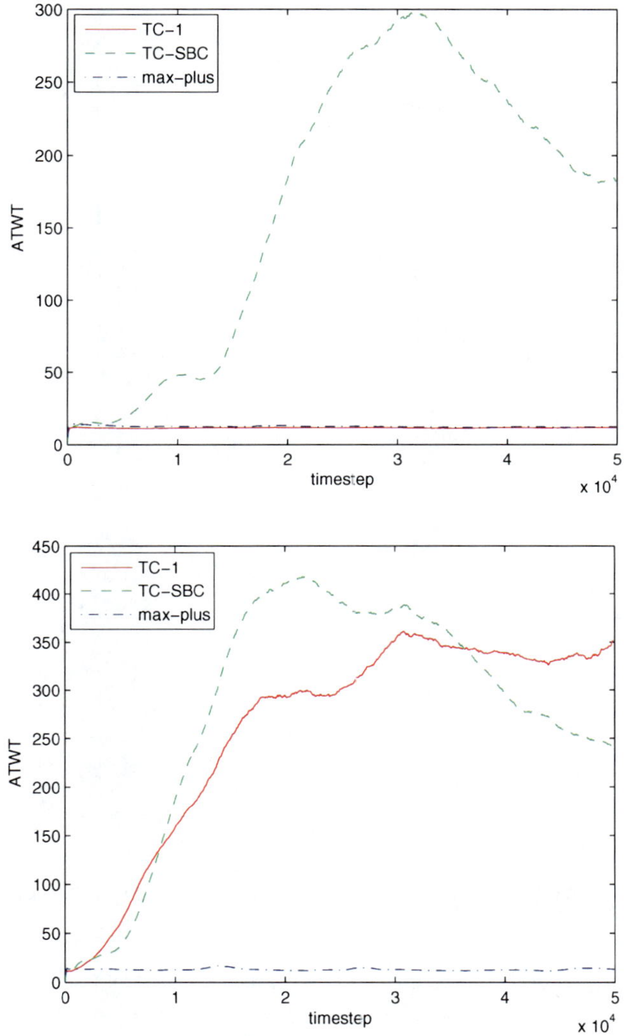

Fig. 16 Average ATWT per timestep for each method in the small network for the baseline (top) and nonuniform destinations (bottom) scenarios

full. As a result, the ATWT is not updated, leading to an artificially low score. At the end of each run, TC-1 had a much higher number cars waiting to enter the network than TC-SBC, and max-plus had none.

6.2.2 Large Networks

We also consider the same three scenarios in larger networks, similar to the Jillesville network we worked with before (see Figure 2), to show that the max-plus

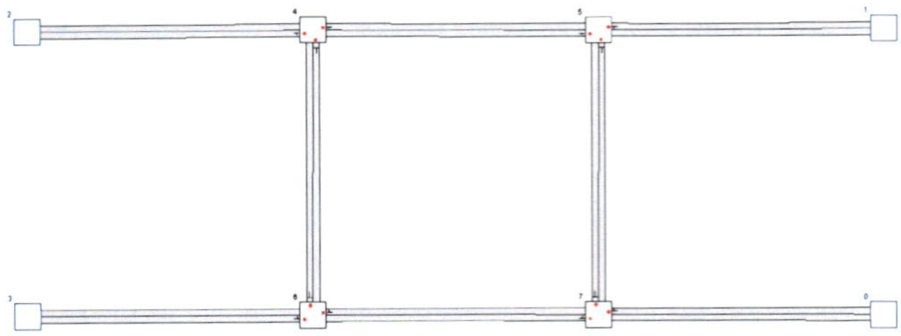

Fig. 17 The small network used in the long routes scenario

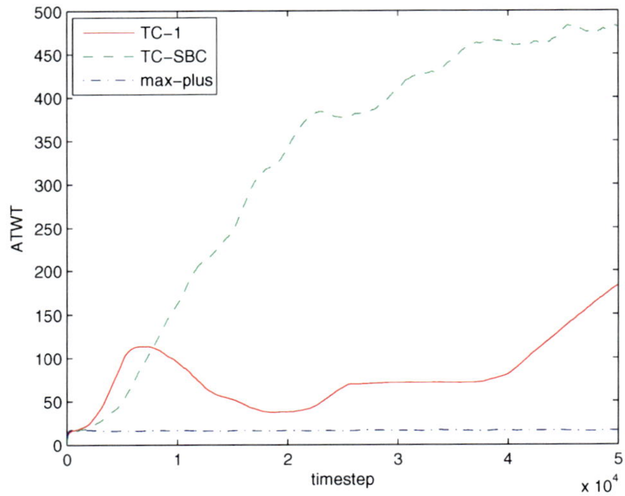

Fig. 18 Average ATWT per timestep in the small network for the long routes scenario

approach scales well and that the qualitative differences between the methods are the same in more realistic scenarios. Figure 19 shows the network used for the baseline and nonuniform destinations scenarios. It includes 15 agents and roads with four lanes. The left side of Figure 20 shows results from the baseline scenario, which has uniform destinations. As with the smaller network, max-plus and TC-1 perform very similarly in this scenario, all though max-plus's coordination results in slightly slower learning. However, TC-SBC no longer performs worse than the other two methods, probably because the network is now large enough to incur substantial traffic congestion. TC-SBC, thanks to its congestion bit, can cope with this occurrence better than TC-1.

The right side of Figure 20 shows results from the nonuniform destinations scenario. In this case, traffic from the top edge nodes travel only to the bottom edge nodes and vice versa. Similarly, traffic from the left edge nodes travels only to right edge nodes and vice versa. As a result, all local traffic is eliminated and max-plus

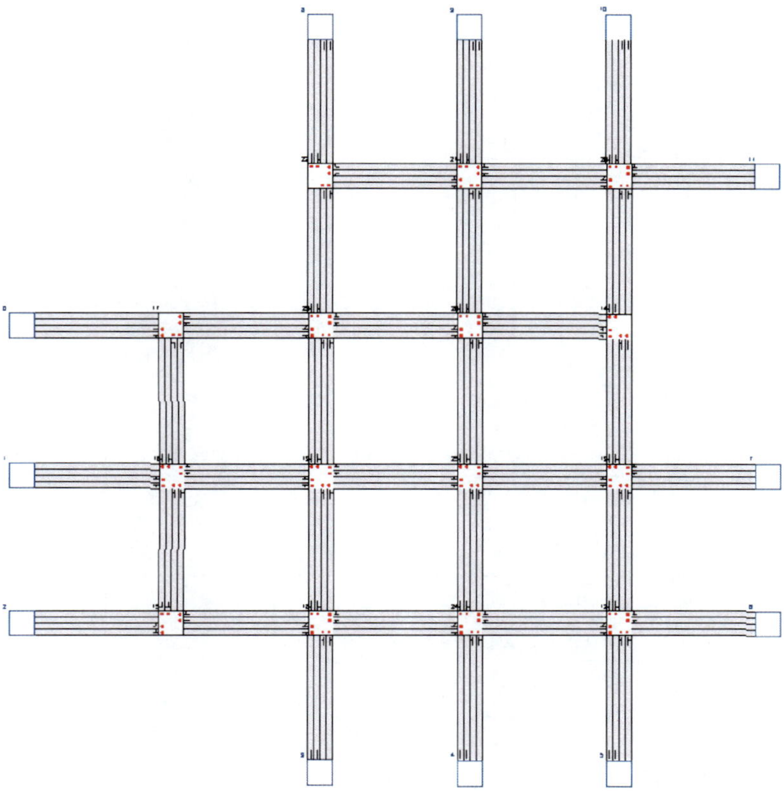

Fig. 19 The large network, similar to Jillesville, used in the baseline and nonuniform destinations scenarios

performs much better than TC-1 and TC-SBC. TC-SBC performs substantially better than TC-1, as the value of its congestion bit is even greater in this scenario.

To implement the long routes scenario, we remove one edge node from the two intersections that have two edge nodes (the top and bottom right nodes in Figure 19). Traffic destinations are uniformly distributed but the new network structure ensures that no local traffic occurs. The results of the long routes scenario are shown in Figure 21. As before, max-plus substantially outperforms the other two methods, confirming that its advantage is due to the absence of local traffic rather than other factors.

6.3 Discussion of Max-Plus Results

The experiments presented above demonstrate a strong correlation between the amount of local traffic and the value of coordinated learning. The max-plus method consistently outperforms both non-coordinated methods in each scenario where local traffic has been eliminated. Hence, these results help explain under what

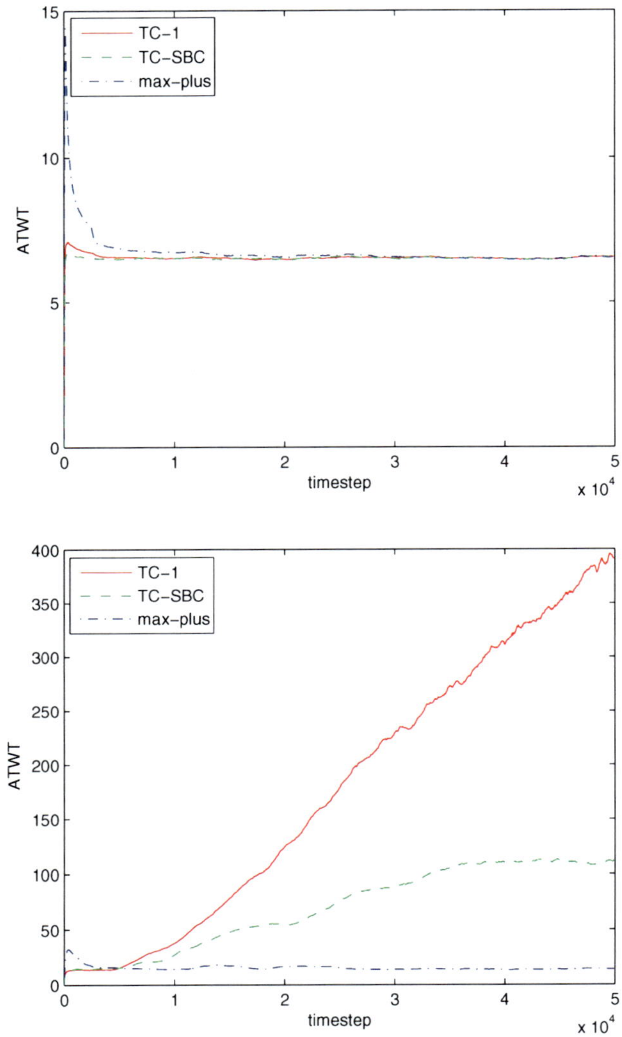

Fig. 20 Average ATWT per timestep for each method in the large network for the baseline (top) and nonuniform destinations (bottom) scenarios

circumstances coordinated methods can be expected to perform better. More specifically, they confirm the hypothesis that, under highly saturated conditions, coordination is beneficial when the amount of local traffic is small.

Even when there is substantial local traffic, the max-plus method achieves the same performance as the alternatives, although it learns more slowly. Hence, this method appears to be substantially more robust, as it can perform well in a much broader range of scenarios.

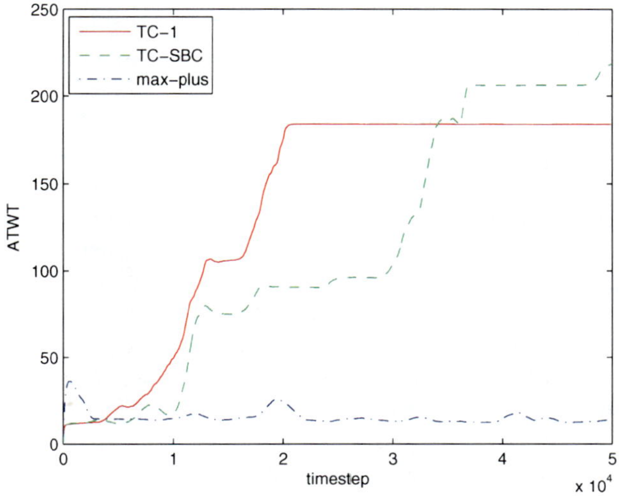

Fig. 21 Average ATWT per timestep for each method in the long routes scenario

By testing both small and large networks, the results also demonstrate that max-plus is practical in realistic settings. While max-plus has succeeded in small applications before [15], this chapter presents its first application to a large-scale problem. In fact, in the scenarios without local traffic, the performance gap between max-plus and the other methods was consistently larger in the big networks than the small ones. In other words, as the number of agents in the system grows, the need for coordination increases. This property makes the max-plus approach particularly attractive for solving large problems with complex networks and numerous agents.

Finally, these results also provide additional confirmation that max-plus can perform well on cyclic graphs. The algorithm has been shown to converge only for tree-structured graphs, although empirical evidence suggests that it also excels on small cyclic graphs [15]. The results presented in this chapter show that this performance also occurs in larger graphs, even if they are not tree-structured.

7 Conclusions

This chapter presented several methods for learning efficient urban traffic controllers by multiagent reinforcement learning. First, the general multiagent reinforcement learning framework used to control traffic lights in this chapter was described. Next, three extensions were described which improve upon the basic framework in various ways: agents (traffic junctions) taking into account congestion information from neighboring agents; handling partial observability of traffic states; and coordinating the behavior of multiple agents by coordination graphs.

In the first extension, letting traffic lights take into account the level of traffic congestion at neighboring traffic lights had a beneficial influence on the performance of the algorithms. The algorithms using this new approach always performed better than the original method if there is traffic congestion, in particular when traffic conditions are highly dynamic such as is the case when there are both quiet times and rush hours.

In the second extension, partial observability of the traffic state was successfully overcome by estimating belief states and combining this with multiagent variants of approximate POMDP solution methods. These variants are the most likely state (MLS) approach and Q-MDP, previously only used in the single case. It was shown that the state transition model and value function could also be estimated (learned) effectively under partial observability.

The third extension extends the MARL approach to traffic control to include explicit coordination between neighboring traffic lights. Coordination is achieved using the max-plus algorithm, which estimates the optimal joint action by sending locally optimized messages among connected agents. This work presents the first application of max-plus to a large-scale problem and thus verifies its efficacy in realistic settings. Empirical results on both large and small traffic networks demonstrate that max-plus performs well on cyclic graphs, although it has been proven to converge only for tree-structured graphs. Furthermore, the results provide a new understanding of the properties a traffic network must have for such coordination to be beneficial and show that max-plus outperforms previous methods on networks that possess those properties.

Acknowledgments. The authors wish to thank Nikos Vlassis, Merlijn Steingröver, Roelant Schouten, Emil Nijhuis, and Lior Kuyer for their contributions to this chapter, and Marco Wiering for his feedback on several parts of the research described here.

References

1. Abdulhai, B., et al.: Reinforcement Learning for True Adaptive Traffic Signal Control. ASCE Journal of Transportation Engineering 129(3), 278–285 (2003)
2. Moore, A.W., Atkenson, C.G.: Prioritized Sweeping: Reinforcement Learning with less data and less time. Machine Learning 13, 103–130 (1993)
3. Bakker, B., Steingrover, M., Schouten, R., Nijhuis, E., Kester, L.: Cooperative multi-agent reinforcement learning of traffic lights. In: Proceedings of the Workshop on Cooperative Multi-Agent Learning, European Conference on Machine Learning, ECML 2005 (2005)
4. Barto, A.G., Bradtke, S.J., Singh, S.P.: Learning to act using real-time dynamic programming. Artificial Intelligence 72, 81–138 (1995)
5. Bellman, R.E.: Dynamic Programming. Princeton University Press, Princeton (1957)
6. Cassandra, T.: Exact and Approximate Algorithms for Partially Observable Markov Decision Processes. PhD thesis, Brown University (1998)
7. Chiu, S.: Adaptive Traffic Signal Control Using Fuzzy Logic. In: Proceedings of the IEEE Intelligent Vehicles Symposium, pp. 98–107 (1992)
8. Crick, C., Pfeffer, A.: Loopy belief propagation as a basis for communication in sensor networks. In: Proceedings of Uncertainty in Artificial Intelligence, UAI (2003)

9. Foy, M.D., Benekohal, R.F., Goldberg, D.E.: Signal timing determination using genetic algorithms. Transportation Research Record No. 1365, pp. 108–115
10. Guestrin, C., Lagoudakis, M.G., Parr, R.: Coordinated reinforcement learning. In: Proceedings Nineteenth International Conference on Machine Learning (ICML), pp. 227–234 (2002)
11. Hauskrecht, M.: Value-function approximations for partially observable markov decision processes. J. of AI Research 13, 33–94 (2000)
12. Jaakkola, T., Singh, S.P., Jordan, M.I.: Monte-carlo reinforcement learning in non-Markovian decision problems. Advances in Neural Information Processing Systems 7 (1995)
13. Kaelbling, L.P., Littman, M.L., Cassandra, A.R.: Planning and acting in partially observable stochastic domains. Artificial Intelligence 101(1-2), 99–134 (1998)
14. Kaelbling, L.P., Littman, M.L., Moore, A.W.: Reinforcement learning: A survey. Journal of Artificial Intelligence Research 4, 237–285 (1996)
15. Kok, J.R., Vlassis, N.: Collaborative Multiagent Reinforcement Learning by Payoff Propagation. J. Mach. Learn. Res. 7, 1789–1828 (2006)
16. Kschischang, F.R., Frey, B.J., Loeliger, H.A.: Factor graphs and the sum-product algorithm. IEEE Transactions on Information Theory 47, 498–519 (2001)
17. Liu, J., Chen, R.: Sequential Monte Carlo methods for dynamic systems. Journal of the American Statistical Association 93 (1998)
18. Mitchell, T.M.: Machine learning. McGraw-Hill, New York (1997)
19. Shoufeng, M., et al.: Agent-based learning control method for urban traffic signal of single intersection. Journal of Systems Engineering 17(6), 526–530 (2002)
20. Smallwood, R.D., Sondik, E.J.: The optimal control of partially observable Markov processes over a finite horizon. Operations Research 21, 1071–1088 (1973)
21. Spaan, M.T.J., Vlassis, N.: A point-based POMDP algorithm for robot planning. In: Proceedings of 2004 IEEE International Conference on Robotics and Automation, ICRA (2004)
22. Spall, J.C., Chin, D.C.: Traffic-Responsive Signal Timing for System-wide Traffic Control. Transportation Research Part C: Emerging Technologies 5(3), 153–163 (1997)
23. Steingröver, M., Schouten, R., Peelen, S., Nijhuis, E., Bakker, B.: Reinforcement learning of traffic light controllers adapting to traffic congestion. In: Proceedings of the Belgium-Netherlands Artificial Intelligence Conference, BNAIC 2005 (2005)
24. Steingröver, M., Schouten, R.: Reinforcement Learning of Traffic Light Controllers under Partial Observability. MSc Thesis, Informatics Institute, Universiteit van Amsterdam (2007)
25. Striebel, C.T.: Sufficient Statistics in the Optimal Control of. Stochastic Systems. Journal of Mathematical Analysis and Applications 12, 576–592 (1965)
26. Sutton, R.S., Barto, A.G.: Reinforcement learning: An introduction. MIT Press, Cambridge (1998)
27. Sutton, R.S.: Integrated architectures for learning, planning, and reacting based on approximating dynamic programming. In: Proceedings of the Seventh International Conference on Machine Learning (ICML). Morgan Kaufmann, Austin (1990)
28. Taale, H., Bäck, T., Preuss, M., Eiben, A.E., de Graaf, J.M., Schippers, C.A.: Optimizing traffic light controllers by means of evolutionary algorithms. In: Proceedings of EUFIT 1998 (1998)
29. Tan, K.K., Khalid, M., Yusof, R.: Intelligent traffic lights control by fuzzy logic. Malaysian Journal of Computer Science 9-2 (1995)

30. Thorpe, T.L., Anderson, C.: Traffic light control using Sarsa with three state representations. Technical report, IBM Cooperation (1996)
31. Vlassis, N.: A Concise Introduction to Multiagent Systems and Distributed Artificial Intelligence. In: Synthesis Lectures in Artificial Intelligence and Machine Learning. Morgan & Claypool Publishers, San Francisco (2007)
32. White, C.C., Scherer, W.T.: Finite memory suboptimal design for partially observed Markov decision processes. Operations Research 42(3), 439–455 (1994)
33. Wiering, M.: Multi-Agent Reinforcement Learning for Traffic Light Control. In: Proc. 17th International Conf. on Machine Learning (ICML), pp. 1151–1158 (2000)
34. Wiering, M., van Veenen, J., Vreeken, J., Koopman, A.: Intelligent traffic light control. Technical report, Dept. of Information and Computing Sciences, Universiteit Utrecht (2004)
35. Wiering, M., Vreeken, J., van Veenen, J., Koopman, A.: Simulation and optimization of traffic in a city. In: IEEE Intelligent Vehicles symposium (IV 2004) (2004)
36. Yedidia, J.S., Freeman, W.T., Weiss, Y.: Understanding belief propagation and its generalizations. In: Exploring Artificial Intelligence in the New Millennium, ch. 8, pp. 239–269 (2003)
37. Zhang, N.L., Poole, D.: Exploiting causal independence in Bayesian network inference. Journal of Artificial Intelligence Research 5, 301–328 (1996)

Fusing Heterogeneous and Unreliable Data from Traffic Sensors

Qing Ou, Hans van Lint, and Serge P. Hoogendoorn

Abstract. Fusing traffic data from a variety of traffic sensors into a coherent, consistent, and reliable picture of the prevailing traffic conditions (e.g. densities, speeds, flows) is a critical and challenging task in any off- or online traffic management or information system which use these data. Recursive Kalman filter-based approaches provide an intuitive and powerful solution for traffic state estimation and data fusion, however, in case the data cannot be straightforwardly aligned over space and time, the equations become unwieldy and computationally expensive. This chapter discusses three alternative data fusion approaches which solve this alignment problem and are tailored to fuse such semantically different traffic sensor data. The so-called PISCIT and FlowResTD methods both fuse spatial data (individual travel times and low-resolution floating car data, respectively) with a prior speed map obtained from either raw data or another estimation method. Both PISCIT and FlowResTD are robust to structural bias in those a priori speeds, which is critically important due to the fact that many real-world local sensors use (arithmetic) time averaging, which induces a significant bias. The extended and generalized Treiber–Helbing filter (EGTF) in turn is able to fuse multiple data sources, as long as for each of these it is possible to estimate under which traffic regime (congested, free flowing) the data were collected. The algorithms are designed such that they can be used in a cascaded setting, each fusing an increasingly accurate posterior speed map with new data, which in the end could be used as input for a model-based/Kalman filter approach for traffic state estimation and prediction.

Q. Ou
Transport & Planning, Faculty of Civil Engineering and Geosciences,
Delft University of Technology
e-mail: Q.Ou@tudelft.nl

Dr. J.W.C. Van Lint
Transport & Planning, Faculty of Civil Engineering and Geosciences,
Delft University of Technology
e-mail: j.w.c.vanlint@tudelft.nl

Prof. Dr. S.P. Hoogendoorn
Transport & Planning, Faculty of Civil Engineering and Geosciences,
Delft University of Technology
e-mail: s.p.hoogendoorn@tudelft.nl

R. Babuška & F.C.A. Groen (Eds.): Interactive Collaborative Information Systems, SCI 281, pp. 511–545.
springerlink.com

1 Introduction

In this section we will motivate the research presented in this chapter, and provide a sketch of both context and background.

1.1 Context and Background: The Need for Reliable Traffic Data Fusion Methods

Traffic data collection and archiving systems are essential tools for online (and real-time) dynamic traffic management (DTM) applications, such as adaptive intersection control, routing, ramp metering, and traffic information services. Put simply, without data from sensors neither traffic management and control nor traffic information services are possible. However, data from a multitude of different traffic sensors do not necessarily mount up to consistent, coherent, and meaningful information on the state of a traffic network, for example, in terms of speed, density, or flow. Traffic state estimation and data fusion techniques are required to translate these available data into a consistent and complete description of the state in a traffic system. Figure 1 illustrates the place of traffic data fusion and state estimation (and prediction) in the context of real-time dynamic traffic management. Later on in this chapter, some of the terms used in this figure (e.g., process and observation model) will be discussed in more detail.

Traffic state estimation and data fusion are also essential for *offline* applications, such as policy/measure evaluation studies and the development, calibration, and validation of the tools necessary to perform these tasks (e.g., traffic simulation

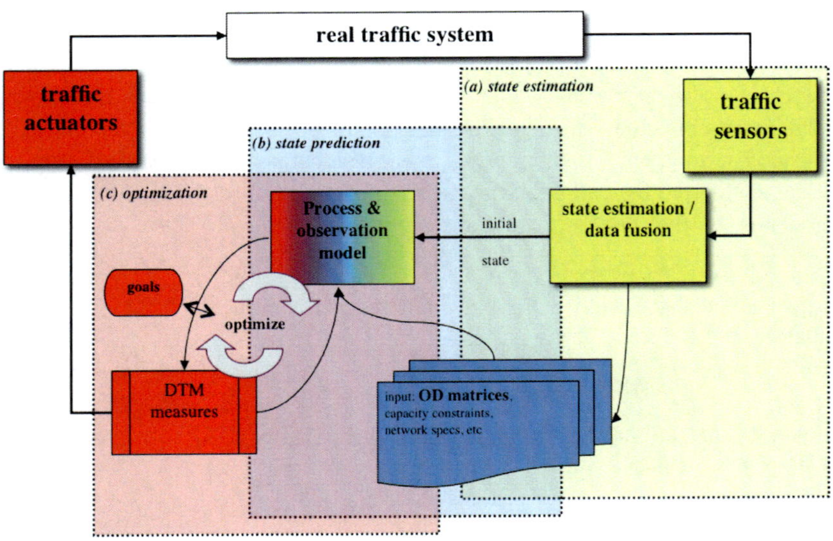

Fig. 1 Schematic representation of traffic state estimation and prediction in the context of (dynamic) traffic management (DTM) and control

models). Finally, the advance of scientific research itself relies heavily on the availability of large amounts of detailed and reliable empirical data and on techniques to extract consistent and reliable information from these. In the last decades, the amount of empirical data becoming available for both online and offline use has steeply increased, particularly in terms of the wide range of sensor technologies developed and applied to collect these data. Traffic sensors may range from inductive loop detectors, radar, microwave, ultrasonic sensors to infrared cameras, and in-vehicle GPS/GSM receivers/transmitters ("floating car data"), to name a few. The motorway network in the Netherlands, for example (around 6600 km), has an inductive loop-based monitoring system installed (with loops about every 500 m); however, this only holds for around 1/3 of the network. Another 1/3 has only limited traffic monitoring, while the other 1/3 has nothing at all. Besides the limited spatial coverage in some areas, a second major issue is that of the available data on average 5–10% is missing or otherwise deemed unreliable, with regular extremes to over 25 or 30%. Also, in the Netherlands, other data sources (than loop detectors) are already available or will become available in the near future. Data from these different sensors (cameras, induction loops, or in-car GPS/GSM devices) are typically characterized by different formats, semantics, temporal and spatial resolution, and accuracy, and also differ in availability and reliability both as a function of location, time and circumstances [21, 23]. Both from technical and methodological points of view, the integration of such heterogeneous data into comprehensive and consistent data warehouses is a complex and challenging task. This chapter focuses predominantly on the second challenge, that is, on methodological tools to fuse heterogeneous traffic data. As we will see in the ensuing, particularly the *semantical* differences over space and time between these data impose strong constraints on the applicability of data fusion techniques.

1.2 Multi-sensor Data Fusion: A Brief Overview

In many fields of science, such as robotics, medical diagnosis, image processing, air traffic control, remote sensing, and ocean surveillance (see e.g. [10, 13, 25, 29, 30]), the de facto method for state estimation is multi-sensor data fusion, a technique by which data from several sensors are combined by means of mathematical and/or statistical models to provide comprehensive and accurate information. A wide variety of approaches have been put forward for multi-sensor data fusion, based on, for instance (extended), Kalman filters, Bayes methods, and Artificial Neural Networks, Dempster-Shaefer theory, or Fuzzy Logic. Which of these is suitable for the problem at hand is governed largely by domain-specific constraints, the characteristics of the data available, and, probably most importantly, by the purpose and application for which the data is (f)used. Using similar arguments as in [3], data fusion generally leads to

- Increased confidence and accuracy and reduced ambiguity;
- increased robustness: one sensor can contribute information where others are unavailable, inoperative, or ineffective;

- enhanced spatial and temporal coverage: one sensor can work when or where another sensor cannot;
- and (more tentatively), decreased costs, because (a) a suite of "average" sensors can achieve the same level of performance as a single, highly-reliable sensor and at a significantly lower cost, and (b) fewer sensors may be required to obtain a (for a particular application) *sufficient* picture of the system state.

With these arguments in mind, data fusion techniques provide an obvious solution for traffic state estimation. However, most approaches to traffic state estimation (e.g. [28]) consider only a *single* source (i.e., minute aggregated or averaged flows and speeds from local inductive loop detectors), whereas of the studies which *do* consider data from various traffic sensors (e.g. [3, 6]) most are concerned with limited traffic networks or corridors (e.g., single traffic links), and not as in the RENAISSANCE approach of Wang and Papageorgiou at comprehensive traffic surveillance and monitoring for entire freeway corridors or networks. As we will elaborate further in the following section, the Kalman Filter (KF) approaches demonstrated in [4, 20, 28] do have other limitations, which relate to the spatiotemporal alignment of the available data. In this chapter, we will present three alternative data fusion approaches, which essentially solve this alignment problem and can be used in conjunction with KF-based state estimators and predictors, for example, in a real-time decision support system for traffic management and control (Fig. 1).

1.3 Chapter Outline

The chapter is structured as follows. In Section 2, first some general characteristics of traffic data and traffic dynamics are recalled. In Section 3, a general framework for traffic state estimation data fusion is presented, and the most common family of approaches (Kalman filters—KF) is discussed. As we will show, when the available data cannot be straightforwardly aligned, such KF approaches may lead to unwieldy computations. To solve this alignment problem, three distinctly different approaches are presented in Sections 4 and 5. Note that these approaches have been published separately in [12, 19]. In Section 6, the methods presented are (qualitatively) compared and critically discussed and several avenues for further research and development are described.

2 Spatiotemporal Characteristics of Traffic Data

2.1 Basic Macroscopic Traffic Variables and Relationships

The basic macroscopic traffic variables used to describe a traffic state include flow q (veh/h), (space) mean speed u (km/h), and vehicular density ρ (veh/km), which are related to one another by

$$q = \rho u \tag{1}$$

Note that Eq. (1) holds only in case u depicts the so-called *space* mean speed u_S, i.e., the arithmetic average of the speeds of vehicles present on the road section r

of interest (with length L_r) at some time instant t. With local detection equipment along r, this speed can be approximated by the local *harmonic* mean speed, that is,

$$u_S \approx u_H = \frac{L_r}{\frac{1}{N}\sum_i \frac{1}{v_i}} \tag{2}$$

The approximation in (2) is exact under the assumption of homogeneity and stationarity on the road section and time period of interest (see e.g. [8]). The local arithmetic (or simply *time*) mean speed $u_T = \frac{1}{N}\sum v_i$ provides a biased approximation of the space mean, due to the fact that in a time sample *faster* observations are over represented. That this bias is significant specifically under congested (low-speed) conditions has been demonstrated for example in [11, 24] for estimating travel times (errors of over 30%) and in [7] for estimating densities, where the resulting errors can mount up to over 100%.

There is a second widely used relationship between the three basic traffic variables, the so-called fundamental diagram. This statistical (not causal) relationship states that vehicles on average maintain a minimum time headway $h = 1/q$ given a certain average distance gap $d = 1/\rho$, that is,

$$q = Q^e(\rho) \tag{3}$$

$Q^e(\rho)$ usually is modeled as a convex function with a slope v_0 (equal to the free or *desired* speed) for $\rho \downarrow 0$, and a maximum flow (capacity flow C) at a critical density ρ_c, which is assumed to discriminate between freely flowing and congested traffic. At a complete standstill, the flow again equals 0 at a maximum density ρ_{jam}. Due to (1), Eq. (3) can also be written as $q = \rho U^e(\rho)$, where $U^e(\rho)$ represents the "equilibrium" speed for a particular density.

2.2 *Dynamics of Traffic: Spatiotemporal Patterns*

The simplest way of describing the propagation of traffic flow over space and time is by considering the continuous conservation law

$$\frac{\partial \rho}{\partial t} + \frac{\partial q}{\partial x} = 0 \tag{4}$$

which combined with (1) and (3) constitutes the well-known (first order) kinematic wave model simultaneously proposed in [9] and [15]. Key to the kinematic wave model is that perturbations in a traffic stream (e.g., a steep increase in density due to a tight platoon of vehicles entering the network or a sudden brake action) propagate along so-called characteristics with a (kinematic wave) speed $c(\rho)$, which is equal to the derivative of the fundamental diagram (3). Substituting (3) in (4) yields

$$\frac{\partial \rho}{\partial t} + c(\rho)\frac{\partial \rho}{\partial x} = 0 \tag{5}$$

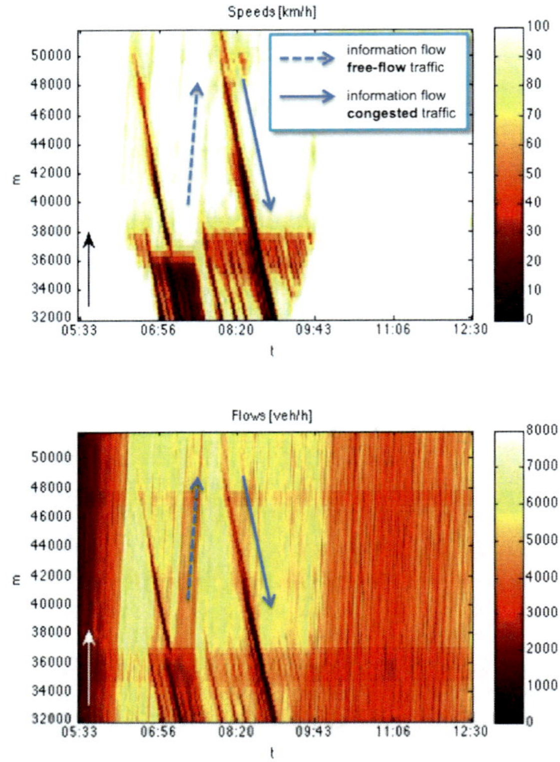

Fig. 2 Speeds (km/h) and flows (veh/h) obtained from inductive loop detectors along a typical densely used 3-lane freeway section in the Netherlands. The arrows bottom left in both graphs indicate the drive direction, and the arrows within both contour maps indicate the two main directions of information flow in free-flowing and congested traffic, respectively.

with

$$c(\rho) = \frac{\partial}{\partial \rho} Q^e(\rho)$$

Although this kinematic wave model has many unrealistic features (dimensionless vehicles, homogeneous traffic flows, instantaneous reaction times, to name just a few), the notion that perturbations in traffic flows move along characteristic curves ($c(\rho)$) is easily illustrated with empirical data. Figure 2 shows some empirical traffic patterns with so-called space–time *contour maps* of the speeds (top graph) and flows (bottom graph) along a 20-km 3-lane freeway section in The Netherlands (the A12 between Gouda and Utrecht). Figure 2 illustrates that there exist two main directions of information flow:

1. Under free-flow conditions with speeds typically around 100 km/h and flows well below capacity (around 4000 veh/h), perturbations (e.g., changes in

average speed or an increase in density) propagate in the same direction of traffic flow, i.e., with positive speeds $c(\rho) = c_{free}$. This is indicated by the striped arrows in Fig. 2.

2. In congested conditions, perturbations propagate in the *opposite* direction of traffic, i.e., with negative speeds $c(\rho) = c_{cong}$ (on freeways typically between -15 and -20 km/h [18]). This characteristic direction of information flow is indicated by the solid arrows.

Note that in congested conditions, a wide variety of patterns can be observed. Loosely using the terminology coined by Kerner [5], we can observe in Fig. 2 (top) regions of *synchronized* traffic ("mild" congestion), where although speed significantly drops, the flow remains relatively high. In contrast, also the so-called *wide moving jams* occur, in which both speed and flow drop to near zero. These wide moving jams often propagate upstream over long distances (more than 15 km in Fig. 2). The debate on how many traffic patterns or phases exist is beyond the scope of this chapter. What is relevant is that information measured with some traffic sensors at one location will travel either up- or downstream with a finite speed depending on the traffic conditions. This implies that for the estimation of traffic quantities between sensor locations one has to resort to traffic theory and models.

2.3 Travel Time and Trajectory Data

Besides flow and speed, many other traffic characteristics can be measured by sensors. The two most common are travel times and (samples) of vehicle trajectories. Travel time can be measured by means of, for example, automated vehicle identification (AVI) systems, which identify vehicles at two consecutive locations A and B at time instants t_A and t_B and deduce the *realized* travel time afterwards with $TT_r = t_B - t_A$. AVI systems may employ camera's and license plate detection technology, or may match vehicles through induction footprints, tolling tags, or otherwise. Methodologically, the most important characteristics of travel time are that

- travel time can only be measured for realized trips, i.e., after a vehicle has finished it. The so-called *actual* travel time TT_a of a vehicle departing at the current moment must hence be predicted per definition (Fig. 3), and
- travel time (or its reciprocal average journey speed $u_r = L_r/TT_r$) is a *scalar* representation of the traffic conditions (e.g., the speed $v(t,x)$) a vehicle encounters during its trip. Figure 3 illustrates this by superimposing vehicle trajectories on a speed contour map. This implies that the relationship between this travel time and the underlying traffic conditions (the speed contour through which a vehicle has "traversed") is $1 : N$. It is possible to estimate travel time from local speeds (e.g. [11, 24]), but conversely, it is not possible to estimate local speeds from travel time, unless other sources of information are available—this is the basis of the piece-wise inverse speed correction algorithm presented in section 4.1.

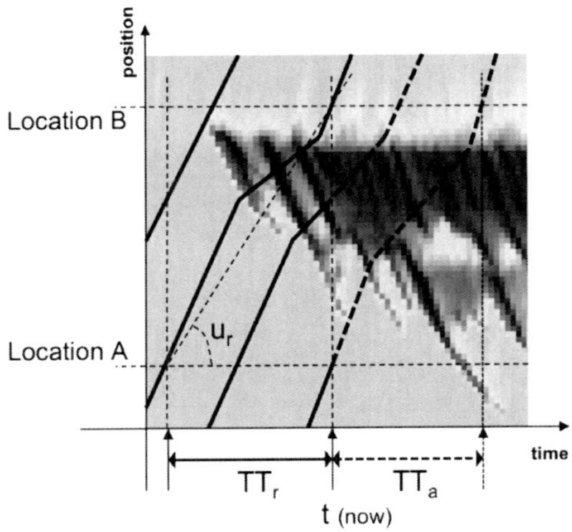

Fig. 3 Relationship between vehicle trajectories (the thick solid and dotted lines), realized and actual travel time (TT_r and TT_a), average journey speed (u_r) and the underlying speeds (represented by a speed contour plot, where dark areas represent low speeds)

By sampling data (location and/or speed) from instrumented vehicles (e.g., through GPS or GSM) at consecutive time instants, also vehicle trajectories can be measured. Clearly, when *all* vehicle trajectories are sampled at high frequencies, the traffic state (prevailing speeds, flows, densities, travel times, etc.) can be completely deduced from these so-called *floating car data* (FCD). However, at penetration rates far below 100%,[1] FCD at best provide a proxy for average speed on the road segments from which these data are collected. Section 4.2 describes a method which is able to fuse these coarse speed estimates with other data sources.

To estimate flows or densities (required for many traffic management purposes such as intersection control, ramp metering, but also for forecasting traffic information), other (local) data sources than travel time or trajectory samples are necessary. Nonetheless, with the in-car ICT revolution, it is reasonable to assume that these penetration rates will rapidly increase in the coming years.

2.4 Summary and Classification of Traffic Data

In conclusion, data from traffic sensors come in many forms and qualities, but can essentially be subdivided along two dimensions. The first relates to their spatiotemporal semantics, that is, do the data represent *local* traffic quantities (speed, time headway(s), etc.) or do the data reflect quantities over space (journey speed, space

[1] It is estimated at the end of 2009 that the penetration rate of real-time traffic information and GPS enabled vehicles which actually transmit their location and speed to their service provider is in the order of one percent or less of the total number of vehicles driving on the Dutch freeways.

Table 1 Classification of data from traffic sensors with some examples

	Event-based	Aggregate
Local	Vehicle passage, speed, length, etc.	Flow, harmonic mean speed, etc.
Spatial	Vehicle travel, time, journey speed, trajectory, etc.	Space mean speed, mean travel time, etc.

mean speed, travel time, trajectories, etc.). The second relates to the degree of aggregation, where data may represent an aggregate or average over fixed time periods (e.g., 1 minute aggregate flows or averaged speeds), or a single event (vehicle passage, travel time, full trajectory, etc.). Table 1 overviews this classification with a few examples. The main consequence for fusing these fundamentally different data is that these data need to be aligned over space and time such that we can employ mathematical and statistical models to average, filter, and combine them into one consistent picture (space–time map) of the traffic conditions. In the next section, we will discuss this in more detail.

3 Traffic State Estimation and Data Fusion Approaches

The generic structure of a traffic state estimation/data fusion approach typically includes three components:

1. Data from whatever sensors available (see e.g., Table 1).
2. Some mathematical model which formalizes the relationship of the data with the underlying dynamic phenomena, such as an analytical traffic flow model or a statistical relationship.
3. Data assimilation techniques, which combine the former two (data + model).

3.1 Recursive State Estimation Approaches (Kalman Filters)

The most widely applied data assimilation technique applied to traffic state estimation problems is the KF and/or its many variations (extended and unscented KF) or simulation-based approaches such as particle filters, which are also known as sequential Monte Carlo methods. For example, Wang et al. [27] and [28] combine a so-called second-order macroscopic traffic flow model with the extended KF to estimate vehicular density and speed from sparse local data from induction loops. Van Lint and co-workers demonstrate a multi-class first-order macroscopic traffic flow model and a dual extended KF for the same purpose [22]. Such KF approaches exploit the fact that analytical traffic flow models (e.g. 5) can be numerically solved and expressed in state-space form, that is

$$\mathbf{x}_{k+1} = \mathbf{f}(\mathbf{x}_k, \mathbf{w}_k, \mathbf{u}_k) + \eta_k \tag{6}$$

$$\mathbf{y}_k = \mathbf{h}(\mathbf{x}_k, \mathbf{v}_k) + \zeta_k \tag{7}$$

In (7), k depicts discrete time steps of duration $t_k - t_{k-1} = \Delta t$) seconds. Equation (6) depicts the process equation, which describes the dynamics of state $\mathbf{x}_{k+1}$ (density and/or speed) as a function of $\mathbf{x}_k$ and external disturbances $\mathbf{u}_k$ (e.g., traffic demand at network boundaries) plus an error term η_k, reflecting errors in the process model (e.g., model misspecification). The function f contains (possibly time-varying) parameters $\mathbf{w}_k$, for example, the kinematic wave speed or road capacity. Equation (7) depicts the observation equation h which relates the system state to (observable) outputs $\mathbf{y}_k$. Also, h may contain (possibly time-varying) parameters $\mathbf{v}_k$. The error term ζ_k depicts errors in either the observation model h and/or the observations themselves. The fundamental diagram of traffic flow $q = Q^e(\rho)$, or $u = U^e(\rho)$, relating speed or flow to density, is a good example of such an observation equation.

Details on KF algorithms can be found in many textbooks (e.g. [16]), let us just make a few remarks on their applicability of fusing semantically different traffic data. The key advantage of KF-based state estimation approaches is that they provide a convenient and principled approach to recursively correct state estimates by balancing the errors (uncertainties) in process and observation model and in the data ($\mathbf{y}_k^{obs}$) with

$$\mathbf{x}_k = \mathbf{G}_k(\mathbf{y}_k^{obs} - \mathbf{y}_k) \tag{8}$$

The so-called Kalman gain $\mathbf{G}_k$ in (8) can be informally understood as

$$\mathbf{G}_k = \frac{\text{uncertainty process model}}{\text{uncertainty observation model \& data}} \times \text{sensitivity obs. model to state variables}$$

This implies that (a) the more uncertain the data are (conceived), the more weight is put on the model predictions and vice versa, and (b) that the KF adjusts $\mathbf{x}_k$ proportional to the sensitivity of the observation model to changes in the state variables. For example, under free-flow conditions the relationship between traffic density and speed is very weak, which would imply only small corrections in state variables ($\mathbf{x}_k$) even if the speeds measured with sensors ($\mathbf{y}_k^{obs}$) differ largely from those predicted by the observation model ($\mathbf{y}_k$). This intuitive structure can be easily explained to traffic operators and professionals using such state estimation tools. Moreover, the same process and observation model can be subsequently used for prediction and control purposes (see Fig. 1), given proper predictions of the boundary conditions (traffic demand, turn fractions, and capacity constraints) and estimates of the model parameters are available.

3.2 The Spatiotemporal Alignment Problem

There is, however, one major drawback of KF approaches, which relates to the spatial and temporal alignment of the data. For every data source used, an (additional) observation equation (7) is required, which relates the data to the traffic state. This is not necessarily a problem, as long as the spatial and temporal discretization of the data (detector locations x_i and time periods of size ΔT) can be aligned with the discretization used in the model (road segments of length Δx and time periods of size Δt). For example, some spatial data can be transformed fairly easily into local measurements, such as sampled floating car data [4, 20]. This, however, is not the case for, e.g., travel time or journey (segment) speeds. These are available after trips are realized, that is, after a time period equal to the measured travel time (or traveled distance divided by the average journey speed). As a result, a (realized) travel time observation equation has the following general form:

$$\mathbf{y}_k = h(\mathbf{x}_k, \mathbf{x}_{k-1}, \cdots, \mathbf{x}_{k-TT^{max}}) + \vartheta_t \tag{9}$$

where the output variable $\mathbf{y}_k$ now depicts (realized) travel time, and TT^{max} is the maximum observed travel time on the route of interest. Using Equation (9) hence necessitates maintaining state variables $\mathbf{x}_k$ (speeds, densities) for as many discrete time instants as required to deduce the longest possible realized travel time. To illustrate this practical and computational problems, consider a traffic corridor constituted of N segments (of typically 500 m) on which state variables (e.g., densities) are estimated, which are corrected every ΔT time units (of typically 1 minute). To facilitate observation equation (9), the number of state variables increases to $N^{aug} = N \times TT^{max}/\Delta T$. On a congested corridor of say 20 km, TT^{max} may become as high as 60 minutes or more, yielding an augmented state vector with 1000 variables or more. Since in the KF procedure also an error covariance matrix is maintained of size $(N^{aug})^2$, this approach would lead to very unwieldy computations for any realistic traffic network or corridor.

We conclude that in case of combining data which cannot be straightforwardly aligned, and/or which require augmenting the state vector with multiple time periods, it makes sense to use a simpler approach for data fusion, that is, one with far less degrees of freedom. The remainder of this chapter will discuss three such alternative state estimation/data fusion techniques. Of course, after fusion, for example, travel times and local sensor data, the resulting (fused) data (i.e., a speed contour map) can in turn be input to a KF-based state estimator. We will return to the combination of approaches in the final section of this chapter.

4 New Traffic Data Fusion Approaches: Methodology

In this section, three state estimation/data fusion approaches are discussed, which were developed (partially) in the ICIS traffic data fusion project. All three methods construct a posterior space–time map of speeds and have the same structure outlined above (fused data = sensor data + (traffic) theory + data assimilation techniques), but differ in their assumptions and in the type of applications and available data for which they are suitable.

- The PISCIT method (*Piece-wise Inverse Speed Correction by using Individual Travel time*, [12]) is based on consistency of route and route-segment travel times and is able to fuse individual travel times with averaged local speeds.

- The FlowResTD method (*Fusion of low resolution topological data*) exploits a simple conservation law and Poisson statistics to fuse low-resolution trajectory data (e.g., from mobile phones) with a priori speed distributions (e.g., from other data sources) into segment-level speeds.
- The EGTF (extended generalized Treiber–Helbing filter) approach [19] finally uses kinematic wave theory and anisotropic spatiotemporal filtering to fuse speed, flow, or other data from which ever traffic sensor available, given each of these also provides quantitative information on the traffic state (free or congested).

For each of these methods, first some general properties and the overall structure of the method are described. Next, the algorithms and associated mathematics are presented. In Section 5, some results and analysis on the basis of synthetic data are given for each of these methods.

4.1 Exploiting Travel Time Consistency: Piece-Wise Inverse Speed Correction by Using Individual Travel Time (PISCIT)

PISCIT is a simple but efficient two-step algorithm to fuse individual travel times with an initially estimated time–space speed contour plot [12]. This initial time–space contour plot of speeds may be the result of a simple interpolation between consecutive detector measurements, or the result of more advanced state estimators described elsewhere in this chapter.

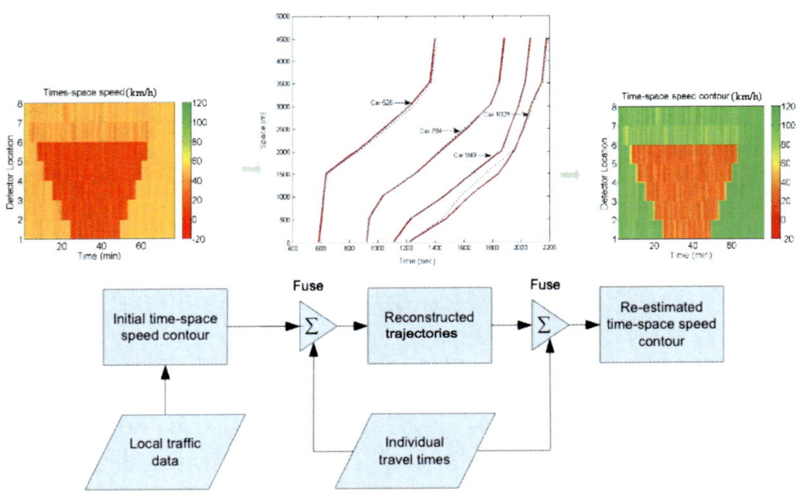

Fig. 4 The PISCIT framework. The two contour plots represent speeds from a simple synthetic example.

4.1.1 Framework of the PISCIT Method

The main assumption underlying the PISCIT algorithm is that travel times measured with cameras (or other AVI systems) have errors which are substantially smaller than travel times estimated from local traffic data such as inductive loop data. This assumption makes sense, given a vehicle is correctly matched at two locations A and B at times t_A and t_B.[2] In that case, the error in the travel time is t_B - t_A. The algorithm (outlined in Fig. 4) consists of two steps:

1. First, a set of individual vehicle trajectories are reconstructed on the basis of an initial time–space speed contour plot, and individual travel times. The *measured* travel times provide (virtually error free) entry and exit times of the approximate vehicle trajectory. These entry and exit times (literally) provide the constraints for each approximated vehicle trajectory over time. The loop data (speeds) provide the information which determines the slope of each trajectory over space and time.
2. All available approximated trajectories from the first step are used to re-estimate (correct) the speeds from the initial (prior) speed contour plot. The result is a posteriori speed contour plot, which fits best with all the estimated trajectories.

The list below provides an overview of the most important notation used in this approach.

i : Road segment index
k : Vehicle index
TT^k : Measured travel time for individual vehicle k on a specified link
tt_i^k / $\hat{tt}_i^k$: Sub-travel time for vehicle k at segment i of the specified link/Estimation
p_i^k / $\hat{p}_i^k$: Period when vehicle k travels at segment i/Estimation
t_i^k / $\hat{t}_i^k$: Moment when vehicle k starts to travel at segment i/Estimation
L_i : Length of segment i
$u(p,i)$: Traffic speed at segment i during period p
$\hat{u}^-(p,i)$: Prior estimated speed at segment i during period p (without fusing travel time information)
$\hat{u}^+(p,i)$: Post estimated speed at segment i during period p (with fusing travel time information)
$\hat{v}^k(p,i)$: Estimated speed for vehicle k at time–space region (p,i)
$\hat{l}^k(p,i)$: Estimated traverse length for vehicle k at time–space region (p,i)
$n(p,i)$: Number of reconstructed trajectories that traverse region (p,i)

[2] Incorrect matches result obviously in erroneous travel times. However, these incorrect matches can be filtered out a priori using outlier removal techniques. Since the goal with PISCIT is to estimate aggregate speeds, these outlier removal techniques may be imposed strictly. An example of such a technique is the following: *Remove all travel times observations in a certain time period further away from the median travel time than* $\alpha \times TT^{75} - TT^{25}$, where TT^{XX} represents the XXth travel time percentile.

4.1.2 Step One: Reconstruction of Individual Vehicle Trajectories

Recall the ingredients for step one:

1. A initial (prior) speed contour plot, and
2. Individual travel times

The main idea behind the approach in step one is to use the so-called proportion-multipliers to repeatedly correct sub-travel times at every segment (based on the prior speed contour plot) such that the sum of the sub-travel times satisfies the total travel time on the whole link (based on the measured travel times). In the ensuing, the following relationships are used (these also hold for the "hatted" symbols $\hat{p}_i^k$ and $\hat{t}_{i+1}^k$):

$$p_i^k = [t_i^k, t_i^k + tt_i^k]$$
$$t_{i+1}^k = t_i^k + tt_i^k$$

As starting a point, we assume that the vehicle is driving with a constant speed over segment i. This yields the following estimation of $\hat{tt}_i^k$:

$$\hat{tt}_i^k = TT^k \times L_i/L \qquad (10)$$
$$\text{where } L = \sum_i L_i$$

that is, the initial estimation of a vehicle trajectory is a straight line. However, based on $\hat{tt}_i^k$ and the entry moment t_1^k, now $\hat{p}_i^k$ can be calculated. Next, we will make use of the prior-estimated time–space speed contour to correct the estimation of $\hat{tt}_i^k$. Since higher speeds imply less sub-travel time, the following intuitive iterative update rule is applied:

$$\hat{tt}_i^k \propto 1/\hat{u}^-(\hat{p}_i^k, i) \qquad (11)$$

with constraint

$$TT^k = \sum_i \hat{tt}_i^k$$

where $i = 1,2,...,n$ and n is the (arbitrary) number of segments divided on the link. When the above equations are iteratively executed, the reconstructed vehicle trajectories provide an approximation of the true vehicle trajectory. Figure 5 schematically outlines this iterative procedure which will stop when the difference between previous values and current ones is smaller than some preset threshold (a small number).

4.1.3 Step Two: Speed Re-estimation

In step two, the estimated vehicle trajectories obtained in the first step are used in turn to correct the speeds in the initial (prior) speed contour plot. To tackle the problem, we introduce a simple and effective linear model with constraints.

First, for each segment (p,i) we collect the estimated speeds $\hat{v}^k(p,i)$, and the approximated traverse lengths $\hat{l}^k(p,i)$ for each vehicle k, which traversed region

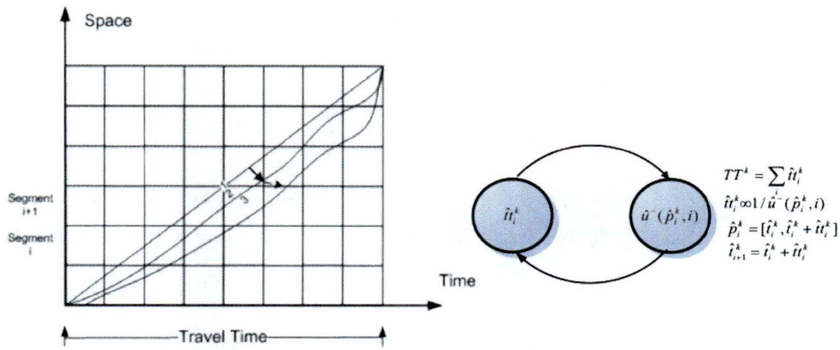

Graph A: Step-by-step approach to the true trajectory for a specified vehicle.

Graph B: Iteratively update estimations of sub-travel times at every segment

Fig. 5 PISCIT step 1: Recursive reconstructions of vehicle trajectories

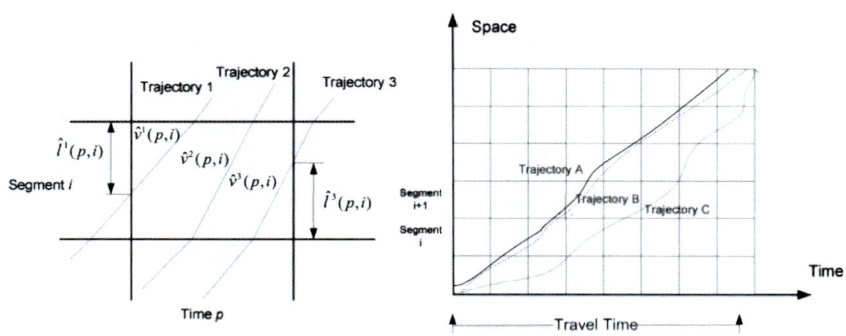

Graph A: Multi-trajectories at the same region

Graph B: Classification of trajectories

Fig. 6 PSCIT step 2: re-estimating segment speeds from estimated individual vehicle trajectories

(p,i), as illustrated in Graph A in Fig. 6. The average speed in each segment (p,i) can be derived by simply averaging these speeds $\hat{v}^k(p,i)$. However, in doing so, this results in a corrected (posteriori) speed contour map which may result in travel times which no longer equal the measured travel times. Taking the measured travel times as constraint, a simple linear model subject to travel time constraint can be established as

$$Y = X\beta + \varepsilon, \text{ with } \varepsilon \sim N(0,\sigma^2 I) \quad (12)$$

where Y is $N \times 1$ (observations), X is $N \times k$ and fixed, at least conditionally, β is $k \times 1$(traffic speeds to be estimated), and ε is an $N \times 1$ vector of Gaussian distributed zero-mean random variables, which reflects the random errors produced by the model. A linear constraint is given due to the travel time constraint, written as

$$H'\beta = h \quad (13)$$

where H' is able to establish the relationship between individual travel times and traffic speeds over time–space. With the knowledge of linear model theory, the least square estimation of β under such a constraint is

$$\hat{\beta} = \tilde{\beta} - (X'X)^{-}H'(H(X'X)^{-}H')^{-1}(H\tilde{\beta} - h) \tag{14}$$

where $\tilde{\beta} = (X'X)^{-}X'Y$. In particular, in our case, X enjoys a very simple form that reads $X = I$. Consequently, the least square estimation $\hat{\beta}$ also enjoys a simple form as

$$\hat{\beta} = Y - H'(HH')^{-1}(HY - h) \tag{15}$$

The main computations stem from the term $(HH')^{-1}$. After all trajectories are classified as shown in Graph B in Fig. 6, the dimensions of HH' are greatly reduced. Finally, $\hat{u}^{+}(p,i)$ can be obtained by

$$\hat{u}^{+}(p,i) = \left(\frac{1}{u(p,i)}\right) \tag{16}$$

where the term on the right-hand side is retrieved from the elements in the estimation $\hat{\beta}$.

4.2 *Conservation Law for Floating Car Data: Fusing Low Resolution Topological Data (FlowResTD)*

Cellular phone location data can be considered a low-resolution type of floating car data. Cellular phones operate using a network made up of multiple radio cells, as shown in Fig. 7. In each cell, a base station provides communication service for phones. Base stations are all interconnected so that an on-call phone can move from one cell to another without losing the connection. All cells are grouped into particular location areas (LA). Typically, a cell ranges in size between 200 and 1000

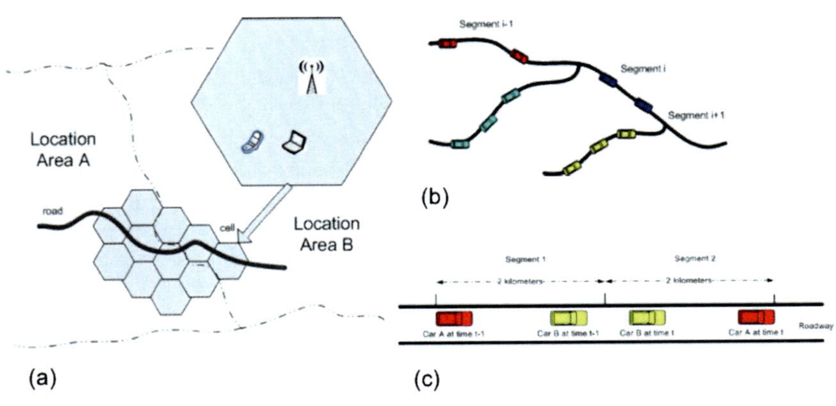

Fig. 7 (a) Example of communication networks; (b) Probe vehicles are identified according to the road segments on which they are; and (c) Illustration of location displacement of vehicles

meters in urban areas, and location areas range between 1 km to several kilometers. In a cellular network, since each cell has its own pre-defined coverage area, the relation between the cellular network and the road network can be pre-matched. The potential availability of GSM-based FCD is large, simply since there are so many cell phones available. On the downside, GSM tracking is limited and restricted by the cell segment sizes and the number of cell phones followed. As seen in Graph (c) (Fig. 7), vehicles A and B would have provided completely different traffic estimation during time t-1 and t. But with segment-level accuracy, the two vehicles will report the same traffic behavior estimation both located on Segment 1 at time t-1 and on Segment 2 at time t. The method proposed in this section is capable of fusing low-resolution positioning data with other data sources, leading to more accurate and reliable speed estimation with less bias. In addition, this method is robust with respect to erroneous measurements.

The following notations are is used in this section:

Δt	: Time interval
$\Psi(x)$	: Number of probe vehicle among all x vehicles
N	: Number of vehicles on a road segment at some time
M	: Number of vehicles from a road segment during interval Δt
n	: Number of all probe vehicles on a road segment
m	: Number of probe vehicles that move to the downstream segment during Δt
v^-	: Observed (initial) speed estimation
$\hat{v}$	: Posterior estimated (corrected) speed in terms of space mean speed

4.2.1 Prototype and Basic Idea

The basic rationale of this method is very similar to the first-order traffic flow theory introduced in Section 2.2, and can be intuitively illustrated with hydrodynamics. Consider a full water container, 1 meter high, and assume it takes 100 seconds to drain the container after the valve is opened. The average rate at which the water level drops can be estimated as 0.01 m/s. Analogously, suppose there are N vehicles ("water") on a road segment (the "container") of length D, and suppose that the upstream traffic is held so that there is no traffic flowing into the downstream segment (see Fig. 8). Suppose it takes Δt time units to "drain" the traffic out of this segment, the space mean speed can be estimated with $v \approx D/\Delta t$. If it is furthermore known that M vehicles enter the downstream segment, the traffic speed can be estimated with

$$v \approx \frac{D \times M}{\Delta t \times N} \tag{17}$$

If the total outflow of the water container is *unknown*, a solution is to track the concentration of indicator solids or isotopes in both the container and the outflow stream, under the assumption that these are uniformly distributed in the water (see Fig. 8). First, a certain amount or number of isotopes are put into the water container and then they are uniformly distributed in the container. When the isotopes flows out with water, one can deduce the amount of the out-flow water based on the amount of

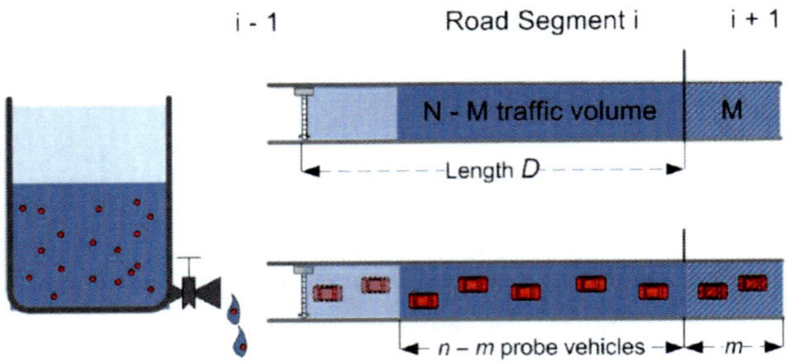

Fig. 8 Model comparison of water container and roadway - probe vehicles may be used to approximate speed in case total flow is unknown

out-flow isotopes, and as a result, the water level drop rate can be further deduced. Extending the method, the same can be done for traffic speed estimation, where probe vehicles with segment level accuracy positioning can serve as such indicators. Suppose that there are n probe vehicles at time t and that m probe vehicles move to the next segment by $t + \Delta t$ (see Fig. 8). The pair (n,m) then becomes characteristic of the space mean traffic speed on this segment during this time period. Analogous to the above, this estimation reads

$$v \approx \frac{D * m}{\Delta t * n} \tag{18}$$

The question is how good this estimate is, i.e., how much confidence we can have in this estimate. For example, the pairs (100, 50) and (10,5) result in the same estimation, but the latter one would be considered less probable. Things become more complicated with prior knowledge of traffic speed (e.g., the speed distribution based on historical data). In the following section, a more accurate estimation method will be given with (and without) prior knowledge of traffic speed combined.

4.2.2 Probe-Vehicle-Count Process and Speed Re-estimation

First, let us define a count process $\{\Psi(x), x \geq 0\}$, where x is the considered vehicle number and $\Psi(x)$ is the number of probe vehicles in this sample (see e.g., Fig. 8). It is easily verified $\Psi(x)$ has the following four attributes:

1. $\Psi(0) = 0$;
2. $\Psi(x+y) - \Psi(x)$ is independent of $\Psi(x)$ for any $x \geq 0, y > 0$;
3. $P(\Psi(x+h) - \Psi(x) = 1) = \lambda h + o(h)$ for small h, where λ is the proportionality factor associated with the percentage of probe vehicles; and
4. $P(\Psi(x+h) - \Psi(x) \geq 2) = o(h)$.

As a result, $\{\Psi(x), x \geq 0\}$ is a Poisson process according to its definition [2]. Now assume we have prior knowledge on the traffic speed distribution $P(v)$, e.g., through other speed data. According to Bayes' rule, we can (re)-estimate the posterior traffic speed distribution with the pair (n,m) via

$$P(v|\Psi(N)=n,\Psi(M)=m) \propto P(\Psi(M)=m|\Psi(N)=n,v) \times P(v|\Psi(N)=n) \quad (19)$$

where $P(v|\Psi(N)=n)=P(v)$ because $\Psi(N)$ is independent of v. However, $P(\Psi(M)=m|\Psi(N)=n,v)$ is not easily deduced, so let us focus on $P(\Psi(M)=m|\Psi(N)=n)$. Since $\{\Psi(x),x\geq 0\}$ is a Poisson process, we have

$$P(\Psi(M)=m|\Psi(N)=n)=\frac{n!}{m!*(n-m)!}\left(\frac{M}{N}\right)^{m}\left(1-\frac{M}{N}\right)^{n-m} \quad (20)$$

Considering that $v\approx\frac{D*M}{\Delta t*N}$ (or $\frac{M}{N}\approx\frac{v\Delta t}{D}$), $P(\Psi(M)=m|\Psi(N)=n,v)$ can be approximately estimated with

$$P(\Psi(M)=m|\Psi(N)=n,v)\approx\frac{n!}{m!*(n-m)!}\left(\frac{v\Delta t}{D}\right)^{m}\left(1-\frac{v\Delta t}{D}\right)^{n-m} \quad (21)$$

Substituting (21) and $P(v|\Psi(N)=n)=P(v)$ into (19) results in

$$P(v|\Psi(N)=n,\Psi(M)=m)\propto\frac{n!}{m!*(n-m)!}\left(\frac{v\Delta t}{D}\right)^{m}\left(1-\frac{v\Delta t}{D}\right)^{n-m}*P(v) \quad (22)$$

In (22), m, n, v, $P(v)$, and D are all known as mentioned above, while unknowns N, M, and λ are not needed. For the best estimation of traffic speed with prior knowledge of the speed distribution, we now have

$$\begin{aligned}\hat{v} &= E(v|\Psi(N)=n,\Psi(M)=m)\\ &= \frac{1}{C}\int\frac{n!}{m!*(n-m)!}\left(\frac{v\Delta t}{D}\right)^{m}\left(1-\frac{v\Delta t}{D}\right)^{n-m}P(v)vdv \end{aligned} \quad (23)$$

where $C=\int\frac{n!}{m!*(n-m)!}\left(\frac{v\Delta t}{D}\right)^{m}\left(1-\frac{v\Delta t}{D}\right)^{n-m}P(v)dv$ is a factor to normalize. In particular, if there is no prior knowledge of actual speed distribution, we simply assume $P(v)$ has a uniform distribution with regard to v.

4.2.3 Physical Explanation and Extended Conclusion

Figure 9 provides a physical explanation for the above estimation equation. The term $\left(\frac{v\Delta t}{D}\right)$ is the outflow traffic (in terms of vehicle number) proportion, and $\left(1-\frac{v\Delta t}{D}\right)$ is the proportion of traffic that remains on this segment. Since speed v is the variable to be estimated, the two proportions are actually unknown. Thus, the pair (n,m) is needed to weigh the two proportions with regard to v. For a fixed v, a larger m means a larger outflow proportion, so it will be more heavily weighted.

Furthermore, when considering the number of vehicles from the *upstream* segment (Fig. 10), more information can be added for estimation. Consider three proportions, that is, the traffic outflow, the remaining traffic and inflow traffic,

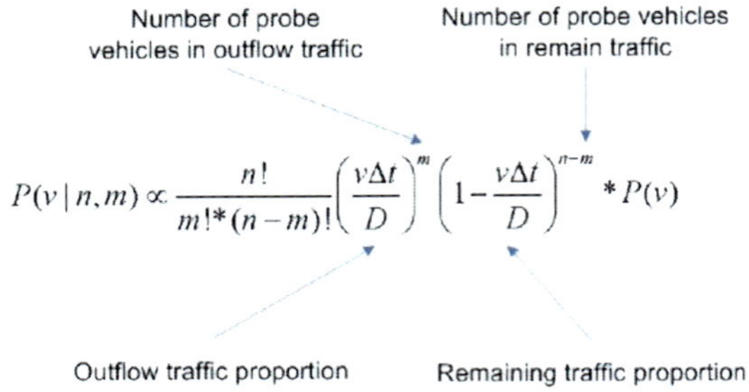

Fig. 9 Physical explanation to estimation equation

proportion, denoted with $\left(\frac{v\Delta t}{D+v\Delta t}\right)$, $\left(\frac{D-v\Delta t}{D+v\Delta t}\right)$, and $\left(\frac{D}{D+v\Delta t}\right)$, respectively. The corresponding weighting factors are m, $n-m$, and $l-n$. Similar to (23) and now considering inflow traffic, we have

$$\begin{aligned}\hat{v} &= E(v|n,m,l)\\ &= \frac{1}{C}\int \frac{l!}{m!(n-m)!(l-n)!}\cdots\\ &\cdots\left(\frac{v\Delta t}{D+v\Delta t}\right)^{m}\left(\frac{D-v\Delta t}{D+v\Delta t}\right)^{n-m}\left(\frac{D}{D+v\Delta t}\right)^{l-n}P(v)\,v\,dv \qquad (24)\end{aligned}$$

where $C=\int \frac{l!}{m!(n-m)!(l-n)!}\left(\frac{v\Delta t}{D+v\Delta t}\right)^{m}\left(\frac{D-v\Delta t}{D+v\Delta t}\right)^{n-m}\left(\frac{D}{D+v\Delta t}\right)^{l-n}P(v)dv$.

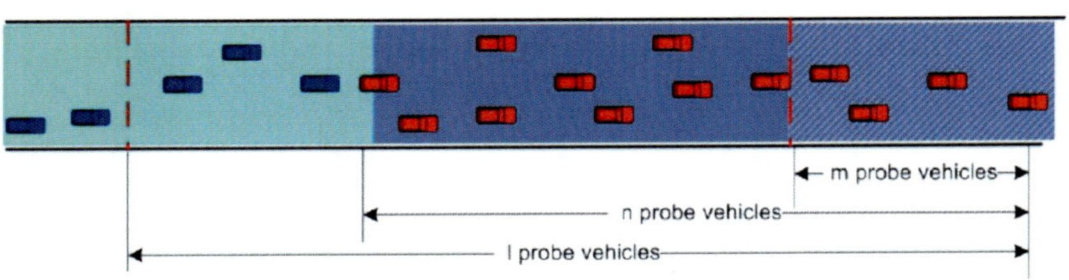

Fig. 10 Probe vehicle inflow and outflow traffic

4.3 *Kinematic Wave Theory: The Extended and Generalized Treiber–Helbing Filter (EGTF)*

This method (EGTF) extends and generalizes the spatiotemporal traffic filter proposed by Treiber and Helbing [18] and is based on the spatiotemporal characteristics discussed in Section 2.2, that is, perturbations in traffic travel along the so-called characteristics (5) with (approximately) constant characteristic speeds c_{cong} in congestion and c_{free} under free flow conditions.

4.3.1 Spatiotemporal Filter for a Single Data Source

Consider a single data source j. For filtering these data over space and time, we use the approach of [18] stating that the reconstructed quantity z at (t,x) is described as follows:

$$z(t,x) = w(t,x)z_{cong}(t,x) + (1 - w(t,x)z_{free}(t,x)) \tag{25}$$

Equation (25) shows that the reconstruction involves a weighted combination using two reconstructions of the signal. The first assumes congested traffic operations (i.e., $z_{cong}(t,x)$) and the second free flow conditions (i.e., $z_{free}(t,x)$). To reconstruct $z(t,x)$ on the basis of data measured at some time and location (t_i,x_i), we compute the time, and space-dependent weights as follows. Let us first define:

$$\begin{aligned} \phi_{cong}(t,x) &\equiv \phi_0\left(t - \frac{x}{c_{cong}}, x\right), \text{ and} \\ \phi_{free}(t,x) &\equiv \phi_0\left(t - \frac{x}{c_{free}}, x\right) \end{aligned} \tag{26}$$

with

$$\phi_0(t,x) = \exp\left(-\frac{|x|}{\sigma} - \frac{|t|}{\tau}\right) \tag{27}$$

where σ and τ are parameters of the filter which describe the width and time window size of the "influence" region around (t_i,x_i). For the value of the weights of a data point (t_i,x_i), we now obtain

$$\begin{aligned} \beta^i_{cong}(t,x) &= \phi_{cong}(x_i - x, t_i - t), \text{ and} \\ \beta^i_{free}(t,x) &= \phi_{free}(x_i - x, t_i - t) \end{aligned} \tag{28}$$

The weights describe the importance of the measurement quantity z_i at the time–space point (t_i,x_i) for the value of the quantity z (to be estimated or reconstructed) at (t,x). Loosely speaking, the weight is determined by the distance between the point (t_i,x_i) and (t,x) considering the speed at which information moves through the flow under free flow or congested conditions.

Let us emphasize that in contrast with [18], we do not assume that the points (t_i,x_i) for which data are available stem from sources that provide data in an equidistant rectangular grid (like, for example, local detectors which produce averaged speeds over fixed time intervals). In fact, the data may also stem from moving data sources, such as trajectory data from moving vehicles (FCD–floating car data), incidental travel time reports, or otherwise.

To determine the value of the quantity $z(t,x)$ on the basis of the congested and the free-flow filter, we use the weights as follows:

$$z_{cong}(t,x) = \frac{\sum_i \beta^i_{cong}(t,x) z_i}{\beta_{cong}(t,x)}, \text{ and}$$
$$z_{free}(t,x) = \frac{\sum_i \beta^i_{free}(t,x) z_i}{\beta_{free}(t,x)} \tag{29}$$

Where the normalization factors are given by

$$\beta_{cong}(t,x) = \sum_i \beta^i_{cong}(t,x), \text{ and}$$
$$\beta_{free}(t,x) = \sum_i \beta^i_{free}(t,x) \tag{30}$$

An important filter design choice is the weight factor $w(t,x)$ used in Eq. (25). This factor describes whether the conditions in (t,x) are dictated by free-flow conditions or by congested conditions or a combination of both. Treiber and Helbing [18] propose to use speed data for this purpose and use the following expression for this weight factor:

$$\begin{aligned} w(t,x) &= \omega(z_{cong}(t,x), z_{free}(t,x)) \\ &= \frac{1}{2}\left[1 + \tanh\left(\frac{V_c - V^*(t,x)}{\Delta V}\right)\right] \end{aligned} \tag{31}$$

with

$$V^*(t,x) = \min\left(V_{cong}(t,x), V_{free}(t,x)\right) \tag{32}$$

where $V_{cong}(t,x)$ and $V_{free}(t,x)$ are calculated with (29), V_c depicts a critical speed marking the transition from free to congested flow, and ΔV a bandwidth around it. Figure 11 provides a graphical explanation of (31) and (32). Note that the functions (31) and (32) are arbitrary filter design choices. Any crisp, smooth, or even fuzzy function which is able to discriminate between free-flowing and congested traffic operations based on whatever data is available (density/occupancy, speed) would in principle do.

4.3.2 Spatiotemporal Filter for Multiple Data Sources

Let $z^{(j)}(t,x)$ denote the considered traffic value as reconstructed from data source j. To fuse data from multiple data sources, we propose the following linear combination:

$$z(t,x) = \frac{\sum_j \alpha^{(j)}(t,x)\phi^{(j)}(t,x) z^{(j)}(t,x)}{\sum_j \alpha^{(j)}(t,x)\phi^{(j)}(t,x)} \tag{33}$$

where the second dynamic weight $\phi^{(j)}(t,x)$ is defined by

$$\begin{aligned} \phi^{(j)}(t,x) &= w^{(j)}(t,x) \cdot \phi^{(j)}_{cong}(t,x) \\ &\quad + (1 - w^{(j)}(t,x)) \cdot \phi^{(j)}_{free}(t,x) \end{aligned} \tag{34}$$

The first (dynamic) weight factor $\alpha^{(j)}(t,x)$ in (33) can be interpreted as a dynamic indicator of the *reliability* of the data from source j at (t,x) and could, for example,

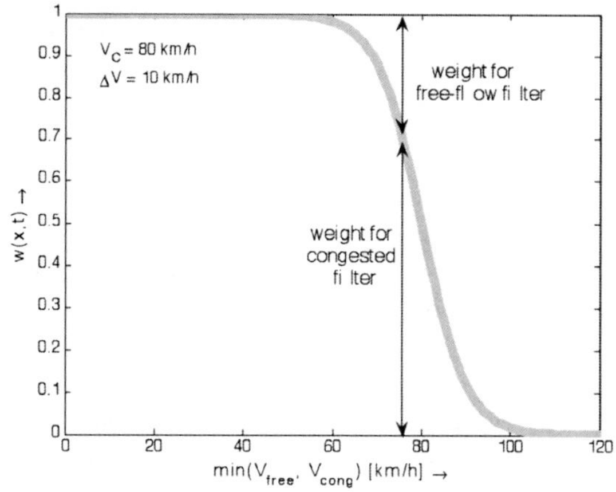

Fig. 11 Calculation of weight for free-flow and congested kernel, adapted from [18]

be determined on the basis of a priori estimates of the measurement accuracy of data source j. For induction loops, where measurements become increasingly unreliable as speeds decrease,[3] it makes sense that $\alpha^{(j)}(t,x)$ is proportional to speed. Although also location tracking equipment (e.g., GPS) is likely to make relative errors proportional to speeds, the reliability of such FCD measurements in terms of speeds would most probably still be higher than that of induction loops.

For the sake of illustration, we will reuse the hyperbolic tangent function in Equation (31) to calculate this weight, which reads

$$\alpha^{(j)}(t,x) = \frac{1}{\Theta_0^{(j)}\left[1+\mu^{(j)}(1-w^{(j)}(t,x))\right]} \tag{35}$$

In (35), $\Theta_0^{(j)}$ represents the standard deviation of the measurement error of data source j at low speeds (under congestion), and $[1+\mu^{(j)}]\Theta_0^{(j)}$ the standard deviation under free-flowing conditions. For induction loops, $\Theta_0^{(j)}$ is typically in the order of 4 km/h, and $\mu^{(j)}$ around $1\frac{1}{2}$ (yielding a standard deviation of around 10 km/h under free-flow conditions).

5 New Traffic Data Fusion Approaches: Results

In this part, the workings of the proposed new approaches will be shown by means of an experiment with synthetic data generated using a microscopic simulation program. The simulation environment used for analyzing PISCIT and FlowResTD is

[3] This is first of all due to increasingly smaller samples per time period (in the limit of zero speed the sample size is also zero), and secondly, caused by the fact that the measurement equipment itself makes larger errors as vehicles pass over at lower speeds.

VISSIM 4.20 [14], and the EGTF method is assessed with FOSIM.[4] The main reason for using a simulation environment is the availability of ground-truth traffic data (speeds).

5.1 Results of Applying the PISCIT Method

In VISSIM, a 4.5-km stretch of freeway was coded (see Fig. 12). At 3750 meters, this freeway (road A) merged with a second freeway stretch (road B), resulting in a possible bottleneck. Road A is equipped with 10 detectors spaced every 500 meters, each of which produces speeds and time stamps when there is a vehicle passing by. At detector locations 1 and 10, two cameras were simulated by Camera A and B, respectively. In such a case, the individual travel times for part of all vehicles can be provided, and the number of camera-captured vehicles takes up about 20% of total number. To make artificial data more realistic, such a traffic demand scenario is set up that traffic demand will exceed traffic supply, producing congestion as a result.

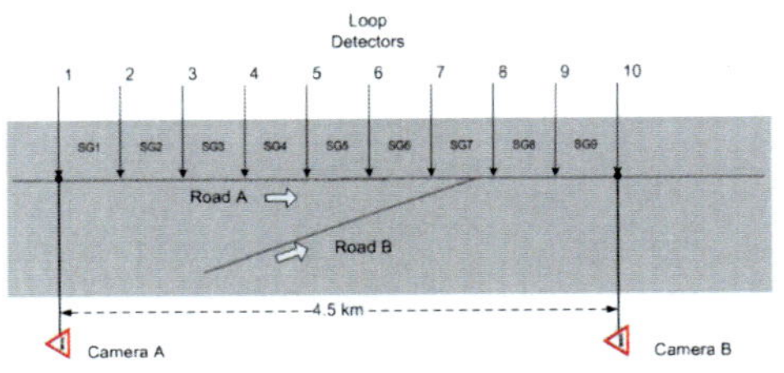

Fig. 12 Experimental setup for PISCIT: layout of freeways and detectors

Suppose that the speeds in time–space plot are non-uniformly overestimated as follows:

$$\text{Deviationrate}(\%) = \begin{cases} \sqrt{(126-u)/u} & \text{if } 0 \le u \le 54(\text{km/h}) \\ (126-u)/u \times 70\% & \text{if } 54 < u \le 126(\text{km/h}) \\ 0\% & \text{otherwise} \end{cases}$$

In the following, quantitative results will be given in terms of the Weighted Relative Error:

$$WRE = \frac{\sum\limits_{p,i} |\hat{u}^{+}(p,i) - u(p,i)|}{\sum\limits_{p,i} u(p,i)} 9 \tag{36}$$

In this scenario, it is assumed that lower speed has relatively more unreliable measures than high speed. So there is more relative deviation with lower speeds.

[4] FOSIM (Freeway Operations Simulation), developed at Delft University of Technology, is a traffic micro-simulation package which has been extensively calibrated and validated for traffic operations on Dutch freeways and is used by the Dutch Ministry of Transport, Public Works and Water Management as a freeway design tool for ex ante freeway capacity estimation [1, 26].

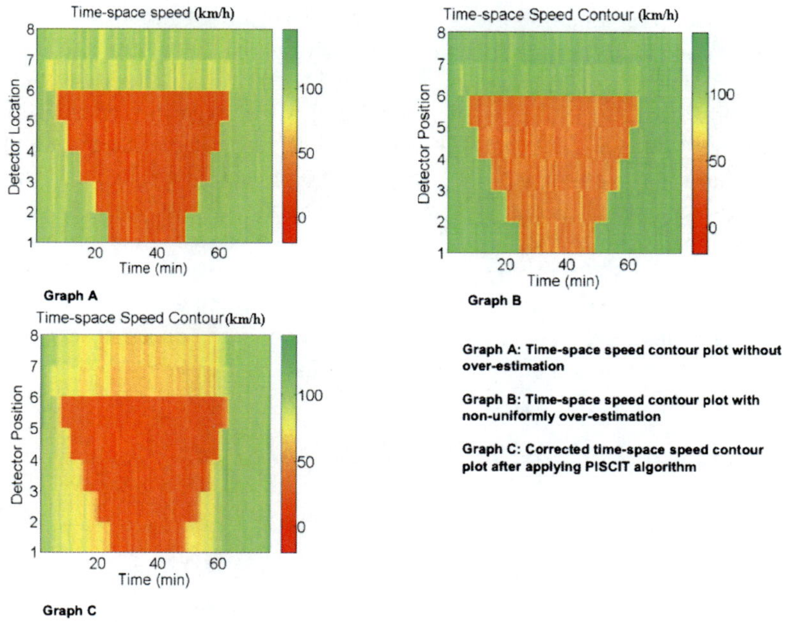

Fig. 13 Comparison of time-space speed contour plots before and after applying PISCIT

Graphs A and B in Fig. 13 are the time–space speed plots with and without deviation, respectively, which share the same color scale from -20 to 120. And as seen, there are significant difference between Graphs A and B (Fig. 13). However, when the algorithm PISCIT is applied, great improvement is made in Graph C compared with A and B. After correction in the two scenarios, both weighted relative errors

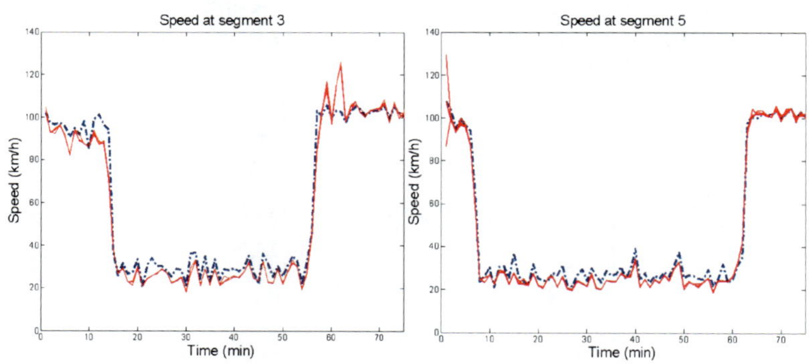

Fig. 14 Robustness of PISCIT with different deviations and random errors. The dashed dot line is ground-truth speeds over time. The four red lines (solid lines) in each graph show the speeds over time under different four cases, which almost overlap each other tightly and only slightly change with the four cases to show the robustness of the algorithm.

are well reduced, resulting in $WRE = 8.23\%$. Similarly, PISCIT can produce a good correction if the speeds are overestimated, as shown in the three graphs in Fig. 13.

Also, the robustness of algorithm PISCIT is shown in Fig. 14, where the estimated speeds through PISCIT at segments 3 and 5 are given in contrast with the true ones represented by blue lines (dashed lines). By zooming in the graphs, it can be found that there are four red lines (solid lines) in each graph, which show the PISICT-estimated speeds under four cases 20% random error, 30% random error in loop detector measurement, 20% random error plus +20% structural deviation, and 20% random error plus -20% structural deviation. The resulting four estimation of speeds by means of PISCIT under the four cases are quite stable and robust. Numerically, at segment 3, $WRE = 8.33\% \pm 0.5\%$ corresponding to different scenarios; at segment 5 $WRE = 4.76\% \pm 0.5\%$ in different scenarios.

5.2 Results of Applying the FlowResTD Method

5.2.1 Experimental Setup

For the purpose of validating this approach, a hypothetical 35.25-km 3-lane highway corridor was coded in VISSIM, made up of 15 segments, each 2.35 km long (Fig. 15). At the 10 km location, there is an off-ramp with a signal control which caused queue spill-back yielding temporary and periodical impacts on the highway. There is a similar situation at the 20 km location. At the 28 km location, highway traffic is interrupted by traffic control across all three lanes, where an intersection is located. All traffic signals are assumed to operate on a fixed-time basis with a 60-second red phase followed by a 140-second green phase. For probe vehicles (10% of the fleet), it is assumed that they reported their segment position every 1 minute. Now, we have $D = 2.35$ km and time interval $\Delta t = 1$ minute.

Although *VISSIM* is able to provide the exact position of each vehicle, we reduced the accurate position to segment level. Since the segment boundaries are predetermined, we know how many probe vehicles remain on a segment and how many

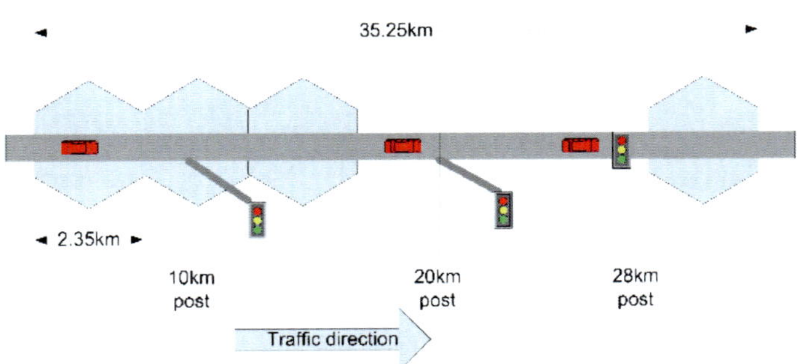

Fig. 15 Experimental setup (FlowResTD): The layout of road infrastructure, segment and scenario settings in VISSIM

flow in and out. In this experiment, the information (m,l,n) on a specific segment is known. To show the effectiveness of this method, we "exaggerate" the errors from observation, causing observed speeds to contain non-uniform structural deviations up to 80% plus random errors based on ground-truth speeds:

$$\text{Structural Deviation rate } (\%) = \left(\frac{0.73 - 1.9}{110} * v + 1.85\right) * 100\%$$

$$\text{Random Error rate } (\%) = N(0{,}0.20^2) \times 100\%$$

where v is the simulated ground-truth speed and $N(0{,}0.20^2)$ is a normal distribution with mean 0 and standard deviation 0.20. This way, the observed speed v^- was fabricated by applying the structural deviation rate and random error rate to the ground truth speed. In particular, the structural deviation rate was primarily dependent on actual speeds. When speeds were higher than 80 km/h, they were underestimated up to 30%. When lower than 80 km/h, they were overestimated even up to 90%. Since there is no extra information available from another data source, a uniform probability distribution with respect to v was assumed to be $P(v) = \frac{1}{1.5v^- - 0.5v^-}$, where $P(v)$ is a generalized probability density function. With more empirical analysis or other data sources, one may assume that v follows other possible distributions such as normal or Poisson or with more accurate parameters. For example, GPS data provide accurate traffic speed information but they may have only sparse time–space coverage. We may make some inference on these GPS data, deducing the most probable prior speed distribution.

By now, we have known vehicle numbers (m,l,n), a prior speed distribution $P(v)$, the segment length D, and the sampling time interval Δt, so that Equation 24 works out with them, leading to posterior speed estimation on a specific road segment. When employing the aforementioned method on each segment, the space mean speeds on all segments of the whole link can be (re)-estimated (corrected).

5.2.2 Overall Results

In Fig. 16, a comparison between Graphs A and B reveals that higher speeds are underestimated and lower speeds are overestimated. But after applying Equation 24, most of the structural deviation is removed and many details which are invisible in Graph B (e.g., more detailed color-gradation, disturbance by signal controls) now emerge.

In Fig. 17, corrections of structural deviation are noticeable regardless of the overestimation of speed or underestimation. The random errors are also diminished. Surprisingly, taking a close look at the right graph in the figure, it is found in the low-speed regions that the corrected speeds exhibit clear signs of periodic traffic control more than the ground-truth speeds do. This reveals that this method may be able to display some critical details in traffic states.

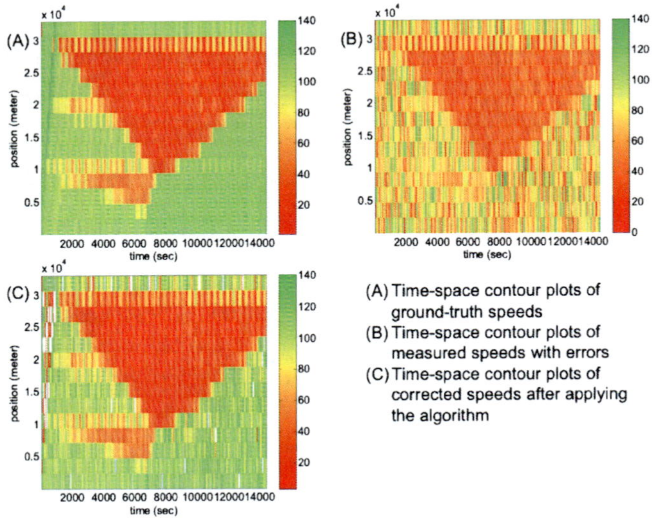

Fig. 16 Comparison between ground-truth, measured, and corrected time–space contour plots

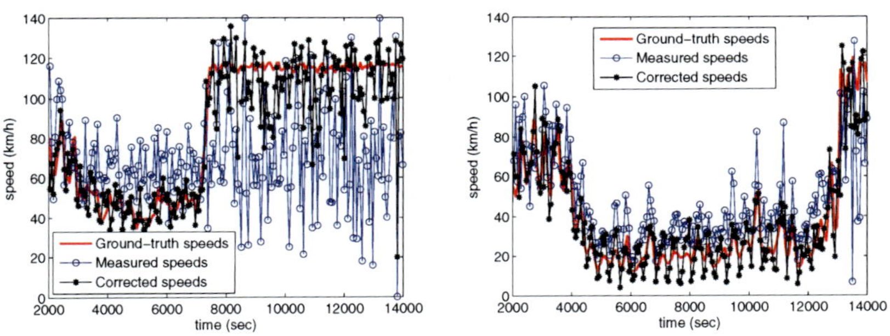

Fig. 17 Speed–time plots on segment 4 from 7.05 km to 9.4 km (left), and segment 9 from 18.8 to 21.5 km (right)

5.2.3 Effects of Different Prior-speed Probability Distributions

In traffic operations, it is difficult to estimate the profile of the prior-speed distribution, which may change with time or circumstances. So the posterior distributions under different prior distributions are analyzed. We assume three prior distributions of substantially different profiles—normal, triangular, and uniform (see Graph A in Fig. 18). Given the fixed pair $(n,m) = (30,15)$ and $D = 2.35$, the corresponding posterior distributions all follow a normal distribution (see Graph B in Fig. 18), which confirms the robustness of this algorithm. It makes sense that the posterior distributions tend to be normal, since this core equation in this approach is mainly made up of the critical expression $\frac{n!}{m!\times(n-m)!}\left(\frac{v\Delta t}{D}\right)^m\left(1-\frac{v\Delta t}{D}\right)^{n-m}$ which is a binomial distribution, a distribution that approximates normal when n,m approach infinity.

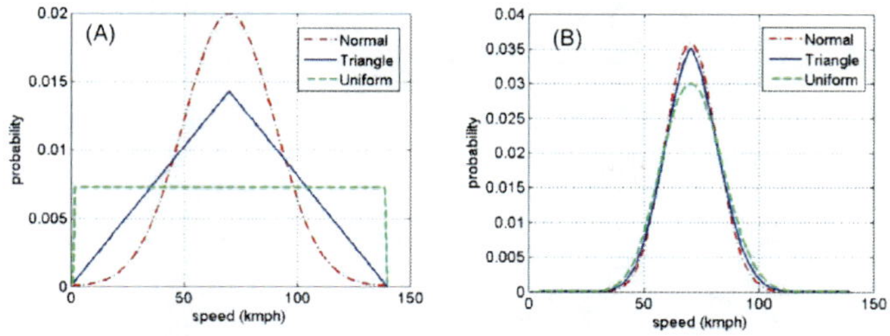

Fig. 18 The influence of the forms of probability distribution

Table 2 Data source weight settings

data source	$\Theta_0^{(j)}$*	$\mu^{(j)}$
u_T^{loop}	4	2
u_H^{loop}	3	1.5
v^{fcd}	1	3

* *in (veh/h)*

5.3 *Results of Applying the EGTF Method*

To generate both ground-truth data (u_S) and all detector data, a FOSIM implementation of the 19-km A13 southbound freeway section between Delft and Rotterdam is used. This corridor contains 6 on ramps, 5 off ramps, and 3 weaving sections (combined on/off ramps). The underlying origin–destination (OD) matrix was estimated on the basis of loop detector data collected on a weekday afternoon peak in 2006

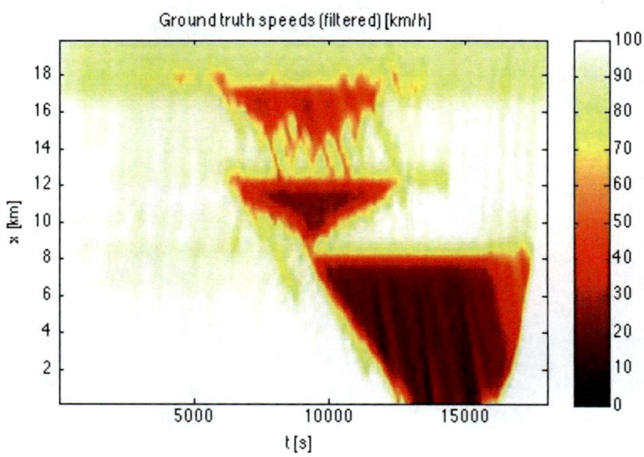

Fig. 19 Space–time contour map of filtered ground-truth speeds

Table 3 MAPE results data fusion

Local data used		Percentage FCD 0%	2%	5%	10%
	none		6.9	5.1	4.4
u_T^{loop}	500m	2.6	3.4	3.5	3.6
	1500m	6.4	5.1	4.4	4.1
u_H^{loop}	500m	1.9	3.5	3.6	3.7
	1500m	6.0	5.2	4.4	4.1

and tuned rudimentary such that a realistic traffic pattern resulted, with congestion occurring at the two bottlenecks where also in reality congestion sets in first.

5.3.1 Experimental Setup

To make a comparison between different scenarios possible, a single simulation was run, from which all data were derived. Since the data fusion/filter mechanism effectively removes the high-frequency fluctuations (smaller than τ (s) and σ) (m) from the data, the reconstructed traffic conditions on the basis of the various data sources are compared against *filtered* ground-truth data (with the same filter settings as will be used for the data fusion itself—see the following paragraph), such that the differences in performance solely reflects error variance due to using different data sources. In our case, the ground truth data reflects the space mean speeds $u_S(k,m)$ over equidistant space time regions (k,m) of size $30(s) \times 100(m)$. Figure 19 shows a speed contour map of filtered ground-truth data.

We consider two data sources here. The first are local traffic sensors spaced 500 or 1500 m, producing either harmonic or time mean speeds (u_H^{loop} and u_T^{loop}) per observation period of 1 minute. The second data source is FCD, where we assume that on average a percentage P (0, 2, 5, and 10%) of all vehicles present report their location and speed every 30 s. The following values for the filter parameters

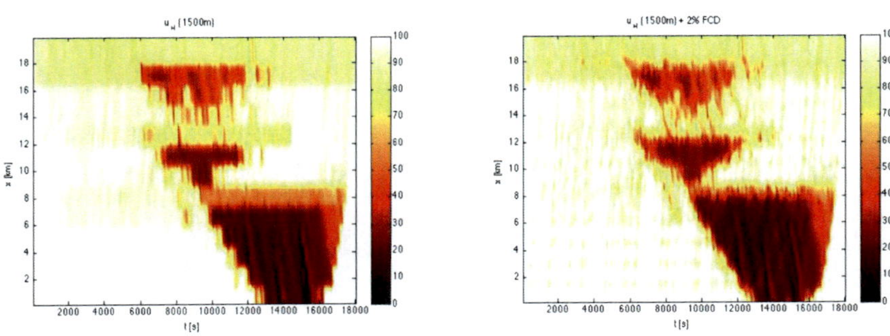

Fig. 20 Space–time contour maps of filtered local harmonic speeds and local harmonic speeds fused with 2% FCD

are used: $c_{cong} = -25$ km/h; $c_{free} = 80$ km/h; $\sigma = 300$ (m); $\tau = 30$ (s); $\Delta V = 10$ (km/h); and $V_c = 80$ (km/h). The data source-specific weights and parameters are listed in Table 2 and were chosen on the basis of rough estimates on a selection of ground-truth data.

5.3.2 Fusing Local Data and FCD

Table 3 shows the mean absolute percentage errors (MAPE) of all possible combinations of data sources. The first row represents results on the basis of just FCD, the first column shows results with solely local traffic data. Harmonic mean speeds from sensors every 500 m clearly outperform all other (combinations of) data sources. However, when spaced at 1500 m, the addition of increasing percentages FCD improves the results significantly. This is even more pronounced in the case time mean speeds are used (which is the quantity collected in the Dutch freeway monitoring system). Vice versa, adding local data to FCD-based estimates improves the quality of the estimate in all cases. Figure 20 illustrates these results and shows that fusion of local and FCD predominantly leads to smoother estimation of the up- and downstream boundaries of congested regions.

6 Discussion and Application Perspectives

In this chapter, we start out with the argument that model-based/KF approaches are (with good reason) the de facto tools for traffic state estimation since they provide a principled approach to balance the uncertainty in different data sources and in our models describing the underlying dynamics. Moreover, such approaches estimate underlying state variables (e.g., traffic densities), which are difficult to measure with sensors but are imperative for many applications (e.g., ramp metering and

Table 4 Overview of state estimation and data fusion approaches in terms of input and output. Note that k here denotes the index of discrete time periods $[t_{k-1}, t_k)$.

	Input	Output
PISCIT	Individual travel times + prior speed map $u^-(t,x)$	posterior (corrected) speed map $u(t,x)$, $t \leq t_k$
FlowResTD	low-resolution probe vehicle data $y(t_j,x_i)$) + prior speed map $u^-(t,x)$	posterior speed map $u(t,x)$, $t \leq t_k$
EGTF	data $y(t_j,x_i)$ (speeds, flow, density, occupancy) from any source + prior speed map $u^-(t,x)$	posterior *smoothed* map of data $y(t,x)$, $t \leq t_k$
Model-based/KF	data $y(t_k,x_i)$ (speeds, flow, density, occupancy) from any source which can be related to state variables $y(t,x) = h(x(t,x))$	posterior vector of state and measurement variables, $x(t,x)$ and $y(t,x)$, with $t \in [t_{k-1}, t_k)$

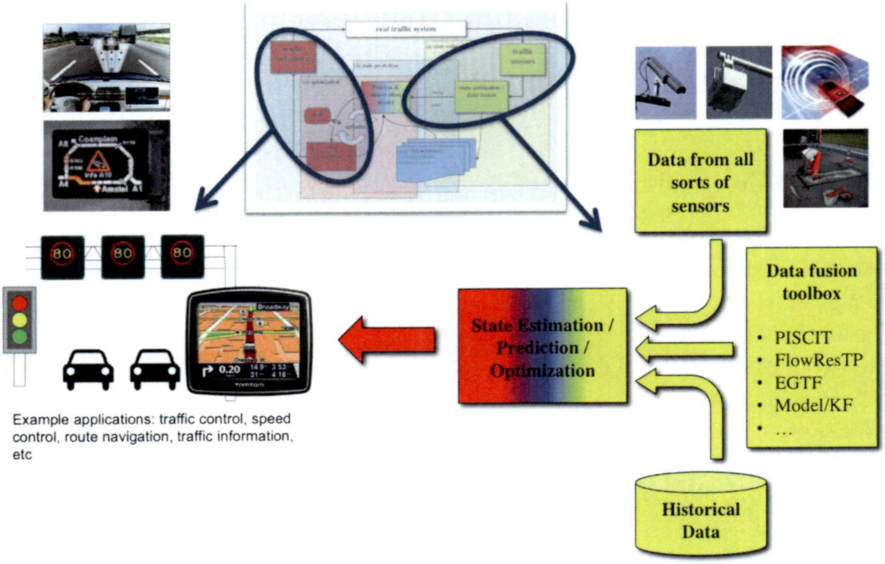

Fig. 21 Stylized structure of an integrated traffic state estimation and data fusion framework on the basis of the traffic control cycle (Fig. 1)

model-based prediction of travel times, queue lengths, etc.). However, in case the available data from different sources (speeds, flows, travel times, sampled vehicle trajectories, etc.) cannot be straightforwardly aligned (in terms of space and time resolution), application of such KF methods becomes cumbersome and computationally expensive (e.g., it results in an explosion of the state space—see (9)).

The three new approaches presented in this chapter solve this alignment problem. However, they can only reconstruct the traffic state in the same representation as observed with sensors, that is, in terms of speeds (PISCIT, FlowResTD, and EGTF) and flows or occupancies (EGTF), or in terms of derived quantities, such as experienced travel times. Table 4 overviews the input–output mapping of the three proposed methods in this chapter. All three take as input a prior space–time map of speeds ($u^{-}(t,x)$) (which can be of arbitrary quality), and as additional input from other—typically not easily aligned—sensor data. This prior speed map could be derived directly from data (given these are provided in an equidistant space–time grid) but could also be the result of any of the other algorithms listed. This implies that these algorithms can be used in a cascaded setting, each fusing an increasingly accurate posterior speed map with new data. Of course, depending on the availability of sensor data, and the requirements set by potential applications (e.g., prediction of departure travel times and optimal routes, optimal speed limits, coordinated ramp metering settings, etc.), not all these algorithms will be required all the time, or at every segment in a traffic network. The three algorithms presented here must be viewed as components in a data fusion toolbox, which can be used to generate reliable inputs for state prediction and control optimization on the basis of heterogeneous and potentially unreliable and inconsistent data. Figure 21 outlines how such a toolbox methods fits in the traffic control cycle outlined in Fig. 1.

Let us illustrate how such a cascaded application could work in practice. For example, in case a prior-speed map obtained from sparsely located loop detectors is biased (e.g., due to wrong averaging), and individual travel times are available, the PISCIT method provides a means to eliminate this speed bias. Particularly for state estimation under severely congested conditions bias correction is very important. Errors in speed of up to 20% may lead to errors in estimated densities of over 100% [17]. In case also low-resolution FCD is available (e.g., obtained from a telco operator), the FlowResTD algorithm can be applied to fuse the (corrected) speed map obtained from PISCIT with GSM-based floating car data, resulting in a second posterior speed map, now incorporating the GSM data. The EGTF method then finally could be used to fuse this second speed map with any other data available (e.g., the average travel times from surveillance camera's), and—most importantly—to align this final posterior speed map with the space–time discretization required by the application (e.g., a coordinated ramp metering algorithm) and/or a model-based/KF state estimator and predictor. This KF estimator/predictor in turn can then be used to estimate underlying state variables (e.g., densities), and employed for state prediction and possibly the optimization of other traffic management measures.

7 Conclusion and Further Research Avenues

The three data fusion approaches discussed in this chapter provide robust solutions for fusing semantically different traffic data into a posterior estimate of traffic speeds. The methods solve the alignment problem associated with recursive data assimilation approaches such as (extended or unscented) Kalman filters. The PISCIT and FlowResTD methods both fuse spatial data (individual travel times and low-resolution floating car data, respectively) with an a priori speed contour map, constructed, for example, with local data. Both methods are robust to structural bias in those a priori speeds, which is critically important due to the fact that many real-world local sensors use (arithmetic) time averaging, which induces a significant bias. The EGTF approach in turn is able to fuse multiple data sources, as long as for each of these it is possible to deduce under which traffic regime (congested, free flowing) the data was collected. The algorithms are designed such that they can be used in a cascaded setting, each fusing an increasingly accurate posterior speed map with new data, which in the end could be used as input for a model-based/Kalman filter approach for traffic state estimation and prediction.

Further research could focus on finding the optimal combinations of these (and possibly other) methods, in terms of their parameter settings and mathematical structure, and preferably automated methods for their calibration and validation. Furthermore, their strengths and weaknesses should be thoroughly tested under different sensor combinations and network characteristics (e.g., freeway and urban traffic networks). Finally, research into the computational complexity in a real-world context (a traffic management center) should be investigated.

Acknowledgments. This work was partially supported by the Dutch Technology Foundation STW, Applied Science Division of NWO and the Technology Program of the Ministry of Economic Affairs under project DCB.7814 "Reversing the traffic jam".

References

1. Fosim (freeway operations simulation) (2008), `http://www.fosim.nl/IndexUK.html`
2. Papoulis, A.: Probability, Random Variables, and Stochastic Processes, 2nd edn., pp. 548–549. McGraw-Hill, New York (1984)
3. Dailey, D.J., Harn, P., Lin, P.J.: Its data fusion. Tech. Rep. Research Project T9903, Task 9, Washington State Transportation Commission, Department of Transportation and the U.S. Department of Transportation, Federal Highway Administration (1996)
4. Herrera, J.C., Bayen, A.M.: Traffic flow reconstruction using mobile sensors and loop detector data. In: Transportation Research Board 87th Annual Meeting, p. 18. Transportation Research Board (2008); 08-1868
5. Kerner, B.S.: The Physics of Traffic: Empirical Freeway Pattern Features, Engineering Applications, and Theory. Springer, Berlin (2004)
6. Kikuchi, S., Miljkovitc, D., van Zuylen, H.: Examination of methods that adjust observed traffic volumes on a network. Transport Research Record 1717, 109–119 (2000)
7. Knoop, V., Hoogendoorn, S., Zuylen, H.V.: Impact of data averaging methods on macroscopic traffic flow characteristics. In: Springer (ed.) Traffic and Granular Flow 2007. Springer, Heidelberg (2007) (to be published)
8. Leutzbach, W.: Introduction to the theory of traffic flow. Springer, Heidelberg (1987)
9. Lighthill, M., Whitham, G.: On kinematic waves ii: A theory of traffic flow on long crowded roads. Proc. R. Soc. A 229(1178), 317–345 (1955)
10. Linn, R.J., Hall, D.L.: A survey of multi-sensor data fusion systems. In: Proceedings of the SPIE - The International Society for Optical Engineering, Orlando, Florida, pp. 13–29 (1991)
11. Ni, D., Wang, H.: Trajectory reconstruction for travel time estimation. Journal of Intelligent Transportation Systems 12(3), 113–125 (2008)
12. Ou, Q., Van Lint, J., Hoogendoorn, S.: Piecewise inverse speed correction by using individual travel times. Transportation Research Record: Journal of the Transportation Research Board 2049(-1), 92–102 (2008); 10.3141/2049-11
13. Piella, G.: A general framework for multiresolution image fusion: from pixels to regions. Information Fusion 4(4), 259–280 (2003)
14. PTV: Vissim traffic microsimulation package (2005), `http://www.ptv.de`
15. Richards, P.: Shock waves on the highway. Operations Research 4, 42–51 (1956)
16. Simon, D.: Optimal State Estimation: Kalman, H-infinity, and Nonlinear Approaches. John Wiley & Sons, New York (2006)
17. Stipdonk, H., Toorenburg, J.v., Postema, M.: Phase diagram distortion from traffic parameter averaging. In: European Transport Conference (ETC), Leiden, The Netherlands (2008)
18. Treiber, M., Helbing, D.: Reconstructing the spatio-temporal traffic dynamics from stationary detector data. Cooper@tive Tr@nsport@tion Dyn@mics 1, 3.1–3.24 (2002)
19. Van Lint, J.W.C., Hoogendoorn, S.: A robust and efficient method for fusing heterogeneous data from traffic sensors on freeways. Computer-Aided Civil and Infrastructure Engineering 24, 1–17 (2009)

20. Van Lint, J., Hoogendoorn, S.P.: The technical and economic benefits of data fusion for real-time monitoring of freeway traffic. In: World Congress of Transportation Research. WCTRS, Berkely, CA, USA (2007)
21. Van Lint, J.W.C.: Reliable real-time framework for short-term freeway travel time prediction. Journal of transportation engineering-asce 132(12), 921–932 (2006)
22. Van Lint, J.W.C., Hoogendoorn, S.P., Hegyi, A.: Dual ekf state and parameter estimation in multi-class first-order traffic flow models. In: Proceedings of the 17th IFAC (International Federation of Automatic Control) World Congress, Seoul, Korea (2008)
23. Van Lint, J.W.C., Hoogendoorn, S.P., Van Zuylen, H.J.: Accurate travel time prediction with state-space neural networks under missing data. Transportation Research Part C: Emerging Technologies 13(5-6), 347–369 (2005)
24. Van Lint, J.W.C., Van der Zijpp, N.J.: Improving a travel time estimation algorithm by using dual loop detectors. Transportation Research Record 1855, 41–48 (2003)
25. Varshney, P.K.: Multisensor data fusion. Electronics and Communications Engineering Journal, 245–253 (December 1997)
26. Vermijs, R.G.M.M., Schuurman, H.: Evaluating capacity of freeway weaving sections and on-ramps using the microscopic simulation model fosim. In: Proceedings of the second international symposium on highway capacity, Sydney, Australia, vol. 2, pp. 651–670 (1994)
27. Wang, Y., Papageorgiou, M.: Real-time freeway traffic state estimation based on extended kalman filter: a general approach. Transportation Research Part B 39, 141–167 (2005)
28. Wang, Y., Papageorgiou, M., Messmer, A.: A real time freeway network traffic surveillance tool. IEEE Transactions on control systems technology 14(1) (2006)
29. Xiong, N., Svensson, P.: Multi-sensor management for information fusion: issues and approaches. Information Fusion 3(2), 163–186 (2002)
30. Yager, R.R.: A framework for multi-source data fusion. Information Sciences 163(1-3), 175–200 (2004)

Bayesian Networks for Expert Systems: Theory and Practical Applications

Wim Wiegerinck, Bert Kappen, and Willem Burgers

Abstract. Bayesian networks are widely accepted as models for reasoning with uncertainty. In this chapter, we focus on models that are created using domain expertise only. After a short review of Bayesian network models and common Bayesian network modeling approaches, we will discuss in more detail three applications of Bayesian networks. With these applications, we aim to illustrate the modeling power and flexibility of the Bayesian networks, which go beyond the standard textbook applications. The first network is applied in a system for medical diagnostic decision support. A distinguishing feature of this network is the large amount of variables in the model. The second one involves an application for petrophysical decision support to determine the mineral content of a well, based on borehole measurements. This model differs from standard Bayesian networks in terms of its continuous variables and nonlinear relations. Finally, we will discuss an application for victim identification by kinship analysis based on DNA profiles. The distinguishing feature in this application is that Bayesian networks are generated and computed on-the-fly based on case information.

1 Introduction

In modeling intelligent systems for real world applications, one inevitably has to deal with uncertainty. This uncertainty is due to the impossibility to model all the

Wim Wiegerinck
SNN Adaptive Intelligence, Geert Grooteplein 21, 6525 EZ Nijmegen, The Netherlands
e-mail: w.wiegerinck@science.ru.nl

Bert Kappen
Radboud University Nijmegen, Donders Institute for Brain, Cognition and Behaviour, Geert Grooteplein 21, 6525 EZ Nijmegen, The Netherlands
e-mail: b.kappen@science.ru.nl

Willem Burgers
SNN Adaptive Intelligence, Geert Grooteplein 21, 6525 EZ Nijmegen, The Netherlands
e-mail: w.burgers@science.ru.nl

R. Babuška & F.C.A. Groen (Eds.): Interactive Collaborative Information Systems, SCI 281, pp. 547–578.
springerlink.com

different conditions and exceptions that can underlie a finite set of observations. Probability theory provides the mathematically consistent framework to quantify and to compute with uncertainty. In principle, a probabilistic model assigns a probability to each of its possible states. In models for real world applications, the number of states is so large that a sparse model representation is inevitable. A general class with a representation that allows modeling with many variables are the Bayesian networks [7, 14, 20].

Bayesian networks are nowadays well established as a modeling tool for expert systems in domains with uncertainty [22]. Reasons are their powerful but conceptual transparent representation for probabilistic models in terms of a network. Their graphical representation, showing the conditional independencies between variables, is easy to understand for humans. On the other hand, since a Bayesian network uniquely defines a joint probability model, inference—drawing conclusions based on observations—is based on the solid rules of probability calculus. This implies that the mathematical consistency and correctness of inference are guaranteed. In other words, all assumptions in the method are contained in model, i.e., the definition of variables, the graphical structure, and the parameters. The method has no hidden assumptions in the inference rules. This is unlike other types of reasoning systems such as e.g., Certainty Factors (CFs) that were used in e.g., MYCIN—a medical expert system developed in the early 1970s [24]. In the CF framework, the model is specified in terms of a number of if-then-else rules with certainty factors. The CF framework provides prescriptions how to invert and/or combine these if-then-else rules to do inference. These prescriptions contain implicit conditional independence assumptions which are not immediately clear from the model specification and have consequences in their application [13].

Probabilistic inference is the problem of computing the posterior probabilities of unobserved model variables given the observations of other model variables. For instance in a model for medical diagnoses, given that the patient has complaints x and y, what is the probability that he/she has disease z? Inference in a probabilistic model involve summations or integrals over possible states in the model. In a realistic application, the number of states to sum over can be very large. In the medical example, the sum is typically over not only all combinations of unobserved factors that could influence the disease probability, such as different patient conditions, risk factors, but also alternative explanations for the complaints, etc. In general these computations are intractable. Fortunately, in Bayesian networks with a sparse graphical structure and with variables that can assume a small number of states, efficient inference algorithms exists such as the junction tree algorithm [7, 14].

The specification of a Bayesian network can be described in two parts, a qualitative and a quantitative part. The qualitative part is the graph structure of the network. The quantitative part consists of specification of the conditional probability tables or distributions. Ideally, both specifications are inferred from data [15]. In practice, however, data are often insufficient even for the quantitative part of the specification. The alternative is to do the specification of both parts by hand, in collaboration with domain experts. Many Bayesian networks are created in this way. Furthermore, Bayesian networks are often developed with the use of software packages such as

Hugin (`www.hugin.com`) or Netica (`www.norsys.com`). These packages typically contain a graphical user interface (GUI) for modeling and an inference engine based on the junction tree algorithm for computation.

Although the networks created in this way can be quite complex, the scope of these software packages obviously has its limitations. In this chapter, we discuss three models in which the standard approach to Bayesian modeling as outlined above was infeasible for different reasons: the large number of variables in the first model, the need to model continuous-valued variables in the second model, and the need to create models on-the-fly from data in the third application.

The first model has been developed for an application for medical diagnostic decision support (Promedas, in collaboration with UMC Utrecht). The main functionality of the application is to list the most probable diseases given the patient-findings (complaints, tests, physical examinations etc.) that are entered. The system is aimed to support diagnosis in general internal medicine, covering a large medical domain with several specializations. However, a considerable level of detail at which the disease areas are modeled is essential for the system to be of practical use. For this application, this means that the model should contain thousands of diseases and a factor 10 more of relations between diseases and findings. With such numbers of variables and relations, the standard modeling approach is infeasible.

The second model has been developed for an application for petrophysical decision support (in collaboration with SHELL E&P). The main function of this application is to provide a probability distribution of the mineral composition of a potential reservoir based on remote borehole measurements. In the underlying model, the number of variables is limited. However, variables are continuous valued. One of them represents the volume fractions of 13 minerals, and is therefore a 13-D continuous variable. Any sensible discretization in a standard Bayesian network approach would lead to a blow up of the state space. Owing to nonlinearities and constraints, a Bayesian network with linear-Gaussian distributions [3] is also not a solution.

Finally, we will discuss an application for victim identification by kinship analysis based on DNA profiles (Bonaparte, in collaboration with NFI). Victims should be matched with missing persons in a pedigree of family members. In this application, the model follows from Mendelian laws of genetic inheritance and from principles in DNA profiling. Inference needs some preprocessing but is otherwise reasonably straightforward. In this application, however, the challenge is that the model structure depends on the family structure of the missing person. This structure will differ from case to case and a standard approach with a static network is obviously insufficient. In this application, modeling is implemented in the engine. The application generates Bayesian networks on-the-fly based on case information. Next, it does the required inferences for the matches.

The chapter is organized as follows. First, we will provide a short review of Bayesian networks in section 2. Next, in sections 3, 4, and 5, we will discuss the three applications. In particular, we will discuss the underlying Bayesian network models and the modeling approaches at a rather detailed level. Furthermore, we will discuss the inference methods that we applied whenever they deviate from the standard junction tree approach. In section 6, we will end with discussion and conclusion.

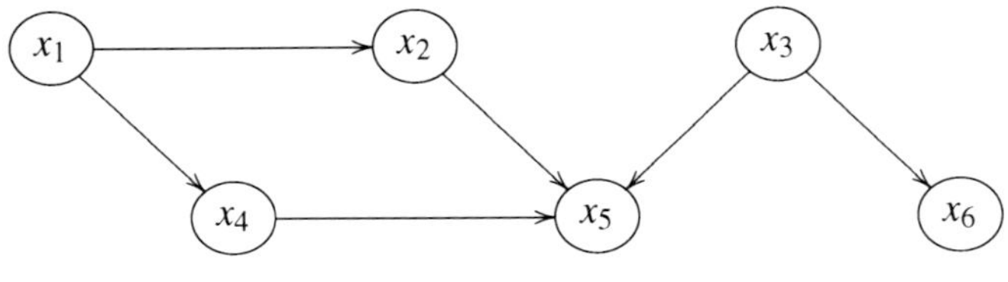

Fig. 1 DAG representing a Bayesian network $P(x_1)P(x_2|x_1)P(x_3)P(x_4|x_1)P(x_5|x_2,x_3,x_4)$ $P(x_6|x_3)$

2 Bayesian Networks

In this section, we first give a short and rather informal review of the theory of Bayesian networks (subsection 2.1). Furthermore, in subsection 2.2, we briefly discuss Bayesian networks modeling techniques, and in particular, the typical approach that is taken in most Bayesian network applications. We briefly discuss pros and cons of this approach, and in particular, why this approach does not work in the applications that we will discuss in the later sections.

2.1 Bayesian Network Theory

In order to introduce notation, we start by considering a joint probability distribution, or probabilistic model, $P(X_1,\ldots,X_n)$ of n stochastic variables $X_1,\ldots,X_n$. Variables X_j can be in state x_j. A state, or value, is a realization of a variable. We use shorthand notation

$$P(X_1 = x_1,\ldots,X_n = x_n) = P(x_1,\ldots,x_n) \tag{1}$$

to denote the probability (in continuous domains: the probability density) of variables X_1 in state x_1, variable X_2 in state x_2 etc.

A Bayesian network is a probabilistic model P on a finite directed acyclic graph (DAG). For each node i in the graph, there is a random variable X_i together with a conditional probability distribution $P(x_i|x_{\pi(i)})$, where $\pi(i)$ are the parents of i in the DAG, see figure 1. The joint probability distribution of the Bayesian network is the product of the conditional probability distributions

$$P(x_1,\ldots,x_n) = \prod_{i=1}^{n} P(x_i|x_{\pi(i)})\,. \tag{2}$$

Since any DAG can be ordered such that $\pi(i) \subseteq 1,\ldots i-1$ and any joint distribution can be written as

$$P(x_1,\ldots,x_n) = \prod_{i=1}^{n} P(x_i|x_{i-1},\ldots,x_1)\,, \tag{3}$$

it can be concluded that a Bayesian network assumes

$$P(x_i|x_{i-1},\ldots,x_1) = P(x_i|x_{\pi(i)}) \,. \tag{4}$$

In other words, the model assumes: given the values of the direct parents of a variable X_i, this variable X_i is independent of all its other predecessing variables in the graph.

Since a Bayesian network is a probabilistic model, one can compute marginal distributions and conditional distributions by applying the standard rules of probability calculus. For instance, in a model with discrete variables, the marginal distribution of variable X_i is given by

$$P(x_i) = \sum_{x_1}\cdots\sum_{x_{i-1}}\sum_{x_{i+1}}\cdots\sum_{x_N} P(x_1,\ldots,x_N) \,. \tag{5}$$

Conditional distributions such as $P(x_i|x_j)$ are obtained by the division of two marginal distributions

$$P(x_i|x_j) = \frac{P(x_i,x_j)}{P(x_j)} \,. \tag{6}$$

The bottleneck in the computation is the sum over combinations of states in (5). The number of combinations is exponential in the number of variables. A straightforward computation of the sum is therefore only feasible in models with a small number of variables. In sparse Bayesian networks with discrete variables, efficient algorithms that exploit the graphical structure, such as the junction tree algorithm [7, 14, 16] can be applied to compute marginal and conditional distributions. In more general models, exact inference is infeasible and approximate methods such as sampling have to be applied [3, 17].

2.2 *Bayesian Network Modeling*

The construction of a Bayesian network consists of deciding about the domain, what are the variables that are to be modeled, and what are the state spaces of each of the variables. Then the relations between the variables have to be modeled. If these are to be determined by hand (rather than by data), then it is a good rule of thumb to construct a Bayesian network from cause to effect. Start with nodes that represent independent root causes, then model the nodes which they influence, and so on until we end at the leaves, i.e., the nodes that have no direct influence on other nodes. Such a procedure often results in sparse network structures that are understandable for humans [22].

Often, models are constructed using Bayesian network software such as the earlier mentioned packages. With the use of a graphical user interface (GUI), nodes can be created. The nodes represent the variables in the system. Typically, variables can assume only values from a finite set. When a node is created, it can be linked to other nodes, under the constraint that there are no directed loops in the network. Finally—or during this process—the table of conditional probabilities are defined, often by educated guesses, and sometimes inferred from data. Many

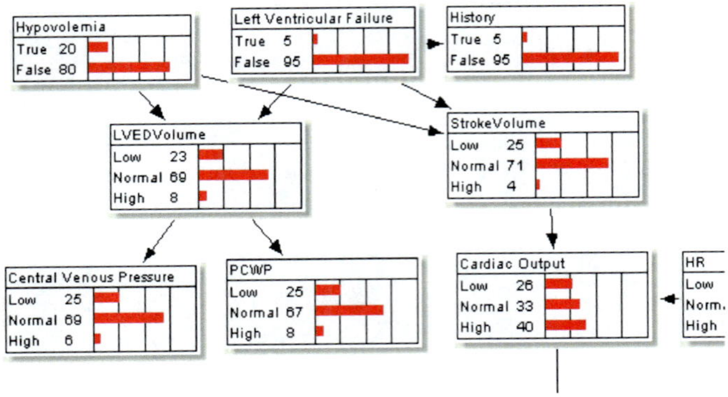

Fig. 2 Screen shot of part of the 'Alarm network' in the BayesBuilder GUI

Bayesian networks that are found in literature fall into this class, see e.g., `www.norsys.com/netlibrary/`. In figure 2, a part of the ALARM network as represented in BayesBuilder (`www.snn.ru.nl/`) is plotted. The ALARM network was originally designed as a network for monitoring patients in intensive care [2]. It consists of 37 variables, each with 2, 3, or 4 states. It can be considered as a relatively large member of this class of models. An advantage of the GUI-based approach is that a small or medium sized Bayesian network, i.e., with up to a few dozen of variables, where each variable can assume a few states, can be developed quickly, without the need of expertise on Bayesian networks modeling or inference algorithms.

In the following sections, we will discuss three Bayesian networks for real world applications that fall outside the class of models that have been built using these modeling tools. The main reason is that the GUI has no added value for these models. The first model is too complex, and would contain too many variables for the GUI. In the second one, the complexity is more in the variables themselves than in the network structure. In the third model, the network consists of a few types of nodes that have simple and well-defined relations among each other. However, for each different case in the application, a different network has to be generated. It does not make sense for this application to try to build these networks beforehand in a GUI.

3 Promedas: A Probabilistic Model for Medical Diagnostic Decision Support

Modern-day medical diagnosis is a very complex process, requiring accurate patient data, a profound understanding of the medical literature and many years of clinical experience. This situation applies particularly to internal medicine, because it covers an enormous range of diagnostic categories. As a result, internal medicine is differentiated in super-specializations.

Diagnosis is a process, by which a doctor searches for the cause (usually a disease) that best explains the symptoms of a patient. The search process is sequential, in the sense that patient symptoms suggest some initial tests to be performed. Based on the outcome of these tests, a tentative hypothesis is formulated about the possible cause(s). Based on this hypothesis, subsequent tests are ordered to confirm or reject this hypothesis. The process may proceed in several iterations until the patient is finally diagnosed with sufficient certainty and the cause of the symptoms is established.

A significant part of the diagnostic process is standardized in the form of protocols which are sets of rules that prescribe which tests to perform and in which order, based on the patient symptoms and previous test results. These rules form a decision tree, whose nodes are intermediate stages in the diagnostic process and whose branches point to additional testing, depending on the current test results. The protocols are defined in each country by a committee of medical experts.

In the majority of the diagnoses that are encountered, the guidelines are sufficiently accurate to make the correct diagnosis. For these "routine" cases, a decision support system is not needed. In 10–20 % of the cases, however, the diagnostic process is more difficult. As a result of the uncertainty about the correct diagnosis and about the next actions to perform, the decisions made by different physicians at different stages of the diagnostic process do not always agree and lack "rationalization." In these cases, normally a particularly specialized colleague or the literature is consulted. For these difficult cases, computer-based decision support may serve as an alternative source of information. In addition, a computer-aided decision support system can be of help by pointing to alternative diagnoses that may be overlooked otherwise. It may thus result in an improved and more rationalized diagnostic process, as well as higher efficiency and cost-effectiveness.

Since 1996, SNN and UMC Utrecht have been developing a clinical diagnostic decision support system for internal medicine, called Promedas. In this system, patient information, such as age and gender, and findings, such as symptoms, results from physical examination, and laboratory tests can be entered. The system then generates patient-specific diagnostic advice in the form of a list of likely diagnoses and suggestions for additional laboratory tests that may be relevant for a selected diagnosis.

The system is intended to support diagnostics in the setting of the outpatient clinic and for educational purposes. Its target users are general internists, super specialists (e.g., endocrinologists, rheumatologists, etc.), interns and residents, medical students, and others working in the hospital environment. Currently, a trial version of the program is installed at department of internal medicine in UMC Utrecht. It contains about 3500 diagnoses and is based on 50000 relations. The program is connected to the electronic patient records, so that physicians can easily consult the program without having to enter all the data manually. A live demo can be found on `www.promedas.nl`

Promedas is based on a Bayesian network. In the remainder of the section, we will describe the model in further detail. We focus on the modeling part, including certain modeling approaches, model choices, and methods to facilitate inference. Medical details of the model are outside the scope of this section.

3.1 Building Large Scale Probabilistic Models

For this application, in which rare diseases play an important role, data are insufficient to train the model. When modeling a Bayesian network by hand, the standard procedure is to specify a network structure of local interactions and to specify those probabilities that are needed to define these interactions quantitatively. For medium sized networks (up to 50 – 100 variables), this is doable using the methodology and Bayesian network software tools such as those discussed in subsection 2.2. However, our aim was to scale up the system to thousands of variables. For larger systems, it is more difficult to keep overview, and not to get lost in the spaghetti of relations and interactions. In addition, available medical knowledge is in general limited to bivariate relations between disease and test in terms of sensitivity and specificity. Therefore, we decided to take a more structured approach in which we assume a generic structure of the model. The general assumption in this structure is that risk factors influence the probabilities of diseases and that diseases influence the probabilities of findings (symptoms, tests etc.). We furthermore restrict to models in which the parameters can be determined from the available medical knowledge of bivariate relations. In order to further facilitate modeling, we have developed a database in which medical specialists can enter their knowledge in a structured and not too complicated way.

In the following, we sketch the structure of the database. Then, we sketch how the Bayesian network is defined and which model choices we have made. Finally, we sketch how a differential diagnosis is computed in this model.

3.1.1 Database Structure

The database contains information from which the structure of the network can be derived as well as its model parameters. In addition, the database contains meta-information, such as information about the structure of Promedas' graphical user interface. This involves mainly the grouping and naming of findings and risk factors into medical relevant categories such as complaints, physical examination, medication, lab results, and subdivisions of these. In addition, descriptions, annotations, and references are included. In the remainder of this subsection, however, we restrict to information that is directly relevant for the computational model.

The database contains three types of variables:

1. *Risk factors* such as occupation, drug use, and past and concurrent diseases; Risk factors are coded binary (true=1/false=0).
2. *Diagnoses* such as current diseases, syndromes, drug side effects, pregnancy; Diagnoses are coded binary (true=1/false=0).
3. *Tests* or findings, such as lab tests, symptoms, physical examination etc. Tests are binary or multinomial (decreased/normal/increased/strongly increased, etc.). When the discretization is not obvious because the test is continuous

by nature, then the discretization is defined in the database with cut-off points according to medical standards where possible. Discretization may depend on gender and age. The state space of the tests is such that there is always one "normal" state. Binary variables are defined such that false is the "normal" state.

Furthermore, the database contains quantifications. These are needed to model the probabilities in the Bayesian network. Quantifications can apply to single variables, and to relations between variables. Relations can be defined between risk factors and diagnoses and between tests and diagnoses. Relations can only be defined for non-normal states, e.g., between diagnosis d being true and test t being in "increased" state. The idea behind this is that relations code positive influences. The absence of the relation between diagnosis d being true and test t in "normal" state implies the assumption that the mere presence of a disease will never make the result of a test more likely to be normal than without the disease being present.

The database contains four types of quantifications:

1. *Priors.* For each diagnosis d there is prior p_d. This is the prior probability of diagnosis d being true in absence of all risk factors.
2. *Leaks.* For each test, there is the so-called leak $p_{t=s}$ of each non-normal test-state. This leak is roughly interpreted as the prior probability of the test being in state $t = s$ in absence of all diagnoses. In an ideal test, the result is normal in absence of diagnoses, so any non-normal state has zero probability. In non-ideal tests, probability is leaked to non-normal test states. Leaks are used e.g., to model the probability of a test being positive without apparent cause.
3. *Mult-factors.* For each risk–diagnosis relation there is a "mult-factor" m_{dr} by which the odds of the prior probability of diagnosis d are multiplied in the presence of the risk factor r.
4. *Senses.* For each test–diagnosis relation, there is one or more "sens" $p_{dt=s}$. A sens relates a diagnosis to a non-normal test-state. This is the probability that the presence of the disease d causes the test t to be in state s (rather than the leak or other diseases). The "'sens" is closely related to sensitivity, the probability of a positive test given the presence of the disease d (regardless the leak or other diseases).

These quantifications can be age and gender dependent.

3.1.2 Network Definition

The global architecture of the diagnostic model is described by a diagnosis-layer that is connected to a layer with tests. The main assumption is that different diagnoses can coexist. Note that there are no nodes for gender, age, and risk-factors. These are assumed to be observed. All other probabilities in the network are conditioned on these observations (as in e.g., (8), below). Default case is a male of 55 with all the risk-factors false. The global architecture of Promedas is similar to the QMR-DT network [25]. QMR stands for Quick Medical Reference, which is a heuristic representation with about 600 diseases and 4000 findings. The QMR-DT network, where DT stands for Decision Theoretic, is a reformulation as a two-layer Bayesian

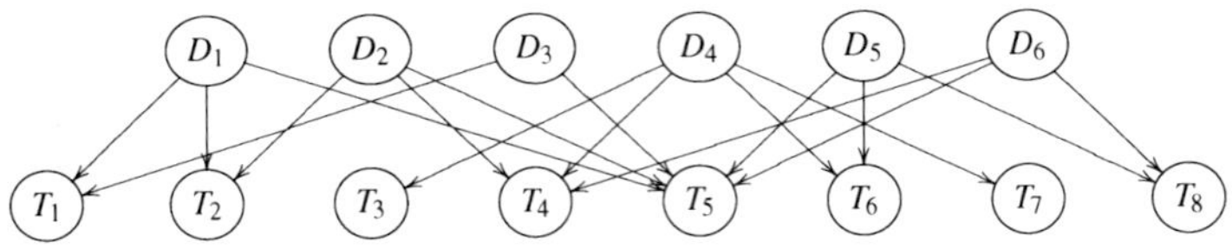

Fig. 3 Network structure in the Promedas model

network. Main differences with Promedas are the absorption of risk factors, and the modeling of multivalued tests in Promedas rather than the binary tests in QMR-DT. Furthermore, Promedas is based on a different knowledge base.

Diagnoses are modeled as a priori independent binary variables. Their prior probabilities (in the absence of risk factors) are read from the database. In the case that a risk factor is set to true, $r = 1$, the prior of a related diagnosis is affected according to a multiplication of prior odds,

$$\frac{P(d=1|r=1)}{P(d=0|r=1)} = m_{dr}\frac{P(d=1|r=0)}{P(d=0|r=0)}\,, \tag{7}$$

where m_{rd} is the "mult-factor" of risk factor r in relation to diagnosis d. This implies, after rearranging terms

$$P(d=1|r=1) = \frac{m_{rd}P(d=1|r=0)}{1+(m_{rd}-1)P(d=1|r=0)}\,. \tag{8}$$

The conditional distributions of tests are modeled using the so-called noisy-OR and noisy-MAX gates [21]. Both will be explained below in more detail. The motivation to use these table parameterizations is that they are convenient to model because there is only one (or a few) parameter(s) for each diagnosis–test relation (rather than exponentially many as in the free form table), while on the other hand, they provide a medically reasonable model that is easy to interpret [25]. Another important reason is that inference is efficient [27] as we will discuss later in this section.

In order to construct the noisy-OR and noisy-MAX, we first consider the deterministic OR-gate $OR(v|u_0,\dots,u_n)$. Here, v and u_i are binary variables.

$$OR(v|u_0,\dots,u_n) = \begin{cases} 1 & \text{if } v = \max(u_0,\dots,u_n) \\ 0 & \text{otherwise} \end{cases}\,. \tag{9}$$

So $v = 1$ (true) if any of the u_i's is 1. Otherwise $v = 0$. Now the noisy-OR gate is modeled as follows (v, u_i and d_i are binary):

$$NoisyOR(v|d_1,\dots d_n) = \sum_{\{u_0,\dots,u_n\}} OR(v|u_0,\dots,u_n)\prod_{i=1}^{n} P(u_i|d_i)P(u_0)\,. \tag{10}$$

The variables $u_0,\dots,u_n$ can be considered as latent or auxiliary variables in this model. Furthermore, the probabilities $P(u_i = 1|d_i = 0)$ are zero in this model. The probability $P(u_0 = 1)$ is often called the "leak". The interpretation is that noisy-OR is a noisy version of the deterministic OR, in which there is a finite probability that (1) although all inputs $d_i = 0$, the outcome is $v = 1$ due to the leak, and (2) although there are inputs $d_i = 1$, the outcome is $v = 0$ due to the fact that $P(u_i = 0|d_i = 1)$ is non-zero. However, the more the inputs $d_i = 1$, the higher the probability that the outcome is $v = 1$. In Promedas, noisy-ORs are applied for binary tests: d_i are the disease states, and v is the test result. The more the diseases are present, the higher the probability of a positive test result. The required probabilities to model the noisy-ORs are read from the database (leaks and senses).

Now we will construct noisy-MAX. The idea is similar as the noisy-OR gate, with an additional a winner-take-all mechanism. The idea is that if some diseases cause a test result to have a slightly increased value, and other diseases cause a test result to have a strongly increased value, then the observed test result will be strongly increased. In order to proceed, we order the states of the test $s_0 < s_1 < \dots s_K$, where "normal" has the lowest order (so $s_0 =$ "normal"). Next, in order to model diseases causing the test result to have a certain value, we define a noisy-OR gate NOR_j for each of the test-values $s_j > s_0$ (except for the "'normal" value, since diagnoses cannot cause values to be normal). The outcome of a noisy-OR gate is either 1 or 0. The outcomes of NOR_j are relabeled ($0 \rightarrow s_0$ and $1 \rightarrow s_j$), and the result is either s_0 or the value s_j.

The winner-take-all mechanism is modeled by the deterministic MAX-gate $MAX(t|v_1,\dots,v_n)$. The variable t can assume all the potential values of its parent variables. The MAX-gate is defined as

$$MAX(t|v_1,\dots,v_n) = \begin{cases} 1 & \text{if } t = \max(v_1,\dots,v_n) \\ 0 & \text{otherwise} \end{cases} \tag{11}$$

Noisy-MAX tables for tests $P(t|d_1,\dots d_n)$ can be represented by NOR_j's for each of the test-values s_j, having subsets $d_{j1},\dots,d_{jn_j}$ of diagnoses that are related to test-state $t = s_j$ as parents, combined with a deterministic MAX-gate for the winner-take-all mechanism (see figure 3),

$$P(t|d_1,\dots d_n) = \sum_{\{v_1,\dots,v_K\}} MAX(t|v_1,\dots,v_K) \prod_{j=1}^{K} NOR_j(v_j|d_{j1},\dots,d_{jn_j}) \,. \tag{12}$$

The interpretation of the noisy-MAX model is as follows. Each of the diseases has a probability to trigger the test to be in a certain state, regardless of the presence or absence of other diseases. If different diseases have a probability to trigger the test to be in the same state, then a combination of them makes this state more likely. If different diseases trigger the test to be in different states, then the strongest state is observed. For instance, if one disease triggers the body temperature to be "increased" and another triggers the temperature to be "strongly increased", then the model assumption is that the "strongly increased" temperature will be observed. A drawback may be that many causes of an "increased" temperature would in reality have an additive effect. Other models could be designed to incorporate such effect. However, such models would lack the crucial computational efficiency of the noisy-MAX

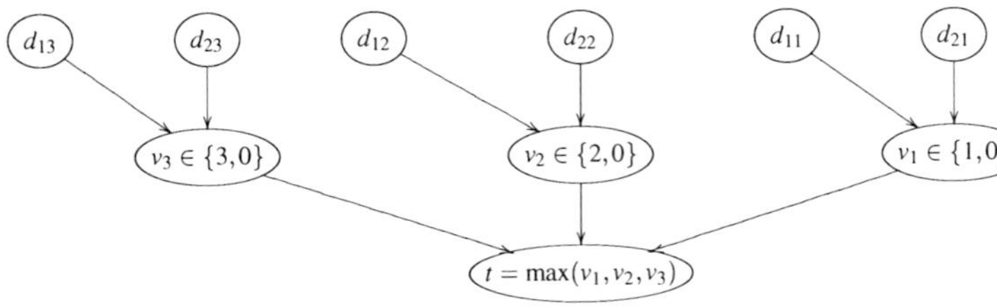

Fig. 4 Test t with ordered states $0 < 1 < 2 < 3$ are modeled as a noisy-MAX, which can itself be modeled as the MAX of the outcomes of three noisy-OR gates. In this example, diagnoses d_{ij} are connected to binary noisy-OR gates NOR_j. The outcome of a noisy-OR gate is either 1 or 0. The outcomes of NOR_j are relabeled $(0/1 \to 0/j)$ and subsequently fed into a MAX gate, which returns the maximum value.

model. Another issue that one could discuss is what to do with tests that have positive and negative states, such as "decreased","normal", "increased". Again, other models could be designed to better incorporate the combination of a "decreased" and an "increased" effect, but this would also be at the expense of computational efficiency. In Promedas, we decided to be pragmatic and enforce an ordering.

3.2 *Inference*

The main inference task in the application is to compute the probabilities of diagnoses given the observed values of tests and risk factors. In general, inference would be computationally infeasible for networks of the size of Promedas. Therefore simplifying assumptions are introduced to make the inference task cheaper. One assumption is that all risk factors are assumed to be observed (in the application, their default value is false). This excludes any uncertainty in these variables. In this way, there will be no correlations between diagnoses through risk factors. Another simplification is to take only diagnoses into account which are connected to at least one test-node that is observed to be in a non-normal state. Other diagnoses are not of interest in the task of supporting the physician.

3.2.1 Efficient Inference in Noisy-MAX

Another assumption is the noisy-MAX model. As we mentioned earlier, one of the reasons to adopt this model is that inference is more efficient. There are a several properties of this model that make inference more efficient than in most other conditional probability models. See e.g., [27] for a more detailed and exposure of a general class of such models.

- *Decoupling of the parents of MAX.* If we apply the max operator over a set of variables v_i, where each v_i can have either value s_0 or s_i, with $s_0 < \ldots < s_K$, then an outcome $\max(v_1, \ldots, v_K) = s_j$ implies that all $v_k = s_0$ for $k > j$. Furthermore $v_j = s_j$ if $s_j > s_0$. The outcome does not contain any information about the

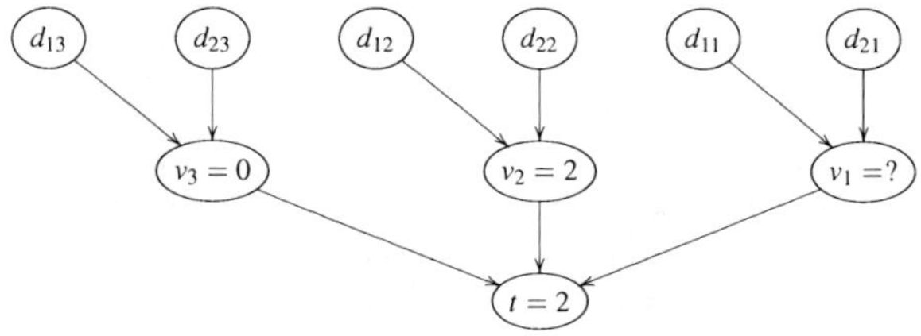

Fig. 5 Inference with noisy-MAX. Observed test value $t = 2$ implies that the outcome of $v_3 = 0$, and $v_2 = 2$. The observed test value does not give any information about v_1.

variables v_k with $k < j$. See Figure 5. This implies that we can take out the factor $MAX(t|v_1, \ldots, v_K)$ and decouple the intermediate variables as follows:

$$P(t = s_j|d_1, \ldots d_n) = \prod_{k=j+1}^{K} NOR_k(v_k = s_0|d_{k1}, \ldots, d_{kn_k})$$
$$\times \; NOR_k(v_j = s_j|d_{j1}, \ldots, d_{jn_j}) \prod_{k=1}^{j-1} \sum_{\{v_k\}} NOR_j(v_k|d_{k1}, \ldots, d_{kn_j}) \qquad (13)$$

- *Decoupling of the parents of OR with outcome "false"*. A related property is that observing that a variable modeled by a noisy-OR gate is equal to be zero, $v = 0$, implies that all states of the intermediate nodes in the noisy-OR $u_0, \ldots, u_n$ are zero. In other words, these can be considered as observed. We can remove the factor $OR(v = 0|u_0, \ldots, u_n)$ and decouple the diagnoses in (10),

$$NoisyOR(v = 0|d_1, \ldots d_n) = \prod_{i=1}^{n} P(u_i = 0|d_i) P(u_0 = 0) \,. \qquad (14)$$

- *Undirected links of OR with outcome "true"*. Straightforward expansion of *OR* leads to

$$OR(v = 1|u_0, \ldots, u_n) = 1 - \prod_{i=0}^{n} \delta_{u_i 0} \,. \qquad (15)$$

In order to rewrite this expression, we define the auxiliary potential ψ

$$\psi(u_0, z = 0) = -\delta_{u_0 0} \,, \qquad (16)$$
$$\psi(u_i, z = 0) = \delta_{u_i 0} \quad \text{for } i > 0, \qquad (17)$$
$$\psi(u_i, z = 1) = 1 \,, \qquad (18)$$

where z is an auxiliary switch variable. Note that $\psi(u_0, z = 0)$ is negative! With these potentials, we can decompose the *OR* as a sum-product,

$$OR(v = 1|u_0, \ldots, u_n) = \sum_{\{z\}} \prod_{i=0}^{n} \psi(u_i, z) \,, \qquad (19)$$

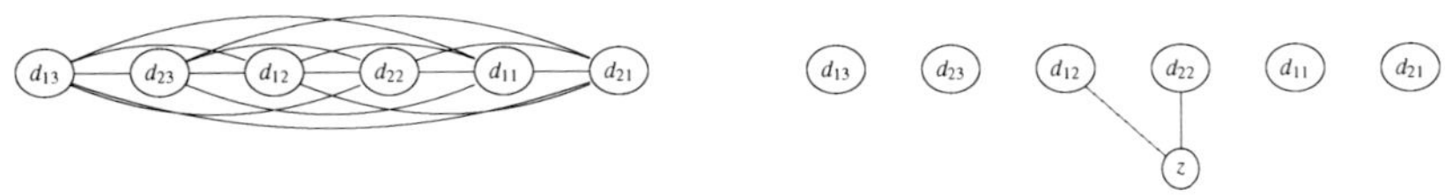

Fig. 6 Inference with noisy-MAX. Graphical structure of the undirected (moral) graph on the diagnoses which results from absorbing the evidence of observed test value $t = 2$. Left: with noisy-MAX modeled as a free form conditional probability table all the parents are connected. Right: exploiting the structure of noisy-MAX results in a much more sparse representation. z is the auxiliary switch variable, see text.

and hence, using now the auxiliary potentials defined by

$$\phi(z=0) = P(u_0 = 1) - 1 , \tag{20}$$

$$\phi(z=1) = 1 , \tag{21}$$

$$\phi(d_i,z=0) = 1 - P(u_i = 1|d_i) , \tag{22}$$

$$\phi(d_i,z=1) = 1 , \tag{23}$$

the noisy-OR decomposes as

$$NoisyOR(v = 1|d_1,\ldots,d_n) = \sum_{\{z\}} \phi(z) \prod_{i=1}^{n} \phi(d_i,z) . \tag{24}$$

The use of these potentials in general lead to a much smaller clique-size in the junction tree algorithm, see figure 6.

Inference in Promedas is now performed as follows. Given a set of test values, the diagnoses nodes that are related to at least one non-normal test value are selected. For these diagnoses, the present risk-factors and the evidences of the test-state-variables v_j are collected. The risk-factors and test-state variables in normal state $v_j = s_0$ are directly absorbed in the priors of diagnoses using the mult factors and the senses in the database. The non-trivial part of the computation are the test-state variables in non-normal state $v_j = s_j$ that are created for each non-normal test value $t = s_j$. For these variables, undirected noisy-OR structures as in (24) are constructed using senses and leaks from the database. Standard junction tree algorithm is applied to the resulting undirected model (note that in undirected graphs, there is no coupling of the parents as preprocessing for the junction tree algorithm. In *directed* graphs, there is. This coupling is known as moralization and leads to larger cliques). The posterior probabilities of the selected diagnosis are computed and reported as the differential diagnosis (a list of the most probable diagnoses) for the case at hand.

3.3 The Current Application

Promedas has been further developed by Promedas B.V. Additional methods to further speed up have been implemented. However, these are outside the scope of this chapter. A live demo can be found on `www.promedas.nl`.

3.4 Summary

Promedas is an application for medical diagnostic decision support. Its primary aim is to find a differential diagnosis based on test results (anamnesis, physical examination, lab-tests, etc.) . Given the large number of variables, a conventional Bayesian network approach is infeasible. We took a knowledge-based approach in which the network is compiled from a database of relations provided by medical experts. In order to make computation feasible, we designed a tractable model parameterization.

4 A Petrophysical Decision Support System

Oil and gas reservoirs are located in the earth's crust at depths of several kilometers, and when located offshore, in water depths of a few meters to a few kilometers. Consequently, the gathering of critical information such as the presence and type of hydrocarbons, size of the reservoir, and the physical properties of the reservoir such as the porosity of the rock and the permeability is a key activity in the oil and gas industry.

Pre-development methods to gather information on the nature of the reservoirs range from gravimetric, 2D, and 3D seismic to the drilling of exploration and appraisal boreholes. Additional information is obtained while a field is developed through data acquisition in new development wells drilled to produce hydrocarbons, time-lapse seismic surveys, and in-well monitoring of how the actual production of hydrocarbons affects physical properties such as the pressure and temperature. The purpose of information gathering is to decide which reservoirs can be developed economically, and how to adapt the means of development best to the particular nature of a reservoir.

The early measurements acquired in exploration, appraisal, and development boreholes are a crucial component of the information gathering process. These measurements are typically obtained from tools on the end of a wireline that are lowered into the borehole to measure the rock and fluid properties of the formation. Their is a vast range of possible measurement tools [23]. Some options are very expensive and may even risk other data acquisition options. In general, acquiring all possible data imposes too great an economic burden on the exploration, appraisal, and development. Hence, data acquisition options must be exercised carefully bearing in mind the learning of already acquired data and general hydrocarbon field knowledge. Also important is a clear understanding of what data can and cannot be acquired later and the consequences of having an incorrect understanding of the nature of a reservoir on the effectiveness of its development.

Making the right data acquisition decisions, as well as the best interpretation of information obtained in boreholes, forms one of the principal tasks of petrophysicists. The efficiency of a petrophysicist executing her/his task is substantially influenced by the ability to gauge her/his experience to the issues at hand. Efficiency is hampered when a petrophysicist's experience level is not yet fully sufficient and by the rather common circumstance that decisions to acquire particular types of information or not must be made in a rush, at high costs and shortly after receiving other information that impact on that very same decision. Mistakes are not entirely uncommon and almost always painful. In some cases, non-essential data are obtained at the expense of extremely high cost, or essential data are not obtained at all, causing development mistakes that can jeopardize the amount of hydrocarbon recoverable from a reservoir and induce significant cost increases.

The overall effectiveness of petrophysicists is expected to improve using a decision support system (DSS). In practice, a DSS can increase the petrophysicists' awareness of low probability but high impact cases and alleviate some of the operational decision pressure.

In cooperation with Shell E&P, SNN has developed a DSS tool based on a Bayesian network and an efficient sampler for inference. The main tasks of the application is the estimation of compositional volume fractions in a reservoir on the basis of measurement data. In addition, it provides insight into the effect of additional measurements. Besides an implementation of the model and the inference, the tool contains GUI in which the user can take different views on the sampled probability distribution and on the effect of additional measurements. The tool is currently under evaluation within Shell E&P.

In the remainder of this section, we will describe the Bayesian network approach for the DSS tool. We focus on our modeling and inference approach. A more detailed description of the model, in particular in relation to the petrophysical relevant quantities will be published elsewhere [5].

4.1 Probabilistic Modeling

The primary aim of the model is to estimate the compositional volume fractions of a reservoir on the basis of borehole measurements. Owing to incomplete knowledge, limited amount of measurements, and noise in the measurements, there will be uncertainty in the volume fractions. We will use Bayesian inference to deal with this uncertainty.

The starting point is a model for the probability distribution $P(\mathbf{v},\mathbf{m})$ of the compositional volume fractions $\mathbf{v}$ and borehole measurements $\mathbf{m}$. A causal argument *"The composition is given by the (unknown) volume fractions, and the volume fractions determine the distribution measurement outcomes of each of the tools"* leads us to a Bayesian network formulation of the probabilistic model,

$$P(\mathbf{v},\mathbf{m}) = \prod_{i=1}^{Z} P(m_i|\mathbf{v})P(\mathbf{v}) \,. \tag{25}$$

In this model, $P(\mathbf{v})$ is the so-called *prior*, the prior probability distribution of volume fractions before having seen any data. In principle, the prior encodes the generic geological and petrophysical knowledge and beliefs [26]. The factor $\prod_{i=1}^{Z} P(m_i|\mathbf{v})$ is the *observation model*. The observation model relates volume fractions $\mathbf{v}$ to measurement outcomes m_i of each of the Z tools i. The observation model assumes that *given* the underlying volume fractions, measurement outcomes of the different tools are independent. Each term in the observation model gives the probability density of observing outcome m_i for tool i given that the composition is $\mathbf{v}$. Now given a set of measurement outcomes $\mathbf{m}^o$ of a subset *Obs* of tools, the probability distribution of the volume fractions can be updated in a principled way by applying *Bayes' rule*,

$$P(\mathbf{v}|\mathbf{m}^o) = \frac{\prod_{i\in Obs} P(m_i^o|\mathbf{v})P(\mathbf{v})}{P(\mathbf{m}^o)} . \tag{26}$$

The updated distribution is called the *posterior* distribution. The constant in the denominator $P(\mathbf{m}^o) = \int_{\mathbf{v}} \prod_{i\in Obs} P(m_i^o|\mathbf{v})P(\mathbf{v})\mathrm{d}\mathbf{v}$ is called the *evidence*.

In our model, $\mathbf{v}$ is a 13 dimensional vector. Each component represents the volume fraction of one of 13 most common minerals and fluids (water, calcite, quartz, oil, etc.). So each component is bounded between zero and one. The components sum up to one. In other words, the volume fractions are confined to a simplex $\mathbb{S}^K = \{\mathbf{v}|0 \le v_j \le 1, \sum_k v_k = 1\}$. There are some additional physical constraints on the distribution of $\mathbf{v}$, for instance that the total amount of fluids should not exceed 40% of the total formation. The presence of more fluids would cause a collapse of the formation.

Each tool measurement gives a 1D continuous value. The relation between composition and measurement outcome is well understood. Based on the physics of the tools, petrophysicists have expressed these relations in terms of deterministic functions $f_j(\mathbf{v})$ that provide the idealized noiseless measurement outcomes of tool j given the composition $\mathbf{v}$ [26]. In general, the functions f_j are nonlinear. For most tools, the noise process is also reasonably well understood—and can be described by either a Gaussian (additive noise) or a log-Gaussian (multiplicative noise) distribution.

A straightforward approach to model a Bayesian network would be to discretize the variables and create conditional probability tables for priors and conditional distributions. However, due to the dimensionality of the volume fraction vector, any reasonable discretization would result in an infeasible large state space of this variable. We therefore decided to remain in the continuous domain.

The remainder of this section describes the prior and observation model, as well as the approximate inference method to obtain the posterior.

4.2 The Prior and the Observation Model

The model has two ingredients: the prior of the volume fractions $P(\mathbf{v})$ and the observation model $P(m_j|\mathbf{v})$.

There is not much detailed domain knowledge available about the prior distribution. Therefore, we decided to model the prior using conveniently parameterized family of distributions. In our case, $\mathbf{v} \in \mathbb{S}^K$, this lead to the Dirichlet distribution [3, 17]

$$Dir(\mathbf{v}|\alpha,\mu) \propto \prod_{j=1}^{K} v_j^{\alpha\mu_j - 1} \delta\left(1 - \sum_{i=1}^{K} v_i\right) . \tag{27}$$

The two parameters $\alpha \in \mathbb{R}_+$ (precision) and $\mu \in \mathbb{S}^K$ (vector of means) can be used to fine-tune the prior to our liking. The delta function—which ensures that the simplex constraint holds—is put here for clarity, but is in fact redundant if the model is constrained to $\mathbf{v} \in \mathbb{S}^K$. Additional information, e.g., the fact that the amount of fluids may not exceed 40% of the volume fraction can be incorporated by multiplying the prior by a likelihood term $\Phi(\mathbf{v})$ expressing this fact. The resulting prior is of the form

$$P(\mathbf{v}) \propto \Phi(\mathbf{v}) Dir(\mathbf{v}|\alpha,\mu) . \tag{28}$$

The other ingredient in the Bayesian network are the observation models. For most tools, the noise process is reasonably well understood and can be reasonably well described by either a Gaussian (additive noise) or a log-Gaussian (multiplicative noise) distribution. In the model, measurements are modeled as a deterministic tool function plus noise,

$$m_j = f_j(\mathbf{v}) + \xi_j , \tag{29}$$

in which the functions f_j are the deterministic tool functions provided by domain experts. For tools where the noise is multiplicative, a log transform is applied to the tool functions f_j and the measurement outcomes m_j. A detailed description of these functions is beyond the scope of this chapter. The noises ξ_j are Gaussian and have a tool-specific variance σ_j^2. These variances have been provided by domain experts. So, the observational probability models can be written as

$$P(m_j|\mathbf{v}) \propto \exp\left(-\frac{(m_j - f_j(\mathbf{v}))^2}{2\sigma_j^2}\right) . \tag{30}$$

4.3 Bayesian Inference

The next step is given a set of observations $\{m_i^o\}$, $i \in \text{Obs}$, to compute the posterior distribution. If we were able to find an expression for the evidence term, i.e., for the marginal distribution of the observations $P(\mathbf{m}^o) = \int_{\mathbf{v}} \prod_{i \in \text{Obs}} P(m_i^o|\mathbf{v}) P(\mathbf{v}) \mathrm{d}\mathbf{v}$, then the posterior distribution (26) could be written in closed form and readily evaluated. Unfortunately, $P(\mathbf{m}^o)$ is intractable, and a closed-form expression does not exist. In order to obtain the desired compositional estimates, we therefore have to resort to approximate inference methods. Pilot studies indicated that sampling methods gave the best performance.

The goal of any sampling procedure is to obtain a set of N samples $\{x_i\}$ that come from a given (but maybe intractable) distribution π. Using these samples, we can approximate expectation values $\langle A \rangle$ of a function $A(x)$ according to

$$\langle A \rangle = \int_x A(x)\pi(x)\mathrm{d}x \approx \frac{1}{N}\sum_{i=1}^{N} A(x_i) . \tag{31}$$

For instance, if we take $A(x) = x$, the approximation of the mean $\langle x \rangle$ is the sample mean $\frac{1}{N}\sum_{i=1}^{N} x_i$.

An important class of sampling methods are the so-called Markov Chain Monte Carlo (MCMC) methods [3, 17]. In MCMC sampling, a Markov chain is defined as that which has an equilibrium distribution π, in such a way that (31) gives a good approximation when applied to a sufficiently long chain $x_1, x_2, \ldots, x_N$. In order to make the chain independent of the initial state x_0, a burn-in period is often taken into account. This means that one ignores the first $M \ll N$ samples that come from intermediate distributions and begins storing the samples once the system has reached the equilibrium distribution π.

In our application, we use the hybrid Monte Carlo (HMC) sampling algorithm [10, 17]. HMC is a powerful class of MCMC methods that are designed for problems with continuous state spaces, such as considered in this section. HMC can in principle be applied to any noise model with a continuous probability density, so that there is no restriction to Gaussian noise models. HMC uses Hamiltonian dynamics in combination with a Metropolis [19] acceptance procedure to find regions of higher probability. This leads to a more efficient sampler than a sampler that relies on random walk for phase space exploration. HMC also tends to mix more rapidly than the standard Metropolis Hastings algorithm. For details of the algorithm, we refer to the literature [10, 17].

In our case, $\pi(\mathbf{v})$ is the posterior distribution $p(\mathbf{v}|m_i^o)$ in (26). The HMC sampler generates samples $\mathbf{v}_1, \mathbf{v}_2, \ldots, \mathbf{v}_N$ from this posterior distribution. Each of the N samples is a full K-dimensional vector of volume fractions constrained to $\mathbb{S}^K$. The number of samples is of the order of $N = 10^5$, which takes a few seconds on a standard PC. Figure 7 shows an example of a chain of 10,000 states generated by the sampler. For visual clarity, only two components of the vectors are plotted (quartz and dolomite). The plot illustrates the multivariate character of the method: for example, the traces show that the volume fractions of the two minerals tend to be mutually exclusive: either 20% quartz, or 20% dolomite, but generally, not both. From the traces, all kind of statistics can be derived. As an example, the resulting 1D marginal distributions of the mineral volume fractions are plotted.

The performance of the method relies heavily on the quality of the sampler. Therefore, we looked at the ability of the system to estimate the composition of a (synthetic) reservoir and the ability to reproduce the results. For this purpose, we set the composition to a certain value $\mathbf{v}^*$. We apply the observation model to generate measurements $\mathbf{m}^o$. Then, we run HMC to obtain samples from the posterior $P(\mathbf{v}|\mathbf{m}^o)$. Consistency is assessed by comparing results of different runs to each other and by comparing them with the "ground truth" $\mathbf{v}^*$. Results of simulations confirm that the sampler generates reproducible results, consistent with the underlying compositional vector [5]. In these simulations, we took the observation model to generate measurement data (the generating model) equal to the observation model that is used to compute the posterior (the inference model). We also performed simulations where they are different, in particular, in their assumed variance. We found that the sampler is robust to cases where the variance of the generating model is smaller than the variance of the inference model. In the cases where the variance of the generating model is bigger, we found that the method is robust up to differences

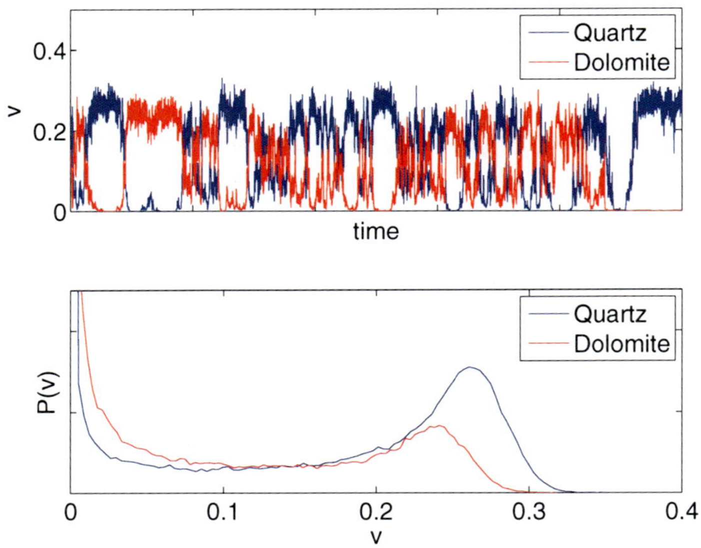

Fig. 7 Diagrams for quartz and dolomite. Top: time traces (10 000 time steps) of the volume fractions of quartz and dolomite. Bottom: Resulting marginal probability distributions of both fractions.

of a factor 10. After that we found that the sampler suffered severely from local minima, leading to irreproducible results.

4.4 Decision Support

Suppose that we have obtained a subset of measurement outcomes $\mathbf{m}^o$, yielding a distribution $P(\mathbf{v}|\mathbf{m}^o)$. One may subsequently ask the question which tool t should be deployed next to gain as much information as possible?

When asking this question, one is often interested in a specific subset of minerals and fluids. Here, we assume this interest is actually in one specific component u. The question then reduces to selecting the most informative tool(s) t for a given mineral u.

We define the informativeness of a tool as the expected decrease of uncertainty in the distribution of v_u after obtaining a measurement with that tool. Usually, entropy is taken as a measure for uncertainty [17], and so a measure of informativeness is the expected entropy of the distribution of v_u after measurement with tool t,

$$\begin{aligned}\langle \mathrm{H}_{u,t}|\mathbf{m}^o\rangle \equiv -\int P(m_t|\mathbf{m}^o)\int P(v_u|m_t,\mathbf{m}^o)\\ \times \log\left(P(v_u|m_t,\mathbf{m}^o)\right)\mathrm{d}v_u\mathrm{d}m_t\ .\end{aligned} \tag{32}$$

Note that the information of a tool depends on the earlier measurement results since the probabilities in (32) are conditioned on $\mathbf{m}^o$.

The most informative tool for mineral u is now identified as that tool t^* which yields in expectation the lowest entropy in the posterior distribution of v_u:

$$t^*_{u|\mathbf{m}^o} = \arg\min_t \langle \mathrm{H}_{u,t} | \mathbf{m}^o \rangle \tag{33}$$

In order to compute the expected conditional entropy using HMC sampling methods, we first rewrite the expected conditional entropy (32) in terms of quantities that are conditioned only on the measurement outcomes $\mathbf{m}^o$,

$$\begin{aligned}\langle \mathrm{H}_{u,t} | \mathbf{m}^o \rangle = -\int\int P(v_u, m_t | \mathbf{m}^o) \\ \times \log\left(P(v_u, m_t | \mathbf{m}^o)\right) \mathrm{d}v_u \mathrm{d}m_t \\ + \int P(m_t | \mathbf{m}^o) \int \log\left(P(m_t | \mathbf{m}^o)\right) \mathrm{d}m_t \ .\end{aligned} \tag{34}$$

Now the HMC run yields a set $V = \{v^j_1, v^j_2, \ldots, v^j_K\}$ of compositional samples (conditioned on $\mathbf{m}^o$). We augment these by a set $M = \{m^j_1 = f_1(\mathbf{v}^j) + \xi^j_1, \ldots, m^j_Z = f_Z(\mathbf{v}^j) + \xi^j_Z\}$ of synthetic tool values generated from these samples (which are indexed by j) by applying equation (29). Subsequently, discretized joint probabilities $P(v_u, m_t | \mathbf{m}^o)$ are obtained via a 2D binning procedure over v_u and m_t for each of the potential tools t. The binned versions of $P(v_u, m_t | \mathbf{m}^o)$ (and $P(m_t | \mathbf{m}^o)$) can be directly used to approximate the expected conditional entropy using a discretized version of equation (34).

The outcome of our implementation of the decision support tool is a ranking of tools according to the expected entropies of their posterior distributions. In this way, the user can select a tool based on a trade-off between expected information and other factors, such as deployment costs and feasibility.

4.5 The Application

The application is implemented in C++ as a stand-alone version with a GUI running on a Windows PC. The application has been validated by petrophysical domain experts from Shell E&P. The further use by Shell of this application is beyond the scope of this chapter.

4.6 Summary

This chapter described a Bayesian network application for petrophysical decision support. The observation models are based on the physics of the measurement tools. The physical variables in this application are continuous-valued. A naive Bayesian network approach with discretized values would fail. We remained in the continuous domain and used the HMC algorithm for inference.

5 Bonaparte: A Bayesian Network for Disaster Victim Identification

Society is increasingly aware of the possibility of a mass disaster. Recent examples are the WTC attacks, the tsunami, and various airplane crashes. In such an event, the recovery and identification of the remains of the victims is of great importance, both for humanitarian as well as legal reasons. Disaster victim identification (DVI), i.e., the identification of victims of a mass disaster, is greatly facilitated by the advent of modern DNA technology. In forensic laboratories, DNA profiles can be recorded from small samples of body remains which may otherwise be unidentifiable. The identification task is the match of the unidentified victim with a reported missing person. This is often complicated by the fact that the match has to be made in an indirect way. This is the case when there is no reliable reference material of the missing person. In such a case, DNA profiles can be taken from relatives. Since their profiles are statistically related to the profile of the missing person (first degree family members share about 50% of their DNA) an indirect match can be made.

In cases with one victim, identification is a reasonable straightforward task for forensic researchers. In the case of a few victims, the puzzle to match the victims and the missing persons is often still doable by hand, using a spread sheet, or with software tools available on the internet [9]. However, large scale DVI is infeasible in this way and an automated routine is almost indispensable for forensic institutes that need to be prepared for DVI.

Bayesian networks are very well suited to model the statistical relations of genetic material of relatives in a pedigree [11]. They can directly be applied in kinship

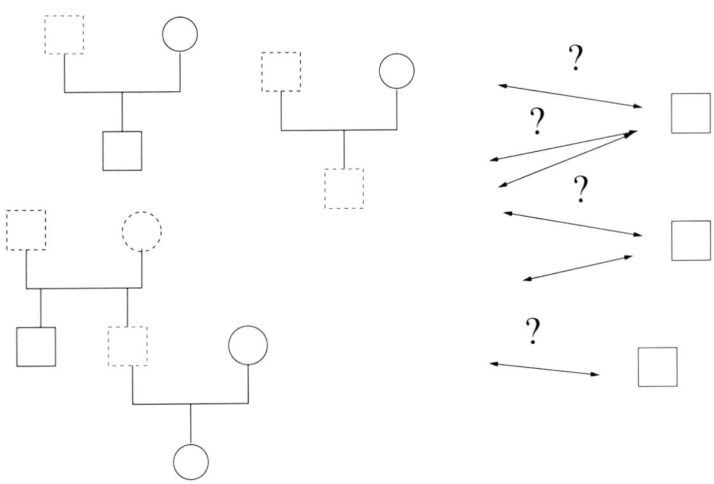

Fig. 8 The matching problem. Match the unidentified victims (blue, right) with reported missing persons (red, left) based on DNA profiles of victims and relatives of missing persons. DNA profiles are available from individuals represented by solid squares (males) and circles (females).

analysis with any type of pedigree of relatives of the missing persons. An additional advantage of a Bayesian network approach is that it makes the analysis tool more transparent and flexible, allowing to incorporate other factors that play a role—such as measurement error probability, missing data, statistics of more advanced genetic markers etc.

Currently, we develop software for DVI, called Bonaparte. This development is in collaboration with NFI (Netherlands Forensic Institute). The computational engine of Bonaparte uses automatically generated Bayesian networks and Bayesian inference methods, enabling to correctly do kinship analysis on the basis of DNA profiles combined with pedigree information. It is designed to handle large-scale events, with hundreds of victims and missing persons. In addition, it has GUI, including a pedigree editor, for forensic analysts. Data-interfaces to other laboratory systems (e.g., for the DNA-data input) will also be implemented.

In the remainder of this section, we will describe the Bayesian model approach that has been taken in the development of the application. We formulate the computational task, which is the computation of the likelihood ratio of two hypotheses. The main ingredient is a probabilistic model P of DNA profiles. Before discussing the model, we will first provide a brief introduction to DNA profiles. In the last part of the section, we describe how P is modeled as a Bayesian network, and how the likelihood ratio is computed.

5.1 Likelihood Ratio of Two Hypotheses

Assume that we have a pedigree with an individual MP who is missing (the Missing Person). In this pedigree, there are some family members that have provided DNA material, yielding the profiles. Furthermore, there is an Unidentified Individual UI, whose DNA is also profiled. The question is, is $UI = MP$? In order to proceed, we assume that we have a probabilistic model P for DNA evidence of family members in a pedigree. In order to compute the probability of this event, we need hypotheses to compare. The common choice is to formulate two hypotheses. The first is the hypothesis H_1 that indeed $UI = MP$. The alternative hypothesis H_0 is that UI is an unrelated person U. In both hypotheses, we have two pedigrees: the first pedigree has MP and family members FAM as members; the second one has only U as member. In order to compare the hypotheses, we compute the likelihoods of the evidence from the DNA profiles under the two hypotheses,

- Under H_p, we assume that $MP = UI$. In this case, MP is observed and U is unobserved. The evidence is $E = \{DNA_{MP} + DNA_{FAM}\}$.
- Under H_d, we assume that $U = UI$. In this case, U is observed and MP is observed. The evidence is $E = \{DNA_U + DNA_{FAM}\}$.

Under the model P, the likelihood ratio of the two hypotheses is

$$LR = \frac{P(E|H_p)}{P(E|H_d)} . \tag{35}$$

If in addition a prior odds $P(H_p)/P(H_d)$ is given, then the posterior odds $P(H_p|E)/P(H_d|E)$ follows directly from multiplication of the prior odds and likelihood ratio,

$$\frac{P(H_p|E)}{P(H_d|E)} = \frac{P(E|H_p)P(H_p)}{P(E|H_d)P(H_d)} . \tag{36}$$

5.2 DNA Profiles

In this subsection, we provide a brief introduction on DNA profiles for kinship analysis. A comprehensive treatise can be found in e.g., [6]. In humans, DNA found in the nucleus of the cell is packed on chromosomes. A normal human cell has 46 chromosomes, which can be organized in 23 pairs. From each pair of chromosomes, one copy is inherited from father and the other copy is inherited from mother. In 22 pairs, chromosomes are homologous, i.e., they have practically the same length and contain in general the same genes (functional elements of DNA). These are called the autosomal chromosomes. The remaining chromosome is the sex-chromosome. Males have an X and a Y chromosome. Females have two X chromosomes.

More than 99% of the DNA of any two humans of the general population is identical. Most DNA is therefore not useful for identification. However, there are well-specified locations on chromosomes where there is variation in DNA among individuals. Such a variation is called a genetic marker. In genetics, the specified locations are called loci. A single location is a locus.

In forensic research, the short tandem repeat (STR) markers are the most currently used. The reason is that they can be reliably determined from small amounts of body tissue. Another advantage is that they have a low mutation rate, which is important for kinship analysis. STR markers is a class of variations that occur when a pattern of two or more nucleotides is repeated. For example,

$$(CATG)_3 = CATGCATGCATG . \tag{37}$$

The number of repeats x (which is 3 in the example) is the variation among the population. Sometimes, there is a fractional repeat, e.g., $CATGCATGCATGCA$, and this would be encoded with repeat number $x = 3.2$, since there are three repeats and two additional nucleotides. The possible values of x and their frequencies are well documented for the loci used in forensic research. These ranges and frequencies vary between loci. To some extent they vary among subpopulations of humans. The STR loci are standardized. The NFI uses CODIS (Combined DNA Index System) standard with 13 specific core STR loci, each on different autosomal chromosomes.

The collection of markers yields the DNA profile. Since chromosomes exist in pairs, a profile will consist of pairs of markers. For example, in the CODIS standard, a full DNA profile will consist of 13 pairs, (the following notation is not common standard)

$$\bar{\mathbf{x}} = ({}^1x^1, {}^1x^2), ({}^2x^1, {}^2x^2), \ldots, ({}^{13}x^1, {}^{13}x^2) , \tag{38}$$

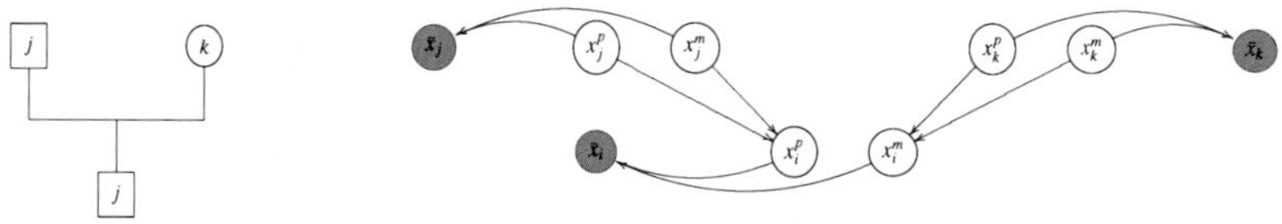

Fig. 9 A basic pedigree with father, mother, and child. Squares represent males, circles represent females. Right: corresponding Bayesian network. Grey nodes are observables. x_j^p and x_j^m represents paternal and maternal allele of individual j. See text.

in which each ${}^{\mu}x^s$ is a number of repeats at a well-defined locus μ. However, since chromosomes exists in pairs, there will be two alleles ${}^{\mu}x^1$ and ${}^{\mu}x^2$ for each location, one paternal—on the chromosome inherited from father—and one maternal. Unfortunately, current DNA analysis methods cannot identify the phase of the alleles, i.e., whether an allele is paternal or maternal. This means that $({}^{\mu}x^1, {}^{\mu}x^2)$ cannot be distinguished from $({}^{\mu}x^2, {}^{\mu}x^1)$. In order to make the notation unique, we order the observed alleles of a locus such that ${}^{\mu}x^1 \leq {}^{\mu}x^2$.

Chromosomes are inherited from parents. Each parent passes one copy of each pair of chromosomes to the child. For autosomal chromosomes, there is no (known) preference which one is transmitted to the child. There is also no (known) correlation between the transmission of chromosomes from different pairs. Since chromosomes are inherited from parents, alleles are inherited from parents as well. However, there is a small probability that an allele is changed or mutated. This mutation probability is about 0.1%.

Finally, in the DNA analysis, sometimes failures occur in the DNA analysis method and an allele at a certain locus drops out. In such a case, the observation is $({}^{\mu}x^1, ?)$, in which "?" is a wild card.

5.3 A Bayesian Network for Kinship Analysis

In this subsection, we will describe the building blocks of a Bayesian network to model probabilities of DNA profiles of individuals in a pedigree. First, we observe that inheritance and observation of alleles at different loci are independent. Therefore, for each locus, we can make an independent model P_{μ}. In the model described below, we will consider a model for a single locus, and we will suppress the μ dependency for notational convenience.

5.3.1 Allele Probabilities

We will consider pedigrees with individuals i. In a pedigree, each individual i has two parents, a father $f(i)$ and a mother $m(i)$. An exception is when a individual is a founder. In that case, it has no parents in the pedigree.

Statistical relations between DNA profiles and alleles of family members can be constructed from the pedigree, combined with models for allele transmission. On the given locus, each individual i has a paternal allele x_i^f and an maternal allele x_i^m. f and m stands for "father" and "mother". The pair of alleles is denoted as $x_i = (x_i^f, x_i^m)$. Sometimes, we use superscript s which can have values $\{f,m\}$. Therefore, each allele in the pedigree is indexed by (i,s), where i runs over individuals and s over phases (f,m). The alleles can assume N values, where N as well as the allele values depend on the locus.

An allele from a founder is called "founder allele". Therefore, a founder in the pedigree has two founder alleles. The simplest model for founder alleles is to assume that they are independent, and each follows a distribution $P(a)$ of population frequencies. This distribution is assumed to be given. In general $P(a)$ will depend on the locus. More advanced models have been proposed in which founder alleles are correlated. For instance, one could assume that founders in a pedigree come from a single but unknown subpopulation [1]. This model assumption yields corrections to the outcomes in models without correlations between founders. A drawback is that these models may lead to a severe increase in required memory and computation time. In this chapter, we will restrict ourself to models with independent founder alleles.

If an individual i has its parents in the pedigree the allele distribution of an individual given the alleles of its parents is as follows,

$$P(x_i|x_{f(i)},x_{m(i)}) = P(x_i^f|x_{f(i)})P(x_i^m|x_{m(i)}) \,, \tag{39}$$

where

$$P(x_i^f|x_{f(i)}) = \frac{1}{2} \sum_{s=f,m} P(x_i^f|x_{f(i)}^s) \,, \tag{40}$$

$$P(x_i^m|x_{m(i)}) = \frac{1}{2} \sum_{s=f,m} P(x_i^m|x_{m(i)}^s) \,. \tag{41}$$

In order to explain (40) in words, individual i obtains its paternal allele x_i^f from its father $f(i)$. However, there is a 50% chance that this allele is the *paternal* allele $x_{f(i)}^f$ of father $f(i)$ and a 50% chance that it is his *maternal* allele $x_{f(i)}^m$. A similar explanation applies to (41).

The probabilities $P(x_i^f|x_{f(i)}^s)$ and $P(x_i^m|x_{m(i)}^s)$ are given by a mutation model $P(a|b)$, which encodes the probability that allele of the child is a while the allele on the parental chromosome that is transmitted is b. The precise mutation mechanisms for the different STR markers are not known. There is evidence that mutations from father to child are in general about 10 times as probable as mutations from mother to child. Gender of each individual is assumed to be known, but for notational convenience, we suppress dependency of parent gender. In general, mutation tends to decrease with the difference in repeat numbers $|a-b|$. Mutation is also locus dependent [4].

Several mutation models have been proposed, see e.g., [8]. As we will see later, however, the inclusion of a detailed mutation model may lead to a severe increase in required memory and computation time. Since mutations are very rare, one could ask whether there is any practical relevance in a detailed mutation model. The simplest mutation model is of course to assume the absence of mutations, $P(a|b) = \delta_{a,b}$. Such model enhances efficient inference. However, any mutation in any single locus would lead to a 100% rejection of the match, even if there is a 100% match in the remaining markers. Mutation models are important to get some model tolerance against such case. The simplest non-trivial mutation model is a uniform mutation model with mutation rate μ (not to be confused with the locus index μ),

$$P(a|a) = 1 - \mu\,, \tag{42}$$

$$P(a|b) = \mu/(N-1) \quad \text{if } a \neq b\,. \tag{43}$$

Mutation rate may depend on locus and gender.

An advantage of this model is that the required memory and computation time increase only slightly compared to the mutation-free model. Note that the population frequency is in general not invariant under this model: the mutation makes the frequency more flat. One could argue that this is a realistic property that introduces diversity in the population. In practical applications in the model, however, the same population frequency is assumed to apply to founders in different generations in a pedigree. This implies that if more unobserved references are included in the pedigree to model ancestors of an individual, then the likelihood ratio will (slightly) change. In other words, formally equivalent pedigrees will give (slightly) different likelihood ratios.

5.3.2 Observations

Observations are denoted as $\bar{x}_i$, or $\bar{x}$ if we do not refer to an individual. The parental origin of an allele can not be observed, so alleles $x^f = a, x^m = b$ yields the same observation as $x^f = b, x^m = a$. We adopt the convention to write the smallest allele first in the observation: $\bar{x} = (a,b) \Leftrightarrow a \leq b$. In the case of an allele loss, we write $\bar{x} = (x,?)$ where "?" stands for a wild card. We assume that the event of an allele loss can be observed (e.g., via the peak hight [6]). This event is modeled by L. With $L = 1$ there is allele loss, and there will be a wild card "?". A full observation is coded as $L = 0$. The case of loss of two alleles is not modeled, since in this particular case we simply have no observation.

The observation model is now straightforwardly written down. Without allele loss ($L = 0$), alleles y results in an observation $\bar{y}$. This is modeled by the deterministic table

$$P(\bar{x}|y, L=0) = \begin{cases} 1 & \text{if } \bar{x} = \bar{y}\,, \\ 0 & \text{otherwise.} \end{cases} \tag{44}$$

Note that for a given y there is only one $\bar{x}$ with $\bar{x} = \bar{y}$.

With allele loss ($L = 1$), we have

$$\begin{cases} P(\bar{x} = (a,F)|(a,b), L=1) = \frac{1}{2} \\ P(\bar{x} = (b,F)|(a,b), L=1) = \frac{1}{2} \end{cases} \quad \text{if } \; a \neq b\,, \tag{45}$$

and

$$P(\bar{x} = (a, F)|(a,a), L = 1) = 1 \; . \tag{46}$$

That is, if one allele is lost, the alleles (a,b) leads to an observation a (then b is lost), or to an observation b (then a is lost). Both events have 50% probability. If both alleles are the same, so the pair is (a,a), then, of course, a is observed with 100% probability.

5.4 *Inference*

By multiplying all allele priors, transmission probabilities and observation models, a Bayesian network of alleles x and DNA profiles of individuals $\bar{x}$ in a given pedigree is obtained. Assume that the pedigree consists of a set of individuals $\mathscr{I} = 1, \ldots, K$ with a subset of founders $\mathscr{F}$, and assume that allele losses L_j are given, then this probability reads

$$P(\{\bar{x},x\}_{\mathscr{I}}) = \prod_j P(\bar{x}_j|x_j,L_j) \prod_{i\in\mathscr{I}\setminus\mathscr{F}} P(x_i|x_{f(i)},x_{m(i)}) \prod_{i\in\mathscr{F}} P(x_i) \; . \tag{47}$$

Under this model, the likelihood of a given set DNA profiles can now be computed. If we have observations $\bar{x}_j$ from a subset of individuals $j \in \mathscr{O}$, then the likelihood of the observations in this pedigree is the marginal distribution $P(\{\bar{x}\}_{\mathscr{O}})$, which is the marginal probability

$$P(\{\bar{x}\}_{\mathscr{O}}) = \sum_{x_1} \cdots \sum_{x_K} \prod_{j\in\mathscr{O}} P(\bar{x}_j|x_j,L_j) \prod_{i\in\mathscr{I}\setminus\mathscr{F}} P(x_i|x_{f(i)},x_{m(i)}) \prod_{i\in\mathscr{F}} P(x_i) \; . \tag{48}$$

This computation involves the sum over all states of allele pairs x_i of all individuals.

In general, the allele-state space can be prohibitively large. This would make even the junction tree algorithm infeasible if it would straightforwardly be applied. Fortunately, a significant reduction in memory requirement can be achieved by "value abstraction": if the observed alleles in the pedigree are all in a subset A of M different allele values, then we can abstract from all the unobserved allele values and consider them as a single state z. If an allele is z, then it means that it has a value that is not in the set of observed values A. We now have a system in which states can assume only $M+1$ values which are generally a lot smaller than N, the number of a priori possible allele values. This procedure is called value abstraction [12]. The procedure is applicable if for any $a \in A$, $L \in \{0,1\}$, and $b_1, b_2, b_3, b_4 \notin A$, the following equalities hold

$$P(a|b_1) = P(a|b_2) \tag{49}$$

$$P(\bar{x}|a,b_1,L) = P(\bar{x}|a,b_2,L) \tag{50}$$

$$P(\bar{x}|b_1,a,L) = P(\bar{x}|b_2,a,L) \tag{51}$$

$$P(\bar{x}|b_1,b_2,L) = P(\bar{x}|b_3,b_4,L) \tag{52}$$

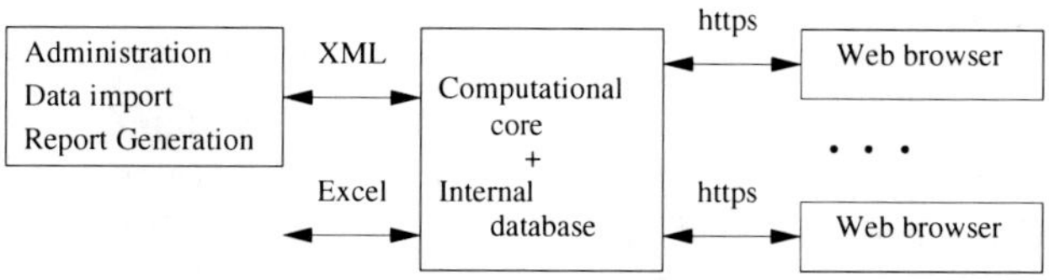

Fig. 10 Bonaparte's basic architecture

If these equalities hold, then we can replace $P(a|b)$ by $P(a|z)$ and $P(\bar{x}|a,b)$ by $P(\bar{x}|a,z)$ etc. in the abstracted state representation. The conditional probability of z then follows from

$$P(z|x) = 1 - \sum_{a \in A} P(a|x) \tag{53}$$

for all x in $A \cup z$. One can also easily verify that the observation probabilities satisfy the condition. The uniform mutation model satisfies condition (49) since $P(a|b) = \mu/(N-1)$ for any $a \in A$ and any $b \notin A$. Note that condition (49) does not necessarily hold for a general mutation model, and so value abstraction could then not be applied.

Using value abstraction as a preprocessing step, a junction tree-based algorithm can straightforwardly be applied to compute the desired likelihood. In this way, likelihoods and likelihood ratios are computed for all loci, and reported to the user.

5.5 The Application

Bonaparte has been designed to facilitate large-scale matching. The application has a multi-user client-server-based architecture, see Fig. 10. Its computational core and the internal database run on a server. All matching results are stored in internal database. Rewind to any point in back in time is possible. Via an XML and secure https interfaces, the server connects to other systems. Users can login via a web browser so that no additional software is needed on the clients. The current version Bonaparte is now under user validation. A live demo version will be made available on www.dnadvi.nl.

5.6 Summary

Bonaparte is an application of Bayesian networks for victim identification by kinship analysis based on DNA profiles. The Bayesian networks are used to model statistical relations between DNA profiles of different individuals in a pedigree. By Bayesian inference, likelihood ratios and posterior odds of hypotheses are computed, which are the quantities of interest for the forensic researcher. The probabilistic relations between variables are based on first principles of genetics. A feature of this application is the automatic, on-the-fly derivation of models from data, i.e., the pedigree structure of a family of a missing person.

6 Discussion

Human decision makers are often confronted with highly complex domains. They have to deal with various sources of information and various sources of uncertainty. The quality of the decision is strongly influenced by the decision makers experience to correctly interpret the data at hand. Computerized decision support can help us to improve the effectiveness of the decision maker by enhancing awareness and alerting the user to uncommon situations that may have high impact. Rationalizing the decision process may alleviate some of the decision pressure.

Bayesian networks are widely accepted as a principled methodology for modeling complex domains with uncertainty, in which different sources of information are to be combined, as needed in intelligent decision support systems. However, many of the examples of Bayesian networks as described in literature—models with a few dozen of variables, each with a few states, and fixed relations—may suggest a limitation in the expressive power of the methodology [18].

In this chapter, we described three Bayesian networks for real-world applications. These models are based on the same principled methodology as standard Bayesian networks, but go beyond the above mentioned limitations. The Promedas model has several orders of magnitudes more variables. The petrophysical model has continuous-valued variables. The Bonaparte model as well as the Promedas model have non-static relations.

Fundamental differences of these models with most standard Bayesian networks are (1) the model development approach and (2) the operational power and flexibility of the applications. Standard Bayesian networks are often developed using off-the-shelf GUI-based software. An advantage of this approach is that small-or medium-sized Bayesian networks can be developed quickly, without the need of expertise on Bayesian networks modeling or inference algorithms. The models described in this chapter, on the other hand, have been developed from scratch, based on first principles and with customized implementations of inference algorithms (junction tree based, or approximate such as the HMC method). This development approach requires more expertise, but it has more flexibility as it is not constrained by the development software and can better handle the various problems posed by the applications, such as the large number of variables, the continuous-valued variables, and on-the-fly model derivation from data, etc.

We have discussed in detail three applications of Bayesian networks. With these applications, we aimed to illustrate the modeling power of the Bayesian networks, which goes beyond the standard textbook applications. The applications domains of the models (medicine, petrophysics, and forensics) demonstrate that Bayesian networks can be applied in a wide variety of domains with different types of domain requirements.

Finally, we would like to stress that the Bayesian network technology is only one side of the model. The other side is the domain knowledge, which is maybe even more important for the model. Therefore, Bayesian network modeling always requires a close collaboration with domain experts. Even then, the model is of course only one of many ingredients of an application, such as user-interface, data-management, user-acceptance etc. which are all essential to make the application a success.

Acknowledgments. The research for the Promedas project has been supported by the Dutch Technology Foundation STW, the Applied Science Division of NWO and the Technology Program of the Ministry of Economic Affairs. We thank Kees Albers and Martijn Leisink (SNN), Jan Neijt (UMC Utrecht), Mirano Spalburg (Shell E & P), Klaas Slooten, and Carla Bruijning (NFI) for their collaboration. Finally, we thank the anonymous reviewers for their useful comments.

References

1. Balding, D., Nichols, R.: DNA profile match probability calculation: how to allow for population stratification, relatedness, database selection and single bands. Forensic Science International 64(2-3), 125–140 (1994)
2. Beinlich, I., Suermondt, H., Chavez, R., Cooper, G., et al.: The ALARM monitoring system: A case study with two probabilistic inference techniques for belief networks. In: Proceedings of the Second European Conference on Artificial Intelligence in Medicine, vol. 256. Springer, Berlin (1989)
3. Bishop, C.: Pattern recognition and machine learning. Springer, Heidelberg (2006)
4. Brinkmann, B., Klintschar, M., Neuhuber, F., Hühne, J., Rolf, B.: Mutation rate in human microsatellites: influence of the structure and length of the tandem repeat. The American Journal of Human Genetics 62(6), 1408–1415 (1998)
5. Burgers, W., Wiegerinck, W., Kappen, H., Spalburg, M.: A Bayesian petrophysical decision support system for estimation of reservoir compositions (submitted)
6. Butler, J.: Forensic DNA typing: biology, technology, and genetics of STR markers. Academic Press, London (2005)
7. Castillo, E., Gutierrez, J.M., Hadi, A.S.: Expert Systems and Probabilistic Network Models. Springer, Heidelberg (1997)
8. Dawid, A., Mortera, J., Pascali, V.: Non-fatherhood or mutation? A probabilistic approach to parental exclusion in paternity testing. Forensic science international 124(1), 55–61 (2001)
9. Drábek, J.: Validation of software for calculating the likelihood ratio for parentage and kinship. Forensic Science International: Genetics 3(2), 112–118 (2009)
10. Duane, S., Kennedy, A., Pendleton, B., Roweth, D.: Hybrid Monte Carlo Algorithm. Phys. Lett. B 195, 216 (1987)
11. Fishelson, M., Geiger, D.: Exact genetic linkage computations for general pedigrees. Bioinformatics 198(Suppl. 1), S189–S198 (2002)
12. Friedman, N., Geiger, D., Lotner, N.: Likelihood computations using value abstraction. In: Proceedings of the Sixteenth Conference on Uncertainty in Artificial Intelligence, pp. 192–200. Morgan Kaufmann Publishers, San Francisco (2000)
13. Heckerman, D.: Probabilistic interpretations for mycin's certainty factors. In: Kanal, L., Lemmer, J. (eds.) Uncertainty in artificial intelligence, pp. 167–196. North-Holland, Amsterdam (1986)
14. Jensen, F.: An Introduction to Bayesian networks. UCL Press (1996)
15. Jordan, M.: Learning in graphical models. Kluwer Academic Publishers, Dordrecht (1998)

16. Lauritzen, S., Spiegelhalter, D.: Local computations with probabilities on graphical structures and their application to expert systems. Journal of the Royal Statistical Society. Series B (Methodological), 157–224 (1988)
17. MacKay, D.: Information theory, inference and learning algorithms. Cambridge University Press, Cambridge (2003)
18. Mahoney, S., Laskey, K.: Network engineering for complex belief networks. In: Proc. 12th Conf. on Uncertainty in Artificial Intelligence, pp. 389–396. Morgan Kaufmann, San Francisco (1996)
19. Metropolis, N., Rosenbluth, A., Rosenbluth, M., Teller, A., Teller, E.: Equation of state calculations by fast computing machines. The journal of chemical physics 21(6), 1087 (1953)
20. Pearl, J.: Probabilistic Reasoning in Intelligent systems: Networks of Plausible Inference. Morgan Kaufmann Publishers, Inc., San Francisco (1988)
21. Pradhan, M., Provan, G., Middleton, B., Henrion, M.: Knowledge engineering for large belief networks. In: Proc. Tenth Conf. on Uncertainty in Artificial Intelligence, pp. 484–490 (1994)
22. Russell, S., Norvig, P., Canny, J., Malik, J., Edwards, D.: Artificial intelligence: a modern approach. Prentice Hall, Englewood Cliffs (2003)
23. Schlumberger: Log Interpretation Principles/Applications. Schlumberger Limited (1991)
24. Shortliffe, E., Buchanan, B.: A model of inexact reasoning in medicine. Mathematical Biosciences 23(3-4), 351–379 (1975)
25. Shwe, M., Middleton, B., Heckerman, D., Henrion, M., Lehman, H., Cooper, G.: Probabilistic Diagnosis Using a Reformulation of the Internist-1/ QMR Knowledge Base. Methods of Information in Medicine 30, 241–255 (1991)
26. Spalburg, M.: Bayesian uncertainty reduction for log evaluation. SPE International (2004); SPE88685
27. Takinawa, M., D'Ambrosio, B.: Multiplicative factorization of noisy-MAX. In: Proceedings of the 15th Conference on Uncertainty in Artificial Intelligence UAI 1999, pp. 622–630 (1999)

Index